Essential
Genetics _and_
Genomics

Give Students the Choice

We're happy to offer you and your students *Essential Genetics and Genomics, Seventh Edition* in varying formats to best meet individual preferences, whether it be print, digital only, or a combination.

Option 1 – Print + Digital

Essential Genetics and Genomics, Seventh Edition WITH Navigate 2 Advantage Access

Each new print copy of *Essential Genetics and Genomics, Seventh Edition* includes Navigate 2 Advantage Access FREE and unlocks a comprehensive and interactive eBook, a *Readiness Assessment* to ensure students are prepared for learning and applying introductory genetics, a study center with links to useful and relevant internet resources, student practice activities and assessments, a full suite of instructor resources, and learning analytics reporting tools. Students can purchase the printed text with digital access at **go.jblearning.com/EssentialGenetics**

ISBN: 978-1-284-15245-6

Option 2 – Digital
50% off the list price of the printed textbook

Standalone Navigate 2 Advantage Access for *Essential Genetics and Genomics, Seventh Edition*

Students can purchase Navigate 2 Advantage Access for *Essential Genetics and Genomics, Seventh Edition* and have full access to the complete and interactive eBook, *Readiness Assessment*, and the Study Center with practice activities and assessments for further independent study. Standalone access can be purchased at **go.jblearning.com/EssentialGeneticsNav2**

ISBN: 978-1-284-15269-2

Option 3 – Custom Print or Digital

When it comes to teaching, one size rarely fits all. That's why we customize our content to create the flexible and affordable solution that best fits your needs. Easily create your own unique textbook aligned with your syllabus and course objectives by adding, removing, or rearranging content. We'll deliver your custom textbook to you and your students as a printed text or an eBook. Visit **go.jblearning.com/Custom** to get started today.

QUESTIONS?

Visit **go.jblearning.com/FindMyRep** or call 1-800-832-0034 to speak to your dedicated account manager.

Essential
Genetics *and*
Genomics

SEVENTH EDITION

Daniel L. Hartl, PhD
Harvard University

JONES & BARTLETT
LEARNING

World Headquarters
Jones & Bartlett Learning
5 Wall Street
Burlington, MA 01803
978-443-5000
info@jblearning.com
www.jblearning.com

Jones & Bartlett Learning books and products are available through most bookstores and online booksellers. To contact Jones & Bartlett Learning directly, call 800-832-0034, fax 978-443-8000, or visit our website, www.jblearning.com.

15275-3

Production Credits
VP, Product Management: Amanda Martin
Director of Product Management: Laura Pagluica
Product Specialist: Audrey Schwinn
Product Assistant: Loren-Marie Durr
Production Manager: Dan Stone
Senior Production Editor, Navigate: Jessica deMartin
Digital Products Manager: Jordan McKenzie
Digital Products Specialist: Angela Dooley
Marketing Manager: Lindsay White

Manufacturing and Inventory Control Supervisor: Amy Bacus
Composition: codeMantra U.S. LLC
Cover Design: Kristin E. Parker
Text Design: Kristin E. Parker
Rights & Media Specialist: John Rusk
Media Development Editor: Troy Liston
Cover Image (Title Page): © ktsdesign/Shutterstock
Printing and Binding: LSC Communications
Cover Printing: LSC Communications

Library of Congress Cataloging-in-Publication Data
Names: Hartl, Daniel L., author.
Title: Essential genetics and genomics / Daniel L. Hartl.
Other titles: Essential genetics
Description: Seventh edition. | Burlington, MA : Jones & Bartlett Learning,
[2018] | Preceded by Essential genetics / Daniel L. Hartl. 6th ed. 2014. |
Includes bibliographical references and index.
Identifiers: LCCN 2018042898 | ISBN 9781284152456 (pbk.) | ISBN 9781284152685
(e-ISBN)
Subjects: | MESH: Genetic Phenomena | Genomics
Classification: LCC QH430 | NLM QU 500 | DDC 572.8/6—dc23
LC record available at https://lccn.loc.gov/2018042898

6048

Printed in the United States of America
22 21 20 19 18   10 9 8 7 6 5 4 3 2 1

Brief Contents

Contents

© ktsdesign/Shutterstock

© Heiko Kueverling/Shutterstock

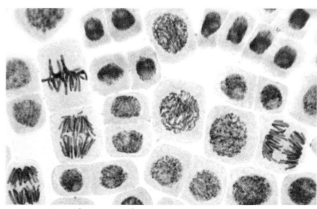

© Dimarion/Shutterstock

© picturepartners/Shutterstock

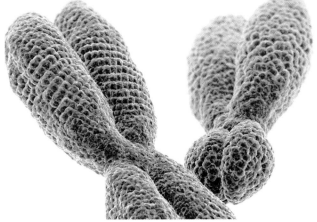

© Iaremenko Sergii/Shutterstock

© H.Tanaka/Shutterstock

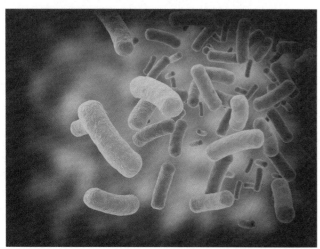

© Jezper/Shutterstock

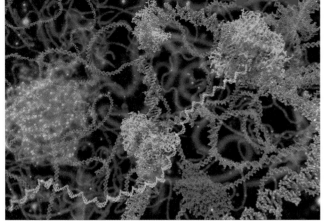

© Juan Gaertner/Shutterstock

© Revers/Shutterstock

Chapter 10   Genomics, Proteomics, and Genetic Engineering   **317**

© Vchal/ShutterStock

Chapter 11   The Genetic Control of Development   **353**

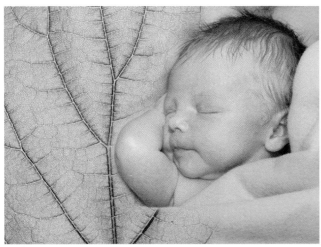

© Valentina Razumova/Shutterstock

© Robert Mcgillivray/Shutterstock

Chapter 13   Molecular Genetics of the Cell Cycle and Cancer   **416**

© Photographee.eu/Shutterstock

© Hobbit/Shutterstock

© Yuganov Konstantin/Shutterstock

Preface

A good teacher aims to *uncover* a subject, not *cover* it. So said a wise former teacher of mine. In revising *Essential Genetics and Genomics, Seventh Edition*, I've tried to heed this advice. To *uncover* a subject means to expose, exhibit, unveil. To help *uncover* genetics, this new edition is:

- **Streamlined**, with *emphasis on concepts* illuminated by vivid example and stripped of extraneous detail

- **Focused,** with *Learning Objectives* stated explicitly at the beginning of each chapter

- **Skills oriented**, with *Stop & Think* problems inserted at strategic points in the text to enhance the reading experience and encourage higher-order, analytical thinking.

The brevity of the text meets the needs of the shorter, less comprehensive introductory course of one semester or quarter. The choice of topics is intended to help students master the following scientific competencies:

- Understand the basic processes of gene transmission, mutation, expression, and regulation.

- Analyze genetic processes using numerical relationships including ratios and proportions.

- Learn to formulate genetic hypotheses in a statistical framework, work out their consequences, and test the results against observed data.

- Develop basic skills in problem solving, including single-concept exercises, those requiring the application of several concepts in logical order, and numerical problems requiring some arithmetic for solution.

- Interpret genetic data and make valid inferences to reveal the underlying causes.

- Acquire an appreciation of current trends in genetics, as well as the social and historical context in which genetics has developed.

Scientific competency is the desired learning outcome of any course in a curriculum in STEM (science, technology, engineering, mathematics). Genetics is an excellent subject for achieving scientific competency. It is integrative over a broad territory, ranging from molecular biology to evolutionary genetics. It is also quantitative, using concepts from probability, statistics, and computational biology. Fortunately, students come to a course in genetics highly motivated because of media reports about the human genome and genetic risk factors for disease, as well as many social and ethical controversies related to genetics such as direct-to-consumer and over-the-counter genetic testing, genetic privacy, cloning, stem-cell research, and genetically modified organisms. The challenges for the instructor are to sustain this motivation and to help students acquire the skills and habits of thought that constitute scientific competency.

What's New in the *Seventh Edition?*

This seventh edition has been completely revised and updated. Each chapter has been thoroughly reworked. Important new methods and findings have been added, including **synthetic bacteria, higher-order chromatin structure, high-throughput genomic sequencing methods, personalized medicine,** and **CRISPR/Cas9 gene editing.** As new material has been added, an equal or greater amount of nonessential or outdated material has been deleted. Several of the chapters have been reorganized to allow smooth integration of the new material. The entire text has been condensed, clarified, and updated.

Major revisions and additions in the *Seventh Edition* include the following:

- Chapter 1 includes a new section emphasizing that most common traits are actually **complex traits** affected by multiple interacting genetic factors as well as environment. This principle includes most **common diseases**, which are influenced by multiple genetic risk factors and lifestyle choices. The section on genomes and proteomes has been updated to include **Syn3.0**, the first living, multiplying bacterial cell whose genome was created entirely by chemical synthesis.

- Chapter 2 includes a new section emphasizing that **genes affect traits at multiple levels** (molecular, cellular, developmental, morphological, and behavioral), and that in many cases the traits that are affected appear to be unrelated until the underlying biology is understood.

- Chapter 3 now includes discussion of the **epigenetic specification of the kinetochore**, in which a specialized histone (CENPA) replaces histone 3 in **centromeric nucleosomes** that helps recruit kinetochore-associated proteins leading to the assembly of the mature kinetochore to which spindle fibers attach.

- Chapter 4 has a much shortened and sharper discussion of the **principal types of genetic variation** with emphasis on single nucleotide polymorphisms (SNPs), copy number variations (CNVs), and simple tandem repeats (STRs). The update also includes a summary of the results of **genome-wide association studies (GWAS)** to detect genetic risk factors for common disorders and genetic factors affecting complex traits.

- Chapter 5 has been significantly shortened and streamlined.

- Chapter 6 includes a completely reorganized and simplified discussion of DNA replication updated to include the **trombone model of replication** showing how synthesis of the leading and lagging strands is coordinated. The section on **massively parallel sequencing** has been expanded and moved to Chapter 10.

- Chapter 7 contains a clearer description of how bacterial cells are brought together in conjugation, and unnecessary detail on genetic fine structure of the rII gene in bacteriophage T4 has been eliminated.

- Chapter 8 makes good use of the **Stop & Think** feature to reinforce fundamental concepts and processes of transcription and translation.

- Chapter 9 includes a major new section on how chromatin is organized into higher-order structures composed of **topologically associating domains (TADs)**, **insulators**, and **compartments**. The section on **RNA interference** and **long noncoding RNAs** has been updated, and the material on programmed DNA rearrangements has been removed because it is less generally applicable than once thought to be.

- A completely revised and reorganized Chapter 10 includes a summary of the latest **high-throughput DNA sequencing technology** including **reversible terminators, ion-torrent**

sequencing, **single-molecule sequencing,** and **nanopore sequencing**. It also includes a major new section on **personalized medicine (precision medicine)** as well as **direct-to-consumer genetic services** and **over-the-counter genetic testing kits**. This section points out the potential values of these approaches but also emphasizes their associated **ethical, legal, and social implications**. Finally, Chapter 10 contains a wholly new section on **CRISPR-Cas9 technology for genetic engineering** showing how CRISPR-Cas9 functions as a sort of immune system in bacteria, how the CRISPR-Cas9 molecules are used to create targeted knockout mutations, and how they are used in DNA editing to change the genome sequence in a predetermined manner. Methods of CRISPR-Cas9 use in insects, mice, and plants are also discussed.

- Chapter 11 contains a slightly expanded section on the use of **epistasis in the analysis of switch-regulation pathways**, and some unnecessary details have been omitted such as minutiae of genetic control of yeast mating type.

- New to Chapter 12 is a major new section on estimates of the **rate of base-substitution mutation in humans** as determined by genome sequencing of parental and offspring genomes. The mutation rate increases steadily with father's age but not with mother's age, and we discuss why this finding is completely consistent with the reproductive biology of males and females. I've also deleted the section on the "ClB method" for detecting mutations in *Drosophila*, as this is mainly of historical interest.

- Chapter 13 puts greater emphasis on the connection between genetic control of the **cell cycle and cancer**, and it has been extensively revised, many of the illustrations simplified, and dispensable details eliminated.

- Chapter 14 contains a shorter, streamlined section on **molecular phylogenetics**.

- What's new in Chapter 15 is a major new section on **genome-wide association studies (GWAS)** focusing on the usually **large number of genes affecting complex traits** and their typically **small individual effects**. Each genetic risk factor for a disease usually increases disease risk by only a small amount. This generalization underlines the importance of proper interpretation of direct-to-consumer (DTC) and over-the-counter (OTC) genetic testing. The chapter also includes a new discussion of **physiological epistasis and statistical epistasis** and explains why genes can exhibit a great deal of physiological epistasis at the molecular, cellular, and organismal levels without showing any substantial statistical epistasis at the population level. A classic method for estimating the number of genes affecting quantitative traits, based on differences between means of inbred lines and the genetic variance, has been deleted because it is obsolete and usually results in absurdly small estimates.

Chapter Organization

Each chapter begins with a set of **Learning Objectives** to orient students toward the knowledge and skills they should focus on. Explicit **Learning Objectives** help students to:

- *Identify* what they should know or be able to do as a result of their study.

- *Focus* on the knowledge they should have acquired from studying the chapter.

- *Guide* the student to identify key concepts and use them at a variety of learning levels including comprehension, application, analysis, and synthesis.

- *Highlight the skills* they should acquire through practice problems of various types.

Each chapter has an opening paragraph that gives an overview of what is to come, illustrates the subject with engaging examples, and shows how the material is connected to genetics as a whole. The section and subsection **Headings** are in the form of complete sentences that encapsulate the main message. The text makes liberal use of **Numbered** and **Bulleted Lists** to aid students in organizing their learning, as well as **Key Concepts** set apart from the main text to

emphasize important principles. A feature called **Stop & Think** recognizes that assessments in real time are critical to reinforce understanding. Each chapter also includes the **Human Connection**. This special feature highlights a research paper in human genetics that reports a key experiment or raises important social, ethical, or legal issues. Each **Human Connection** has a brief introduction of its own, explaining the importance of the experiment and the context in which it

was carried out. At the end of each chapter is a complete **Chapter Summary** in the form of bullet points highlighting the most important concepts.

Each chapter also includes several different types and levels of **Problems**, including concept, synthesis,

---

LEARNING OBJECTIVES

- To understand how genetic information is stored in the base sequence of DNA. For a given sequence of bases in a transcribed strand of protein-coding DNA, you will specify the sequence of bases in the corresponding region of messenger RNA and the sequence of amino acids in the protein. For a mutation in which a specified base is replaced with another, you will deduce the resulting mRNA and protein sequence.

- To realize that enzymes work in sequence in a metabolic pathway. Given a linear metabolic pathway for an essential nutrient, you will deduce which intermediates will restore the ability to grow mutant strains that are defective for any of the enzymes in the pathway. Conversely, using data that specify which intermediates in a linear metabolic pathway restore the ability of mutants to grow, you will infer the order of the enzymes and intermediates in the pathway.

- To learn that genetic complementation is the operational definition of a gene. Given data on the complementation or lack of complementation among all pairs of a set of mutations affecting a biological process, you will sort the mutations into complementation groups, each corresponding to a different gene.

---

and discussion questions in the form of **Issues and Ideas**, a guide to problem solving called **Solutions: Step By Step**, and application and analysis problems designated **Concepts In Action**.

At the end of the book are **Answers** to even-numbered problems, a complete **Glossary** of key terms in genetics, and a compilation of frequently used **Word Roots** that will help students to understand key genetic terms and make them part of their vocabulary. Answers to odd-numbered problems will be available to instructors.

Contents

The organization and number of chapters in the *Seventh Edition* have been retained because they appeal to the majority of instructors who teach genetics. An important feature is the presence of an introductory chapter providing a broad overview of the gene: what it is, what it does, how it changes, how

---

KEY CONCEPT

At any position on the paired strands of a DNA molecule, if one strand has an A, then the partner strand has a T; and if one strand has a G, then the partner strand has a C.

---

it evolves. Today, most students learn about DNA in grade school or high school. In my teaching, I have found it rather artificial to pretend that DNA does not exist until the middle of the term. The introductory chapter, therefore, serves to connect the more advanced concepts that students are about to learn with what they already know. It also serves to provide each student with a solid framework for integrating the material that comes later. Throughout each chapter, there is a balance between challenge and motivation, between observation and theory, and between principle and concrete example. Molecular and classical genetics are integrated throughout, and the principles of human genetics are interwoven with the entire fabric of the book. On the other hand, the book is also liberally supplied with examples from animals and plants, especially model organisms. Several points related to organization and coverage should be noted:

CHAPTER SUMMARY

- Inherited traits are affected by genes.
- Genes are composed of the chemical deoxyribonucleic acid (DNA).
- DNA replicates to form (usually identical) copies of itself.
- DNA contains a code specifying what types of enzymes and other proteins are made in cells.
- DNA occasionally mutates, and the mutant forms specify altered proteins.
- A mutant enzyme is an "inborn error of metabolism" that blocks one step in a biochemical pathway for the metabolism of small molecules.
- Genetic analysis of mutants of the fungus *Neurospora* unable to synthesize an essential nutrient led to the one gene–one enzyme hypothesis.

- Different mutations in the same gene can be identified by means of a complementation test, in which the mutants are brought together in the same cell or organism. Mutations in the same gene fail to complement one another, whereas mutations in different genes show complementation.
- Most traits are complex traits affected by multiple genes as well as by environmental factors.
- Organisms change genetically through generations in the process of biological evolution.
- Because of their common descent, organisms share many features of their genetics and biochemistry.

Chapter Summary: Summary of overall concepts discussed in chapter.

ISSUES AND IDEAS

- What special feature of the structure of DNA allows each strand to be replicated without regard to the other?
- What does it mean to say that a strand of DNA specifies the structure of a molecule of RNA?
- What types of RNA participate in protein synthesis, and what is the role of each type of RNA?
- What is meant by the phrase *genetic code*, and how is the genetic code relevant to the translation of a

polypeptide chain from a molecule of messenger RNA?
- What is meant by the term *genetic analysis*, and how is genetic analysis exemplified by the work of Beadle and Tatum using *Neurospora*?
- What is a complementation test, and what is it used for in genetic analysis?

Issues and Ideas: Questions asking for genetic principles to be restated in the student's own words.

Solutions: Step by Step: A section that demonstrates problems worked in full, explaining step by step a path of logical reasoning that can be followed to analyze the problem.

SOLUTIONS: STEP BY STEP

PROBLEM 1 In the human gene for the beta chain of hemoglobin, the oxygen-carrying protein in the red blood cells, the first 30 nucleotides in the protein-coding region are as shown here.

3'-TACCACGTGGACTGAGGACTCCTCTTCAGA-5'

(a) What is the sequence of the partner strand?
(b) If the DNA duplex of this gene were transcribed from left to right, what is the base sequence of the RNA across this part of the coding region?
(c) What is the sequence of amino acids in this part of the beta-globin polypeptide chain?
(d) In the mutation responsible for sickle-cell anemia, the red T indicated is replaced with an A. The mutant is present at relatively high frequency in some human populations because carriers of the gene are more resistant to falciparum malaria than are noncarriers. What is the amino acid replacement associated with this mutation?

SOLUTION. (a) The partner strand is deduced from the rule that A pairs with T and G pairs with C; however, keep in mind that the paired DNA strands have opposite polarity (that is, their 5'-to-3' orientations are reversed). (b) The RNA strand is synthesized in the 5'-to-3'

direction, which means that the template DNA strand is transcribed in the 3'-to-5' direction, which happens to be the same left-to-right orientation of the strand shown above. The base sequence is deduced from the usual base-pairing rules, except that A in DNA pairs with U in RNA. (c) The polypeptide chain is translated in successive groups of three nucleotides (each group constituting a codon), starting at the 5' end of the coding sequence in the RNA and moving in the 5'-to-3' direction. The amino acid corresponding to each codon can be found in the genetic code table. (d) The change from T to A in the transcribed strand alters a GAG codon into a GUG codon in the RNA transcript, resulting in the replacement of the normal glutamic acid (GAG) with valine (V). The nonmutant duplex, the RNA transcript, and the amino acid sequence are as shown below. The amino acid that is replaced in the sickle-cell mutant is indicated in red.

3'-TACCACGTGGACTGAGGACTCCTCTTCAGA-5'
5'-ATGGTGCACCTGACTCCTGAGGAGAAGTCT-3'

5'-AUGGUGCACCUGACUCCUGAGGAGAAGUCU-3'
　MetValHisLeuThrProGluGluLysSer

CONCEPTS IN ACTION: PROBLEMS FOR SOLUTION

1.1 Prior to the Avery, MacLeod, and McCarty experiment, what features of cells and chromosomes were already known that could have been interpreted as evidence that DNA is an important constituent of the genetic material?

1.2 In the early years of the twentieth century, why did most biologists and biochemists believe that proteins were probably the genetic material?

1.3 From their examination of the structure of DNA, what were Watson and Crick able to infer about the probable mechanisms of DNA replication, coding capability, and mutation?

1.4 What are three principal structural differences between RNA and DNA?

1.5 A region along an RNA transcript contains no U. What base will be missing in the corresponding region of the template strand of DNA?

1.6 When the base composition of a DNA sample from the bacterium *Salinicoccus roseus* was determined, 23.6 percent of the bases were found to be guanine. The DNA of this organism is known to be double stranded. What is the percentage of adenine in its DNA?

1.7 DNA extracted from a certain virus has the following base composition: 15 percent adenine, 25 percent thymine, 20 percent guanine, and 40 percent cytosine. How would you interpret this result in terms of the structure of the viral DNA?

1.8 A duplex DNA molecule contains 532 occurrences of the dinucleotide 5'-GT-3' in one or the other of the paired strands. What other dinucleotide is also present exactly 532 times?

1.9 A repeating polymer with the sequence

5'-GAUGAUGAUGAU . . .-3'

was found to produce only two types of polypeptides in a translation system that uses cellular components but not living cells (called an *in vitro*

translation system). One polypeptide consisted of repeating Asp and the other of repeating Met. How can you explain this result?

1.10 If one strand of a DNA duplex has the sequence 5'-GTCAT-3', what is the sequence of the complementary strand? (Write the answer with the 5' end at the left.)

1.11 Consider a region along one strand of a double-stranded DNA molecule consists of tandem repeats of the trinucleotide 5'-CTA-3', so that the sequence in this strand is 5'-CTACTACTACTA . . .-3'. What is the sequence in the other strand? (Write the answer with the 5' end at the left.)

1.12 Part of the protein-coding region in a gene has the base sequence 3'-ACAGCATAAACGTTC-5'. What is the sequence of the partner DNA strand?

1.13 If the DNA sequence in Problem 1.12 is the template strand that is transcribed in the synthesis of messenger RNA, would it be transcribed from left to right or from right to left? What base sequence would this region of the RNA contain?

1.14 What amino acid sequence would be synthesized from the messenger RNA region in Problem 1.12?

1.15 If a mutation occurs in the DNA sequence in Problem 1.12 in which the red C is replaced with a T, what amino acid sequence would result?

1.16 A polymer is made that has a random sequence consisting of 25 percent U's and 75 percent C's. Among the amino acids in the polypeptide chains resulting from in vitro translation, what is the expected frequency of Pro? Of Phe?

1.17 With *in vitro* translation of an RNA into a polypeptide chain, the translation can begin anywhere along the RNA molecule. A synthetic RNA molecule has the sequence

5'-CGCUUACCACAUGUCGCGAAC-3'

Concepts in Action: Problems for Solution: Problems that require the student to reason using genetic concepts. The problems make use of a variety of formats, and many require some numerical calculation.

Chapter 1 is an overview of genetics designed to bring students with disparate backgrounds to a common level of understanding. This chapter enables classical genetics, molecular genetics, evolutionary genetics, and genomics to be integrated throughout the rest of the book. Included in Chapter 1 are the basic concepts of genetics: genes as regions of DNA that function through transcription and translation, that change by mutation, and that affect organisms through inborn errors of metabolism. Chapter 1 also explains that most traits are actually complex traits affected by multiple genetic and environmental factors, and it introduces genomics and proteomics.

Chapters 2 through 5 are the core of Mendelian genetics, including segregation and independent assortment, the chromosome theory of heredity, mitosis and meiosis, linkage and chromosome mapping, tetrad analysis in fungi, and chromosome mechanics. An important principle of genetics, too often ignored or given inadequate treatment, is that of the complementation test and how complementation differs from segregation or other genetic principles. Chapter 4 expands on the use of molecular markers in genetics, because these are the principal types of genetic markers in use today.

Chapter 6 deals with DNA, including the details of DNA structure and replication. It also discusses how basic research that revealed the molecular mechanisms of DNA replication ultimately led to such important practical applications as DNA hybridization analysis, DNA sequencing, and the polymerase chain reaction. These examples illustrate the value of basic research in leading, often quite unpredictably, to practical applications.

Chapter 7 deals with the principles of genetics in prokaryotes, beginning with the genetics of mobile DNA, plasmids, and integrons, and their relationships to the evolution of multiple antibiotic resistance. There is a thorough discussion of mechanisms of genetic recombination in microbes, including transformation, conjugation, and transduction, as well as a discussion of temperate and virulent bacteriophages.

Chapters 8 through 12 focus on molecular genetics in the strict sense. Chapter 8 examines the details of gene expression, including transcription, RNA processing, and translation.

Chapter 9 is an integrative chapter that deals with genetic mechanisms of regulation, with examples of mechanisms of gene regulation in prokaryotes as well as eukaryotes. Broader aspects of gene regulation that are topics of much current research, such as higher-order chromatin organization, imprinting, and RNAi are included.

Chapter 10 deals with high-throughput genome sequencing and its implications for personalized medicine and the ethical, legal, and social implications of this technology. It also includes basic methods of recombinant DNA, and there is a major new section of CRISPR/Cas9 in DNA editing and its application to genetic engineering.

Chapter 11 examines the genetic control of development with emphasis on models in *C. elegans*, *D. melanogaster*, and *A. thaliana*.

Chapter 12 focuses on mechanisms of mutation and DNA repair, including chemical mutagens.

Chapter 13 stresses cancer from the standpoint of the genetic control of the cell cycle, with emphasis on the checkpoints that, in normal cells, result either in inhibition of cell division or in programmed cell death (apoptosis). Cancer results from a series of successive mutations, usually in somatic cells, which overcome the normal checkpoints that control cellular proliferation.

Chapters 14 and 15 deal with molecular evolution and population genetics. The discussion includes gene trees and species trees and the population genetics of the CCR5 receptor mutation that confers resistance to infection by HIV. It also includes DNA typing in criminal investigations, paternity testing, the effects of inbreeding, and the evolutionary mechanisms that drive changes in allele frequency. The approach to quantitative genetics includes a discussion of how particular genes influencing quantitative traits (QTLs, or quantitative-trait loci) may be identified and mapped by linkage analysis. There is also a section on what has been learned from genome-wide association studies of complex traits in humans, including the identification of QTLs through genetic mapping or studies of candidate genes.

The Student Experience

Stop & Think

A unique feature of this book is found in boxes called **Stop & Think**. These are problems that ask

STOP & THINK 4.3

Yeast cells of genotype *A b* are crossed with those of genotype *a B*. Among 170 unordered tetrads that were analyzed, the following numbers were observed of each type.

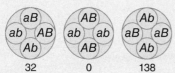

aB ab AB Ab	AB ab ab AB	Ab aB aB Ab
32	0	138

Based on these data, what is the map distance between the genes?

a student to pause and think about a concept and apply it to an actual situation. Often these problems use the results of classical experiments to help the student transform a concept from abstract to concrete and carry it from thought to action. Current pedagogy places great emphasis on assessments, and assessments are especially important in genetics because there are many different concepts to be mastered individually as well as in combination. Today's students (and their teachers, too) are often distracted by neighbors, background noise, text messages, email, and all the other disturbances and diversions of the modern world. Three to five **Stop & Think** pauses at strategic points in each chapter encourage students to verify their own understanding of a concept just explained and how to use it to solve an actual problem. The answers are provided at the end of each chapter.

The Human Connection

the human connection in each chapter is our way of connecting to the world of human genetics outside the classroom. All the connections include short excerpts from the original literature of genetics, usually papers, each introduced with a short explanatory

THE HUMAN CONNECTION

Double Trouble

Andrew Fire,[1] SiQun Xu,[1] Mary K. Montgomery,[1] Steven A. Kostas,[1] Samuel E. Driver,[2] and Craig C. Mello[2] (1998)
[1]Carnegie Institution of Washington, Baltimore, Maryland; [2]University of Massachusetts Medical School, Worcester, Massachusetts.

Potent and Specific Genetic Interference by Double-Stranded RNA in Caenorhabditis elegans

Weird and unexpected results began to be reported as soon as it became possible to introduce engineered RNA molecules into organisms. In extreme cases, the engineered RNA prevented the expression of endogenous host genes with sequence homology. At first, it seemed possible that the engineered RNA acted as an antisense inhibitor, in which the introduced RNA undergoes base pairing with the endogenous transcripts and interferes with their function. If this were true, the inhibitory effect of the introduced RNA should be strongly concentration dependent. In this path-breaking paper, the authors show that introduced double-stranded RNA (dsRNA) mediates the inhibitory effects, and that only a few molecules per cell are required. The nematode worm *C. elegans* proved to be ideal for these experiments because, in contrast to some other organisms, dsRNA can be transported from cell to cell and from parent to offspring.

Experimental introduction of RNA into cells can be used in certain biological systems to interfere with the function of an endogenous gene.... Here we investigate the requirements for structure and delivery of the interfering RNA. To our surprise, we found that double-stranded RNA was substantially more effective at producing interference than was either strand individually.... Only a few molecules of injected double-stranded RNA were required per affected cell, ... suggesting that there could be a catalytic or amplification component of the interference process....

> **To our surprise, we found that double-stranded RNA was substantially more effective at producing interference than was either strand individually.**

Fire and his colleagues looked more closely at this phenomenon, concentrating on the *unc-22* (uncoordinated-22) gene, loss-of-function mutations of which cause severe twitching in the worms. When they injected single-stranded RNA either identical or complementary to *unc-22* mRNA, only minimal interference was observed.

In contrast, a sense–antisense mixture produced highly effective interference with endogenous gene activity. The mixture was at least two orders of magnitude more effective than either strand alone. . . . The potent interfering activity of the sense–antisense mixture could reflect the formation of double-stranded RNA (dsRNA) or, conceivably, some other synergy between the strands....

The phenotype induced by the introduced RNA was identical to that of conventional loss-of-function mutations of *unc-22*. They concluded by suggesting that RNA interference might be a more general phenomenon.

Double-stranded RNA could conceivably mediate interference more generally in other nematodes, in other invertebrates, and, potentially, in vertebrates. RNA interference might also operate in plants. . . . Genetic interference by dsRNA could be used by the organism for physiological gene silencing.

A. Fire, et al. *Nature* 391(1998): 806–810.

passage. Many of the connections are excerpts from classic materials, such as Allison's work on the sickle-cell trait and resistance to malaria, but by no means are all the "classic" papers old papers. The pieces are called **the human connection** because each connects the material to something that broadens or enriches its implications for human beings. Some of the connections raise issues of ethics in the application of genetic knowledge, social issues that need to be addressed, or issues related to laboratory animals.

They illustrate other things as well. Because each connection names the place where the research was carried out, the student will see that great science is done in many universities and research institutions throughout the world. In papers that use outmoded or unfamiliar terminology, or archaic gene symbols, I have substituted the modern equivalent to make the material more accessible to the student.

Solutions Step by Step

Each chapter contains a section titled **Solutions: Step by Step** that demonstrates problems worked in full, explaining step by step a path of logical reasoning that can be followed to analyze the problem. The **Solutions: Step by Step** serve as another level of review of the important concepts used in working problems. The solutions also emphasize some of the most common mistakes made by beginning students and give pointers on how students can avoid falling into these conceptual traps.

Levels and Types of Problems

Each chapter provides numerous problems for solution, graded in difficulty, so students can test their understanding. The problems are of two different types:

- *Issues and Ideas* ask for genetic principles to be restated in the student's own words; some are matters of definition or call for the application of elementary principles.

- *Concepts in Action* are problems that require the student to reason using genetic concepts. The problems make use of a variety of formats, and many require some numerical calculation. The level of mathematics is that of arithmetic and elementary probability as it pertains to genetics. None of the problems uses mathematics beyond elementary algebra. The problems range in difficulty from easy to hard. They are primarily at **Bloom's higher order cognitive level**, and most require **analyzing** data, **evaluating** evidence, or **creating** hypotheses or experiments.

Answers to Problems

The answers to the even-numbered **Concepts in Action** are included in the **Answer** section at the end of the book. The answers are complete; they explain the logical foundation of the solution and lay out the methods. The answers to the remainder of the **Concepts in Action** problems are available with the online instructor's resources and in the optional online Study Guide and Solutions Manual.

5.6 The observation is quite unexpected, because a 47, XXX female would be expected to produce many XX-bearing eggs, and a 47, XYY male would be expected to produce many XY-bearing sperm. Apparently, the extra X chromosome in 47, XXX females, and the extra Y chromosome in 47, XYY males, are eliminated from the nucleus prior to

Word Roots and Glossary

I have included a compilation of **Word Roots** that students find helpful in interpreting and remembering the meaning of technical terms. This precedes the **Glossary** of key words.

ante-	*preceding, before*	antedate, preceding a date
apo-	*former, from*	aporepressor, precursor to repressor
aut-, auto-	*self*	autogenous, self-generated
bi-	*two*	bidirectional, going in two directions

Illustrations

Every chapter is richly illustrated with beautiful graphics in which color is used functionally to enhance the value of each illustration as a learning aid. The illustrations are also heavily annotated with "process boxes" explaining step by step what is happening at each level of the illustration. These labels make the art user friendly, inviting, and maximally informative.

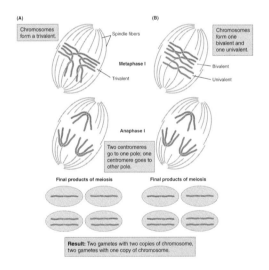

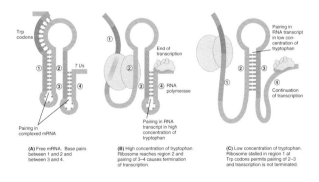

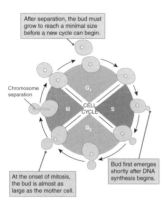

After separation, the bud must grow to reach a minimal size before a new cycle can begin.

Chromosome separation

CELL CYCLE

G_1
S
G_2
M

At the onset of mitosis, the bud is almost as large as the mother cell.

Bud first emerges shortly after DNA synthesis begins.

Teaching Tools

Adaptability and Flexibility

There is no compelling reason to start at the beginning and proceed straight to the end of this text. Each chapter is a self-contained unit that stands on its own. This feature gives the book the flexibility to be used in a variety of course formats. Throughout the book, we have integrated classical and molecular principles, so you can begin a course with almost any of the chapters. Most teachers will prefer starting with the overview in Chapter 1, possibly as suggested reading, because it brings every student to the same basic level of understanding. Teachers preferring to cover Mendel early should continue with Chapter 2; those preferring to teach the details of DNA early should continue with Chapter 6. Some teachers are partial to a chromosomes-early format, which would suggest continuing with Chapter 3, followed by Chapters 2 and 4. A novel approach would put genomics first, which could be implemented by continuing with Chapter 10. The writing and illustration programs were designed to accommodate a variety of formats, and we encourage teachers to take advantage of this flexibility to meet their own needs.

Instructor Resources

An unprecedented offering of traditional and interactive multimedia supplements is available to assist instructors and aid students in mastering genetics. Additional information and review copies of any of the following items are available through your Jones & Bartlett Learning sales representative.

The *Image Bank in PowerPoint format* provides all the illustrations and photos (to which Jones & Bartlett Learning owns the copyright or has permission to reprint digitally), inserted into PowerPoint slides. With the Microsoft® PowerPoint program you can quickly and easily copy individual image slides into your existing lecture slides.

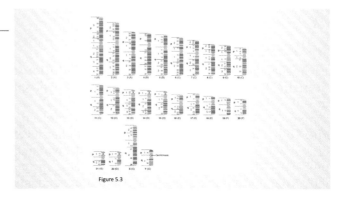

Figure 5.3

A *Table Bank* provides images of all of the tables (to which Jones & Bartlett Learning owns the copyright or has permission to reprint digitally), in a PDF file for easy use.

TABLE 9.1 Characteristics of Partial Diploids Containing Several Combinations of *lacI*, *lacO*, and *lacP* Alleles

Genotype	Synthesis of *lac* mRNA	Lac phenotype
1. F' *lacO* *lacZ*⁺/*lacO*⁺ *lacZ*⁺	Constitutive	+
2. F' *lacO*⁺ *lacZ*⁺/*lacO* *lacZ*⁺	Constitutive	+
3. F' *lacI*⁻ *lacZ*⁺/*lacI*⁺ *lacZ*⁺	Inducible	+
4. F' *lacI*⁺ *lacZ*⁺/*lacI*⁻ *lacZ*⁺	Inducible	+
5. F' *lacO* *lacZ*⁻/*lacO*⁺ *lacZ*⁺	Inducible	+
6. F' *lacO* *lacZ*⁺/*lacO*⁺ *lacZ*⁻	Constitutive	+
7. F' *lacI* *lacZ*⁺/*lacI*⁺ *lacZ*⁺	Uninducible	−
8. F' *lacI*⁺ *lacZ*⁺/*lacI* *lacZ*⁺	Uninducible	−
9. F' *lacP*⁻ *lacZ*⁺/*lacP*⁺ *lacZ*⁺	Inducible	+
10. F' *lacP*⁺ *lacZ*⁺/*lacP*⁻ *lacZ*⁺	Inducible	+
11. F' *lacP*⁺ *lacZ*⁻/*lacP*⁻ *lacZ*⁺	Uninducible	−
12. F' *lacP*⁺ *lacZ*⁺/*lacP*⁻ *lacZ*⁻	Inducible	+

A set of *Lecture Outlines in PowerPoint format* provides outline summaries of each chapter. The slide set can be customized to meet your classroom needs.

DNA is a double helix

- DNA backbone forms right-handed helix

- Each DNA strand has polarity = directionality

- The paired strands are oriented in opposite directions = antiparallel

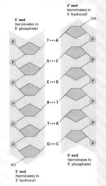

Figure 06.06: A segment of a DNA molecule showing the antiparallel orientation of the complementary strands.

© Eky Studio/Shutterstock. Copyright © 2019 by Jones & Bartlett Learning, LLC an Ascend Learning Company www.jblearning.com

The *Test Bank* contains over 700 test items. There is a mix of factual, descriptive, analytical, and quantitative question types. A typical chapter file contains 20 multiple-choice objective questions, 15 fill-in-the-blank questions, and 15 quantitative problems. Versions easily compatible with most course management software are available to adopting instructors upon request.

5. The R-type strain of *S. pneumoniae* does not cause pneumonia because bacterial cells are
A. Surrounded by a polysaccharide capsule
B. Unable to synthesize a polysaccharide capsule
C. Unable to form colonies
D. Undetectable by the immune system

Ans: B

6. A DNA strand consists of any sequence of four kinds of nucleotides. Suppose there were only 16 different amino acids instead of 20. Which of the following statements would be correct descriptions of the minimal number of nucleotides necessary to create a genetic code?
A. 1
B. 2, provided that chain termination does not require a special codon
C. 3, provided that chain termination does require a special codon
D. 2, no matter how chain termination is accomplished
E. Both B and C

Ans: E

7. tNRA is
A. The major structural material making up ribosomes
B. A molecule that incorporates a specific amino acid into the growing protein when it recognizes a specific group of three bases
C. The major structural component of chromosomes
D. The molecule that carries the genetic information from DNA and is used as a template for protein synthesis
E. The major building block of proteins

Ans: B

8. rRNA is
A. The major structural material making up ribosomes
B. The molecule that carries the genetic information from DNA and is used as a template for protein synthesis

2

Readiness Assessment and Readiness Review

How ready are you to learn introductory genetics? For the first time we are offering a Readiness Assessment for you to find out. Simply answer the online questions found within Navigate 2 (accessible via the access code in the front of the text*). Once complete, you will be given a score and directed to the color-coded in-text Readiness Review section(s) that will most help you prepare (found at the back of the text). Never has learning genetics been so easy and fun!

STEP 1: Redeem your code.

STEP 2: Take the quiz.

STEP 3: Get your score.

STEP 4: Learn, review, and practice!

Problems

Problem 2.1 Calculate the percentage and ratios of each continent's population to that of the world.

Continent	Proportion (percent)	Calculation	Ratio (Continent : World)	Calculation
Asia				
Africa				
Europe				
North America	0.077 (7.7%)		1:13	
South America				
Australia/ Oceania				
Antarctica				

Problem 2.2 What is the ratio of the population of Asia to North America?

_____

Problem 2.3 What is the ratio of the area of Asia to North America?

_____

Problem 2.4 What is the ratio of the population of Africa to Europe?

_____

Problem 2.5 What is the ratio of the area of Africa to Europe?

_____

STEP 5: Retake the quiz to check for improvement.

*Access can also be purchased separately. Visit go.jblearning.com/hartl7e to learn more.

Acknowledgments

I am indebted to my colleagues whose advice and thoughts were immensely helpful throughout the preparation of the five editions of this book. These colleagues range from specialists in various aspects of genetics who checked for accuracy or suggested improvement to instructors who evaluated the material for suitability in teaching or sent me comments on the text as they used it in their courses.

Laura Adamkewitz, George Mason University
Jeremy C. Ahouse, Brandeis University
Mary Alleman, Duquesne University
Jennifer Armstrong, Scripps, Pitzer, and Claremont McKenna Colleges
Peter D. Ayling, University of Hull
John C. Bauer, Stratagene, Inc., La Jolla, CA
Anna C. Berkowitz, Purdue University
Mary K. B. Berlyn, Yale University
Thomas A. Bobik, University of Florida
Carrie Baker Brachmann, University of California, Irvine
Jessica Brzyski, Seton Hill University
Colin G. Brooks, The Medical School, Newcastle
Mirella Vasquez Brooks, University of Hawaii at Manoa
Jill A. Buettner, Richland College
Jeffrey J. Byrd, St. Mary's College of Maryland
Susan L. Carney, Hood College
Pierre Carol, Université Joseph Fourier
Domenico Carputo, University of Naples
Sean Carroll, University of Wisconsin
Chris Caton, University of Birmingham
John Celenza, Boston University
Richard W. Cheney, Jr., Christopher Newport University
Alan C. Christiensen, University of Nebraska, Lincoln
Michael J. Christoffers, North Dakota State University
Erin Cram, Northeastern University
Christoph Cremer, University of Heidelberg
Marion Cremer, Ludwig Maximilians University
Thomas Cremer, Ludwig Maximilians University
Leslie Dendy, University of New Mexico
John W. Drake, National Institute of Environmental Health Sciences, Research Triangle Park, NC

Stephen J. D'Surney, University of Mississippi
Kathleen Dunn, Boston College
Chris Easton, State University of New York
David Eisenmann, University of Maryland, Baltimore County
Wolfgang Epstein, University of Chicago
Silviu Faitar, D'Youville College
Brian E. Fee, Manhattan College
Gyula Ficsor, Western Michigan University
Robert G. Fowler, San Jose State University
David W. Francis, University of Delaware
Gail Gasparich, Towson University
Elliott S. Goldstein, Arizona State University
Ruth Grene, Virginia Tech
Patrick Guilfoile, Bemidji State University
Jeffrey C. Hall, Brandeis University
Mark L. Hammond, Campbell University
Randall K. Harris, William Carey University
Steven Henikoff, Fred Hutchinson Cancer Research Center, Seattle, WA
Charles Hoffman, Boston College
Ivan Huber, Fairleigh Dickinson University
Kerry Hull, Bishop's University
Lynn A. Hunter, University of Pittsburgh
Richard Imberski, University of Maryland
Bradley J. Isler, Ferris State University
Diana Ivankovic, Anderson University
Joyce Katich, Monsanto, Inc., St. Louis, MO
Jeane M. Kennedy, Monsanto, Inc., St. Louis, MO
Jeffrey King, University of Berne
Tobias A. Knoch, German Cancer Research Center, Heidelberg, Germany
Laszlo Kovacs, Missouri State University
Yan B. Linhart, University of Colorado
K. Brooks Low, Yale University
Sally A. MacKenzie, Purdue University
Gustavo Maroni, University of North Carolina
Jeffrey Mitton, University of Colorado
Sara Morris, Biology MPCC, North Platte, NE
Robert K. Mortimer, University of California
Gisela Mosig, Vanderbilt University
John R. Nambu, Florida Atlantic University
Steve O'Brien, National Cancer Institute

Kevin O'Hare, Imperial College, London

Michael V. Osier, Rochester Institute of Technology

Catherine A. Palmer, Portland State University

Ronald L. Phillips, University of Minnesota

Jennifer R. Powell, Gettysburg College

Robert Pruitt, Purdue University

Peggy Redshaw, Austin College

Pamela Reinagel, California Institute of Technology

Susanne Renner, University of Missouri

Lynn S. Ripley, University of Medicine and Dentistry of New Jersey

Andrew J. Roger, Dalhousie University, Halifax

Moira E. Royston, St. Joseph's College, NY

Kenneth E. Rudd, National Library of Medicine, Bethesda, MD

Mary Russell, Kent State University at Trumbull

Thomas F. Savage, Oregon State University

Joseph Schlammadinger, University of Debrecen

Brian W. Schwartz, Columbus State University

David Shepard, University of Delaware

Alastair G. B. Simpson, Dalhousie University

Leslie Smith, National Institute of Environmental Health Sciences, Research Triangle Park, NC

Charles Staben, University of Kentucky

Julie Dangremond Stanton, Washington State University

Johan H. Stuy, Florida State University

David T. Sullivan, Syracuse University

Jeanne Sullivan, West Virginia Wesleyan College

Millard Susman, University of Wisconsin

Fusheng Tang, University of Arkansas, Little Rock

Irwin Tessman, Purdue University

James H. Thomas, University of Washington

Michael Thomas, Idaho State University

Jan Trybula, State University of New York, College at Potsdam

Michael Tully, University of Bath

L.K. Tuominen, John Carroll University

David Ussery, The Technical University of Denmark

George von Dassow, Friday Harbor Laboratories, Friday Harbor, WA

Denise Wallack, Muhlenberg College

Kenneth E. Weber, University of Southern Maine

Tamara Western, Okanagan University College

Taek H. You, Campbell University

Finally, a very special thanks goes to geneticist Elena R. Lozovsky of Harvard University, who contributed substantially to the instructional materials at the end of each chapter, as well as to the instructor and student supplements. Elena's help and support is very gratefully acknowledged. I also wish to acknowledge the superb art, production, and editorial staff at Jones & Bartlett Learning who helped make this book possible: Matt Kane, Audrey Schwinn, Loren-Marie Durr, John Rusk, Troy Liston, Dan Stone, and Kristin Parker. I am also grateful to the many people, acknowledged in the legends of the figures, who contributed photographs, drawings, and micrographs from their own research and publications. Every effort has been made to obtain permission to use copyrighted material and to make full disclosure of its source. We are grateful to the authors, journal editors, and publishers for their cooperation. Any errors or omissions are wholly inadvertent and will be corrected at the first opportunity.

We would also like to thank and acknowledge Dr. Bruce Cochrane for his work updating the instructor's resources and Dr. Rebecca Reiss for her work on the assessments, including the online Readiness Assessment and the Readiness Review found in the back of the book.

Dan Hartl

About the Author

Daniel L. Hartl is Higgins Professor of Biology at Harvard University, a Professor of Immunology and Infectious Diseases at the Harvard T. H. Chan School of Public Health, and a Senior Associate Member of the Broad Institute of M.I.T. and Harvard. He is a member of the National Academy of Sciences and the American Academy of Arts and Sciences. Hartl received his B.S. degree and Ph.D. from the University of Wisconsin and carried out postdoctoral research at the University of California at Berkeley. His research interests include molecular genetics, genomics, molecular evolution, and population genetics.

For the Student

Special features designed to help in mastering the material are emphasized in the **Preface.** In my experience, students who struggle in genetics and genomics do so for two reasons. They may lack effective reading and study habits, or they may fail to self-assess. By self-assessment I mean asking yourself whether you understand a concept well enough to express it in your own words, and whether you understand it well enough to use in solving problems. Here are some pointers for improving both study skills and problem solving.

Tips for Learning Concepts

- Go to class, take notes by hand (in telegraphic style, abbreviating as needed), and copy your notes in complete sentences and legible handwriting as soon as possible thereafter. Words written by hand are retained in memory better than words typed on a keyboard.

- Plan 30–35 minute reading sessions, but only when you are not tired or distracted. Most people find that their attention begins to wane after 30–35 minutes of intense concentration.

- Read attentively. Find a quiet, clean, well-lighted place and turn off your laptop and smartphone.

- Start by skimming what you think you can cover in your reading session, including a preliminary look at the illustrations; this is your drone's eye view of the terrain that will help keep you oriented.

- Look up unfamiliar words in the glossary.

- Reread difficult sections and make handwritten notes of the key points.

- Highlight, underline, or better yet summarize the key concepts in your own words. The textbook is designed to help in recognizing these: key terms are in **boldface,** key points are highlighted with **bulleted lists,** and key concepts are set off and labeled **Key Concept.**

- Take a break from reading to solve the problems in **Stop & Think.** They are designed to help you assess whether you have understood the concepts you've just read well enough to apply them.

Tips for Problem Solving

- Make use of the **Solutions: Step By Step.** These guide you through the reasoning used to solve the major types of problems arising from the concepts in each chapter.

- Don't start working a problem until you're sure you understand what is being asked.

- Use the glossary if necessary to understand the key terms in a problem.

- Start with some easy problems to gain self-confidence.

- Once you know how to solve a certain type of problem, don't spend time on similar ones that you already know how to solve. It's problems that you don't immediately know how to solve that you really learn from.

- Don't rush. Haste makes waste, as the saying goes—and when you rush you are more likely to make stupid mistakes.

- Break a complex problem into smaller parts that you can attack individually, and use the parts you understand as leverage to get at the more difficult parts.

- Don't give up! Never, never, never! You may have to attack a problem from two or three different angles before you find yourself on a productive track.

- Don't work backwards from the answer. There's an old adage that "if you know where you're going you can find a way to get there." The problem is that the way you find to "get there" may use completely messed-up logic. What's worse, you will have trained your neurons to use the wrong logic. Learning a concept is hard enough, unlearning one that you misunderstand is harder still.

- Work in small groups with other students if you can, but be sure you understand the reasoning behind any answer your group comes up with. Sometimes a fellow student can explain a concept more clearly than your instructor.

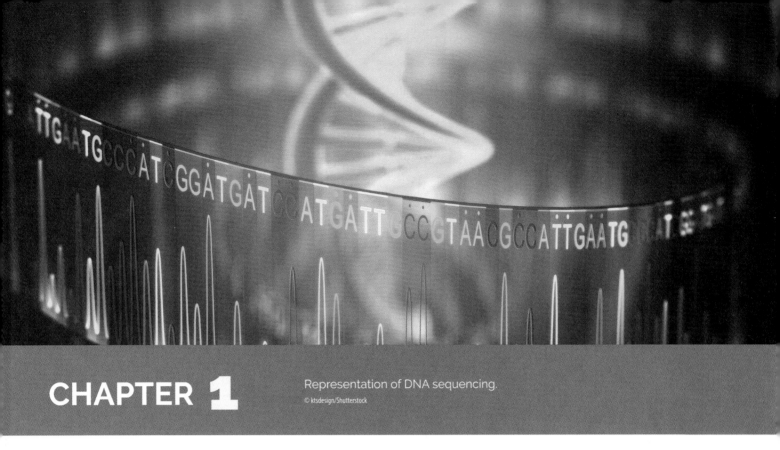

CHAPTER 1

The Genetic Code of Genes and Genomes

LEARNING OBJECTIVES

- To understand how genetic information is stored in the base sequence of DNA. For a given sequence of bases in a transcribed strand of protein-coding DNA, you will specify the sequence of bases in the corresponding region of messenger RNA and the sequence of amino acids in the protein. For a mutation in which a specified base is replaced with another, you will deduce the resulting mRNA and protein sequence.

- To realize that enzymes work in sequence in a metabolic pathway. Given a linear metabolic pathway for an essential nutrient, you will deduce which intermediates will restore the ability to grow mutant strains that are defective for any of the enzymes in the pathway. Conversely, using data that specify which intermediates in a linear metabolic pathway restore the ability of mutants to grow, you will infer the order of the enzymes and intermediates in the pathway.

- To learn that genetic complementation is the operational definition of a gene. Given data on the complementation or lack of complementation among all pairs of a set of mutations affecting a biological process, you will sort the mutations into complementation groups, each corresponding to a different gene.

Genetics is worth studying for many reasons, ranging from applications in medicine, agriculture, public health, and conservation biology to ongoing research in cell biology, development, neuroscience, and evolution. Genetics also deals with some of the great philosophical themes of human life and individual differences:

> A spermatozoon plunges headlong into an ovum, and immediately a long-term project is set in motion. The cells proliferate at a dizzying rate, clustering, diversifying. Out of that initial, infinitesimal particle will grow a beating heart, hands, fingernails, hair, glands, and a brain with the power to think of itself … But from time to time nature, too, gets things wrong, and so you'll have six fingers on one hand or one leg shorter than the other, or else she may construct a brain incapable of understanding the simplest things.

From *Death in August* by Marco Vichi, translated by Stephen Sartarelli (2011, Pegasus Books, p. 123)

There is indeed a developmental plan encoded in our DNA that makes each of us a member of the human species. But still we differ from one another. Besides rare anomalies like six fingers, we differ from one another in many everyday, observable characteristics, or traits, like hair color, eye color, skin color, height, weight, and personality. Some of these traits differ because of heredity, others because of culture. The color of your eyes results from biological inheritance, but the native language you learned as a child results from cultural inheritance. Many traits are influenced jointly by biological inheritance and environmental factors or lifestyle choices. How much you weigh is determined in part by your inheritance but also in part by how much food you eat, its nutritional content, and your exercise habits.

Genetics is the study of biologically inherited traits, including traits that are influenced in part by the environment. **Genomics** is the study of all the genes in an organism to understand their molecular organization, function, interaction, and evolutionary history. The fundamental concept of genetics and genomics is that:

KEY CONCEPT

Inherited traits are determined by **genes**—the elements of heredity that are transmitted from parents to offspring in reproduction.

The existence of genes and the rules governing their transmission from generation to generation were first articulated by Gregor Mendel in 1866. Mendel's formulation of inheritance was in terms of the abstract rules by which genes (he called them "factors") are transmitted from parents to offspring. His objects of study were garden peas, with variable traits like pea color and plant height. The foundation of genetics as a molecular science also dates back to the 1860s when Friedrich Miescher discovered a new type of weak acid, abundant in the nuclei of white blood cells, which turned out to be what we now call **DNA (deoxyribonucleic acid)**. For many years, the biological function of DNA was unknown, and no role in heredity was ascribed to it.

In this book, you will learn a lot about genes and genomes. You will learn what constitutes a gene and how it works in physiology and development, in health and disease. You will also learn how genomes are organized and the activities of different genes coordinated in space and time. If you know nothing about genetics, you will be brought up to speed. And if you already know something, you will see it in a different light. There are lots of details, but try not to get so tangled up in them that you lose sight of how genetics can help you understand the great themes—birth, consciousness, death—that make the details worth knowing.

1.1 DNA is the molecule of heredity.

The importance of the cell nucleus in inheritance became clear in the 1870s, when the nuclei of the male and female reproductive cells were observed to fuse in the process of fertilization. The next major advance was the discovery of **chromosomes**, threadlike objects inside the nucleus that become visible in the light microscope when stained with certain dyes. Chromosomes exhibit a characteristic "splitting" behavior, in which each daughter cell formed by cell division receives an identical complement of chromosomes. More evidence for the importance of chromosomes was provided by the observation that, whereas the number of chromosomes in each cell differs from one biological species to the next, the number of chromosomes is nearly always constant within the cells of any particular species. These features of chromosomes were well understood by about 1900, and they made it seem likely that chromosomes were the carriers of the genes.

By the 1920s, several lines of indirect evidence suggested a close relationship between chromosomes, and DNA. Microscopic studies with special stains showed that DNA is present in chromosomes. Various types of proteins are present in chromosomes, too. But whereas most of the DNA in cells of higher organisms is present in chromosomes, and the amount of DNA per cell is constant, the amount and kinds of proteins and other large molecules differ greatly from one type of cell to another. The indirect evidence for DNA as the genetic material was unconvincing, because crude chemical analyses had suggested (erroneously, as it turned out) that DNA lacked the chemical diversity needed in a genetic substance. The favored candidate for the genetic material was protein, because proteins were known to be an exceedingly diverse collection of molecules. Proteins therefore became widely accepted as the genetic material, and DNA was thought to provide only the structural framework of chromosomes. Any researcher who hoped to demonstrate that DNA was the genetic material had a double handicap. Such experiments had to demonstrate not only that DNA *is* the genetic material but also that proteins are *not* the genetic material. Some of the experiments regarded as decisive in implicating DNA are described in this section.

Genetic traits can be altered by treatment with pure DNA.

One type of bacterial pneumonia in mammals is caused by strains of *Streptococcus pneumoniae* able to synthesize a gelatinous capsule composed of polysaccharide (complex carbohydrate). This capsule surrounds the bacterium and protects it from the defense mechanisms of the infected animal; thus it enables the bacterium to cause disease. When a bacterial cell is grown on solid medium, it undergoes repeated cell divisions to form a visible clump of cells called a **colony**. The enveloping capsule makes the size of each colony large and gives it a glistening or smooth (S) appearance (**FIGURE 1.1**). Certain strains of *S. pneumoniae*, however, are unable to synthesize the capsular polysaccharide, and they form small colonies that have a rough (R) surface. The R strains do not cause pneumonia;

lacking the capsule, these bacteria are inactivated by the immune system of the host. Both types of bacteria "breed true" in the sense that the progeny formed by cell division have the capsular type of the parent, either S or R.

When mice are injected with living R cells or with S cells that have been killed with extreme heat, the animals remain healthy. However, in 1928 Frederick Griffith showed that when mice are injected with a *mixture* of living R cells and heat-killed S cells, they often die of pneumonia (**FIGURE 1.2**). Bacteria isolated from blood samples of the dead mice produce S cultures with a capsule typical of the injected S cells, even though the injected S cells had been killed by heat. Evidently, the injected material from the dead S cells includes a substance that can enter living R bacterial cells and give them the ability to synthesize the S-type capsule. In other words, the R bacteria can be changed—or undergo **transformation**—into S bacteria, and the new characteristics are inherited by descendants of the transformed bacteria.

Griffith's transformation of *Streptococcus* was not in itself definitive, but in 1944 the chemical substance responsible for changing the R cells into S cells was identified as DNA. In a milestone experiment, Oswald Avery, Colin MacLeod, and Maclyn McCarty showed that the substance causing the transformation of R cells into S cells was DNA. In preparation for the experiment, they had to develop chemical procedures for obtaining DNA in almost pure form from bacterial cells, which had not been done before. When they added DNA isolated from S cells to growing cultures of R cells, they observed that a few S-type cells were produced. Although the DNA preparations contained traces of protein and RNA (ribonucleic acid, an abundant cellular macromolecule chemically related to DNA), the transforming activity was not altered by treatments that destroy either protein or RNA. However, treatments that destroy DNA eliminated

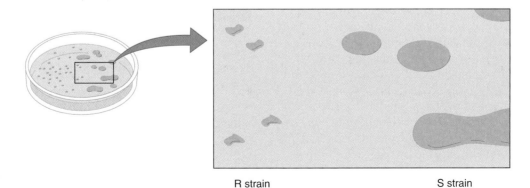

R strain   S strain

FIGURE 1.1 Colonies of *Streptococcus pneumoniae*. The small colonies on the left are from a rough (R) strain, and the large colonies on the right are from a smooth (S) strain. The S colonies are larger because of the capsule on the S cells.

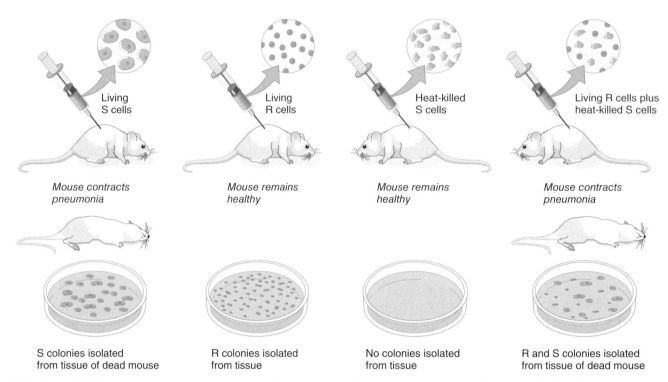

FIGURE 1.2 Griffith's experiment demonstrating bacterial transformation. A mouse remains healthy if injected with either the nonvirulent R strain of *S. pneumoniae* or heat-killed cell fragments of the usually virulent S strain. R cells in the presence of heat-killed S cells are transformed into the virulent strain, causing pneumonia in the mouse.

the transforming activity (**FIGURE 1.3**). These experiments implied that the substance responsible for genetic transformation was the DNA of the cell—and hence that DNA is the genetic material.

Transmission of DNA is the link between generations.

A second pivotal finding was reported by Alfred Hershey and Martha Chase in 1952. They studied cells of the intestinal bacterium *Escherichia coli* after infection by the virus T2. A virus that attacks bacterial cells is called a **bacteriophage**, often shortened to **phage**. (*Bacteriophage* means "bacteria eater.") The T2 particle is exceedingly small, yet it has a complex structure composed of a head containing the phage DNA, a tail, and tail fibers. (The head of a human sperm is about 30–50 times larger in both length and width than the head of T2.) Hershey and Chase were already aware that T2 infection proceeds via the attachment of a phage particle by the tip of its tail to the bacterial cell wall, entry of phage material into the cell, multiplication of this material to form a hundred or more progeny phage, and release of the progeny phage by bursting (lysis) of the bacterial host cell. They also knew that T2 particles are composed of DNA and protein in approximately equal amounts.

The intricate color patterns on butterfly wings demonstrate the complexity that can evolve in developmental processes.

© Ervin Monn/Shutterstock.

Because DNA contains phosphorus but no sulfur, whereas most proteins contain sulfur but no phosphorus, it is possible to label DNA and proteins differentially by the use of radioactive isotopes of the two elements. Hershey and Chase produced particles containing radioactive DNA by infecting *E. coli* cells that had been grown for several generations in a medium that included ^{32}P (a radioactive isotope of phosphorus)

(A) The transforming activity in S cells is not destroyed by heat.

S cells killed by heat

S cell extract (contains mostly DNA with a little protein and RNA)

(B) The transforming activity is not destroyed by either protease or RNase.

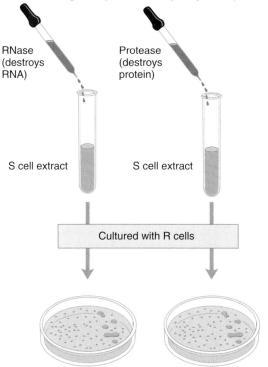

RNase (destroys RNA)

Protease (destroys protein)

S cell extract

S cell extract

Cultured with R cells

In both cases, progeny of R cells produce R colonies and a few S colonies.

Conclusion: Transforming activity is not protein or RNA.

(C) The transforming activity is destroyed by DNase.

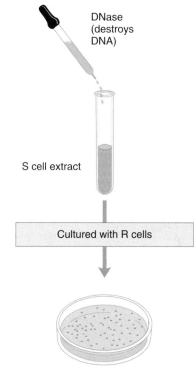

DNase (destroys DNA)

S cell extract

Cultured with R cells

Progeny of R cells produce R colonies only.

Conclusion: Transforming activity is most likely DNA.

FIGURE 1.3 A diagram of the experiment demonstrating that DNA is the active material in bacterial transformation. (A) Purified DNA extracted from heat-killed S cells can convert some living R cells into S cells, but the extract may still contain undetectable traces of protein and/or RNA. (B) The transforming activity is not destroyed by either protease or RNase. (C) The transforming activity is destroyed by DNase and so probably consists of DNA.

and then collecting the phage progeny. They obtained other particles containing labeled proteins in the same way, using medium that included ^{35}S (a radioactive isotope of sulfur).

In the experiments summarized in **FIGURE 1.4**, nonradioactive *E. coli* cells were infected with phage labeled with *either* ^{32}P (part A) *or* ^{35}S (part B) in order to follow the DNA and proteins separately. Infected cells were separated from unattached phage particles by centrifugation, resuspended in fresh medium, and then swirled violently in a kitchen blender to shear attached phage material from the cell surfaces.

This treatment was found to have no effect on the subsequent course of the infection, which implies that the genetic material must enter the infected cells very soon after phage attachment. The kitchen blender turned out to be the critical piece of equipment. Other methods had been tried to tear the phage heads from the bacterial cell surface, but nothing had worked reliably. Hershey later explained, "We tried various grinding arrangements, with results that weren't very encouraging. When Margaret McDonald loaned us her kitchen blender, the experiment promptly succeeded."

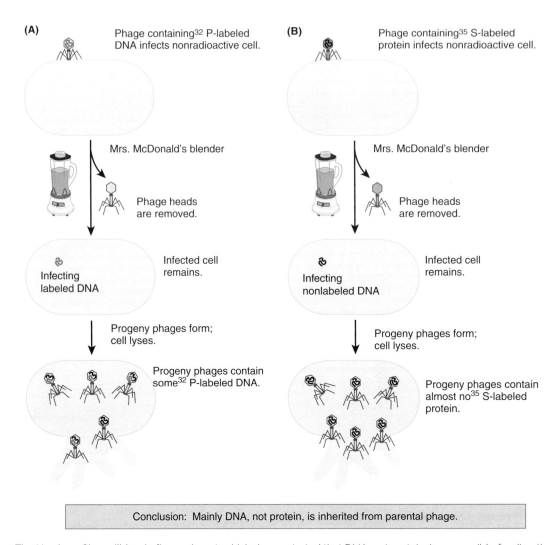

Conclusion: Mainly DNA, not protein, is inherited from parental phage.

FIGURE 1.4 The Hershey–Chase ("blender") experiment, which demonstrated that DNA, not protein, is responsible for directing the reproduction of phage T2 in infected *E. coli* cells. (A) Radioactive DNA is transmitted to progeny phage in substantial amounts. (B) Radioactive protein is transmitted to progeny phage in negligible amounts.

After the phage heads were removed by blending, the infected bacteria were examined. Most of the radioactivity from ^{32}P-labeled phage was found to be associated with the bacteria, whereas only a small fraction of the ^{35}S radioactivity was present in the infected cells. The retention of most of the labeled DNA, contrasted with the loss of most of the labeled protein, implied that a T2 phage transfers most of its DNA, but very little of its protein, to the cell it infects. The critical finding (Figure 1.4) was that about 50 percent of the transferred ^{32}P-labeled DNA, but less than 1 percent of the transferred ^{35}S-labeled protein, was inherited by the *progeny* phage particles. Hershey and Chase interpreted this result to mean that the genetic material in T2 phage is DNA.

The transformation experiment and the Hershey–Chase experiment are regarded as classic demonstration

that genes consist of DNA. At the present time, the equivalent of the transformation experiment is carried out daily in many research laboratories throughout the world, usually with bacteria, yeast, or animal or plant cells grown in culture. These experiments indicate that DNA is the genetic material in these organisms as well as in phage T2.

KEY CONCEPT

There are no known exceptions to the generalization that DNA is the genetic material in all cellular organisms.

It is worth noting, however, that in a few types of viruses, the genetic material consists of the other type of nucleic acid called RNA.

STOP & THINK 1.1

In this diagram of the Hershey–Chase experiment, G_0 represents the original population of bacteriophage with radioactive DNA. The progeny bacteriophage (G_1) showed half the amount of radioactivity in their DNA as the G_0 bacteriophage did.

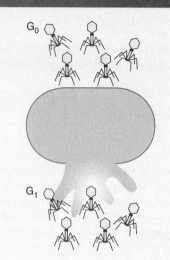

(a) If the G_1 bacteriophage were used to infect bacteria, what fraction of the original G_0 radioactivity would be present in their progeny (the G_2 bacteriophage)?
(b) Suppose that after each cycle of infection, the progeny bacteriophage are used to initiate the next cycle of infection. What fraction of the original G_0 radioactivity would be present in the G_5 bacteriophage?

1.2   The structure of DNA is a double helix composed of two intertwined strands.

Even after it was shown that genes consist of DNA, many questions remained. How is the DNA in a gene duplicated when a cell divides? How does the DNA in a gene control a hereditary trait? What happens to the DNA when a mutation (a change in the DNA) takes place in a gene? Important clues to the answers to these questions emerged from the discovery of the three-dimensional structure of the DNA molecule itself. This structure is discussed next.

A central feature of double-stranded DNA is complementary base pairing.

In the early 1950s, a number of researchers began to try to understand the detailed molecular structure of DNA. The first essentially correct three-dimensional structure of the DNA molecule was proposed in 1953 by James Watson and Francis Crick at Cambridge University. The structure was dazzling in its elegance and revolutionary in suggesting how DNA duplicates itself, controls hereditary traits, and undergoes mutation. Even while their tin sheet and wire model of the DNA molecule was still incomplete, Crick announced in his favorite pub that, "We have discovered the secret of life."

In the Watson–Crick structure, DNA consists of two long chains of subunits twisted around one another to form a double-stranded helix. The double helix is right-handed, which means that as one looks along the barrel, each chain follows a clockwise path as it progresses. You can see the right-handed coiling in part A of **FIGURE 1.5** if you imagine yourself

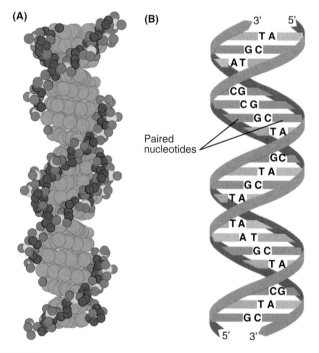

FIGURE 1.5 Molecular structure of a DNA double helix. (A) A "space-filling" model, in which each atom is depicted as a sphere. (B) A diagram highlighting the helical backbones on the outside of the molecule and the stacked A—T and G—C base pairs inside.

looking up into the structure from the bottom: The smaller spheres outline the "backbone" of each individual strand, and they coil in a clockwise direction. The subunits of each strand are **nucleotides**, each of which contains any one of four chemical constituents called **bases**. The four bases in DNA are

Adenine (A)	Guanine (G)
Thymine (T)	Cytosine (C)

The chemical structures of the nucleotides and bases need not concern us at this point. A key point for our present purposes is that the bases in the double helix are paired as shown in Figure 1.5, part B. That is,

KEY CONCEPT

At any position on the paired strands of a DNA molecule, if one strand has an A, then the partner strand has a T; and if one strand has a G, then the partner strand has a C.

The base pairing between A and T and between G and C is said to be **complementary base pairing**; the complement of A is T, and the complement of G is C. The complementary pairing in the duplex molecule means that each base along one strand of the DNA is matched with a base in the opposite position on the other strand. Furthermore,

KEY CONCEPT

Nothing restricts the sequence of bases in a single strand, so any sequence could be present along one strand.

This principle explains how only four bases in DNA can code for the huge amount of information needed to make an organism. It is the linear order or *sequence* of bases along the DNA that encodes the genetic information, and the sequence is completely unrestricted.

The complementary pairing is also called *Watson–Crick base pairing*. In the three-dimensional structure (Figure 1.5, part A), the base pairs are represented by the larger spheres filling the interior of the double helix. The base pairs lie almost flat, stacked on top of one another perpendicular to the long axis of the double helix, like pennies in a roll. When discussing a DNA molecule, biologists frequently refer to the individual strands as **single-stranded DNA** and to the double helix as **double-stranded DNA** or **duplex DNA**.

Each DNA strand has a **polarity**, or directionality, like a chain of circus elephants linked trunk to tail. In this analogy, each elephant corresponds to one nucleotide along the DNA strand. The polarity is determined by the direction in which the nucleotides are pointing. The "trunk" end of the strand is called the 5′ *end* of the strand, and the "tail" end is called the 3′ *end*. In double-stranded DNA, the paired strands are oriented in opposite directions: The 5′ end of one strand is aligned with the 3′ end of the other. The oppositely oriented strands are said to be **antiparallel**. In illustrating DNA molecules, we use an arrow-like ribbon to represent the backbone, and we use tabs jutting off the ribbon to represent the nucleotides. The polarity of a DNA strand is indicated by the direction of the

arrow-like ribbon. The tail of the arrow represents the 5′ end of the DNA strand, the head the 3′ end.

Beyond the most optimistic hopes, knowledge of the structure of DNA immediately gave clues to its function:

1. The sequence of bases in DNA could be copied by using each of the separate "partner" strands as a pattern for the creation of a new partner strand with a complementary sequence of bases.

2. The DNA could contain genetic information in coded form in the sequence of bases, analogous to letters printed on a strip of paper.

3. Changes in genetic information (mutations) could result from errors in copying in which the base sequence of the DNA became altered.

In the remainder of this chapter, we discuss some of the implications of these clues.

In replication, each parental DNA strand directs the synthesis of a new partner strand.

"It has not escaped our notice," wrote Watson and Crick, "that the specific base pairing we have postulated immediately suggests a copying mechanism for the genetic material." The copying process in which a single DNA molecule becomes two identical molecules is called **replication**. The replication mechanism that Watson and Crick had in mind is illustrated in **FIGURE 1.6**. The strands of the original (parent) duplex separate, and each individual strand serves as a pattern, or **template**, for the synthesis of a new strand (replica). The replica strands are synthesized by the addition of successive nucleotides in such a way that each base in the replica is complementary (in the Watson–Crick pairing sense) to the base across the way in the template strand. Although the mechanism in Figure 1.6 is simple in principle, it is a complex process that is fraught with geometric problems and requires a variety of enzymes and other proteins. The end result of replication is that a single double-stranded molecule becomes replicated into two copies with identical sequences:

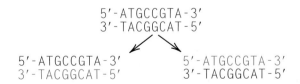

Here the bases in the newly synthesized strands are shown in red. In the duplex on the left, the top strand is the template from the parental molecule and the bottom strand is newly synthesized; in the duplex on the right, the bottom strand is the template from the parental molecule and the top strand is newly synthesized.

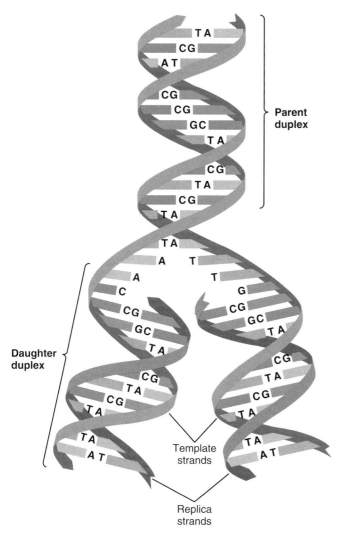

FIGURE 1.6 Replication in a long DNA duplex as originally proposed by Watson and Crick. The parental strands separate, and each parental strand serves as a template for the formation of a new daughter strand by means of A—T and G—C base pairing.

🧠 STOP & THINK 1.2

Shown here is part of the base sequence in one strand in a DNA duplex undergoing replication.

5'-TAGCAAAAATAGC-3'

What is the base sequence in the daughter strand?

1.3   Genes affect organisms through the action of proteins.

One of the important principles of molecular genetics is that genes exert their effects on organisms indirectly. For most genes, the genetic information contained in the nucleotide sequence specifies a particular type of *protein*. Proteins control the chemical and physical processes of cells known as **metabolism**. Many proteins are **enzymes**, a term introduced in 1878 to refer to the biological catalysts that accelerate biochemical reactions. Enzymes are essential for the breakdown of organic molecules, generating the chemical energy needed for cellular activities; they are also essential for the synthesis of small molecules and for their assembly into larger molecules and complex cellular structures.

Although the fundamental connection between genes and proteins was not widely appreciated until the 1940s, the first evidence for a relationship came much earlier. The pioneering observations were made by Archibald Garrod, a British physician, who studied genetic diseases caused by inherited defects in metabolism. He concluded that an inherited defect in metabolism results from an inherited defect in an enzyme. The key observations on which Garrod based this conclusion are summarized in the following sections.

Enzyme defects result in inborn errors of metabolism.

In 1908 Garrod gave a series of lectures in which he proposed this fundamental hypothesis about the relationship between enzymes and disease:

KEY CONCEPT

Any hereditary disease in which cellular metabolism is abnormal results from an inherited defect in an enzyme.

Such diseases became known as **inborn errors of metabolism**, a term still in use today.

Garrod studied a number of inborn errors of metabolism in which the patients excreted abnormal substances in the urine. One of these was **alkaptonuria**. In this case, the abnormal substance excreted is **homogentisic acid**:

This is a conventional chemical representation in which each corner of the hexagon represents a carbon atom, and hydrogen atoms attached to the ring are not shown. The six-carbon ring is called a *phenyl* ring. An early name for homogentisic acid was *alkapton*—hence the name *alkaptonuria*. Even though alkaptonuria is rare, with an incidence of about one in 200,000 people, it was well known even before Garrod studied it. The disease itself is relatively mild, but it has one striking symptom: The urine of the patient turns black because

FIGURE 1.7 Urine from a person with alkaptonuria turns black because of the oxidation of the homogentisic acid that it contains.

Courtesy Daniel De Aguiar.

of the oxidation of homogentisic acid (**FIGURE 1.7**). This is why alkaptonuria is also called *black urine disease*. The passing of black urine can hardly escape being noticed. One case was described in the year 1649:

> The patient was a boy who passed black urine and who, at the age of fourteen years, was submitted to a drastic course of treatment that had for its aim the subduing of the fiery heat of his viscera, which was supposed to bring about the condition in question by charring and blackening his bile. Among the measures prescribed were bleedings, purgation, baths, a cold and watery diet, and drugs galore. None of these had any obvious effect, and eventually the patient, who tired of the futile and superfluous therapy, resolved to let things take their natural course. None of the predicted evils ensued. He married, begat a large family, and lived a long and healthy life, always passing urine black as ink.

(Quotation from Garrod, 1908.)

Garrod was primarily interested in the biochemistry of alkaptonuria, but he took note of family studies that indicated that the disease was inherited as though it were due to a defect in a single gene. As to the biochemistry, he deduced that the problem in alkaptonuria was the patients' inability to break down the phenyl ring of six carbons that is present in homogentisic acid. Where does this ring come from? Mammals are unable to synthesize it and must obtain it from their diet. Garrod proposed that homogentisic acid originates as a breakdown product of two amino acids, phenylalanine and tyrosine, which also contain a phenyl ring. An **amino acid** is one of the "building blocks" from which proteins are made. Phenylalanine and tyrosine are constituents of normal proteins. The scheme that illustrates the relationship between the molecules is shown in **FIGURE 1.8**. Any such sequence of biochemical

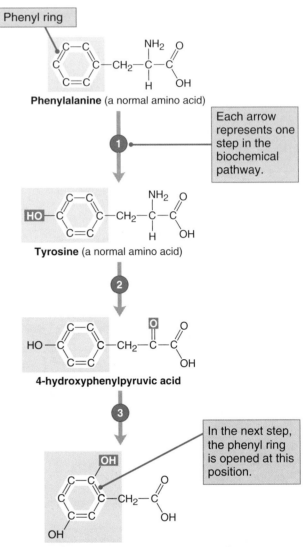

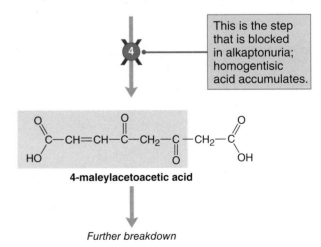

FIGURE 1.8 Metabolic pathway for the breakdown of phenylalanine and tyrosine. Each step in the pathway, represented by an arrow, requires a particular enzyme to catalyze the reaction. The key step in the breakdown of homogentisic acid is the breaking open of the phenyl ring.

reactions is called a **biochemical pathway** or a **metabolic pathway**. Each arrow in the pathway represents a single step depicting the transition from the "input" or **substrate molecule**, shown at the tail of the arrow, to the "output" or **product molecule**, shown at the tip. Biochemical pathways are usually oriented either vertically with the arrows pointing down, as in Figure 1.8, or horizontally, with the arrows pointing from left to right. Garrod did not know all of the details of the pathway in Figure 1.8, but he did understand that the key step in the breakdown of homogentisic acid is the breaking open of the phenyl ring and that the phenyl ring in homogentisic acid comes from dietary phenylalanine and tyrosine.

What allows each step in a biochemical pathway to occur? Garrod's insight was to see that each step requires a specific enzyme to catalyze the reaction and allow the chemical transformation to take place. Persons with an inborn error of metabolism, such as alkaptonuria, have a defect in one step of a metabolic pathway because they lack a functional enzyme for that step. When an enzyme in a pathway is defective, the pathway is said to have a **block** at that step. One frequent result of a blocked pathway is that the substrate of the defective enzyme accumulates. Observing the accumulation of homogentisic acid in patients with alkaptonuria, Garrod proposed that there must be an enzyme whose function is to open the phenyl ring of homogentisic acid and that this enzyme is missing in these patients. Discovery of all the enzymes in the pathway in Figure 1.8 took a long time. The enzyme that opens the phenyl ring of homogentisic acid was not actually isolated until 50 years after Garrod's lectures. In normal people it is found in cells of the liver. Just as Garrod had predicted, the enzyme is defective in patients with alkaptonuria.

The pathway for the breakdown of phenylalanine and tyrosine, as it is understood today, is shown in **FIGURE 1.9**. In this figure the emphasis is on the enzymes rather than on the structures of the **metabolites**, or small molecules, on which the enzymes act. As Garrod would have predicted, each step in the pathway requires the presence of a particular enzyme that catalyzes that step. Although Garrod knew only about alkaptonuria, in which the defective enzyme is homogentisic acid 1,2-dioxygenase, we now know the clinical consequences of defects in the other enzymes. Unlike alkaptonuria, which is a relatively benign inherited disease, the others are very serious. The condition known as **phenylketonuria (PKU)** results from the absence of (or a defect in) the enzyme **phenylalanine hydroxylase (PAH)**. When this step in the pathway is blocked, phenylalanine accumulates. The excess phenylalanine is broken down into harmful metabolites that cause defects in myelin formation that damage a child's developing nervous system and lead to severe mental retardation.

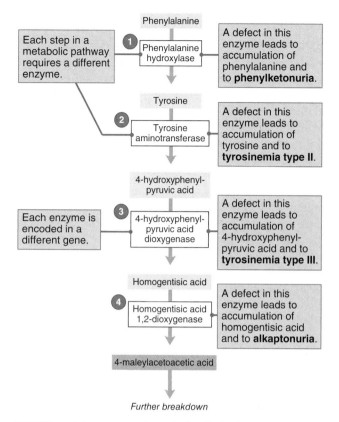

FIGURE 1.9 Inborn errors of metabolism in the breakdown of phenylalanine and tyrosine. A different inherited disease results when each of the enzymes is missing or defective. Alkaptonuria results from a defective homogentisic acid 1,2-dioxygenase, phenylketonuria from a defective phenylalanine hydroxylase.

If PKU is diagnosed in children soon enough after birth, they can be placed on a specially formulated diet low in phenylalanine. The child is allowed only as much phenylalanine as can be used in the synthesis of proteins, so excess phenylalanine does not accumulate. The special diet is very strict. It excludes meat, poultry, fish, eggs, milk and milk products, legumes, nuts, and bakery goods manufactured with regular flour. These foods are replaced by a synthetic formula that is very expensive. With the special diet, however, the detrimental effects of excess phenylalanine on mental development can largely be avoided. In many countries, including the United States, all newborn babies have their blood tested for chemical signs of PKU. Routine screening is cost effective because PKU is relatively common. In the United States, the incidence is about one in 8000 among Caucasian births. The disease is less common in other ethnic groups.

In the metabolic pathway in Figure 1.9, defects in the breakdown of tyrosine or of 4-hydroxyphenylpyruvic acid lead to types of tyrosinemia. These are also severe diseases. Type II is associated with skin lesions and mental retardation, type III with severe liver dysfunction.

A defective enzyme results from a mutant gene.

It follows from Garrod's work that a defective enzyme results from a mutant gene. How does a mutant gene result in a defective enzyme? Garrod did not speculate. For all he knew, genes *were* enzymes. This would have been a logical hypothesis at the time. We now know that the relationship between genes and enzymes is somewhat indirect. With a few exceptions, each enzyme is *encoded* in a particular sequence of nucleotides present in a region of DNA. The DNA region that codes for the enzyme, as well as adjacent regions that regulate when and in which cells the enzyme is produced, make up the "gene" that encodes the enzyme.

The genes for the enzymes in the biochemical pathway in Figure 1.9 have all been identified and the nucleotide sequence of the DNA determined. In the following list, and throughout this text, we use the typographical convention that the names of *genes* are printed in *italic* type, whereas gene products are printed in regular type. In Figure 1.9 the numbers 1 through 4 correspond to the following genes and enzymes:

1. The gene *PAH* on the long arm of chromosome 12 encodes phenylalanine hydroxylase (PAH).

2. The gene *TAT* on the long arm of chromosome 16 encodes tyrosine aminotransferase (TAT).

3. The gene *HPD* on the long arm of chromosome 12 encodes 4-hydroxyphenylpyruvic acid dioxygenase (HPD).

4. The gene *HGD* on the long arm of chromosome 3 encodes homogentisic acid 1,2-dioxygenase (HGD).

Genetic analysis led to the one gene–one enzyme hypothesis.

Garrod's thinking was far ahead of his time, and his conclusions about inborn errors of metabolism were largely ignored. The influential experiments connecting genes with enzymes were carried out in the 1940s by George W. Beadle and Edward L. Tatum using a filamentous fungus *Neurospora crassa*, commonly called red bread mold, an organism they chose because both genetic and biochemical analysis could be done with ease. In these experiments they identified new mutations that each caused a block in the metabolic pathway for the synthesis of some needed nutrient and showed that each of these blocks corresponded to a defective enzyme needed for one step in the pathway. The experimental approach, now called **genetic analysis**, was important because it solidified the link between genetics and biochemistry. Equally as

important, the experimental approach is widely applicable to understanding any complex biological process, ranging from the genetic control of the cell cycle or cancer to that of development or behavior. For this reason the methods of genetic analysis warrant a closer examination.

N. crassa grows in the form of filaments on a great variety of substrates including laboratory medium containing only inorganic salts, a sugar, and one vitamin. Such a medium is known as a **minimal medium** because it contains only the nutrients that are essential for growth of the organism. The filaments consist of a mass of branched threads separated into interconnected, multinucleate compartments allowing free interchange of nuclei and cytoplasm. Each nucleus contains a single set of seven chromosomes. Beadle and Tatum recognized that the ability of *Neurospora* to grow in minimal medium implied that the organism must be able to synthesize all of the other small molecules needed for growth, such as amino acids. If the biosynthetic pathways needed for growth are controlled by genes, then a mutation in a gene responsible for synthesizing an essential nutrient would be expected to render a strain unable to grow unless the strain were provided with the nutrient.

These ideas were tested in the following way. Spores of nonmutant *Neurospora* were irradiated with either x-rays or ultraviolet light to produce mutant strains with various nutritional requirements. The isolation of a set of mutants affecting any biological process, in this case metabolism, is called a **mutant screen**. In the initial step for identifying mutants, summarized in **FIGURE 1.10**, the irradiated spores (purple) were used in crosses with an untreated strain (green). Ascospores produced by the sexual cycle in fruiting bodies were individually germinated in **complete medium**, a complex medium enriched with a variety of amino acids, vitamins, and other substances expected to be essential metabolites whose synthesis could be blocked by a mutation. Even those ascospores containing a new mutation affecting synthesis of an essential nutrient would be expected to germinate and grow in complete medium.

To identify which of the irradiated ascospores contained a new mutation affecting the synthesis of an essential nutrient, spores from each culture were transferred to minimal medium (**FIGURE 1.11**, Part A). The vast majority of cultures yielded spores that could grow on minimal medium; these cultures lacked any new mutation of the desired type and were discarded. The cultures that were kept were the small number producing spores unable to grow on minimal medium, because these were mutant cultures that contained a new mutation blocking the synthesis of some essential nutrient.

THE HUMAN CONNECTION

One Gene, One Enzyme

George W. Beadle and Edward L. Tatum (1941)
Stanford University, Stanford, California

Genetic Control of Biochemical Reactions in Neurospora

How do genes control metabolic processes? The suggestion that genes control enzymes was made very early in the history of genetics, most notably by the British physician Archibald Garrod in his 1903 book, *Inborn Errors of Metabolism*. Nevertheless, the precise relationship between genes and enzymes was still uncertain. Perhaps each enzyme is controlled by more than one gene, or perhaps each gene contributes to the control of several enzymes. The classic experiments of Beadle and Tatum showed that the relationship is usually remarkably simple: One gene codes for one enzyme. Their pioneering experiments united genetics and biochemistry, and for the "one gene, one enzyme" concept, Beadle and Tatum were awarded a Nobel Prize in 1958 (Joshua Lederberg shared the prize for his contributions to microbial genetics). Because we now know that some enzymes contain polypeptide chains encoded by two (or occasionally more) different genes, a more accurate statement of the principle is "one gene, one polypeptide." Beadle and Tatum's experiments also demonstrate the importance of choosing the right organism. *Neurospora* had been introduced as a genetic organism only a few years earlier, and Beadle and Tatum realized that they could take advantage of this organism's ability to grow on a simple medium composed of known substances.

Beadle and Tatum's work was published in 1941 and can be found at the reference at the end of this feature. In it, they point out the limitations of starting with the physiological basis of a trait (such as black urine disease) and

> **❝These preliminary results appear to us to indicate that the approach may offer considerable promise as a method of learning more about how genes regulate development and function. ❞**

attempting to determine its genetic basis. First, these analyses are limited to traits in which the variants are nonlethal. Second, the variants must have visible effects. To get around these problems, Beadle and Tatum turned the problem on its head.

[These limitations] have led us to investigate the general problem of the genetic control of development and metabolic reactions by reversing the ordinary procedure . . . [by setting out] to determine if and how genes control known biochemical reactions . . . If the organism must be able to carry out a certain chemical reaction to survive on a given medium, a mutant unable to do this will obviously be lethal on this medium. . . . [It can be] studied, however, if it will grow on a medium to which has been added the essential product of the genetically blocked reaction. . . .

Thus, rather than starting with observed differences in traits among individuals, Beadle and Tatum started by generating mutations (in their case, mutations resulting from x-irradiation of *Neurospora* cells), then identified the mutations that were lethal on minimal medium but not on medium supplemented with the normal product of the mutated gene. This experimental approach ranks among the most important experimental tool of genetic analysis.

G. W. Beadle and E. L. Tatum, Genetic control of biochemical reactions in *Neurospora. Proc. Natl. Acad. Sci. USA* 27 (1941): 499–506.

Spores from each mutant culture were then transferred to a series of media to determine whether the mutation results in a requirement for a vitamin, an amino acid, or some other substance. In the example illustrated in Figure 1.11, Part B, the mutant strain requires one (or possibly more than one) amino acid, because a mixture of all amino acids added to the minimal medium allows growth. Because the proportion of irradiated cultures with new mutations was very small, only a negligible number of cultures would contain two or more new mutations that had occurred simultaneously.

For nutritional mutants requiring amino acids, further experiments testing each of the amino acids

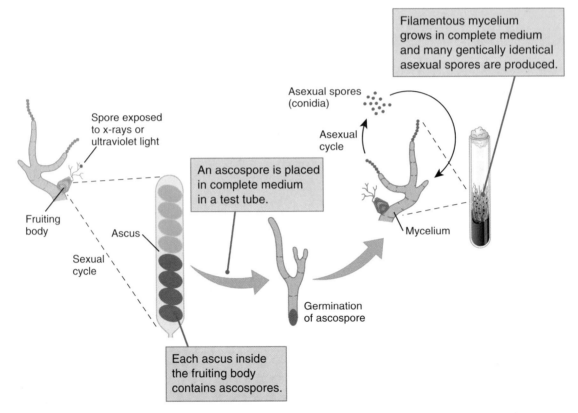

FIGURE 1.10 Beadle and Tatum obtained mutants of the filamentous fungus *Neurospora crassa* by exposing asexual spores to x-rays or ultraviolet light. The treated spores were used to start the sexual cycle in fruiting bodies. After any pair of cells and their nuclei undergo fusion, meiosis takes place almost immediately and results in eight sexual spores (ascospores) included in a single ascus. These are removed individually and cultured in complete medium. Ascospores that carry new nutritional mutants are identified later by their inability to grow in minimal medium.

individually usually revealed that only one amino acid was required to be added to minimal medium to support growth. In Figure 1.11, Part C, the mutant strain requires the amino acid arginine. Even in the 1940s some of the possible intermediates in amino acid biosynthesis had been identified. These were recognized by their chemical resemblance to the amino acid and by being present at low levels in the cells of organisms. In the case of arginine, two candidates were ornithine and citrulline. All mutants requiring arginine were, therefore, tested in medium supplemented with either ornithine alone or citrulline alone (Figure 1.11, Part D). One class of arginine-requiring mutants, designated Class I, was able to grow in minimal medium supplemented with either ornithine, citrulline, or arginine. Other mutants, designated Class II, were able to grow in minimal medium supplemented with either citrulline or arginine but not ornithine. A third class, Class III, was able to grow only in minimal medium supplemented with arginine.

The types of arginine-requiring mutants illustrate the principle of genetic analysis as applied to metabolic pathways. The basic principle is that

KEY CONCEPT

If a strain with a mutant enzyme that blocks a particular step in a linear metabolic pathway can grow when an intermediate is added to the growth medium, it means that the location of the intermediate in the pathway is *downstream* of the enzymatic step that is blocked.

This principle makes intuitive sense because, if the intermediate were upstream of the metabolic block, then adding the intermediate to the growth medium would not allow growth, because conversion of the intermediate would still be blocked at the point of the mutant enzyme.

Application of the principle to the linear pathway for arginine biosynthesis is shown in **FIGURE 1.12**, where arginine is the end product starting with some precursor metabolite, and ornithine and citrulline are intermediates in the pathway. The mutants imply the order of the intermediates shown because

- Mutants in Class I are able to grow in the presence of either ornithine or citrulline, which means that

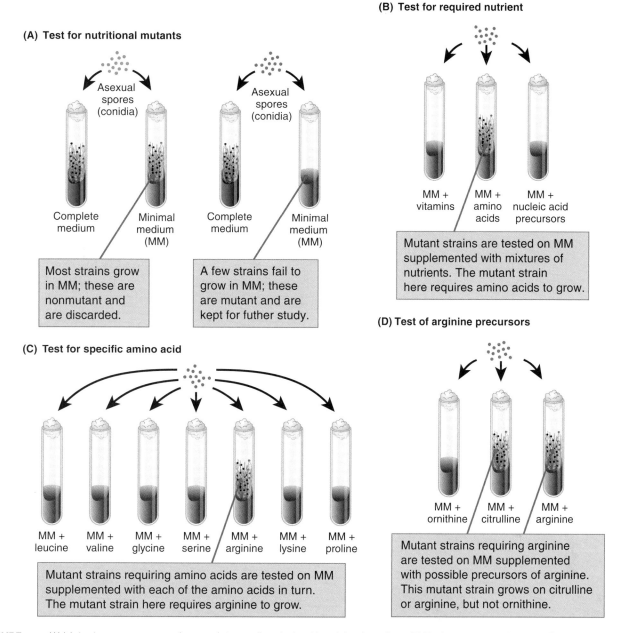

(A) Test for nutritional mutants

Asexual spores (conidia) → Complete medium / Minimal medium (MM)

Asexual spores (conidia) → Complete medium / Minimal medium (MM)

Most strains grow in MM; these are nonmutant and are discarded.

A few strains fail to grow in MM; these are mutant and are kept for futher study.

(B) Test for required nutrient

MM + vitamins / MM + amino acids / MM + nucleic acid precursors

Mutant strains are tested on MM supplemented with mixtures of nutrients. The mutant strain here requires amino acids to grow.

(C) Test for specific amino acid

MM + leucine / MM + valine / MM + glycine / MM + serine / MM + arginine / MM + lysine / MM + proline

Mutant strains requiring amino acids are tested on MM supplemented with each of the amino acids in turn. The mutant strain here requires arginine to grow.

(D) Test of arginine precursors

MM + ornithine / MM + citrulline / MM + arginine

Mutant strains requiring arginine are tested on MM supplemented with possible precursors of arginine. This mutant strain grows on citrulline or arginine, but not ornithine.

FIGURE 1.11 (A) Mutant spores can grow in complete medium but not in minimal medium. (B) Each new mutant is tested for growth in minimal medium supplemented with a mixture of nutrients. (C) Mutants that can grow on minimal medium supplemented with amino acid are tested with each amino acid individually. (D) Mutants unable to grow in the absence of arginine are tested with likely precursors of arginine.

both ornithine and citrulline are downstream of any of the enzymes blocked in Class I mutants.

■ Mutants in Class II are able to grow in the presence of citrulline but not ornithine, which means that citrulline is located downstream of the enzymatic block in Class II mutants and that ornithine is upstream of the metabolic block in Class II mutants.

■ Mutants in Class III are unable to grow in the presence of either citrulline or ornithine, which

means that these intermediates are upstream of any of the enzymatic steps blocked in Class III mutants.

The structure of the pathway in Figure 1.12 was further confirmed by the observations that Class III mutants accumulate citrulline and Class II mutants accumulate ornithine. Ultimately, direct biochemical experiments demonstrated that the inferred enzymes were actually present in nonmutant strains but defective in mutant strains.

STOP & THINK 1.3

Suppose you do a mutant screen for *Neurospora* mutants unable to grow on minimal medium unless they are supplemented with an amino acid we'll call "A." Based on their molecular structures, you surmise that two molecules, X and Y, are intermediates in the biochemical pathway for the synthesis of A, but you are unsure which of the following pathways may be correct:

(a) X ⟶ Y ⟶ A

(b) Y ⟶ X ⟶ A

You find two classes of mutants that require A for growth.

Class 1 grows on minimal medium supplemented with A but not with X or Y.

Class 2 grows on minimal medium supplemented with A or X but not with Y.

Which of the pathways **(a)** or **(b)** do these data support?

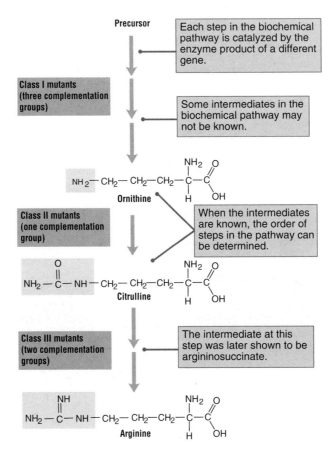

FIGURE 1.12 Metabolic pathway for arginine biosynthesis inferred from genetic analysis of *Neurospora* mutants.

Mutant screens sometimes isolate different mutations in the same gene.

Beadle and Tatum were fortunate to study metabolic pathways in a relatively simple organism in which each gene specifies a single enzyme, a relation often called the **one gene–one enzyme hypothesis**. In such a situation, genetic analysis of the mutants reveals a great deal more about the metabolic pathway than merely the order of the intermediates. By classifying each mutation according to the particular gene it is in and grouping all the mutations in each gene together, each set of mutations and, therefore, each individual gene, correspond to one enzymatic step in the metabolic pathway. In Figure 1.12, for example, Class I includes mutations in any of three different genes, which implies that there are three steps in the pathway between the precursor and ornithine. Similarly, Class III comprises mutations in either of two different genes, which implies that there are two steps in the pathway between the citrulline and arginine. However, Class II consists of mutations in only one gene, which implies only one step in the pathway between ornithine and citrulline.

Mutations that have defects in the same gene are identified by means of a **complementation test**, in which two mutations are brought together into the same cell. In most multicellular organisms (and even some sexual unicellular organisms), the usual way to do this is by means of a mating. When two parents, each carrying one of the two mutations, are crossed, fertilization brings the reproductive cells containing the two mutations together, and through ordinary cell division each cell in the resulting offspring carries one copy of each mutant gene. In *Neurospora* this procedure does not work because nuclear fusion is followed almost immediately by the formation of ascospores, each of which has only one set of chromosomes.

Complementation tests are nevertheless possible in *Neurospora* owing to the multinucleate nature of the filaments. Certain strains, including those studied by Beadle and Tatum, have the property that when the filaments from two mutant organisms come into physical contact, the filaments fuse and the new filament contains multiple nuclei from both of the participating partners. This sort of hybrid filament is called a **heterokaryon**, and it contains mutant forms of both genes. The word roots of the term *heterokaryon* mean "different nuclei." (A list of the most common word roots used in genetics can be found at the end of the book.)

When a heterokaryon formed from two nutritional mutants is inoculated into minimal medium, it may grow or it may fail to grow. If it grows in minimal medium, the mutant genes are said to undergo **complementation**, and this result indicates that the mutations are in different genes. On the other hand, if the heterokaryon fails to grow in minimal medium, the result indicates **noncomplementation**, and the two mutations are inferred to be in the same gene.

The inferences from complementation or noncomplementation emerge from the logic illustrated in **FIGURE 1.13**. Here the multinucleate filament is shown, and the mutant nuclei are color coded according to which of two different genes (red or purple) is mutant. The thick red and purple horizontal lines represent the proteins encoded in the mutant nuclei, and the × represents a defect in the protein resulting from a mutation in the corresponding gene.

Part A depicts the situation in which the mutant strains have mutations in different genes. In the heterokaryon, the red nuclei produce mutant forms of the red protein and normal forms of the purple protein, whereas the purple nuclei produce mutant forms of the purple protein and normal forms of the red protein. The result is that the red/purple heterokaryon has normal forms of both the red and purple proteins. It also has mutant forms of both proteins, but these do not matter. What matters is that the normal proteins allow the heterokaryon to grow on minimal medium because all needed nutrients can be synthesized. In other words, the normal purple gene in the red nucleus complements the defective purple gene in the purple nucleus, and the other way around. The logic of complementation is captured in the ancient nursery rhyme "Jack Sprat could eat no fat / His wife could eat no lean / And so between the two of them / They licked the platter clean," because each partner makes up for the defect in the other.

Part B in Figure 1.13 shows a heterokaryon formed between mutants with defects in the same gene, in this case purple. Both of the purple nuclei encode a normal form of the red protein, but each purple nucleus encodes a defective purple protein. When the nuclei are together, two different mutant forms of the purple protein are produced, and so the biosynthetic pathway that requires the purple protein is still blocked, and the heterokaryon is unable to grow in minimal medium. In other words, the mutants 2 and 3 in Figure 1.14 fail to complement, and so they are judged to have mutations in the same gene.

The following principle underlies the complementation test.

KEY CONCEPT

The Principle of Complementation: A complementation test brings two mutant genes together in the same cell or organism. If this cell or organism is nonmutant, the mutations are said to *complement* one another and it means that the parental strains have mutations in *different* genes. If the cell or organism is mutant, the mutations fail to complement one another, and it means that the parental mutations are in the *same* gene.

A complementation test identifies mutations in the same gene.

In the mutant screen for *Neurospora* mutants requiring arginine, Beadle and Tatum found that mutants in different classes (Class I, Class II, and Class III in Figure 1.12) always complemented one another. This result makes sense, because the genes in each class encode enzymes that act at different levels between the known intermediates. However, some of the mutants

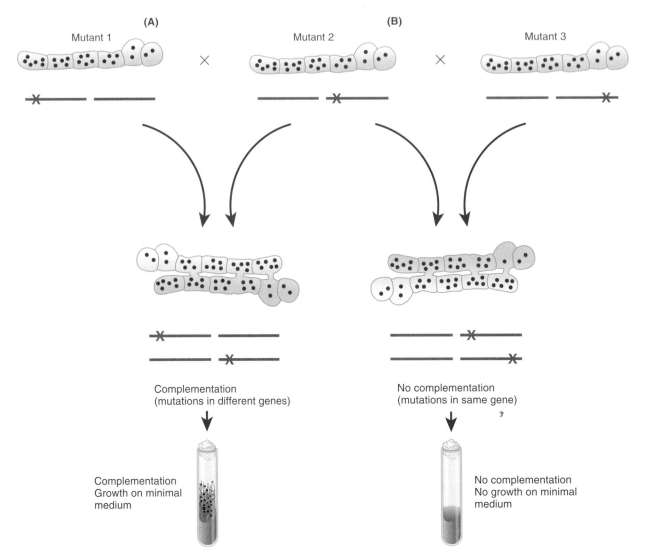

FIGURE 1.13 Molecular interpretation of a complementation test using heterokaryons to determine whether two mutant strains have mutations in different genes (A) or mutations in the same gene (B). In (A) each nucleus contributes a nonmutant form of one or the other polypeptide chain, and so the heterokaryon is able to grow in minimal medium. In (B) both nuclei contribute a mutant form of the same polypeptide chain; hence, no nonmutant form of that polypeptide can be synthesized and the heterokaryon is unable to grow in minimal medium.

in Class I failed to complement others in Class I, and some in Class III failed to complement others in Class III. These results allow the number of genes in each class to be identified.

To illustrate this aspect of genetic analysis, we consider six mutant strains in Class III. These strains were taken in pairs to form heterokaryons and their growth on minimal medium assessed. The data are shown in **FIGURE 1.14**, Part A. The mutant genes in the six strains are denoted *x*1, *x*2, and so forth, and the data are presented in the form of a matrix in which + indicates growth in minimal medium (complementation) and − indicates lack of growth in minimal medium (lack of complementation). The diagonal entries are all −, which reflects the fact that two copies of the identical mutation cannot show complementation. The pattern of + and − signs in

the matrix indicate that mutations *x*1 and *x*5 fail to complement one another; hence, *x*1 and *x*5 are mutations in the same gene. Likewise, mutations *x*2, *x*3, *x*4, and *x*6 fail to complement one another in all possible pairs; hence, *x*2, *x*3, *x*4, and *x*6 are all mutations in the same gene (but a different gene from that represented by *x*1 and *x*5).

Data in a complementation matrix can conveniently be analyzed by arranging the mutant genes in the form of a circle as shown in Figure 1.14, Part B. Then, for each possible pair of mutations, connect the pair by a straight line if the mutations *fail* to complement (− signs in part A). According to the principle of complementation, these lines connect mutations that are in the same gene. Each of the groups of noncomplementing mutations is called a **complementation group**. As we have seen, each

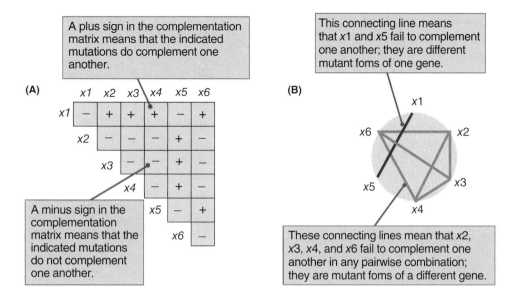

FIGURE 1.14 (A) Results of complementation tests. (B) To interpret the results, arrange the mutations in a circle. Connect by a straight line any pair of mutations that fail to complement—that is, that yield a mutant heterokaryon. Any pair of mutations connected by a straight line are mutations in the same gene, and are more than likely mutations at different nucleotide sites in the gene. This example shows two complementation groups, each of which represents a single gene needed for arginine biosynthesis.

complementation group defines a gene, so the complementation test actually provides the geneticist's operational definition:

KEY CONCEPT

A *gene* is defined experimentally as a set of mutations that make up a single complementation group. Any pair of mutations within a complementation group fail to complement one another.

The mutations in Figure 1.14, therefore, represent two genes, mutation of any one of which results in the inability of the strain to convert citrulline to

arginine. On the basis of the one gene–one enzyme hypothesis, which is largely true for metabolic enzymes in *Neurospora*, the pathway from citrulline to arginine in Figure 1.12 must comprise two steps with an unknown intermediate in between. This intermediate was later found to be argininosuccinate. Likewise, Class I mutants defined three complementation groups; hence, there are three enzymatic steps from the precursor to ornithine. These intermediates were also soon identified. Finally, Class II mutants all failed to complement one another, and the finding of only one complementation group means that there is but a single enzymatic step that converts ornithine to citrulline.

🧠 STOP & THINK 1.4

Among mutations affecting a metabolic pathway in *Neurospora*, one class of mutants blocks the conversion of W into Z. These mutants can grow on minimal medium when supplemented with Z, but they can't when supplemented with W. You carry out complementation tests with six such mutants (*m1, m2, …, m6*) and find the complementation matrix shown here.

(a) How many different genes are indicated by these results?
(b) If each gene codes for a different enzyme in the pathway, how many enzymatic steps are there in the conversion of W into Z?

	m1	m2	m3	m4	m5	m6
m1	−	+	+	−	+	+
m2		−	−	+	+	−
m3			−	+	+	−
m4				−	+	+
m5					−	+
m6						−

Genetic analysis can be applied to the study of any complex biological process.

The type of genetic analysis pioneered by Beadle and Tatum is immensely powerful for identifying the genetic control of complex biological processes. Their approach lays out a systematic path—a sort of recipe—for gene discovery. First, decide what process you want to study. Next figure out what characteristics mutant organisms with a disruption in that process would display. Then do a mutant screen for mutants showing these characteristics. Carry out complementation tests to find out how many different genes that you have identified. And finally, find out what the products of those genes are, what they do, how they interact with each other, and in what order they function.

Beadle and Tatum themselves analyzed many metabolic pathways for a wide variety of essential nutrients, but their experiments were especially important in deciphering pathways of amino acid biosynthesis. Their findings over just a few years are said to have "contributed more knowledge of amino acid biosynthetic pathways than had been accumulated during decades of traditional study." They were awarded the 1958 Nobel Prize in Physiology or Medicine for their research, and in the intervening years at least nine more Nobel Prizes in Physiology or Medicine were awarded in which genetic analysis carried out along the lines of Beadle and Tatum played a significant role. Here is a list, with quotations from the official citations of the Nobel Foundation.

- 1958—George Beadle and Edward Tatum "for their discovery that genes act by regulating definite chemical events." (It was in doing literature research for his Nobel Prize Lecture that Beadle discovered Garrod's earlier work and brought it to the world's attention.)

- 1965—François Jacob, André Lwoff, and Jacques Monod "for their discoveries concerning genetic control of enzyme and virus synthesis"

- 1995—Edward B. Lewis, Christiane Nüsslein-Volhard, and Eric F. Wieschaus "for their discoveries concerning the genetic control of early embryonic development"

- 2000—Leland H. Hartwell, Tim Hunt, and Sir Paul Nurse "for their discoveries of key regulators of the cell cycle"

- 2002—Sydney Brenner, H. Robert Horvitz, and John E. Sulston "for their discoveries concerning genetic regulation of organ development and programmed cell death"

- 2007—Mario R. Capecchi, Martin J. Evans, and Oliver Smithies "for their discoveries of principles for introducing specific gene modifications in mice by the use of embryonic stem cells"

- 2009—Elizabeth H. Blackburn, Carol W. Greider, and Jack W. Szostak "for the discovery of how chromosomes are protected by telomeres and the enzyme telomerase"

- 2013—James E. Rothman, Randy W. Schekman, and Thomas C. Südhof "for their discoveries of machinery regulating vesicle traffic, a major transport system in our cells"

- 2015—Tomas Lindahl, Paul Modrich, and Aziz Sancar "for mechanistic studies of DNA repair"

- 2017—Jeffrey C. Hall, Michael Rosbash, and Michael W. Young "for their discoveries of molecular mechanisms controlling the circadian rhythm"

1.4   Genes specify proteins by means of a genetic code.

The Beadle and Tatum experiments established that a gene specifies the structure of an enzyme but left open the issue of how this happens. We now know that the relationship between genes and proteins is indirect. The genetic information that specifies a protein is actually contained in the sequence of bases in DNA in a manner analogous to letters printed on a strip of paper. In a region of DNA that directs the synthesis of a protein, the genetic code for the protein is contained in only one strand, and it is decoded in a linear order. The result of protein synthesis is a polypeptide chain, which consists of a linear sequence of amino acids connected end to end. Each polypeptide chain folds into a characteristic three-dimensional configuration that is determined by its particular sequence of amino acids. A typical protein is made up of one or more polypeptide chains. Many proteins function as enzymes that participate in metabolic processes such as amino acid biosynthesis.

One of the DNA strands directs the synthesis of a molecule of RNA.

The details of how genes code for proteins were not understood until the 1960s, and an outline of the process is shown in **FIGURE 1.15**. The decoding of the genetic information takes place in two distinct steps known as *transcription* and *translation*. The indirect route of information transfer

$$DNA \rightarrow RNA \rightarrow Protein$$

is known as the **central dogma** of molecular genetics. The term *dogma* means "set of beliefs"; it dates from

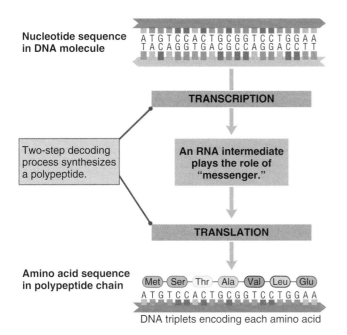

FIGURE 1.15 DNA sequence coding for the first seven amino acids in a polypeptide chain. The DNA sequence specifies the amino acid sequence through a molecule of RNA that serves as an intermediary "messenger." Although the decoding process is indirect, the net result is that each amino acid in the polypeptide chain is specified by a group of three adjacent bases in the DNA. In this example, the polypeptide chain is that of phenylalanine hydroxylase (PAH).

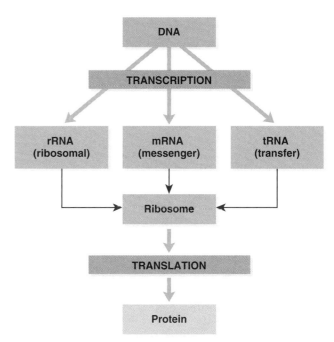

FIGURE 1.16 The "central dogma" of molecular genetics: DNA codes for RNA, and RNA codes for proteins. The DNA → RNA step is transcription, and the RNA → protein step is translation.

the time the idea was first put forward as hypothesis. Since then the "dogma" has been confirmed experimentally, but the term persists. The main concept in the central dogma is that DNA does not code for protein directly but rather acts through an intermediary molecule of **ribonucleic acid (RNA)**. The structure of RNA is similar to, but not identical with, that of DNA. There is a difference in the sugar (RNA contains the sugar **ribose** instead of deoxyribose), RNA is usually single stranded (not a duplex), and RNA contains the base **uracil (U)** instead of thymine (T), which is present in DNA. Three types of RNA take part in the synthesis of proteins:

- A molecule of **messenger RNA (mRNA)**, which carries the genetic information from DNA and is used as a template for polypeptide synthesis. In most mRNA molecules, a relatively high proportion of the nucleotides actually code for amino acids. For example, the mRNA for phenylalanine hydroxylase is 2400 nucleotides in length and codes for a polypeptide of 452 amino acids; in this case, more than 50 percent of the length of the mRNA codes for amino acids.

- Three types of **ribosomal RNA (rRNA)**, which are major constituents of the cellular particles called **ribosomes** on which polypeptide synthesis takes place.

- A set of about 45 **transfer RNA (tRNA)** molecules, each of which carries a particular amino acid as well as a three-base recognition region that base-pairs with a group of three adjacent bases in the mRNA. As each tRNA participates in translation, its amino acid becomes the terminal subunit of the growing polypeptide chain. A tRNA that carries methionine is denoted tRNA^{Met}, one that carries serine is denoted tRNA^{Ser}, and so forth. (Because there are more than 20 different tRNAs, but only 20 amino acids, some amino acids can be attached to any of several tRNAs.)

The central dogma illustrated in **FIGURE 1.16** is the fundamental principle of molecular genetics because it summarizes how the genetic information in DNA becomes expressed in the amino acid sequence in a polypeptide chain.

KEY CONCEPT

The sequence of nucleotides in a gene specifies the sequence of nucleotides in a molecule of messenger RNA; in turn, the sequence of nucleotides in the messenger RNA specifies the sequence of amino acids in the polypeptide chain.

The manner in which genetic information is transferred from DNA to RNA is shown in **FIGURE 1.17.**

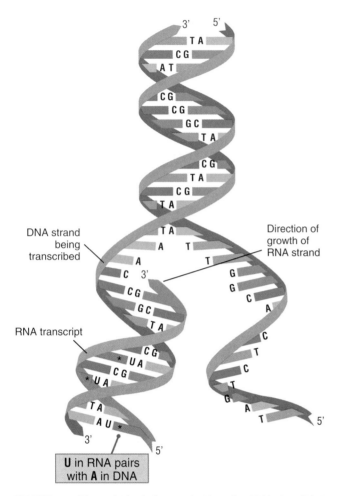

DNA strand being transcribed

Direction of growth of RNA strand

RNA transcript

U in RNA pairs with A in DNA

FIGURE 1.17 Transcription is the production of an RNA strand that is complementary in base sequence to a DNA strand. In this example, a DNA strand is being transcribed into an RNA strand at the bottom left. Note that in an RNA molecule, the base U (uracil) plays the role of T (thymine) in that it pairs with A (adenine). Each A–U pair is marked with an asterisk.

The DNA opens up, and one of the strands is used as a template for the synthesis of a complementary strand of RNA. The process of making an RNA strand from a DNA template is **transcription**, and the RNA molecule that is made is the **transcript**. The base sequence in the RNA is complementary (in the Watson–Crick pairing sense) to that in the DNA template, except that U (which pairs with A) is present in the RNA in place of T. The rules of base pairing between DNA and RNA are summarized below.

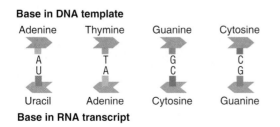

Base in DNA template

Adenine	Thymine	Guanine	Cytosine
A	T	G	C
U	A	C	G
Uracil	Adenine	Cytosine	Guanine

Base in RNA transcript

Like DNA, an RNA strand also exhibits polarity; its 5′ and 3′ ends are determined by the orientation of the nucleotides. The 5′ end of the RNA transcript is synthesized first, and in the RNA–DNA duplex formed in transcription, the polarity of the RNA strand is opposite to that of the DNA strand. Each gene includes particular nucleotide sequences that initiate and terminate transcription. The RNA transcript made from any gene begins at an initiation site in the template strand, which is located "upstream" from the amino acid coding region, and ends at a termination site, which is located "downstream" from the amino acid coding region. For any gene, the length of the RNA transcript is very much smaller than the length of the DNA in the entire chromosome. For example, the transcript of the *PAH* gene for phenylalanine hydroxylase is 90,000 nucleotides in length, but the DNA in all of chromosome 12 is about 130,000,000 nucleotide pairs. In this case, the length of the *PAH* transcript is less than 0.1 percent of the length of the DNA in the chromosome. A different gene in chromosome 12 would be transcribed from a different region of the DNA molecule in chromosome 12, but the transcribed region would again be small in comparison with the total length of the DNA in the chromosome.

A molecule of RNA directs the synthesis of a polypeptide chain.

The synthesis of a polypeptide under the direction of an mRNA molecule is known as **translation**. Although the sequence of bases in the mRNA codes for the sequence of amino acids in a polypeptide, the molecules that actually do the "translating" are the tRNA molecules. The mRNA molecule is translated in nonoverlapping groups of three bases called **codons**. For each codon in the mRNA that specifies an amino acid, there is one tRNA molecule containing a complementary group of three adjacent bases that can pair with the bases in the codon. The correct amino acid is attached to the other end of the tRNA, and when this tRNA comes into line, the amino acid attached to it becomes the new terminal end of the growing polypeptide chain.

The role of tRNA in translation is illustrated in **FIGURE 1.18** and can be described as follows:

KEY CONCEPT

The mRNA is read codon by codon. Each codon that specifies an amino acid matches with a complementary group of three adjacent bases in a single tRNA molecule. One end of the tRNA is attached to the correct amino acid, so the correct amino acid is brought into line.

The tRNA molecules used in translation do not line up along the mRNA simultaneously as shown in Figure 1.18. The process of translation takes place on a ribosome, which combines with a single mRNA and

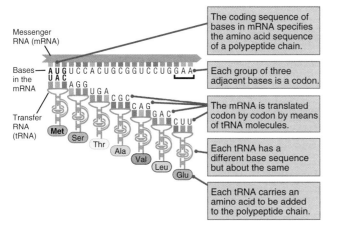

FIGURE 1.18 The role of messenger RNA in translation is to carry the information contained in a sequence of DNA bases to a ribosome, where it is translated into a polypeptide chain. Translation is mediated by transfer RNA (tRNA) molecules, each of which can base-pair with a group of three adjacent bases in the mRNA. Each tRNA also carries an amino acid; when it is brought to the ribosome by base pairing, its amino acid becomes the growing end of the polypeptide chain.

moves along it in steps, three nucleotides at a time (codon by codon). As each new codon comes into place, the next tRNA binds with the ribosome, and the growing end of the polypeptide chain becomes attached to the amino acid on the tRNA. In this way, each tRNA in turn serves temporarily to hold the polypeptide chain as it is being synthesized. As the polypeptide chain is transferred from each tRNA to the next in line, the tRNA that previously held the polypeptide is released from the ribosome. The polypeptide chain elongates one amino acid at a step until any one of three particular codons specifying "stop" is encountered. At this point, synthesis of the chain of amino acids is finished, and the polypeptide chain is released from the ribosome.

The genetic code is a triplet code.

Figure 1.18 indicates that the mRNA codon AUG specifies methionine (Met) in the polypeptide chain, UCC specifies Ser (serine), ACU specifies Thr (threonine), and so on. The complete decoding table is called the **genetic code**, and it is shown in **TABLE 1.1**. For any codon, the column on the left corresponds to the first nucleotide in the codon (reading from the 5′ end), the row across the top corresponds to the second nucleotide, and the column on the right corresponds

TABLE 1.1 The Standard Genetic Code

		Second Nucleotide in Codon			
		U	C	A	G
U		UUU Phe F *Phenylalanine*	UCU Ser S *Serine*	UAU Tyr Y *Tyrosine*	UGU Cys C *Cysteine*	U
		UUC Phe F *Phenylalanine*	UCC Ser S *Serine*	UAC Tyr Y *Tyrosine*	UGC Cys C *Cysteine*	C
		UUA Leu L *Leucine*	UCA Ser S *Serine*	UAA Termination	UGA Termination	A
		UUG Leu L *Leucine*	UCG Ser S *Serine*	UAG Termination	UGG Trp W *Tryptophan*	G
C		CUU Leu L *Leucine*	CCU Pro P *Proline*	CAU His H *Histidine*	CGU Arg R *Arginine*	U
		CUC Leu L *Leucine*	CCC Pro P *Proline*	CAC His H *Histidine*	CGC Arg R *Arginine*	C
		CUA Leu L *Leucine*	CCA Pro P *Proline*	CAA Gln Q *Glutamine*	CGA Arg R *Arginine*	A
		CUG Leu L *Leucine*	CCG Pro P *Proline*	CAG Gln Q *Glutamine*	CGG Arg R *Arginine*	G
A		AUU Ile I *Isoleucine*	ACU Thr T *Threonine*	AAU Asn N *Asparagine*	AGU Ser S *Serine*	U
		AUC Ile I *Isoleucine*	ACC Thr T *Threonine*	AAC Asn N *Asparagine*	AGC Ser S *Serine*	C
		AUA Ile I *Isoleucine*	ACA Thr T *Threonine*	AAA Lys K *Lysine*	AGA Arg R *Arginine*	A
		AUG Met M *Methionine*	ACG Thr T *Threonine*	AAG Lys K *Lysine*	AGG Arg R *Arginine*	G
G		GUU Val V *Valine*	GCU Ala A *Alanine*	GAU Asp D *Aspartic acid*	GGU Gly G *Glycine*	U
		GUC Val V *Valine*	GCC Ala A *Alanine*	GAC Asp D *Aspartic acid*	GGC Gly G *Glycine*	C
		GUA Val V *Valine*	GCA Ala A *Alanine*	GAA Glu E *Glutamic acid*	GGA Gly G *Glycine*	A
		GUG Val V *Valine*	GCG Ala A *Alanine*	GAG Glu E *Glutamic acid*	GGG Gly G *Glycine*	G

First nucleotide in codon (5′ end) — Third nucleotide in codon (3′ end)

Codon — Three-letter and single-letter abbreviations

to the third nucleotide. The complete codon is given in the body of the table, along with the amino acid (or "stop") that the codon specifies. Each amino acid is designated by its full name as well as by a three-letter abbreviation and a single-letter abbreviation. Both types of abbreviations are used in molecular genetics. The code in Table 1.1 is the "standard" genetic code used in translation in the cells of nearly all organisms.

In addition to the 61 codons that code only for amino acids, there are 4 codons that have specialized functions:

- The codon AUG, which specifies Met (methionine), is also the "start" codon for polypeptide synthesis. The positioning of a tRNA^{Met} bound to AUG is one of the first steps in the initiation of polypeptide synthesis, so all polypeptide chains begin with Met. In most organisms, the tRNA^{Met} used for initiation of translation is the same tRNA^{Met} used to specify methionine at internal positions in a polypeptide chain.

- The codons UAA, UAG, and UGA, each of which is a "stop," specify the termination of translation and result in release of the completed polypeptide chain from the ribosome. These codons do not have tRNA molecules that recognize them but are instead recognized by protein factors that terminate translation.

How the genetic code table is used to infer the amino acid sequence of a polypeptide chain may be illustrated using phenylalanine hydroxylase again, in particular the DNA sequence coding for amino acid numbers 1 through 7. The DNA sequence is

```
5'-ATGTCCACTGCGGTCCTGGAA-3'
3'-TACAGGTGACGCCAGGACCTT-5'
```

This region is transcribed into RNA in a left-to-right direction, and because RNA grows by the addition of successive nucleotides to the 3' end (Figure 1.17), it is the bottom strand that is transcribed. The nucleotide sequence of the RNA is that of the top strand of the DNA, except that U replaces T, so the mRNA for amino acids 1 through 7 is

```
5'-AUGUCCACUGCGGUCCUGGAA-3'
```

The codons are read from left to right according to the genetic code shown in Table 1.1. Codon AUG codes for Met (methionine), UCC codes for Ser (serine), and so on. Altogether, the amino acid sequence of this region of the polypeptide is

```
5'-AUGUCCACUGCGGUCCUGGAA-3'
MetSerThrAlaValLeuGlu
```

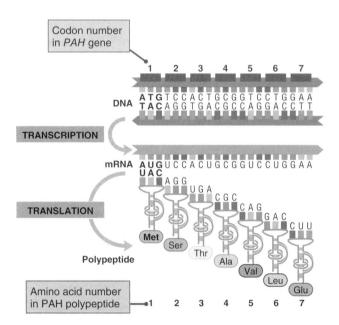

FIGURE 1.19 The central dogma in action. The DNA that encodes PAH serves as a template for the production of a messenger RNA, and the mRNA, in turn, serves to specify the sequence of amino acids in the PAH polypeptide chain through interactions with the tRNA molecules.

or, in terms of the single-letter abbreviations,

```
5'-AUGUCCACUGCGGUCCUGGAA-3'
   M  S  T  A  V  L  E
```

The full decoding operation for this region of the *PAH* gene is shown in **FIGURE 1.19**. In this figure, the initiation codon AUG is highlighted because some patients with PKU have a mutation in this particular codon. As might be expected from the fact that AUG is the initiation codon for polypeptide synthesis, cells in patients with this particular mutation fail to produce any of the PAH polypeptide. Mutation and its consequences are considered next.

1.5   Genes change by mutation.

The term **mutation** refers to any heritable change in a gene (or, more generally, in the genetic material); the term also refers to the process by which such a change takes place. One type of mutation results in a change in the sequence of bases in DNA. The change may be simple, such as the substitution of one pair of bases in a duplex molecule for a different pair of bases. For example, a C—G pair in a duplex molecule may mutate to T—A, A—T, or G—C. The change in base sequence may also be more complex, such as the deletion or addition of base pairs. Geneticists also use the term **mutant**, which refers to the result of a mutation. A mutation yields a mutant gene, which in turn produces a mutant mRNA, a mutant protein, and finally a mutant

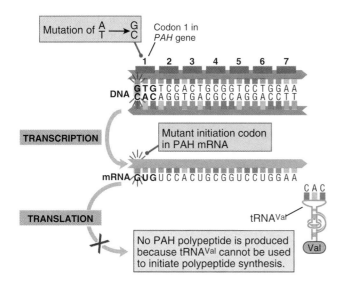

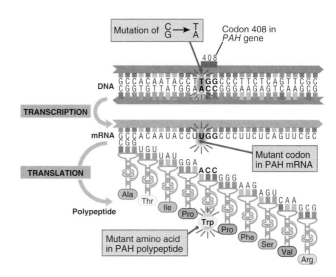

FIGURE 1.20 The M1V mutant in the *PAH* gene. The methionine codon needed for initiation mutates to a codon for valine. Translation cannot be initiated, and no PAH polypeptide is produced.

FIGURE 1.21 The R408W mutant in the *PAH* gene. Codon 408 for arginine (R) is mutated into a codon for tryptophan (W). The result is that position 408 in the mutant PAH polypeptide is occupied by tryptophan rather than by arginine. The mutant protein has no PAH enzyme activity.

organism that exhibits the effects of the mutation—for example, an inborn error of metabolism.

DNA from patients from all over the world who have phenylketonuria has been studied to determine what types of mutations are responsible for the inborn error. There are a large variety of mutant types. More than 400 different mutations have been described. In some cases part of the gene is missing, so the genetic information to make a complete PAH enzyme is absent. In other cases the genetic defect is more subtle, but the result is still either the failure to produce a PAH protein or the production of a PAH protein that is inactive. In the mutation shown in **FIGURE 1.20**, substitution of a G—C base pair for the normal A—T base pair at the very first position in the coding sequence changes the normal codon AUG (Met) used for the initiation of translation into the codon GUG, which normally specifies valine (Val) and cannot be used as a "start" codon. The result is that translation of the PAH mRNA cannot occur, so no PAH polypeptide is made. This mutant is designated M1V because the codon for M (methionine) at amino acid position 1 in the PAH polypeptide has been changed to a codon for V (valine). Although the M1V mutant is quite rare worldwide, it is common in some localities, such as in Québec province in Canada.

One PAH mutant that is quite common is designated R408W, which means that codon 408 in the PAH polypeptide chain has been changed from one coding for arginine (R) to one coding for tryptophan (W). This mutant is one of the four most common in cases of PKU among European Caucasians. The molecular basis of the mutation is shown in **FIGURE 1.21**. In this case, the first base pair in

codon 408 is changed from a C—G base pair into a T—A base pair. The result is that the PAH mRNA has a mutant codon at position 408; specifically, it has UGG instead of CGG. Translation does occur in this mutant because everything else about the mRNA is normal, but the result is that the mutant PAH carries a tryptophan (Trp) instead of an arginine (Arg) at position 408 in the polypeptide chain. The consequence of the seemingly minor change of one amino acid is very drastic, because the mutant PAH has no enzyme activity and so is unable to catalyze its metabolic reaction. In other words, the mutant PAH protein is complete but inactive. With PAH, as with other proteins, some amino acid replacements result in a polypeptide chain that is unable to fold properly. The incorrectly folded polypeptides are digested by proteases in the cell, which recycles the amino acids for use in the synthesis of other proteins.

STOP & THINK 1.5

Suppose you discover a novel mutant form of the enzyme phenylalanine hydroxylase (PAH) with the amino acid replacement R408Q—that is, the amino acid arginine (R) normally found at position 408 in the protein is replaced in the mutant by the amino acid glutamine (G). What single nucleotide substitution in the DNA coding for arginine at position 408 would result in R408Q?

1.6 Most traits are complex traits affected by multiple genetic and environmental factors.

Inborn errors of metabolism illustrate the general principle that genes code for proteins and that mutant genes code for mutant proteins that can result in inherited diseases such as phenylketonuria. But few people have even met anyone with phenylketonuria because its frequency is only about 1 in 10,000 individuals. An inherited trait like phenylketonuria, which is due to mutations in a single gene, is called a **simple Mendelian trait** because it occurs in families according to simple genetic ratios first discovered by Gregor Mendel (Chapter 2). Although about 2000 simple Mendelian diseases have been described, each of them is quite rare in the human population as a whole (although some individual diseases are more common in particular subgroups).

Most traits that you will encounter in everyday life are not simple Mendelian traits. The most commonly encountered diseases include heart disease, diabetes, kidney disease, autism, and bipolar disorder. Almost everyone knows somebody who is affected with one or more of these conditions, often a family member. Each of these common conditions occurs in about 1/100 individuals—at least 100 times more frequent than a typical simple Mendelian disorder.

These common diseases are examples of **complex traits** because their causation is a complex interplay between multiple genetic and environmental factors. A disease in which causation is complex is affected by genetic factors, but each genetic factor, acting alone, does not determine presence of the disease. Each genetic factor is a **risk factor** that increases the chance that an individual carrying the gene will manifest the disease. Each risk factor may have a relatively small effect, but the risk factors are cumulative.

Complex traits are also affected by environmental factors and lifestyle choices. An individual may have multiple genetic risk factors for heart disease, for example, but still delay the onset of the disease, minimize its severity, or avoid it altogether with lifestyle choices like eating a healthy diet, getting regular exercise, and not smoking. Conversely, an individual with few genetic rick factors for heart disease may nevertheless come down with the disease owing to poor lifestyle choices in diet, exercise, and tobacco use.

Not all complex traits are diseases. Most commonly observed differences among individuals are due to variation in complex traits. Height and weight are two prominent examples. Both traits are affected by multiple genetic and environmental factors acting together. In the case of weight, obvious environmental factors are diet and exercise, but weight is also affected by the cumulative impact of at least 700 known genetic factors, each of small effect.

1.7 Evolution means continuity of life with change.

The pathway for the breakdown and excretion of phenylalanine is by no means unique to human beings. One of the remarkable generalizations to have emerged from molecular genetics is that organisms that are very distinct—for example, plants and animals—share many features in their genetics and biochemistry. These similarities indicate a fundamental "unity of life":

KEY CONCEPT

All creatures on Earth share many features of the genetic apparatus, including genetic information encoded in the sequence of bases in DNA, transcription into RNA, and translation into protein on ribosomes with the use of transfer RNAs. All creatures also share certain characteristics in their biochemistry, including many enzymes and other proteins that are similar in amino acid sequence, three-dimensional structure, and function.

Groups of related organisms descend from a common ancestor.

Organisms share a common set of similar genes and proteins because they evolved by descent from a common ancestor. The process of **evolution** takes place when a population of organisms gradually changes in genetic composition through time. Evolutionary changes in genes and proteins result in differences in metabolism, development, and behavior among organisms, which allows them to become progressively better adapted to their environments. From an evolutionary perspective, the unity of fundamental molecular processes in organisms alive today reflects inheritance from a distant common ancestor in which the molecular mechanisms were already in place.

Not only the unity of life but also many other features of living organisms become comprehensible from an evolutionary perspective. For example, the interposition of an RNA intermediate in the basic flow of genetic information from DNA to RNA to protein makes sense if the earliest forms of life used RNA for both genetic information and enzyme catalysis. The importance of the evolutionary perspective in understanding aspects of biology that seem pointless or needlessly complex is summed up in a famous aphorism of the evolutionary biologist Theodosius Dobzhansky: "Nothing in biology makes sense except in the light of evolution."

Biologists distinguish three major kingdoms of organisms:

1. **Bacteria** This group includes most bacteria and cyanobacteria (formerly called blue-green algae). Cells of these organisms lack a membrane-bounded nucleus and mitochondria, are surrounded by a cell wall, and divide by binary fission.

2. **Archaea** This group was initially discovered among microorganisms that produce methane gas or that live in extreme environments, such as hot springs or pools with high salt concentrations. They are widely distributed in more normal environments as well. Superficially resembling bacteria, the cells of archaeans show important differences in the manner in which their membrane lipids are chemically linked. The machinery for DNA replication and transcription in archaeans resembles that of eukaryans, whereas their metabolism strongly resembles that of bacteria. DNA sequence analysis indicates that about half of the genes found in the kingdom Archaea are unique to this group.

3. **Eukarya** This group includes all organisms whose cells contain an elaborate network of internal membranes, a membrane-bounded nucleus, and mitochondria. Their DNA is present in the form of linear molecules organized into true chromosomes, and cell division takes place by means of mitosis. The eukaryotes include plants and animals as well as fungi and many single-celled organisms, such as amoebae and ciliated protozoa.

The members of the groups Bacteria and Archaea are often grouped together into a larger assemblage called **prokaryotes**, which literally means "before [the evolution of] the nucleus." This terminology is convenient for designating prokaryotes as a group in contrast with **eukaryotes**, which literally means "good [well-formed] nucleus."

The molecular unity of life is seen in comparisons of genomes.

The totality of DNA in a cell, nucleus, or organelle is called its **genome**. When used with reference to a species of organism, for example in phrases such as "the human genome," the term genome is defined as the DNA present in a normal reproductive cell.

Modern methods for sequencing DNA are so rapid and efficient that the complete DNA sequence is

known for hundreds of different species of organisms. These include the genomes of multiple representatives of many groups of organisms, including extinct human ancestors sequenced from DNA extracted from fossil bones. **TABLE 1.2** shows a small sample of sequenced genomes. Genome size is given in megabases (Mb), or millions of base pairs.

The organism denoted syn3.0 is very special in that it is not a naturally occurring organism, and we will discuss it later. Among the naturally occurring organisms in Table 1.2 are two bacteria: *Mycoplasma mycoides* is notable for its small genome and limited number of genes; *Escherichia coli* is more typical of bacteria in size and gene number. Both bacteria have

TABLE 1.2 Comparison of Genomes

Organism	Genome size, Mb[a] (approximate)	Number of genes (approximate)
syn3.0 (synthetic DNA bacterium)	0.5	473
Mycoplasma mycoides (causes bovine pneumonia)	1.2	985
Escherichia coli (common colon bacterium)	4.6	4000
Saccharomyces cerevisiae (baker's yeast)	12	6000
Caenorhabditis elegans (soil nematode)	100	20,000
Drosophila melanogaster (fruit fly)	180	16,000
Arabidopsis thaliana (mouse-ear cress)	135	28,000
Mus musculus (laboratory mouse)	2500	25,000
Homo sapiens (human being)	3000	25,000

[a]Millions of base pairs.

about the same density of genes—about one gene per Mb of DNA.

Baker's yeast (*Saccharomyces cerevisiae*) is a single-celled eukaryote. It has a genome size of 12 Mb organized into 16 chromosomes containing about 6000 genes. The gene density is about one gene per 2 Mb—twice that of typical bacteria.

Caenorhabditis elegans, *Drosophila melanogaster*, and the diminutive flowering plant *Arabidopsis thaliana* are complex, multicellular eukaryotes notable for their relatively small genome size, which is still substantially larger than that of baker's yeast. Their gene number is also larger than that of yeast, but only by a factor of 3–4, with an average gene density of one gene per 5–10 Mb. (Not all insects have a genome as small as that of *Drosophila*; the genome size of the mountain grasshopper, *Podisma pedestris*, is about 100 times larger than that of *Drosophila*, but it has about the same number of genes. Such paradoxes of genome size are discussed in Chapter 6.)

The genomes of mouse and human are much larger than those of the other multicellular eukaryotes in Table 1.2, yet they have about the same number of genes. This means that the gene density in mouse and human is much reduced, to roughly one gene per 100 Mb. The reduced gene density is reflected in the fact that only about 1.5 percent of the human genome sequence codes for protein. (About 27 percent of the human genome is present in protein-coding genes, but much of the DNA sequence present in such genes does not actually code for amino acids.)

Which brings us back to syn3.0, the world's first synthetic organism. It was created to identify the minimal set of genes that would enable a bacterial cell to multiply in growth medium containing amino acids and other small-molecule nutrients. Starting with cells of *Mycoplasma mycoides*, researchers at the J. Craig Venter Institute in La Jolla, California systematically knocked out each of the 985 genes to determine which genes were essential for growth. They then chemically synthesized a 0.5-Mb DNA molecule that contained only essential genes. Synthesis of such a large piece of DNA is technically extremely difficult, because long molecules of DNA in solution are fragile and break easily due to mechanical shear. In practice, large molecules have to be synthesized in smaller pieces that must then be assembled in proper order. In the case of syn3.0, the researchers made clever use of living yeast cells to combine the synthetic pieces in the right order and to faithfully replicate the molecule.

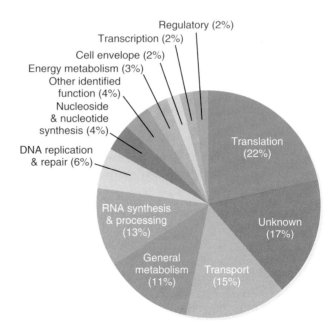

FIGURE 1.22 Functions of the 473 genes in the synthetic bacterium syn3.0.

Data from C. A. Hutchison III et al. *Science* 2016 Mar 25;351(6280):aad6253. doi: 10.1126/science.aad6253.

The completed genome was then tested for viability by being introduced into living *Mycoplasma* cells in which the original DNA had been destroyed.

The project was not without its surprises. There are, for example, more than a few cases in which each of two genes appears to be nonessential, but eliminating both results in cells that grow extremely slowly or not at all. A good analogy is the engines on a twin-engine jet like the Boeing 767: It can fly when one engine is disabled, but not both.

After much work and some trial and error, syn3.0—a viable cell with a synthetic minimal genome comprising 473 genes—was created. The functions of these 473 genes are summarized in **FIGURE 1.22**. As might be expected, the largest numbers of genes function in small-molecule transport, general metabolism, or synthesis and processing of macromolecules. Remarkably, 17 percent of the genes in the minimal genome have no identified function ("unknown"), which means a great deal of biology remains to be discovered. It is also worth noting that a substantial fraction of the genes in syn3.0 have recognizable counterparts in the other organisms listed in Table 1.2, attesting to the molecular unity of life on Earth.

CHAPTER SUMMARY

- Inherited traits are affected by genes.
- Genes are composed of the chemical deoxyribonucleic acid (DNA).
- DNA replicates to form (usually identical) copies of itself.
- DNA contains a code specifying what types of enzymes and other proteins are made in cells.
- DNA occasionally mutates, and the mutant forms specify altered proteins.
- A mutant enzyme is an "inborn error of metabolism" that blocks one step in a biochemical pathway for the metabolism of small molecules.
- Genetic analysis of mutants of the fungus *Neurospora* unable to synthesize an essential nutrient led to the one gene–one enzyme hypothesis.
- Different mutations in the same gene can be identified by means of a complementation test, in which the mutants are brought together in the same cell or organism. Mutations in the same gene fail to complement one another, whereas mutations in different genes show complementation.
- Most traits are complex traits affected by multiple genes as well as by environmental factors.
- Organisms change genetically through generations in the process of biological evolution.
- Because of their common descent, organisms share many features of their genetics and biochemistry.

ISSUES AND IDEAS

- What special feature of the structure of DNA allows each strand to be replicated without regard to the other?
- What does it mean to say that a strand of DNA specifies the structure of a molecule of RNA?
- What types of RNA participate in protein synthesis, and what is the role of each type of RNA?
- What is meant by the phrase *genetic code*, and how is the genetic code relevant to the translation of a polypeptide chain from a molecule of messenger RNA?
- What is meant by the term *genetic analysis*, and how is genetic analysis exemplified by the work of Beadle and Tatum using *Neurospora*?
- What is a complementation test, and what is it used for in genetic analysis?

SOLUTIONS: STEP BY STEP

PROBLEM 1 In the human gene for the beta chain of hemoglobin, the oxygen-carrying protein in the red blood cells, the first 30 nucleotides in the protein-coding region are as shown here.

```
3'-TACCACGTGGACTGAGGACTCCTCTTCAGA-5'
```

(a) What is the sequence of the partner strand?
(b) If the DNA duplex of this gene were transcribed from left to right, what is the base sequence of the RNA across this part of the coding region?
(c) What is the sequence of amino acids in this part of the beta-globin polypeptide chain?
(d) In the mutation responsible for sickle-cell anemia, the red T indicated is replaced with an A. The mutant is present at relatively high frequency in some human populations because carriers of the gene are more resistant to falciparum malaria than are noncarriers. What is the amino acid replacement associated with this mutation?

SOLUTION. (a) The partner strand is deduced from the rule that A pairs with T and G pairs with C; however, keep in mind that the paired DNA strands have opposite polarity (that is, their 5'-to-3' orientations are reversed). (b) The RNA strand is synthesized in the 5'-to-3' direction, which means that the template DNA strand is transcribed in the 3'-to-5' direction, which happens to be the same left-to-right orientation of the strand shown above. The base sequence is deduced from the usual base-pairing rules, except that A in DNA pairs with U in RNA. (c) The polypeptide chain is translated in successive groups of three nucleotides (each group constituting a codon), starting at the 5' end of the coding sequence in the RNA and moving in the 5'-to-3' direction. The amino acid corresponding to each codon can be found in the genetic code table. (d) The change from T to A in the transcribed strand alters a GAG codon into a GUG codon in the RNA transcript, resulting in the replacement of the normal glutamic acid (GAG) with valine (V). The nonmutant duplex, the RNA transcript, and the amino acid sequence are as shown below. The amino acid that is replaced in the sickle-cell mutant is indicated in red.

```
3'-TACCACGTGGACTGAGGACTCCTCTTCAGA-5'
5'-ATGGTGCACCTGACTCCTGAGGAGAAGTCT-3'

5'-AUGGUGCACCUGACUCCUGAGGAGAAGUCU-3'
   MetValHisLeuThrProGluGluLysSer
```

PROBLEM 2 The accompanying diagram shows a linear bioynthetic pathway for an essential nutrient designated F in an organism, such as *Neurospora*, able to grow in a minimal medium. Each red letter indicates one intermediate in the pathway, and each blue number indicates a mutant that blocks one step in the pathway.

$$A \xrightarrow{1} B \xrightarrow{2} C \xrightarrow{3} D \xrightarrow{4} E \xrightarrow{5} F$$

Make a table in which the columns correspond to the intermediates, arranged in alphabetical order, and the rows correspond to the mutants, arranged in numerical order. In the body of the table, insert a plus sign if the mutant will grow on minimal medium supplemented with the nutrient and a minus sign if the mutant will not grow under these conditions. Assume that all intermediates can be transported into the cell from the growth medium.

SOLUTION. This is a classic type of genetic analysis pioneered by Beadle and Tatum. The principle is that a mutant will grow on any intermediate whose position in the pathway is *downstream* of the metabolic block. Hence, mutant 1 will grow on any intermediate except A, mutant 2 will grow on any intermediate except A or B, and so forth. The complete matrix is as shown. It looks exceptionally simple because both the rows (mutants) and columns (intermediates) are arranged in the same order as their constituents appear in the pathway. Normally this will not be the case.

	A	B	C	D	E	F
1	−	+	+	+	+	+
2	−	−	+	+	+	+
3	−	−	−	+	+	+
4	−	−	−	−	+	+
5	−	−	−	−	−	+

PROBLEM 3 A complementation test is used to sort a set of mutants into groups, each group corresponding to a subset of the mutants that have defects in the same gene. Shown here are the genes (1–5) from the previous problem and 10 mutants (a–j) grouped according to the gene they affect.

1	2	3	4	5
a	e, g	c, b, h, j	f	d, i

Gene 1 is represented by mutant *a* only, gene 2 by mutants *e* and *g*, and so forth.

(a) Prepare a square complementation matrix of data, with the rows and columns representing the mutants in alphabetical order. Each entry in the matrix should be a plus sign if the row mutant and the column mutant do show complementation (that is, if they are mutants of different genes) or a minus sign if they do not show complementation (that is, if they are mutants of the same gene).
(b) What is special about the principal diagonal of the matrix? (The principal diagonal is the diagonal that runs from upper left to lower right.) What does this result mean biologically?
(c) What is special about the triangular parts of the matrix above and below the diagonal? What does this result mean biologically?
(d) Prepare a circular diagram of the mutants as discussed in the text, showing which of the mutants form complementation groups.

SOLUTION. (a) The complementation matrix is as shown here. (b) The principal diagonal consists exclusively of minus signs; biologically, this means that a mutant cannot undergo complementation with itself, because two copies of the identical mutation must be in the same gene. (c) The upper and lower triangular matrices are symmetrical, mirror images of one another; biologically, this means that the parent of origin of the mutant makes no difference to whether the mutants undergo complementation. Because of the symmetry of the data matrix, complementation data are often presented only in the form of the upper diagonal. (d) The circular type of the complementation test is also shown. It indicates that the complementation groups are {*a*}, {*b, c, h, j*}, {*d, i*}, {*e, g*}, and {*f*}. The complementation groups are not informative about where the product of each gene acts in the pathway; this information must come from the type of analysis illustrated in the previous problem.

(a–c)

	a	b	c	d	e	f	g	h	i	j
a	−	+	+	+	+	+	+	+	+	+
b	+	−	−	+	+	+	+	−	+	−
c	+	−	−	+	+	+	+	−	+	−
d	+	+	+	−	+	+	+	+	−	+
e	+	+	+	+	−	+	−	+	+	+
f	+	+	+	+	+	−	+	+	+	+
g	+	+	+	+	−	+	−	+	+	+
h	+	−	−	+	+	+	+	−	+	−
i	+	+	+	−	+	+	+	+	−	+
j	+	−	−	+	+	+	+	−	+	−

(d)

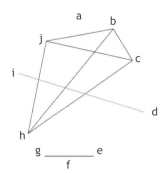

CONCEPTS IN ACTION: PROBLEMS FOR SOLUTION

1.1 Prior to the Avery, MacLeod, and McCarty experiment, what features of cells and chromosomes were already known that could have been interpreted as evidence that DNA is an important constituent of the genetic material?

1.2 In the early years of the twentieth century, why did most biologists and biochemists believe that proteins were probably the genetic material?

1.3 From their examination of the structure of DNA, what were Watson and Crick able to infer about the probable mechanisms of DNA replication, coding capability, and mutation?

1.4 What are three principal structural differences between RNA and DNA?

1.5 A region along an RNA transcript contains no U. What base will be missing in the corresponding region of the template strand of DNA?

1.6 When the base composition of a DNA sample from the bacterium *Salinicoccus roseus* was determined, 23.6 percent of the bases were found to be guanine. The DNA of this organism is known to be double stranded. What is the percentage of adenine in its DNA?

1.7 DNA extracted from a certain virus has the following base composition: 15 percent adenine, 25 percent thymine, 20 percent guanine, and 40 percent cytosine. How would you interpret this result in terms of the structure of the viral DNA?

1.8 A duplex DNA molecule contains 532 occurrences of the dinucleotide 5'-GT-3' in one or the other of the paired strands. What other dinucleotide is also present exactly 532 times?

1.9 A repeating polymer with the sequence

    5'-GAUGAUGAUGAU . . .-3'

was found to produce only two types of polypeptides in a translation system that uses cellular components but not living cells (called an *in vitro*

translation system). One polypeptide consisted of repeating Asp and the other of repeating Met. How can you explain this result?

1.10 If one strand of a DNA duplex has the sequence 5'-GTCAT-3', what is the sequence of the complementary strand. (Write the answer with the 5' end at the left.)

1.11 Consider a region along one strand of a double-stranded DNA molecule consists of tandem repeats of the trinucleotide 5'-CTA-3', so that the sequence in this strand is 5'-CTACTACTACTA . . .-3'. What is the sequence in the other strand? (Write the answer with the 5' end at the left.)

1.12 Part of the protein-coding region in a gene has the base sequence 3'-ACAGCATAAACGTTC-5'. What is the sequence of the partner DNA strand?

1.13 If the DNA sequence in Problem 1.12 is the template strand that is transcribed in the synthesis of messenger RNA, would it be transcribed from left to right or from right to left? What base sequence would this region of the RNA contain?

1.14 What amino acid sequence would be synthesized from the messenger RNA region in Problem 1.12?

1.15 If a mutation occurs in the DNA sequence in Problem 1.12 in which the red C is replaced with T, what amino acid sequence would result?

1.16 A polymer is made that has a random sequence consisting of 25 percent U's and 75 percent C's. Among the amino acids in the polypeptide chains resulting from in vitro translation, what is the expected frequency of Pro? Of Phe?

1.17 With *in vitro* translation of an RNA into a polypeptide chain, the translation can begin anywhere along the RNA molecule. A synthetic RNA molecule has the sequence

    5'-CGCUUACCACAUGUCGCGAAC-3'

How many reading frames are possible if this molecule is translated *in vitro*? How many reading frames are possible if this molecule is translated *in vivo*, in which translation starts with the codon AUG?

1.18 The coding sequence in the messenger RNA for amino acids 1 through 10 of human phenylalanine hydroxylase is

5′-AUGUCCACUGCGGUCCUGGAAAACCCAGGC-3′

  (a) What are the first 10 amino acids?
  (b) What sequence would result from a mutant RNA in which the red A was changed to G?
  (c) What sequence would result from a mutant RNA in which the red C was changed to G?
  (d) What sequence would result from a mutant RNA in which the red U was changed to C?
  (e) What sequence would result from a mutant RNA in which the red G was changed to U?

1.19 How is it possible for a gene with a mutation in the coding region to encode a polypeptide with the same amino acid sequence as the nonmutant gene?

1.20 Shown here is part of a metabolic pathway in a bacterium in which a substrate metabolite (small molecule) X is converted into a final product metabolite W through a sequence of three steps catalyzed by the enzymes A, B, and C. Each of the enzymes is the product of a different gene.

$$X \xrightarrow{A} Y \xrightarrow{B} Z \xrightarrow{C} W$$

Which metabolites would be expected to be missing, and which present in excess, in cells that are mutant for:

  (a) Enzyme A?
  (b) Enzyme B?
  (c) Enzyme C?

1.21 A mutant is isolated with a defect in one of the enzymes in the metabolic pathway in Problem 20, but it is not known which step (A, B, or C) is blocked. The final product W of the pathway is essential for growth. When mutant cells are placed in cultures lacking W, they cannot grow; but when W is added to the medium, they can grow. Experiments are carried out to determine whether any of the intermediates can substitute for W in supporting growth. The mutant cells are found to grow in the presence of Z but not in the presence of X or Y. Deduce from these data what step in the pathway is blocked in the mutant.

 STOP & THINK ANSWERS

ANSWER TO STOP & THINK **1.1**

(a) Because the G_1 bacteriophage have half the radioactivity of the G_0 bacteria, the G_2 would have half of the G_1 or $1/2 \times 1/2 = 1/4$ of the G_0. **(b)** With halving of the radioactivity in each bacteriophage generation, the G_5 would have $(1/2)^5 = 1/32$ of the radioactivity of original G_0 radioactivity.

ANSWER TO STOP & THINK **1.2**

3′-ATCGTTTTTATCG-5′

ANSWER TO STOP & THINK **1.3**

The data support **(b)** Y → X → A because a mutation in the X → A step would allow growth on A but not on X or Y, and a mutation in the Y → X step would allow growth on A or X but not on Y. These are the observed classes of mutants.

ANSWER TO STOP & THINK **1.4**

(a) The complementation data indicate three complementation groups: One group consists of mutants *m1* and *m4*, another of mutants *m2*, *m3*, and *m6*, and the third only of mutant *m5*. **(b)** If each complementation group represents one enzymatic step in the conversion of W into Z, then three enzymatic steps are involved.

ANSWER TO STOP & THINK **1.5**

The possible codons for arginine are CGU, CGC, CGA, CGG, AGA, or AGG; the possible codons for glutamine are CAA or CAG. The actual codon for arginine must be either CGA or CGG because, in either case, a change in the second position from A to G would result in substituting arginine with glutamine. Because a change in a codon (RNA) from A to G corresponds to a change in the transcribed strand of DNA from T to C, the single nucleotide substitution in DNA would be from a T–A nucleotide pair to a C–G nucleotide pair.

Design Credits: Stop & Think icon made by Darius Dan from www.flaticon.com; The Human Connection icon made by Daniel Bruce from www.flaticon.com; Elephant image: © NickBiemans/GettyImages.

CHAPTER 2

Transmission Genetics: Heritage from Mendel

LEARNING OBJECTIVES

- To understand the inheritance of genotypes and phenotypes of a single-gene trait and apply the principle of segregation to predict the types and expected proportions of the progeny from a mating.

- To use the principle of independent assortment to predict the possible types of progeny and their expected proportions with respect to two or more traits.

- To predict how the expected types and proportions of progeny are modified by different types of epistasis.

- To recognize the inheritance patterns of simple Mendelian dominant or recessive traits in human pedigrees.

- To make inferences about the genotypes of individuals in pedigrees based on their own phenotypes and those of their close relatives.

We learned in Chapter 1 how mutations in a single gene can result in severe genetic disorders like phenylketonuria; however, most single-gene disorders are rare in the population as a whole. Disorders like heart disease and diabetes, which are relatively common, are complex traits determined by many genes and environmental factors acting together. From a genetic point of view, the main difference between single-gene, simple Mendelian traits and complex traits is the number of genetic factors involved in their inheritance. A simple Mendelian disorder results from one genetic factor with a large effect, transmitted from generation to generation. A complex disease results from multiple genetic risk factors, each with a small (but not necessarily equal) effects.

The commonality between simple Mendelian traits and complex traits is that each genetic factor, whether a major gene for phenylketonuria or genetic risk factor for heart disease, is transmitted according to the same principles of inheritance. These principles underlie **transmission genetics**. They were discovered by Gregor Mendel, a monk who worked in a monastery in the city of Brno in the Czech Republic. Mendel also taught physics and natural history at a local secondary school. His teaching was said to be "clear, logical, and well suited to the needs of his students."

Mendel's most important biological experiments were his studies of crosses of the common garden pea (*Pisum sativum*). These were carried out from 1856 to 1863 in a small garden plot nestled in a corner of the monastery grounds. He reported his experiments to a local natural history society, published the results and his interpretation in its scientific journal in 1866, and began exchanging letters with one of the leading botanists of the time. At the time, no one understood the true significance of Mendel's findings, and his now-famous paper was ignored until, 16 years after his death, it was finally recognized as pathbreaking. Mendel's breakthrough experiments and concepts are the subject of this chapter.

2.1 Mendel took a distinctly modern view of transmission genetics.

Mendel's name will forever be associated with peas: round or wrinkled, yellow or green, tall or short. But it was not only his choice of experimental organism and his choice of traits that made Mendel's success possible. The basic premise underlying Mendel's experiments represented an important shift in approach. Although he didn't know about DNA or chromosomes, he came to realize that each parent contributed to its progeny a number of separate and distinct elements of heredity ("factors" as he called them—in modern terms, genes). More important still, he realized that each of these

parental factors remained unchanged as it was passed from one generation to the next.

Given the unchanging nature of these factors, Mendel set out to track their movement through generations of pea plants by observing the appearance of the traits associated with them, such as round or wrinkled seeds. He thought in quantitative, numerical terms. Mendel did not ask merely "What types of peas are present?" in the progeny of a cross, but also "What are their numerical ratios?" He proceeded by carrying out simple crossing experiments and then looked for statistical regularities that might identify general rules. In his own words, he wanted to "determine the number of different forms in which hybrid progeny appear" and, among these, to "ascertain their numerical interrelationships."

Mendel selected peas for his experiments for two reasons. First, he had access to varieties that differed in observable alternative characteristics, such as round versus wrinkled seeds, or yellow versus green seeds. Second, his preliminary studies had indicated that peas normally reproduce by self-fertilization, in which pollen produced in a flower is used to fertilize the eggs in the same flower. Left alone, pea flowers always self-fertilize. Carrying out a cross between two different varieties is actually very tedious. One must open the keel petal (which encloses the reproductive structures), remove the immature anthers (the pollen-producing structures) before they shed pollen, and dust the stigma (part of the female structure) with mature pollen taken from a flower on a different plant (**FIGURE 2.1**).

Mendel was careful in his choice of traits.

Mendel recognized the need to study traits that were uniform within any given variety of peas but different between varieties. For this reason, at the beginning of his experiments, he established **true-breeding** varieties, in which the plants produced only progeny like themselves when allowed to self-fertilize. For example, one true-breeding variety always yielded round seeds, whereas another true-breeding variety always yielded wrinkled seeds. For his experiments, Mendel chose seven pairs of varieties, each of which was true-breeding for a different trait. The contrasting traits affected

- seed shape (round versus wrinkled)
- seed color (yellow versus green)
- flower color (purple versus white)
- pod shape (smooth versus constricted)
- pod color (green versus yellow)
- flower and pod position (axial versus terminal)
- stem length (standard versus dwarf)

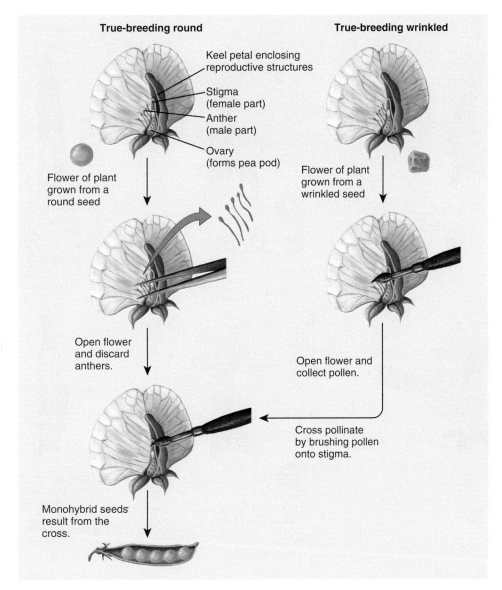

True-breeding round

True-breeding wrinkled

Keel petal enclosing reproductive structures

Stigma (female part)

Anther (male part)

Ovary (forms pea pod)

Flower of plant grown from a round seed

Flower of plant grown from a wrinkled seed

Open flower and discard anthers.

Open flower and collect pollen.

Cross pollinate by brushing pollen onto stigma.

Monohybrid seeds result from the cross.

FIGURE 2.1 Crossing pea plants requires some minor surgery in which the anthers of a flower are removed before they produce pollen. The stigma, the female part of the flower, is not removed. It is fertilized by brushing with mature pollen grains taken from another plant. Each pollinated flower has a single ovary that develops into the seed pod. The ovary contains as many as 10 ovules, which develop into seeds upon fertilization. These seeds represent the second generation; in this instance, a hybrid.

When two varieties that differ in one or more traits are crossed, the progeny constitute a **hybrid** between the parental varieties. Crosses in which the parental varieties differ in one, two, or three traits of interest are called *monohybrid, dihybrid,* or *trihybrid* crosses, respectively. Unless a trait is relevant to the experiment under consideration, it is normally ignored even if the parental varieties happen to differ in regard to this trait.

Reciprocal crosses yield the same types of offspring.

It is worthwhile to examine a few of Mendel's original experiments to learn what his methods were and how he interpreted his results. One pair of traits that he studied was round versus wrinkled seeds. When pollen from a variety of plants with wrinkled seeds was used to cross-pollinate plants from a variety with round seeds, all of the resulting hybrid seeds were round (**FIGURE 2.2**, cross A). Geneticists call the true-breeding parents the **P₁ generation** and the hybrid

KEY CONCEPT

The outcome of a genetic cross does not depend on which trait is present in the male and which is present in the female; reciprocal crosses yield the same result.

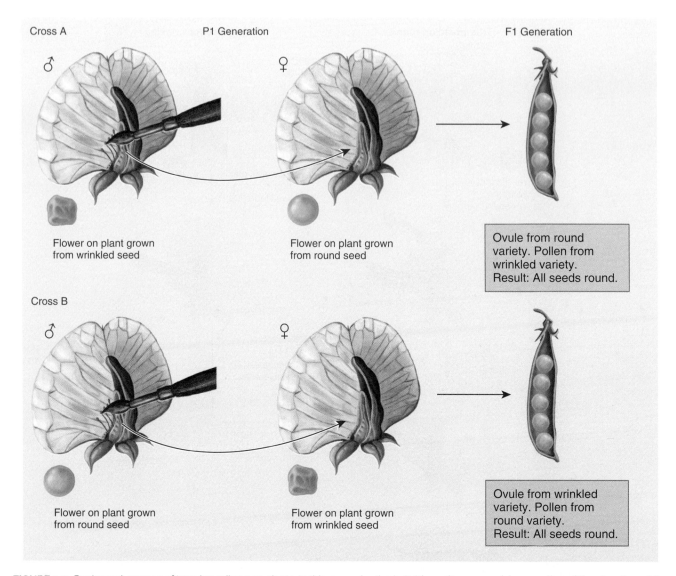

Cross A · P1 Generation · F1 Generation

Flower on plant grown
from wrinkled seed ♂

Flower on plant grown
from round seed ♀

Ovule from round
variety. Pollen from
wrinkled variety.
Result: All seeds round.

Cross B

Flower on plant grown
from round seed ♂

Flower on plant grown
from wrinkled seed ♀

Ovule from wrinkled
variety. Pollen from
round variety.
Result: All seeds round.

FIGURE 2.2 Reciprocal crosses of true-breeding pea plants. In this example, the hybrid seeds are round, irrespective of the direction of the cross.

filial seeds or plants the **F₁ generation**. Mendel also performed the **reciprocal cross** (Figure 2.2, cross B), in which plants from the variety with round seeds were used as the pollen parents and those from the variety with wrinkled seeds as the female parents. As before, all of the F₁ seeds were round. The reciprocal crosses in Figure 2.2 illustrate the principle that, in most cases,

Equivalent results were obtained when Mendel made crosses between plants that differed in any of the pairs of alternative characteristics. In each case, all of the F₁ progeny exhibited only one of the parental traits, and the other trait was absent. The trait expressed in the F₁ generation in each of the monohybrid crosses is shown in **FIGURE 2.3**. The trait expressed in the hybrids Mendel called the *dominant* trait; the trait not expressed in the hybrids he called *recessive*.

The wrinkled mutation causes an inborn error in starch synthesis.

Let us now consider Mendel's round and wrinkled seeds in the context of modern methods of genetic analysis and what we know today. Although most of Mendel's original experimental material has been lost, a strain of peas bearing what is thought to be the original wrinkled mutation was perpetuated by seed dealers in Eastern Europe. Analysis of this mutation using modern methods has revealed the function of the normal gene and the molecular basis of the wrinkled mutation. The normal gene encodes an enzyme, starch-branching enzyme I (SBEI), required to synthesize a branched-chain form of starch known as *amylopectin*. As pea seeds dry, they lose water and shrink. Round seeds contain amylopectin and shrink uniformly; wrinkled seeds lack amylopectin and

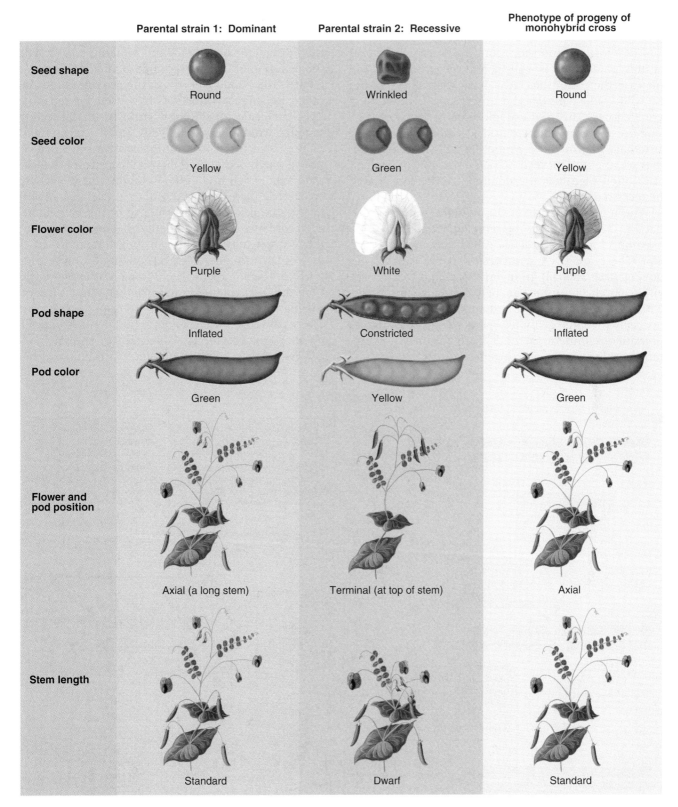

FIGURE 2.3 The seven character differences in peas studied by Mendel. The character considered dominant is the trait that appears in the hybrid produced by crossing. Which of each of the pairs of contrasting characters is dominant is revealed only after the F$_1$ progeny are formed.

shrink irregularly. In other words, wrinkled peas have an inborn error in starch metabolism.

The most common form of a trait occurring in a natural population is considered the **wildtype**, in this case the round pea. Any form that differs from the wildtype is considered a mutant, in this case the wrinkled pea. The wildtype and mutant forms of the gene are represented as *W* and *w*, respectively. (It is customary to print gene symbols in italic type. Geneticists use the mutant form to name the trait. A capital letter often identifies the dominant form, lowercase the recessive.)

The molecular basis of the wrinkled mutation is that the *SBEI* gene has become interrupted by the insertion of a DNA sequence called a *transposable element*. These are DNA sequences that are capable of moving (*transposition*) from one location to another within a chromosome or between chromosomes. Transposable elements are present in most genomes, especially the large genomes of eukaryotes, and many spontaneous mutations result from the insertion of transposable elements into a gene.

FIGURE 2.4, part A, is a simplified diagram of the DNA structure of the wildtype (nonmutant) form of the *SBEI* gene, along with the mutant form showing the insertion of the transposable element. One way to identify the *W* and *w* forms of the gene is a procedure called **gel electrophoresis**. It is used for separating DNA molecules of different sizes. In this procedure, samples containing relatively small fragments of duplex DNA are placed into slots near one edge of a slab of a jelly-like material (usually agarose), which is then submerged in a buffer solution and subjected to an electric field (Figure 2.4, part B). DNA fragments in the samples move in response to the electric field in accordance with their lengths. Shorter fragments move faster and farther than long fragments. In the case of DNA fragments corresponding to the *W* and *w* forms of the *SBEI* gene, the *W* fragment moves farther than the *w* fragment because the *w* fragment is larger (owing to the insertion of the transposable element). The separation of the *W* and *w* fragments is indicated by the dark rectangles, called *bands*, shown in the gel. As noted, a sample containing a mixture of both

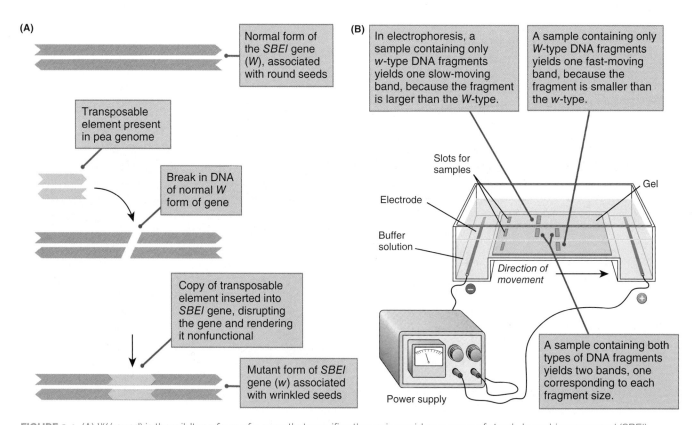

FIGURE 2.4 (A) *W* (round) is the wildtype form of a gene that specifies the amino acid sequence of starch-branching enzyme I (SBEI). The allele *w* (wrinkled) encodes an inactive form of the enzyme, inactive because its DNA sequence is interrupted by the insertion of a transposable element. (B) The molecular difference between *W* and *w* can be detected using electrophoresis. The DNA molecules are separated by size in an electric field. Each distinct size of DNA molecule produces a band at a characteristic position in the gel. A DNA molecule from the *w* gene, because it includes the transposable element, is larger than a molecule from the *W* gene and will migrate more slowly in the gel. A DNA sample containing both types of molecules will yield two bands in the gel.

W and *w* fragments yields two bands, one corresponding to *W* and the other to *w*.

Analysis of DNA puts Mendel's experiments in a modern context.

In discussing Mendel's results with round and wrinkled peas from a modern point of view, we must be careful to specify how the trait is examined. To avoid confusion, we use the terms morphological trait and molecular trait. A *morphological trait* is one that is manifest, plainly shown, and readily perceived by the senses. A *molecular trait* is one that can be perceived only by means of special methods, such as gel electrophoresis, that enable differences between molecules to be visualized. Classical geneticists studied primarily morphological traits (although their observations were sometimes aided by instruments such as the microscope). Modern geneticists study morphological traits too, but they usually supplement this with molecular analysis using techniques such as gel electrophoresis and DNA sequencing. With regard to round and wrinkled peas, the morphological trait corresponds to whether the shape of a seed is manifestly round or wrinkled. The molecular trait corresponds to the pattern of bands in an electrophoresis gel: whether the DNA extracted from a seed yields one rapidly migrating band, one slowly migrating band, or two bands.

Morphological traits are frequently dominant or recessive, but this is not necessarily true of molecular traits. In Figure 2.4, part B, for example, consider the molecular trait defined by the distance traveled by each DNA band from its starting position in the gel. The true-breeding strain with round seeds has a single rapidly migrating band, the true-breeding strain with wrinkled seeds has a single slowly migrating band, and the progeny of the cross (which has round seeds) exhibit both bands.

WW × ww → Ww

In other words, the progeny of the cross between the true-breeding strains show the molecular trait associated with both forms of the gene (in this case, a rapidly migrating DNA band along with a slowly migrating DNA band). In situations in which alternative forms of a gene (in this case, *W* and *w*) can both be detected when they are present in a cell or organism, we say that the forms of the gene are **codominant**. Molecular traits are often (but by no means always) codominant.

In the next section we will use the gel icons to show the molecular traits whose existence Mendel could only infer as he followed the morphological

traits of pea plants through many generations. This approach puts Mendel's experiments in the context of modern molecular genetics.

2.2 Genes come in pairs, separate in gametes, and join randomly in fertilization.

The prevailing concept of heredity in Mendel's time was that the traits of the parents became blended in the hybrid, as though the hereditary material consisted of fluids that became permanently mixed when combined. Following this logic to its natural conclusion, one would expect to see successive generations of offspring move toward a set of shared traits, with little to distinguish one individual from another. This did not happen with Mendel's monohybrid peas. In the first generation of hybrids, the recessive visible trait "disappeared," only to reappear in the next generation, after the hybrid progeny were allowed to undergo self-fertilization. For example, when the round hybrid seeds from the round × wrinkled cross—the F_1 generation—were grown into plants and allowed to self-fertilize, some of the resulting seeds were round and others wrinkled (**FIGURE 2.5**). The progeny seeds produced by self-fertilization of the F_1 generation constitute the **F_2 generation**. Mendel found that the dominant and recessive traits appear in the F_2 progeny in the proportions 3 round : 1 wrinkled.

Similar results were obtained in the F_2 generation of crosses between plants that differed in any of the pairs of alternative characteristics (**TABLE 2.1**). Note that the first two traits (round versus wrinkled seeds and yellow versus green seeds) have many more observations than any of the other traits; this is because seed shape and color can be classified directly in the seeds, whereas the other traits can be classified only in the mature plants. Relative to the inheritance

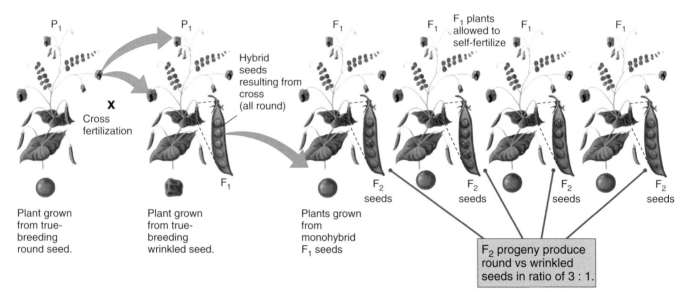

FIGURE 2.5 Some of Mendel's traits (such as flower color and plant height) are visible only in the mature plants grown from seeds, but other traits (such as seed shape and seed color) are visible in the seeds themselves.

TABLE 2.1 Results of Mendel's Monohybrid Experiments

Parental traits	F_1 trait	Number of F_2 progeny	F_2 ratio
round × wrinkled (seeds)	round	5474 round 1850 wrinkled	2.96 : 1
yellow × green (seeds)	yellow	6022 yellow 2001 green	3.01 : 1
purple × white (flowers)	purple	705 purple 224 white	3.15 : 1
inflated × constricted (pods)	inflated	882 inflated 299 constricted	2.95 : 1
green × yellow (unripe pods)	green	428 green 152 yellow	2.82 : 1
axial × terminal (flower position)	axial	651 axial 207 terminal	3.14 : 1
long × short (stems)	long	787 long 277 short	2.84 : 1

of visible traits, the principal conclusions from the data in Table 2.1 were as follows:

- The F_1 hybrids express only the dominant trait.

- In the F_2 generation, some plants show the dominant trait and others show the recessive trait.

- In the F_2 generation, there are approximately three times as many plants with the dominant

trait as plants with the recessive trait. In other words, the F_2 ratio of dominant : recessive is approximately 3 : 1.

In the remainder of this section, we will see how Mendel deduced from these basic observations his hypothesis of discrete genetic units and the principles governing their inheritance. We shall also see how he used statistical analysis to support it.

Genes are physical entities that come in pairs.

Important to Mendel's formulation of his hypothesis was the fact that in his monohybrid crosses, the recessive trait that seemingly disappeared in the F_1 generation reappeared again in the F_2 generation. Not only did the recessive trait reappear, it was in no way different from the trait present in the recessive P_1 plants. In a letter describing this finding, Mendel noted that in the F_2 generation, "the two parental traits appear, separated and unchanged, and there is nothing to indicate that one of them has either inherited or taken over anything from the other." From this finding, Mendel concluded that the hereditary determinants for the traits in the parental lines were transmitted as two different elements that retain their purity in the hybrids. In other words, the hereditary determinants do not "mix" or "contaminate each other." The implication of this conclusion is that a plant with the dominant trait might carry, in unchanged form, a hereditary determinant for the recessive trait.

The hypothesis of genetic transmission that Mendel developed to explain the reappearance of the recessive trait is outlined in **FIGURE 2.6**. The first element of the hypothesis is that each reproductive cell, or **gamete**, contains one representative of each kind

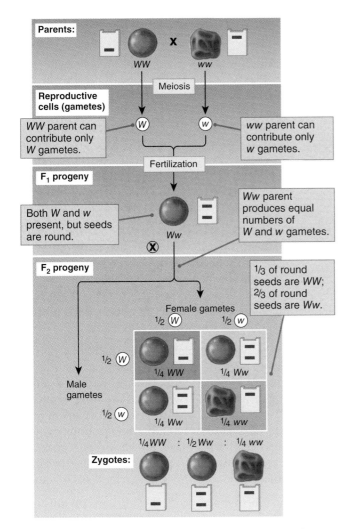

FIGURE 2.6 A diagrammatic explanation of the 3 : 1 ratio of dominant : recessive visible traits observed in the F$_2$ generation of a monohybrid cross. While a 3 : 1 ratio of visible traits is observed, the ratio of molecular traits (*WW* : *Ww* : *ww*) in the F$_2$ generation is 1 : 2 : 1, as depicted by the bands in the DNA gels.

of hereditary determinant in the plant. Mendel proposed that in the true-breeding variety with round seeds, all of the reproductive cells would contain the "round factor" (*W*) and that in the true-breeding variety with wrinkled seeds, all of the reproductive cells would contain the "wrinkled factor" (*w*). When the varieties are crossed, the F$_1$ hybrid should receive one each of *W* and *w* and so have the genetic constitution *Ww* (Figure 2.6). Because, with respect to seed shape, round (*W*) is dominant to wrinkled (*w*), the presence of *w* in the F$_1$ seeds is concealed, and so the seeds are round. Although the mutant *w* form of the gene is concealed with regard to the visible trait, it is not concealed with regard to the molecular trait. This is signified by the gel icon for the F$_1$ progeny in Figure 2.6, in which the DNA band corresponding to the mutant *w* form of the gene is clearly present. Whereas Mendel had to infer the presence of *w* from the progeny of crosses involving the F$_1$ progeny, the analysis of

DNA allows the presence of *w* in the F$_1$ progeny to be detected directly.

The paired genes separate (segregate) in the formation of reproductive cells.

The second key feature of Mendel's hypothesis in Figure 2.6 is that when an F$_1$ plant is self-fertilized (denoted by the encircled cross sign), the *W* and *w* determinants separate from one another and are included in the gametes in equal numbers. This separation of the hereditary elements is the heart of Mendelian genetics. The principle is called **segregation**.

KEY CONCEPT

The Principle of Segregation: In the formation of gametes, the paired hereditary determinants (genes) separate (segregate) in such a way that each gamete is equally likely to contain either member of the pair.

The principle of segregation implies not only that the hereditary determinants separate in the formation of gametes but also that, when separated, the hereditary determinants are completely unaltered by their having been paired in the previous generation. In Mendel's words, neither of them has "inherited or taken over anything from the other."

Gametes unite at random in fertilization.

The third key feature of Mendel's hypothesis is that the gametes produced by segregation come together in pairs *at random* to yield the progeny of the next generation. The assumption of random fertilization means that the result of self-fertilization of the F$_1$ plants in Figure 2.6 can be deduced by cross-multiplication in a square grid, as shown for the F$_2$ progeny. Each square within the grid represents the result of fertilization combining one type of pollen with one type of egg. When a *W*-bearing pollen fertilizes a *W*-bearing egg, the result is a *WW* fertilized egg, or **zygote**. Similarly, when a *W*-bearing pollen fertilizes a *w*-bearing egg, the result is a *Ww* zygote. The consequence of fertilization by *w*-bearing pollen is either a *Ww* zygote (if the egg carries *W*) or a *ww* zygote (if the egg carries *w*). These possibilities are shown in the grid in Figure 2.6.

The critical point is that the outcome of fertilization is a chance—or random—event. The probability of a particular gene combination occurring in the zygote is directly related to the frequency with which a particular gamete occurs. In this instance there is a 1-in-2 chance that a gamete will bear either the *W* or *w* gene (written as a probability of 1/2 along the top and left side of the grid).

♂

	$1/2\ w$	$1/2\ W$
♀ $1/2\ w$	$1/4\ ww$	$1/4\ Ww$
$1/2\ W$	$1/4\ Ww$	$1/4\ WW$

The probability of two such chance events occurring together—one particular female gamete combining with a particular male gamete—is calculated as the product of their individual probabilities. In this case, each zygote combination within the grid has a probability of $1/2 \times 1/2 = 1/4$ (a 1-in-4 chance of occurring). Because, in a Ww seed, it does not matter whether the W came through the pollen or the egg, random combinations of the gametes result in an F_2 generation with the genetic composition $1/4$ WW, $1/2$ Ww, and $1/4$ ww. This is the ratio of genetic types that would be observed via electrophoretic analysis of the DNA in the seeds, owing to the codominance of W and w at the molecular level. However, because of dominance at the level of the visible trait, the underlying $1/4$ WW : $1/2$ Ww : $1/4$ ww ratio is concealed, and instead one observes a ratio of $3/4$ round : $1/4$ wrinkled. (The $3/4$ comes from the fact that round seeds include $1/4$ WW + $1/2$ Ww = $3/4$ altogether.)

In summary, Mendel's key observation and the inference he made from it are as follows:

KEY CONCEPT

In the F_2 progeny of a monohybrid cross, the observed ratio of visible traits is $3/4$ dominant : $1/4$ recessive (or 3 : 1), but the dominance expressed at the level of the visible trait conceals the fact that the underlying ratio of genetic types is

$$1/4 : 1/2 : 1/4\ (\text{or}\ 1 : 2 : 1)$$

For example, $1/4$ WW : $1/2$ Ww : $1/4$ ww.

Genotype means genetic endowment; phenotype means observed trait.

The genetic hypothesis outlined in Figure 2.6 also illustrates another of Mendel's important deductions: Two plants with the same outward appearance—for example, with round seeds—might nevertheless differ in their hereditary makeup. One of the handicaps under which Mendel wrote was the absence of an established vocabulary of terms suitable for describing his concepts. Hence he made a number of seemingly elementary mistakes, such as occasionally confusing the outward appearance of an organism with its hereditary constitution. The necessary vocabulary was developed only after Mendel's work was rediscovered, and it includes the following essential terms.

1. A hereditary determinant of a trait is called a **gene**.

2. The different forms of a particular gene are called **alleles**. In Figure 2.6, the alleles of the gene for seed shape are W for round seeds and w for wrinkled seeds. W and w are alleles because they are alternative forms of the gene for seed shape. Alternative alleles are typically represented by the same letter or combination of letters, distinguished either by upper case versus lower case or by means of superscripts or subscripts or some other typographic identifier.

3. The **genotype** is the genetic constitution of an organism or cell—its molecular makeup. With respect to seed shape in peas, WW, Ww, and ww are examples of the possible genotypes for the W and w alleles. Because gametes contain only one allele of each gene, W and w are examples of genotypes of gametes.

4. A genotype in which the members of a pair of alleles are different, as in the Ww hybrids in Figure 2.6, is said to be **heterozygous**; a genotype in which the two alleles are alike is said to be **homozygous**. A homozygous organism may be homozygous dominant (WW) or homozygous recessive (ww). The terms *homozygous* and *heterozygous* cannot apply to gametes, because gametes contain only one allele of each gene.

5. The observable properties of an organism—including its visible traits—constitute its **phenotype**. Round seeds and wrinkled seeds are phenotypes. So are yellow seeds and green seeds. The phenotype of an organism does not necessarily imply anything about its genotype. For example, a seed with the phenotype "round" could have either the genotype WW or the genotype Ww.

6. A **dominant trait** is that expressed in the phenotype when the genotype is either heterozygous or homozygous. A **recessive trait** is that expressed in the phenotype when a genotype is homozygous for the alternative allele. The presence of a dominant trait masks a recessive trait.

The progeny of the F_2 generation support Mendel's hypothesis.

Mendel realized that the key to proving the genetic hypothesis outlined in Figure 2.6 lay with the round seeds in the F_2 generation. If his hypothesis was correct, then **1/3** of the round seeds should have the genetic composition WW and **2/3** of the round seeds should have the genetic composition Ww. The reason for the **1 : 2** ratio is shown in **FIGURE 2.7**. The ratio of WW : Ww : ww in the F_2 generation is 1 : 2 : 1, but if we disregard the *ww* seeds, then the ratio of *WW* : *Ww* is 1 : 2. In other words, 1/3 of the round seeds are *WW* and 2/3 are *Ww*. These ratios are apparent from the molecular analysis of the round seeds, but Mendel had to identify the genotypes of the seeds on the basis of the breeding behavior of the plants that grew out of them. He realized that, upon self-fertilization, the *WW* genotypes should be true-breeding for round seeds. He also realized that the *Ww* genotypes should produce progeny seeds that are round or wrinkled in the ratio 3:1. Furthermore, among the wrinkled seeds in

the F₂ generation, all should have the genetic composition *ww*, and so, upon self-fertilization, they should be true-breeding for wrinkled seeds.

For several of his traits, Mendel carried out self-fertilization of the F₂ plants in order to test these predictions. His results for round versus wrinkled seeds are summarized in **FIGURE 2.8**. As predicted from Mendel's genetic hypothesis, the plants grown from F₂ wrinkled seeds were true-breeding for wrinkled seeds. They produced only wrinkled seeds in the F₃ generation. Moreover, among 565 plants grown from F₂ round seeds, 193 were true-breeding, producing only round seeds in the F₃ generation, whereas the other 372 plants produced both round and wrinkled seeds in a proportion very close to 3 : 1. The ratio 193 : 372 equals 1 : 1.93, which is very close to the ratio 1 : 2 of *WW* : *Ww* genotypes predicted theoretically from the genetic hypothesis in Figure 2.7. Overall, taking all of the F₂ plants into account, the ratio of genotypes observed was very close to the predicted 1 : 2 : 1 of *WW* : *Ww* : *ww* expected from Figure 2.7.

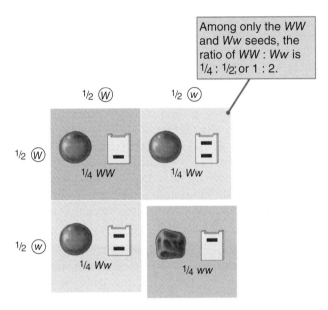

FIGURE 2.7 In the F2 generation, the ratio of *WW* : *Ww* : *ww* is 1 : 2 : 1. However, *among those seeds that are round*, the ratio of *WW* : *Ww* is 1 : 2, hence 1/3 of the round seeds are *WW* and 2/3 are *Ww*.

The progeny of testcrosses also support Mendel's hypothesis.

Mendel devised a second way to test the genetic makeup of the F₁ seeds, lending further support to his hypothesis (Figure 2.6). By crossing the plants grown from F₁ seeds with plants that were homozygous recessive, the genotype of the F₁ seeds would be revealed. Such a cross, between an organism of dominant phenotype (genotype unknown) and an organism of recessive phenotype (genotype known to be homozygous recessive), is called a **testcross**. If the parent with

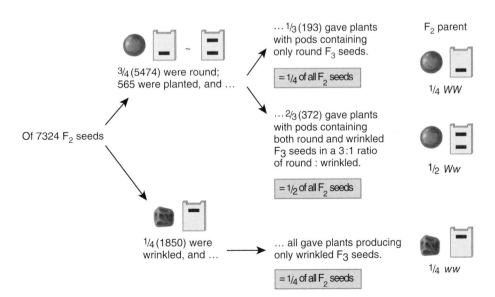

FIGURE 2.8 Mendel's results of self-fertilization of the F2 progeny of a cross between plants with round seeds and plants with wrinkled seeds. When he self-fertilized F2 plants grown from round seeds (the dominant trait), 1/3 of the progeny consisted of all round seeds and 2/3 of the progeny consisted of round : wrinkled seeds in the ratio 3 : 1. The result shows that the F2 seeds with the dominant trait (round) include two genetic types, *WW* and *Ww*, in a ratio of 1 : 2.

the dominant phenotype is homozygous, then the cross will produce progeny with the dominant phenotype. If the parent with the dominant phenotype is heterozygous (for example, *Ww*), then the result of the testcross will be progeny with both dominant and recessive phenotypes, as shown in **FIGURE 2.9**. Because of segregation, the heterozygous parent is expected to produce *W* and *w* gametes in equal numbers. When these gametes combine at random with the *w*-bearing gametes produced by the homozygous recessive parent, the expected progeny are 1/2 with the genotype *Ww* and 1/2 with the genotype *ww*. The former have the dominant visible phenotype (round, because *W* is dominant to *w*), whereas the latter have the recessive visible phenotype (wrinkled), and so the expected ratio of dominant to recessive phenotypes is 1 : 1. If the organism being tested in a testcross is heterozygous, the ratio of visible phenotypes will be the same as the ratio of molecular phenotypes. Both ratios are 1 : 1, as indicated by the gel icons in Figure 2.9. This is why a testcross is often extremely useful in genetic analysis.

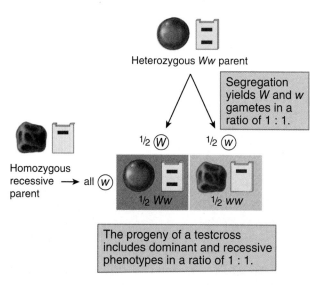

FIGURE 2.9 A testcross shows the result of segregation directly in the phenotypes of the progeny. This example illustrates a testcross of a *Ww* heterozygous parent with a *ww* homozygous recessive. The expected progeny are *Ww* and *ww* in a ratio of 1 : 1.

Mendel carried out a series of testcrosses with the genes for round versus wrinkled seeds, yellow versus green seeds, purple versus white flowers, and long versus short stems. The results are shown in **TABLE 2.2**. In all cases, the ratio of phenotypes among the progeny is very close to the 1 : 1 ratio expected from segregation of the alleles in the heterozygous parent.

Another valuable type of cross is a **backcross**, in which hybrid organisms are crossed with one of the parental genotypes. Backcrosses are commonly used by geneticists and by plant and animal breeders, as we will see in later chapters. Note that the testcrosses in Table 2.2 are also, in effect, backcrosses because the F_1 heterozygous parent that came from a cross between the homozygous dominant and the homozygous recessive is backcrossed with a homozygous recessive.

TABLE 2.2 Results of Mendel's Testcross Experiments

Testcross (F_1 heterozygote × homozygous recessive)	Progeny from testcross	Ratio
round × wrinkled seeds	193 round 192 wrinkled	1.01 : 1
yellow × green seeds	196 yellow 189 green	1.04 : 1
purple × white flowers	85 purple 81 white	1.05 : 1
long × short stems	85 long 79 short	1.01 : 1

STOP & THINK 2.2

In the matings shown here, suppose that the *A* allele is dominant to the *a* allele.

1. *AA × AA*   2. *AA × Aa*   3. *AA × aa*
4. *Aa × Aa*   5. *Aa × aa*   6. *aa × aa*

Which of the matings is expected to produce only progeny with the dominant phenotype? Only progeny with the recessive phenotype? Dominant and recessive phenotypes in a ratio of 3 : 1? Dominant and recessive phenotypes in a ratio of 1 : 1?

2.3 The alleles of different genes segregate independently.

Mendel also paid special attention to seed color—yellow *versus* green—because, like seed shape, seed color can be classified directly in the seeds. The green color is due to a defect in an enzyme necessary to break down the green pigment chlorophyll. Homozygous mutant seeds cannot break down their chlorophyll and, therefore, remain green, whereas wildtype seeds do break down their chlorophyll and turn yellow (like the leaves of certain trees in the autumn). In experiments in which plants homozygous for wrinkled and green (genotype *ww gg*) were crossed with those homozygous for round and yellow (genotype *WW GG*), Mendel made another important discovery. From this cross, the F_1 seeds are doubly heterozygous *Ww Gg* and show a phenotype of round, yellow. When he cultivated the F_1 plants and allowed self-fertilization to take place, among the seeds of the F_2 generation he observed four types of seed phenotypes in the following numbers:

round, yellow	315
round, green	108
wrinkled, yellow	101
wrinkled, green	32
Total	556

In these data, Mendel noted the presence of the expected monohybrid 3 : 1 ratio for each trait separately. With respect to each trait, the progeny were

round : wrinkled
$$= (315 + 108) : (101 + 32)$$
$$= 423 : 133$$
$$= 3.18 : 1$$

yellow : green
$$= (315 + 101) : (108 + 32)$$
$$= 416 : 140$$
$$= 2.97 : 1$$

Furthermore, in the F_2 progeny of the dihybrid cross, the separate 3 : 1 ratios for the two traits were combined at random, as shown in **FIGURE 2.10**. That is, among the 3/4 of the progeny that are round, 3/4 are yellow and 1/4 green; similarly, among the 1/4 of the progeny that are wrinkled, 3/4 are yellow and 1/4 green. The overall proportions of round–yellow to round–green to wrinkled–yellow to wrinkled–green are therefore expected to be

3/4 × 3/4 to 3/4 × 1/4 to 1/4 × 3/4 to 1/4 × 1/4

or

9/16 : 3/16 : 3/16 : 1/16

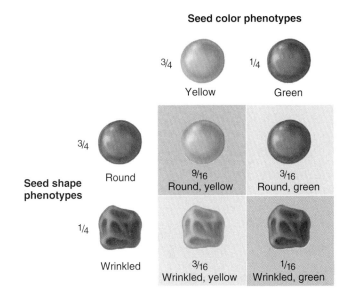

Seed color phenotypes

The observed ratio of 315 : 108 : 101 : 32 equals 9.84 : 3.38 : 3.16 : 1, which is reasonably close to the 9 : 3 : 3 : 1 ratio expected from the cross multiplication of the separate 3 : 1 ratios in Figure 2.10.

Ratio of phenotypes in the F_2 progeny of a dihybrid cross is 9 : 3 : 3 : 1.

FIGURE 2.10 The 3 : 1 ratio of round : wrinkled, when combined at random with the 3 : 1 ratio of yellow : green, yields the 9 : 3 : 3 : 1 ratio that Mendel observed in the F_2 progeny of the dihybrid cross.

The F_2 genotypes in a dihybrid cross conform to Mendel's prediction.

Mendel carried out similar experiments with other combinations of traits. For each pair of traits, he consistently observed the 9 : 3 : 3 : 1 ratio. He also deduced the biological reason for the observation. To illustrate his explanation using the dihybrid round × wrinkled cross, we can represent the dominant and recessive alleles of the pair affecting seed shape as *W* and *w*, respectively, and the allelic pair affecting seed color as *G* (yellow) and *g* (green). Mendel proposed that the underlying reason for the 9 : 3 : 3 : 1 ratio in the F_2 generation is that the segregation of the alleles *W* and *w* for round or wrinkled seeds has no effect on the segregation of the alleles *G* and *g* for yellow or green seeds. Each pair of alleles undergoes segregation into the gametes independently of the segregation of the other pair of alleles. The parental genotypes in the P_1 generation are *WW GG* (round, yellow seeds) and *ww gg* (wrinkled, green seeds). When these are crossed, the genotype of the F_1 hybrid is the double heterozygote *Ww Gg*.

The result of independent segregation in the F_1 plants is that the *W* allele is just as likely to be included in a gamete with *G* as with *g*, and the *w* allele is just as

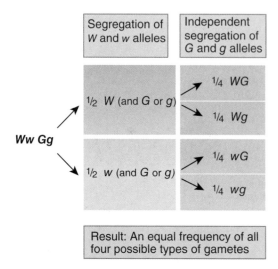

FIGURE 2.11 Independent segregation of the *Ww* and *Gg* allele pairs means that among each of the *W* and *w* classes, the ratio of *G* : *g* is 1 : 1. Likewise, among each of the *G* and *g* classes, the ratio of *W* : *w* is 1 : 1.

"Identical twins" really are genetically identical, but the environment matters, too. In this case, the girl on the right has spent more time in the sun, resulting in a darker skin and bleached hair color.
© sf2301420max/Shutterstock.

likely to be included in a gamete with *G* as with *g*. The independent segregation is illustrated in **FIGURE 2.11**. The independent segregation of the *W*, *w* and the *G*, *g* allele pairs implies that the gametes produced by the double heterozygote *Ww Gg* are

$$1/4 \; W\,G \quad 1/4 \; W\,g \quad 1/4 \; w\,G \quad 1/4 \; w\,g$$

When the four types of gametes combine at random to form the zygotes of the next generation, the result of independent segregation is as shown in **FIGURE 2.12**. Again, we use cross multiplication to show how the F_1 female and male gametes combine at random to produce the F_2 genotypes. This format is called a **Punnett square**. In the Punnett square, the combinations of seed shape and color phenotypes of the F_2 progeny are indicated. Note that the ratio of phenotypes is 9 : 3 : 3 : 1 for round yellow : wrinkled yellow : round green : wrinkled green.

The Punnett square in Figure 2.12 also shows that the ratio of *genotypes* in the F_2 generation is not 9 : 3 : 3 : 1. With independent segregation, the ratio of genotypes in the F_2 generation is

$$1 : 2 : 1 : 2 : 4 : 2 : 1 : 2 : 1$$

The reason for this ratio is shown in **FIGURE 2.13**. Among seeds with the *WW* genotype, the ratio of *GG* : *Gg* : *gg* is 1 : 2 : 1. Among seeds with the *Ww* genotype, the ratio is 2 : 4 : 2 (the 1 : 2 : 1 is multiplied by 2 because there are twice as many *Ww* genotypes as either *WW* or *ww*). And among seeds with the *ww* genotype, the ratio of *GG* : *Gg* : *gg* is 1 : 2

the genotypes. The combined ratio of phenotypes is 9 : 3 : 3 : 1.

Mendel tested the hypothesis of independent segregation by ascertaining whether the predicted genotypes were actually present in the expected proportions. He did the tests by growing plants from the F_2 seeds and obtaining F_3 progeny by self-pollination. To illustrate the tests, consider one series of crosses in which he grew plants from F_2 seeds that were round, green. Note in Figures 2.12 and 2.13 that round, green F_2 seeds are expected to have either the genotype *Ww gg* or the genotype *WW gg* in the ratio 2 : 1. Mendel grew 102 plants from such seeds and found that 67 of them produced pods containing both round, green and wrinkled, green seeds (indicating that the parental plants must have been *Ww gg*) and 35 of them produced pods containing only round, green seeds (indicating that the parental genotype was *WW gg*). The ratio 67 : 35 is in good agreement with the expected 2 : 1 ratio of genotypes.

Mendel's observation of independent segregation of two pairs of alleles has come to be known as the principle of **independent assortment**:

KEY CONCEPT

The Principle of Independent Assortment:
Segregation of the members of any pair of alleles is independent of the segregation of other pairs in the formation of reproductive cells.

Although the principle of independent assortment is fundamental to Mendelian genetics, there are important exceptions when genes are sufficiently close together in the same chromosome.

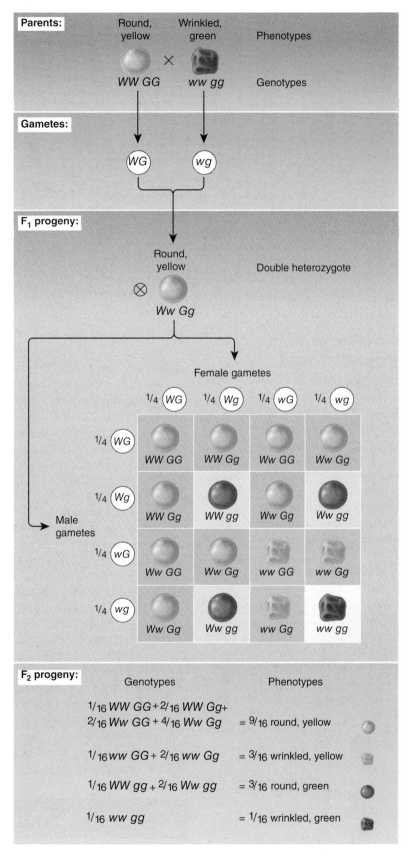

FIGURE 2.12 Diagram showing the basis for the 9 : 3 : 3 : 1 ratio of F_2 phenotypes resulting from a cross in which the parents differ in two traits determined by genes that undergo independent segregation.

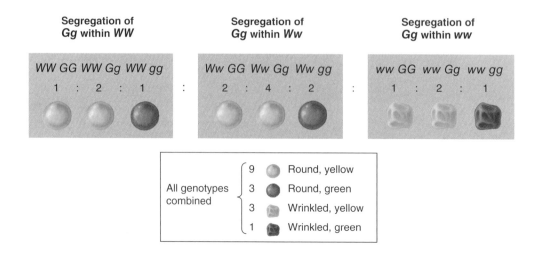

FIGURE 2.13 In the F$_2$ progeny of the dihybrid cross for seed shape and seed color, in any of the genotypes for one of the allele pairs, the ratio of homozygous dominant, heterozygous, and homozygous recessive genotypes for the other allele pair is 1 : 2 : 1.

The progeny of testcrosses show the result of independent assortment.

A second way in which Mendel tested the hypothesis of independent assortment was by carrying out a testcross with the F$_1$ genotypes that were heterozygous for both genes (*Ww Gg*). In this testcross, one parental genotype has to be multiply homozygous recessive—in this case, *ww gg*. As shown in **FIGURE 2.14**, the double heterozygotes produce four types of gametes—*W G*, *W g*, *w G*, and *w g*—in equal proportions, whereas the *ww gg* plants produce only *w g* gametes. Thus the progeny phenotypes are expected to consist of round yellow, round green, wrinkled yellow, and wrinkled green in a ratio of 1 : 1 : 1 : 1. As in a testcross of a monohybrid, the ratio of phenotypes in the progeny is

a direct demonstration of the ratio of gametes produced by the heterozygous parent, because no dominant alleles are contributed by the homozygous recessive parent to obscure the results. In the actual cross, Mendel obtained 55 round yellow, 51 round green, 49 wrinkled yellow, and 53 wrinkled green, which is in good agreement with the predicted 1 : 1 : 1 : 1 ratio. The results were the same in the reciprocal cross with *Ww Gg* as the female parent and *ww gg* as the male parent. This observation confirmed Mendel's assumption that the gametes of both sexes included all possible genotypes in approximately equal proportions.

An interesting historical note: Mendel's paper does not explicitly state either the principle of segregation (sometimes called Mendel's first law) or the principle of independent assortment (sometimes called Mendel's second law). On this basis, one could argue that Mendel did not discover Mendel's laws! On the other hand, Mendel did seem to have a pretty clear idea of what was going on. Six times in his relatively short paper, he repeated what he evidently thought was the main message: "Pea hybrids form germinal and pollen cells that in their composition correspond in equal numbers to all the constant [true-breeding] forms resulting from the combination of traits united through fertilization." One could not make this statement without invoking both segregation and independent assortment.

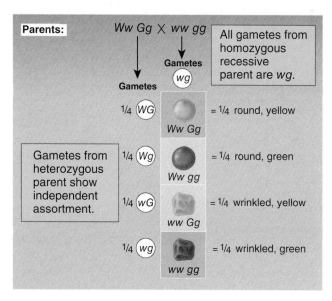

FIGURE 2.14 Genotypes and phenotypes resulting from a testcross of a *Ww Gg* double heterozygote.

⚙ STOP & THINK 2.3

If genes *A* and *B* undergo independent assortment, what is the expected proportion of *aa Bb* offspring from the mating *Aa bb* × *Aa Bb*?

2.4   Chance plays a central role in Mendelian genetics.

As we have seen, chance plays a central role in Mendelian genetics. In the formation and fertilization of gametes, the particular combination of alleles that occurs is random and subject to chance variation. In a genetic cross, the proportions of the different types of offspring obtained are the cumulative result of numerous individual events of fertilization. It is for this reason that a working knowledge of the rules of probability is basic to understanding the transmission of hereditary characteristics.

In the analysis of genetic crosses, the probability of a particular outcome may be considered equivalent to the number of times that an outcome is expected to occur over a large number of repeated trials. This number, expressed as a ratio, is also considered equivalent to the *probability* that this particular outcome will occur in a single trial. For example, in the F_2 generation of the hybrid between pea varieties with round seeds and those with wrinkled seeds, Mendel observed 5474 round seeds and 1850 wrinkled seeds (Table 2.1). In this case, the proportion of wrinkled seeds was $1850/(1850 + 5474) = 1/3.96$, or very nearly 1/4. We may therefore regard 1/4 as the approximate proportion of wrinkled seeds to be expected among a large number of progeny from this cross. Completely equivalently, we can regard 1/4 as the probability that any particular seed chosen at random will be wrinkled.

Evaluating the probability of any possible outcome of a genetic cross usually requires an understanding of the mechanism of inheritance and knowledge of the particular cross. For example, in evaluating the probability of obtaining a round seed from a particular cross, one needs to know that there are two alleles, *W* and *w*, with *W* dominant to *w*. One also needs to know the particular cross, because the probability of round seeds is determined by whether the cross is

$WW \times ww$, in which case all the progeny seeds are expected to be round,
$Ww \times Ww$, in which case 3/4 of the progeny seeds are expected to be round, or
$Ww \times ww$, in which case 1/2 of the progeny seeds are expected to be round.

The addition rule applies to mutually exclusive possibilities.

Sometimes an outcome of interest can be expressed in terms of two or more possibilities. For example, a seed with the phenotype "round" may have either of two genotypes, *WW* or *Ww*. A seed that is round cannot have both genotypes at the same time. Only one possibility, such as the presence of the *WW* or the

Ww genotype, can be realized in any one organism, and the realization of one such possibility precludes the realization of others. In this example, the realization of the genotype *WW* in a seed precludes the realization of the genotype *Ww* in the same seed, and the other way around. Outcomes that exclude each other in this manner are said to be *mutually exclusive*. When the possible outcomes are mutually exclusive, their probabilities are combined according to the addition rule.

KEY CONCEPT

Addition Rule: The probability of the realization of one or the other of two mutually exclusive possibilities, A or B, is the sum of their separate probabilities.

In symbols, where Prob is used to mean *probability*, the addition rule is written

$$\text{Prob } \{A \text{ or } B\} = \text{Prob } \{A\} + \text{Prob } \{B\}$$

The addition rule can be applied to determine the proportion of round seeds expected from the cross $Ww \times Ww$, which is illustrated in Figure 2.7. The round-seed phenotype results from the expression of either of two genotypes, *WW* and *Ww*, and these possibilities are mutually exclusive. In any particular progeny seed, the probability of genotype *WW* is 1/4 and that of *Ww* is 1/2. Hence the overall probability of either *WW* or *Ww* is

$$\text{Prob } \{WW \text{ or } Ww\} = \text{Prob } \{WW\} + \text{Prob } \{Ww\}$$
$$= 1/4 + 1/2 = 3/4$$

Because 3/4 is the probability of an individual seed being round, it is also the expected proportion of round seeds among a large number of progeny.

The multiplication rule applies to independent possibilities.

Possible outcomes that are not mutually exclusive may be *independent*, which means that the realization of one outcome has no influence on the possible realization of any others. For example, in Mendel's crosses for seed shape and color, the two traits are independent, and the proportions of phenotypes in the F_2 generation are expected to be 9/16 round yellow, 3/16 round green, 3/16 wrinkled yellow, and 1/16 wrinkled green. These proportions can be obtained by considering the traits separately, because they are independent. Considering only seed shape, we can expect the F_2 generation to consist of 3/4 round and 1/4 wrinkled seeds. Considering only seed color, we can expect the F_2 generation to consist of 3/4 yellow and 1/4 green. Because the traits are inherited independently, among the 3/4 of the seeds that are round, there should be 3/4 that are yellow, and so the overall proportion of round yellow seeds is expected to be 3/4 × 3/4 = 9/16 (Figure 2.10). Likewise, among the 3/4 of the seeds that are round, there should be 1/4 green, yielding 3/4 × 1/4 = 3/16 as the expected proportion of round, green seeds. The proportions of the other phenotypic classes can be deduced in a similar way using the cross multiplication method illustrated in Figure 2.10. The principle is that when outcomes are independent, the probability that they are realized together is obtained by multiplication.

Successive offspring from a cross are also independent outcomes, which means that the genotypes of early progeny have no influence on the relative proportions of genotypes in later progeny. The independence of successive offspring contradicts the widespread belief that in each human family, the ratio of girls to boys must "even out" at approximately 1 : 1 such that if parents already have, say, four girls, then they are somehow more likely to have a boy the next time around. But this belief is not supported by theory, and it is also contradicted by actual data on the sex ratios in human sibships. (The term **sibship** refers to a group of offspring from the same parents.) The data indicate that parents are no more likely to have a girl on the next birth if they already have five boys than if they already have five girls. The statistical reason is that although the sex ratios tend to balance out when they are averaged across a large number of sibships, they do not need to balance within individual sibships. Thus, among families in which there are five children, the sibships consisting of five boys balance those consisting of five girls, for an overall sex ratio of 1 : 1. However, both of these sibships are unusual in their sex distribution.

When the possible outcomes are independent (such as independent traits or successive offspring from a cross), the probabilities are combined by means of the multiplication rule.

KEY CONCEPT

Multiplication Rule: The probability of two independent possibilities, A and B, being realized simultaneously is given by the product of their separate probabilities.

In symbols, the multiplication rule is

Prob {A and B} = Prob {A} · Prob {B}

The multiplication rule can be used to answer questions like the following: Of two offspring from the mating *Aa* × *Aa*, what is the probability that both have the dominant phenotype? Because the mating is

Aa × *Aa*, the probability that any particular offspring has the dominant phenotype equals 3/4. Using the multiplication rule, we find that the probability that both of two offspring have the dominant phenotype is 3/4 × 3/4 = 9/16.

Here is a typical genetic question that can be answered by using the addition and multiplication rules together: Of two offspring from the mating *Aa* × *Aa*, what is the probability of one dominant phenotype (probability of 3/4) and one recessive (probability of 1/4)? Sibships of one dominant phenotype and one recessive can come about in two different ways, with the dominant born first or with the dominant born second, and these outcomes are mutually exclusive. The probability of the first case is 3/4 × 1/4 and that of the second is 1/4 × 3/4; because the outcomes are mutually exclusive, the probabilities are added. The answer is, therefore,

$$(3/4 \times 1/4) + (1/4 \times 3/4) = 2(3/16) = 3/8$$

The addition and multiplication rules are very powerful tools for calculating the probabilities of genetic events. **FIGURE 2.15** shows how the rules are applied to determine the expected proportions of the nine different genotypes possible among the F_2 progeny produced by self-pollination of a *Ww Gg* dihybrid.

In genetics, independence applies not only to the successive offspring formed by a mating but also to genes that segregate according to the principle of independent assortment (**FIGURE 2.16A**). The independence means that the multiplication rule can be used to determine the probability of the various types of progeny from a cross in which there is independent assortment among numerous pairs of alleles. This principle is the theoretical basis for the expected progeny types from the dihybrid cross shown in Figure 2.15. One can also use the multiplication rule to calculate the probability of a specific genotype among the progeny of a cross. For example, if a quadruple heterozygote of genotype *Aa Bb Cc Dd* is self-fertilized, the probability of a quadruple heterozygote *Aa Bb Cc Dd* offspring is

$$(1/2)(1/2)(1/2)(1/2) = (1/2)^4 = 1/16$$

assuming independent assortment of all four pairs of alleles.

STOP & THINK 2.4

If genes *A* and *B* undergo independent assortment, what is the expected proportion of homozygous offspring from the mating *Aa bb* × *Aa Bb*?

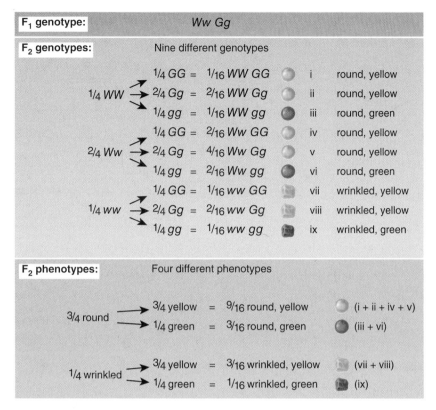

F₁ genotype:	*Ww Gg*

FIGURE 2.15 Example of the use of the addition and multiplication rules to determine the probabilities of the nine genotypes and four phenotypes in the F_2 progeny obtained from self-pollination of a dihybrid F_1. The roman numerals are arbitrary labels identifying the F_2 genotypes.

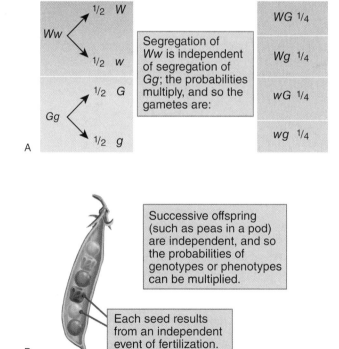

FIGURE 2.16 In genetics, two important types of independence are independent segregation of alleles that show independent assortment (A) and independent fertilizations resulting in successive offspring (B). In these cases, the probabilities of each of the individual outcomes of segregation or fertilization are multiplied to obtain the overall probability.

2.5 The results of segregation can be observed in human pedigrees.

Determining the genetic basis of a trait from the kinds of crosses that we have considered requires controlled matings and large numbers of offspring. The analysis of segregation by this method is not possible in human families, and it is usually not feasible for traits in large domestic animals. However, the mode of inheritance of a trait can sometimes be determined by examining the appearance of the phenotypes that reflect the segregation of alleles in several generations of related individuals. This is typically done with a family tree that shows the phenotype of each individual; such a diagram is called a **pedigree**. An important application of probability in genetics is its use in pedigree analysis.

FIGURE 2.17 depicts most of the standard symbols used in drawing a human pedigree. Females are represented by circles and males by squares. (A diamond is used if the sex of an individual is unknown.) Persons with the phenotype of interest are indicated by colored or shaded symbols. A mating between a female and a male is indicated by joining their symbols with a horizontal line, which is connected vertically to a second horizontal line running beneath that connects

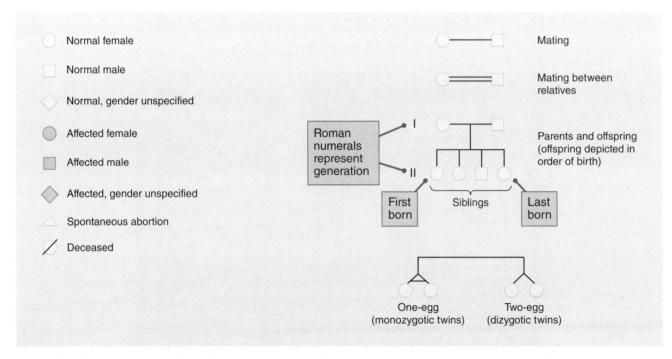

FIGURE 2.17 Conventional symbols used in depicting human pedigrees.

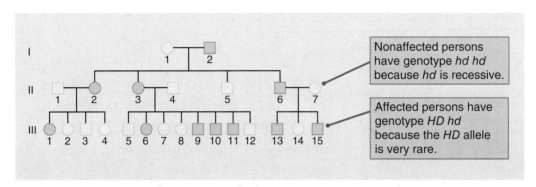

FIGURE 2.18 Pedigree of a human family showing the inheritance of the dominant gene for Huntington's disease, a dominant genetic disorder. The condition does not typically exhibit itself until individuals are 30 years of age or older—after they may have had children.

the symbols for their offspring. The offspring within a sibship, called **siblings** or **sibs** regardless of sex, are represented from left to right in order of their birth.

A typical pedigree for a trait due to a dominant allele is shown in **FIGURE 2.18**. In this example the trait is **Huntington's disease**, which is a progressive nerve degeneration, usually beginning about middle age, that results in severe physical and mental disability and ultimately in death. The numbers in the pedigree are added for convenience in referring to particular persons. The successive generations are designated by Roman numerals. Within any generation, all of the persons are numbered consecutively from left to

right. The pedigree starts with the woman I-1 and the affected man I-2. The pedigree shows the characteristic features of inheritance due to a simple Mendelian dominant allele:

- The trait affects both sexes.
- Every affected person has an affected parent.
- Approximately 1/2 of the offspring of affected persons are affected.

Because the dominant allele, *HD*, that causes Huntington's disease is very rare, all affected persons in the pedigree have the heterozygous genotype *HD hd*.

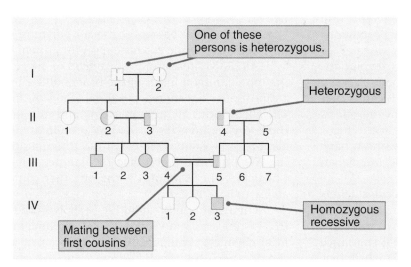

FIGURE 2.19 Pedigree of albinism, a recessive genetic disorder that is characterized by a lack of the pigment melanin. The girl with albinism (right) has very light skin and white (actually colorless) hair. With recessive inheritance, affected persons (red symbols) often have unaffected parents. The double horizontal line indicates a mating between relatives—in this case, first cousins.

Nonaffected persons have the homozygous normal genotype *hd hd*.

A typical pedigree pattern for a trait due to a homozygous recessive allele is shown in **FIGURE 2.19**. The trait is **albinism**, absence of pigment in the skin, hair, and iris of the eyes. The pedigree characteristics of recessive inheritance are as follows:

■ The trait affects both sexes.

■ Most affected persons have parents who are not themselves affected; the parents are heterozygous for the recessive allele and are called **carriers**.

■ Approximately 1/4 of the children of heterozygous parents are affected.

■ The parents of affected individuals are often relatives.

The reason for the 1/4 ratio is that in a mating between carriers (*Aa* × *Aa*), each offspring has a 1/4 chance of being homozygous *aa* and hence being affected. The reason why mating between relatives is important, particularly with traits due to rare recessive alleles, is that when a recessive allele is rare, it is more likely to become homozygous through inheritance from a common ancestor than from parents who are completely unrelated. When a common ancestor of an individual's parents is a carrier, the recessive allele may, by chance, be transmitted down both sides of the pedigree to the parents of the individual. That allele then has a 1/4 chance of becoming homozygous when the relatives mate. Another term for mating between relatives is *inbreeding*.

STOP & THINK 2.5

In the pedigree of recessive inheritance in Figure 2.19, what are the genotypes of individuals III-4 and III-5? Taking into account the fact that individual IV-1 is not affected, what is the probability that IV-1 is heterozygous?

Most differences in human genes are not harmful.

Before the advent of molecular methods, there were many practical obstacles to the study of human genetics. With the exception of traits such as the ABO and other blood groups, few traits showing simple Mendelian inheritance were known. Most of these were associated with genetic diseases, and these presented special challenges.

■ Most genes that cause simple Mendelian genetic diseases are rare, so they are observed in only a small number of families.

■ Many genes for simple Mendelian diseases are recessive, so they are not detected in heterozygous genotypes.

■ The number of offspring per human family is relatively small, so segregation cannot usually be detected in single sibships.

■ The human geneticist cannot perform testcrosses, backcrosses, or other experimental matings.

The result was that geneticists tended to focus on mutations that had major effect on phenotype, and these were usually associated with serious conditions such as phenylketonuria, Huntington's disease, or albinism. Modern molecular genetics has revolutionized the study of human genetics and made it possible to study differences in single nucleotides and other subtle changes, most of which have no obvious effects on phenotype. Human geneticists have come to focus on such subtle changes because, while most have no harmful effects, they do show simple Mendelian inheritance.

But even before the advent of molecular genetics, a few seemingly harmless simple-Mendelian traits had been discovered. One of the best known was associated with the ability to taste a chemical substance known as **phenylthiocarbamide (PTC)**, which has the molecular structure shown here.

The taste polymorphism was discovered in the early 1930s when an industrial chemist was studying PTC and one day carelessly released a cloud of it into the air. The PTC powder didn't bother the chemist at all, but his lab mate loudly complained about the bitter taste it left in his mouth. Out of curiosity the chemist started to test family and friends for their ability to taste PTC, and recruited a geneticist who began to study the situation. It was soon shown from family studies that the ability to taste PTC is a trait inherited as a simple Mendelian dominant. In European populations, about 70 percent of the people are tasters and 30 percent are nontasters, but these proportions differ greatly among ethnic groups. Among people of African or Asian origin, the frequency of tasters is about 90 percent, whereas among Australian aborigines it is only about 50 percent.

The ability to taste PTC is quantitative, however. The most sensitive tasters can taste concentrations as low as 0.001 millimolar (mM) whereas the most insensitive nontasters fail to detect concentrations as high as 10 mM. For classifying individuals as "tasters" or "nontasters," an arbitrary cutoff is employed, typically at a concentration of 0.2 mM PTC. Most of the variation in tasting ability between tasters and nontasters is due to the major taster polymorphism, but there are also differences due to other genes, sex, and probably environmental factors. The result of the other variables is that about 5 percent of the heterozygous tasters get classified as "nontaster" and at least 5

percent of the homozygous nontasters become classified as "tasters."

The molecular basis of the taster polymorphism is now known to reside in a taste receptor protein known as hTAS2R38. There are several alleles of the gene, but the most common forms of the protein differ by three amino acids at scattered positions along the protein. The allelic forms are known as *PAV* and *AVI*, since the three key amino acids in the PAV protein are proline, alanine, and valine whereas these three positions in the AVI protein are occupied by alanine, valine, and isoleucine. The PAV form is the one that confers the ability to taste PTC.

When you think about it, a polymorphism in PTC tasting makes no sense. PTC is a completely artificial chemical synthesized in the laboratory, so why should there be a polymorphism in the ability to taste it? A clue comes from the observation that the chemical structure of PTC resembles a large and heterogeneous class of molecules called *glucosinolates*. These are distasteful compounds synthesized by some plants, including some human food plants, and their synthesis likely evolved as a chemical defense against plant-eating insects. Among the plants that produce glucosinolates is one singled out by former President George H. W. Bush, who in 1989 took broccoli off the White House menu, proclaiming, "I do not like broccoli. And I haven't liked it since I was a little kid and my mother made me eat it. And I'm President of the United States and I'm not going to eat any more broccoli!" (In good-humored protest, broccoli growers throughout the country sent him tons of the stuff.) The irony is that, 17 years after Bush's broccoli boycott, new studies showed that individuals carrying the PAV form of the hTAS2R38 taste receptor do, in fact, find broccoli to be significantly more bitter-tasting than individuals homozygous for the allele encoding AVI form. Tasters also report greater perceived bitterness for collard greens, turnip, rutabaga, and horseradish.

2.6   Simple dominance is not always observed.

In Mendel's experiments, all visible traits had clear dominant–recessive patterns. This was fortunate, because otherwise he might not have made his discoveries. However, departures from strict dominance are frequently observed. In fact, even for such a classic trait as round versus wrinkled seeds in peas, it is an oversimplification to say that round is dominant. At the level of whether a seed is round or wrinkled, round is dominant in the sense that the genotypes *WW* and *Ww* cannot be distinguished by the outward appearance of the seeds. However, every gene potentially affects many traits. It often happens that the same pair of alleles show

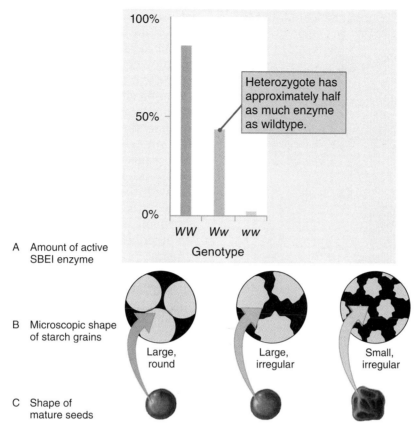

FIGURE 2.20 Two attributes of phenotype affected by Mendel's alleles *W* and *w*, which determine round versus wrinkled seeds. (A) Relative amounts of starch-branching enzyme I (SBEI); the enzyme level in the heterozygous genotype is about halfway between the levels in the homozygous genotypes. (B) Size and shape of the microscopic starch grains; the heterozygote is intermediate. (C) Effect on shape of mature seeds; for seed shape, *W is dominant over w*.

complete dominance for one trait but not complete dominance for another trait. For example, in the case of round versus wrinkled seeds, the biochemical defect in wrinkled seeds is the absence of the active form of the enzyme starch-branching enzyme I (SBEI), which is needed for the synthesis of amylopectin, a branched-chain form of starch. Seeds that are heterozygous *Ww* have only half as much SBEI as homozygous *WW* seeds, and seeds that are homozygous *ww* have virtually none (**FIGURE 2.20**, part A). In addition, homozygous *WW* peas contain large, well-rounded starch grains. As a result, the seeds retain water and shrink uniformly as they ripen, and so they do not become wrinkled. In homozygous *ww* seeds, the starch grains lack amylopectin; they are irregular in shape. When these seeds ripen, they lose water too rapidly and shrink unevenly, resulting in the wrinkled phenotype.

The *w* allele also affects the shape of the starch grains in *Ww* heterozygotes. In heterozygous seeds, the starch grains are intermediate in shape (Figure 2.20, part B). Nevertheless, their amylopectin content is high enough to result in uniform shrinking of the seeds and no wrinkling (Figure 2.20, part C). Thus there is an apparent

paradox of dominance. If we consider only the overall shape of the seeds, round is dominant over wrinkled. If we examine the shape of the starch grains with a microscope, however, all three genotypes can be distinguished from each other: large rounded starch grains in *WW*, large irregular grains in *Ww*, and small irregular grains in *ww*.

The example in Figure 2.20 makes it clear that "dominance" is not simply a property of a particular pair of alleles independent of which aspect of the phenotype is observed. When a gene affects multiple traits (as most genes do), a particular pair of alleles might show complete dominance for some traits but not others. The general principle illustrated in Figure 2.20 is that

KEY CONCEPT

The total phenotype of an organism consists of many different physical and biochemical attributes, and dominance may be observed for some of these attributes and not for others; thus dominance is a property of a pair of alleles in relation to a particular attribute of phenotype.

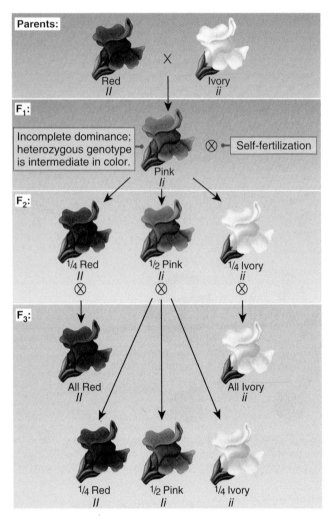

Parents:

Red
II

X

Ivory
ii

F₁:

Incomplete dominance;
heterozygous genotype
is intermediate in color.

⊗ — Self-fertilization

Pink
Ii

F₂:

1/4 Red
II
⊗

1/2 Pink
Ii
⊗

1/4 Ivory
ii
⊗

F₃:

All Red
II

All Ivory
ii

1/4 Red
II

1/2 Pink
Ii

1/4 Ivory
ii

FIGURE 2.21 Incomplete dominance in the inheritance of flower
color in snapdragons.

Flower color in snapdragons illustrates incomplete dominance.

When the phenotype of the heterozygous genotype is intermediate between the phenotypes of the homozygous genotypes, there is said to be **incomplete dominance**. As shown in Figure 2.20, the *W* and *w* alleles for SBEI show incomplete dominance for the traits "amount of active SBEI enzyme" and "microscopic shape of starch grains." Incomplete dominance sometimes occurs for visible traits, too. An example concerns flower color in the snapdragon *Antirrhinum* (**FIGURE 2.21**). In wildtype flowers, a red type of anthocyanin pigment is formed by a sequence of enzymatic reactions. A wildtype enzyme, encoded by the *I* allele, affects the rate of the overall reaction—the more enzyme available, the more red pigment produced. The alternative *i* allele codes for an inactive enzyme, and *ii* flowers, which have no red pigment, are ivory in color. Because the amount of the critical enzyme is reduced in *Ii* heterozygotes, the amount of

red pigment in the flowers is reduced also, and the effect of the dilution is to make the flowers pink.

The result of Mendelian segregation is observed directly when snapdragons that differ in flower color are crossed. For example, a cross between plants from a true-breeding red-flowered variety and a true-breeding ivory-flowered variety results in F₁ plants with pink flowers. In the F₂ progeny obtained via self-fertilization of the F₁ hybrids, one experiment resulted in 22 plants with red flowers, 52 with pink flowers, and 23 with ivory flowers. These numbers agree fairly well with the Mendelian ratio of 1 dominant homozygote : 2 heterozygotes : 1 recessive homozygote. In agreement with the predictions from simple Mendelian inheritance, when self-fertilized, the red-flowered F₂ plants produced only red-flowered progeny; the ivory-flowered plants produced only ivory-flowered progeny; and the pink-flowered plants produced red, pink, and ivory progeny in the proportions 1/4 red : 1/2 pink : 1/4 ivory.

The human ABO blood groups illustrate both dominance and codominance.

Beginning students are often confused by the difference between incomplete dominance and codominance. Incomplete dominance means that the phenotype of the heterozygous genotype is *intermediate* between those of the homozygous genotypes. Incomplete dominance is more frequent for morphological traits than for molecular traits. For example, in the snapdragon in Figure 2.21, the color pink is intermediate between red and white. Codominance means that the heterozygous genotype exhibits the traits associated with *both* homozygous genotypes. Codominance is more frequent for molecular traits than for morphological traits. In Figure 2.7, for example, the gel pattern of the heterozygous *Ww* genotype shows both the small DNA fragment associated with *WW* and the large DNA fragment associated with *ww*, and therefore *W* and *w* are regarded as codominant.

An example illustrating both dominance and codominance is found in the familiar A, B, AB, and O human blood groups determined by polysaccharides (polymers of sugars) present on the surface of red blood cells. Both the A and B polysaccharides are formed from a precursor substance that is modified by the enzyme product of either the *Iᴬ* or the *Iᴮ* allele. The gene products are transferase enzymes that attach either of two types of sugar units to the precursor (**FIGURE 2.22**). People of genotype *IᴬIᴬ* produce red blood cells having only the A polysaccharide and are said to have blood type A. Those of genotype *IᴮIᴮ* have red blood cells with only the B polysaccharide and have blood type B. Heterozygous *IᴬIᴮ* people have red cells with both the A and the B polysaccharide and have blood type AB. The *IᴬIᴮ* genotype illustrates

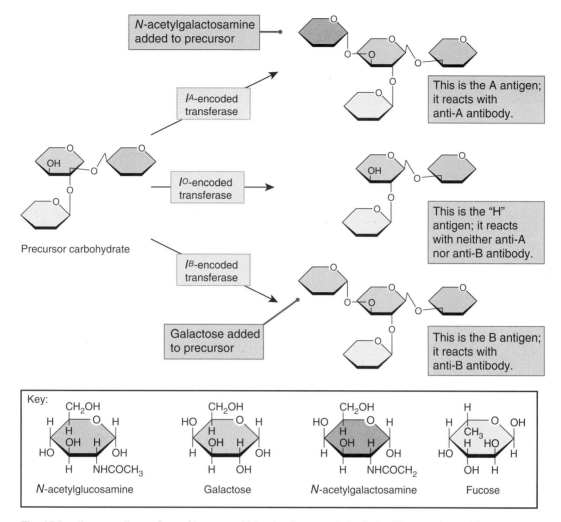

FIGURE 2.22 The ABO antigens on the surface of human red blood cells are carbohydrates. They are formed from a precursor carbohydrate by the action of transferase enzymes encoded by alleles of the *I* gene. Allele *I*^{O} codes for an inactive enzyme and leaves the precursor unmodified. The unmodified form is called the H substance. The *I*^{A} allele encodes an enzyme that adds *N*-acetylgalactosamine (purple) to the precursor. The *I*^{B} allele encodes an enzyme that adds galactose (green) to the precursor. The other colored sugar units are *N*-acetylglucosamine (orange) and fucose (yellow). The sugar rings also have side groups attached to one or more of their carbon atoms; these are shown in the detailed structures inside the box.

codominance, because the heterozygous genotype has the characteristic of both homozygous genotypes—in this case, the presence of both the A and the B carbohydrate on the red blood cells. Although the polypeptides encoded by *I*^{A} and *I*^{B} differ in only 4 out of 355 amino acids, these differences are at strategic positions in the molecules and change their substrate specificity.

Both *I*^{A} and *I*^{B} are dominant to the recessive allele *I*^{O}. The *I*^{O} allele has a single-base deletion in codon 86 that shifts the translational reading frame of the mRNA, resulting in an incomplete, inactive enzyme. The precursor substrate remains unchanged, and neither the A nor the B type of polysaccharide is produced. Homozygous *I*^{O}*I*^{O} persons therefore lack both the A and the B polysaccharide; they are said to have blood type O. In *I*^{A}*I*^{O} heterozygotes, presence of the *I*^{A} allele results in production of the A polysaccharide; and in *I*^{B}*I*^{O} heterozygotes, presence of the *I*^{B} allele results in production

of the B polysaccharide. The result is that *I*^{A}*I*^{O} persons have blood type A and *I*^{B}*I*^{O} persons have blood type B, and so *I*^{O} is recessive to both *I*^{A} and *I*^{B}. The genotypes and phenotypes of the ABO blood-group system are summarized in the first three columns of **TABLE 2.3**.

The ABO blood groups are critical in medicine because of the frequent need for blood transfusions. An important feature of the ABO system is that most human blood contains antibodies to either the A or the B polysaccharide. An **antibody** is a protein that is made by the immune system in response to a stimulating molecule called an **antigen** and is capable of binding to the antigen. An antibody is usually specific in that it recognizes only one antigen. Some antibodies combine with antigen and form large molecular aggregates that may precipitate.

Antibodies act in the body's defense against invading viruses and bacteria, as well as other cells, and

TABLE 2.3 Genetic Control of the Human ABO Blood Groups

Genotype	Antigens present on red blood cells	ABO blood group phenotype	Antibodies present in blood fluid	Blood types that can be tolerated in transfusion	Blood types that can accept blood for transfusion
$I^A I^A$	A	Type A	Anti-B	A & O	A & AB
$I^A I^O$	A	Type A	Anti-B	A & O	A & AB
$I^B I^B$	B	Type B	Anti-A	B & O	B & AB
$I^B I^O$	B	Type B	Anti-A	B & O	B & AB
$I^A I^B$	A & B	Type AB	Neither anti-A nor anti-B	A, B, AB, & O	AB only
$I^O I^O$	Neither A nor B	Type O	Anti-A & anti-B	O only	A, B, AB, & O

help remove such invaders from the body. Although antibodies do not normally form without prior stimulation by the antigen, people capable of producing anti-A and anti-B antibodies do produce them. Production of these antibodies may be stimulated by antigens similar to polysaccharides A and B present on the surfaces of many common bacteria. However, a mechanism called *tolerance* prevents an organism from producing antibodies against its own antigens. This mechanism ensures that A-antigen or B-antigen elicits antibody production only in people whose own red blood cells do not contain A or B, respectively. The end result is:

KEY CONCEPT

People of blood type O make both anti-A and anti-B antibodies: Those of blood type A make anti-B antibodies, those of blood type B make anti-A antibodies, and those of blood type AB make neither type of antibody.

The antibodies found in the blood fluid of people with each of the ABO blood types are shown in the fourth column in Table 2.3. The clinical significance of the ABO blood groups is that transfusion of blood containing A or B red-cell antigens into persons who make antibodies against these antigens results in an agglutination reaction in which the donor red blood cells are clumped. In this reaction, the anti-A antibody will agglutinate red blood cells of either blood type A or blood type AB, because both carry the A antigen. Similarly, anti-B antibody will agglutinate red blood cells of either blood type B or blood type

AB. When the blood cells agglutinate, many blood vessels are blocked, and the recipient of the transfusion goes into shock and may die. Incompatibility in the other direction, in which the donor blood contains antibodies against the recipient's red blood cells, is usually acceptable because the donor's antibodies are diluted so rapidly that clumping is avoided. The types of compatible blood transfusions are shown in the last two columns of Table 2.3. Note that a person of blood type AB can receive blood from a person of any other ABO type; type AB is called a *universal recipient*. Conversely, a person of blood type O can donate blood to a person of any ABO type; type O is called a *universal donor*.

A mutant gene can affect more than one trait.

We've already emphasized that most easily observed phenotypic differences among individuals (for example, height and weight) and most common diseases (for example, heart disease and diabetes) are complex traits affected by multiple genetic risk factors. In these cases, a single trait is affected by multiple genes.

The converse is also true: A single gene can affect multiple traits. The effects on different traits may result from examining the phenotype at different levels (molecular, cytological, morphological), or they may result from secondary or indirect effects of the gene. The various, sometimes seemingly unrelated effects of a mutant gene are called **pleiotropic effects**, and the phenomenon itself is known as **pleiotropy**.

An example of different traits being affected at different levels of phenotype is found in the wildtype (*W*) and mutant (*w*) alleles of the gene for starch-branching

FIGURE 2.23 Among cats with white fur and blue eyes, about 40 percent are born deaf.

© Medioimages/Alamy Images.

enzyme I (SBEI) in peas already discussed. At the molecular level, the homozygous mutant (*ww*) has no detectable enzyme and the heterozygous mutant (*Ww*) has half the amount observed in homozygous wildtype (*WW*). At the cytological level, the shape of the starch grains (Figure 2.20) ranges from large and round (*WW*) to large and irregular (*Ww*) to small and irregular (*ww*). And at the level of morphology, seed shape is either round (*WW* or *Ww*) or wrinkled (*ww*).

Although one can understand how a mutation affecting SBEI can affect enzyme level, starch grain morphology, and seed shape, the biology underlying pleiotropic effects is often much more obscure. An example is seen in cats with white fur and blue eyes, of which about 40 percent are born deaf (**FIGURE 2.23**). This form of deafness can be regarded as a pleiotropic effect of white fur and blue eyes. In this case, the biology underlying the connection is known. The connection is that, during embryonic development, pigment cells derived from the neural crest migrate to various tissues, including hair follicles and the eyes, where their function is to form pigment, as well as to the middle ear, where their function is essential for hearing. This is why defective pigment cells resulting in white fur and blue eyes can also lead to deafness, which may be regarded as a pleiotropic effect of white fur and blue eyes.

A mutant gene is not always expressed in exactly the same way.

Simple Mendelian ratios are not always observed even when a trait is determined by the alleles of a single gene. The reason is that the same genotype may be expressed in different individuals in different ways. Variation in the phenotypic expression of a particular genotype may happen because other genes modify the phenotype or because the biological processes that produce the phenotype are sensitive to environmental conditions.

The types of variable gene expression are usually grouped into two categories:

- **Variable expressivity** refers to genes that are expressed to different degrees in different organisms. For example, inherited genetic diseases in human beings are often variable in expression from one person to the next. One patient may be very sick, whereas another with the same disease is less severely affected. Variable expressivity means that the same mutant gene can cause a severe form of the disease in one person but a mild form in another. The different degrees of expression often form a continuous series from full expression to almost no expression of the expected phenotypic characteristics.

- **Penetrance** refers to the proportion of organisms whose phenotype matches their genotype for a given trait. A genotype that is always expressed has a penetrance of 100 percent. A penetrance of less than 100 percent (*incomplete penetrance*) is the extreme of variable expressivity in which the genotype is not expressed to any detectable degree in some individuals. For example, people with a genetic predisposition to lung cancer may not get the disease if they don't smoke tobacco. A lack of gene expression may result from environmental conditions, such as in the example of not smoking, or from the effects of other genes.

2.7 Epistasis can affect the observed ratios of phenotypes.

In genetic crosses in which two mutations that affect different steps in a single biochemical pathway are both segregating, the typical F_2 dihybrid ratio of 9 : 3 : 3 : 1 is not observed. One example is found in the interaction of two recessive mutations, each in a different gene, that affect flower coloration in peas. Plants of genotypes *CC* and *Cc* have purple flower color, which is the normal or wildtype expression of the trait, whereas homozygous *cc* plants have white flowers. For the other gene, plants of genotypes *PP* and *Pp* have wildtype purple flowers, whereas homozygous *pp* plants have white flowers. Geneticists often use a dash to indicate an allele whose identity is not specified; for example, the symbol *C*− means that in this genotype, one allele is known to be *C* and the other (unspecified) allele, indicated by the dash, may be either *C* or *c*. The symbol *C*− is therefore a shorthand designation meaning "either *CC* or *Cc*." Using this type of symbolism, we could

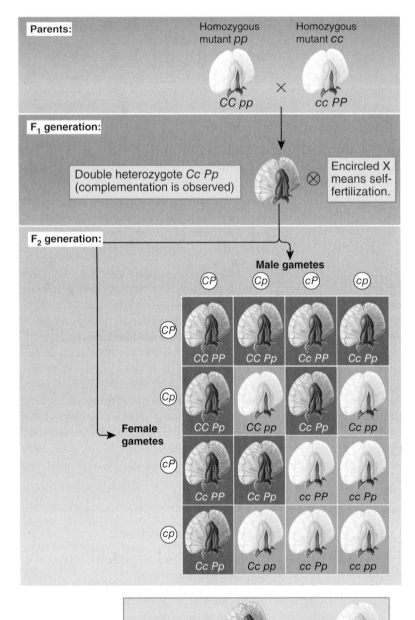

FIGURE 2.24 A cross showing epistasis in the determination of flower color in peas. Formation of the purple pigment requires the dominant allele of both the C and P genes. With this type of epistasis, the dihybrid F$_2$ ratio is modified to 9 purple : 7 white.

say that genotypes *C–* and *P–* have wildtype purple flowers, whereas genotypes *cc* and *pp* have white flowers. Homozygous recessive *cc* or *pp* plants have white flowers regardless of the genotype of the other gene.

FIGURE 2.24 shows a cross between the homozygous recessive genotypes *CC pp* and *cc PP*. The phenotype of the flowers in the F$_1$ generation is the wildtype purple. Why? The reason is that each wildtype allele codes for a different enzyme needed in the biochemical

pathway for the synthesis of purple pigment. Suppose, for example, that the product of the *C* gene (Enzyme C) acts earlier than the product of the *P* gene (Enzyme P). Then the pathway would be

$$
\begin{array}{ccc}
& \text{Enzyme} & \text{Enzyme} \\
& \text{C} & \text{P} \\
\text{Precursor} \rightarrow & \text{Intermediate} \rightarrow & \text{Purple} \\
& & \text{pigment}
\end{array}
$$

where each step may also require other enzymes not shown. The genotype of the F$_1$ plant from the cross *CC pp* × *cc PP* is a double heterozygote *Cc Pp* and has purple flowers owing to the fact that the *C* allele produces a wildtype Enzyme C and the *P* allele produces a wildtype Enzyme P. A cross between recessive mutants with similar phenotypes (in this case white flowers) constitutes a **complementation test**, and it is typically carried out to determine whether the recessive mutations are alleles of the same gene. When the phenotype of the F$_1$ offspring is wildtype (as in this case, with purple flowers), the result means that the recessive mutations are in different genes. The observation of purple flowers in the F$_1$ progeny may at first seem strange and unexpected, because the original cross involved two homozygous recessive mutants, each with white flowers. But once the result of the complementation test tells us that the mutant *c* and the mutant *p* alleles are in different genes, then the phenotype of purple flowers in the F$_1$ generation makes perfect sense because it implies that the genotype of the F$_1$ offspring is the doubly heterozygous *Cc Pp*.

In the F$_2$ generation of the cross in Figure 2.24, independent assortment results in a ratio of genotypes of 9 *C– P–* : 3 *C– pp* : 3 *cc P–* : 1 *cc pp*. The ratio of F$_2$ *phenotypes* is not 9 : 3 : 3 : 1, however, but rather 9 purple : 7 white because the genotypes *C– pp*, *cc P–*, and *cc pp* all have white flowers (Figure 2.24). In other words, the 9 : 7 ratio of purple : white flowers is a modified form of the 9 : 3 : 3 : 1 ratio in which the "9" class has purple flowers and the "3 : 3 : 1" classes all have white flowers. This is an example of **epistasis**, a term that refers to any type of gene interaction that results in the F$_2$ dihybrid ratio of 9 : 3 : 3 : 1 being modified into some other ratio. In a more general sense, what this means is that one gene is masking the expression of the other.

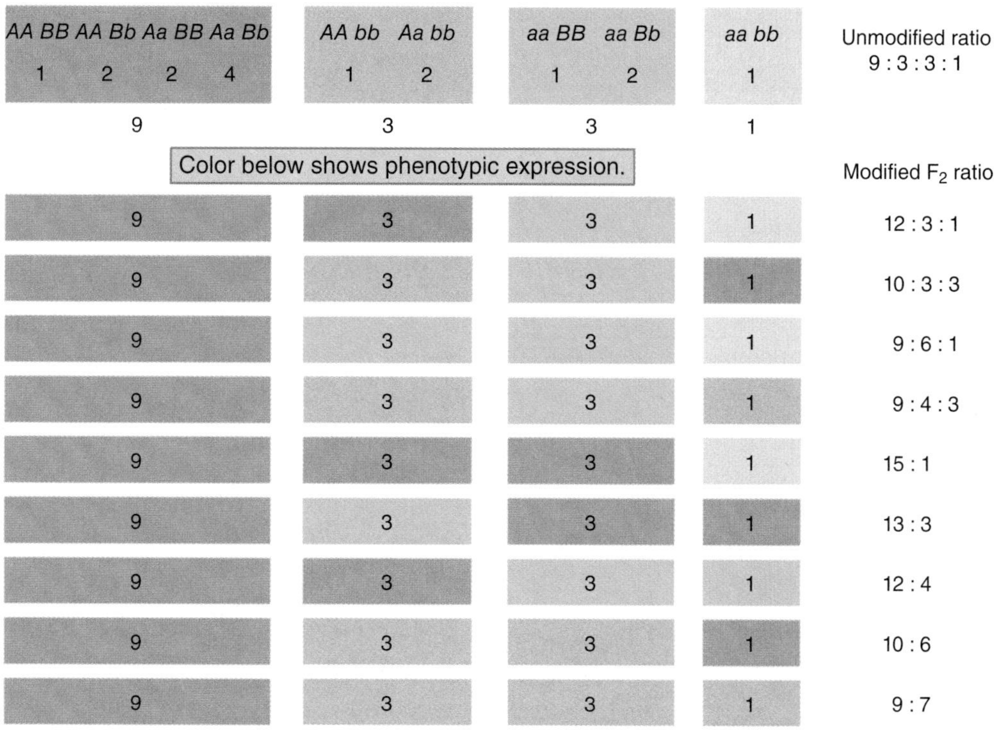

AA BB AA Bb Aa BB Aa Bb	AA bb Aa bb	aa BB aa Bb	aa bb	Unmodified ratio 9 : 3 : 3 : 1
1 2 2 4	1 2	1 2	1	
9	3	3	1	

Color below shows phenotypic expression.   Modified F$_2$ ratio

9	3	3	1	12 : 3 : 1
9	3	3	1	10 : 3 : 3
9	3	3	1	9 : 6 : 1
9	3	3	1	9 : 4 : 3
9	3	3	1	15 : 1
9	3	3	1	13 : 3
9	3	3	1	12 : 4
9	3	3	1	10 : 6
9	3	3	1	9 : 7

FIGURE 2.25 Modified F$_2$ dihybrid ratios. In each row, different colors indicate different phenotypes.

For a trait determined by the interaction of two genes, each with a dominant allele, there are only a limited number of ways in which the 9 : 3 : 3 : 1 dihybrid ratio can be modified. The possibilities are illustrated in **FIGURE 2.25**. The rectangles across the top row shows the 9 : 3 : 3 : 1 ratio expected of independent assortment when there is no epistasis. Each of the other diagrams shows a different modification of the 9 : 3 : 3 : 1 ratio, depending on which genotypes have the same phenotype (indicated by rectangles of the same color).

Taking all the possible modified ratios in Figure 2.25 together, there are nine possible dihybrid ratios when both genes show complete dominance. Examples of each of the modified ratios are known. Some of the most frequently encountered modified ratios are illustrated in the following examples, which are taken from a variety of organisms. Other examples can be found in the problems at the end of the chapter.

9:7 This is the ratio observed when a homozygous recessive mutation in either or both of two different genes results in the same mutant phenotype. It is exemplified by the segregation of purple and white flowers in Figure 2.24. Genotypes that are C– for the C gene and P– for the P gene have purple flowers; all other genotypes have white flowers. Recall that the dash in C– means that the unspecified allele could be

either C or c, and so C– means "either CC or Cc." Similarly, the dash in P– means that the unspecified allele could be either P or p.

12:3:1 A modified dihybrid ratio of the 12 : 3 : 1 variety results when the presence of a dominant allele of one gene masks the genotype of a different gene. For example, if the A– genotype renders the B– and bb genotypes indistinguishable, then the dihybrid ratio is 12 : 3 : 1 because the A– B– and A– bb genotypes are expressed as the same phenotype.

In a genetic study of the color of the hull in oat seeds, a variety having white hulls was crossed with a variety having black hulls. The F$_1$ hybrid seeds had black hulls. Among 560 progeny in the F$_2$ generation produced by self-fertilization of the F$_1$, the following seed phenotypes were observed in the indicated numbers:

418 black hulls   106 gray hulls   36 white hulls

The observed ratio of phenotypes is 11.6 : 2.9 : 1, or very nearly 12 : 3 : 1. These results can be explained by a genetic hypothesis in which the black-hull phenotype results from the presence of a dominant allele (say, A) and the gray-hull phenotype results from another dominant allele (say, B) whose effect is apparent only in the aa homozygotes. On the basis of this

THE HUMAN CONNECTION

Blood Feud

Karl Landsteiner 1901

Anatomical Institute, Vienna, Austria

On Agglutination Phenomena in Normal Blood

Early blood transfusions were extremely hazardous. The patient receiving the blood often went into shock and died. This outcome was caused by massive clumping (agglutination) of red blood cells in the recipient, leading to blockage of the oxygen supply to many vital organs. In this paper, Landsteiner demonstrates that the clumping reaction can be observed in the test tube and that blood cells from each person can be classified as type A, type B, or type O, according to whether the cells are agglutinated by blood sera from other persons. Landsteiner found that blood serum of normal human beings is often capable of agglutinating red blood cells of other healthy individuals. His key results are shown in the table, where the + sign denotes agglutination. The blood samples were taken from volunteers at the Institute at which Landsteiner worked ("Person 1," "Person 2," and so forth). "Person 1" is Landsteiner himself. In this excerpt, we have preserved the terms corpuscle (red blood cell) as in the original but have replaced the blood type that Landsteiner called type C with its modern equivalent, type O. (Blood type AB was not found in these experiments,

> ❝ *The observations allow us to explain the variable results in therapeutic transfusions of human blood.* ❞

because the number of persons tested was too small.) Landsteiner's discovery led quickly to the matching of donor and recipient for the ABO blood groups in blood transfusions, and the disastrous incompatibility reactions were almost completely eliminated. In addition to its value in blood transfusions, this work was important in showing that individuals could in some cases be distinguished by their ABO blood type, and Landsteiner pointed out that in some cases, the agglutination reactions would be suitable for identification in forensic investigation. As an interesting exercise, you may wish to deduce the blood type of "Person 1."

In group A the serum reacted on the corpuscles of group B, but not on those of group A, whereas the A corpuscles are influenced in the same manner by serum of group B. In the O group the serum aggregates the corpuscles of A and B, while the O corpuscles are not affected by sera of A and B. ... The observations allow us to explain the variable results in therapeutic transfusions of human blood.

Sera	Blood corpuscles of					
	Person 1	Person 2	Person 3	Person 4	Person 5	Person 6
Person 1	–	+	+	–	+	+
Person 2	–	–	+	–	–	+
Person 3	–	+	–	–	+	–
Person 4	–	+	+	–	+	+
Person 5	–	–	+	–	–	+
Person 6	–	+	–	–	+	–

Wiener Klinische Wochenschrift 14: 1132–1134. Original in German. Excerpt from translation in S. H. Boyer, IV. 1963. *Papers on Human Genetics*. Englewood Cliffs, NJ: Prentice-Hall, pp. 27–31.

hypothesis, the original true-breeding varieties must have had genotypes *aa bb* (white) and *AA BB* (black). The F₁ has genotype *Aa Bb* (black). If the *A, a* allele pair and the *B, b* allele pair undergo independent assortment, then the F₂ generation is expected to have the following composition of genotypes:

9/16	*A– B–*	(black hull)
3/16	*A– bb*	(black hull)
3/16	*aa B–*	(gray hull)
1/16	*aa bb*	(white hull)

This type of epistasis accounts for the 12 : 3 : 1 ratio.

13:3 This type of epistasis is illustrated by the difference between White Leghorn chickens (genotype *CC II*) and White Wyandotte chickens (genotype *cc ii*). The *C* allele is responsible for colored feathers but in White Leghorns the *I* allele is a dominant inhibitor of feather coloration. The F₁ generation of a dihybrid cross between these breeds has the genotype *Cc Ii*, which results in the presence of white feathers because of the inhibitory effects of the *I* allele. In the F₂ generation, only the *C– ii* genotype has colored feathers; hence there is a 13 : 3 ratio of white : colored.

9:4:3 This dihybrid ratio (often stated as 9 : 3 : 4) is observed when homozygosity for a recessive allele with respect to one gene masks the expression of the genotype of a different gene. For example, if the *aa* genotype has the same phenotype regardless of whether the genotype is *B–* or *bb*, then the 9 : 4 : 3 ratio results.

In the mouse, the grayish coat color called agouti is produced by the presence of a horizontal band of yellow pigment just beneath the tip of each hair. The agouti pattern results from the presence of a dominant allele *A*, and in *aa* animals the coat color is black. A second dominant allele, *C*, is necessary for the formation of hair pigments of any kind, and *cc* animals are albino (white fur). In a cross of *AA CC* (agouti) × *aa cc* (albino), the F₁ progeny are *Aa Cc* and agouti. Crosses between F₁ males and females produce F₂ progeny in the following proportions:

9/16	*A– C–*	(agouti)
3/16	*A– cc*	(albino)
3/16	*aa C–*	(black)
1/16	*aa cc*	(albino)

The dihybrid ratio is therefore 9 agouti : 4 albino : 3 black.

9:6:1 This dihybrid ratio is observed when homozygosity for a recessive allele of either of two genes results in the same phenotype but the phenotype of the double homozygote is distinct. For example, red coat color in Duroc–Jersey pigs requires the presence of two dominant alleles *R* and *S*. Pigs of genotype *R– ss* and *rr S–* have sandy-colored coats, and *rr ss* pigs are white. The F₂ dihybrid ratio is therefore

9/16	*R– S–*	(red)
3/16	*R– ss*	(sandy)
3/16	*rr S–*	(sandy)
1/16	*rr ss*	(white)

The 9 : 6 : 1 ratio results from the fact that both single recessives have the same phenotype.

STOP & THINK 2.6

In summer squash (varieties of *Cucurbita pepo*), a gene we may call *A* has a dominant allele that results in white fruit. Another gene we may call *B* has a dominant allele resulting in yellow fruit, and in the homozygous *bb* genotype, the fruit is green. The genes undergo independent assortment. In a cross *Aa Bb* (white) × *Aa Bb* (white), what phenotypes are expected among the progeny fruit, and in what proportions?

CHAPTER SUMMARY

- Inherited traits are determined by the genes present in the reproductive cells united in fertilization.

- Genes are usually inherited in pairs, one from the mother and one from the father.

- The genes in a pair may differ in DNA sequence and in their effect on the expression of a particular inherited trait.

- The maternally and paternally inherited genes are not changed by being together in the same organism.

- In the formation of reproductive cells, the paired genes separate again into different cells.

- Random combinations of reproductive cells containing different genes result in Mendel's ratios of traits appearing among the progeny.

- Simple Mendelian inheritance results in characteristic patterns in human pedigrees for both dominant and recessive traits.

- When two possible outcomes of a cross are mutually exclusive, they cannot occur together. In this case, the probability that either one or the other outcome occurs is given by the sum of their respective probabilities (the addition rule).

- When two possible outcomes of a cross are independent, then knowledge that one has occurred provides no information whether the other has occurred. In this case, the probability that both outcomes occur together is given by the product of their respective probabilities (the multiplication rule).

- The ratios actually observed for any traits are determined by the types of dominance and gene interaction (epistasis).

ISSUES AND IDEAS

- What constitutes the genotype of an organism? What constitutes the phenotype? Why is it important in genetics that genotype and phenotype be distinguished?

- What is the difference between a gene and an allele? How can a gene have more than two alleles? Give an example of multiple alleles of a gene.

- What is the principle of segregation, and how is this principle demonstrated in the results of Mendel's monohybrid crosses?

- What is the principle of independent assortment, and how is this principle demonstrated in the results of Mendel's dihybrid crosses?

- Explain why random union of male and female gametes is necessary for Mendelian segregation and independent assortment to occur.

- What is the difference between mutually exclusive possibilities and independent possibilities? How are the probabilities of these two types of possible outcomes combined? Give two examples of genetic possibilities that are mutually exclusive and two examples of genetic possibilities that are independent.

- When two pairs of alleles show independent assortment, under what conditions will a $9:3:3:1$ ratio of phenotypes in the F_2 generation not be observed?

SOLUTIONS: STEP BY STEP

PROBLEM 1 In the diagram of an electrophoresis gel shown here, the bands *a*, *b*, and *c* each originate from a gene. Two of the bands represent alleles of the same gene, and the other represents a band from a gene in a different chromosome. The lane labeled F_1 shows the banding pattern observed in a hybrid, and those labeled F_2 are the three types of progeny observed in the F_2 generation. The F_1 hybrid is, therefore, heterozygous for the alleles represented by two of the bands and homozygous for the allele represented by the remaining band.

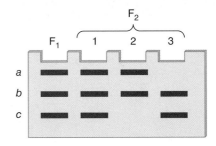

(a) Which of the bands represent alleles of the same gene?
(b) What is the genotype of the F_1 hybrid?
(c) What are the genotypes 1, 2, and 3 in the F_2 generation?
(d) What is the expected ratio of progeny types 1, 2, and 3 in the F_2 generation?

SOLUTION. (a) The key to understanding which bands represent alleles of the same gene is to see if they undergo segregation. The bands designated *a* and *c* do segregate: Among the F_2 progeny, type 1 shows both bands, type 2 shows only *a*, and type 3 shows only *c*. This result is consistent with the F_1 hybrid having genotype *a*/*c*, where *a* and *c* are alleles. This inference implies that band *b* represents the homozygous allele of another gene. (b) Based on the inferences in part (a), the genotype of the F_1 hybrid is *a*/*c*; *b*/*b*. (c) The genotype of type 1 is *a*/*c*; *b*/*b*, that of type 2 is *a*/*a*; *b*/*b*, and that of type 3 is *c*/*c*; *b*/*b*. (d) The expected ratio of types 1 : 2 : 3 is 1/2 : 1/4 : 1/4 or 2 : 1 : 1, because the progeny of the cross *a*/*c* × *a*/*c* are expected to be *a*/*a* (type 2), *a*/*c* (type 1), and *c*/*c* (type 3) in the ratio of 1 : 2 : 1.

PROBLEM 2 Suppose that the genes *A* and *B* undergo independent assortment. In the cross *A*/*a* *B*/*b* × *A*/*a* *B*/*b*, what fraction of the progeny are expected to be *A*/*a* *B*/− (that is, either *A*/*a* *B*/*B* or *A*/*a* *B*/*b*). Solve this problem without using a Punnett square.

SOLUTION. Because *A* and *B* are independent, the genes can be considered separately and the answers multiplied. For the cross *A*/*a* × *A*/*a*, the fraction of progeny of genotype *A*/*a* is 1/2. For the cross *B*/*b* × *B*/*b*, the fraction of progeny that are either *B*/*B* or *B*/*b* is 1/4 + 1/2 = 3/4.

Altogether, the expected fraction of *A*/*a* *B*/− progeny is 1/2 × 3/4 = 3/8. For this type of problem, drawing a Punnett square for a dihybrid cross is unnecessary (and, during an examination, takes too much time).

PROBLEM 3 The pretty purple in the petals of *Primula vulgaris* shown here results from the pigment *malvidin*. Synthesis of malvidin requires a dominant allele *K*, and in *kk* flowers the corresponding parts of the petals are white. A dominant allele *D* of another gene suppresses the expression of the *K* allele, whereas in *dd* genotypes the effect of *K* is expressed. The two genes undergo independent assortment. In the cross *K*/*k* *D*/*d* × *K*/*k* *D*/*d*, what is the expected ratio of purple : white flowers?

© Kalina Iwaszko/Shutterstock.   © Christophe Rolland/Shutterstock.

SOLUTION. Because of independent assortment, drawing a Punnett square for the dihybrid cross is unnecessary. The expected ratio of the genotypes *K*− *D*− : *K*− *dd* : *kk* *D*− : *kk* *dd* is 9 : 3 : 3 : 1. Flowers of genotype *K*− *D*− are white owing to the presence of *D*, those of genotype *K*− *dd* are purple, and those of genotypes *kk* *D*− and *kk* *dd* are white because of the absence of *K*. The overall expected ratio of purple : white is, therefore, 3 : 13.

CONCEPTS IN ACTION: PROBLEMS FOR SOLUTION

2.1 A round pea seed is germinated, and the mature plant is allowed to self-fertilize. It produces some wrinkled seeds. What was the genotype of the original seed? What is the expected proportion of wrinkled seeds produced by the mature plant?

2.2 A monohybrid cross is carried out between pea plants with round and wrinkled seeds. A single round seed is chosen at random from the F_2 generation, and its DNA examined by electrophoresis as described in the text. What is the probability that the gel pattern appears as shown below?

2.3 A woman is affected with a trait due to a dominant gene that shows 75 percent penetrance. What is the probability that, if she has a child, it will be affected?

2.4 The recurrence risk of a genetic disorder is the probability that the next child born into a sibship will be affected, given that one or more previous children is affected. What is the recurrence risk for:

(a) A rare dominant trait in which one parent is affected?

(b) A rare recessive trait in which neither parent is affected?

(c) A rare recessive trait in which one parent is affected?

2.5 With independent assortment, how many different types of gametes are possible from the genotype K/k; L/l; M/m; P/p, and in what proportions are they expected?

2.6 The accompanying diagram shows an electrophoresis gel in which DNA samples are placed ("loaded") in the depressions ("wells") at the top of the gel, and electrophoresis is carried out such that the DNA fragments move in the downward direction. The dashed lines on the right denote the positions to which DNA fragments of various sizes would migrate. The fragment sizes are given in kilobase (kb) pairs; 1 kb refers to a duplex DNA molecule 1000 base pairs in length. Also shown is the position of a DNA fragment corresponding to part of the coding region of a gene in DNA extracted from a homozygous wildtype (AA) organism. Assuming that a_1 is a mutant allele that has a 2-kb insertion of DNA into the wildtype fragment, and that a_2 is a mutant allele that has a 1-kb deletion within the wildtype fragment, show the positions at which DNA bands would be expected in each of the other genotypes shown.

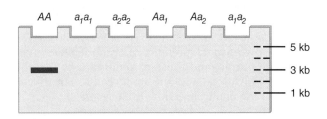

2.7 A woman who is homozygous recessive for a mutation that results in deafness marries an unrelated man who is also deaf because he is homozygous for a recessive mutation. They have a child whose hearing is normal. Explain how this can happen. What genetic principle does this situation exemplify?

2.8 The pedigree illustrated here shows individual II-2 affected with a rare recessive trait. Let A and a represent the dominant and recessive alleles.

(a) What is the genotype of II-2 ?

(b) What are the genotypes of I-1 and I-2?

(c) What are the possible genotypes of II-1 and II-3?

(d) What is the probability that II-3 is a heterozygous "carrier" of the a allele?

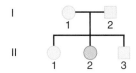

2.9 In Problem 2.8, what is the probability that both II-1 and II-3 are carriers? That neither is a carrier? That at least one is a carrier?

2.10 The pedigree below is for a rare autosomal recessive trait with complete penetrance. What is the probability that at least one of IV-1, IV-2, and IV-4 is a carrier?

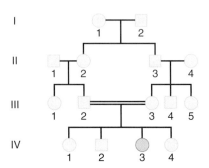

2.11 In *Drosophila*, the dominant allele Cy (Curly) results in curly wings. The cross $Cy/+ \times Cy/+$, where $+$ represents the wildtype allele of Cy, results in a ratio of 2 curly : 1 wildtype among the progeny. When the curly progeny in this generation are crossed with one another, their offspring also show a ratio of 2 curly : 1 wildtype. How can this result be explained?

2.12 Complementation tests of the recessive mutants a through f produced the data in the accompanying matrix. The circles represent missing data. Assuming that all of the missing mutant combinations would yield data consistent with the entries that are known, complete the table by filling each circle with a $+$ or $-$ as needed.

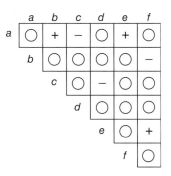

2.13 In the shepherd's purse, *Capsella bursapastoris*, the capsule containing the seeds can be either triangular or ovoid. A cross between certain true-breeding strains with triangular capsules yielded an F_1 generation with triangular capsules. The observed F_2 ratio was 15 triangular : 1 ovoid. What genetic hypothesis can explain these results? What crosses would you carry out to test this hypothesis?

2.14 A woman is heterozygous for two harmful recessive alleles in different chromosomes, one for phenylketonuria (PKU) and the other for cystic fibrosis (CF). If she has a daughter, what is the probability that the child will carry neither of the recessive alleles? Exactly one? Both? If she has a granddaughter, what is the probability that the child will carry both recessives?

2.15 In the summer squash, *Curcurbita pepo,* the shape of the fruit in wildtype genotypes is a Frisbee-like flattened shape known as disc. A homozygous recessive mutation in either of two different genes results in spherical fruit. The fruit of the double recessive homozygous genotype is elongated, like an American football. Assuming independent assortment, what ratio of disc : sphere : elongate would be expected in the F₂ generation of a cross between homozygous disc and homozygous elongate?

2.16 A modernized version of Mendel's monohybrid cross between true-breeding round (*WW*) and true-breeding wrinkled (angular) (*ww*) is shown below. Self-fertilization of the F₁ hybrid plants results in the F₂ progeny. Among 400 F₂ progeny, what are the expected numbers (a)–(f) of each of the seed-shape and DNA-band phenotypes? (Hint: One or more of the expected numbers could be 0.)

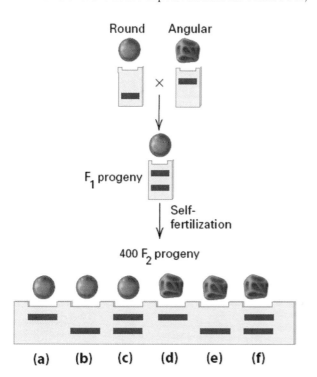

2.17 Huntington's disease is a rare neurological degenerative disorder resulting from a rare dominant mutant allele designated *HD*. The allele has

complete penetrance, but the disorder is usually manifested after the age of 45. A young man has learned that his father has developed the disease.

(a) What is the probability that the young man will later develop the disorder?

(b) What is the probability that a child of the young man carries the *HD* allele?

2.18 Consider human sibships with four children, and assume that each birth is equally likely to result in a boy (B) or a girl (G).

(a) What fraction is expected to include at least one boy?

(b) What fraction is expected to have two girls and two boys?

(c) What fraction is expected to have the birth order GBBG?

2.19 What is the probability that a sibship of five children will include at least one boy and at least one girl?

2.20 In plants, certain mutant genes are known that affect the ability of gametes to participate in fertilization. Suppose that an allele *A* is such a mutation, and that pollen cells bearing the *A* allele are only half as likely to survive and participate in fertilization as pollen cells bearing the *a* allele. What is the expected ratio of *AA* : *Aa* : *aa* plants in the F₂ generation of a monohybrid cross? (Hint: Use a Punnett square.)

2.21 Assume that the trait in the accompanying pedigree results from simple Mendelian inheritance.

(a) Is the trait likely to be due to a dominant allele or a recessive allele? Explain.
(b) What is the meaning of the double horizontal line connecting III-1 with III-2?
(c) What is the biological relationship between III-1 and III-2?
(d) If the allele responsible for the condition is rare, what are the most likely genotypes of all of the persons in the pedigree in generations I, II, and III? (Use *A* and *a* for the dominant and recessive alleles, respectively.)

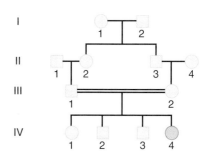

⊕ STOP & THINK ANSWERS

ANSWER TO STOP & THINK 2.1

The offspring genotypes are (1) *AA*, (2) *aa*, (3) *Aa*, (4) *Aa*, (5) *Aa*, (6) *aa*, (7) *Aa*, and (8) *AA*.

ANSWER TO STOP & THINK 2.2

Matings 1, 2, and 3 produce only progeny with the dominant phenotype; mating 6 produces only progeny with the recessive phenotype; mating 4 produces dominant : recessive in the ratio 3 : 1; mating 5 produces dominant : recessive in the ratio 1 : 1.

ANSWER TO STOP & THINK 2.3

Aa × *Aa* produces 1/4 *aa* progeny and *bb* × *Bb* produces 1/2 *Bb* progeny; with independent assortment, the expected proportion of *aa Bb* progeny is 1/4 × 1/2 = 1/8.

ANSWER TO STOP & THINK 2.4

For the *A* gene, the *Aa* × *Aa* genotypes segregate to produce 1/4 *AA*, 1/2 *Aa*, and 1/4 *aa* offspring; according to the addition rule, the proportion of homozygous offspring is therefore 1/4 + 1/4 = 1/2.

For the *B* gene, the *bb* × *Bb* genotypes produce 1/2 homozygous *bb* and 1/2 heterozygous *Bb* offspring. Using the multiplication rule, the overall probability of an offspring being homozygous for both genes is 1/2 × 1/2 = 1/4.

ANSWER TO STOP & THINK 2.5

Neither III-4 nor III-5 is affected, but they have an affected child and therefore III-4 and III-5 must both be heterozygous. Because IV-1 is not affected, the genotype must be either homozygous dominant or heterozygous, which are expected in the relative proportions 1/4 : 1/2. The probability that IV-1 is heterozygous is therefore (1/2)/(1/4 + 1/2) = 2/3.

ANSWER TO STOP & THINK 2.6

Using a dash (–) as wild card to indicate either allele of a gene, the expected genotypes of the offspring are 9/16 *A– B–*, 3/16 *A– bb*, 3/16 *aa B–*, 1/16 *aa bb*. The genotypes *A– B–* and *A– bb* have white fruit, *aa B–* have yellow, and *aa bb* have green. The expected phenotypes are therefore white : yellow : green in the proportions 12/16 : 3/16 : 1:16 (or a ratio of 12 : 3 : 1).

Design Credits: Stop & Think icon made by Darius Dan from www.flaticon.com; The Human Connection icon made by Daniel Bruce from www.flaticon.com; Elephant image: © NickBiemans/GettyImages.

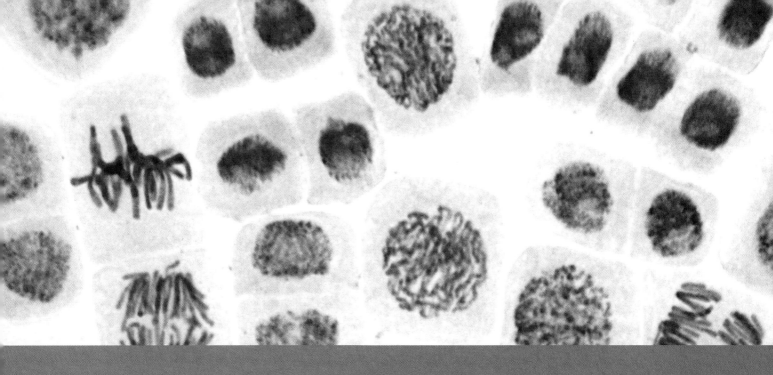

CHAPTER 3

The Chromosomal Basis of Heredity

LEARNING OBJECTIVES

- To predict what products of mitosis or meiosis would result when a chromosome undergoes normal separation or nondisjunction.

- To recognize the characteristic pattern of *X*-linked inheritance in a pedigree.

- To describe the structure of the nucleosome and explain why physical accessibility of the histone tails to chemical modification is important.

- To define the function of the centromere.

- To explain how telomerase restores the missing material that cannot be copied in DNA replication.

- For any specified genetic cross, to use the binomial distribution to calculate the probability of any particular combination of genotypes or phenotypes among the progeny.

- To be able to formulate a genetic hypothesis, use it to predict the expected results of a cross, compare the expected results with observed results by means of a chi-square test for goodness of fit, and interpret the *P* value of the test as to whether or not the hypothesis should be rejected.

When Gregor Mendel carried out his experiments in 1860s, the biological basis of the transmission of the hereditary factors from one generation to the next was a mystery. Neither the role of the nucleus in reproduction nor the details of cell division had been discovered. Once these phenomena were understood, and when microscopy had improved enough that the chromosomes could be observed and were finally realized to be the bearers of the genes, new understanding came at a rapid pace. This chapter examines the mechanism of chromosome segregation in cell division and the relationship between DNA and chromosomes.

3.1 Each species has a characteristic set of chromosomes.

The importance of the cell nucleus and its contents was suggested as early as the 1840s when it was noted that, in dividing cells, the nucleus divided first. By the 1870s it had become clear that nuclear division is a universal attribute of cell division. The importance of the nucleus in inheritance was reinforced by the nearly simultaneous discovery that the nuclei of two gametes fuse in the process of fertilization. The next major advance came in the 1880s with the discovery of **chromosomes**, which had been made visible by light microscopy when stained with basic dyes. A few years later, chromosomes were found to segregate by an orderly process into the daughter cells formed by cell division, as well as into the gametes formed by the division of reproductive cells. Finally, three important regularities were observed about the **chromosome complement** (the complete set of chromosomes) of plants and animals:

1. The nucleus of each **somatic cell** (a cell of the body, in contrast with a **germ cell**, or gamete) contains a fixed number of chromosomes typical of the particular species. However, the numbers vary tremendously among species and have little relationship to the complexity of the organism (**TABLE 3.1**).

2. The chromosomes in the nuclei of somatic cells are usually present in pairs. For example, the 46 chromosomes of human beings consist of 23 pairs (**FIGURE 3.1**). Similarly, the 14 chromosomes of peas consist of 7 pairs. Cells with nuclei of this sort, containing two similar sets of chromosomes, are called **diploid**. The chromosomes are present in pairs because one chromosome of each pair derives from the maternal parent of the organism and the other from its paternal parent.

TABLE 3.1 Somatic Chromosome Numbers of Some Plant and Animal Species

Organism	Chromosome number	Organism	Chromosome number
Field horsetail	216	Yeast (*Saccharomyces cerevisiae*)	32
Bracken fern	116	Fruit fly (*Drosophila melanogaster*)	8
Giant sequoia	22	Nematode (*Caenorhabditis elegans*)	11 ♀, 12 ♂
Macaroni wheat	28	House fly	12
Bread wheat	42	Scorpion	4
Fava bean	12	Geometrid moth	224
Garden pea	14	Common toad	22
Mustard cress (*Arabidopsis thaliana*)	10	Chicken	78
Corn (*Zea mays*)	20	Mouse	40
Lily	24	Gibbon	44
Snapdragon	16	Human being	46

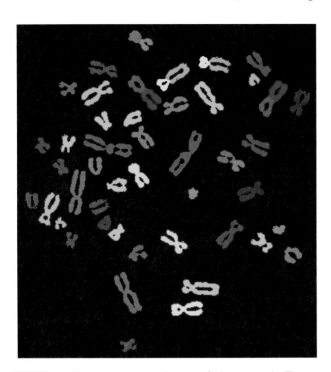

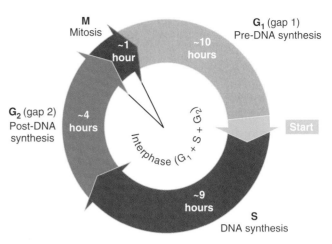

FIGURE 3.2 The cell cycle of a typical mammalian cell growing in tissue culture with a generation time of 24 hours.

FIGURE 3.1 Chromosome complement of a human male. There are 46 chromosomes, present in 23 pairs. At the stage of the division cycle in which these chromosomes were observed, each chromosome consists of two identical halves lying side by side longitudinally. Except for the members of one chromosome pair (the XY pair that determines sex), the members of all of the chromosome pairs are the same color because they contain DNA molecules that were labeled with the same mixture of fluorescent dyes. The colors differ from one pair to the next because the dye mixtures differ in color. In some cases, the long and the short arm have been labeled with different colors.

Courtesy of Michael R. Speicher, Institute of Genome Genetics, Medical University of Graz.

3. The germ cells, or gametes, contain only one set of chromosomes, consisting of one member of each of the pairs. The gamete nuclei are said to be **haploid**. The haploid gametes unite in fertilization to produce the diploid state of somatic cells.

In a multicellular organism, which develops from a single fertilized egg, the presence of the diploid chromosome number in somatic cells and the haploid chromosome number in germ cells indicates that there are *two* different processes of nuclear division. One of these, mitosis, maintains chromosome number, whereas the other, meiosis, halves the number. These two processes are examined in the following sections.

3.2 The daughter cells of mitosis have identical chromosomes.

Mitosis is a precise process of nuclear division that ensures that each of two daughter cells receives a diploid complement of chromosomes identical with the diploid complement of the parent cell. Mitosis is usually accompanied by **cytokinesis**, the process in which the cell itself divides to yield two daughter

cells. The essential details of mitosis are the same in all organisms, and the basic process is remarkably uniform:

1. Each chromosome is already present as a duplicated structure at the beginning of nuclear division. (The duplication of each chromosome coincides with the replication of the DNA molecule contained within it.)

2. Each chromosome divides longitudinally into identical halves that become separated from each other.

3. The separated chromosome halves move in opposite directions, and each becomes included in one of the two daughter nuclei that are formed.

In a cell that is not undergoing mitosis, the chromosomes are not visible with a light microscope. This stage of the cell cycle is called **interphase**. In preparation for mitosis, the genetic material (DNA) in the chromosomes is replicated during a period of interphase called **S** (**FIGURE 3.2**). (The S stands for *synthesis* of DNA.) DNA replication is accompanied by chromosome duplication. Before and after S, there are periods, called G_1 and G_2, respectively, in which DNA replication does not take place. The **cell cycle**, or the life cycle of a cell, is commonly described in terms of these three interphase periods followed by mitosis, **M**. The order of events is therefore $G_1 \rightarrow S \rightarrow G_2 \rightarrow M$, as shown in Figure 3.2. In this representation, the M period includes cytokinesis, which is the division of the cytoplasm into two approximately equal parts, each containing one daughter nucleus. The length of time required for a complete life cycle varies with cell type. In higher eukaryotes, the majority of cells require from 18 to 24 hours. The relative duration of the different periods in the cycle also varies considerably with cell type. Mitosis is usually the shortest period, requiring from 1/2 hour to 2 hours.

The cell cycle is an actively regulated process controlled by mechanisms that are essentially identical in all eukaryotes. The transitions from G_1 into S and from G_2 into M are called **checkpoints** because the transitions are delayed unless key processes have been completed (Figure 3.2). For example, for DNA replication to be initiated at the G_1/S checkpoint, some cell types require that sufficient time must have elapsed since the preceding mitosis, whereas other cell types require that the cell must have attained a particular size. Similarly, for the M phase to begin at the G_2/M checkpoint, DNA replication and repair of any DNA damage must be completed.

In mitosis, the replicated chromosomes align on the spindle, and the sister chromatids pull apart.

The diagram in **FIGURE 3.3** shows the essential features of chromosome behavior in mitosis. The process is conventionally divided into four stages: **prophase**, **metaphase**, **anaphase**, and **telophase**. The stages have the following characteristics.

1. Prophase   In interphase, the chromosomes have the form of extended filaments and cannot be seen as discrete bodies with a light microscope. Except for the presence of one or more conspicuous dark bodies, each called a **nucleolus**, the nucleus has a diffuse, granular appearance. The beginning of prophase is marked by the

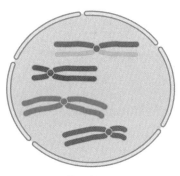

condensation of chromosomes to form visibly distinct, thin threads within the nucleus. Each chromosome is already longitudinally double, consisting of two closely associated subunits called **chromatids**. The longitudinally bipartite nature of each chromosome is readily seen later in prophase. Each pair of chromatids is the product of the duplication of one chromosome in the S period of interphase. The chromatids in a pair are held together at a specific region of the chromosome called the **centromere**. As prophase progresses, the chromosomes become shorter and thicker, as a result of further coiling. At the end of prophase, the nucleoli disappear and the nuclear envelope, a membrane surrounding the nucleus, abruptly disintegrates.

Prophase

2. Metaphase   At the beginning of metaphase, the **mitotic spindle** forms. The spindle is an elongated, football-shaped array of spindle fibers consisting primarily of microtubules formed by polymerization of the protein tubulin. Many other proteins and at least one

RNA-protein complex regulate tubulin polymerization and microtubule organization. The ends or *poles* of the spindle, where the microtubules converge, mark the locations of the *centrosomes*, which are the microtubule organizing centers where tubulin polymerization is initiated. Each pair of centrosomes results from the duplication of a single centrosome that takes place in interphase, followed by migration of the daughter centrosomes to opposite sides of the nuclear envelope.

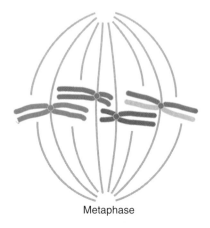

Metaphase

The spindle features three types of microtubules: (1) those that anchor the centrosome to the cell membrane, (2) those that arch between the centrosomes, and (3) those that become attached to the chromosomes. In spindle formation, microtubules grow out from the spindle poles in essentially random directions as new tubulin subunits are added to the growing end of the polymer. For the microtubules that become attached to the chromosomes, the site of attachment is a structure technically known as the **kinetochore**, which coincides with the position of the centromere. Once spindle fibers have attached, each chromosome is moved to a position near the center of the cell where its kinetochore lies on an imaginary plane approximately equidistant from the spindle poles. This imaginary plane is called the metaphase plate. Aligned on the metaphase plate, the chromosomes reach their maximum condensation and are easiest to count and examine for differences in shape and appearance.

3. Anaphase   In anaphase, the proteins holding the centromeres together are degraded. The centromeres become separated, and the two **sister chromatids** of each chromosome move toward opposite poles of the spindle. Once the centromeres separate, each sister chromatid is regarded as a separate chromosome in its own right. Chromosome

Anaphase

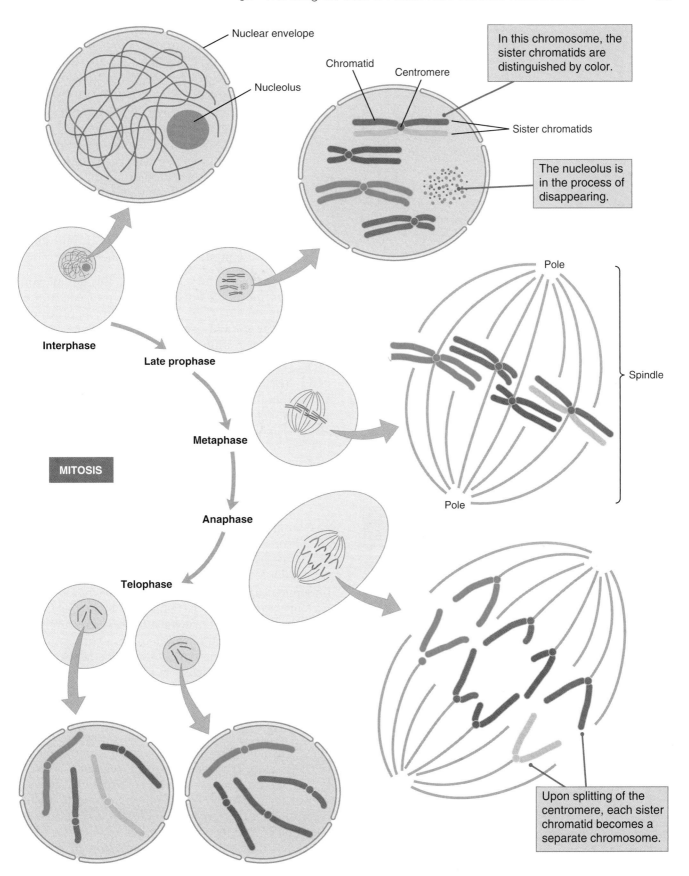

FIGURE 3.3 Chromosome behavior during mitosis in an organism with two pairs of chromosomes (red/rose versus green/blue). At each stage, the smaller, inner diagram represents the entire cell, and the larger diagram is an exploded view showing the chromosomes at that stage.

Telophase

movement results in part from progressive shortening of the spindle fibers attached to the centromeres, which pulls the chromosomes in opposite directions toward the poles. At the completion of anaphase, the chromosomes lie in two groups near opposite poles of the spindle. Each group contains the same number of chromosomes that was present in the original interphase nucleus.

4. Telophase In telophase, a nuclear envelope forms around each compact group of chromosomes, nucleoli are formed, and the spindle disappears. The chromosomes undergo a reversal of condensation until they are no longer visible as discrete entities. The two daughter nuclei slowly assume a typical interphase appearance as the cytoplasm of the cell divides in two by means of a gradually deepening furrow around the periphery. (In plants, a new cell wall is synthesized between the daughter cells and separates them.)

⊶◉⊷ STOP & THINK 3.1

For the terms describing the mitotic cell cycle (interphase, prophase, metaphase, anaphase, and telophase), fill in the term that corresponds to the main event that occurs during that stage:

_____ Chromosomes become visible

_____ Cytokinesis

_____ Chromosomes align on metaphase plate

_____ DNA replication

_____ Sister chromatids move apart

3.3 Meiosis results in gametes that differ genetically.

Meiosis is a mode of cell division in which cells are created that contain only one member of each pair of chromosomes present in the premeiotic cell. When a diploid cell with two sets of chromosomes undergoes meiosis, the result is four daughter cells, each genetically different and each containing one haploid set of chromosomes.

Meiosis consists of two successive nuclear divisions. The essentials of chromosome behavior during meiosis are outlined in **FIGURE 3.4**. This outline affords an overview of meiosis as well as an introduction to the process as it takes place in a cellular context.

1. Prior to the first nuclear division, the members of each pair of chromosomes become closely associated along their length (part A). The chromosomes that pair with each other are said to be **homologous** chromosomes. Because each member of a pair of homologous chromosomes is already replicated, it consists of a duplex of two sister chromatids joined at the centromere. The pairing of the homologous chromosomes therefore produces a four-stranded structure.

2. In the first nuclear division, the homologous chromosomes are separated from each other, one member of each pair going to opposite poles of the spindle (part B). Two nuclei are formed, each containing a haploid set of duplex chromosomes (part C).

3. The second nuclear division loosely resembles a mitotic division, *but there is no DNA replication*. At metaphase, the chromosomes align on the metaphase plate, and at anaphase, the chromatids of each chromosome are separated into opposite daughter nuclei (part D). The net effect of the two divisions in meiosis is the creation of four haploid daughter nuclei, each containing the equivalent of a single sister chromatid from each pair of homologous chromosomes (part E).

Figure 3.4 does not show that at the time of chromosome pairing, the homologous chromosomes can exchange genes. The exchanges result in the formation of chromosomes that consist of segments from one homologous chromosome intermixed with segments from the other. In Figure 3.4, the exchanged chromosomes would be depicted as segments of alternating color. The exchange process is one of the critical feature of meiosis, and it will be examined in the next section.

In animals, meiosis takes place in specific cells called *meiocytes*, a general term for the primary oocytes and spermatocytes in the gamete-forming tissues (**FIGURE 3.5**). The oocytes form egg cells and the spermatocytes form sperm cells. Although the process of meiosis is similar in all sexually reproducing organisms, in the female of both animals and plants, only one of the four products develops into a functional cell (the other three disintegrate). In animals, the products of meiosis form gametes (sperm or eggs).

In plants, the situation is slightly more complicated:

1. The products of meiosis typically form *spores*, which undergo one or more mitotic divisions to produce a haploid *gametophyte* organism.

First (reductional) division

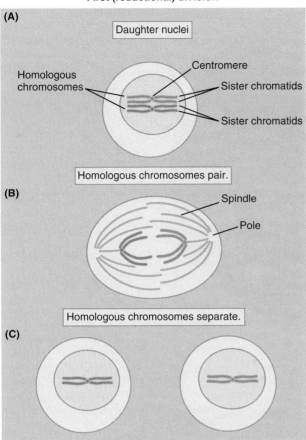

(A)

Daughter nuclei

Homologous chromosomes

Centromere

Sister chromatids

Sister chromatids

Homologous chromosomes pair.

(B)

Spindle

Pole

Homologous chromosomes separate.

(C)

Second (equational) division

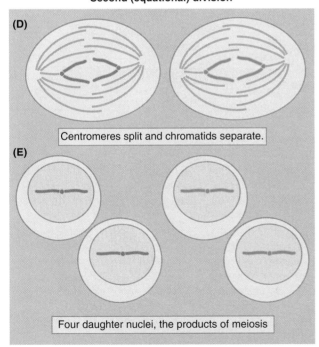

(D)

Centromeres split and chromatids separate.

(E)

Four daughter nuclei, the products of meiosis

FIGURE 3.4 Overview of the behavior of a single pair of homologous chromosomes in meiosis. The key differences from mitosis are the pairing of homologous chromosomes (A) and the two successive nuclear divisions (B and D) that reduce the chromosome number by half. For clarity, this diagram does not incorporate crossing-over, an interchange of chromosome segments that takes place at the stage depicted in part A.

The gametophyte produces gametes by mitotic division of a haploid nucleus (**FIGURE 3.6**).

2. Fusion of haploid gametes creates a diploid zygote that develops into the *sporophyte* plant, which undergoes meiosis to produce spores and so restarts the cycle.

Meiosis is a more complex and considerably longer process than mitosis and usually requires days or even weeks. The entire process of meiosis is illustrated in its cellular context in **FIGURE 3.7**. The essence is that *meiosis consists of two divisions of the nucleus but only one replication of the chromosomes*. The nuclear divisions—called the *first meiotic division* and the *second meiotic division*—can be separated into a sequence of stages similar to those used to describe mitosis. The distinctive events of this important process occur during the first division of the nucleus; these events are described in the following section.

The first meiotic division reduces the chromosome number by half.

The first meiotic division (meiosis I) is sometimes called the **reductional division** because it divides the chromosome number in half. By analogy with mitosis, the first meiotic division can be split into the four stages of **prophase I**, **metaphase I**, **anaphase I**, and **telophase I**. These stages are generally more complex than their counterparts in mitosis. The stages and substages can be visualized with reference to Figure 3.7 and **FIGURE 3.8**.

1. Prophase I   This long stage lasts several days in most higher organisms and is commonly divided into five substages: *leptotene*, *zygotene*, *pachytene*, *diplotene*, and *diakinesis*. These are descriptive terms that indicate the appearance of the chromosomes at each substage.

In **leptotene**, which literally means "thin thread," the chromosomes first become visible as long, thread-like structures. The pairs of sister chromatids can be distinguished by electron microscopy. In this initial phase of condensation of the chromosomes, numerous dense granules appear at irregular intervals along their length. These localized contractions, called *chromomeres*, have a characteristic number, size, and position in a given chromosome (Figure 3.8, part A).

Leptotene

The **zygotene** ("paired thread") period is marked by the lateral pairing, or **synapsis**, of homologous chromosomes, beginning at the chromosome tips. As the pairing

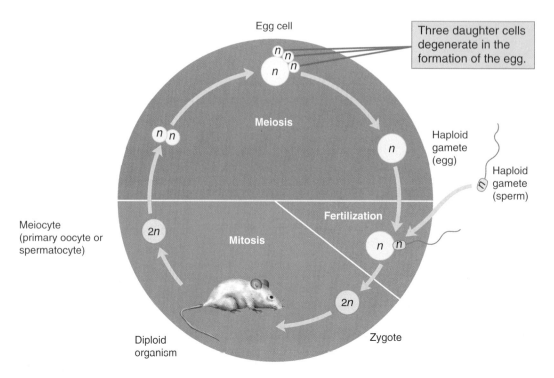

FIGURE 3.5 The life cycle of a typical animal. The number *n* is the number of chromosomes in the haploid chromosome complement. In males, the four products of meiosis develop into functional sperm; in females, only one of the four products develops into an egg.

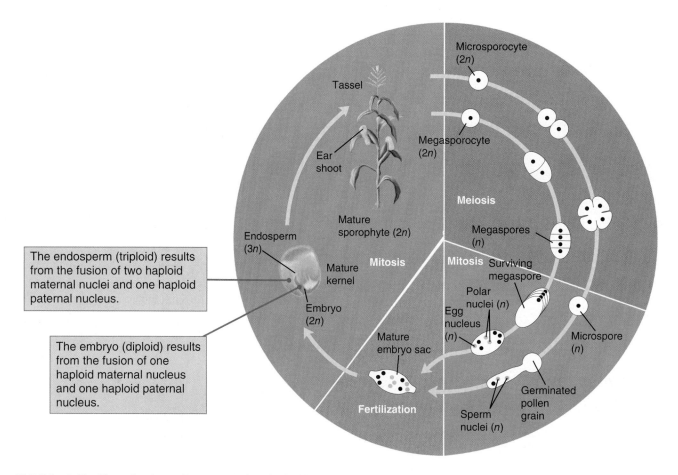

FIGURE 3.6 The life cycle of corn, *Zea mays*. As is typical in higher plants, the diploid spore-producing (sporophyte) generation is conspicuous, whereas the gamete-producing (gametophyte) generation is microscopic. The egg-producing spore is the *megaspore* and the sperm-producing spore is the *microspore*. Nuclei participating in meiosis and fertilization are shown in yellow and green.

process proceeds along the length of the chromosomes, it results in a precise chromomere-by-chromomere association (Figure 3.8, part B and part F). Synapsis is accompanied by synthesis of the *synaptonemal complex*, a protein structure that helps to hold the aligned homologous chromosomes together. Each pair of synapsed homologous chromosomes is referred to as a **bivalent**.

Zygotene

During **pachytene** (Figure 3.8, part C and part D), condensation of the chromosomes continues. The term literally means "thick thread," and throughout this period, the chromosomes continue to shorten and thicken (Figure 3.7). By late pachytene, it can sometimes be seen that each bivalent (that is, each set of paired chromosomes) actually consists of a **tetrad** of four chromatids, but the two sister chromatids of each chromosome are usually juxtaposed very tightly. The important event of genetic exchange, **crossing-over**, takes place during pachytene, but crossing-over does not become apparent until the transition to diplotene. In Figure 3.7, the sites of exchange are indicated by the points where chromatids of different colors cross over each other.

Pachytene

At the onset of **diplotene** ("double thread"), the synaptonemal complex breaks down and the synapsed chromosomes begin to separate, and the diplotene chromosomes become obviously double (Figure 3.8, part E). However, the homologous chromosomes remain held together at intervals along their length by cross-connections resulting from crossing-over. Each cross-connection, called a **chiasma** (plural, *chiasmata*), is formed by a breakage and rejoining between nonsister chromatids. As shown in the chromosome and diagram in **FIGURE 3.9**, *a chiasma results from physical exchange between chromatids of homologous chromosomes*. In normal meiosis, each bivalent usually has at least one chiasma, and bivalents of long chromosomes often have three or more.

Diplotene

The final period of prophase I is **diakinesis**, in which the homologous chromosomes seem to repel each other and the segments not connected by chiasmata move apart. (Diakinesis means "moving apart.") It is at this substage that the chromosomes attain their maximum condensation. The homologous chromosomes in a bivalent remain connected by at least one chiasma, which persists until the first meiotic anaphase. Near the end of diakinesis, the formation of a spindle is initiated and the nuclear envelope breaks down.

Diakinesis

KEY CONCEPT

Genes on different chromosomes undergo independent assortment because nonhomologous chromosomes align at random on the metaphase plate in meiosis I.

2. Metaphase I The bivalents become positioned with the centromeres of the two homologous chromosomes on opposite sides of the metaphase plate (part A, **FIGURE 3.10**). As each bivalent moves onto the metaphase plate, its centromeres are oriented at random with respect to the poles of the spindle. As shown in **FIGURE 3.11**, the bivalents formed from nonhomologous pairs of

Metaphase I

chromosomes can be oriented on the metaphase plate in either of two ways. The orientation of the centromeres determines which member of each bivalent will subsequently move to each pole. If each of the nonhomologous chromosomes is heterozygous for a pair of alleles, one type of alignment results in *A B* and *a b* gametes, and the other type results in *A b* and *a B* gametes (Figure 3.11). Because the metaphase alignment takes place at random, the two types of alignment—and therefore the four types of gametes—are equally frequent. The ratio of the four types of gametes is 1 : 1 : 1 : 1, which means that the *A, a* and *B, b* pairs of alleles undergo independent assortment. In other words,

The experimental demonstration of this principle in 1913 gave strong support to the idea, already accepted by many geneticists, that the chromosomes were the cellular objects that contained the genetic material. These studies were carried out by Eleanor Carothers working with a species of grasshopper.

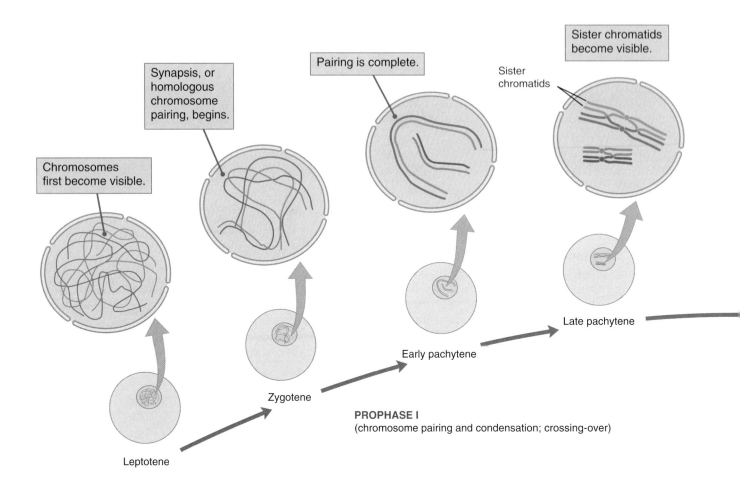

Chromosomes first become visible.

Synapsis, or homologous chromosome pairing, begins.

Pairing is complete.

Sister chromatids become visible.

Sister chromatids

Leptotene

Zygotene

Early pachytene

Late pachytene

PROPHASE I
(chromosome pairing and condensation; crossing-over)

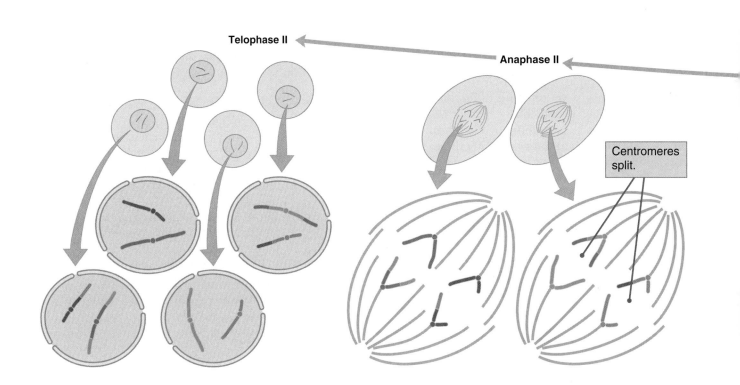

Telophase II

Anaphase II

Centromeres split.

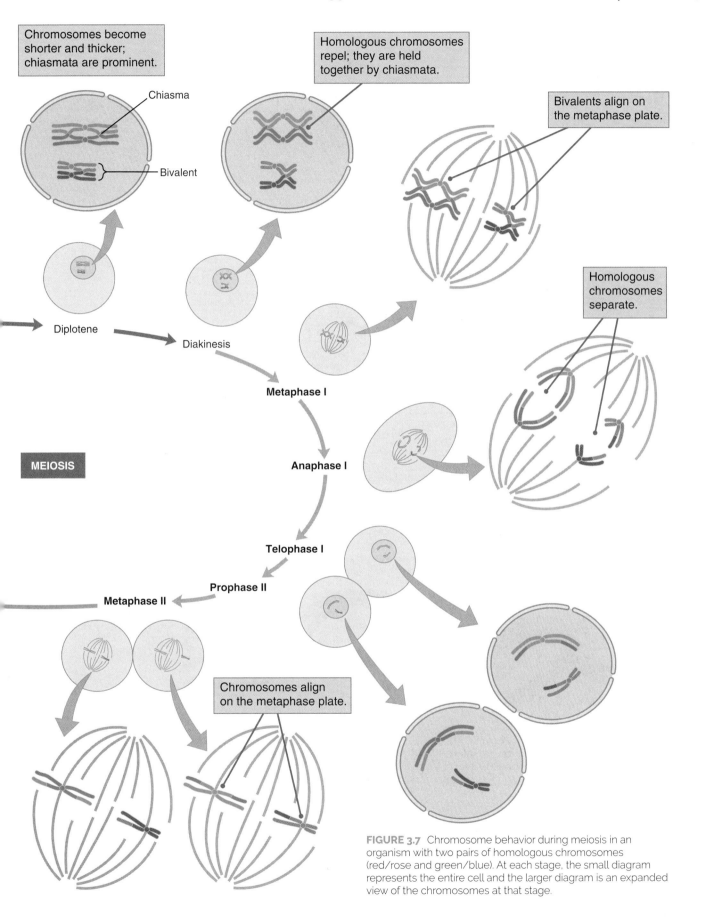

Chromosomes become shorter and thicker; chiasmata are prominent.

Chiasma

Bivalent

Homologous chromosomes repel; they are held together by chiasmata.

Bivalents align on the metaphase plate.

Homologous chromosomes separate.

Diplotene

Diakinesis

MEIOSIS

Metaphase I

Anaphase I

Telophase I

Prophase II

Metaphase II

Chromosomes align on the metaphase plate.

FIGURE 3.7 Chromosome behavior during meiosis in an organism with two pairs of homologous chromosomes (red/rose and green/blue). At each stage, the small diagram represents the entire cell and the larger diagram is an expanded view of the chromosomes at that stage.

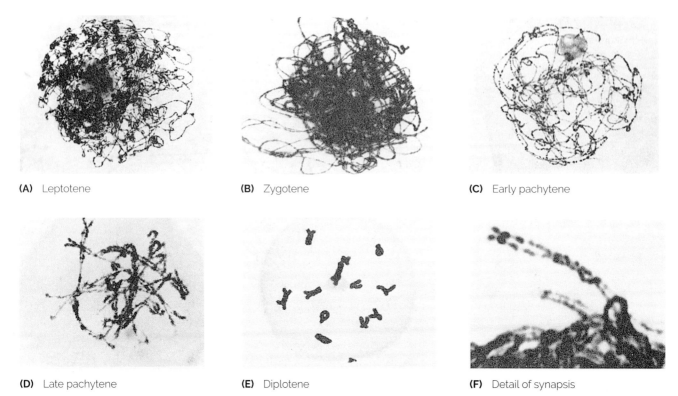

(A) Leptotene   **(B)** Zygotene   **(C)** Early pachytene

(D) Late pachytene   **(E)** Diplotene   **(F)** Detail of synapsis

FIGURE 3.8 Substages of prophase of the first meiotic division in microsporocytes of a lily (*Lilium longiflorum*). (A) Leptotene, in which condensation of the chromosomes is initiated and bead-like chromomeres are visible along the length of the chromosomes. (B) Zygotene, in which pairing (synapsis) of homologous chromosomes occurs (paired and unpaired regions can be seen particularly at the lower left in this photograph). (C) Early pachytene, in which synapsis is completed and crossing-over between homologous chromosomes occurs. (D) Late pachytene, showing the continued shortening and thickening of the chromosomes. (E) Diplotene, characterized by mutual repulsion of the paired homologous chromosomes, which remain held together at one or more cross points (chiasmata) along their length; diplotene is followed by diakinesis (not shown), in which the chromosomes reach their maximum contraction. (F) Zygotene (at higher magnification in another cell) showing paired homologs and matching of chromomeres during synapsis.

Parts A, B, C, E, and F courtesy of Marta Walters and Santa Barbara Botanic Gardens, Santa Barbara, California. Part D courtesy of Herbert Stern. Used with permission.

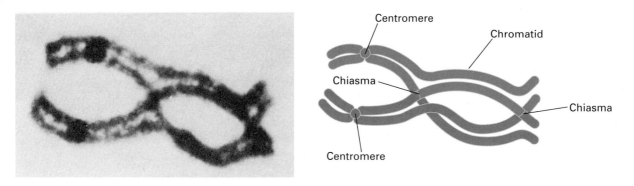

FIGURE 3.9 Light micrograph (A) and interpretative drawing (B) of a bivalent consisting of a pair of homologous chromosomes. This bivalent was photographed at late diplotene in a spermatocyte of the salamander *Oedipina poelzi*. It shows two chiasmata where the chromatids of the homologous chromosomes appear to exchange pairing partners.

Part A courtesy of Dr. James Kezer. Used with permission by Dr. Stanley K. Sessions.

Anaphase I

3. Anaphase I In this stage, homologous chromosomes, each composed of two chromatids joined at an undivided centromere, separate from one another and move to opposite poles of the spindle (Figure 3.10, part B). Chromosome separation at anaphase is the cellular basis of the segregation of alleles.

> **KEY CONCEPT**
>
> The physical separation of homologous chromosomes in anaphase is the physical basis of Mendel's principle of segregation.

Note, however, that the centromeres of the sister chromatids are stuck together tightly and behave as a

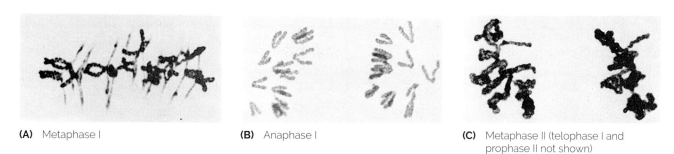

(A) Metaphase I   **(B)** Anaphase I   **(C)** Metaphase II (telophase I and prophase II not shown)

(D) Anaphase II   **(E)** Telophase II

FIGURE 3.10 Later meiotic stages in microsporocytes of the lily *Lilium longiflorum*. (A) Metaphase I. (B) Anaphase I. (C) Metaphase II. (D) Anaphase II. (E) Telophase II. Cell walls have begun to form in telophase, which will lead to the formation of four pollen grains.

Courtesy of Herbert Stern. Used with permission.

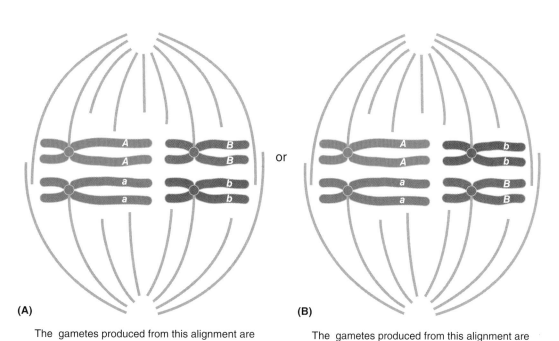

or

(A)

The gametes produced from this alignment are

A B : A B : a b : a b

(B)

The gametes produced from this alignment are

A b : A b : a B : a B

> Because the alignments are equally likely, the overall ratio of gametes is
>
> *A B : A b : a B : a b* = 1:1:1:1
>
> This ratio is characteristic of independent assortment.

FIGURE 3.11 Random alignment of nonhomologous chromosomes at metaphase I results in the independent assortment of genes on nonhomologous chromosomes.

single unit, owing to the presence of a protein "glue" that holds them together until the onset of anaphase II, when these glue proteins are broken down.

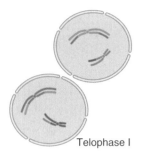

Telophase I

4. Telophase I At the completion of anaphase I, a haploid set of chromosomes consisting of one homolog from each bivalent is located near each pole of the spindle (Figure 3.9). In telophase, the spindle breaks down, and, depending on the species, either a nuclear envelope briefly forms around each group of chromosomes or the chromosomes enter the second meiotic division after only a limited uncoiling.

The second meiotic division is equational.

The second meiotic division (meiosis II) is sometimes called the **equational division** because the chromosome number remains the same in each cell before and after the second division. In some species, the chromosomes pass directly from telophase I to **prophase II** without loss of condensation; in others, there is a brief pause between the two meiotic divisions and the chromosomes may "decondense" (uncoil) somewhat. *Chromosome replication never takes place between the two divisions*; the chromosomes present at the beginning of the second division are identical with those present at the end of the first division.

After a short prophase (prophase II) and the formation of second-division spindles, the centromeres of the chromosomes in each nucleus become aligned on the central plane of the spindle at **metaphase II** (Figure 3.10, part C). In **anaphase II**, the centromeres divide longitudinally and the chromatids of each chromosome move to opposite poles of the spindle (Figure 3.10, part D). Once the centromere has split at anaphase II, each chromatid is considered a separate chromosome.

Telophase II (Figure 3.10, part E) is marked by a transition to the interphase condition of the chromosomes in the four haploid nuclei, accompanied by division of the cytoplasm. Thus the second meiotic division superficially resembles a mitotic division. However, there is an important difference:

KEY CONCEPT

The chromatids of a chromosome are usually not genetically identical along their entire length because of crossing-over associated with the formation of chiasmata during prophase of the first division.

STOP & THINK 3.2

For the terms describing the substages of prophase I of meiosis (leptotene, zygotene, pachytene, diplotene, and diakinesis), fill in the term that corresponds to the main event that occurs during that stage:

_____ Bivalents are maximally condensed
_____ Crossing-over takes place
_____ Chromosome threads first become
                        visible
_____ Homologous chromosomes pair
_____ Chiasmata become prominent

3.4 Eukaryotic chromosomes are highly coiled complexes of DNA and protein.

Each eukaryotic chromosome contains a single, uninterrupted molecule of DNA that can be extremely long. The average length of a chromosome in the human genome is 130 million base pairs (Mb), ranging from 250 Mb (chromosome 1, the longest) to 47 Mb (chromosome 21, the shortest). If the DNA in chromosome 1 were extended to its full length, it would exceed the diameter of the nucleus of an average human cell by a factor of about 15,000. In this section, we take a closer look at how molecules of such enormous length are packaged to fit into the nucleus.

The nucleosome is the basic structural unit of chromatin.

The DNA of all eukaryotic chromosomes is associated with numerous protein molecules in a stable ordered aggregate called **chromatin**. Some of the proteins present in chromatin determine chromosome structure and the changes in structure that occur during the division cycle of the cell. Other chromatin proteins appear to have important roles in regulating chromosome functions.

The simplest form of chromatin is present in nondividing eukaryotic cells, when chromosomes are not sufficiently condensed to be visible by light microscopy. Chromatin isolated from such cells is a complex aggregate of DNA and proteins. The major class of chromosomal proteins are the **histone** proteins. Histones are largely responsible for the structure of chromatin. Five major types—H1, H2A, H2B, H3, and H4—are present in the chromatin of nearly all eukaryotes in amounts about equal in mass to that of the DNA. Histones are small proteins (100 to 200 amino acids) that differ from most other proteins in that from 20 to 30 percent of the amino acids are lysine and arginine, both of which have a positive charge. (Only a few percent of the amino acids of a typical protein are lysine and arginine.) The positive charges enable histone molecules to bind to DNA, primarily by electrostatic attraction to the negatively charged phosphate groups in the sugar–phosphate backbone of DNA. Placing chromatin in a solution with a high salt concentration (for example, 2 molar NaCl) to eliminate the electrostatic attraction causes the histones to dissociate from the DNA. Histones also bind tightly to each other; both DNA–histone and histone–histone binding are important for chromatin structure.

The histone molecules from different organisms are remarkably similar to one another, with the exception of H1. In fact, the amino acid sequences of H3 molecules from widely different species are almost identical. For example, the sequences of H3 of cow chromatin and pea chromatin differ by only 4 of 135 amino acids. The H4 proteins of all organisms also are quite similar; cow and pea H4 differ by only 2 of 102 amino acids. There are few other proteins whose amino acid sequences vary so little from one species to the next. When the variation is very small between organisms, one says that the sequence is highly *conserved*. The extraordinary conservation in histone composition through hundreds of millions of years of evolutionary divergence is consistent with the important role of these proteins in the structural organization of eukaryotic chromosomes.

In the electron microscope, chromatin resembles a regularly beaded thread formed into a coiled fiber, known as the **30-nm chromatin fiber** (**FIGURE 3.12**), with an average diameter ranging from 300 to 350 angstrom (Å). The beadlike units within the 30-nm chromatin fiber are called **nucleosomes.** The molecular composition of nucleosomes is illustrated in part A of **FIGURE 3.13**. Each unit has a definite composition, consisting of two molecules each of H2A, H2B, H3, and H4; a segment of DNA containing about 200 nucleotide pairs; and one molecule of histone H1. The complex of two subunits each of H2A, H2B, H3, and H4, as well as part of the DNA, forms each "bead," and the remaining DNA bridges between the beads. Histone H1 also appears to play a role in bridging between the beads, but it is not shown in Figure 3.13, part A.

Brief treatment of chromatin with certain DNases yields a collection of small particles of quite uniform size consisting only of histones and DNA. The DNA fragments in these particles are of lengths equal to about 200 nucleotide pairs or small multiples of that unit size (the precise size varies with species and tissue). These particles result from cleavage of the linker DNA segments between the beads (Figure 3.13, part B). More extensive treatment with DNase results in loss of the H1 histone and digestion of all the DNA except that protected by the histones in the bead. The resulting structure is called a **core particle**, which consists of an octamer of pairs of H2A, H2B, H3, and H4, around

which the remaining DNA, approximately 145 base pairs, is wound in about one and three-fourths turns. Each nucleosome is composed of a core particle, additional DNA called *linker DNA* that links adjacent core particles (the linker DNA is removed by extensive nuclease digestion), and one molecule of H1 that binds to the histone octamer and to the linker DNA.

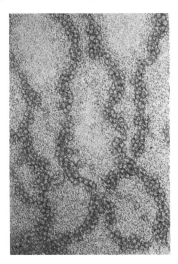

FIGURE 3.12 Electron micrograph of the 30-nm chromatin fiber in mouse chromosomes.

Courtesy of Barbara A. Hamkalo, University of California, Irvine.

The amino ends of the histone proteins, which constitute about 25 percent of the total length, are known as **histone tails** because they are accessible to enzymes that modify particular amino acid residues such as by the addition of one or more acetyl ($-COCH_3$), methyl ($-CH_3$), or phosphate [$-OP(=O)(OH)_2$] groups to the amino acid. The modifications resulting from these histone acetylase, methylase, or phosphorylase enzymes are reversible by the corresponding histone deacetylase, demethylase, or dephosphorylase enzymes. Modifications of the histone tails are important to gene activity. Acetylated histones tend to bind DNA more loosely and usually render chromatin more accessible to transcription, whereas methylated histones can either promote or impede transcription depending on the particular histone residue that is modified. Histone modifications are thought to be important features of *chromatin remodeling* that takes place in the regulation of gene activity.

Chromatin fibers form discrete chromosome territories in the nucleus.

In forming the 30-nm chromatin fiber in Figure 3.12, the string of nucleosomes forms a series of stacked, right-handed coils. Viewed from the side (**FIGURE 3.14**, part A), each nucleosome is attached to its neighbor by linker DNA that stretches nearly linearly across to the opposite side of the coil. Looking down at the 30-nm fiber from the top (Figure 3.14, part B), one can trace the path of the linker DNA as it travels down the length of the fiber. In each revolution around the fiber axis, the path of the linker DNA closely approximates the shape of a seven-pointed star.

In the nucleus of a nondividing cell, the 30-nm chromatin fiber is organized into higher order structures that can be visualized using modern methods of optical sectioning and image reconstruction. **FIGURE 3.15**

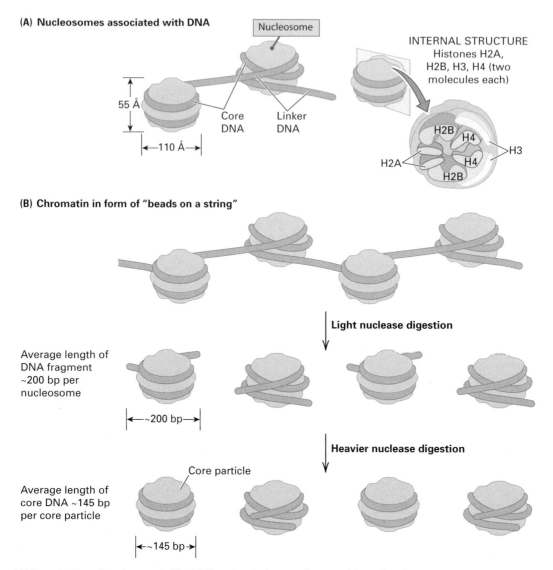

FIGURE 3.13 (A) Organization of nucleosomes. The DNA molecule is wound one and three-fourths turns around a histone octamer called the core particle. If H1 were present, it would bind to the octamer surface and to the linkers, causing the linkers to cross. (B) Effect of treatment with micrococcal nuclease. Brief treatment cleaves the DNA between the nucleosomes and results in core particles associated with histone H1 and approximately 200 base pairs of DNA. More extensive treatment results in loss of H1 and digestion of all but 145 base pairs of DNA in intimate contact with each core particle.

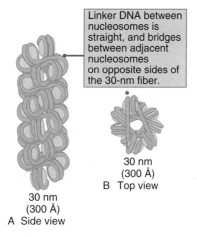

FIGURE 3.14 Model of how nucleosomes are packed into the 30-nm chromatin fiber.

Modified from J. T. Finch and A. Klug, *Proc. Natl. Acad. Sci. USA* 73 (1976): 1897–1901.

shows a computer-generated image of 30-nm chromatin fibers within the nucleus of a nondividing cell. The chromatin fibers are folded into small *chromatin loops* with a DNA content of approximately 100 kb each, and these are further organized into *chromatin domains* with a DNA content of approximately 1 Mb each. Each chromosome arm occupies a discrete **chromosome territory**, denoted by the different colors. In cells cycling through mitosis, the chromosome territories are disrupted when the chromosomes condense and the cell divides, but they are reconstituted again in the next interphase. However, the chromosome territories may differ in position in different cell types as well as in the same cell type at different times in development.

The pattern of chromatin folding in Figure 3.18 looks at first glance like a tangled mess, but in reality the chromatin is carefully folded in such a way as

FIGURE 3.15 Computer-generated image of chromosome territories formed by 30-nm chromatin fibers within the nucleus of a nondividing cell.

Courtesy of Tobias A. Knoch, Erasmus MC, Rotterdam, and Kirchhoff-Institute for Physics, Ruperto-Carola University, Heidelberg.

to avoid knots and tangles. Evidence for this picture comes from experiments in which stretches of DNA that are near each other are chemically crosslinked, isolated, and sequenced to determine which parts of which chromosomes are in close proximity. The results indicate that folded interphase chromatin is in the form of a so-called **fractal globule**, in which the chromatin thread follows a path that allows dense packing without knots or tangles. Much like a sailor carefully coils a rope so that it will play out at great speed without forming knots or tangles, the coils of interphase chromatin allow easy access and unwinding. The fractal globule type of folding enables regions of chromatin or individual genes to undergo unfolding or reorganization with minimal disturbance to nearby regions or genes.

Chromosome territories are correlated with gene densities. The territories of chromatin domains containing relatively few genes tend to be located near the periphery of the nucleus or near the nucleolus, whereas the territories of chromosome domains that are relatively gene rich tend to be located toward the interior of the nucleus. For example, human chromosome 18 (85 Mb in size) is relatively gene poor whereas chromosome 19 (67 Mb in size) is relatively gene rich. In the nucleus, chromosome 18 territories tend to be at the nuclear periphery whereas those of chromosome 19 tend to be in the interior.

The spaces between the chromatin domains form a network of channels, like the holes permeating through a sponge. The channels are large enough to allow passage of the molecular machinery for replication, transcription, and RNA processing. Evidence suggests that these molecules gain access to chromatin by means of passive diffusion. Replication, transcription, and RNA processing are all ordered processes. DNA replication takes place in small discrete regions that exhibit a reproducible temporal and spatial pattern, and transcription takes place in a few hundred discrete locations. However, many important details are still unknown about the organization of chromatin in the nucleus and how chromosome territories function in the coordination of the central molecular processes of replication, transcription, and RNA processing.

The metaphase chromosome is a hierarchy of coiled coils.

The hierarchical nature of chromosome structure is illustrated in **FIGURE 3.16**. Assembly of DNA and histones is the first level, resulting in a sevenfold reduction in length of the DNA and the formation of a beaded flexible fiber 110 Å (11 nm) wide (part B), roughly five times the width of free DNA (part A). The structure of chromatin varies with the concentration of salts, and the 110-Å fiber is present only when the salt concentration is quite low. In the living cell, this is usually compacted into the 30-nm chromatin fiber (part C), which in the interphase nucleus is folded into 100-kb chromatin loops that are organized into 1-Mb chromatin domains that form the chromosome territories.

In cells cycling through mitosis, the interphase chromatin organization is replaced by a more compact organization in which the 30-nm chromatin fiber condenses into a chromatid of the metaphase chromosome (Figure 3.16, parts D through F). Chromosome condensation is an ordered, energy-consuming process orchestrated by a protein complex called *condensin* that works to actively coil the chromatin. Although the structures of some condensin proteins are known, the details of chromatin condensation are still largely unknown, and there is no strong evidence supporting any of the particular coiled structures greater than the 30-nm chromatin fiber depicted in Figure 3.16.

In electron micrographs of isolated metaphase chromosomes from which histones have been removed, the partly unfolded DNA has the form of an enormous number of loops that seem to extend from a central core or **scaffold** (**FIGURE 3.17**), which is composed of a number of nonhistone chromosomal proteins. Electron microscopic studies of chromosome condensation in mitosis and meiosis suggest that the scaffold extends along the chromatid and that the 30-nm fiber becomes arranged into a helix of loops radiating from the scaffold. Details are not known about the additional folding that is required of the fiber in each loop to produce the fully condensed metaphase chromosome.

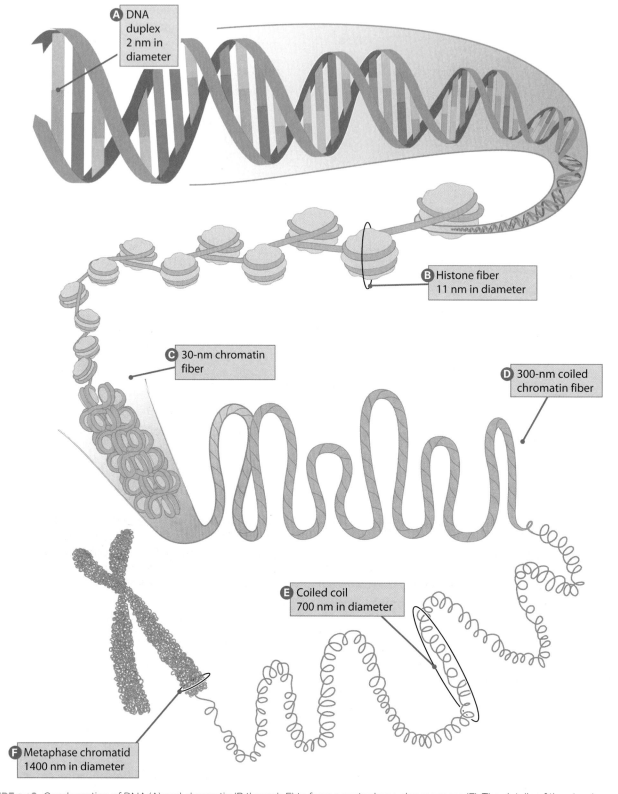

FIGURE 3.16 Condensation of DNA (A) and chromatin (B through E) to form a metaphase chromosome (F). The details of the structures in D–F are hypothetical.

The genetic significance of the compaction of DNA and protein into chromatin and ultimately into the chromosome is that it greatly facilitates the movement of the genetic material during nuclear division. Relative to a fully extended DNA molecule, the length of a metaphase chromosome is reduced by a factor of approximately 10^4 because of chromosome condensation. Without chromosome condensation, the chromosomes would become so entangled that there would be many more abnormalities in the distribution of genetic material into daughter cells.

An analogy may be helpful in appreciating the prodigious feat of packaging that chromosome condensation represents. If the 250-Mb DNA molecule in human chromosome 1 were a cooked spaghetti noodle 1 mm in diameter, it would stretch for 25 miles; in chromosome condensation, this noodle is gathered together, coil upon coil, until at metaphase it is a canoe-sized tangle of spaghetti 16 feet long and 2 feet wide. After cell division, the noodle is unwound again.

Heterochromatin is rich in satellite DNA and low in gene content.

Certain regions of the chromosome have a dense, compact structure in interphase and are darkly stainable

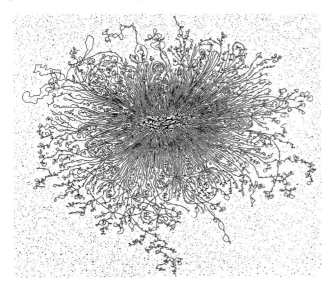

FIGURE 3.17 Electron micrograph of a partially disrupted anaphase chromosome of the milkweed bug *Oncopeltus fasciatus*, showing multiple loops of 30-nm chromatin at the periphery.

Courtesy of Bruno Zimm and Ruth Kavenoff. Used with permission of Georgianna Zimm, University of California, San Diego.

by many standard dyes used to make chromosomes visible. Regions of chromatin that are compact and heavily stained in interphase are known as **heterochromatin**. The rest of the chromatin, which becomes visible only after chromosome condensation in mitosis or meiosis, is called **euchromatin**. Sometimes the heterochromatin remains highly condensed throughout the cell cycle and can be distinguished even at metaphase. The major heterochromatic regions are adjacent to the centromere; smaller blocks are present at the ends of the chromosome arms (the telomeres) and interspersed with the euchromatin (**FIGURE 3.18**). At the DNA level, a substantial part of the heterochromatin consists of long tracts of relatively short base sequences, typically from 5 to 500 base pairs in length, each repeated in tandem. The highly repeated sequences are often called **satellite DNA** for reasons related to the original method of their isolation. Each satellite sequence has its own distinctive distribution in the heterochromatin. In many species, an entire chromosome—such as the Y chromosome in *Drosophila*—is almost completely heterochromatic.

The genetic content of heterochromatin is summarized in the following generalization:

KEY CONCEPT

The number of genes located in heterochromatin is small relative to the number in euchromatin.

The relatively small number of genes means that many large blocks of heterochromatin are genetically almost inert, or devoid of function. Indeed, heterochromatic blocks can often be rearranged in the genome, duplicated, or even deleted without major phenotypic consequences.

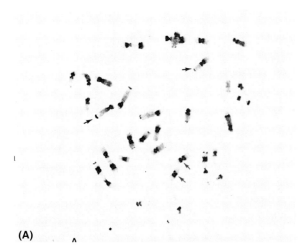

(A)

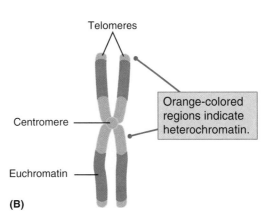

(B)

FIGURE 3.18 (A) Metaphase chromosomes of the ground squirrel *Ammospermophilus harrissi*, stained to show the heterochromatic regions near the centromere of most chromosomes (red arrows) and the telomeres of some chromosomes (black arrows). (B) An interpretive drawing.

Part A courtesy of T.C. Hsu, Ph.D., and used with permission of Sen Pathak, Ph.D., Anderson Cancer Center, University of Texas.

3.5 The centromere and telomere are essential parts of chromosomes.

Eukaryotic chromosomes contain regions specialized for maneuvering the chromosomes in cell division and for capping the ends. These regions are discussed next.

The centromere is essential for chromosome segregation.

The *centromere* is a specific region of the eukaryotic chromosome that becomes visible as a distinct morphological entity along the chromosome during condensation. It serves as the point of assembly of the *kinetochore*, the complex of DNA and proteins to which the spindle fibers attach and move the chromosomes in both mitosis and meiosis. The kinetochore is also the site at which the spindle fibers shorten, causing the chromosomes to move toward the poles. In higher eukaryotes, each centromeric region encompasses 1 million base pairs or more. These regions of heterochromatin contain various kinds of repetitive DNA sequences, as well as a patchwork of DNA sequences derived from duplicated regions from elsewhere in the genome.

Common to all human centromeres are hundreds of thousands of copies of a 170-bp DNA sequence called **alpha satellite** (**FIGURE 3.19**). The blocks of alpha-satellite DNA are associated with chromatin composed of nucleosomes that contain a specialized histone variant called CENPA that replaces the normal histone 3 (**FIGURE 3.20**, part A). The presence of CENPA nucleosomes helps in the recruitment of more than a dozen kinetochore proteins, resulting in a fully assembled kinetochore (Figure 3.20, part B). Because alpha-satellite sequences can differ from one

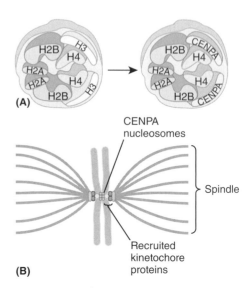

FIGURE 3.20 Assembly of the kinetochore around the centromere. (A) Centromeric nucleosomes are modified by replacement of histone 3 with CENPA protein. (B) Modified centromeric nucleosomes recruit kinetochore proteins, and the assembled kinetochore serves as an attachment site for spindle microtubules.

another in as much as 50 percent of their sequence, the replacement of histone 3 with CENPA in the nucleosomes is not directly specified by the sequence of the alpha-satellite array. Kinetochore assembly is therefore said to be **epigenetic**, a term that refers to persistent changes in chromatin structure or gene expression that are not directly specified by the DNA sequence.

The telomere is essential for the stability of the chromosome tips.

Each end of a linear chromosome is composed of a special DNA–protein structure called a **telomere** that is essential for chromosome stability. Genetic and microscopic observations first indicated that telomeres are special structures. In *Drosophila*, chromosomes without ends formed by x-ray breakage cannot be recovered; in maize, broken chromosome ends frequently fuse with one another and form new chromosomes with abnormal structures (often having two centromeres). The process of DNA replication cannot begin precisely at the 3′ end of a template strand, so the 3′ end of a replicated duplex must terminate in a short stretch in which the DNA is single stranded. This single-stranded overhang is subject to degradation by nucleases. Without some mechanism to restore the digested end, the DNA molecule in a chromosome would become slightly shorter with each replication. There is such a mechanism, and in mutant cells in which the mechanism is defective, each chromosome end does become shorter in each replication until, eventually, there is so much degradation that the cell dies.

The mechanism of restoring the ends of a DNA molecule in a chromosome relies on an enzyme called **telomerase**. This enzyme works by adding tandem

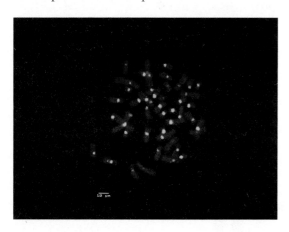

FIGURE 3.19 Hybridization of human metaphase chromosomes (red) with alpha-satellite DNA. The yellow areas result from hybridization with the labeled DNA. The sites of hybridization of the alpha satellite coincide with the centromeric regions of all 46 chromosomes.

Courtesy of Paula Coelho and Claudio E. Sunkel Cariola, IBMC.

repeats of a simple sequence to the 3′ end of a DNA strand. In the ciliated protozoan *Tetrahymena*, in which the enzyme was first discovered, the simple repeating sequence is −TTGGGG-3′, and in humans and other vertebrate organisms, it is −TTAGGG-3′. The tandem repeats of these sequences constitute the telomere. As the repeating telomere sequence is being elongated, DNA replication occurs to synthesize a partner strand, and so, for example, the telomere sequence of the right-hand end of any *Tetrahymena* chromosome would be a DNA duplex of the form

```
-TTGGGGTTGGGGTTGGGGTTGGGGTTGGGGTTGGGGTTGGGG-3'
-AACCCCAACCCCAACCCCAACCCC-5'
```

with a single-stranded overhang at the 3′ end that can be elongated further by the telomerase.

The role of telomerase in the replication of chromosomal DNA is illustrated in FIGURE 3.21. Part A represents the duplex DNA in a chromosome, with the telomere sequences shown in red. Because DNA replication cannot start precisely at the 3′ end, the 5′ end of each daughter strand in part B is a little shorter than the template strand from which it was replicated. The unreplicated part of the telomere sequence is subject to degradation by nucleases. The 3′ end of each daughter molecule also has a short telomere, because this end is replicated from the underhanging 5′ end of the telomere in the parental strand. The shortened telomere remaining at each 3′ end is the substrate of the telomerase, which elongates each 3′ end by the addition of more repeating telomere units (in the case of *Tetrahymena*, −TTGGGG-3′). Telomere elongation restores the structure of the original parental chromosome in which each end has a larger number of telomere repeats at the 3′ end and a smaller number of repeats at the 5′ end.

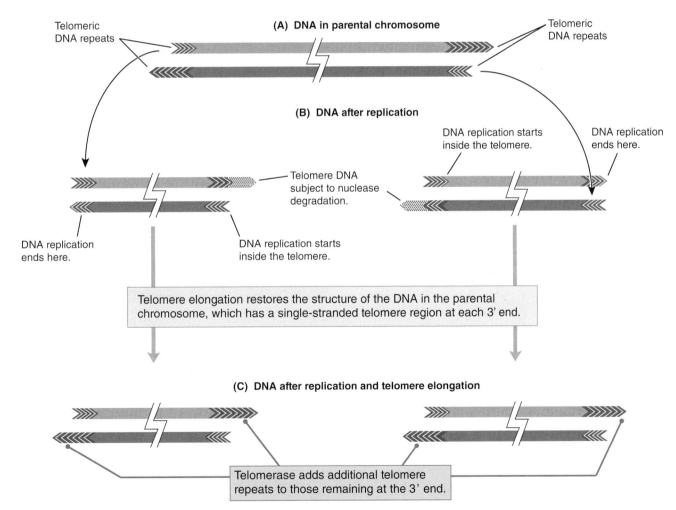

FIGURE 3.21 The function of telomerase. (A) Chromosomal DNA is double stranded. Each end of each strand terminates in a set of telomere repeats, but the 3′ end of each strand is longer as a result of telomerase action after the previous replication. (B) In the replication of each parental DNA strand, the new daughter DNA strand is initiated within the telomere repeat at the 3′ end; the telomere at the 5′ end in the new strand is shorter than that in the parental strand. The unreplicated 3′ end of the parental strand is vulnerable to digestion by nucleases. (C) In the daughter DNA duplexes formed by replication, the 3′ strand at each end is elongated via the addition of telomere repeats by the telomerase. The length and 3′ overhang of the telomeres are restored to the state that was present in the original parental molecule.

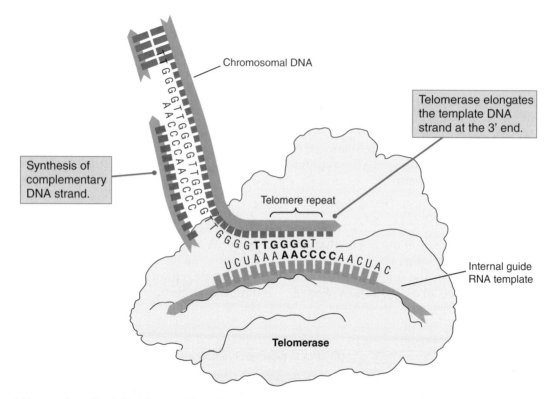

FIGURE 3.22 Telomere formation in *Tetrahymena*. The telomerase enzyme contains an internal guide RNA with a sequence complementary to the telomere repeat. The RNA undergoes base pairing with the telomere repeat and serves as a template for telomere elongation. The newly forming DNA strand is produced by DNA polymerase.

Relatively few copies of the telomere repeat are necessary to prime the telomerase to add additional copies and form a telomere. Remarkably, the telomerase enzyme incorporates an essential RNA molecule, called a *guide RNA*, that contains sequences complementary to the telomere repeat and that serves as a template for telomere synthesis and elongation. For example, the *Tetrahymena* guide RNA contains the sequence 3'-AACCCCAAC-5'. The guide RNA undergoes base pairing with the telomere repeat and serves as a template for telomere elongation by the addition of more repeating units (**FIGURE 3.22**). The complementary DNA strand of the telomere is synthesized by cellular DNA replication enzymes. In the telomeric regions of most eukaryotic chromosomes, there are also longer, moderately repetitive DNA sequences just preceding the terminal repeats. These sequences differ among organisms and even among different chromosomes in the same organism.

Telomere length limits the number of cell doublings.

Normal human cells in culture undergo only a limited number of divisions and then stop. The cells are still alive and metabolically active, but their entry into the S phase of DNA replication and subsequently into mitosis is blocked. Human cells in culture exposed to oxygen in amounts comparable to those in the body undergo about 70 divisions. The reason they stop doubling is due in part to telomere length. Normal cells have numerous proteins that can sense mishaps in the cell cycle and trigger one of several cell-cycle **checkpoints** that stop the process

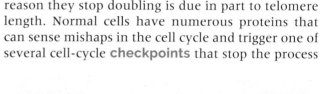

STOP & THINK 3.3

Imagine a chromosome in which the double-stranded DNA molecule has strands that are exactly the same length, as shown here.

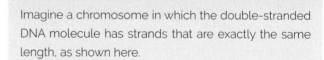

The chromosome undergoes one round of DNA replication in a eukaryotic cell in the presence of an inhibitor of telomerase. After this one round of replication, which end or ends (A, B, C, or D) are shorter than when replication began, and why?

until either the dysfunction is corrected or else cellular self-destruction is initiated. Among the types of damage detected is the presence of double-stranded breaks in DNA. In normal cells, proteins that detect double-stranded breaks are associated with telomeres but are not activated as long as telomere length is restored in successive cell divisions. In most cells in the body, however, the amount of telomerase is limited, so telomeres become a bit shorter with each division. When the telomeres reach a length of about 5 kb (which means about 800 copies of the telomeric repeat 5'-TTAGGG-3'), the proteins that recognize double-stranded breaks are activated and the ensuing checkpoint stops the cell cycle. The signal for shortened telomeres to be recognized as double-stranded breaks appears to be the addition of two methyl groups to a lysine residue at position 20 in histone H4 in the telomere-associated nucleosomes (forming a histone-designated H4K20me2).

In adult humans, most cells have telomeres of a length that permits only a few divisions until they are arrested by the DNA double-stranded break checkpoint. Only a minority of cells, known as **progenitor cells**, have telomeres of about 10 kb, and these are the cells that can undergo about 70 divisions before they stop. A loss of 5 kb of telomeric DNA in 70 divisions represents a loss of about 70 bp of telomere in each cell cycle. Progenitor cells lose telomeric DNA in each cell cycle even when they divide in the body. Progenitor cells from children divide in culture more times than those from middle-age adults, and those from middle-age adults more times than those from the elderly. The progressive loss of ability to divide explains in part why the healing process takes longer in the elderly.

Two types of cells that are not restricted in their number of divisions should be mentioned. One type is the **embryonic stem cell**, which exists in early embryos and has relatively high telomerase activity. These cells can undergo many cycles of cell division and can differentiate into many different types of specialized cells. These properties explain why embryonic stem cells have been of great interest in research, although their origin from embryos has raised moral qualms in some quarters.

The other type of cell that is unrestricted in number of divisions is a *cancer cell*. The mechanism of unlimited division in this case is that cancer cells have certain mutations that reactivate the telomerase gene and other mutations that override normal controls over cell division, such as the DNA double-stranded break checkpoint. The molecular mechanisms of cancer origination and progression are important in themselves, and because the analysis of mutations in cancer cells has yielded deep insights into normal processes in the cell cycle.

3.6 Genes are located in chromosomes.

Not long after the rediscovery of Mendel's paper in 1900, it became widely assumed that genes were physically located in the chromosomes. The strongest evidence was that Mendel's principles of segregation and independent assortment paralleled the behavior of chromosomes in meiosis. But the first indisputable proof that genes are parts of chromosomes was obtained in experiments concerned with the pattern of transmission of the **sex chromosomes**, the chromosomes responsible for determination of the separate sexes in some plants and in almost all higher animals. We will examine these results in this section.

Special chromosomes determine sex in many organisms.

The sex chromosomes are an exception to the rule that all chromosomes of diploid organisms are present in pairs of morphologically similar homologs. As early as 1891, microscopic analysis had shown that one of the chromosomes in males of some insects, such as grasshoppers, does not have a homolog. This unpaired chromosome was called the **X chromosome**, and it was present in all somatic cells of the males but in only half the sperm cells. The biological significance of these observations became clear when females of the same species were shown to have two X chromosomes.

In other species in which the females have two X chromosomes, the male has one X chromosome along with a morphologically different chromosome. This different chromosome is referred to as the **Y chromosome**, and it pairs with the X chromosome during meiosis in males because the X and Y share a small region of homology. The difference in the chromosomal constitution of males and females is a chromosomal mechanism for determining sex at the time of fertilization (**FIGURE 3.23**). Whereas every egg cell contains an X chromosome, half the sperm cells contain an X chromosome and the rest contain a Y chromosome. Fertilization of an X-bearing egg by an X-bearing sperm results in an XX zygote, which normally develops into a female; and fertilization by a Y-bearing sperm results in an XY zygote, which normally develops into a male. The result is a criss-cross pattern of inheritance of the X chromosome in which a male receives his X chromosome from his mother and transmits it only to his daughters. The XX-XY type of chromosomal sex determination is found in mammals, including human beings, in many insects, and in other animals, as well as in some flowering plants.

The X and Y chromosomes together constitute the sex chromosomes; this term distinguishes them from other pairs of chromosomes, which are called

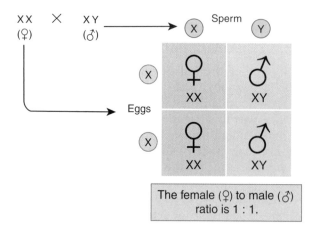

The female (♀) to male (♂) ratio is 1 : 1.

FIGURE 3.23 In chromosomal sex determination as found in humans and many other animals, each son gets his X chromosome from his mother and his Y chromosome from his father.

FIGURE 3.24 The photographs show the eye-color phenotypes of a wildtype, red-eyed male and a mutant, white-eyed male of the fruit fly, *Drosophila melanogaster.*

Courtesy of E. R. Lozovsky.

autosomes. Although the sex chromosomes control the developmental switch that determines the earliest stages of female or male development, the developmental process itself requires many genes scattered throughout the chromosome complement, including genes on the autosomes. The X chromosome also contains many genes with functions unrelated to sexual differentiation. In most organisms, the Y chromosome carries few genes other than those related to male determination. In human beings, for example, the Y chromosome is about 51 Mb in length and contains many nonfunctional genes thought to be remnants of genes whose functional counterparts are in the X chromosome.

X-linked genes are inherited according to sex.

The compelling evidence that genes are in chromosomes came from the study of a *Drosophila* gene for white eyes, which proved to be present in the X chromosome. In Mendel's crosses, reciprocal crosses gave the same result; it did not matter which trait was present in the male parent and which in the female parent. One of the earliest exceptions to this rule was found by Thomas Hunt Morgan in 1910, in an early study of a mutation in the fruit fly *Drosophila melanogaster* that had white eyes. The wildtype eye color is a brick-red combination of red and brown pigments (**FIGURE 3.24**). Although white eyes can result from certain combinations of autosomal genes that eliminate the pigments individually, the white-eye mutation that Morgan studied results in a metabolic block that knocks out both pigments simultaneously.

Morgan's study started with a single male with white eyes that appeared in a wildtype laboratory population that had been maintained for many generations. In a mating of this male with wildtype females

(cross A, **FIGURE 3.25**), all of the F$_1$ progeny of both sexes had red eyes, showing that the allele for white eyes is recessive. In the F$_2$ progeny from the mating of F$_1$ males and females, Morgan observed 2459 red-eyed females, 1011 red-eyed males, and 782 white-eyed males. The white-eyed phenotype was somehow connected with sex, because all of the white-eyed flies were males.

On the other hand, white eyes were not restricted to males. For example, when red-eyed F$_1$ females from the cross of wildtype ♀♀ × white ♂♂ were backcrossed with their white-eyed fathers, the progeny consisted of both red-eyed and white-eyed females and red-eyed and white-eyed males in approximately equal numbers.

A key observation came from the mating of white-eyed females with wildtype males (cross B, Figure 3.25). All of the female progeny had wildtype eyes, but all of the male progeny had white eyes. This is the reciprocal of the cross A of wildtype ♀♀ × white ♂♂, which had yielded only wildtype females and wildtype males, and so the reciprocal crosses gave different results.

Morgan realized that reciprocal crosses would yield different results if the allele for white eyes were present in the X chromosome. This is because the X chromosome is transmitted in a different pattern by males and females, and the Y chromosome does not contain a counterpart of the *white* gene. Figure 3.25 shows that a male transmits his X chromosome only to his daughters, whereas a female transmits one of her X chromosomes to the offspring of both sexes. A gene located in the X chromosome is said to be **X-linked**.

Hemophilia is a classic example of human X-linked inheritance.

A classic example of a human trait with an X-linked pattern of inheritance is **hemophilia A**, a severe disorder of blood clotting determined by a recessive allele. Affected persons lack a blood-clotting protein called

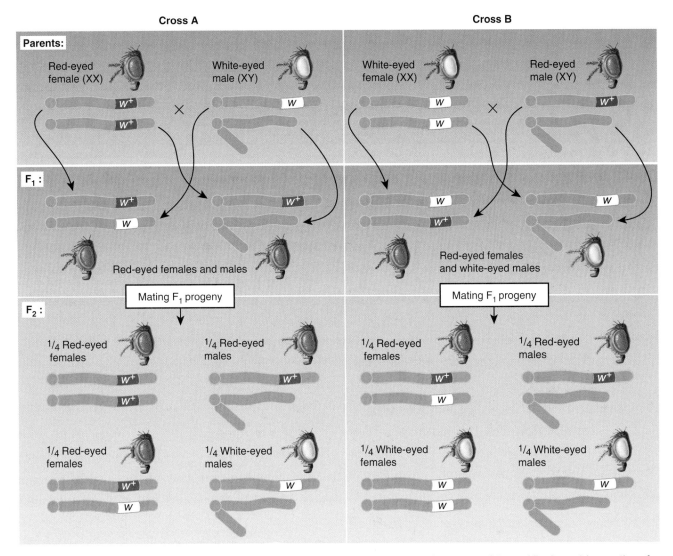

FIGURE 3.25 A chromosomal interpretation of the results obtained in F$_1$ and F$_2$ progenies in crosses of *Drosophila*. Cross A is a mating of a wildtype (red-eyed) female with a white-eyed male. Cross B is the reciprocal mating of a white-eyed female with a red-eyed male. In the X chromosome, the wildtype w^1 allele is shown in red and the mutant w allele in white. The Y chromosome does not carry either allele of the w gene.

factor VIII that is needed for normal clotting, and they suffer excessive, often life-threatening bleeding after injury. A famous pedigree of hemophilia starts with Queen Victoria of England (**FIGURE 3.26**). One of her sons, Leopold, was hemophilic, and three of her daughters were heterozygous carriers of the gene. Two of Victoria's granddaughters were also carriers, and by marriage they introduced the gene into the royal families of Russia and Spain. The heir to the Russian throne of the Romanoffs, Tsarevich Alexis, was afflicted with the condition. He inherited the gene from his mother, Tsarina Alix, one of Victoria's granddaughters. The Tsar, the Tsarina, Alexis, and his four sisters were all executed by the Bolsheviks in the 1918 Russian revolution. The present royal family of England is descended from a normal son of Victoria and is free of the disease.

X-linked inheritance in human pedigrees shows several characteristics that distinguish it from other modes of genetic transmission:

1. For any rare trait due to an X-linked recessive allele, the affected individuals are exclusively, or almost exclusively, male. There is an excess of affected males because females carrying the rare X-linked recessive allele are almost exclusively heterozygous and so do not express the mutant phenotype.

2. Affected males who reproduce have normal sons. This follows from the fact that a male transmits his X chromosome only to his daughters.

3. A woman whose father was affected has normal sons and affected sons in the ratio 1 : 1.

THE HUMAN CONNECTION

Sick of Telomeres

William C. Hahn,[1,2] **Christopher M. Counter,**[3] **Ante S. Lundberg,**[1,2] **Roderick L. Beijersbergen,**[1] **Mary W. Brooks,**[1] **and Robert A. Weinberg,**[1] 1999

[1]*Massachusetts Institute of Technology, Cambridge, Massachusetts;* [2]*Harvard Medical School, Boston, Massachusetts;* [3]*Duke University Medical Center, Durham, North Carolina*

Creation of Human Tumor Cells with Defined Genetic Elements

Some years ago, the head of the Department of Medicine at Washington University School of Medicine confronted the head of the Department of Genetics (me) and said, "Your Professor Blank studies telomeres. Who cares about telomeres? Nobody ever gets sick because of their telomeres!" I recount this story to emphasize that the directions of basic research that prove most important in the long run cannot usually be predicted, even by experts. As this important paper shows, lots of people get sick—very sick—because of their telomeres. The researchers noted that rodent cells in culture easily become converted to cancer cells by mutations in genes that limit cell division, but this is not the case with human cells. They suspected that telomere biology might play a role in the difference. Unlike adult human cells, rodent cells maintain telomerase activity, and their chromosomes have much longer telomeres than human chromosomes. With each cell division, human cells progressively erode their telomeric DNA and cease to divide

> **❝** *...it is now highly likely that telomere maintenance contributes directly to oncogenesis [cancer].* **❞**

when the telomeres become too short. The researchers engineered human cells to express the hTERT gene for telomerase and also introduced genes known to be associated with cancer. They found that the human cells now behaved in culture like the cancerous rodent cells. The human cells with excess telomerase also caused tumors in mice.

When these cells were introduced into [immunologically deficient] mice, rapidly growing tumors were repeatedly observed with high efficiency. . . . We conclude that ectopic expression of a defined set of genes . . . suffices to convert normal human cells into tumorigenic cells. . . . It is now highly likely that telomere maintenance contributes directly to oncogenesis by allowing pre-cancerous cells to proliferate beyond the number of replicative doublings allotted to their normal precursors.

Nature 400: 464–468.

This is true because any daughter of an affected male must be heterozygous for the recessive allele.

In birds, moths, and butterflies, the sex chromosomes are reversed.

In some organisms, sex is determined by sex chromosomes, but the mammalian situation is reversed: The males are XX and the females are XY. This type of sex determination is found in birds, in some reptiles and fish, and in moths and butterflies. The reversal of XX and XY in the sexes results in an opposite pattern of nonreciprocal inheritance of X-linked genes. For example, some breeds of chickens have feathers with alternating transverse bands of light and dark color, resulting in a phenotype referred to as barred. The feathers are uniformly colored in the nonbarred

phenotypes of other breeds. Reciprocal crosses between true-breeding barred and nonbarred types give the results shown in **FIGURE 3.27**. The results indicate that the gene determining barring is in the chicken X chromosome and is dominant. To distinguish sex determination in birds, butterflies, and moths from the usual XX-XY mechanism, in these organisms the sex chromosome constitutions are usually designated WZ for the female and ZZ for the male.

Experimental proof of the chromosome theory came from nondisjunction.

The parallel between the inheritance of the *Drosophila white* mutation and the genetic transmission of the X chromosome supported the chromosome theory of heredity that genes are parts of chromosomes. Other

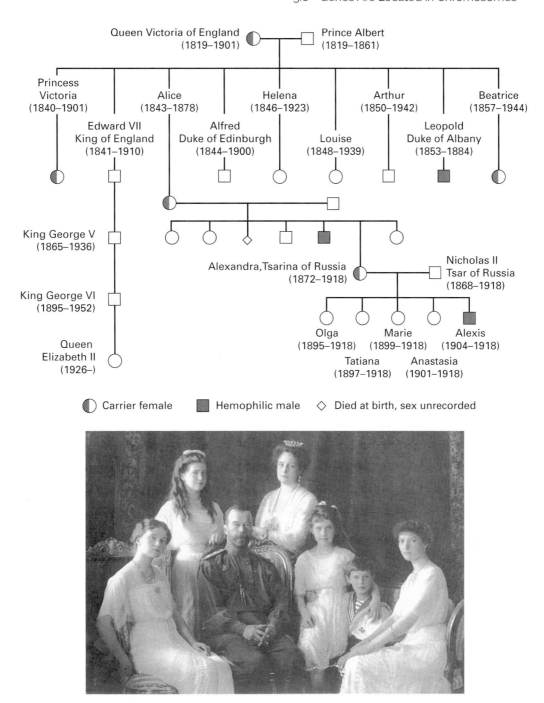

FIGURE 3.26 Genetic transmission of hemophilia A among the descendants of Queen Victoria of England, including her granddaughter, Tsarina Alexandra of Russia, and Alexandra's five children. The photograph is that of Tsar Nicholas II, Tsarina Alexandra, and their children. Tsarevich Alexis was afflicted with hemophilia.

Photo courtesy of Boston Public Library, Print Department.

experiments with *Drosophila* provided the definitive proof.

One of Morgan's students, Calvin Bridges, discovered rare exceptions to the expected pattern of inheritance in crosses with several X-linked genes. To understand these experiments, it is necessary to know that *Drosophila* is unusual among organisms with an XX-XY type of sex determination in that the Y chromosome, although it is associated with

maleness, is not male determining. In *Drosophila*, XXY embryos develop into morphologically normal, fertile females, whereas XO embryos develop into morphologically normal, but sterile, males. (The O is written in the formula XO to emphasize that a sex chromosome is missing.) The sterility of XO males shows that the Y chromosome, though not necessary for male development, is essential for male fertility.

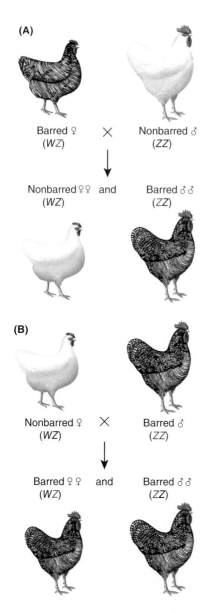

FIGURE 3.27 In sex determination in birds, the female has the unmatched sex chromosomes (called W and Z), whereas the male has the matched sex chromosomes (ZZ). The recessive mutant gene for nonbarred feathers is in the Z chromosome. (A) A cross of barred females with nonbarred males yields nonbarred female and barred male progeny. (B) A cross of nonbarred females with barred males yields barred female and barred male progeny. These results are the opposite of those observed with the white-eye *Drosophila* mutant in Figure 3.25.

When Bridges crossed white-eyed *Drosophila* females with red-eyed males, most of the progeny consisted of the expected red-eyed females and white-eyed males. However, about 1 in every 2000 F$_1$ flies was an exception: either a white-eyed female or a red-eyed male. Bridges showed that these rare exceptional offspring resulted from occasional failure of the two X chromosomes in the mother to separate from each other during meiosis—a phenomenon called **nondisjunction**. The consequence of nondisjunction of the X chromosomes is the formation of some eggs

with two X chromosomes and others with none. Four classes of zygotes are expected from the fertilization of these abnormal eggs (FIGURE 3.28). Animals with no X chromosome are not detected because embryos that lack an X chromosome die early in development; likewise, most progeny with three X chromosomes are not viable. Microscopic examination of the chromosomes of the exceptional progeny from the cross white ♀♀ × wildtype ♂♂ showed that the exceptional white-eyed females had two X chromosomes *plus* a Y chromosome and that the exceptional red-eyed males had a single X but were *lacking* a Y. The latter were sterile XO males.

These and related experiments demonstrated conclusively the validity of the chromosome theory of heredity.

KEY CONCEPT

Chromosome Theory of Heredity: Genes are contained in the chromosomes.

Bridges's evidence for the chromosome theory was that exceptional behavior of chromosomes is precisely paralleled by exceptional inheritance of their genes. This proof of the chromosome theory ranks among the most important and elegant experiments in genetics.

🧠 STOP & THINK 3.4

In the pedigree for *X*-linked hemophilia shown here, what is the probability that the individual denoted III-1 is heterozygous for the mutant gene?

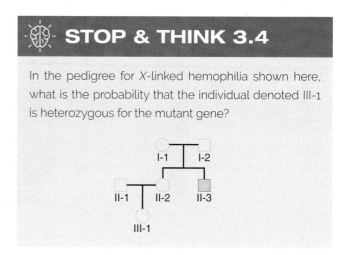

3.7 Genetic data analysis makes use of probability and statistics.

Genetic transmission includes a large component of chance. A particular gamete from an *Aa* organism might or might not include the *A* allele, depending on chance. A particular gamete from an *Aa Bb* organism might or might not include both the *A* and *B* alleles, depending on the chance orientation of the

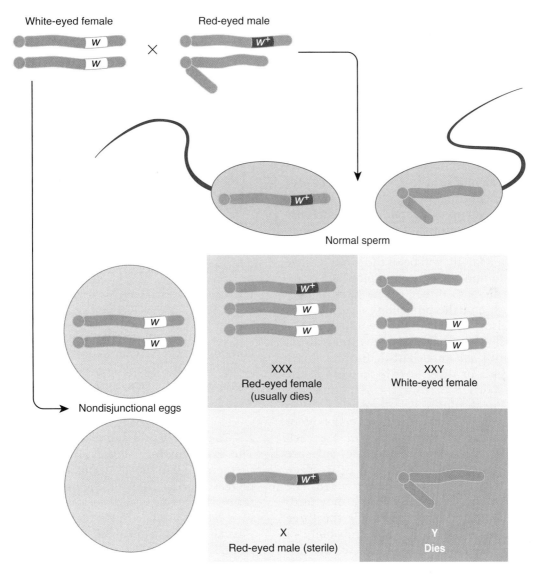

White-eyed female   Red-eyed male

Normal sperm

Nondisjunctional eggs

XXX
Red-eyed female
(usually dies)

XXY
White-eyed female

X
Red-eyed male (sterile)

Y
Dies

FIGURE 3.28 The results of meiotic nondisjunction of the X chromosomes in a female *Drosophila*.

chromosomes on the metaphase I plate. Genetic ratios result not only from the chance assortment of genes into gametes but also from the chance combination of gametes into zygotes. Although exact predictions are not possible for any particular event, it is possible to determine the probability that a particular event might be realized. In this section, we consider some of the probability methods used in interpreting genetic data.

Progeny of crosses are predicted by the binomial probability formula.

The addition rule of probability deals with possible outcomes of a genetic cross that are *mutually exclusive*. Outcomes are mutually exclusive if they are incompatible in the sense that they cannot occur at the same time. For example, there are four mutually exclusive

outcomes of the sex distribution of sibships with three children—namely, the inclusion of 0, 1, 2, or 3 girls. These have the probabilities 1/8, 3/8, 3/8, and 1/8, respectively. The addition rule states that the overall probability of any combination of mutually exclusive outcomes is equal to the sum of the probabilities of the outcomes taken separately. For example, the probability that a sibship of size 3 contains *at least one girl* includes the outcomes 1, 2, and 3 girls, and so the overall probability of at least one girl equals 3/8 + 3/8 + 1/8 = 7/8.

The multiplication rule of probability deals with outcomes of a genetic cross that are independent. Any two outcomes are independent if the knowledge that one outcome is actually realized provides no information about whether the other is realized also. For example, in a sequence of births, the sex of any one child is not affected by the sex distribution of any

children born earlier and has no influence whatso-
ever on the sex distribution of any siblings born later.
Each successive birth is independent of all the others. When possible outcomes are independent, the
multiplication rule states that the probability of any
combination of outcomes being realized equals the
product of the probabilities of each of the outcomes
taken separately. For example, the probability that
a sibship of three children will consist of three girls
equals $1/2 \times 1/2 \times 1/2$, because the probability of
each birth resulting in a girl is $1/2$, and the successive
births are independent.

Probability calculations in genetics frequently use
the addition and the multiplication rules together.
For example, to find the probability that each of
three children in a family will be of the same sex, we
use both the addition and the multiplication rules.
The probability that all three will be girls is $(1/2)$
$(1/2)(1/2) = 1/8$, and the probability that all three
will be boys is also $1/8$. Because these outcomes are
mutually exclusive (a sibship of size three cannot
include three boys *and* three girls), the probability
of either three girls or three boys is the sum of the
two probabilities, or $1/8 + 1/8 = 1/4$. The other possible outcomes for sibships of size three are that two
of the children will be girls and the other a boy, or
that two will be boys and the other a girl. For each
of these outcomes, three different orders of birth are
possible—for example, GGB, GBG, and BGG—each
having a probability of $1/2 \times 1/2 \times 1/2 = 1/8$. The
probability of two girls and a boy, disregarding birth
order, is the sum of the probabilities for the three
possible orders, or $3/8$; likewise, the probability of
two boys and a girl is also $3/8$. Therefore, the distribution of probabilities for the sex ratio in families
with three children is

GGG   BGG   GBB   BBB
               GBG   BGB
               GGB   BBG
$(1/2)^3 + 3(1/2)^2(1/2)^1 + 3(1/2)^1(1/2)^2 + (1/2)^3$
$1/8   +   3/8   +   3/8   +   1/8   = 1$

The sex-ratio probabilities can be obtained by
expanding the binomial expression $(p + q)^n$, in which
p is the probability of the birth of a girl $(1/2)$, q the
probability of the birth of a boy $(1/2)$, and n the number of children. In the present example,

$$(p + q)^3 = 1p^3 + 3p^2q + 3pq^2 + 1q^3$$

in which the red numerals are the possible number
of birth orders for each sex distribution. Similarly, the
binomial distribution of probabilities for the sex ratios
in families of five children is

$$(p + q)^5 = 1p^5 + 5p^4q + 10p^3q^2 + 10p^2q^3 + 5pq^4 + 1q^5$$

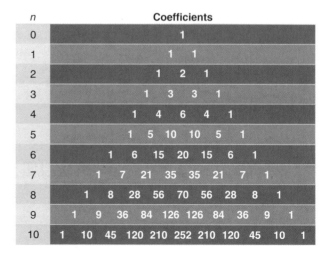

FIGURE 3.29 Pascal's triangle. The numbers are the coefficients
of the terms obtained by multiplying out the expression $(p + q)^n$
for successive values of n from 0 through 10.

Each term tells us the probability of a particular
combination. For example, the third term is the
probability of three girls (p^3) and two boys (q^2) in a
family having five children:

$$10(1/2)^3(1/2)^2 = 10/32 = 5/16$$

There are $n + 1$ terms in a binomial expansion.
The exponents of p decrease by one from n in the
first term to 0 in the last term, and the exponents of q
increase by one from 0 in the first term to n in the last
term. The coefficients generated by successive values
of n can be arranged in a regular triangle known as
Pascal's triangle (**FIGURE 3.29**). Note that the horizontal rows of the triangle are symmetrical and that
each number is the sum of the two numbers on either
side of it in the row above.

In general, if the probability of a possible outcome
A is p and that of B is q, and the two events are independent and mutually exclusive, then the probability
that A will be realized four times and B two times—
in a specific order—is p^4q^2, by the multiplication rule.
However, suppose that we are interested in the combination of events "four of A and two of B," irrespective
of order. In that case, we multiply the probability that
the combination 4A : 2B will be realized in any one
specific order by the number of possible orders. The
number of different combinations of six things, four of
one kind and two of another, is

$$\frac{6!}{4!2!} = \frac{1 \times 2 \times 3 \times 4 \times 5 \times 6}{(1 \times 2 \times 3 \times 4) \times (1 \times 2)} = 15$$

The symbol ! stands for **factorial**, or the product
of all positive integers from 1 through a given number.
Except for $n = 0$, the formula for factorial is $n! = 1 \times$
$2 \times 3 \times 4 \times \ldots \times n - 1 \times n$. The case $n = 0$ is an

TABLE 3.2 Factorials			
n	*n!*	*n*	*n!*
0	1	8	40,320
1	1	9	362,880
2	2	10	3,628,800
3	6	11	39,916,800
4	24	12	479,001,600
5	120	13	6,227,020,800
6	720	14	87,178,291,200
7	5040	15	1,307,674,368,000

exception because 0! is defined as equal to 1. The values of the first few factorials are given in **TABLE 3.2**. The value of *n!* increases very rapidly as *n* increases; 15! is more than a trillion.

The factorial formula $6!/(4! \times 2!) = 720/(24 \times 2) = 15$ is the coefficient of the term $p^4 q^2$ in the expansion of the binomial $(p + q)^6$. Therefore, the probability that outcome A will be realized four times and outcome B two times is $15 p^4 q^2$.

The general rule for repeated trials of events with constant probabilities is as follows:

KEY CONCEPT

If the probability of possibility A is *p* and the probability of the alternative possibility B is *q*, then the probability that, in *n* trials, A is realized *s* times and B is realized *t* times is

$$\frac{n!}{s!t!} p^s q^t \qquad (3.1)$$

in which $s + t = n$ and $p + q = 1$. Equation (3.1) applies even when either *s* or *t* equals 0, because 0! is defined to equal 1. (Remember also that any number raised to the zero power equals 1; for example, $2^0 = 1$.) Any individual term in the expansion of the binomial $(p + q)^n$ is given by Equation (3.1) for the appropriate values of *s* and *t*.

In Equation (3.1), $n!/(s! \, t!)$ enumerates all possible ways in which *s* elements of one kind and *t* elements of another kind can be arranged in order, provided that the *s* elements and the *t* elements are not distinguished among themselves. A specific example might include *s* yellow peas and *t* green peas. Although the yellow peas and the green peas can be distinguished from each other

because they have different colors, the yellow peas are not distinguishable from one another (because they are all yellow), and the green peas are not distinguishable from one another (because they are all green). Altogether there are $n!/(s! \, t!)$ different orders in which the yellow and green peas can be arranged in a row.

Let us use Equation (3.1) to calculate the probability that a mating between two heterozygous parents yields exactly the expected 3 : 1 ratio of the dominant and recessive traits among sibships of a particular size. The probability *p* of a child showing the dominant trait is 3/4, and the probability *q* of a child showing the recessive trait is 1/4. Suppose that we wanted to know how often families with eight children would contain exactly six children with the dominant phenotype and two with the recessive phenotype. This is the "expected" Mendelian ratio. In this case, $n = 8$, $s = 6$, $t = 2$, and the probability of this combination of events is

$$\frac{8!}{6!2!} p^6 q^2 = \frac{6! \times 7 \times 8}{6! \times 2!} (3/4)^6 (1/4)^2 = 0.31$$

That is, in only 31 percent of the families with eight children would the offspring exhibit the expected 3 : 1 phenotypic ratio; the other sibships would deviate in one direction or the other because of chance variation. The importance of this example is in demonstrating that although a 3 : 1 ratio is the "expected" outcome (and also the single most probable outcome), the majority of the families (69 percent) actually have a distribution of offspring different from 3 : 1.

Chi-square tests goodness of fit of observed to expected numbers.

Geneticists often need to decide whether an observed ratio is in satisfactory agreement with a theoretical prediction. Mere inspection of the data is unsatisfactory because different investigators may disagree. Suppose, for example, that we crossed a plant having purple flowers with a plant having white flowers and, among the progeny, observed 14 plants with purple flowers and 6 with white flowers. Is this result close enough to be accepted as a 1 : 1 ratio? What if we observed 15 plants with purple flowers and 5 with white flowers? Is this result consistent with a 1 : 1 ratio? There is

STOP & THINK 3.5

Suppose that the mating *Aa* × *Aa*, where *a* is a recessive allele, produces eight offspring. What is the probability that the ratio of dominant to recessive offspring equals 1 : 1?

bound to be statistical variation in the observed results from one experiment to the next. Who is to say what results are consistent with a particular genetic hypothesis? In this section, we describe a test of whether observed results deviate too far from a theoretical expectation. The test is called a test for **goodness of fit**, where the word *fit* means how closely the observed results "fit," or agree with, the expected results.

A conventional measure of goodness of fit is a value called **chi-square** (symbol, χ^2), which is calculated from the number of progeny observed in each of various classes, compared with the number expected in each of the classes on the basis of some genetic hypothesis. For example, in a cross between plants with purple flowers and those with white flowers, we may be interested in testing the hypothesis that the parent with purple flowers is heterozygous for a pair of alleles determining flower color and that the parent with white flowers is homozygous recessive. Suppose further that we examine 20 progeny plants from the mating and find that 14 are purple and 6 are white. The procedure to be followed in testing this genetic hypothesis (or any other genetic hypothesis) by means of the chi-square method is as follows:

1. *State the genetic hypothesis in detail, specifying the genotypes and phenotypes of the parents and the possible progeny.* In the example using flower color, the genetic hypothesis implies that the genotypes in the cross purple × white could be represented as *Pp* × *pp*. The possible progeny genotypes are either *Pp* or *pp*.

2. *Use the rules of probability to make explicit predictions of the types and proportions of progeny that should be observed if the genetic hypothesis is true. Convert the proportions to numbers of progeny (percentages are not allowed in a χ^2 test).* If the hypothesis about the flower-color cross is true, then we expect the progeny genotypes *Pp* and *pp* in a ratio of 1 : 1. Because the hypothesis is that *Pp* flowers are purple and *pp* flowers are white, we expect the phenotypes of the progeny to be purple or white in the ratio 1 : 1. Among 20 progeny, the expected numbers are 10 purple and 10 white.

3. *For each class of progeny in turn, subtract the expected number from the observed number. Square this difference and divide the result by the expected number.* In our example, the calculation for the purple progeny is $(14 - 10)^2/10 = 1.6$, and that for the white progeny is $(6 - 10)^2/10 = 1.6$.

4. *Sum the result of the numbers calculated in step 3 for all classes of progeny. The summation is the value of χ^2 for these data.* The sum for the purple and white classes of progeny is $1.6 + 1.6 = 3.2$, and this is the value of χ^2 for the experiment, calculated on the assumption that our genetic hypothesis is correct.

In symbols, the calculation of χ^2 can be represented by the expression

$$\chi^2 = \sum \frac{(observed - expected)^2}{expected}$$

in which Σ means the summation over all the classes of progeny. Note that χ^2 is calculated using the observed and expected *numbers*, not the proportions, ratios, or percentages. Using something other than the actual numbers is the beginner's most common mistake in applying the χ^2 method. The χ^2 value is reasonable as a measure of goodness of fit, because the closer the observed numbers are to the expected numbers, the smaller the value of χ^2. A value of $\chi^2 = 0$ means that the observed numbers fit the expected numbers perfectly.

As another example of the calculation of χ^2, suppose that the progeny of an $F_1 \times F_1$ cross include two contrasting phenotypes observed in the numbers 99 and 45. The genetic hypothesis might be that the trait is determined by a pair of alleles of a single gene, in which case the expected ratio of dominant : recessive phenotypes among the F_2 progeny is 3 : 1. Considering the data, the question is whether the observed ratio of 99 : 45 is in satisfactory agreement with the expected 3 : 1. Calculation of the value of χ^2 is illustrated in **TABLE 3.3**. The total number of progeny is $99 + 45 = 144$. The *expected* numbers in the two classes, on the basis of the genetic hypothesis that the

TABLE 3.3 Calculation of χ^2 for a Monohybrid Ratio				
Phenotype (class)	Observed number	Expected number	Deviation from expected $(obs - exp)^2$	$\dfrac{(obs - exp)^2}{exp}$
Wildtype	99	108	−9	0.75
Mutant	45	36	+9	2.25
Total	144	144		$\chi^2 = 3.00$

true ratio is 3 : 1, are calculated as $(3/4) \times 144 = 108$ and $(1/4) \times 144 = 36$. Because there are two classes of data, there are two terms in the χ^2:

$$\chi^2 = \frac{(99 - 108)^2}{108} + \frac{(45 - 36)^2}{36}$$
$$= 0.75 + 2.25 = 3.00$$

Once the χ^2 value has been calculated, the next step is to interpret whether this value represents a good fit or a bad fit to the expected numbers. This assessment is done with the aid of the graphs in **FIGURE 3.30**. The x-axis gives the χ^2 values measuring goodness of fit, and the y-axis gives the probability P that a worse fit (or one equally bad) would be obtained by chance, assuming that the genetic hypothesis is true. If the

genetic hypothesis is true, then the observed numbers should be reasonably close to the expected numbers. Suppose that the observed χ^2 is so large that the probability of a fit as bad or worse is very small. Then the observed results do *not* fit the theoretical expectations. This means that the genetic hypothesis used to calculate the expected numbers of progeny must be rejected, because the observed numbers of progeny deviate too much from the expected numbers.

In practice, the critical values of P are conventionally chosen as 0.05 (the 5 percent level) and 0.01 (the 1 percent level). For P values ranging from 0.01 to 0.05, the probability that chance alone would lead to a fit as bad or worse is between 1 in 20 experiments and 1 in 100. This is the purple region in Figure 3.30; if the P value falls in this range, the correctness of the genetic

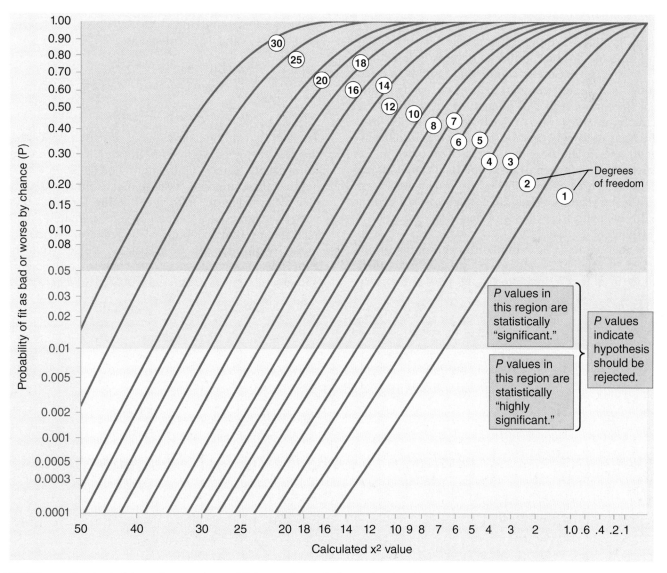

FIGURE 3.30 Graphs for interpreting goodness of fit to genetic predictions using the chi-square test. For any calculated value of χ^2 along the x-axis, the y-axis corresponding to the curve with the appropriate number of degrees of freedom gives the probability P that chance alone would produce a fit as bad as or worse than that actually observed, when the genetic predictions are correct. Tests with P in the purple region (less than 5 percent) or in the green region (less than 1 percent) are regarded as statistically significant and normally require rejection of the genetic hypothesis that led to the prediction.

hypothesis is considered very doubtful. The result is said to be **statistically significant** at the 5 percent level. For P values smaller than 0.01, the probability that chance alone would lead to a fit as bad or worse is less than 1 in 100 experiments. This is the green region in Figure 3.30; in this case, the result is said to be **statistically highly significant** at the 1 percent level, and the genetic hypothesis is rejected outright. If the terminology of statistical significance seems backwards, it is because the term *significant* refers to the magnitude of the difference between the observed and the expected numbers; in a result that is statistically significant, there is a large ("significant") difference between what is observed and what is expected.

To use Figure 3.30 to determine the P value corresponding to a calculated χ^2, we need the number of **degrees of freedom** of the particular χ^2 test. For the type of χ^2 test illustrated in Table 3.3, the number of degrees of freedom equals the number of classes of data minus 1. Table 3.3 contains two classes of data (wildtype and mutant), so the number of degrees of freedom is $2 - 1 = 1$. The reason for subtracting 1 is that, in calculating the expected numbers of progeny, we make sure that the total number of progeny is the same as that actually observed. For this reason, one of the classes of data is not really "free" to contain any number we might specify. Because the expected number in one class must be adjusted to make the total come out correctly, 1 "degree of freedom" is lost. Analogous χ^2 tests with three classes of data have 2 degrees of freedom, and those with four classes of data have 3 degrees of freedom.

Once we have determined the appropriate number of degrees of freedom, we can interpret the χ^2 value in Table 3.3. Refer to Figure 3.30, and observe that each curve is labeled with its degrees of freedom. To determine the P value for the data in Table 3.3, in which the χ^2 value is 3.00, first find the location of $\chi^2 = 3.00$ along the x-axis in Figure 3.30. Trace vertically from 3.00 until you intersect the curve with 1 degree of freedom. Then trace horizontally to the left until you intersect the y-axis and read the P value—in this case, $P = 0.08$. This means that chance alone would produce a χ^2 value as great as or greater than 3 in about 8 percent of experiments of the type in Table 3.3; because the P value is in the blue region, the goodness of fit to the hypothesis of a 3 : 1 ratio of wildtype : mutant is judged to be satisfactory.

As a second illustration of the χ^2 test, we will determine the goodness of fit of Mendel's round-versus-wrinkled data to the expected 3 : 1 ratio. Among the 7324 seeds that he observed, 5474 were round and 1850 were wrinkled. The expected numbers are (3/4) × 7324 = 5493 round and (1/4) × 7324 = 1831 wrinkled. The χ^2 value is calculated as

$$\chi^2 = \frac{(5474 - 5493)^2}{5493} + \frac{(1850 - 1831)^2}{1831}$$
$$= 0.26$$

The fact that the χ^2 is less than 1 already implies that the fit is very good. To find out how good, note that the number of degrees of freedom equals $2 - 1 = 1$ because there are two classes of data (round and wrinkled). From Figure 3.30, the P value for $\chi^2 = 0.26$ with 1 degree of freedom is approximately 0.65. This means that in about 65 percent of all experiments of this type, a fit as bad or worse would be expected simply because of chance; only about 35 percent of all experiments would yield a better fit.

CHAPTER SUMMARY

- Chromosomes in eukaryotic cells are usually present in pairs.

- The chromosomes of each pair separate in meiosis, one going to each gamete.

- In meiosis, the chromosomes of different pairs undergo independent assortment.

- Chromosomes consist largely of DNA combined with histone proteins in a compact, highly coiled configuration.

- In many animals including humans, sex is determined by a special pair of chromosomes, the

X and Y. Females are chromosomally XX and males XY.

- X-linked genes result in characteristic patterns of inheritance in human pedigrees.

- Irregularities in the inheritance of an X-linked gene in *Drosophila* gave experimental proof of the chromosomal theory of heredity.

- The progeny of genetic crosses follow the binomial probability formula.

- The chi-square statistical test is used to determine how well-observed genetic data agree with expectations from a hypothesis.

ISSUES AND IDEAS

- For a sexual species to maintain a constant chromosome number from generation to generation, why is it important that gametes contain half of the chromosome complement present in somatic cells?

- The term *mitosis* derives from the Greek *mitos*, which means "thread." The term *meiosis* derives from the Greek *meioun*, which means "to make smaller." What feature, or features, of these types of nuclear division might have led to the choice of these terms?

- Explain the meaning of the terms *reductional division* and *equational division*. What is "reduced" or "kept equal"? To which nuclear divisions do these terms refer?

- How is independent assortment of genes on different chromosomes related physically to the

process of chromosome alignment on the metaphase plate in meiosis I?

- What are some of the important differences between the first meiotic division and the second meiotic division?

- Why is X-linked inheritance often called "criss-cross inheritance"? How can this term be misleading in regard to the genetic transmission of the X chromosome?

- In what ways is the inheritance of Y-linked genes different from that of X-linked genes?

- How did nondisjunction "prove" the chromosome theory of heredity?

- Why is a statistical test necessary to determine whether an observed set of data yields an acceptable fit to the result expected from a particular genetic hypothesis? What statistical test is conventionally used for this purpose?

SOLUTIONS: STEP BY STEP

PROBLEM 1 The accompanying diagrams show the appearance of a pair of homologous chromosomes in prophase I of meiosis. Arrange the diagrams in chronological order, and identify each stage as leptotene, zygotene, pachytene, diplotene, or diakinesis.

SOLUTION The terms themselves help to distinguish one stage of prophase I from the next. *Leptotene* literally means "thin thread," when each chromosome is in an extended, threadlike condition prior to synapsis; this stage corresponds to diagram (B). *Zygotene* means "paired threads," and the pairing begins at the chromosome tips; this is configuration (D). *Pachytene* means "thick thread"; it commences when pairing is completed and the homologous chromosomes still appear to be single, which corresponds to diagram (E). *Diplotene* means "double thread"; at this time each homologous chromosome clearly consists of two sister chromatids, and chiasmata

are apparent, which is shown in diagram (A). *Diakinesis* means "moving apart," and in this stage the synapsed homologous chromosomes begin to repel one another,

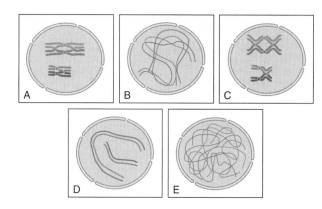

being held together by the chiasmata, producing configuration (C). Therefore, the order of the stages is B–D–E–A–C; that is, leptotene (B), zygotene (D), pachytene (E), diplotene (A), and diakinesis (C).

PROBLEM 2 Most color blindness in people is due to relatively common X-linked recessive alleles. A woman with normal color vision whose father was color blind marries a normal man. What types of color vision are expected in the offspring, and in what frequencies?

SOLUTION. In these kinds of problems it is helpful to draw a pedigree, showing the information given, and to identify the genotypes of the persons in the pedigree insofar as possible. In this case the pedigree is as shown here. The woman whose father was color blind is number II-1. (Her father's genotype must be as shown, because he was color blind.) We are told nothing about her mother's genotype, but because II-1 has normal color vision, the mother I-1 must have at least one nonmutant allele (designated cb^+). The normal male II-2 must have a nonmutant allele in his X chromosome, as shown.

I cb^+ X/ –   ⚪——▇ cb X/Y
                1   2

II cb^+ X/cb X   ⚪——▢ cb^+ X/Y
                      1   2

III   ?

The progeny in question are those in generation III, and their expected composition is shown in the Punnett square that follows. The expected offspring are 1/2 normal females, 1/4 normal males, and 1/4 color-blind males. Half of the female offspring are carriers of the recessive allele (heterozygous).

	Chromosome from father	
	1/2 cb^+ X	1/2 Y
1/2 cb^+ X	1/4 cb^+ X/cb^+ X **Normal female (noncarrier)**	1/4 cb^+ X/Y **Normal male**
1/2 cb X	1/4 cb X/cb^+ X **Normal female (carrier)**	1/4 cb X/Y **Color-blind male**

(X chromosome from mother — row labels at left)

PROBLEM 3 Suppose that a cell undergoes meiosis in a normal human male, and consider the possibility that nondisjunction of the sex chromosomes takes place. Determine what chromosome constitution would be present in sperm formed under the following conditions:

(a) Meiosis and chromosomal disjunction take place normally.
(b) Nondisjunction takes place in meiosis I.
(c) Nondisjunction happens to the X chromosome in meiosis II.
(d) Nondisjunction happens to the Y chromosome in meiosis II.

SOLUTION. In approaching problems like this, it is essential to draw diagrams of the meiotic divisions, showing the postulated events of nondisjunction. The consequences then become clear. The diagrams shown here illustrate the normal situation (A), along with the types of nondisjunction stipulated in the problem. The X chromosome is indicated in red, the Y in blue. At metaphase I each chromosome consists of two chromatids attached to a single centromere aligned on the metaphase plate. First-division disjunction, in which the homologous chromosomes separate from each other, takes place in anaphase I; and second-division disjunction, in which the sister chromatids separate from each other, takes place in anaphase II. (a) In a normal meiosis (part A), half the sperm contain an X chromosome and the other half contain a Y chromosome. For (**b**) – (**c**), the consequences of the nondisjunction events are clear from the diagrams. In each case the red arrows indicate nondisjunction, and the abnormal gametes are indicated by the salmon color. (b) The abnormal sperm resulting from XY nondisjunction in meiosis I (part B) carry either no sex chromosome ("nullo-X") or both an X and a Y. (c) The abnormal gametes resulting from X nondisjunction in meiosis II (part C) are either nullo-X or XX. (d) The abnormal gametes resulting from Y nondisjunction in meiosis II (part D) are either nullo-Y or YY.

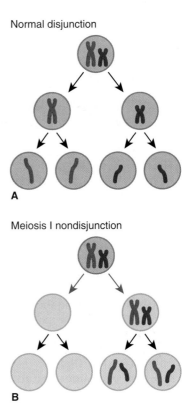

Normal disjunction

A

Meiosis I nondisjunction

B

Meiosis II nondisjunction (X)

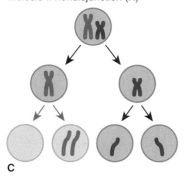

C

Meiosis II nondisjunction (Y)

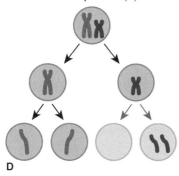

D

PROBLEM 4 A single female *Drosophila* fly collected in an apple orchard is allowed to lay eggs in bottle of culture medium in the laboratory. After larval development and metamorphosis, 144 adults emerge, among which 83 are females and 61 are males. Do these data give any reason to reject the hypothesis that the sex ratio is 1 : 1?

SOLUTION. First you need to convert the expected ratio of 1 : 1 into expected *numbers* of individuals of each sex, which equal 72 females and 72 males. The chi-square value is given by $\Sigma(observed - expected)^2/expected$, where the summation is over all classes of data, in this case females and males. Hence

$$\chi^2 = \frac{(83-72)^2}{72} + \frac{(61-72)^2}{72} = 1.681$$
$$+\ 1.681 + 1.671 = 3.36$$

Since there are two classes of data (females and males), there is 1 degree of freedom. The *P* value from Figure 3.30 is about 0.07, which means that there is about a 7 percent chance of obtaining a fit as bad or worse than 83 : 61. Based on these data alone, therefore, there is not statistically significant evidence against a 1 : 1 sex ratio, and, hence, the hypothesis cannot be rejected.

CONCEPTS IN ACTION: PROBLEMS FOR SOLUTION

3.1 A somatic cell has 46 chromosomes aligned at metaphase. How many chromosomes are present at anaphase, immediately after the centromeres have split?

3.2 The chemical colchicine is a "spindle poison" that interferes with the organization of the spindle. Somatic cells undergoing division in the presence of colchicine arrest at metaphase. Eventually the splitting of the centromeres that is characteristic of anaphase occurs, but cell division does not take place. If a cat cell with a normal diploid complement of 38 chromosomes undergoes one round of the cell cycle in the presence of colchicine, what is the expected number of chromosomes in the resulting cell?

3.3 Emmer wheat (*Triticum dicoccum*) has a somatic chromosome number of 28, and rye (*Secale cereale*) has a somatic chromosome number of 14. Hybrids produced by crossing these cereal grasses have many characteristics intermediate between the parental species, but they are nearly sterile and unable to reproduce. How many chromosomes do the hybrids possess?

3.4 Albinism refers to a total lack of skin pigment due to a rare recessive gene. What is the probability that a mating produces an albino child if:

(a) Both parents are normally pigmented, as are the grandparents, but each parent has an albino sibling.

(b) The father is albino and the mother has no family history of albinism.

(c) The mother is albino and the father has normal pigmentation although his father was albino.

3.5 What would you expect to happen to the chromosomes in successive cell cycles in a cell lineage that had a nonfunctional telomerase? Explain your answer.

3.6 Coat color in domestic pigs is determined by the alleles of two genes that undergo independent assortment. Genotypes of the form *R– S–* have red coat color (the dash is a "wild card" meaning that any allele can be present), genotypes *rr S–* and *R– ss* have sandy coats, and *rr ss* are white. In a cross of *Rr Ss × Rr Ss*, what is expected ratio of red : sandy : white?

3.7 It is often advantageous to be able to determine the sex of newborn chickens from their plumage. How could this be done by using the Z-linked dominant allele *S* for silver plumage and the recessive allele *s* for gold plumage? (Remember that, in chickens, the homogametic and heterogametic sexes are the reverse of those in mammals: Females are WZ and males are ZZ.)

3.8 The diagrams shown here depict anaphase in cell division in a cell of a hypothetical organism with two pairs of chromosomes. Identify the panels as being anaphase of mitosis, anaphase I of meiosis, or anaphase II of meiosis, stating on what basis you reached your conclusions.

3.9 A cytogeneticist examines cells in stamen hairs of *Tradescantia* in an attempt to estimate the duration of the various stages in mitosis and the cell cycle as a whole. She examines 2000 cells and finds 295 cells in prophase, 148 cells in metaphase, 78 cells in anaphase, and 109 cells in telophase. Assuming that the cells are sampled in proportion to the duration of each stage in the cell cycle, what conclusion can be drawn about the relative length of each stage of the cell cycle, including the time spent in interphase (G_1, S, and G_2)? Express each answer as a percentage of the total cell-cycle time.

3.10 Fruit flies with the chromosome constitution XXY are fertile females. Random segregation of the sex chromosomes results in eggs of the following chromosomal types and frequencies: XX (1/6), XY (1/3), X (1/3), Y (1/6). If an XXY female is crossed with a normal XY male, what is the expected fraction of fertilized eggs that will survive? (Note: In this cross, fertilized eggs with sex chromosome constitutions XXX and YY die; all others survive.)

3.11 The most common form of color blindness in humans results from X-linked recessive alleles. One type of allele, call it cb^r, results in defective red perception, whereas another type of allele, call it cb^g, results in defective green perception. A woman who is heterozygous cb^r/cb^g and a normal male produce a son whose chromosome constitution is XXY. What are the possible genotypes of this child under each of the following circumstances?

(a) The nondisjunction took place in meiosis I in the mother.

(b) The nondisjunction took place in the cb^r-bearing chromosome in meiosis II in the mother.

(c) The nondisjunction took place in the cb^g-bearing chromosome in meiosis II in the mother.

3.12 A recessive mutation of an X-linked gene in humans results in hemophilia, a condition marked by a prolonged increase in the time needed for blood clotting. Suppose that a mating between phenotypically normal parents produces two normal daughters and one son affected with hemophilia.

(a) What is the probability that at least one of the daughters is a heterozygous carrier?

(b) If one of the daughters and a normal male produce a son, what is the probability that the son will be affected?

3.13 People with the chromosome constitution 47,XXY are phenotypically male. A normal woman whose father had hemophilia mates with a normal man and produces an XXY son who also has hemophilia. What kind of nondisjunction can explain this result?

3.14 The trait represented by the filled symbols in the accompanying pedigree is a rare inherited trait with complete penetrance. What mode of inheritance does the pedigree suggest? Choose among autosomal dominant, autosomal recessive, X-linked, Y-linked, or mitochondrial.

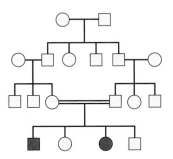

3.15 Duchenne-type muscular dystrophy is an inherited disease of muscle due to a mutant form of a protein called dystrophin. The pattern of inheritance of the disease has these characteristics: (1) affected males have unaffected children, (2) the unaffected sisters of affected males often have affected sons, and (3) the unaffected brothers of affected males have unaffected children. What type of inheritance do these findings suggest? Explain your reasoning.

3.16 Interpret the P-value in a χ^2 test. Which of the following is correct?

(a) The P-value is the probability that the hypothesis is true.

(b) The P-value is the probability that the hypothesis is false.

(c) The P-value is the probability of a chi-square value as great or greater than that observed, given that the hypothesis is true.

(d) The P-value is the probability of a chi-square value as small or smaller than that observed, given that the hypothesis is false.

3.17 In the pedigree shown here, the male I-2 is affected with color blindness owing to an X-linked recessive mutation. What is the probability that male IV-1 is color blind? (Assume that the only possible source of the color blindness mutation in the pedigree is from male I-2.)

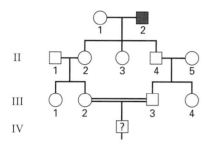

3.18 What are the values of a chi-square that yield *P* values of 5 percent (statistically significant) when there are 1, 2, 3, 4, and 5 degrees of freedom? For the goodness-of-fit type of chi-square test illustrated in this chapter, how many classes of data do these degrees of freedom represent? Because the chi-square values regarded as significant increase with the number of degrees of freedom, does this mean that it becomes increasingly "hard" (less likely) to obtain a statistically significant chi-square value when the genetic hypothesis is true?

3.19 In *Drosophila pseudoobscura*, the alleles *P* and *p* determine red versus pink eyes, and *Na* and *na* determine wide (wildtype) versus narrow wings. A dihybrid cross was carried out to produce flies homozygous for both *p* and *na*. The following phenotypes were obtained in the F2 generation:

red eyes, wide wings	583
red eyes, narrow wings	193
pink eyes, wide wings	168
pink eyes, narrow wings	56

Test these data for agreement with the 9 : 3 : 3 : 1 ratio expected under the hypothesis that the two pairs of alleles undergo independent assortment.

3.20 In a sibship of eight children, how many different birth orders will result in three boys and five girls? One such birth order would be BBBGGGGG, for example.

3.21 The accompanying pedigree and gel diagram show the molecular phenotypes obtained from genomic DNA samples. The bands are characteristic DNA fragments that distinguish two alleles of a single gene. What mode of inheritance does the pedigree suggest? On the basis of this hypothesis, and using *A1* to represent the allele associated with the 4-kb band and *A2* to represent the allele associated with the 9-kb band, deduce the genotype of each individual in the pedigree.

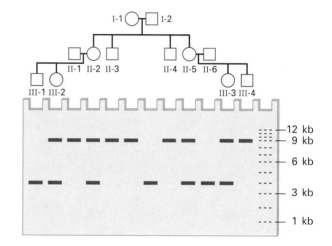

 STOP & THINK ANSWERS

ANSWER TO STOP & THINK **3.1**

Interphase (DNA replication), prophase (Chromosomes become visible), metaphase (Chromosomes align on metaphase plate), anaphase (Sister chromatids move apart), telophase (Cytokinesis).

ANSWER TO STOP & THINK **3.2**

Leptotene (Chromosome threads first become visible), zygotene (Homologous chromosomes pair), pachytene (Crossing-over takes place), diplotene (Chiasmata become prominent), diakinesis (Bivalents are maximally condensed).

ANSWER TO STOP & THINK **3.3**

The ends labeled B and C will be shorter. Replication of the top strand proceeds from right to left because the daughter DNA strand can only be elongated at the 3' end. Since replication of the top strand cannot start exactly at its 3' end, the 3' end labeled B will be shortened. Likewise, replication of the bottom strand proceeds right to left, and therefore its 3' end, labeled C, will be shortened. Normally, the shortening of the B and C ends would be restored by telomerase.

ANSWER TO STOP & THINK 3.4

The fact that II-3 is affected implies that I-1 must be heterozygous for the mutant gene. This being the case, the probability that II-2 inherits the mutant gene is 1/2; and if II-2 is heterozygous, then the probability that she transmits the mutant gene to III-1 is 1/2. Putting the two events together, the probability that III-1 is heterozygous is $1/2 \times 1/2 = 1/4$.

ANSWER TO STOP & THINK 3.5

For any one offspring, the probability of the dominant phenotype is 3/4 and that of the recessive phenotype is 1/4. We are given that $n = 8$ and asked what is the probability that the ratio of dominant to recessive is 1 : 1 (which in this case means 4 dominant and 4 recessive offspring). Use Equation (3.1) with $p = 3/4$, $q = 1/4$, $n = 8$, $s = 4$, and $t = 4$. The result is a

probability of $\dfrac{8!}{4!4!}(3/4)^4(1/4)^4 = 0.087$, or a little

less than 10 percent.

ANSWER TO STOP & THINK 3.6

The expected number of AA is $2/3 \times 200 = 133.3$ and that of Aa is $1/3 \times 200 = 66.7$. The chi-square value is

$$\frac{(125 - 133.3)^2}{133.3} + \frac{(75 - 66.7)^2}{66.7} = 1.55$$

This chi-square has 1 degree of freedom (because there are two classes of data), and Figure 3.30 implies that the P value is about 0.20 (the exact value is $P = 0.21$). Because $P > 0.05$, the chi-square test gives no reason to reject the hypothesis that 125 : 75 shows any greater deviation from 2 : 1 than would be expected by chance alone.

Multicolored varieties of Zea mays.
© picturepartners/Shutterstock

CHAPTER 4

Gene Linkage and Genetic Mapping

LEARNING OBJECTIVES

- To predict, for a specified genetic map with two genes, the kinds and relative frequencies of gametes that would be produced by an individual of a given genotype.

- To analyze the results of a genetic cross with three linked genes to deduce the genotypes of the parents, the order of the genes along the chromosome, the map distances between the genes, and the degree of interference between crossovers.

- To distinguish between single-nucleotide polymorphism (SNP) and copy-number variation (CNV).

- To explain how linkage is detected between two genes in an organism with unordered tetrads and estimate the distance in the genetic map between the genes.

- To explain how linkage between a gene and its centromere is detected in an organism with ordered tetrads and estimate the map distance between the gene and its centromere.

Genetic mapping means determining the relative positions of genes along a chromosome. It is one of the main experimental tools in genetics. This may seem odd in organisms in which the DNA sequence of the genome has been determined. If every gene in an organism is already sequenced, then what is the point of genetic mapping? The answer is that a gene's sequence does not always reveal its function, nor does a genomic DNA sequence reveal which genes interact in a complex biological process. When a new mutant gene is discovered, the first step in genetic analysis is usually genetic mapping to determine its position in the genome. It is at this point that the genomic sequence, if known, becomes useful, because in some cases the position of the mutant gene coincides with a gene whose sequence suggests a role in the biological process being investigated. For example, in the case of flower color, a new mutation may map to a region containing a gene whose sequence suggests that it encodes an enzyme in anthocyanin synthesis. But the function of a gene is not always revealed by its DNA sequence, and so in some cases, further genetic or molecular analysis is necessary to sort out which one of the genes in a sequenced region corresponds to a mutant gene mapped to that region. In human genetics, genetic mapping is important because it enables genes associated with hereditary diseases, such as those that predispose to breast cancer, to be localized and correlated with the genomic sequence in the region.

4.1 Linked alleles tend to stay together in meiosis.

In meiosis, homologous chromosomes form pairs in prophase I by undergoing synapsis; the individual members of each pair separate from one another at anaphase I. Genes that are close enough together in the same chromosome might therefore be expected to be transmitted together. Thomas Hunt Morgan examined this issue using two genes present in the X chromosome of *Drosophila*. One was a mutation for white eyes, the other a mutation for miniature wings. Morgan found that the *white* and *miniature* alleles present in each X chromosome of a female do tend to remain together in inheritance, a phenomenon known as **linkage**. Nevertheless, the linkage is incomplete. Some gametes are produced that have different combinations of the *white* and *miniature* alleles than those in the parental chromosomes. The new combinations are produced because homologous chromosomes can exchange segments when they are paired. This process (crossing-over) results in **recombination** of alleles between the homologous chromosomes. The probability of recombination between any two genes serves as a measure of genetic distance between the genes

and allows the construction of a **genetic map**, which is a diagram of a chromosome showing the relative positions of the genes. The linear order of genes along a genetic map is consistent with the conclusion that each gene occupies a well-defined position, or **locus**, in the chromosome, with the alleles of a gene in a heterozygote occupying corresponding locations in the pair of homologous chromosomes.

In discussing linked genes, it is necessary to distinguish which alleles are present together in the parental chromosomes. This is done by means of a slash ("/"). The alleles in one chromosome are depicted to the left of the slash, and those in the homologous chromosome are depicted to the right of the slash. For example, in the cross *AA BB* × *aa bb*, the genotype of the doubly heterozygous progeny is denoted *A B/a b* because the *A* and *B* alleles were inherited in one parental chromosome and the alleles *a* and *b* were inherited in the other parental chromosome. In this genotype the *A* and *B* alleles are said to be in the **coupling** or *cis* **configuration**; likewise, the *a* and *b* alleles are in coupling. Among the four possible types of gametes, the *A B* and *a b* types are called **parental combinations** because the alleles are in the same configuration as in the parental chromosomes, and the *A b* and *a B* types are called **recombinants** (**FIGURE 4.1**, part A).

Another possible configuration of the *A, a* and *B, b* allele pairs is *A b/a B*. In this case the *A* and *B* alleles are said to be in the **repulsion** or *trans* **configuration**. Now the parental and recombinant gametic types are reversed (Figure 4.1, part B). The *A b* and *a B* types are the parental combinations, and the *A B* and *a b* types are the recombinants.

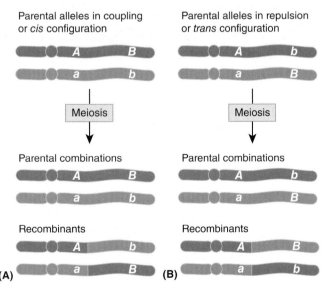

FIGURE 4.1 For any pair of alleles, the gametes produced through meiosis have the alleles either in a parental configuration or in a recombinant configuration. Which types are parental and which are recombinant depends on whether the configuration of the alleles in the parent is (A) coupling or (B) repulsion.

The degree of linkage is measured by the frequency of recombination.

In his early experiments with *Drosophila*, Morgan found mutations in each of several X-linked genes that provided ideal materials for studying linkage. One of these genes, with alleles w^+ and w, determines normal red eye color versus white eyes; another such gene, with the alleles m^+ and m, determines whether the size of the wings is normal or miniature. The initial cross is shown as Cross 1 in **FIGURE 4.2**. It was a cross between females with white eyes and normal wings and males with red eyes and miniature wings:

$$\frac{w\ m^+}{w\ m^+}\ ♀♀ \times \frac{w^+\ m}{Y}\ ♂♂$$

In this way of writing the genotypes, the horizontal line replaces the slash. Alleles written above the line are present in one chromosome, and those written below the line are present in the homologous chromosome. In the females, both X chromosomes carry w and m^+. In males, the X chromosome carries the alleles w^+ and m. (The Y written below the line denotes the Y chromosome in the male.) Figure 4.2 illustrates a simplified symbolism, commonly used in *Drosophila* genetics, in which a wildtype allele is denoted by a + sign in the appropriate position. The + symbolism is unambiguous because the linked genes in a chromosome are always written in the same order. Using the + notation,

$$\frac{w\ +}{w\ +}\quad \text{means}\quad \frac{w\ m^+}{w\ m^+}$$

and

$$\frac{+\ m}{Y}\quad \text{means}\quad \frac{w^+\ m}{Y}$$

The resulting F_1 female progeny from Cross 1 have the genotype $w\ +/+\ m$ (or, equivalently, $w\ m^+/w^+\ m$). In this genotype, the w^+ and m^+ alleles are in repulsion. When these females were mated with $w\ m$/Y males, the offspring denoted as Progeny 1 in Figure 4.2 were obtained. In each class of progeny, the gamete from

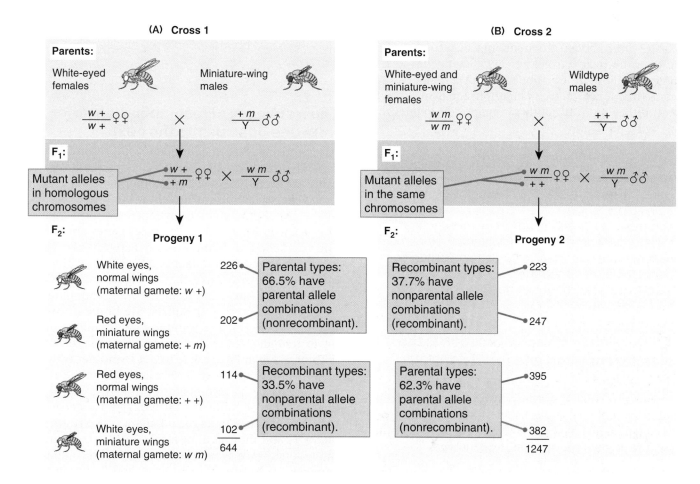

FIGURE 4.2 An experiment demonstrating that the frequency of recombination between two mutant alleles is independent of whether they are present in the same chromosome or in homologous chromosomes. (A) Cross 1 produces F_1 females with the genotype $w\ +/+\ m$, and the $w - m$ recombination frequency is 33.5 percent. (B) Cross 2 produces F_1 females with the genotype $w\ m/+\ +$, and the $w - m$ recombination frequency is 37.7 percent. These values are within the range of variation expected to occur by chance.

the female parent is shown in the column at the left, and the gamete from the male parent carries either *w m* or the Y chromosome. The cross is equivalent to a testcross, and so the phenotype of each class of progeny reveals the alleles present in the gamete from the mother.

The results of Cross 1 show a great departure from the 1 : 1 : 1 : 1 ratio of the four male phenotypes that is expected with independent assortment. If genes in the same chromosome tended to remain together in inheritance but were not completely linked, this pattern of deviation might be observed. In this case, the combinations of phenotypic traits in the parents of the original cross (parental phenotypes) were present in 428/644 (66.5 percent) of the F_2 males, and non-parental combinations (recombinant phenotypes) of the traits were present in 216/644 (33.5 percent). The 33.5 percent recombinant X chromosomes is called the **frequency of recombination**, and it should be contrasted with the 50 percent recombination expected with independent assortment.

The recombinant X chromosomes $w^+ m^+$ and *w m* result from crossing-over in meiosis in F_1 females. In this example, the frequency of recombination between the linked *w* and *m* genes was 33.5 percent. With other pairs of linked genes, the frequency of recombination ranges from near 0 to 50 percent. Even genes in the same chromosome can undergo independent assortment (frequency of recombination equal to 50 percent) if they are sufficiently far apart. This implies the following principle:

KEY CONCEPT

Genes with recombination frequencies smaller than 50 percent are present in the same chromosome (linked). Two genes that undergo independent assortment, indicated by a recombination frequency equal to 50 percent, either are in nonhomologous chromosomes or are located far apart in a single chromosome.

The frequency of recombination is the same for coupling and repulsion heterozygotes.

Morgan also studied progeny from the coupling configuration of the w^+ and m^+ alleles, which results from the mating designated as Cross 2 in Figure 4.2. In this case, the original parents had the genotypes

$$\frac{w\ m}{w\ m}\,\female\female \times \frac{+\ +}{Y}\,\male\male$$

The resulting F_1 female progeny from Cross 2 have the genotype *w m*/+ + (equivalently, *w m*/*w⁺ m⁺*). In this

case the wildtype alleles are in the same chromosome. When these F_1 female progeny were crossed with *w m*/Y males, they yielded the types of progeny tabulated as Progeny 2 in Figure 4.2.

Because the alleles in Cross 2 are in the coupling configuration, the parental-type gametes carry either *w m* or + +, and the recombinant gametes carry either *w* + or + *m*. The types of gametes are the same as those observed in Cross 1, but the parental and recombinant types are opposite. Yet the frequency of recombination is approximately the same: 37.7 percent versus 33.5 percent. The difference is within the range expected to result from random variation from experiment to experiment. The consistent finding of equal recombination frequencies in experiments in which the mutant alleles are in the *trans* or the *cis* configuration leads to the following conclusion:

KEY CONCEPT

Recombination between linked genes takes place with the same frequency whether the alleles of the genes are in the repulsion (*trans*) configuration or in the coupling (*cis*) configuration; it is the same no matter how the alleles are arranged.

The frequency of recombination differs from one gene pair to the next.

The principle that the frequency of recombination depends on the particular pair of genes may be illustrated using the recessive allele *y* of another X-linked gene in *Drosophila*, which results in yellow body color instead of the usual gray color determined by the y^+ allele. The *yellow body (y)* and *white eye (w)* genes are linked. The frequency of recombination between the genes is demonstrated in the data in **FIGURE 4.3**. The layout of the crosses is like that in Figure 4.2. In Cross 1, the female has *y* and *w* in the *trans* configuration (+ *w*/*y* +); in Cross 2, the alleles are in the *cis* configuration (*y w*/+ +). The *y* and *w* genes exhibit a much lower frequency of recombination than that observed with *w* and *m* in Figure 4.2. To put it another way, the genes *y* and *w* are more closely linked than are *w* and *m*. In Cross 1, the recombinant progeny are + + and *y w*, and they account for 130/9027 = 1.4 percent of the total. In Cross 2, the recombinant progeny are + *w* and *y* +, and they account for 94/7838 = 1.2 percent of the total. Once again, the parental and recombinant gametes are reversed in Crosses 1 and 2, because the configuration of alleles in the female parent is *trans* in Cross 1 but *cis* in Cross 2, yet the frequency of recombination between the genes is within the range expected with experimental error.

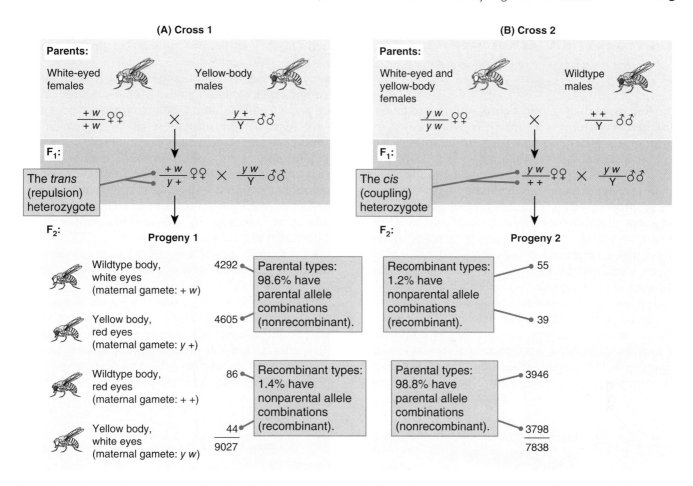

FIGURE 4.3 An experiment demonstrating that the frequency of recombination between two genes depends on the genes. The frequency of recombination between *w* and *y* is much less than that between *w* and *m* in Figure 4.2. The *y−w* experiment also confirms the equal frequency of recombination in *trans* and *cis* heterozygous genotypes. (A) The *trans* heterozygous females, + *w/y* +, yield 1.4 percent recombination. (B) The *cis* heterozygous females, *y w/* + +, yield 1.2 percent recombination.

The results of these and other experiments give support to two general principles of recombination:

- The recombination frequency is a characteristic of a particular pair of genes.

- Recombination frequencies are the same in *cis* (coupling) and *trans* (repulsion) heterozygotes.

Recombination does not occur in *Drosophila* males.

Early experiments in *Drosophila* genetics also indicated that the organism is unusual in that recombination does not take place in males. The absence of recombination in *Drosophila* males means that all alleles located in a particular chromosome show complete linkage in the male. For example, the genes *cn* (cinnabar eyes) and *bw* (brown eyes) are both in chromosome 2, but they are so far apart that in females, they show 50 percent recombination. Because the genes exhibit 50 percent recombination, the cross

$$\frac{cn\ bw}{++}\ ♀♀\ \times\ \frac{cn\ bw}{cn\ bw}\ ♂♂$$

yields progeny of genotype *cn bw/cn bw* and + +/*cn bw* (the nonrecombinant types) as well as *cn* +/*cn bw* and + *bw/cn bw* (the recombinant types) in the proportions 1 : 1 : 1 : 1. The outcome of the reciprocal cross is different. Because no crossing-over occurs in males, the reciprocal cross

$$\frac{cn\ bw}{cn\ bw}\ ♀♀\ \times\ \frac{cn\ bw}{++}\ ♂♂$$

yields progeny only of the nonrecombinant genotypes *cn bw/cn bw* and ++/*cn bw* in equal proportions. The absence of recombination in *Drosophila* males is a convenience often exploited in experimental design; as shown in the case of *cn* and *bw*, all the alleles present in any chromosome in a male must be transmitted as a group, without being recombined with alleles present in the homologous chromosome. The absence of crossing-over in *Drosophila* males is atypical; in most other animals and plants, recombination takes place in both sexes, though not necessarily with the same frequency.

4.2 Recombination results from crossing-over between linked alleles.

The linkage of the genes in a chromosome can be represented in the form of a *genetic map*, which shows the linear order of the genes along the chromosome spaced so that the distances between adjacent genes is proportional to the frequency of recombination between them. A genetic map is also called a **linkage map** or a **chromosome map**. The concept of genetic mapping was first developed by Morgan's student Alfred H. Sturtevant in 1913. The early geneticists understood that recombination between genes takes place by an exchange of segments between homologous chromosomes in the process now called crossing-over. Each crossover is manifested physically as a chiasma, or cross-shaped configuration, between homologous chromosomes; chiasmata are observed in prophase I of meiosis. Each chiasma results from the breaking and rejoining of chromatids during meiosis, with the result that there is an exchange of corresponding segments

between them. The theory of crossing-over is that each chiasma results in a new association of genetic markers. This process is illustrated in **FIGURE 4.4**. When there is no crossing-over (part A), the alleles present in each homologous chromosome remain in the same combination. When a crossover does take place (part B), the outermost alleles in two of the chromatids are interchanged (recombined).

The unit of distance in a genetic map is called a **map unit**; one map unit is equal to 1 percent recombination. For example, two genes that recombine with a frequency of 3.1 percent are said to be located 3.1 map units apart. One map unit is also called a **centimorgan**, abbreviated cM, in honor of T. H. Morgan. A distance of 3.1 map units therefore equals 3.1 centimorgans and indicates 3.1 percent recombination between the genes. An example is shown in part A of **FIGURE 4.5**, which deals with the *Drosophila* mutants *w* for white eyes and *dm (diminutive)* for small body. The female parent in the testcross is the *trans* heterozygote, but as we have seen, this configuration is equivalent in frequency of recombination to the *cis* heterozygote. Among 1000 progeny there are 31 recombinants. Using this estimate, we can

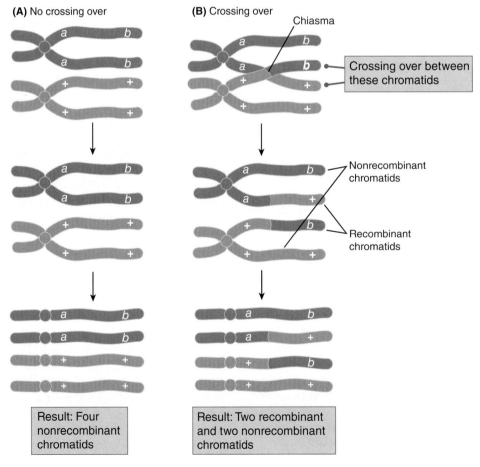

(A) No crossing over

(B) Crossing over

Chiasma

Crossing over between these chromatids

Nonrecombinant chromatids

Recombinant chromatids

Result: Four nonrecombinant chromatids

Result: Two recombinant and two nonrecombinant chromatids

FIGURE 4.4 Diagram illustrating crossing-over between two genes. (A) When there is no crossover between two genes, the alleles are not recombined. (B) When there is a crossover between them, the result is two recombinant and two nonrecombinant products, because the exchange takes place between only two of the four chromatids.

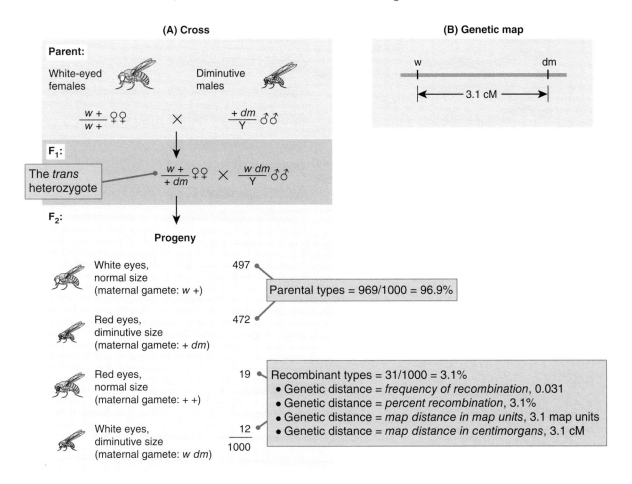

FIGURE 4.5 An experiment illustrating how the frequency of recombination is used to construct a genetic map. (A) There is 3.1 percent recombination between the genes *w* and *dm*. (B) A genetic map with *w* and *dm* positioned 3.1 map units (3.1 centimorgans, cM) apart, corresponding to 3.1 percent recombination. The map distance equals frequency of recombination only when the frequency of recombination is sufficiently small.

express the genetic distance between *w* and *dm* in four completely equivalent ways:

- As the *frequency of recombination*—in this case 0.031

- As the *percent recombination*, or 3.1 percent

- As the distance in *map units*—in this example, 3.1 map units

- As the distance in *centimorgans*, or 3.1 cM

A genetic map based on these data is shown in Figure 4.5, part B. The chromosome is represented as a horizontal line, and each gene is assigned a position on the line according to its genetic distance from other genes. In this example, there are only two genes, *w* and *dm*, and they are separated by a distance of 3.1 cM, or 3.1 map units. Genetic maps are usually truncated to show only the genes of interest. The full genetic map of the *Drosophila* X chromosome extends considerably farther in both directions than indicated in this figure.

Physically, one map unit corresponds to a length of the chromosome in which, on the average, one crossover is formed in every 50 cells undergoing meiosis.

This principle is illustrated in **FIGURE 4.6**. If one meiotic cell in 50 has a crossover, the frequency of *crossing-over* equals 1/50, or 2 percent. Yet the frequency of *recombination* between the genes is 1 percent. The correspondence of 1 percent recombination with 2 percent crossing-over is a little confusing until you consider that a crossover results in two recombinant chromatids and two nonrecombinant chromatids (Figure 4.6). A crossover frequency of 2 percent means that of the 200 chromosomes that result from meiosis in 50 cells, exactly 2 chromosomes (those involved in the crossover) are recombinant for genetic markers spanning the particular chromosome segment. To put the matter in another way, 2 percent crossing-over corresponds to 1 percent recombination because only half of the chromatids in each cell with a crossover are actually recombinant.

In situations in which there are **genetic markers** along the chromosome, such as the *A, a* and *B, b* pairs of alleles in Figure 4.6, recombination between the marker genes takes place only when a crossover occurs *between* the genes. **FIGURE 4.7** illustrates a case in which a crossover takes place between the gene *A* and

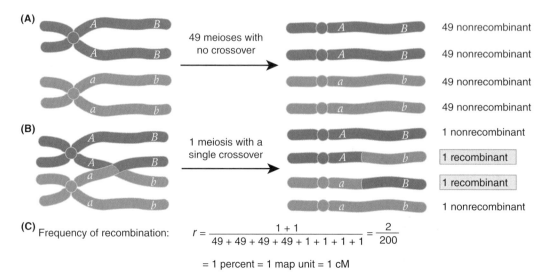

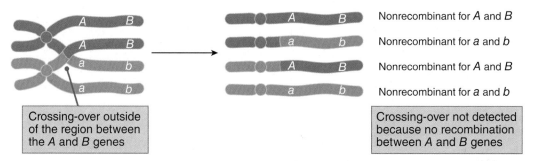

FIGURE 4.6 Diagram of chromosomal configurations in 50 meiotic cells, in which 1 has a crossover between 2 genes. (A) The 49 cells without a crossover result in 98 *A B* and 98 *a b* chromosomes; these are all nonrecombinant. (B) The cell with a crossover yields chromosomes that are *A B*, *A b*, *a B*, and *a b*, of which the middle two types are recombinant chromosomes. (C) The recombination frequency equals 2/200, or 1 percent, also called 1 map unit or 1 cM. Hence, 1 percent recombination means that 1 meiotic cell in 50 has a crossover in the region between the genes.

FIGURE 4.7 Crossing-over outside the region between two genes is not detectable through recombination. Although a segment of chromosome is exchanged, the genetic markers outside the region of the crossovers stay in the nonrecombinant configurations, in this case *A B* and *a b*.

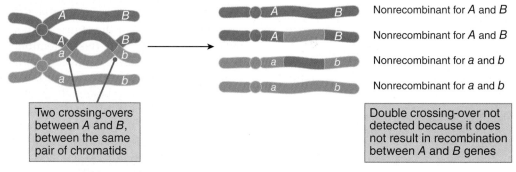

FIGURE 4.8 If two crossovers take place between marker genes *A* and *B*, and both involve the same pair of chromatids, then neither crossover is detected because all of the resulting chromosomes are nonrecombinant *A B* or *a b*.

the centromere, rather than between the genes *A* and *B*. The crossover does result in the physical exchange of segments between the innermost chromatids. However, because it is located outside the region between *A* and *B*, all of the resulting gametes must carry either the *A B* or the *a b* allele combination. These are nonrecombinant chromosomes. The presence of the crossover is

undetected because it is not in the region between the genetic markers.

In some cases, the region between genetic markers is large enough that two (or even more) crossovers can be formed in a single meiotic cell. One possible configuration for two crossovers is shown in **FIGURE 4.8**. In this example, both crossovers are between the same pair of chromatids. The result is that there is a physical exchange of a segment of chromosome between the marker genes, but the double crossover remains undetected because the markers themselves are not recombined. The absence of recombination results from the fact that the second crossover reverses the effect of the first, insofar as recombination between *A* and *B* is concerned. The resulting chromosomes are either *A B* or *a b*, both of which are nonrecombinant.

Because double crossovers in a region between two genes can remain undetected (this happens when they do not result in recombinant chromosomes), there is an important distinction between the distance between two genes as measured by the recombination frequency and as measured in map units:

■ The *map distance* between two genes equals one-half of the average number of crossovers that take place in the region per meiotic cell; it is a measure of crossing-over.

■ The *recombination frequency* between two genes indicates how much recombination is actually observed in a particular experiment; it is a measure of recombination.

The difference between map distance and recombination frequency arises because double crossovers that do not yield recombinant gametes, like the one depicted in Figure 4.8, *do* contribute to the map distance but *do not* contribute to the recombination frequency.

The distinction is important only when the region in question is large enough for double crossing-over to occur. If the region between the genes is short enough that no more than one crossover can occur in the region in any one meiosis, then map units and recombination frequencies are the same (because there are no multiple crossovers that can undo each other). This is the basis for defining a map unit as being equal to 1 percent recombination:

KEY CONCEPT

Over an interval so short that multiple crossovers are precluded (typically yielding 10 percent recombination or less), the map distance equals the recombination frequency because all crossovers result in recombinant gametes.

Furthermore, when adjacent chromosome regions separating linked genes are so short that multiple crossovers are not formed, the recombination frequencies (and hence the map distances) between the genes are additive. This important feature of recombination, as well as the logic used in genetic mapping, is illustrated by the example in **FIGURE 4.9**. The genes are located in the X chromosome of *Drosophila*—*y* for yellow body, *rb* for ruby eye color, and *cv* for shortened wing crossvein. The experimentally measured

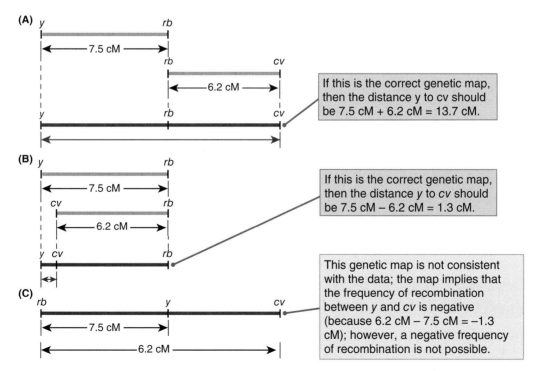

FIGURE 4.9 In *Drosophila*, the genes *y* (yellow body) and *rb* (ruby eyes) have a recombination frequency of 7.5 percent, and *rb* and *cv* (shortened wing crossvein) have a recombination frequency of 6.2 percent. There are three possible genetic maps, depending on whether *rb* is in the middle (part A), *cv* is in the middle (part B), or *y* is in the middle (part C). Map (C) can be excluded because it implies that *rb* and *y* should be closer than *rb* and *cv*, whereas the observed recombination frequency between *rb* and *y* is actually greater than that between *rb* and *cv*. Maps (A) and (B) are compatible with the data given.

recombination frequency between genes *y* and *rb* is 7.5 percent, and that between *rb* and *cv* is 6.2 percent. The genetic map might be any one of three possibilities, depending on which gene is in the middle (*y*, *cv*, or *rb*). Map C, which has *y* in the middle, can be excluded because it implies that the recombination frequency between *rb* and *cv* should be greater than that between *rb* and *y*, and this contradicts the observed data. In other words, map C can be excluded because it implies that the frequency of recombination between *y* and *cv* must be negative.

Maps A and B are both consistent with the observed recombination frequencies. They differ in their predictions regarding the recombination frequency between *y* and *cv*. Using the principle of additivity of map distances, the predicted *y*−*cv* map distance in A is 13.7 map units, whereas the predicted *y*−*cv* map distance in B is 1.3 map units. In fact, the observed recombination frequency between *y* and *cv* is 13.3 percent. Map A is therefore correct. However, there are actually two genetic maps corresponding to map A. They differ only in whether *y* is placed at the left or at the right. One map is *y*−*rb*−*cv*, which is the one shown in Figure 4.9; the other is *cv*−*rb*−*y*. The two ways of depicting the genetic map are completely equivalent because there is no way of knowing from the recombination data whether *y* or *cv* is closer to the telomere. (Other data indicate that *y* is, in fact, near the telomere.)

A linkage group is a genetic map of the genes in a chromosome.

A genetic map can be expanded by the type of reasoning shown in Figure 4.9 to include all the known genes in a chromosome; these genes constitute a linkage group. The number of linkage groups is the same as the haploid number of chromosomes of the species. For example, cultivated corn (*Zea mays*) has 10 pairs of chromosomes and 10 linkage groups. A partial genetic map of chromosome 10 is shown in FIGURE 4.10, along with the dramatic phenotypes caused by some of the mutations. The ears of corn shown in parts C and F demonstrate the result of Mendelian segregation. The ear in part C shows a 3 : 1 segregation of yellow : orange kernels produced by the recessive *orange pericarp-2* (*orp-2*) allele in a cross between two heterozygous genotypes.

Courtesy of M. G. Neuffer, College of Agriculture, Food, and Natural Resources, University of Missouri.

The ear in part F shows a 1 : 1 segregation of marbled : white kernels produced by the dominant allele *R1-mb* in a cross between a heterozygous genotype and a homozygous wildtype.

Courtesy of M. G. Neuffer, College of Agriculture, Food, and Natural Resources, University of Missouri.

One feature of the human genetic map is that, across all 23 pairs of chromosomes, the length of the genetic map in females is longer than the genetic map in males. The female genetic map is about 4400 cM, whereas the male genetic map is about 2700 cM, so there is about 60 percent more recombination in females than in males. Averaged over both sexes, the length of the human genetic map for all 23 pairs of chromosomes is about 3500 cM. Because the total DNA content per haploid set of human chromosomes is about 3000 Mb, there is, very roughly, 1 cM per million base pairs in the human genome.

Physical distance is often—but not always—correlated with map distance.

Generally speaking, the greater the physical separation between genes along a chromosome, the greater the map distance between them. Physical distance and genetic map distance are usually correlated, because a greater distance between genetic markers affords a greater chance for a crossover to take place; crossing-over is a physical exchange between the chromatids of paired homologous chromosomes.

STOP & THINK 4.1

Individuals heterozygous for dominant and recessive alleles of two linked genes are crossed with others who are homozygous recessive for both genes. The classes and number of offspring observed among 640 progeny were as follows:

| *A B*/*a b* 95 | *A b*/*a b* 216 | *a B*/*a b* 232 | *a b*/*a b* 97 |

What is the frequency of recombination between these genes? Were the recessive alleles in the doubly heterozygous parent in coupling or repulsion?

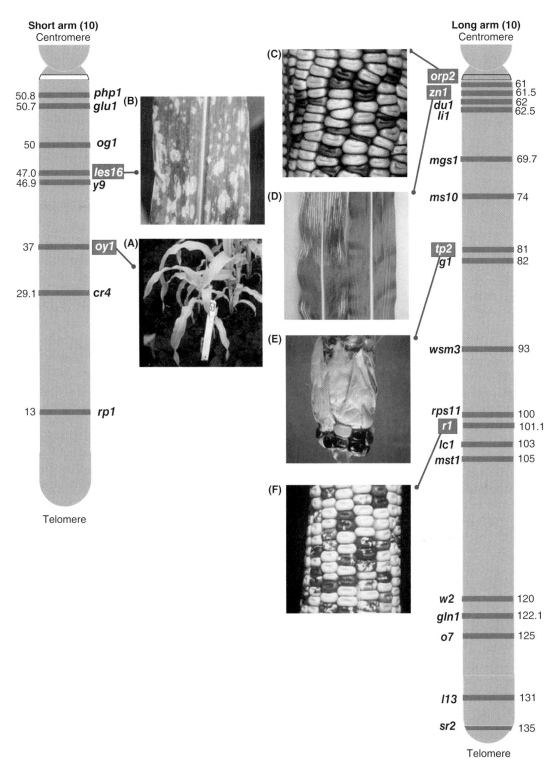

FIGURE 4.10 Genetic map of chromosome 10 of corn, *Zea mays*. The map distance to each gene is given in standard map units (centimorgans) relative to a position 0 for the telomere of the short arm (lower left). (A) Mutations in the gene *oil yellow-1 (oy1)* result in a yellow-green plant. The plant in the foreground is heterozygous for the dominant allele *Oy1*; behind is a normal plant. (B) Mutations in the gene *lesion-16 (les16)* result in many small to medium-sized, irregularly spaced, discolored spots on the leaf blade and sheath. The photograph shows the phenotype of a heterozygote for *Les16*, a dominant allele. (C) The *orp2* allele is a recessive expressed as orange pericarp, a maternal tissue that surrounds the kernels. The ear shows the segregation of *orp2* in a cross between two heterozygous genotypes, yielding a 3 : 1 ratio of yellow : orange seeds. (D) The gene *zn1 is zebra necrotic-1*, in which dying tissue appears in longitudinal leaf bands. The leaf on the left is homozygous *zn1*, that on the right is wildtype. (E) Mutations in the gene *teopod-2 (tp2)* result in many small, partially podded ears and a simple tassel. An ear from a plant heterozygous for the dominant allele *Tp2* is shown. (F) The mutation *R1-mb* is an allele of the *r1* gene, resulting in red or purple color in the aleurone layer of the seed. Note the marbled color in kernels of an ear segregating for *R1-mb*.

Data from E.H. Coe. Photos courtesy of M. G. Neuffer, College of Agriculture, Food, and Natural Resources, University of Missouri.

On the other hand, the general correlation between physical distance and genetic map distance is by no means absolute. We have already noted that the frequency of recombination between genes may differ in males and females. An unequal frequency of recombination means that the sexes can have different map distances in their genetic maps, although the physical chromosomes of the two sexes are the same and the genes must have the same linear order. For example, because there is no recombination in male *Drosophila*, the map distance between any pair of genes located in the same chromosome, when measured in the male, is 0. (On the other hand, genes on different chromosomes do undergo independent assortment in males.)

The general correlation between physical distance and genetic map distance can even break down in a single chromosome. For example, crossing-over is much less frequent in heterochromatin, which consists primarily of gene-poor regions near the centromeres, than in euchromatin. Consequently, a given length of heterochromatin will appear much shorter in the genetic map than an equal length of euchromatin. In heterochromatic regions, therefore, the genetic map gives a distorted picture of the physical map. An example of such distortion is illustrated in **FIGURE 4.11**, which compares the physical map and the genetic map of chromosome 2 in *Drosophila*. The physical map depicts the appearance of the chromosome in metaphase of mitosis. Two genes near the tips and two near the euchromatin–heterochromatin junction are indicated in the genetic map. The map distances across the euchromatic arms are 54.5 and 49.5 map units, respectively, for a total euchromatic map distance of 104.0 map units. However, the heterochromatin,

which constitutes approximately 25 percent of the entire chromosome, has a genetic length in map units of only 3.0 percent. The distorted length of the heterochromatin in the genetic map results from the reduced frequency of crossing-over in the heterochromatin. In spite of the distortion of the genetic map across the heterochromatin, in the regions of euchromatin there is a good correlation between the physical distance between genes and their distance, in map units, in the genetic map.

One crossover can undo the effects of another.

When two genes are located far apart along a chromosome, more than one crossover can be formed between them in a single meiosis, and this complicates the interpretation of recombination data. The probability of multiple crossovers increases with the distance between the genes. Multiple crossing-over complicates genetic mapping because map distance is based on the number of physical exchanges that are formed, and some of the multiple exchanges between two genes do not result in recombination of the genes and hence are not detected. As we saw in Figure 4.8, the effect of one crossover can be canceled by another crossover further along the way. If two exchanges between the same two chromatids take place between the genes *A* and *B*, then their net effect will be that all chromosomes are nonrecombinant, either *A B* or *a b*. Two of the products of this meiosis have an interchange of their middle segments, but the chromosomes are not recombinant for the genetic markers and so are genetically indistinguishable from noncrossover chromosomes. The possibility of such canceling events means that the observed recombination value is an *underestimate* of the true exchange frequency and the map distance between the genes. In higher organisms, double crossing-over is effectively precluded in chromosome segments that are sufficiently short, usually about 10 map units or less. Therefore, multiple crossovers that cancel each other's effects can be avoided by using recombination data for closely linked genes to build up genetic linkage maps.

The minimum recombination frequency between two genes is 0. The recombination frequency also has a maximum:

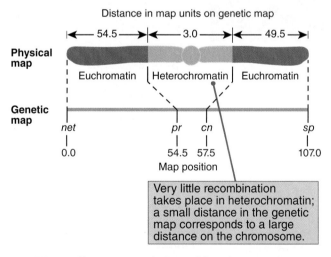

FIGURE 4.11 Chromosome 2 in *Drosophila* as it appears in metaphase of mitosis (physical map, top) and in the genetic map (bottom). The genes *pr* and *cn* are actually in euchromatin but are located near the junction with heterochromatin. The total map length is 54.5 + 49.5 + 3.0 = 107.0 map units. The heterochromatin accounts for 3.0/107.0 = 2.8 percent of the total map length but constitutes approximately 25 percent of the physical length of the metaphase chromosome.

KEY CONCEPT

No matter how far apart two genes may be, the maximum frequency of recombination between any two genes is 50 percent.

Fifty percent recombination is the same value that would be observed if the genes were on nonhomologous chromosomes and assorted independently.

The maximum frequency of recombination is observed when the genes are so far apart in the chromosome that at least one crossover is almost always formed between them. In part B of Figure 4.6, it can be seen that a single exchange in every meiosis would result in half of the products having parental combinations and the other half having recombinant combinations of the genes. The occurrence of two exchanges between two genes has the same effect, as shown in **FIGURE 4.12**. Part A shows a two-strand double crossover, in which the same chromatids participate in both exchanges; no recombination of the marker genes is detectable. When the two exchanges have one chromatid in common (three-strand double crossover, parts B and C), the result is indistinguishable from that of a single exchange; two products with parental combinations and two with recombinant combinations are produced. Note that there are two types of three-strand doubles, depending on which three chromatids participate. The final possibility is

that the second exchange connects the chromatids that did not participate in the first exchange (four-strand double crossover, part D), in which case all four products are recombinant.

In most organisms, when double crossovers are formed, the chromatids that take part in the two exchange events are selected at random. In this case, the expected proportions of the three types of double exchanges are 1/4 four-strand doubles, 1/2 three-strand doubles, and 1/4 two-strand doubles. This means that on the average, $(1/4)(0) + (1/2)(2) + (1/4)(4) = 2$ recombinant chromatids will be found among the 4 chromatids produced from meioses with two exchanges between a pair of genes. This is the same proportion obtained with a single exchange between the genes. Moreover, a maximum of 50 percent recombination is obtained for any number of exchanges.

Double crossing-over is detectable in recombination experiments that employ **three-point crosses**,

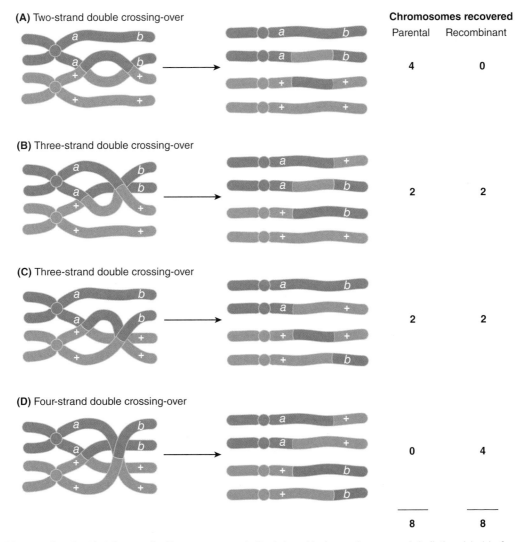

FIGURE 4.12 Diagram showing that the result of two crossovers in the interval between two genes is indistinguishable from independent assortment of the genes, provided that the chromatids participate at random in the crossovers. (A) A two-strand double crossover. (B) and (C) The two types of three-strand double crossovers. (D) A four-strand double crossover.

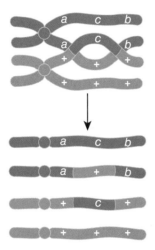

FIGURE 4.13 Diagram showing that two crossovers that occur between the same chromatids and span the middle pair of alleles in a triple heterozygote will result in a reciprocal exchange of the middle pair of alleles between the two chromatids.

which include three pairs of alleles. If a third pair of alleles, c^+ and c, is located between the outermost genetic markers (**FIGURE 4.13**), double exchanges in the region can be detected when the crossovers flank the c gene. The two crossovers, which in this example take place between the same pair of chromatids, would result in a reciprocal exchange of the c^+ and c alleles between the chromatids. A three-point cross is an efficient way to obtain recombination data; it is also a simple method for determining the order of the three genes, as we will see in the next section.

4.3   Double crossovers are revealed in three-point crosses.

The data in **TABLE 4.1** result from a testcross in corn with three genes in a single chromosome. The analysis illustrates the approach to interpreting a three-point cross. The recessive alleles of the genes in this cross are lz (for lazy or prostrate growth habit), gl (for glossy leaf), and su (for sugary endosperm), and the multiply heterozygous parent in the cross had the genotype

$$\frac{Lz \quad Gl \quad Su}{lz \quad gl \quad su}$$

where each symbol with an initial capital letter represents the dominant allele. (The use of this type of symbolism is customary in corn genetics.) The two classes of progeny that inherit noncrossover (parental-type) gametes are therefore the wildtype plants and those with the lazy-glossy-sugary phenotype. The number of progeny in these classes is far larger than the number in any of the crossover classes. Because the frequency of recombination is never greater than 50 percent, the very fact that these progeny are the most numerous indicates that the gametes that gave rise to them have the parental allele configurations, in this case $Lz\ Gl\ Su$ and $lz\ gl\ su$. Using this principle, we could have inferred the genotype of the heterozygous parent even if the genotype had not been stated. This is a point important enough to state more generally:

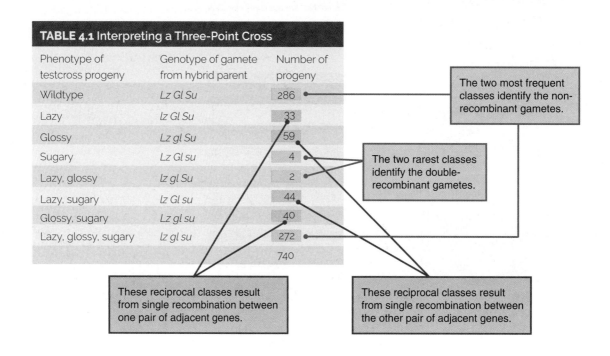

Phenotype of testcross progeny	Genotype of gamete from hybrid parent	Number of progeny
TABLE 4.1 Interpreting a Three-Point Cross		
Wildtype	*Lz Gl Su*	286
Lazy	*lz Gl Su*	33
Glossy	*Lz gl Su*	59
Sugary	*Lz Gl su*	4
Lazy, glossy	*lz gl Su*	2
Lazy, sugary	*lz Gl su*	44
Glossy, sugary	*Lz gl su*	40
Lazy, glossy, sugary	*lz gl su*	272
		740

The two most frequent classes identify the non-recombinant gametes.

The two rarest classes identify the double-recombinant gametes.

These reciprocal classes result from single recombination between one pair of adjacent genes.

These reciprocal classes result from single recombination between the other pair of adjacent genes.

KEY CONCEPT

In any genetic cross, no matter how complex, the two most frequent types of gametes with respect to any pair of genes are *nonrecombinant*; these provide the linkage phase (*cis* versus *trans*) of the alleles of the genes in the multiply heterozygous parent.

In mapping experiments, the gene sequence is usually not known. In this example, the order in which the three genes are shown is entirely arbitrary. However, there is an easy way to determine the correct order from three-point data. The gene order can be deduced by identifying the genotypes of the double-crossover gametes produced by the heterozygous parent and comparing these with the nonrecombinant gametes. Because the probability of two simultaneous exchanges is considerably smaller than that of either single exchange, the double-crossover gametes will be the least frequent types. It is clear in Table 4.1 that the classes composed of four plants with the sugary phenotype and two plants with the lazy-glossy phenotype (products of the *Lz Gl su* and *lz gl Su* gametes, respectively) are the least frequent and therefore constitute the double-crossover progeny. Now we apply another principle:

KEY CONCEPT

The effect of double crossing-over is to interchange the members of the *middle* pair of alleles between the chromosomes.

This principle is illustrated in **FIGURE 4.14**. With three genes there are three possible orders, depending on which gene is in the middle. If *gl* were in the middle (part A), the double-recombinant gametes would be *Lz gl Su* and *lz Gl su*, which is inconsistent with the data. Likewise, if *lz* were in the middle (part C), the double-recombinant gametes would be *Gl lz Su* and *gl Lz su*, which is also inconsistent with the data.

The correct order of the genes, *lz−su−gl*, is given in part B, because in this case, the double-recombinant gametes are *Lz su Gl* and *lz Su gl*, which Table 4.1 indicates is actually the case. Although one can always infer which gene is in the middle by going through all three possibilities, there is a shortcut. Each double-recombinant gamete will always match one of the parental gametes in two of the alleles. In Table 4.1, for example, the double-recombinant gamete *Lz Gl su* matches the parental gamete *Lz Gl Su* except for the allele *su*. Similarly, the double-recombinant gamete

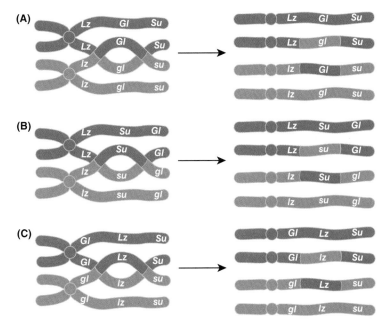

FIGURE 4.14 The order of genes in a three-point testcross may be deduced from the principle that double recombination interchanges the middle pair of alleles. For the genes *Lu*, *Gl*, and *Su*, there are three possible orders (parts A, B, and C), each of which predicts a different pair of gametes as the result of double recombination. Only the order in part B is consistent with the finding that *Lz Gl su* and *lz gl Su* are the double-recombinant gametes.

lz gl Su matches the parental gamete *lz gl su* except for the allele *Su*. The middle gene can be identified because the "odd man out" in the comparisons—in this case, the alleles of *Su*—is always the gene in the middle. The reason is that only the middle pair of alleles is interchanged by double crossing-over.

Taking the correct gene order into account, the genotype of the heterozygous parent in the cross yielding the progeny in Table 4.1 should be written as

$$\frac{Lz \quad Su \quad Gl}{lz \quad su \quad gl}$$

The consequences of single crossing-over in this genotype are shown in **FIGURE 4.15**. A single crossover in the *lz−su* region (part A) yields the reciprocal recombinants *Lz su gl* and *lz Su Gl*, and a single crossover in the *su−gl* region (part B) yields the reciprocal recombinants *Lz Su gl* and *lz su Gl*. The consequences of double crossing-over are illustrated in **FIGURE 4.16**. There are four different types of double crossovers: a two-strand double (part A), two types of three-strand doubles (parts B and C), and a four-strand double (part D). These types were illustrated earlier in Figure 4.12, where the main point was that with two genetic markers flanking the crossovers, the occurrence of double crossovers cannot be detected genetically. The difference in the present case is that, here, the genetic marker *su* is located in the middle between

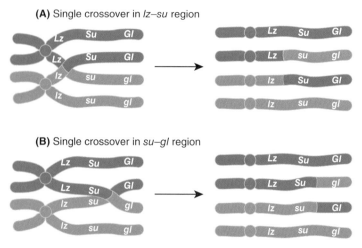

(A) Single crossover in *lz–su* region

(B) Single crossover in *su–gl* region

FIGURE 4.15 Result of single crossovers in a triple heterozygote, using the *Lz—Su—Gl* region as an example. (A) A crossover between *Lz* and *Su* results in two gametes that show recombination between *Lz* and *Su* and two gametes that are nonrecombinant. (B) A crossover between *Su* and *Gl* results in two gametes that show recombination between *Su* and *Gl* and two gametes that are nonrecombinant.

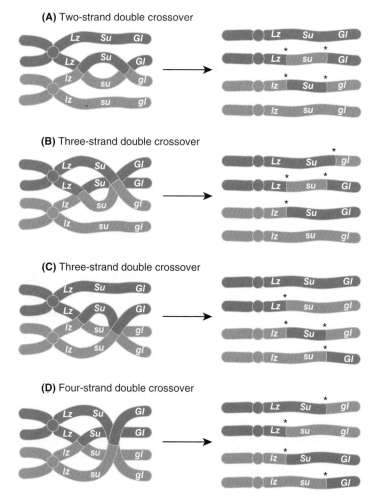

(A) Two-strand double crossover

(B) Three-strand double crossover

(C) Three-strand double crossover

(D) Four-strand double crossover

FIGURE 4.16 Result of double crossovers in a triple heterozygote, using the *Lz—Su—Gl* region as an example. Note that chromosomes showing double recombination derive from the two-strand double crossover (A) or from either type of three-strand double crossover (B and C). The four-strand double crossover (D) results only in single-recombinant chromosomes.

the two crossovers, so some of the double crossovers can be detected genetically. On the right in Figure 4.16, the asterisks mark the sites of crossing-over between nonsister chromatids. In terms of recombination, the result is that

- A two-strand double crossover (part A) yields the reciprocal double-recombinant products *Lz su Gl* and *lz Su gl*.

- One three-strand double crossover (part B) yields the double-recombinant product *Lz su Gl* and two single-recombinant products, *Lz Su gl* and *lz Su Gl*.

- The other three-strand double crossover (part C) yields the double-recombinant product *lz Su gl* and two single-recombinant products, *Lz su gl* and *lz su Gl*.

- The four-strand double crossover (part D) yields reciprocal single recombinants in the *lz—su* region, namely *Lz su gl* and *lz Su Gl*, and reciprocal single recombinants in the *su—gl* region, namely *Lz Su gl* and *lz su Gl*.

Note that the products of recombination in the three-strand double crossovers (parts B and C) are the reciprocals of each other. Because these two types of double crossovers are equally frequent, the reciprocal products of recombination are expected to appear in equal numbers.

We can now summarize the data in Table 4.1 in a more informative way by writing the genes in the correct order and grouping reciprocal gametic genotypes together. This grouping is shown in **TABLE 4.2**. Note that each class of single recombinants consists of two reciprocal products and that these are found in approximately equal frequencies (40 versus 33 and 59 versus 44). This observation illustrates an important principle:

KEY CONCEPT

The two reciprocal products resulting from any crossover, or any combination of crossovers, are expected to appear in approximately equal frequencies among the progeny.

In calculating the frequency of recombination from the data, remember that the double-recombinant chromosomes result from *two* exchanges, one in each of the chromosome regions defined by the three genes. Therefore, chromosomes that are recombinant

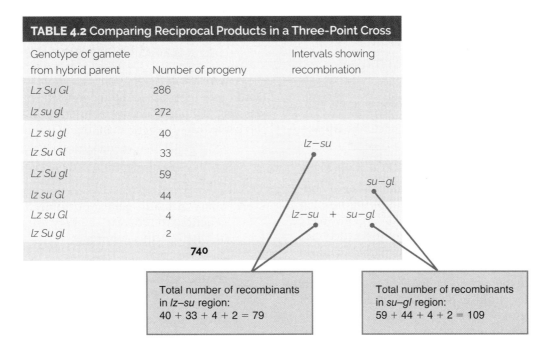

TABLE 4.2 Comparing Reciprocal Products in a Three-Point Cross

Genotype of gamete from hybrid parent	Number of progeny	Intervals showing recombination
Lz Su Gl	286	
lz su gl	272	
Lz su gl	40	lz–su
lz Su Gl	33	
Lz Su gl	59	su–gl
lz su Gl	44	
Lz su Gl	4	lz–su + su–gl
lz Su gl	2	
	740	

Total number of recombinants in *lz–su* region:
40 + 33 + 4 + 2 = 79

Total number of recombinants in *su–gl* region:
59 + 44 + 4 + 2 = 109

between *lz* and *su* are represented by the following chromosome types:

Lz su gl	40
lz Su Gl	33
Lz su Gl	4
lz Su gl	2
	79

The total implies that 79/740, or 10.7 percent, of the chromosomes recovered in the progeny are recombinant between the *lz* and *su* genes, so the map distance between these genes is 10.7 map units, or 10.7 cM. Similarly, the chromosomes that are recombinant between *su* and *gl* are represented by

Lz Su gl	59
lz su Gl	44
Lz su Gl	4
lz Su gl	2
	109

In this case the recombination frequency between *su* and *gl* is 109/740, or 14.7 percent, so the map distance between these genes is 14.7 map units, or 14.7 cM. The genetic map of the chromosome segment in which the three genes are located is therefore

lz   su   gl
|◄—10.7 map units—►|◄———— 14.7 map units ————►|

The most common error in learning how to interpret three-point crosses is to forget to include the double recombinants when calculating the recombination frequency between adjacent genes. You can keep from falling into this trap by remembering that the

double-recombinant chromosomes have single recombination in *both* regions.

Interference decreases the chance of multiple crossing-over.

The detection of double crossing-over makes it possible to determine whether exchanges in two different regions of a pair of chromosomes are formed independently of each other. Using the information from the example with corn, we know from the recombination frequencies that the probability of recombination is 0.107 between *lz* and *su* and 0.147 between *su* and *gl*. If recombination is independent in the two regions (which means that the formation of one crossover does not alter the probability of the second crossover), the probability of a single recombination in both regions is the product of these separate probabilities, or $0.107 \times 0.147 = 0.0157$ (1.57 percent). This implies that in a sample of 740 gametes, the expected number of double recombinants would be 740×0.0157, or 11.6, whereas the number actually observed was only 6 (Table 4.2). Such deficiencies in the observed number of double recombinants are common; they reflect a phenomenon called chromosome **interference**, in which a crossover in one region of a chromosome reduces the probability of a second crossover in a nearby region. Over short genetic distances, chromosome interference is nearly complete.

The **coefficient of coincidence** is the observed number of double-recombinant chromosomes divided by the expected number. Its value provides a quantitative measure of the degree of interference, which is defined as

i = Interference

= 1 − (Coefficient of coincidence)

From the data in the corn example, the coefficient of coincidence is calculated as follows:

- Observed number of double recombinants = 6

- Expected number of double recombinants = 0.107 × 0.147 × 740 = 11.6

- Coefficient of coincidence = 6/11.6 = 0.52

The 0.52 means that the observed number of double recombinants was only about half of the number expected if crossing-over in the two regions were independent. The value of the interference depends on the distance between the genetic markers and on the species. In some species, the interference increases as the distance between the two outside markers becomes smaller, until a point is reached at which double crossing-over is eliminated; that is, no double recombinants are found, and the coefficient of coincidence equals 0 (or, to say the same thing, the interference equals 1). In *Drosophila* this distance is about 10 map units.

Whatever the pattern of interference, as long as each crossover involves a randomly chosen pair of nonsister chromatids, the frequency of recombination between any two genes increases with distance to a maximum of 50 percent. This principle can be inferred from Figure 4.16, which shows that, averaged across all classes of double crossovers, the overall frequency of recombination between *Lz* and *Gl* is 1/2. You can see this in Figure 4.16 by counting: Among the 16 possible products of meiosis in which a double crossover occurs between the genes *Lz* and *Gl*, 8 products show recombination between the genes. This is a general principle no matter how many crossovers take place between the genes. The only effect of interference is that, with less interference, the maximum of 50 percent recombination is approached more slowly as distance along the chromosome increases.

STOP & THINK 4.2

Crosses between individuals of genotype *A B C / a b c* and those of genotype *a b c / a b c* yielded 1000 progeny, among which 95 were recombinant in the *A–B* interval but not in the *B–C* interval, 195 were recombinant in the *B–C* interval but not the *A–B* interval, and 5 were recombinant in both intervals. What is the frequency of recombination between *A* and *B*? Between *B* and *C*? What are the values of the coefficient of coincidence and the interference based on these data?

4.4 Polymorphic DNA sequences are used in human genetic mapping.

Until quite recently, mapping genes in human beings was very tedious and slow. Numerous practical obstacles complicated genetic mapping in human pedigrees:

1. Most genes that cause genetic diseases are rare, so they are observed in only a small number of families.

2. Many mutant genes of interest in human genetics are recessive, so they are not detected in heterozygous genotypes.

3. The number of offspring per human family is relatively small, so segregation cannot usually be detected in single sibships.

4. The human geneticist cannot perform testcrosses or backcrosses, because human matings are not dictated by an experimenter.

Human genetics has been revolutionized by the use of techniques for manipulating DNA, especially large-scale, automated DNA sequencing. These techniques have enabled investigators to carry out genetic mapping in human pedigrees primarily by using genetic markers present in the DNA itself, rather than through the phenotypes produced by mutant genes. There are many differences in DNA sequence from one person to the next. On the average, the DNA sequences at corresponding positions in any two chromosomes, taken from any two people, differ at approximately 1 in every 1000 base pairs. A genetic difference that is relatively common in a population is called a **polymorphism**. Most polymorphisms in DNA sequence are not associated with any inherited disease or disability; many occur in DNA sequences that do not code for proteins. Nevertheless, each of the polymorphisms serves as a convenient genetic marker, and those genetically linked to genes that cause hereditary diseases are particularly important.

Single-nucleotide polymorphisms (SNPs) are abundant in the human genome.

The simplest type of polymorphisms detected by DNA sequencing is one in which the identity of the nucleotide pair present at a given position along the DNA differs among individuals. A position in the DNA showing this kind of variation is known as a **single-nucleotide polymorphism** (**SNP**, pronounced "snip"). An example is shown in **FIGURE 4.17**. In this case, the SNP is defined by the presence of a T–A base pair at a particular site in some DNA molecules and by the presence of a C–G base pair at the same site in other DNA molecules. The SNP defines two "alleles" for which there

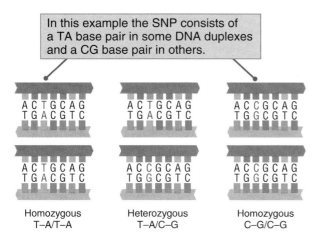

In this example the SNP consists of a TA base pair in some DNA duplexes and a CG base pair in others.

```
A C T G C A G      A C T G C A G      A C C G C A G
T G A C G T C      T G A C G T C      T G G C G T C

A C T G C A G      A C C G C A G      A C C G C A G
T G A C G T C      T G G C G T C      T G G C G T C
```

Homozygous   Heterozygous   Homozygous
T–A/T–A   T–A/C–G   C–G/C–G

FIGURE 4.17 A single-nucleotide polymorphism (SNP) with either a T–A base pair or a C–G base pair at a particular nucleotide site in the DNA. Three genotypes are possible: homozygous T–A/T–A; heterozygous T–A/C–G; or homozygous C–G/C–G.

could be three genotypes among individuals in the population (Figure 4.17): homozygous with T–A at the corresponding site in both homologous chromosomes, heterozygous with T–A in one chromosome and C–G in the homologous chromosome, or homozygous with C–G at the corresponding site in both homologous chromosomes. The word *allele* is in quotation marks above because the SNP need not be in a coding sequence, or even in a gene. In the human genome, any two randomly chosen DNA molecules are likely to differ at one SNP site about every 1000 bp in noncoding DNA and at about one SNP site every 2000 bp in protein-coding DNA. Because most of the DNA in the human genome does not code for protein, one SNP every 1000 bp in a genome of 3 billion bp implies that any two randomly chosen genomes are likely to differ at about 3 million nucleotide sites. While this is an impressive amount of genetic variation, it is nevertheless true that any two randomly chosen genomes are 99.9 percent identical!

SNPs are important for several reasons. At the species level, SNPs are important because each SNP represents a mutation that adds genetic diversity to the population. Most of the SNPs in human populations have no detectable effect on health or well-being, such as those associated with the ability to taste certain chemicals that impart a disagreeable bitterness to vegetables like broccoli or cauliflower. These SNPs are neutral in their effects, or nearly so. Many SNPs are harmful. Some have major deleterious effects like those that cause simple Mendelian disorders such as phenylketonuria, while others have milder deleterious effects as risk factors for complex diseases such as type 2 diabetes or heart disease. A few SNPs are beneficial, like the ones associated with increased production of an enzyme for lactose digestion that allows adults to drink milk without unpleasant side effects. SNPs that are beneficial are important in the long run,

since they are among the most important sources of genetic variation that allows organisms to evolve and adapt to their ever-changing environments.

At the population level, SNPs are important because they provide a rich source of genetic markers scattered throughout the genome for use in genome-wide association studies of complex diseases. In a typical study of this type, thousands of individuals are genotyped for hundreds of thousands of SNPs to ascertain which of the SNPs, if any, are significantly overrepresented in individuals manifesting a complex disease or are significantly correlated with some quantitative measurement such as blood pressure. Although multiple genetic factors affect complex traits and their effects are usually small, their effects can nevertheless be detected in studies of a sufficiently large number of individuals. Genome-wide association studies have already been carried out for more than 500 complex traits and common diseases, and more than 6000 genetic factors affecting these traits have been identified. Considering that most of the leading causes of death have a genetic component, virtually everyone carries multiple risk factors for one or more complex diseases.

Finally, at the individual level, SNPs are important because they make each of us genetically unique. Except for identical twins, each us of has a unique combination of SNPs that occurs in no one else, has never occurred before, and will never occur again. Our unique combination of SNPs allows us to be identified by our DNA sequence, affects our maturation and behavior, contains information about our ancestral history, influences our risk of complex diseases and how we react to drugs used to treat them, and determines what kind of hereditary endowment we transmit to our children.

Gene dosage can differ owing to copy-number variation (CNV).

Another important type of genetic variation consists of regions of the genome that are present in differing copy number from one chromosome to the next. As illustrated in **FIGURE 4.18**, such genomic regions are normally present in a single copy, but in some chromosomes, they may be present in two or more copies (duplicated), and in other chromosomes, they may be absent (deleted). This type of genetic variation is known as copy-number variation (CNV), and the copies are usually adjacent in the chromosome.

By definition, CNVs are greater than 1 kb in size, but many are much larger. All CNVs are submicroscopic. The length of the CNV region in Figure 4.18 has been exaggerated for clarity. The average length of a CNV is 200 to 300 kb. Each particular CNV is rare, and most individuals are heterozygous for only 0 to 4 CNVs (in contrast to being heterozygous for several

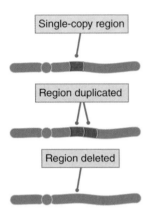

FIGURE 4.18 Copy-number variation (CNV), in which some chromosomes have a single copy of a region, other chromosomes have two (or sometimes more) copies of the region, and still other chromosomes may have a deletion of the region.

million SNPs). Nevertheless, in the aggregate, averaged across the whole genome in a large number of individuals, CNVs encompass 300 to 450 Mbp, or 10 to 15 percent of the entire genome. Many CNVs are located in regions near known mutant genes associated with hereditary diseases. For example, CNVs in alpha- and beta-hemoglobin genes are known to be associated with resistance to malaria, and CNVs in an HIV-1 receptor gene CCL3 are associated with resistance to AIDS.

Most copy-number variants, particularly those that are large enough to include one or more complete genes, are deleterious. This is why each particular CNV is quite rare, because to a greater or lesser extent it impairs the survival or reproduction of its carriers, depending on the size of the CNV and the genes included. In Chapter 5, we will discuss examples of CNVs that are risk factors for schizophrenia and autism. Because deleterious CVNs are rare, they almost occur in heterozygous form with a normal chromosome, which to some extent can ameliorate their harmful effects.

On the other hand, a few CNVs are beneficial to their carriers, and through evolutionary time these CNVs can increase in frequency in the population owing to the action of natural selection. One example is the subject of the next section.

Copy-number variation has helped human populations adapt to a high-starch diet.

An excellent example of a beneficial CNV in human evolution is copy-number variation in the gene for the starch-degrading enzyme amylase, which is produced in the salivary gland and pancreas.

As anatomically modern humans left their original homes in Southern Africa about 200,000 years ago and migrated first north to populate all of Africa and then, about 60,000 years ago, into the Middle East and then other parts of the world, the migrating populations had to cope with different climates, soil types, water availability, and native plants and animals. Some of the novel conditions were accommodated by changes in diet. Traditional hunter-gatherer populations in humid climates such as tropical forests typically eat a diet rich in meat or fish and relatively low in starch. Pastoralists, who follow herds like reindeer or bison as they migrate during the year, also have diets rich in meat and relatively low in starch. In contrast, hunter-gatherer populations in arid climates rely more on roots and tubers for food and increase their intake of starch. Populations that switch to agriculture also have a dramatic increase in dietary starch because cultivated staple foods like wheat, rice, potatoes, and corn are rich in starch.

One consequence of copy-number variation is that the amount of protein produced from any gene included in the region is usually proportional to the copy number. Increased gene copy number usually implies increased protein level. In human populations with high-starch diets, it might be expected that increased amylase is beneficial because more calories can be absorbed as the starch is digested. In human populations, increased amylase results from copy-number variation. The results of one study of amylase gene copy number are shown in **FIGURE 4.19**. Seven populations were assayed for amylase gene copy number, four with low-starch diets (blue) and three with high-starch diets (red). Individuals in populations with a low-starch diet had an average of 5.4 copies of the amylase gene, whereas those with a high-starch diet had an average of 6.7 copies of the gene. The comparisons exclude geography as an explanation for the difference. Note, for example, the contrasting populations in Tanzania in East Africa at the lower right and the contrasting populations in Siberia and Japan at the upper right. Our nearest primate relatives, chimpanzees and gorillas, have only one copy of the amylase gene; therefore, amylase gene copy number apparently has increased generally in human evolution, perhaps beginning at a time when our ancestors left the humid forests for the arid savannahs and began to eat more starchy roots and tubers.

Short tandem repeats (STRs) often differ in copy number.

In addition to copy-number variation due to genomic regions larger than 1 kb, another type of DNA polymorphism results from differences in the number of copies of a short DNA sequence that may be repeated many times in tandem at a particular site in a chromosome. In a particular chromosome, the tandem repeats may contain any number of copies, typically ranging from tens to hundreds, depending on the length of

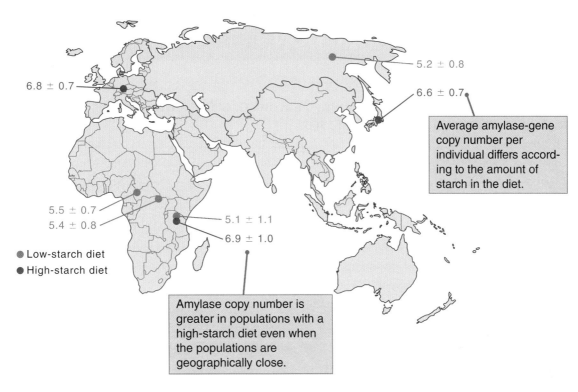

6.8 ± 0.7

5.2 ± 0.8

6.6 ± 0.7

Average amylase-gene copy number per individual differs according to the amount of starch in the diet.

5.5 ± 0.7
5.4 ± 0.8

5.1 ± 1.1
6.9 ± 1.0

● Low-starch diet
● High-starch diet

Amylase copy number is greater in populations with a high-starch diet even when the populations are geographically close.

FIGURE 4.19 Amylase copy number varies with the amount of starch in human diets.

Data from G.H. Perry, et al. *Nat. Genet.* 39 (2007): 1256–1260.

the repeat. When a DNA molecule is cleaved with an enzyme that cuts the DNA at sites flanking the tandem repeat, the size of the DNA fragment produced is determined by the number of repeats present in the molecule. When cleaved and separated in a gel, a DNA fragment cleaved from chromosome with fewer repeats is shorter than that from a chromosome with more repeats. A genetic polymorphism resulting from a tandemly repeated short DNA sequence is called a **short tandem repeat (STR)**. An example of an STR is the repeating sequence

5'-...TGATGATGATGATGATGA...-3'

and the polymorphism consists of differences in the number of TGA repeats. A particular "allele" of this STR is defined by the number of TGA repeats it includes. One source of the utility of STRs in human genetic mapping is the high density of STRs across the genome. There is an average of one STR per 2 kb of human DNA. The second utility of STRs in genetic mapping is the large number of alleles that can be present in any human population. The large number of alleles also implies that most people will be heterozygous, and so their DNA will yield two bands upon cleavage with the appropriate enzyme. Because of their high degree of variation among individuals, DNA polymorphisms are also widely used in DNA typing in criminal investigations (Chapter 14).

In genetic mapping, the phenotype of a person with respect to an STR polymorphism is a pattern of bands in a gel. As with any other type of genetic marker, the genotype of a person with respect to the polymorphism is inferred, insofar as it is possible, from the phenotype. Linkage between different polymorphic loci is detected through lack of independent assortment of the alleles in pedigrees, and recombination and genetic mapping are carried out using the same principles as apply in other organisms, except that in humans, because of the small family size, different pedigrees are pooled for analysis. Primarily through the use of DNA polymorphisms, genetic mapping in humans has progressed rapidly.

To give an example of the type of data used in human genetic mapping, a three-generation pedigree of a family segregating for several alleles of an STR is illustrated in **FIGURE 4.20**. In this example, each of the parents is heterozygous, as are all of the children. Yet every person can be assigned his or her genotype because the STR alleles are codominant. STR polymorphisms are an important type of genetic marker used in genetic mapping in human pedigrees because STR polymorphisms are prevalent, are located in virtually all regions of the chromosome set, and have multiple alleles and so yield a high proportion of heterozygous genotypes. Furthermore, only a small amount of biological material is needed to perform the necessary tests.

THE HUMAN CONNECTION

Starch Contrast

George H. Perry[1,2], Nathaniel J. Dominy[3], Katrina G. Claw[1], Arthur S. Lee[2], Heike Fiegler[4], Richard Redon[4], John Werner[1], et al., 2007

[1]*Arizona State University, Tempe, Arizona;* [2]*Brigham and Women's Hospital, Boston, Massachusetts;* [3]*University of California, Santa Cruz, California;* [4]*The Wellcome Trust Sanger Institute, Hinxton, United Kingdom.*

Diet and the Evolution of Human Amylase Gene Copy Number Variation

Evolution of increased enzyme activity can occur through regulatory mutations that increase transcription of a single gene or through increases in gene copy number. This study reports a strong correlation between the amount of starch in the diets of human populations and the number of copies of a gene encoding salivary amylase, a starch-degrading enzyme. Humans are not alone among primates in having evolved higher amylase activity. A group of Old World monkeys called cercopithecines, which includes macaques and mangabeys, produces even more salivary amylase than humans. Cercopithecines are unique among primates in storing starchy foods, such as the seeds of unripe fruits, in a cheek pouch, and it is a plausible hypothesis that the increased amylase facilitates digestion of the starch. It is not known whether the increased amylase production in cercopithecines is due to copy number variation or to some other mechanism.

First, we can consider the variation in starch content in diets of different human populations:

> A distinction can be made between "high-starch" populations for which starchy food resources comprise a substantial portion of the diet and "low-starch" populations with traditional diets that incorporate relatively few starchy foods [but] instead emphasize proteinaceous resources (for example, meats and blood) and simple saccharides (for example, from fruit, honey, and milk). . . .

Then the genomes of individuals from those populations can be analyzed to determine the copy number of the *AMY1* gene, which encodes salivary amylase.

We estimated *AMY1* copy number in three high-starch and four low-starch population samples. . . . Notably, the proportion of individuals from the combined high-starch sample with at least six *AMY1* copies (70 percent) was nearly two times greater than that for low-starch populations (37 percent). . . .

One again, however, the scientific maxim that "correlation does not necessarily imply causation" applies here. Are these differences truly the result of evolutionary adaptation to starch in the diet, or could something like genetic drift be responsible? The authors note that *AMY1* copy number correlates more highly with dietary starch than with the geographic distribution of the populations study. They conclude:

> We favor a model in which *AMY1* copy number has been subject to positive or directional selection in at least some high-starch populations but has evolved neutrally (that is, through genetic drift) in low-starch populations. . . . Comparisons with other great apes suggest that *AMY1* copy number was probably gained in the human lineage. . . . The initial human specific increase in *AMY1* copy number may have been coincident with a dietary shift early in hominin evolutionary history.

66 We favor a model in which AMY1 copy number has been subject to positive or directional selection in at least some high-starch populations. . . 99

Green, R. E., et al. (2010). *Science*, 328 (5979):710–722.

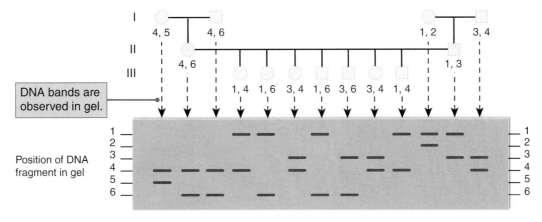

FIGURE 4.20 Human pedigree showing segregation of STR alleles. Six alleles (1–6) are present in the pedigree, but any one person can have only one allele (if homozygous) or two alleles (if heterozygous).

4.5 Tetrads contain all four products of meiosis.

In some species of fungi, each meiotic tetrad is contained in a sac-like structure, called an **ascus**, and can be recovered as an intact group. Each product of meiosis is included in a reproductive cell called an **ascospore**, and all of the ascospores formed from one meiotic cell remain together in the ascus (**FIGURE 4.21**). The advantage of these organisms for the study of recombination is the potential for analyzing all of the products from each meiotic division. For example, one can see immediately from the diagram in Figure 4.21 that a tetrad containing the products of a single meiosis in a heterozygous *Aa* organism contains 2 *A* ascospores and 2 *a* ascospores. The 2 *A* : 2 *a* segregation means that the Mendelian ratio of 1 : 1 is realized in the products of each individual meiotic division and is not merely an average over a large number of meioses.

Two other features of ascus-producing organisms are especially useful for genetic analysis: (1) They are haploid, so dominance is not a complicating factor because the genotype is expressed directly in the phenotype. (2) They produce very large numbers of progeny, making it possible to detect rare events and to estimate their frequencies accurately. Furthermore, the life cycles of the organisms tend to be short. The only diploid stage is the zygote, which undergoes meiosis soon after it is formed; the resulting haploid meiotic products (which form the ascospores) germinate to regenerate the vegetative stage (**FIGURE 4.22**). In most of the organisms, the meiotic products, or their derivatives, are not arranged in any particular order in the ascus. However, bread molds of the genus *Neurospora* and related organisms have the useful characteristic that the meiotic products are arranged in a definite order directly related to the planes of the meiotic divisions.

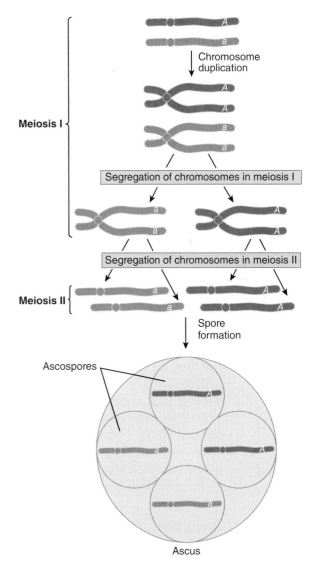

FIGURE 4.21 Formation of an ascus containing all of the four products of a single meiosis. Each ascospore present in the ascus is a reproductive cell formed from one of the products of meiosis.

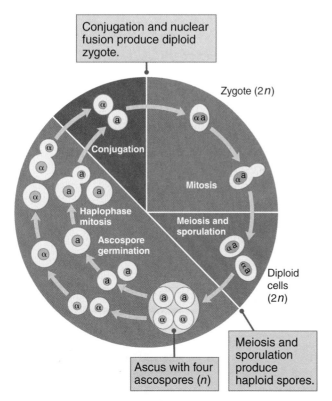

FIGURE 4.22 Life cycle of the yeast *Saccharomyces cerevisiae*. Mating type is determined by the alleles a and α. Both haploid and diploid cells normally multiply by mitosis (budding). Depletion of nutrients in the growth medium induces meiosis and sporulation of cells in the diploid state. Diploid nuclei are shown in red, haploid nuclei in yellow.

Unordered tetrads have no relation to the geometry of meiosis.

In the tetrads when two pairs of alleles are segregating, three patterns of segregation are possible. For example, in the cross $A B \times a b$, the three types of tetrads are

(AB) (AB) (ab) (ab)   referred to as **parental ditype**, or **PD**. Only two genotypes are represented, and their alleles have the same combinations found in the parents.

(Ab) (Ab) (aB) (aB)   referred to as **nonparental ditype**, or **NPD**. Only two genotypes are represented, but their alleles have nonparental combinations.

(AB) (Ab) (aB) (ab)   referred to as **tetratype**, or **TT**. All four of the possible genotypes are present.

Tetratype tetrads demonstrate that crossing-over takes place at the four-strand stage of meiosis and is reciprocal.

We noted earlier that tetrads from heterozygous organisms regularly contain 2 *A* and 2 *a* ascospores, which implies that Mendelian segregation takes place in each meiosis. The existence of tetratype tetrads for linked genes demonstrates two features about crossing-over that we have assumed, so far without proof.

1. The exchange of segments between parental chromatids takes place in the first meiotic prophase, *after the chromosomes have duplicated.* Tetratype tetrads demonstrate this assertion because only two of the four products of meiosis show recombination. This would not be possible unless crossing-over took place at the four-strand stage.

2. The exchange process consists of the breaking and rejoining of the two chromatids, resulting in the *reciprocal* exchange of equal and corresponding segments. Tetratype tetrads demonstrate this point because they contain the reciprocal products (*A b* and *a B* if the parental alleles were in coupling, *A B* and *a b* if they were in repulsion).

Tetrad analysis affords a convenient test for linkage.

Tetrad analysis is an effective way to determine whether two genes are linked, because

KEY CONCEPT

When genes are *unlinked*, the parental ditype tetrads and the nonparental ditype tetrads are expected in equal frequencies (PD = NPD).

The reason for the equality PD = NPD for unlinked genes is shown in part A of **FIGURE 4.23**, where the two pairs of alleles *A, a* and *B, b* are located in different chromosomes. In the absence of crossing-over between either gene and its centromere, the two chromosomal configurations are equally likely at metaphase I, and so PD = NPD. When there is a crossover between either gene and its centromere (Figure 4.23, part B), a tetratype tetrad results, but this does not change the fact that PD = NPD.

In contrast, when genes are linked, parental ditypes are far more frequent than nonparental ditypes. To see why, assume that the genes are linked and consider the events required for the production of the three types of tetrads. **FIGURE 4.24** shows that when no crossing-over takes place between the genes, a PD tetrad is formed. Single crossover between the genes results in a TT tetrad. The formation of a two-strand, three-strand, or four-strand double crossover results in a PD, TT, or NPD tetrad, respectively. With linked genes, meiotic cells with no crossovers always outnumber those with four-strand double crossovers. Therefore,

KEY CONCEPT

Linkage is indicated when nonparental ditype tetrads appear with a much lower frequency than parental ditype tetrads (NPD << PD).

(A) No crossing over

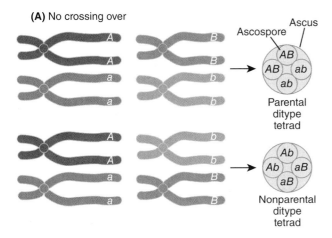
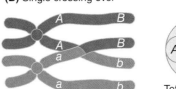

Parental ditype tetrad

Nonparental ditype tetrad

(B) Crossing over between one of the genes and its centromere

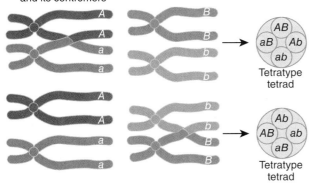

Tetratype tetrad

Tetratype tetrad

FIGURE 4.23 Types of unordered asci produced with two genes in different chromosomes. (A) In the absence of a crossover, random arrangement of chromosome pairs at metaphase I results in two different combinations of chromatids, one yielding PD tetrads and the other NPD tetrads. (B) When a crossover takes place between one gene and its centromere, the two chromosome arrangements yield TT tetrads. If both genes are closely linked to their centromeres (so that crossing-over is rare), few TT tetrads are produced.

The relative frequencies of the different types of tetrads can be used to determine the map distance between two linked genes. The simplest case is one in which the genes are sufficiently close that double and higher levels of crossing-over can be neglected. In this case, tetratype tetrads arise only from meiotic cells in which a single crossover occurs between the genes (Figure 4.24, part A and part B). As we saw in Figure 4.6, the genetic map distance across an interval is defined as one-half the proportion of cells with a crossover in the interval, so the map distance implied by the tetrads is given by

Map distance =

$$\frac{1}{2} \times \frac{\text{Number of tetratype tetrads}}{\text{Total number of tetrads}} \times 100 \quad (4.1)$$

To take a specific example, suppose 100 tetrads are analyzed from the cross $A\ B \times a\ b$, and the result is that 91 are PD and 9 TT. The finding that NPD $\ll$ PD means that the genes are linked, and the fact that

(A) No crossing over

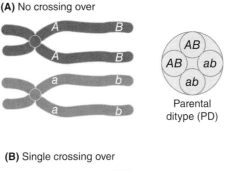

Parental ditype (PD)

(B) Single crossing over

Tetratype (TT)

(C) 2-strand double crossing over

Parental ditype (PD)

(D) 3-strand double crossing over

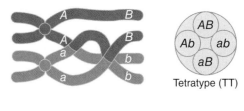

Tetratype (TT)

(E) 3-strand double crossing over

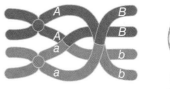

Tetratype (TT)

(F) 4-strand double crossing over

Nonparental ditype (NPD)

FIGURE 4.24 Types of tetrads produced with two linked genes. (A) In the absence of a crossover, a PD tetrad is produced. (B) With a single crossover between the genes, a TT tetrad is produced. (C–F) Among the four possible types of double crossovers between the genes, only the four-strand double crossover in part F yields an NPD tetrad.

NPD = 0 means that the genes are so closely linked that double crossing-over does not occur between them. The map distance between *A* and *B* is calculated as follows:

$$\text{Map distance } \frac{1}{2} \times \frac{9}{100} \times 100 = 4.5 \, \text{cM}$$

We must emphasize that Equation (4.1) is valid only when NPD = 0, so that interference across the region prevents the occurrence of double crossing-over. When double crossovers do take place in the interval, then NPD ≠ 0, and the formula for map distance has to be modified to take the double crossovers into account.

The mapping procedure using tetrads differs from that presented earlier in the chapter in that the map distance is not calculated directly from the number of recombinant and nonrecombinant chromatids. Instead, the map distance is calculated directly from the tetrads and the inferred crossovers that give rise to each type of tetrad. However, it is not necessary to carry out a full tetrad analysis for estimating linkage. The alternative is to examine spores chosen at random after allowing the tetrads to break open and disseminate their spores. This procedure is called *random-spore analysis*, and the linkage relationships are determined exactly as described earlier for *Drosophila* and corn. In particular, the frequency of recombination equals the number of spores that are recombinant for the genetic markers divided by the total number of spores.

STOP & THINK 4.3

Yeast cells of genotype *A b* are crossed with those of genotype *a B*. Among 170 unordered tetrads that were analyzed, the following numbers were observed of each type.

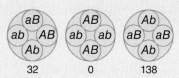

32	0	138

Based on these data, what is the map distance between the genes?

The geometry of meiosis is revealed in ordered tetrads.

In the bread mold *Neurospora crassa*, the products of meiosis are contained in an *ordered* array of ascospores (FIGURE 4.25). A zygote nucleus, contained in a sac-like ascus, undergoes meiosis almost immediately after it is formed. The four nuclei produced by meiosis are in a linear, ordered sequence in the ascus; each of them undergoes a mitotic division to form two genetically identical and adjacent ascospores. Each mature ascus contains eight ascospores arranged in four pairs, each pair derived from one of the products of meiosis. The ascospores can be removed one by one from an ascus and each germinated in a culture tube to determine its genotype.

Ordered asci also can be classified as PD, NPD, or TT with respect to two pairs of alleles, which makes it possible to assess the degree of linkage between the genes. The fact that the arrangement of meiotic products is ordered also makes it possible to determine the recombination frequency between any particular gene and its centromere. The logic of the mapping technique is based on the feature of meiosis shown in FIGURE 4.26.

KEY CONCEPT

Homologous centromeres of parental chromosomes separate at the first meiotic division; the centromeres of sister chromatids separate at the second meiotic division.

Thus, in the absence of crossing-over between a gene and its centromere, the alleles of the gene (for example, *A* and *a*) must separate in the first meiotic division; this separation is called **first-division segregation**. If, instead, a crossover is formed between the gene and its centromere, the *A* and *a* alleles do not become separated until the second meiotic division; this separation is called **second-division segregation**. The distinction between first-division and second-division segregation is shown in Figure 4.26. As shown in part A, only two possible arrangements of the products of meiosis can yield first-division segregation—*A A a a* or *a a A A*. However, four patterns of second-division segregation are possible because of the random arrangement of homologous chromosomes at metaphase I and of the chromatids at metaphase II. These four arrangements, which are shown in part B, are

$$A\,a\,A\,a \quad a\,A\,a\,A \quad A\,a\,a\,A \quad a\,A\,A\,a$$

The percentage of asci with second-division segregation patterns for a gene can be used to map the gene with respect to its centromere. For example, let us assume that 30 percent of a sample of asci from a cross have a second-division segregation pattern for the *A* and *a* alleles. This means that 30 percent of the cells undergoing meiosis had a crossover between the *A* gene and its centromere. Because the map distance between two genes is, by definition, equal to one-half times the proportion of cells with a crossover between the genes, the map distance between a gene and its centromere is given by the equation

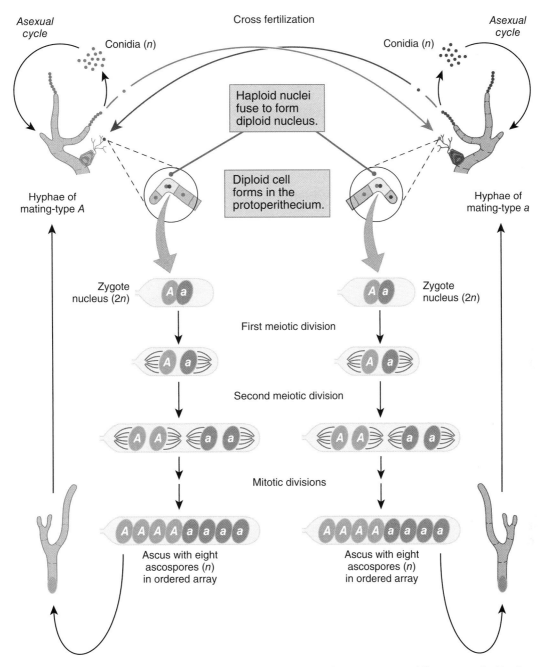

FIGURE 4.25 The life cycle of *Neurospora crassa*. The vegetative body consists of partly segmented filaments called hyphae. Conidia are asexual spores that function in the fertilization of organisms of the opposite mating type. A protoperithecium develops into a structure in which numerous cells undergo meiosis.

$$\text{Map distance} = \frac{1}{2} \times \frac{\text{Number of asci with second division segregation}}{\text{Total number of asci}} \times 100 \quad (4.2)$$

Equation (4.2) is valid as long as the gene is close enough to the centromere for us to neglect multiple crossovers. Reliable linkage values are best determined for genes that are near the centromere. The location of more distant genes is then accomplished by mapping these genes relative to genes nearer the centromere.

If a gene is far from its centromere, crossing-over between the gene and its centromere will be so frequent that the *A* and *a* alleles become randomized with respect to the four chromatids. The result is that the six possible spore arrangements shown in Figure 4.26 are all equally frequent. Therefore, when the chromatids participating in each crossover are chosen at random,

KEY CONCEPT

The maximum frequency of second-division segregation asci is 2/3.

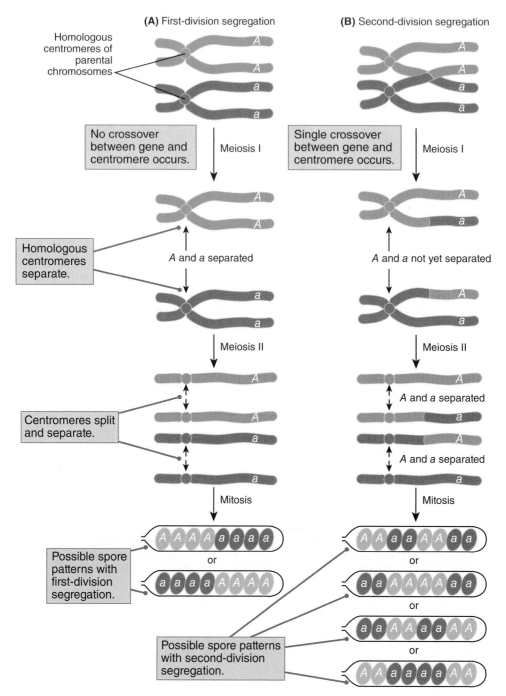

(A) First-division segregation

(B) Second-division segregation

Homologous centromeres of parental chromosomes

No crossover between gene and centromere occurs.

Meiosis I

Single crossover between gene and centromere occurs.

Meiosis I

Homologous centromeres separate.

A and *a* separated

A and *a* not yet separated

Meiosis II

Meiosis II

Centromeres split and separate.

A and *a* separated

A and *a* separated

Mitosis

Mitosis

Possible spore patterns with first-division segregation.

or

or

or

or

Possible spore patterns with second-division segregation.

FIGURE 4.26 First- and second-division segregation in *Neurospora*. (A) First-division segregation patterns are found in the ascus when a crossover between the gene and centromere does not take place. The alleles separate (segregate) in meiosis I. Two spore patterns are possible, depending on the orientation of the pair of chromosomes on the first-division spindle. The orientation shown results in the pattern in the upper ascus. (B) Second-division segregation patterns are found in the ascus when a crossover between the gene and the centromere delays separation of *A* from *a* until meiosis II. Four patterns of spores are possible, depending on the orientation of the pair of chromosomes on the first-division spindle and that of the chromatids of each chromosome on the second-division spindle. The orientation shown results in the pattern in the top ascus.

Gene conversion suggests a molecular mechanism of recombination.

Genetic recombination may be regarded as a process of breakage and repair between two DNA molecules. In eukaryotes, the process takes place early in meiosis after each molecule has replicated, and with respect to genetic markers, it results in two molecules of the parental type and two recombinants. For genetic studies of recombination, fungi such as yeast or *Neurospora* are particularly useful, because all four products of

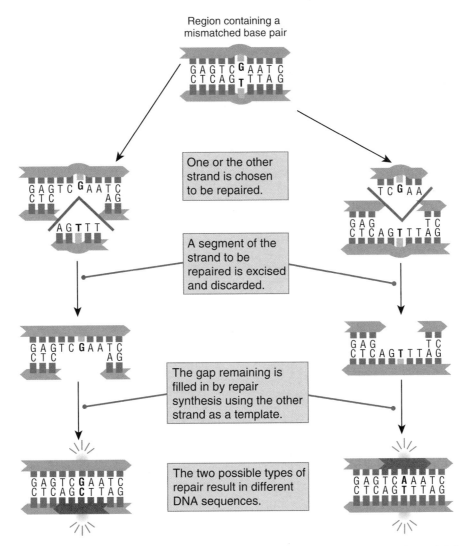

FIGURE 4.27 Mismatch repair consists of the excision of a segment of a DNA strand containing a base mismatch followed by repair synthesis. Either strand can be excised and corrected. In this example, the G−T mismatch is corrected to either G−C (left) or A−T (right).

🧠 STOP & THINK 4.4

Ordered tetrads from a strain of *Neurospora* of genotype *Aa* were analyzed. Among 120 tetrads analyzed, 80 were found to have all of their *A*-bearing spores adjacent to one another at either the top or bottom end of the ascus. Based on these data, what is the map distance between the gene and its centromere?

any meiosis can be recovered in a four-spore (yeast) or eight-spore (*Neurospora*) ascus. As we have noted, most asci from heterozygous *Aa* diploids contain ratios of

2 *A* : 2 *a* in four-spored asci, or
4 *A* : 4 *a* in eight-spored asci.

demonstrating normal Mendelian segregation. Occasionally, however, aberrant ratios are also found, such as

3 *A* : 1 *a* or 1 *A* : 3 *a* in four-spored asci, and
5 *A* : 3 *a* or 3 *A* : 5 *a* in eight-spored asci

Different types of aberrant ratios can also occur. The aberrant asci are said to result from **gene conversion** because it appears as if one allele has "converted" the other allele into a form like itself. Gene conversion is frequently accompanied by recombination between genetic markers on either side of the conversion event, even when the flanking markers are tightly linked. This implies that gene conversion can be one consequence of the recombination process.

Gene conversion results from a normal DNA repair process in the cell known as **mismatch repair**. In this process, an enzyme recognizes any base pair in a DNA duplex in which the paired bases are mismatched—for example, G paired with T, or A paired with C. When such a mismatch is found in a molecule of duplex DNA, a small segment of one strand is excised and replaced with a new segment synthesized using the remaining

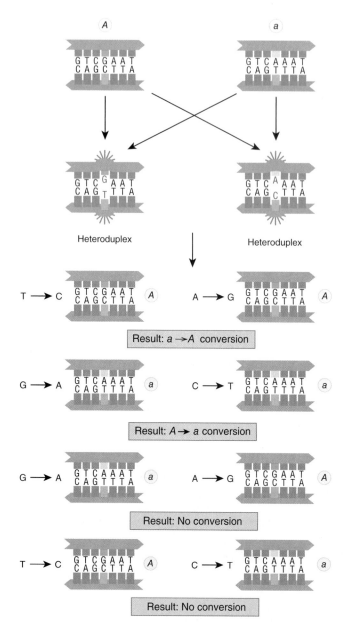

FIGURE 4.28 Mismatch repair resulting in gene conversion. Only a small part of the heteroduplex region is shown.

strand as a template. In this manner the mismatched base pair is replaced. **FIGURE 4.27** shows an example in which a mismatched G—T pair is being repaired. The strand that is excised could be either the strand containing T or the one containing G, and the newly synthesized (repaired) segment, shown in red, would contain either a C or an A, respectively. The two possible products of repair differ in DNA sequence.

The role of mismatch repair in gene conversion is illustrated in **FIGURE 4.28**. The pair of DNA duplexes across the top represents the DNA molecules of two alleles in a cell undergoing meiosis. One duplex contains a G—C base pair highlighted in color; this corresponds to the *A* allele. The other duplex contains an A—T base pair at the same position, which corresponds to the *a* allele. In the process of recombination, the participating

DNA duplexes can exchange pairing partners. The result is shown in the second row. The exchange of pairing partners creates a **heteroduplex** region in which any bases that are not identical in the parental duplexes become mismatched. In this example, one heteroduplex contains a G—T base pair and the other contains an A—C base pair. At this point, the mismatch repair system comes into play and corrects the mismatches. Each mismatch can be repaired in either of two ways, so there are four possible ways in which the mismatches can be repaired. One type of repair results in gene conversion of *a* to *A*, another results in gene conversion of *A* to *a*, and the remaining two restore the sequences of the original duplexes and so do not result in gene conversion.

4.6 Recombination is initiated by a double-stranded break in DNA.

In prophase I of meiosis, chiasmata are the physical manifestations of crossing-over between DNA molecules. These structures bridge between pairs of sister chromatids in a bivalent and are important in the proper alignment of the bivalent at the metaphase plate in preparation for anaphase I. Bivalents that lack chiasmata to help hold them together are prone to undergo nondisjunction.

The crossovers needed for the chiasmata to form are initiated by programmed double-stranded breaks in DNA (**FIGURE 4.29**, part A). In a double-stranded break, the size of the gap is usually increased by nuclease digestion of the broken ends, with greater degradation of the 5' ends leaving overhanging 3' ends as shown in the illustration. These gaps are repaired using the unbroken homologous DNA molecule as a template, but in meiosis the repair process can result in crossovers that yield chiasmata between nonsister chromatids. These crossovers are also the physical basis of what is observed genetically as recombination.

A double-stranded break does not necessarily result in a crossover, however. Repair of the double-stranded break by the noncrossover pathway is illustrated in Figure 4.29, part B. The first step in repair is that a broken 3' end invades the homologous unbroken DNA duplex, forming a short heteroduplex region with one strand and a looped-out region of the other strand called a **D loop**. (Specific proteins are required to mediate strand invasion; in *Escherichia coli* the strand-invasion protein is known as RecA.) In the illustration, the heteroduplex region is the region where the light blue strand is paired with the red strand. Because it is a heteroduplex, any base-pair mismatches in this region could be corrected by mismatch repair in such a way as to result in gene conversion. Such heteroduplex regions are typically only a few hundred base pairs in length. They are much shorter than a gene and vastly shorter than a chromosome, and

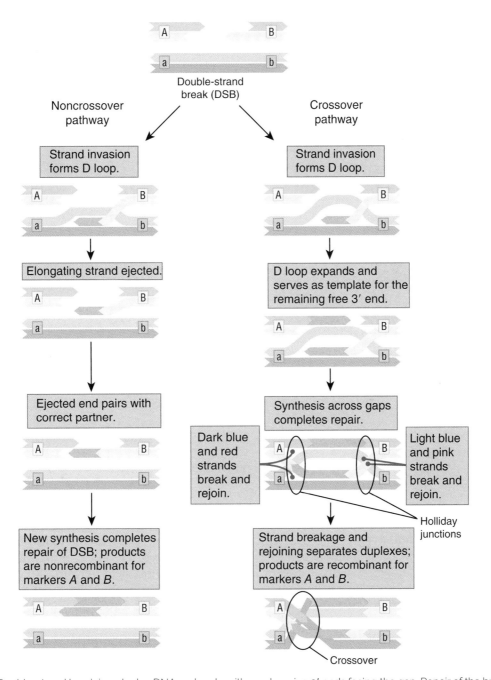

FIGURE 4.29 (A) Double-strand break in a duplex DNA molecule with overhanging 3′ ends facing the gap. Repair of the break makes use of the nonbroken homologous duplex. (B) Repair pathway that does not result in crossing-over, although the heteroduplex regions can undergo gene conversion. (C) Repair pathway that does result in crossing-over, also with possible gene conversion in heteroduplex regions.

Model from D. K. Bishop and D. Zickler, *Cell* 117 (2004): 9–15.

so gene conversions are rare events except for short regions very near the site of a double-stranded break.

At one end of the heteroduplex, the free 3′ end of the broken DNA strand is extended (brown), but after a time it is ejected from the template, and the strands of the unbroken duplex are able to come together again. At this point, the extension of the 3′ end is long enough that pairing can take place with the complementary strand in the broken duplex. At the same time, this pairing provides a template for the 3′ end of the other broken strand. Extension of the 3′ ends

across the remaining gaps completes the repair of the double-stranded break. Note that although gene conversion can occur in the noncrossover pathway, the resulting duplex DNA molecules are nonrecombinant.

The crossover pathway for repairing a double-stranded break is illustrated in Figure 4.29, part C. Again invasion of the unbroken duplex forms a D loop and a short heteroduplex region in which gene conversion can occur. As in the noncrossover pathway, the free 3′ end of the broken DNA strand is extended (brown), but in this case it continues until it displaces the partner strand

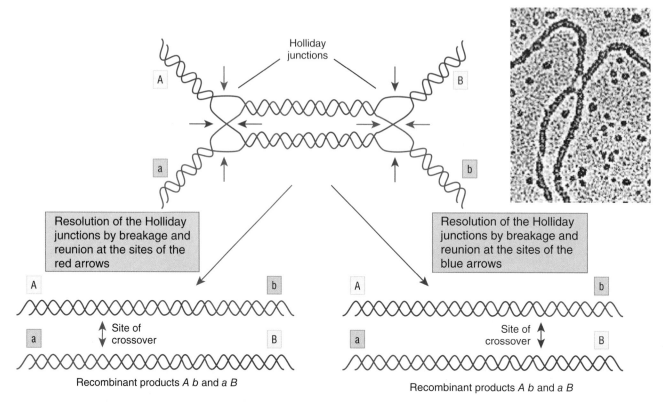

FIGURE 4.30 (A) Two Holliday junctions in a pair of DNA molecules undergoing recombination (the parental chromosomes are AB and ab); (B and C) two modes of resolution depending on which strands are broken and rejoined. Part D is an electron micrograph showing a single Holliday junction between a pair of DNA molecules.

Illustration modified from B. Alberts. *Essential Cell Biology*. Garland Science, 1997. Illustration reproduced with permission of Huntington Potter, Johnnie B. Byrd Sr., Alzheimer's Center & Research Institute.

(pink) of the template strand (red). The displaced strand can then serve as a template for the elongation of the 3′ end of the other broken strand. Eventually, the extensions of the broken strands become long enough that they can be attached to the broken 5′ ends. This completes the repair of the double-stranded break, but note that the resulting structure includes two places where the strands have exchanged pairing partners. Each of the structures where pairing partners are switched is called a **Holliday junction**, named after Robin Holliday, who first predicted that such structures would be involved in recombination.

The problem with Holliday junctions is that they are places where DNA strands from different duplex molecules are interconnected. How the strands are interconnected is shown for the DNA double helices in **FIGURE 4.30**, part A. Resolution of the Holliday junctions is necessary for the DNA molecules to become free of one another. This requires breakage and rejoining of one pair of DNA strands at each Holliday junction. The breakage and rejoining is an enzymatic function carried out by an enzyme called the **Holliday junction-resolving enzyme**.

Parts B and C in Figure 4.30 show two ways in which the Holliday structures can be resolved. Breakage and rejoining of the strands indicated by the red arrows results in a crossover at the site of the left-hand

Holliday junction, whereas breakage and rejoining of the strands indicated by the blue arrows results in a crossover at the site of the right-hand Holliday junction. In both cases, the resulting DNA molecules have a crossover that yields reciprocal recombinant *A b* and *a B* products. (In principle, resolution could also take place at the red arrows in one Holliday junction and the blue arrows in the other, but these resolutions result in noncrossover products. It is unclear how often these noncrossover types of resolution take place.)

It is important to understand double-strand break repair, because the process has far-reaching practical applications in a powerful method of genetic engineering that allows genes to be duplicated, deleted, or their sequences changed at will. The method, known as CRISPR, makes use of certain molecules originally isolated from bacteria that allow a double-strand break to be targeted to a specific site in the genome. Details of the CRISPR method of gene editing are discussed in Chapter 10.

Recombination tends to take place at preferred positions in the genome.

In some organisms, including humans and other mammals, programmed double-stranded DNA breaks are much more likely to occur at certain positions in the genome than others. Crossovers resulting in recombination are much more likely to occur at these positions, which are referred to as **hotspots of recombination**. The human genome contains about 30 thousand hotspots of recombination, spaced an average of 100 kb apart. These hotspots are often located in the spaces between genes, and they differ greatly from one to the next in the likelihood of a double-stranded break. One particular protein has been implicated in about 40 percent of hotspots across the human genome. The protein is known as PRDM9, and it is known to bind with double-stranded DNA at sites that match or nearly match the 13-mer sequence

$$5'\text{-CCGCCGTMWCCWC-}3'$$

where M means A or C, and W means T or A. The protein does not cause the double-stranded break directly. PRDM9 is actually a methyl transferase that attaches methyl ($-CH_3$) groups to the amino acid lysine in histone H3, which predisposes the DNA to undergo a double-stranded break. More than 30 alleles of the gene encoding PRDM9 have been identified, which differ in their propensity to recognize the 13-mer target and to methylate histone H3. Variation in the alleles encoding PRDM9 accounts for the differing efficiency of PRDM9-associated hotspots.

Although recombination tends to be initiated by means of double-stranded breaks at hotspots of recombination, it should be emphasized that there are so many hotspots relative to number of crossover events that the sites of recombination in any particular meiotic cell show a great deal of randomness. In the human genome, for example, there are about 30 thousand hotspots of recombination and about 60 crossover events per meiosis. Roughly speaking, each crossover could take place at any of about 500 hotspots in the vicinity. The relatively small number of crossovers per meiosis means that crossovers occur essentially at random sites chosen from among the very large number of hotspots that occur across the genome.

CHAPTER SUMMARY

- Genes that are located in the same chromosome and that do not show independent assortment are said to be linked.

- The alleles of linked genes present together in the same chromosome tend to be inherited as a group.

- Crossing-over between homologous chromosomes results in recombination, which breaks up combinations of linked alleles.

- The frequency of recombination serves as a measure of distance between linked genes along a chromosome, providing a genetic map of the relative positions of the genes.

- The map distance between genes in a genetic map is related to the frequency of crossing-over between the genes in meiosis.

- Physical distance along a chromosome is often, but not always, correlated with map distance.

- Variations in DNA sequence among individuals (polymorphisms) serve as genetic markers along the genome that are used for genetic mapping, tracing the genetic ancestry of individuals, and many other purposes.

- Two major classes of DNA polymorphism are single-nucleotide polymorphisms (SNPs) and copy-number polymorphisms (CNVs). SNPs are common across the genome, but CNVs are rare. Polymorphisms due to varying numbers of short-tandem repeats (STRs) are also abundant across the genome and are used in genetic mapping and DNA typing.

- Tetrads are sensitive indicators of linkage because they include all the products of meiosis.

- At the DNA level, recombination is initiated by a double-stranded break in a DNA molecule. Use of the homologous DNA molecule as a template for repair can result in a crossover, in which both strands of the participating DNA molecules are broken and rejoined.

ISSUES AND IDEAS

- Distinguish between genetic recombination and genetic complementation. Is it possible for two mutant genes to show complementation but not recombination? Is it possible for two mutant genes to show recombination but not complementation?

- In genetic analysis, why is it important to know the position of a gene along a chromosome?

- What is the maximum frequency of recombination between two genes? Is there a maximum map distance between two genes?

- Why is the frequency of recombination over a long interval of a chromosome always smaller than the map distance over the same interval?

- What is meant by the term *chromosome interference*?

- In human genetics, why are molecular variations in DNA sequence, rather than phenotypes such as eye color or blood-group differences, used for genetic analysis?

- In genetic analysis, what is so special about the ability to examine tetrads in certain fungi?

■ Explain how tetratype tetrads demonstrate that recombination takes place at the four-strand stage of meiosis and is reciprocal.

■ Explain why the observation PD >> NPD with respect to tetrads is a sensitive indicator of linkage.

SOLUTIONS: STEP BY STEP

PROBLEM 1 In *Drosophila*, the recessive mutant allele *spineless* (*ss*) results in thin bristles, *cinnabar* (*cn*) results in bright red eyes, and *ebony* (*e*) results in a black body color. A cross is carried out between females of genotype *ss cn e / + + +* and males of genotype *ss cn e / ss cn e*. In this type of symbolism, each + denotes the nonmutant allele of the gene written at the corresponding position. From this cross, the following 1000 progeny were obtained:

ss	cn	e	/	ss	cn	e		241
ss	+	e	/	ss	cn	e		223
+	cn	+	/	ss	cn	e		202
+	+	+	/	ss	cn	e		212
ss	cn	+	/	ss	cn	e		25
ss	+	+	/	ss	cn	e		31
+	cn	e	/	ss	cn	e		31
+	+	e	/	ss	cn	e		35

Determine which, if any, of the genes are linked, and for those that are linked, estimate the frequency of recombination between the genes.

SOLUTION. To determine which, if any, genes are linked, consider the mutants in pairs, and sum the data to find the total number of parental types and recombinant types for each pair. For *ss* and *cn*, the parental types are 241 + 212 + 25 + 35 = 513 and the recombinant types are 223 + 202 + 31 + 31 = 487. These numbers are close enough to 500 : 500 that one may infer that *ss* and *cn* are unlinked. Similarly, for *cn* and *e*, the parental types sum to 515 and the recombinant types to 485, hence *cn* and *e* are unlinked. For *ss* and *e*, however, the parental types sum to 878 and the recombinant types to 122, and this result implies that *ss* and *e* are linked. The estimated frequency of recombination is 122/1000 = 0.122 or 12.2 percent.

PROBLEM 2 In *Drosophila*, the recessive mutant allele *cinnabar* (*cn*) results in bright red eyes, *curved* (*c*) results in curved wings, and *plexus* (*px*) results in extra wing veins. All three genes are linked. In a cross between *cn c px / + + +* females and *cn c px / cn c px* males, the following progeny were counted:

cn	c	px	/	cn	c	px	296
cn	c	+	/	cn	c	px	63
cn	+	+	/	cn	c	px	119
cn	+	px	/	cn	c	px	10
+	c	px	/	cn	c	px	86
+	c	+	/	cn	c	px	15
+	+	+	/	cn	c	px	329
+	+	px	/	cn	c	px	82
		Total					1000

(a) What is the frequency of recombination between *cn* and *c*?
(b) What is the frequency of recombination between *c* and *px*?
(c) What is the frequency of recombination between *cn* and *px*?
(d) Why is the frequency of recombination between *cn* and *px* smaller than the sum of that between *cn* and *c* and that between *c* and *px*?
(e) What is the coefficient of coincidence across this region? What is the value of the interference?
(f) Draw a genetic map of the region, showing the locations of *cn*, *c*, and *px* and the map distances between the genes.

SOLUTION. Do not try to hurry through linkage problems! You will be rewarded by taking time to organize the information in the optimal manner. First, group the progeny types into reciprocal pairs—*cn c px* with + + +, *cn c +* with + + *px*, and so forth—and make a new list organized as shown here. (Ignore the *cn c px* chromosome from the father because it contributes no information about recombination.)

cn	c	px	296	625
+	+	+	329	
cn	c	+	63	145
+	+	px	82	
cn	+	+	119	205
+	c	px	86	
cn	+	px	10	25
+	c	+	15	
	Total			1000

In this tabulation, a space has been inserted between the pairs of reciprocal products in order to keep the groups separate. The number next to each brace is the total number of chromosomes in the group. The most numerous group of reciprocal chromosomes (in this case, *cn c px* and + + +) consists of the nonrecombinants, and the least numerous group of reciprocal chromosomes (in this case, *cn + px* and + *c* +) consists of the double recombinants. Rearrange the order of the groups, if necessary, so that the nonrecombinants are at the top of the list and the double recombinants are at the bottom. (In the present example, rearrangement is not necessary.) At this point, also make sure that the order of the genes is correct as given, by comparing the genotypes of the double recombinants with those of the nonrecombinants. If the gene order is correct, then it will require two recombination events (one in each interval) to derive the double-recombinant chromosomes from the nonrecombinants. If this is not the case, rearrange the order of the genes. (The "odd man out" in comparing the double recombinants with the nonrecombinants is always the gene in the middle.) In this particular example, the gene order is correct as given. Finally, with this preliminary bookkeeping done, we can proceed to tackle the questions. **(a)** The frequency of recombination between *cn* and *c* is given by the totals of all classes of progeny showing recombination in the *cn* − *c* interval, in this case (205 + 25)/1000 = 0.23. **(b)** The frequency of recombination between *c* and *px* equals (145 + 25)/1000 = 0.17. **(c)** The frequency of recombination between *cn* and *px* equals (145 + 205)/1000 = 0.35. (Note that the double recombinants are not included in this total, because the double recombinants are not recombined for *cn* and *px*; their allele combinations for *cn* and *px* are the same as in the parents.) **(d)** The frequency of recombination between *cn* and *px* (0.35) is smaller than the sum of that between *cn* and *c* and that between *c* and *px* (0.23 + 0.17 = 0.40) because of double crossovers. **(e)** The coefficient of coincidence equals the observed number of double recombinants divided by the expected number. The observed number is 25 and the expected number is 0.23 × 0.17 × 1000 = 39.1; the coefficient of coincidence therefore equals 25/39.1 = 0.64. The interference equals 1 − coefficient of coincidence, and so the interference equals 1 − 0.64 = 0.36. **(f)** The genetic map is shown in the accompanying diagram. The distances are in map units (centimorgans). However, the map distances of

23 and 17 map units are based on the 23 percent and 17 percent recombination observed between *cn* and *c* and between *c* and *px*, respectively; the actual distances in map units are probably a little greater than these estimates because of a small amount of double recombination within each of the intervals.

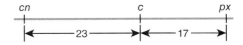

PROBLEM 3 In *Neurospora*, the gene *arg12* encodes the enzyme needed to convert ornithine to citrulline in the pathway of arginine biosynthesis. The gene was discovered in the experiments of Beadle and Tatum discussed in Chapter 1. In a cross between *arg12* and *ARG12* strains, where *arg12* denotes the mutant allele and *ARG12* the nonmutant allele, two-thirds of the resulting asci show second-division segregation. What does this observation imply about the map distance between *arg12* and the centromere?

SOLUTION. Two-thirds is the maximum proportion of second-division segregation that can occur. This value is observed for any gene that is so far from the centromere that one or more crossovers are almost certain to take place between the gene and the centromere. Hence, we can deduce that *arg12* is at least 50 map units from the centromere. This map distance is a minimum, and the true map distance could be greater.

CONCEPTS IN ACTION: PROBLEMS FOR SOLUTION

4.1 A double heterozygote has the repulsion configuration *A b / a B* of two linked genes that have a frequency of recombination of 0.20. If a randomly chosen gamete carries *A*, what is the probability that it also carries *B*?

4.2 What gametes, and in what frequencies, are produced by a female *Drosophila* of genotype *A B / a b* when the genes are present in the same chromosome and the frequency of recombination between the genes is 8 percent? What gametes, and in what frequencies, are produced by a male of the same genotype?

4.3 A cell undergoing meiosis in an organism with unordered tetrads undergoes a double crossover between two markers. If the ratio of 2-strand : 3-strand : 4-strand doubles is 1 : 2 : 1, what is the ratio of PD : TT : NPD tetrads? (PD stands for parental ditype tetrad, TT for tetratype, and NPD for nonparental ditype.)

4.4 A coefficient of coincidence of 0.36 implies which one or more of the following statements are true:

(a) The frequency of double crossovers was 36 percent.

(b) The frequency of double crossovers was 36 percent of the number that would be expected if there were no interference.

(c) There were 0.36 times as many single crossovers as double crossovers.

(d) There were 0.36 times as many single crossovers in one region as there were in an adjacent region.

(e) There were 0.36 times as many parental as recombinant progeny.

4.5 A gene in *Neurospora*, a fungus with ordered tetrads, shows 10 percent second-division segregation. What is the map distance between the gene and the centromere?

4.6 In *Drosophila pseudoobscura*, the eye-color mutation *purple (pr)* and the wing mutation *crossveinless (cv)* are located in chromosome 3 at a distance of 18 map units. What phenotypes, and in what proportions, would you expect in the progeny from the mating of $pr^+ cv^+ / pr\ cv$ females with $pr\ cv / pr\ cv$ males?

4.7 Construct a genetic map of a chromosome from the following recombination frequencies between individual pairs of genes: $r-c$, 10; $c-p$, 12; $p-r$, 3; $s-c$, 16; $s-r$, 8. You will discover that the distances are not strictly additive. Why aren't they?

4.8 A *Drosophila* cross is carried out with a female that is heterozygous for both the *y* (yellow body) and *bb* (bobbed bristles) mutations. Both genes are located in the X chromosome. Among 200 male progeny, there were 49 wildtype for both traits, 51 with yellow body, 41 with bobbed bristles, and 59 mutant for both genes. Do these genes show evidence for linkage? [Note: The appropriate chi-square test is a test for a 1 : 1 ratio of parental : recombinant gametes.]

4.9 Two genes in chromosome 7 of corn are identified by the recessive alleles *gl* (glossy), determining glossy leaves, and *ra* (ramosa), determining branching of ears. When a plant heterozygous for each of these alleles was crossed with a homozygous recessive plant, the progeny consisted of the following genotypes with the numbers of each indicated:

Gl ra/gl ra 98   *gl Ra/gl ra* 91
Gl Ra/gl ra 7   *gl ra/gl ra* 4

Calculate the frequency of recombination between these genes.

4.10 In the yellow-fever mosquito, *Aedes aegypti*, a dominant gene *DDT* for DDT resistance (DDT is dichlorodiphenyltrichloroethane, a long-lasting insecticide) and a dominant gene *Dl* for dieldrin resistance (dieldrin is another long-lasting insecticide) are known to be in the same chromosome. A cross was carried out between a DDT-resistant strain and a dieldrin-resistant strain, and female progeny resistant to both insecticides were testcrossed with wildtype males. Among the progeny, 99 were resistant to both insecticides, 88 were resistant to DDT only, 89 were resistant to dieldrin only, and 106 were sensitive to both insecticides.

(a) Are *DDT* and *Dl* alleles of the same gene? How can you tell?

(b) Are *DDT* and *Dl* linked?

(c) What can you deduce about the genetic positions of *DDT* and *Dl* along the chromosome?

4.11 Two true-breeding strains of mice are crossed to produce F_1 mice that are heterozygous for three linked genes with alleles *Aa*, *Bb*, and *Dd*. Numerous triply heterozygous F_1 mice are testcrossed,

and the genotypes and numbers of the resulting progeny are as follows:

A–	B–	D–	10
A–	B–	dd	350
A–	bb	D–	100
A–	bb	dd	40
aa	B–	D–	60
aa	B–	dd	120
aa	bb	D–	320
Total			1000

(a) Which gene is in the middle?

(b) Specify the genotype of the F_1 triple heterozygote as completely as possible, with the genes in the correct order and the correct alleles on each chromosome.

(c) Which two genes are closest together?

(d) What is the map distance between the two closest genes? Assume that interference is complete between these two genes.

(e) If interference is not complete, how will the true map distance differ from the value you calculated in part **(d)**? Briefly explain why this is so.

4.12 In corn, the genes *v* (virescent seedlings), *pr* (red aleurone), and *bm* (brown midrib) are all on chromosome 5, but not necessarily in the order given. The cross

$$v^+\ pr\ bm/v\ pr^+\ bm^+ \times v\ pr\ bm/v\ pr\ bm$$

produces 1000 progeny with the following phenotypes:

v^+	pr	bm	226
v	pr^+	bm^+	229
v^+	pr	bm^+	153
v	pr^+	bm	185
v^+	pr^+	bm	59
v	pr	bm^+	71
v^+	pr^+	bm^+	36
v	pr	bm	41
Total			1000

(a) Determine the gene order, the recombination frequencies between adjacent genes, the coefficient of coincidence, and the interference.

(b) Explain why, in this example, the recombination frequencies are not good estimates of map distance.

4.13 The male I-2 in the accompanying pedigree is affected with Huntington disease, a type of neuromuscular degeneration caused by a rare autosomal dominant mutation *HD* with complete penetrance. The wildtype, nonmutant allele is denoted *hd*. The woman II-1 is also affected.

A gene with alleles *A* and *a* is linked to the Huntington locus with a recombination frequency of 10 percent. The bands in the gel labeled *A* and *a* distinguish between the alleles.

(a) Is the genotype of II-1 *HD A/hd a* or is it *HD a/hd A*?

(b) Given the pattern of bands in the gel, what is the probability that III-1 will be affected?

(c) Given the pattern of bands in the gel, what is the probability that III-2 will be affected?

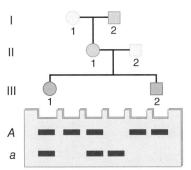

4.14 A human geneticist discovers the molecular variation in DNA sequence illustrated in the accompanying diagrams of electrophoresis gels. In the human population as a whole, she finds any of four phenotypes, shown in panel A. She believes that this may be a simple genetic polymorphism with three alleles, like the ABO blood groups. There are two alleles that yield DNA fragments of different sizes, fast (*F*) or slow (*S*) migration, and a "null" allele (*O*) in which the DNA fragment is deleted. The genotypes in panel A would therefore be, from left to right, *FF* or *FO*, *SS* or *SO*, *FS*, and *OO*. In the population as a whole, the putative *OO* genotype is extremely common, and the *FS* genotype is quite rare. To investigate this hypothesis further, the geneticist studies offspring of matings between parents who have the putative FS genotype (panel B). The types of progeny, and their numbers, are shown in panel C.

(a) What result would be expected from the three-allele hypothesis?

(b) Are the observed data consistent with this result? Why or why not?

(c) Suggest a genetic hypothesis that can explain the data in panel C.

(d) Are the data consistent with your hypothesis?

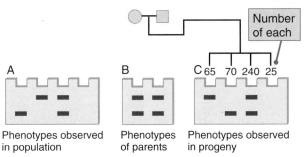

Phenotypes observed in population   Phenotypes of parents   Phenotypes observed in progeny

4.15 The following classes and frequencies of ordered tetrads were obtained from the cross $a^+ b^+ \times a\ b$ in *Neurospora*. (Only one member of each pair of spores is shown.) What is the order of the genes in relation to the centromere?

1-2	3-4	5-6	7-8	Number of asci
$a^+ b^+$	$a^+ b^+$	$a\ b$	$a\ b$	1766
$a^+ b^+$	$a\ b$	$a^+ b^+$	$a\ b$	220
$a^+ b^+$	$a\ b^+$	$a^+ b$	$a\ b$	14

4.16 A portion of the linkage map of chromosome 2 in the tomato is illustrated here. The oblate phenotype is a flattened fruit, the peach phenotype is hairy fruit (like a peach), and compound influorescence means clustered flowers.

Among 1000 gametes produced by a plant of genotype *o ci +/+ + p*, what types of gametes would be expected, and what number would be expected of each? Assume that the chromosome interference across this region is 80 percent but that interference within each region is complete.

4.17 The yeast *Saccharomyces cerevisiae* has unordered tetrads. In a cross carried out to study the linkage relationships among three genes, the tetrads in the accompanying table were obtained. The cross was between a strain of genotype *+ b c* and one of genotype *a + +*.

(a) From these data determine which, if any, of the genes are linked.

(b) For any linked genes, determine the map distances.

Tetrad type	Genotypes of spores in tetrads				Number of tetrads
1	a + +	a + +	+ b c	+ b c	132
2	a b +	a b +	+ + c	+ + c	124
3	a + +	a + c	+ b +	+ b c	64
4	a b +	a b c	+ + +	+ + c	80
Total					400

4.18 A small portion of the genetic map of *Neurospora crassa* chromosome VI is illustrated here. The *cys-1* mutation blocks cysteine synthesis, and the *pan-2* mutation blocks pantothenic acid synthesis. Assuming complete chromosome interference, determine the expected frequencies of the following types of asci in a cross of *cys-1 pan-2* × *CYS-1 PAN-2*.

(a) First-division segregation of *cys-1* and first-division segregation of *pan-2*

(b) First-division segregation of *cys-1* and second-division segregation of *pan-2*

(c) Second-division segregation of *cys-1* and first-division segregation of *pan-2*

(d) Second-division segregation of *cys-1* and second-division segregation of *pan-2*

(e) Parental ditype, tetratype, and nonparental ditype tetrads

cys-1   pan-2

|← —————— 7 cM —————— →|← 3 cM →|

4.19 The accompanying gel diagram shows the positions of DNA bands associated with the *A, a* and *B, b* allele pairs for two linked genes. On the left are the phenotypes of the parents, and on the right are the phenotypes of the progeny and the number of each observed. Is the linkage phase of *A* and *B* in the doubly heterozygous parent coupling or repulsion? What is the frequency of recombination between these genes?

```
              Progeny
Parents   ┌──────────────────────┐
          155    44    36    165

   A      ▬▬   ▬▬   ▬▬   ▬▬   ▬▬
   a      ▬▬   ▬▬   ▬▬        ▬▬
   B      ▬▬   ▬▬        ▬▬   ▬▬
   b      ▬▬   ▬▬   ▬▬   ▬▬   ▬▬
```

4.20 Janet is performing a testcross to determine the linkage relationships between three *Drosophila* genes, *dpy*, *unc*, and *dor*. Her entire grade depends on her getting this right! She testcrosses females heterozygous for the recessive alleles:

> *dpy* (dumpy body) / *dpy*⁺ (normal body)
> *unc* (uncoordinated) / *unc*⁺ (coordinated)
> *dor* (deep orange eye) / *dor*⁺ (red eye).

The cross yields the following results:

normal body, red eye, coordinated	75
normal body, red eye, uncoordinated	348
normal body, deep orange eye, uncoordinated	96
dumpy body, red eye, coordinated	110
dumpy body, deep orange eye, coordinated	306
dumpy body, deep orange eye, uncoordinated	65

(a) What is the genotype of the F_1 heterozygous female?

(b) Construct a map of the region indicating the order of the genes and the distances (in map units) between them.

(c) What is the interference in this region?

⚙ STOP & THINK ANSWERS

ANSWER TO STOP & THINK 4.1

The frequency of recombination is $(95 + 97)/640 = 0.30$. [It is not $(216 + 232)/640 = 0.70$ because the frequency of recombination is always less than 0.50.] Since *A B* and *a b* are the recombinant chromosomes, the parental chromosomes must have been *A b* and *a B*, so the doubly heterozygous parent had the genotype *A b / a B*, in which the recessive alleles are in repulsion.

ANSWER TO STOP & THINK 4.2

The frequency of recombination in the *A−B* interval is $(95 + 5)/1000 = 0.10$, and that in the *B−C* interval is $(195 + 5)/1000 = 0.20$. The double recombinants must be added in both cases because they have recombination in both *A−B* and *B−C*. The expected number of double recombinants is therefore $0.1 \times 0.2 \times 1000 = 20$, and the observed number of 5.

The coefficient of coincidence equals $5/20 = 0.25$ and the interference equals 1 − coefficient of coincidence $= 1 − 0.25 = 0.75$.

ANSWER TO STOP & THINK 4.3

From left to right, the types of asci are tetratype, nonparental ditype, and parental ditype. The map distance between *A* and *B* is therefore $(1/2) \times (32/170) \times 100 = 9.4$ map units.

ANSWER TO STOP & THINK 4.4

Asci in which spores with the same allele are clustered at either one or the other end of the ascus show first-division segregation. Hence $120 − 80 = 40$ of the asci shows second-division segregation. The map distance between the gene and its centromere is therefore $(1/2) \times (40/120) \times 100 = 16.7$ map units.

CHAPTER 5

Human Chromosomes and Chromosome Behavior

LEARNING OBJECTIVES

- To describe the normal human chromosome complement and the implications of extra chromosomes (for example, Down syndrome) or missing chromosomes, including the X and Y sex chromosomes.

- To define dosage compensation with respect to the X chromosome, to cite a classic example of the single-active-X principle observed at the level of phenotype, and to explain how dosage compensation helps ameliorate the effects of extra X chromosomes.

- To diagram ectopic crossing-over between repeated sequences in the same chromosome arm and to show the resulting crossover products.

- To predict the result of crossing-over in the inversion loop of a paracentric inversion or of a pericentric inversion.

- To illustrate what types of gametes would be expected from meiosis in an individual that is heterozygous for a reciprocal translocation.

Organisms with an extra chromosome or a missing chromosome usually have developmental or other types of abnormalities. The abnormalities result from the increase or decrease in copy number (*dosage*) of the genes in this chromosome. Some organisms, usually rare, are found to have a variation in chromosome structure. The abnormal chromosome may have a particular segment missing, duplicated, reversed in orientation, or attached to a different chromosome. Each of these structural abnormalities has different genetic implications. In this chapter we consider the human chromosome complement and some of the major chromosomal abnormalities encountered in human populations. We also examine chromosome abnormalities in other organisms. Generally speaking, animals are much less tolerant of chromosomal changes than are plants. As we shall see, the acquisition of entire extra sets of chromosomes is not necessarily harmful, especially in plants. In some lineages of plants, the duplication of entire chromosome sets has figured prominently in genome evolution and the origin of species.

5.1 Humans have 46 chromosomes in 23 pairs.

The normal chromosome complement of a cell in mitotic metaphase from a human male is illustrated in **FIGURE 5.1**. The chromosomes have been labeled via a technique called **chromosome painting**, in which different colors are "painted" on each chromosome by

hybridization (formation of duplex molecules) with DNA strands labeled with different fluorescent dyes. Individual chromosomes are first isolated by any of a variety of techniques, and then the chromosome-specific DNA samples are labeled with fluorescence. A mixture of differently labeled strands from all the chromosomes is used in hybridization with metaphase chromosomes squashed onto a glass slide, allowing the fluorescent strands to hybridize with complementary strands present in the chromosomes. Unhybridized DNA is washed from the slide, and the preparation is examined through a confocal microscope to read the fluorescent signals for conversion into visible colors. (A confocal microscope produces images of a single region in a single focal plane, because it is able to reject scattered and extraneous light.)

The standard human karyotype consists of 22 pairs of autosomes and two sex chromosomes.

Chromosome painting dramatically identifies the pairs of homologous chromosomes. The presentation shown in part A of Figure 5.1 is a *metaphase spread*, in which the chromosomes are arranged just as they appear in the cytological preparation. A more conventional representation, called a **karyotype**, is shown in part B of Figure 5.1. In a karyotype, the autosomes in the metaphase spread are rearranged systematically in pairs, from longest to shortest, and numbered from

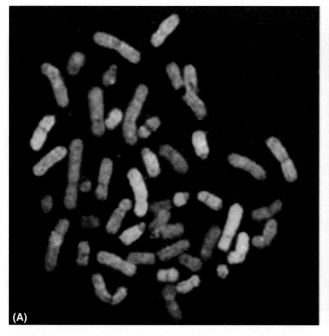
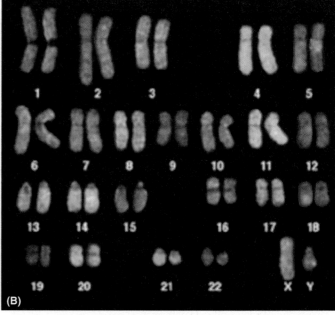

(A)   (B)

FIGURE 5.1 Human chromosome painting, in which each pair of chromosomes is labeled by hybridization with a different fluorescent probe. (A) Metaphase spread showing the chromosomes in a random arrangement as they were squashed onto the slide. (B) A karyotype, in which the chromosomes have been grouped in pairs and arranged in conventional order. Chromosomes 1–20 are arranged in order of decreasing size, but for historical reasons, chromosome 21 precedes chromosome 22, even though chromosome 21 is smaller.

Courtesy of Johannes Wienberg, Ludwig-Maximillians-University, and Thomas Ried, National Institutes of Health.

1 (the longest) through 22. In this example, the sex chromosomes are set off at the bottom right. The single X and Y chromosomes are evident. The karyotype of a normal human female has a pair of X chromosomes, instead of an X and a Y, in addition to the 22 pairs of autosomes. Chromosome painting is of considerable utility in human cytogenetics because even complex chromosome rearrangements can be detected rapidly and easily.

Another, less colorful metaphase spread and karyotype are shown in **FIGURE 5.2**. In this case the chromosomes have been treated with a staining reagent called Giemsa, which causes the chromosomes to exhibit transverse bands (*G-bands*). The bands form in large regions in which the base composition of the DNA has a relatively low abundance of G−C base pairs, and the banding pattern is specific for each pair of homologs. These bands permit smaller segments of each chromosome arm to be identified. The chromosomes are grouped into seven sets denoted by the letters A through G. (The X chromosome is included in group C, the Y in group G.) These conventional groupings date from a time prior to G-banding and chromosome painting, when the chromosomes could be sorted only by size and centromere position.

The nomenclature of the banding patterns in human chromosomes is shown in **FIGURE 5.3**, where the red letter beneath each chromosome indicates its group. For each chromosome, the short arm is designated with the letter p, which stands for "petite," and the long arm by the letter q, which stands for "not p." Within each arm, the regions are numbered according to standard conventions. Some familiar genetic landmarks in the human genome are the ABO blood-group locus at 9q34; the red–green color-blindness genes at Xq28, and the male-determining gene on the Y chromosome, called *SRY* (sex-determining region, Y) at Yp11.3.

FIGURE 5.4 shows a chromosome painting of a human chromosome complement at metaphase of mitosis. Only one of each pair of homologous autosomes is shown, along with the X and Y chromosome. Below each chromosome is the amount of DNA in the chromosome in megabase pairs (Mb), an estimate of the number of genes from the human genome sequence, and the approximate gene density. Gene density is not highly correlated with chromosome size, and it can differ greatly from one chromosome to the next. Two of the smallest chromosomes, 19 and 22, have the highest gene densities (27 and 23 genes per Mb, respectively), and two of the largest chromosomes (4 and 5) have among the lowest gene densities (8 and 9 genes per Mb, respectively).

Chromosomes with no centromere, or with two centromeres, are genetically unstable.

As is true in nearly all eukaryotic organisms, each human chromosome is linear and has a single centromere. Chromosomes are often classified according to the relative position of their centromeres, which determines the appearance of the daughter chromosomes as they separate from each other in anaphase (**FIGURE 5.5**). A chromosome with its centromere about in the middle is a *metacentric chromosome*; the arms

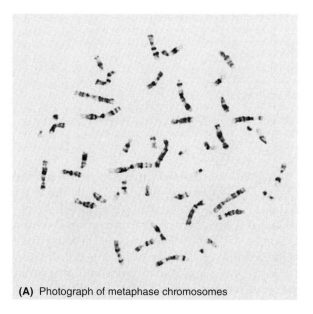

(A) Photograph of metaphase chromosomes

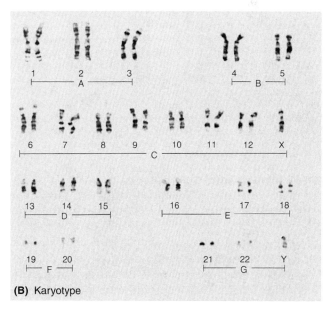

(B) Karyotype

FIGURE 5.2 A karyotype of a normal human male. Blood cells arrested in metaphase were stained with Giemsa and photographed with a microscope. (A) The chromosomes as seen in the cell by microscopy. (B) The chromosomes have been cut out of the photograph and paired with their homologs.

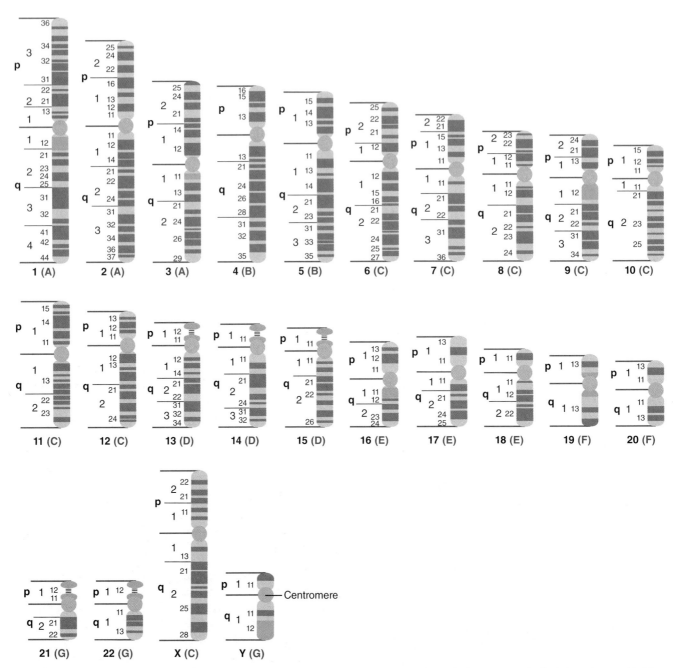

FIGURE 5.3 Designations of the bands and interbands in the human karyotype. Beneath each chromosome is the lettered group (A–G) to which it belongs.

are of approximately equal length, and each daughter chromosome forms a V shape at anaphase. When the centromere is somewhat off center, the chromosome is a *submetacentric chromosome*, and each daughter chromosome forms a J shape at anaphase. A chromosome with the centromere very close to one end appears I-shaped at anaphase because the arms are grossly unequal in length; such a chromosome is *acrocentric*.

Chromosomes with a single centromere are usually the only ones that are reliably transmitted from parental cells to daughter cells or from parental organisms to their progeny. When a cell divides, spindle

fibers attach to the kinetochore associated with the centromere of each chromosome and pulls the sister chromatids to opposite poles. A chromosome that lacks a centromere is an **acentric chromosome**. Acentric chromosomes are genetically unstable because they cannot be maneuvered properly during cell division and are lost. Occasionally, a chromosome arises that has two centromeres and is said to be **dicentric**. A dicentric chromosome is also genetically unstable because it is not transmitted in a predictable fashion. The dicentric chromosome is frequently lost from a cell when the two centromeres proceed to opposite

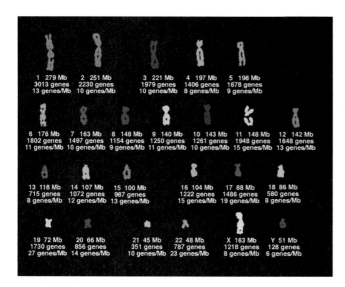

FIGURE 5.4 The human chromosome complement at metaphase of mitosis showing the amount of DNA in each chromosome, the estimated number of genes, and the approximate gene density. For the autosomes, only one of each homologous pair is shown.

Sequence data from International Human Genome Sequencing Consortium, *Nature* 409 (2001): 860-921, and J.C. Venter, et al., *Science* 291 (2001): 1304-1351. Chromosome image courtesy of Michael R. Speicher. Institute of Human Genetics, Medical University of Graz.

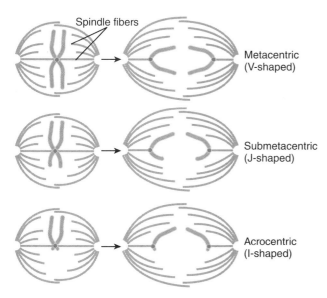

FIGURE 5.5 Three possible shapes of monocentric chromosomes in anaphase as determined by the position of the centromere (shown in dark blue).

poles in the course of cell division; in this case, the chromosome is stretched and forms a *bridge* between the daughter cells. This bridge may not be included in either daughter nucleus, or it may break, with the result that each daughter nucleus receives a broken chromosome. We will consider one mechanism by which dicentric and acentric chromosomes are formed when we discuss inversions.

Although most dicentric chromosomes are genetically unstable, if the two centromeres are close enough

together, they can frequently behave as a single unit and be transmitted normally. This principle was important in the evolution of human chromosome 2. Among higher primates, chimpanzees and human beings have 23 pairs of chromosomes that are similar in morphology and G-banding pattern, but chimpanzees have no obvious homolog of human chromosome 2, a large metacentric chromosome. Instead, chimpanzees have two medium-sized acrocentric chromosomes not found in the human genome. The cause of this situation is shown in **FIGURE 5.6**. The G-banding patterns indicate that human chromosome 2 was formed by fusion of the telomeres between the short arms of two acrocentric chromosomes that, in chimpanzees, remain acrocentrics. The chromosome fusion reduced the chromosome number in the human lineage from 48, which is characteristic of the great apes (chimpanzee, gorilla, and orangutan), to the number 46.

Dosage compensation adjusts the activity of X-linked genes in females.

For all organisms with XX–XY sex determination, there is a problem of the dosage of genes on the X chromosome, because females have two copies of this chromosome whereas males have only one. (There is less of a problem with Y-linked genes, because the Y chromosome is largely heterochromatic and carries relatively few genes.) In most organisms, a mechanism of **dosage compensation** has evolved in which the unequal dosage in the sexes is corrected either by increasing the activity of genes in the X chromosome in males or by reducing the activity of genes in the X chromosome in females.

The mechanism of dosage compensation in mammals is seemingly simple. In the early cleavage divisions of the embryo, at roughly the 64 to 128 cell stage, one and only one X chromosome in each cell, chosen at random, remains genetically active, and any other X chromosomes that may be present in the cell undergo a process of **X inactivation**. Any X chromosome that is inactivated in a particular somatic cell remains inactive in all the descendants of that cell (**FIGURE 5.7**); hence, the inactive state of an X chromosome is inherited from parental cell to daughter cell.

The process of X-chromosome inactivation takes place in all embryos with two or more X chromosomes, including normal XX females. The inactivation process is one of chromosome condensation initiated at a site called *XIC* (for *X-inactivation center*) near the centromere on the long arm between Xq11.2 and Xq21.1. The *XIC* includes a transcribed region in band Xq13 designated *Xist* (for *X-inactivation–specific transcript*). Transcription of *Xist* is the earliest event observed in X inactivation, and *Xist* transcription defines which X chromosome will be the inactive X chromosome. Remarkably, the spliced transcript of *Xist* does not

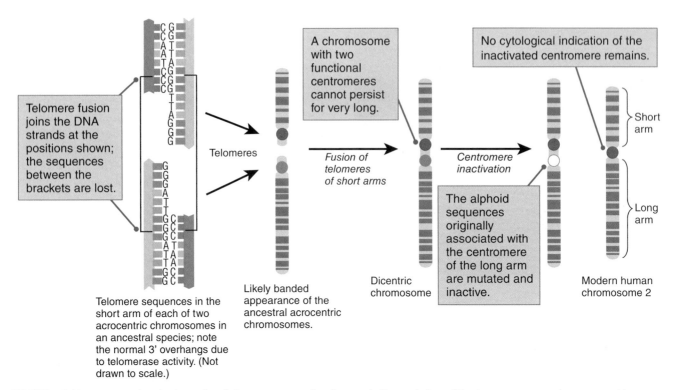

Telomere fusion joins the DNA strands at the positions shown; the sequences between the brackets are lost.

Telomeres

Telomere sequences in the short arm of each of two acrocentric chromosomes in an ancestral species; note the normal 3' overhangs due to telomerase activity. (Not drawn to scale.)

Likely banded appearance of the ancestral acrocentric chromosomes.

A chromosome with two functional centromeres cannot persist for very long.

Fusion of telomeres of short arms

Dicentric chromosome

Centromere inactivation

The alphoid sequences originally associated with the centromere of the long arm are mutated and inactive.

No cytological indication of the inactivated centromere remains.

Short arm

Long arm

Modern human chromosome 2

FIGURE 5.6 Human ancestors had 24 pairs of chromosomes rather than 23. In the evolution of the human genome, two acrocentric chromosomes fused to create human chromosome 2.

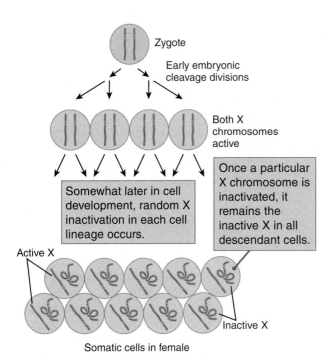

Zygote

Early embryonic cleavage divisions

Both X chromosomes active

Somewhat later in cell development, random X inactivation in each cell lineage occurs.

Once a particular X chromosome is inactivated, it remains the inactive X in all descendant cells.

Active X

Inactive X

Somatic cells in female

FIGURE 5.7 Schematic diagram of somatic cells of a normal female showing that the female is a mosaic for X-linked genes. The two X chromosomes are shown in red and blue. An active X is depicted as a straight chromosome, an inactive X as a tangle. Each cell has just one active X, but the particular X that remains active is a matter of chance. In human beings, the inactivation includes all but a few genes in the tip of the short arm.

contain an open reading frame encoding a protein. It appears to function as a noncoding RNA, and as transcription of *Xist* continues, the spliced transcript progressively coats the inactive X chromosome, spreading outward from the *XIC*. Thereafter, other molecular changes take place along the inactive X chromosome that are typically associated with gene silencing. In mouse embryos in which the *Xist* homolog has been disrupted, X inactivation does not take place, which demonstrates that *Xist* is essential for inactivation.

On the other hand, a study of the level of transcription of 624 genes along the human X chromosome has shown that about 15 percent of the X-linked genes escape inactivation to some degree. The *inactive X* is, therefore, not completely silenced. The transcribed genes occur in large blocks that tend to be located in the distal portions of the arms, especially the short arm, a pattern suggesting that escape from X inactivation may be correlated with distance from *Xist*. In any event, most of the genes that escape inactivation have levels of transcription that range from 15 to 50 percent of those observed for their homologs in the active X chromosome; hence, the level of activity is robust. The number of genes that escape complete X inactivation, and their levels of transcription, could readily account for some of the differences in expression of traits between males and females, for phenotypic variation among females heterozygous for such X-linked mutations as hemophilia A, and for phenotypic variation among individuals with abnormal numbers of X chromosomes.

X-chromosome inactivation has two consequences. First, it results in dosage compensation. It equalizes the number of active copies of X-linked genes in females and males. Although a female has two X chromosomes and a male has only one, because of inactivation of one X chromosome in each of the somatic cells of a female, both sexes have the same number of active X chromosomes. The mechanism of dosage compensation by means of X inactivation was originally proposed by Mary Lyon and is called the **single-active-X principle**.

The second consequence of X-chromosome inactivation is that a normal female is a mosaic for the expression of X-linked genes. A genetic *mosaic* is an individual that contains cells of two or more different genotypes. A normal female is a mosaic for gene expression, because the X chromosome that is genetically active can differ from one cell to the next. The mosaicism can be observed directly in females that are heterozygous for X-linked alleles that determine different forms of an enzyme, A and B: When cells from a heterozygous female are individually cultured in the laboratory, half of the clones are found to produce only the A form of the enzyme and the other half to produce only the B form. Mosaicism can also be observed directly in women who are heterozygous for an X-linked recessive mutation that results in the absence of sweat glands; these women exhibit large patches of skin in which sweat glands are present (these patches are derived from embryonic cells in which the normal X chromosome remained active and the mutant X was inactivated) and other large patches of skin in which sweat glands are absent (these patches are derived from embryonic cells in which the normal X chromosome was inactivated and the mutant X remained active.)

The calico cat shows visible evidence of X-chromosome inactivation.

In some cases, the result of random X inactivation in females can be observed in the external phenotype. One example is the "calico" pattern of coat coloration in female cats. Two alleles affecting coat color are present in the X chromosome in cats. One allele results in an orange coat color, the other in a black coat color. Because a normal male has only one X chromosome, he has either the orange or the black allele. A female can be heterozygous for orange and black, and in this case the coat color is "calico"—a mosaic of orange and black patches mixed with patches of white. **FIGURE 5.8** is a photograph of a female cat with the

FIGURE 5.8 Ginger is a female cat heterozygous for the orange and black coat color alleles. She shows the classic "calico" pattern of patches of orange and black.

classic calico pattern. The orange and black patches result from X-chromosome inactivation. In cell lineages in which the X chromosome bearing the orange allele is inactivated, the X chromosome with the black allele is active and so the fur is black. In cell lineages in which the X chromosome with the black allele is inactivated, the orange allele in the active X chromosome results in orange fur. (The white patches are due to an unrelated autosomal dominant mutation.)

STOP & THINK 5.1

In this pedigree, the woman I-1 is heterozygous for X-linked hemophilia A, which results from inadequate levels of blood clotting factor VIII. The normal level of factor VIII in the plasma is 120 nanomolar (nM), but any level greater than 60 nM is still associated with normal blood clotting.

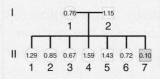

The number inside each symbol in the pedigree is the level of factor VIII observed in the plasma of that individual. Which of the females in the pedigree are likely to be heterozygous for the mutation? How can you tell? Why is there a difference between homozygous and heterozygous females?

Some genes in the X chromosome are also present in the Y chromosome.

The silencing of genes in the inactive X chromosome evolved gradually as the X and Y chromosomes progressively diverged from their ancestral chromosomes and the Y chromosome began to lose the function of most of its genes. The gene inactivation in the inactive X chromosome therefore affects individual genes and blocks of genes, and some genes in the inactive X are not silenced. Some of the genes that escape X inactivation have functional homologs in the Y chromosome, whereas others do not.

Two continuous regions that escape X inactivation are found at the tips of the arms. These are regions in which the Y chromosome does retain homologous genes that are functional. These regions of homology enable the X and Y chromosomes to synapse in spermatogenesis, and a crossover takes place that holds the chromosomes together to ensure their proper segregation during anaphase I. The regions of shared X–Y homology define the **pseudoautosomal regions**: *PARp* is a 2.7-Mb region at the terminus of the short arms and *PARq* is a 0.3-Mb region at the terminus of the long arms. Because crossing-over regularly takes place at least in *PARp*, the rate of recombination per nucleotide pair is at least 20-fold greater in the *PARp* than in the autosomes.

The pedigree patterns of inheritance of genes in the pseudoautosomal regions are indistinguishable from patterns characteristic of autosomal inheritance. The reason is that a mutant allele in a pseudoautosomal region is neither completely X linked nor completely Y linked but can move back and forth between the X and Y chromosomes because of recombination in the pseudoautosomal region. A gene that shows an autosome-like pattern of inheritance, but that is known from molecular studies to reside in a pseudoautosomal region, is said to show *pseudoautosomal inheritance*.

The pseudoautosomal region of the X and Y chromosomes has gotten progressively shorter in evolutionary time.

Comparative cytogenetic and molecular studies suggest that the X and Y chromosomes began their existence as a pair of ordinary autosomes in the common ancestor of modern mammals and birds. They started to diverge in DNA sequence and gene content at about the same time that the evolutionary lineage of mammals diverged from that of birds, some 300–350 million years (MY) ago. One must assume that prior to this time, recombination took place at normal levels throughout the entire proto-X and proto-Y chromosomes and that their gene contents were identical.

In the human genome as it exists today, the Y chromosome includes a small number of genes that are important for male fertility. The Y chromosome is also populated with many repeat sequences, some extremely long, which can undergo gene conversion and serve as sites for homologous recombination. One of the key genes in the Y chromosome is the master sex-determining gene *SRY*, located in the short arm near, but not included in, the pseudoautosomal region. The gene *SRY* codes for a protein transcription factor, the **testis-determining factor (TDF)**. When TDF is present, it stimulates transcription of its target genes, and their products in turn direct embryonic development toward the male sex by inducing the undifferentiated embryonic genital ridge (the precursor of the gonad) to develop as a testis.

Once *SRY* had evolved as a sex-determining mechanism, the Y chromosome began to diverge in DNA sequence from the X chromosome, and the region of possible X–Y recombination became progressively restricted to the telomeric regions. In regions with no X–Y recombination, there is a steady selection pressure for genes in the Y chromosome to undergo mutational degeneration into nonfunctional states. This results from the forced heterozygosity of the Y chromosome, which allows multiple deleterious mutations to accumulate through time because there is no opportunity for recombination to regenerate Y chromosomes that are free of deleterious mutations. Hence any Y-linked gene whose function is nonessential will tend to degenerate gradually because of the accumulation of mutations, and at the same time there will be selection pressure for dosage compensation of the homologous gene in the X chromosome. Eventually, only the dosage-compensated X-linked gene will remain functional.

Apparently blocks of genes were removed from the region of X–Y recombination in large chunks. Molecular evidence for this conclusion is summarized in **FIGURE 5.9**. Shown at the left are the locations of some protein-coding sequences in the short arm of the modern X chromosome from band Xp11 to the telomere. All of these genes have homologous sequences in the modern Y chromosome, as shown at the right. The amount of sequence divergence between the X and Y homologs shows a remarkable pattern. In the positions of the codons where a nucleotide substitution can occur without changing the encoded amino acid, the proportion of nucleotide differences between the X and Y homologs for the genes *GYG2–AMELX* is 0.07–0.11, for *TB4X–UTX* it is 0.23–0.36, for *SMCX* it is 0.52, and for other genes outside the region shown it is >0.94. Because the evolutionary rate of nucleotide substitutions at synonymous sites in mammalian genes is known, we can say that these levels of divergence correspond to divergence times of 30–50 MY for *GYG2–AMELX*, 80–130 MY for *TB4X UTX*, 130–170 MY

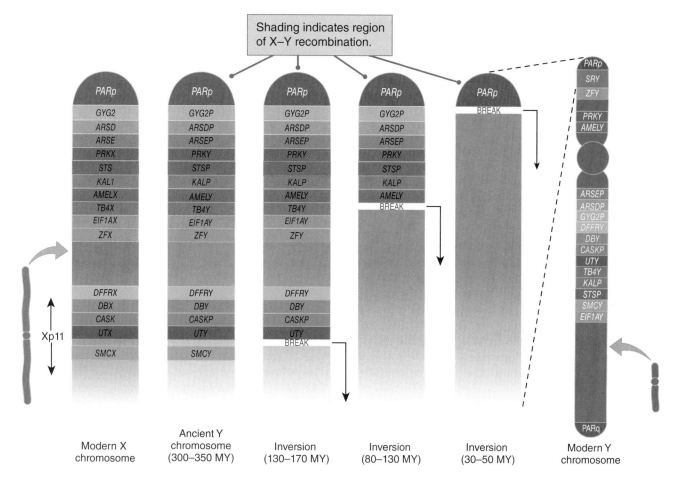

FIGURE 5.9 Progressive shortening of the mammalian X–Y pseudoautosomal region through time due to inversions in the Y chromosome, inferred from DNA sequence data. The arrows denote the distal (nearest the telomere) breakpoint of each successive inversion interrupting the pseudoautosomal region.

Data from B.T. Lahn and D.C. Page, *Science* 286 (1999): 964–967.

for *SMCX*, and 300–350 MY for other genes outside the region shown. The simplest explanation is that these were the times when successive blocks of genes were removed from the region of X–Y recombination by chromosome rearrangements—namely, inversions—in which a region somewhere in the interior of the Y chromosome becomes reversed in orientation. One way in which such an inversion can take place is by means of homologous recombination between two repeated DNA sequences that are present in opposite orientations at different locations in the Y chromosome, a mechanism that is examined in greater detail later in this chapter. Whether the inversions actually took place via homologous recombination or were generated by some other process of breakage and reunion is not known. However the inversions happened, the evolutionary reconstruction in Figure 5.9 shows that the 130–170 MY inversion breakpoint was adjacent to *UTY*, the 80–130 MY inversion breakpoint adjacent to *AMELY*, and the 30–50 MY year breakpoint adjacent to the present-day pseudoautosomal region on the short arm. As each of these inversions was fixed in the evolving Y chromosome, it removed the corresponding block of genes from the region of X–Y recombination, so that the rate of sequence divergence between the X and Y homologs accelerated. Other rearrangements in the Y chromosome, which it is not possible to trace from these data, led to some additional scrambling of the gene order in the Y chromosome.

The history of human populations can be traced through studies of the Y chromosome.

Because the Y chromosome does not undergo recombination along most of its length, genetic markers in the Y are completely linked and so remain together as the chromosome is transmitted from generation to generation. Therefore, the genetic relation between Y chromosomes can be traced, because chromosomes that are closely related will share more alleles along their length than will more distantly related chromosomes. The set of alleles at two or more loci present in a particular chromosome is called a **haplotype**. For many genealogical studies of the Y chromosome, short tandem repeat (STR) polymorphisms are convenient because of their relatively high rate of mutation and the large number of alleles. The logic is that Y chromosomes with haplotypes that share alleles at each of 20–30 STRs across the chromosome must have descended from the same ancestral Y chromosome in the very recent past. For haplotypes differing at a single locus the genetic relationship is less close, for those differing at two loci it is still less close, and so forth. This simple logic is the basis of tracing population history through Y-chromosome polymorphisms. Haplotypes that share many alleles have a more recent common ancestral Y chromosome than haplotypes that share fewer alleles. Furthermore, because the rate of STR mutation can be estimated, the time at which the ancestral chromosome existed can be deduced. This reasoning forms the basis of the estimate that the most recent common ancestor of all extant human Y chromosomes existed 50–150 thousand years ago. Such estimates are not highly precise, and there are many assumptions that must be made. Much can be learned about human population history through studies of the Y chromosome. The following discussion highlights three specific examples.

A Legacy of Genghis Khan

At its maximum extent stretching from China to Russia through to the Middle East and then into Eastern Europe, the Mongol Empire of the thirteenth century comprised the largest land empire that history has known. The founder was a man originally called Temujin, born in 1162. As a young man he organized a confederation of tribes, who around 1200 took to their small Mongolian ponies equipped with high wooden saddles and stirrups, and armed with bows and arrows began to conquer their neighbors. Soon thereafter, Temujin adopted the name Genghis Khan, which means Universal Ruler. He was often merciless, exterminating the men and boys of rebellious cities and kidnapping the women and girls. In answer to a question about the source of happiness, he is reputed to have said, "The greatest happiness is to vanquish your enemies, to chase them before you, to rob them of their wealth, to see those dear to them bathed in tears, to clasp to your bosom their wives and daughters." Through their multiple wives, concubines, and innumerable unrecorded sexual conquests, Genghis Khan and his descendants were very prolific. His eldest son Tushi had 40 acknowledged sons, and his grandson Kubilai Khan (under whom the Mongol Empire reached its maximum extent) had 22 acknowledged sons.

Although the legacy of Genghis Khan is well recorded in history, it was hardly expected that it would show up in studies of the Y chromosome. But the genotypes of 32 markers along the Y chromosome

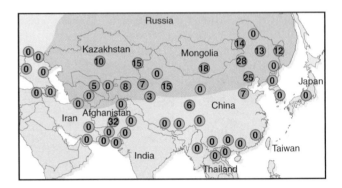

FIGURE 5.10 Percent of men with Y-chromosome haplotypes (red numbers) thought to have descended from Genghis Khan or his close male relatives. The populations sampled are near and bordering the ancient Mongol Empire. Sample sizes range from 30 to more than 60.

Data from T. Zerjal, *Am. J. Hum. Genet.* 72 (2003): 717–721.

of 2123 men sampled from throughout a large region of Asia yielded the remarkable result in **FIGURE 5.10**. Each circle represents a population sample, with its area proportion to the sample size. The red numbers denote the percentages of a group of nearly identical Y-chromosome haplotypes. The most recent common ancestor of these closely related haplotypes is estimated as existing 1000 ± 300 years ago. Furthermore, the geographical region in which the closely related haplotypes cluster is included largely within the Mongol Empire (shading). The sole exception is composed of the ethnic Hazara of Pakistan. This provides a clue to the origin of the closely related Y chromosomes, because the Hazara consider themselves to be of Mongol origin, and many claim to be direct male-line descendants of Genghis Khan. Whatever their origin, the closely related Y chromosomes are found in about 8 percent of the males throughout a large region of Asia. Direct proof of the connection with Genghis Khan could, in principle, be obtained by determining the haplotype of the Y chromosome in material recovered from his grave. He died in 1227 from injuries sustained in a fall from a horse, but his burial place is unknown.

A Legacy of the Cohanim

The Lemba are a group of about 50,000 Bantu-speaking people living predominantly in South Africa and Zimbabwe. They drew attention about 100 years ago because of their vaguely Jewish customs including dietary restrictions and male circumcision, and especially because of their oral history of their ancestors arriving by boat from a city called Sena, variously placed in Yemen, Judea, Egypt, or Ethiopia. Studies of 12 polymorphic Y-chromosome markers among 136 Lemba males from six clans has shed some light on the situation. The Y chromosomes from the Lemba derive from one of two lineages. One is closely related to the Bantu and the other is clearly Semitic. About 50 percent of the Y chromosomes of one Lemba clan (the Buba) have haplotypes closely related to a haplotype of Judaic origin called the *Cohen modal haplotype*, because it occurs primarily in the Cohanim (the plural of Cohen), the priestly lineage said to be descended from Moses's brother Aaron. Although the Cohen model haplotype has a frequency at least 50 percent in the Cohanim, it is rare in other Semitic groups. This finding affords some support for the Lemba's oral history, and the estimated time for the most recent common ancestor of the Lemba and Cohanim Y chromosomes is roughly 3000–5000 years. The earliest of these dates would be consistent with the time when the Assyrian King Shalmaneser V sent the 10 tribes of Israel into exile. Sometimes known as the "black Jews of south Africa," the Lemba are technically not Jewish. Judaism is transmitted through the maternal lineage, and Lemba tradition holds that only men survived the perilous voyage from Sena.

Origin of European Gypsies

Arriving in Eastern Europe about 1000 years ago, the Roma (Gypsies) were persecuted for centuries. They were held and bartered as slaves until the 1860s, and they were the only ethnic group besides Jews to be singled out for extermination in the Nazi death camps. Today they number more than 12 million people located in many countries around the world. Their origin has been disputed. The term "Gypsy" reflects a legend that they originated in Egypt, but their language (Romanes) has some similarities to languages of the Indian subcontinent.

Studies of the Y chromosome have clarified this situation, too. A group of closely related haplotypes was found among men in all of 14 Romani populations studied and accounted for 44.8 percent of all the Romani Y chromosomes. Elsewhere in the world, this haplotype is frequent only in the Indian subcontinent. In this study, mitochondrial DNA haplotypes were also examined. Mitochondrial DNA is also convenient for tracing population history because it does not undergo recombination and is transmitted through the female. A particular group of mitochondrial DNA haplotypes was found in 26.5 percent of the female lineages among the Romani populations. This haplotype, too, derives from the Indian subcontinent. The origin of the Y-chromosomal and mitochondrial DNA haplotypes, and the relatively high frequency of a small number of haplotypes among the Roma, are consistent with a small group of founders originating in the Indian subcontinent. Given the time of their appearance in Eastern Europe, it has been suggested that their migration was actually a flight from the armies of Mahmud of Ghazni invading from what is now Afghanistan about 1000 years ago.

5.2 Chromosome abnormalities are frequent in spontaneous abortions.

Approximately 15 percent of all recognized pregnancies in human beings terminate in spontaneous abortion, and in about half of all spontaneous abortions, the fetus has a major chromosome abnormality. **TABLE 5.1** summarizes the average rates of

TABLE 5.1 Chromosome Abnormalities per 100,000 Recognized Human Pregnancies		
Chromosome constitution	Number among spontaneously aborted fetuses	Number among live births
Normal	7500	84,450
Trisomy		
13	128	17
18	223	13
21	350	113
Other autosomes	3176	0
Sex chromosomes		
47,XYY	4	46
47,XXY	4	44
45,X	1350	8
47,XXX	21	44
Translocations		
Balanced (euploid)	14	164
Unbalanced (aneuploid)	225	52
Polyploid		
Triploid	1275	0
Tetraploid	450	0
Others (mosaics, etc.)	280	49
Total	15,000	85,000

chromosome abnormality found per 100,000 recognized pregnancies in several studies. The term **trisomic** refers to an otherwise diploid organism that has an extra copy of an individual chromosome. Many of the spontaneously aborted fetuses have trisomy of one of the autosomes. Triploids, which have three sets of chromosomes (total count 69), and tetraploids, which have four sets of chromosomes (total count 92), are also common in spontaneous abortions. Triploids and tetraploids are examples of **euploid** conditions, because they have the same relative gene dosage as found in the diploid. In contrast, relative gene dosage is upset in a trisomic, because three copies of the genes located in the trisomic chromosome are present, whereas two copies of the genes in the other chromosomes are present. Such unbalanced chromosome complements are said to be **aneuploid**. Although it is not apparent in the data in Table 5.1, in most organisms, euploid abnormalities generally have less severe phenotypic effects than aneuploid abnormalities. In Table 5.1 the term *balanced translocation* refers to a euploid condition in which nonhomologous chromosomes have an interchange of parts, but all of the parts are present; the term *unbalanced translocation* refers to an aneuploid condition in which some part of the genome is missing. The much greater survivorship of the balanced translocation indicates that a euploid chromosomal abnormality is generally less harmful than an aneuploid chromosome abnormality.

When an otherwise diploid organism has a missing copy of an individual chromosome, the condition is known as **monosomy**. In most organisms, chromosome loss (resulting in monosomy) is a more frequent event than chromosome gain (resulting in trisomy). However, monosomies are conspicuously absent in the data on spontaneous abortions in Table 5.1. Their absence is undoubtedly due to another feature of monosomy:

KEY CONCEPT

A missing copy of a chromosome (monosomy) usually results in more harmful effects than an extra copy of the same chromosome (trisomy).

In human fertilizations, monosomic zygotes are probably created in even greater numbers than trisomic zygotes, but monosomy is not found among aborted fetuses in Table 5.1 because the abortions take place so early in development that the pregnancy goes unrecognized by the mother. Data relevant to very early abortions come from medical records of women attempting to become pregnant who, while trying to conceive, undergo a pregnancy hormone test every day. The hormone assayed is human chorionic gonadotropin, a glycoprotein first produced by the embryo soon after conception at about the time of implantation in the

uterine wall. The results are that most such women conceive every month, but in 50 to 60 percent of the cases, implantation fails to occur or the embryo undergoes spontaneous abortion shortly thereafter. Given the high level of chromosomal abnormalities in the late spontaneous abortions in Table 5.1, the majority of these early spontaneous abortions are likely to have chromosomal abnormalities, primarily monosomy. These data imply a huge fetal wastage, but this serves the important biological function of eliminating many fetuses that would be grossly abnormal in their physical and mental development because of major chromosomal abnormalities.

Down syndrome results from three copies of chromosome 21.

Table 5.1 demonstrates that monosomy or trisomy of most human autosomes is incompatible with life. There are three exceptions: trisomy 13, trisomy 18, and trisomy 21. The first two are rare conditions associated with major developmental abnormalities, and the affected infants can survive for only a few days or weeks.

Trisomy 21 is **Down syndrome** (or *Down's syndrome*), which occurs in about 1 in 750 live-born children. Its major symptom is intellectual disability, but there can also be multiple physical abnormalities, such as heart defects. Affected children are small in stature because of delayed maturation of the skeletal system; their muscle tone is poor, resulting in a characteristic facial appearance; and they have a shortened life span of usually less than 50 years. Nevertheless, for a major chromosomal abnormality, the symptoms are relatively mild, and most children with Down syndrome can relate well to other people.

> Children with Down syndrome usually take great pleasure in their surroundings, their families, their toys, their playmates. Happiness comes easily, and throughout life they usually maintain a childlike good humor. They are not burdened with the grown-up cares that come to most people with adolescence and adulthood. Life is simpler and less complex. The emotions that others feel seem to be less intense for them. They are sometimes sad, happy, angry, or irritable, like everyone else, but their moods are generally not so profound and they blow away more quickly.... Children with Down syndrome, though slow, are still very responsive to their environment, to those around them, and to the affection and encouragement they receive from others.

> Quoted from D. W. Smith and A. A. Wilson. *The Child with Down's Syndrome.* (Philadelphia, PA: Saunders, 1973.)

Most cases of Down syndrome are caused by nondisjunction, which means the failure of homologous chromosomes to separate in meiosis. The result of chromosome-21 nondisjunction is one gamete that contains two copies of chromosome 21 and one that contains none. If the gamete with two copies participates in fertilization, then a zygote with trisomy 21 is produced. The gamete with no copy may also participate in fertilization, but zygotes with monosomy 21 do not survive even through the first few days or weeks of pregnancy. About three-fourths of the trisomy-21 fetuses also undergo spontaneous abortion (Table 5.1). If this were not the case, and all trisomy-21 fetuses survived to birth, the incidence of Down syndrome would rise to 1 in 250, approximately a threefold increase over the incidence actually observed.

For unknown reasons, nondisjunction of chromosome 21 is more likely to happen in oogenesis than in spermatogenesis, and so the abnormal gamete in Down syndrome is usually the egg. Furthermore, the risk of nondisjunction of chromosome 21 increases dramatically with the age of the mother, resulting in a risk of Down syndrome that approaches 4 percent in mothers age 45 and older (**FIGURE 5.11**). For this reason, many physicians recommend that older women who are pregnant have cells from the fetus tested in order to detect Down syndrome prenatally.

About 3 percent of all cases of Down syndrome are due not to simple nondisjunction but to an abnormality in chromosome structure. In these cases the risk of recurrence of the syndrome in subsequent children is very high—up to 20 percent of births. The high risk is caused by a chromosomal translocation in one of the parents, in which chromosome 21 has been broken and become attached to another chromosome.

Trisomic chromosomes undergo abnormal segregation.

In a trisomic organism, the segregation of chromosomes in meiosis is upset because the trisomic chromosome has two pairing partners instead of one.

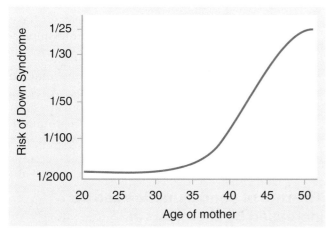

FIGURE 5.11 Risk of Down syndrome in the absence of prenatal screening as related to mother's age. The scale on the vertical axis is logarithmic. The graph is based on 9479 Down syndrome cases among 6,008,450 births in England and Wales over the period 1989–1998.

Data from J. K. Morris, D. E. Mutton, and E. Alberman, *J. Med. Screen.* 9 (2002): 2–6.

The behavior of the chromosomes in meiosis depends on the manner in which the homologous chromosome arms pair and on the chiasmata formed between them. In some cells, the three chromosomes form a **trivalent** in which distinct parts of one chromosome are paired with homologous parts of each of the others (**FIGURE 5.12**, part A). In metaphase, the trivalent is usually oriented with two centromeres pointing toward one pole and the other centromere pointing toward the other pole. The result is that at the end of both meiotic divisions, one pair of gametes contains two copies of the trisomic chromosome, and the other pair of gametes contains only a single copy. Alternatively, the trisomic chromosome can form one normal bivalent and one **univalent**, or unpaired chromosome, as shown in Figure 5.12, part B. In anaphase I, the bivalent disjoins normally and the univalent usually proceeds randomly to one pole or the other. Again, the end result is the formation of two products of meiosis that contain two copies of the trisomic chromosome and two products of meiosis that contain one copy. To state the matter in another way, a trisomic organism with three copies of a chromosome (say, C C C) will produce gametes half of which contain two copies (C C) and half of which contain one copy (C). When mated with a chromosomally normal individual, a trisomic is therefore expected to produce trisomic and normal progeny in a ratio of 1 : 1. This theoretical expectation is borne out in experimental organisms.

An extra X or Y chromosome usually has a relatively mild effect.

Many types of sex-chromosome trisomies, as a group, are even more frequent among newborns than is trisomy 21 (Table 5.1). There are two reasons why extra sex chromosomes have phenotypic effects that are relatively mild compared with those of autosomal trisomies. First, the single-active-X principle results in the silencing of most X-linked genes in all but one X chromosome in each somatic cell. Second, the Y chromosome contains relatively few functional genes.

The four most common types of sex-chromosome abnormalities are described below. The karyotypes are given in the conventional fashion, with the total number of chromosomes listed first, followed by the sex chromosomes that are present. For example, in the designation 47,XXX the number 47 refers to the total number of chromosomes, and XXX indicates that the person has three X chromosomes.

- **47,XXX** This condition is often called **trisomy-X syndrome.** People with the karyotype 47,XXX are female. Many are phenotypically normal or nearly normal, though the frequency of mild mental disability is somewhat greater than it is among 46,XX females.

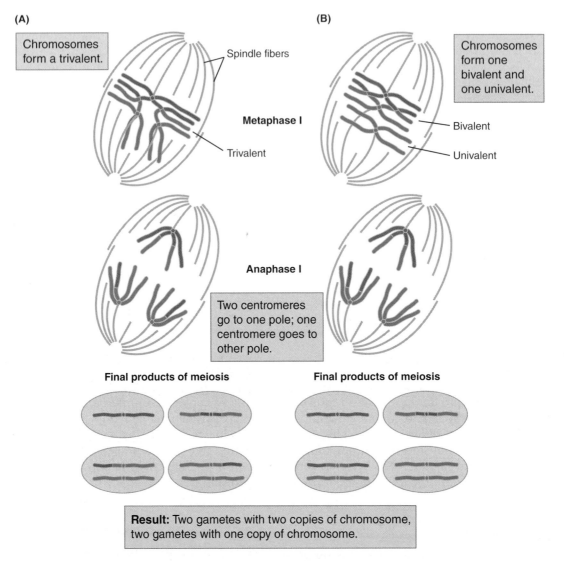

(A)

Chromosomes form a trivalent.

Spindle fibers

Metaphase I

Trivalent

(B)

Chromosomes form one bivalent and one univalent.

Bivalent

Univalent

Anaphase I

Two centromeres go to one pole; one centromere goes to other pole.

Final products of meiosis

Final products of meiosis

Result: Two gametes with two copies of chromosome, two gametes with one copy of chromosome.

FIGURE 5.12 Meiotic synapsis in a trisomic. (A) Formation of a trivalent. (B) Formation of a bivalent and a univalent. Both types of synapsis result in one pair of gametes containing two copies of the trisomic chromosome and the other pair of gametes containing one copy of the trisomic chromosome.

- **47,XYY** This condition is often called **double-Y syndrome**. These people are male and tend to be tall, but they are otherwise phenotypically normal. At one time it was thought that 47,XYY males developed severe personality disorders and were at a high risk of committing crimes of violence—a belief based on an elevated incidence of 47,XYY among violent criminals. Further study indicated that most 47,XYY males have slightly impaired mental function and that, although their rate of criminality is higher than that of normal males, the crimes are mainly nonviolent petty crimes such as theft. The majority of 47,XYY males are phenotypically and psychologically normal, have mental capabilities in the normal range, and have no criminal convictions.

- **47,XXY** This condition is called **Klinefelter syndrome**. Affected persons are male. They tend to be tall, do not undergo normal sexual

maturation, are sterile, and in some cases have enlargement of the breasts. Mild mental impairment is common.

- **45,X** Monosomy of the X chromosome in females is called **Turner syndrome**. Affected persons are phenotypically female but are short in stature and do not exhibit sexual maturation. Mental abilities are typically within the normal range.

The rate of nondisjunction can be increased by chemicals in the environment.

Because a large fraction of aneuploid zygotes terminate in miscarriage or result in congenital defects or intellectual disability, the identification of environmental hazards that may increase the incidence of meiotic errors is of great importance. Environmental risk

THE HUMAN CONNECTION

Catch 21

Jerome Lejeune, Marthe Gautier, and Raymond Turpin (1959)
National Center for Scientific Research, Paris, France

Study of the Somatic Chromosomes of Nine Down Syndrome Children [original in French]

Down syndrome had been one of the greatest mysteries in human genetics. One of the most common forms of intellectual disability, the syndrome did not follow any pattern of Mendelian inheritance. Yet, some families had two or more children with Down syndrome. (Many of these cases are now known to be due to a translocation involving chromosome 21.) This paper marked a turning point in human genetics by demonstrating that Down syndrome actually results from the presence of an extra chromosome. It was the first identified chromosomal disorder. The excerpt uses the term "telocentric," which means a chromosome that has its centromere very near one end. In the human genome,

> **66** *Analysis of the chromosome set of the "perfect" cells reveals the presence of six small telocentric chromosomes in Down syndrome boys (instead of five in the standard man) and five small telocentric chromosomes in Down syndrome girls (instead of four in the standard woman).* **99**

the smallest chromosomes are three very small, telocentric chromosomes—chromosomes 21, 22, and Y. A typical male has five small telocentrics (21, 21, 22, 22, and Y); a typical female has four (21, 21, 22, and 22). (The X chromosome is a medium-sized chromosome with its centromere somewhat off center.) In the table that follows, note the variation in chromosome counts in the "doubtful" cells. The methods for counting chromosomes were then very difficult, and many errors were made either by counting two nearby chromosomes as one or by including in the count of one nucleus a chromosome that actually belonged to a nearby nucleus. Lejeune and collaborators wisely chose

		Number of Chromosomes					
		"Doubtful" Cells			"Perfect" Cells		
		46	47	48	46	47	48
Boys	1	6	10	2	—	11	—
	2	—	2	1	—	9	—
	3	—	1	1	—	7	—
	4	—	3	—	—	1	—
	5	—	—	—	—	8	—
Girls	1	1	6	1	—	5	—
	2	1	2	—	—	8	—
	3	1	2	1	—	4	—
	4	1	1	2	—	4	—

J. Lejeune, M. Cautier, and R. Turpin, *Comptes Rendus Hebd. Seances Acad. Sci.* 248 (1959): 1721–1722.

(continues)

to ignore these doubtful counts and based their conclusion only on the "perfect" cells. Sometimes good science is a matter of knowing which data to ignore.

The culture of fibroblast cells from nine Down syndrome children reveals the presence of 47 chromosomes, the supernumerary chromosome being a small telocentric one. The hypothesis of the chromosomal determination of Down syndrome is considered. . . . The observations made in these nine cases (five boys and four girls) are recorded in the [accompanying] table.

. . . . It therefore seems legitimate to conclude that there exists in Down syndrome children a small supernumerary telocentric chromosome, accounting for the abnormal figure of 47. To explain these observations, the hypothesis of nondisjunction of a pair of small telocentric chromosomes at the time of meiosis can be considered. . . . It is, however, not possible to say that the supernumerary small telocentric chromosome is indeed a normal chromosome and at the present time the possibility cannot be discarded that a fragment resulting from another type of aberration is involved.

These results were persuasive, but they also pointed to the uncertainty involved in karyotype analysis—in some cases, the karyotypes of individual cells may be difficult or impossible to accurately determine. It is for this reason that, in modern diagnostic tests of embryonic cells, the karyotypes of multiple cells (typically 20) are usually determined.

J. Lejeune, M. Cautier, and R. Turpin, *Comptes Rendus Hebd. Seances Acad. Sci.* (1959) 248:1721–1722.

factors that have been suggested include: radiation, smoking, alcohol consumption, oral contraceptives, fertility drugs, environmental pollutants, pesticides, among others. When significant effects have been found, they are usually small and not always reproducible, due in part to confounding effects of other factors such as maternal age. In view of the maternal-age effect, the female sex hormone estrogen and molecules resembling estrogen have long been under suspicion. With this background in mind, it was no great surprise to learn that modest concentrations of a common estrogen mimic known as bisphenol A [technical name 2,2-(4,4-dihydroxy-diphenol) propane] caused about an eightfold increase in the incidence of aneuploidy in mice. Bisphenol A is the basic subunit of polycarbonate plastic products widely used as a can liner in the food and beverage industry. In its polymerized form it may be completely harmless, but the monomers can leach out of plastic products under certain conditions. It is noteworthy that detectable levels of the chemical are found in the urine of 90 percent of the U.S. population.

5.3 Chromosome rearrangements can have important genetic effects.

This section deals with abnormalities in chromosome structure. Each of the principal types of structural aberrations has characteristic genetic effects.

Chromosome aberrations were initially discovered through their genetic effects, which, though confusing at first, were eventually understood as resulting from abnormal chromosome structure. This was later confirmed directly by microscopic observations.

A chromosome with a deletion has genes missing.

A chromosome sometimes arises in which a segment is missing. Such a chromosome is said to have a **deletion** or a **deficiency**. Deletions are generally harmful to the organism, and the usual rule is the larger the deletion, the greater the harm. Very large deletions are usually lethal, even when heterozygous with a normal chromosome. Small deletions are often viable when they are heterozygous with a structurally normal homolog, because the normal homolog supplies gene products that are necessary for survival. However, even small deletions are usually homozygous-lethal (when both members of a pair of homologous chromosomes carry the deletion).

Among the *copy-number variations (CNVs)* observed in the human genome (Chapter 4), deletions account for a significant proportion. Deletions can be formed in two major ways. One is by chromosome breakage and reunion. Chromosome breaks result from double-stranded breaks in the DNA backbone. Chromosome breaks occur spontaneously at a low rate, but they can also be induced by x-rays and certain

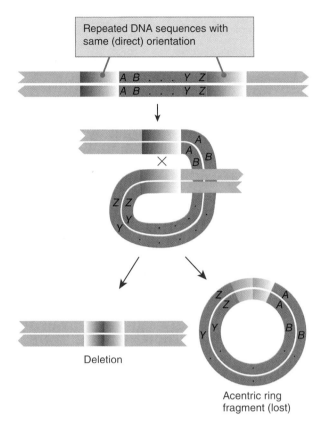

FIGURE 5.13 Ectopic recombination between direct repeats in the same DNA molecule results in deletion of the material between the repeats.

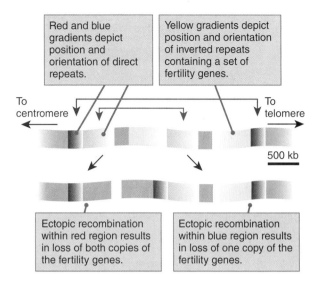

FIGURE 5.14 Ectopic recombination in the human Y chromosome results in deletion of genes affecting male fertility.

chemicals that cause double-stranded breaks in DNA. A deletion is created when a chromosome arm is broken in two places, when the broken ends bearing the centromere and the telomere fuse and the part left out remains as an acentric fragment that is lost.

Deletions can also be created by homologous recombination between repeated DNA sequences present at different sites along the DNA, a process known as **ectopic recombination**. An example is shown in **FIGURE 5.13**. In this case, each copy of the repeated DNA sequence is indicated by a color gradient. The gradient runs from left to right in both copies, which indicates that both copies of the repeated DNA sequence have the same orientation along the DNA, a configuration known as **direct repeats**. If the direct repeats undergo pairing and homologous recombination, the result is a deletion of the material between the direct repeats, because the small circular acentric fragment containing this material is lost.

Examples of deletions caused by ectopic recombination in the human Y chromosome are shown in **FIGURE 5.14**, which depicts a region of the Y chromosome that includes several large repeated sequences, shown here as gradients. The red and blue gradients indicate direct repeats, whereas the yellow gradients indicate **inverted repeats**, in which the repeated sequences are in reverse orientation. The yellow inverted repeats include genes important for male

fertility. To give an idea of scale, the red repeats are each 229 kb in length, and the region between the repeats is 3.5 Mb. As shown in the diagram, homologous recombination in the red repeats results in loss of both sets of copies of the male-fertility genes. About one in 4000 males has this deletion, which causes complete sterility. A deletion with less drastic effects results from recombination within the blue repeats, which yields a deletion of only one set of the fertility genes. Although loss of these genes does not result in complete sterility, it does impair spermatogenesis. Nevertheless about 1 percent of males have a Y chromosome with this deletion.

Deletions can be detected genetically by making use of the fact that a chromosome with a deletion no longer carries the wildtype alleles of the genes that have been eliminated. For example, in *Drosophila*, many *Notch* deletions are large enough to remove the nearby wildtype allele of *white* also. When these deleted chromosomes are heterozygous with a structurally normal chromosome carrying the recessive *w* allele, the fly has white eyes because the wildtype *w⁺* allele is no longer present in the deleted *Notch* chromosome. This "uncovering" of the recessive allele implies that the corresponding wildtype allele of *white* has also been deleted. Once a deletion has been identified, its size can be assessed genetically by determining which recessive mutations in the region are uncovered by the deletion. This method is illustrated in **FIGURE 5.15**.

Rearrangements are apparent in giant polytene chromosomes.

In the nuclei of cells in the larval salivary glands and certain other tissues of *Drosophila* and other two-winged (dipteran) flies, there are giant chromosomes, called **polytene chromosomes**, that

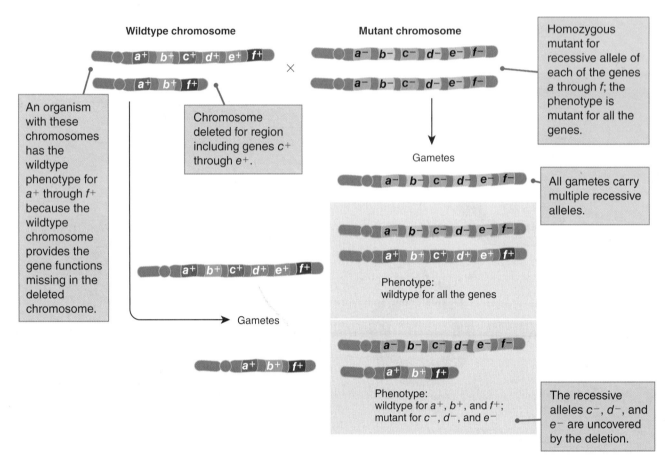

Wildtype chromosome

Mutant chromosome

> Homozygous mutant for recessive allele of each of the genes *a* through *f*; the phenotype is mutant for all the genes.

X

> An organism with these chromosomes has the wildtype phenotype for *a*+ through *f*+ because the wildtype chromosome provides the gene functions missing in the deleted chromosome.

> Chromosome deleted for region including genes *c*+ through *e*+.

Gametes

> All gametes carry multiple recessive alleles.

Gametes

Phenotype: wildtype for all the genes

Phenotype: wildtype for *a*+, *b*+, and *f*+; mutant for *c*−, *d*−, and *e*−

> The recessive alleles *c*−, *d*−, and *e*− are uncovered by the deletion.

FIGURE 5.15 Mapping of a deletion by testcrosses. The F$_1$ heterozygotes with the deletion express the recessive phenotype of all deleted genes. The expressed recessive alleles are said to be "uncovered" by the deletion.

contain about 1000 DNA molecules laterally aligned (**FIGURE 5.16**). Each of these chromosomes has a volume many times greater than that of the corresponding chromosome at mitotic metaphase in ordinary somatic cells, as well as a constant and distinctive pattern of transverse banding. The polytene structures are formed by repeated replication of the DNA in a closely synapsed pair of homologous chromosomes without separation of the replicated chromatin strands or of the two chromosomes. Polytene chromosomes are atypical chromosomes and are formed in "terminal" cells; that is, the larval cells containing them do not divide further during the development of the fly and are later eliminated in the formation of the pupa. However, the polytene chromosomes have been especially valuable in the genetics of *Drosophila* and are ideal for the study of chromosome rearrangements.

About 5000 darkly staining transverse bands have been identified in the polytene chromosomes of *D. melanogaster*. The linear array of bands, which has a pattern that is constant and characteristic for each species, provides a finely detailed **cytological map** of the chromosomes. The banding pattern is such that short regions in any of the chromosomes can be identified. Because of their large size and finely detailed morphology, polytene chromosomes are exceedingly useful for the study of deletions and other chromosome aberrations. For example, all the *Notch* deletions cause particular bands to be missing in the salivary chromosomes. Physical mapping of deletions also allows particular genes, otherwise known only from genetic studies, to be assigned to specific bands or regions in the salivary chromosomes.

Physical mapping of genes in part of the *Drosophila* X chromosome is illustrated in Figure 5.16. The banded chromosome is shown, and beneath it are the designations of the individual bands. On the average, each band contains about 20 kb of DNA, but there is considerable variation in DNA content from band to band. The mutant X chromosomes labeled I through VI in the figure have deletions. The deleted part of each chromosome is shown with dashes. These deletions define regions along the chromosome, some of which correspond to specific bands. For example, the deleted region in both chromosomes I and II that is

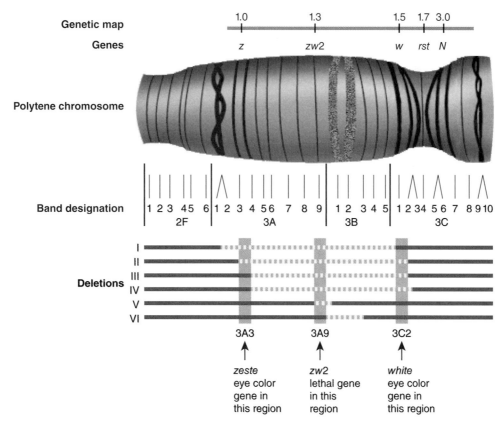

Genetic map

Genes

Polytene chromosome

Band designation

Deletions

3A3
↑
zeste
eye color
gene in
this region

3A9
↑
zw2
lethal gene
in this
region

3C2
↑
white
eye color
gene in
this region

FIGURE 5.16 Part of the X chromosome in polytene salivary gland nuclei of *Drosophila melanogaster* and the extent of six deletions (I–VI) in a set of chromosomes. Any recessive allele that is uncovered by a deletion must be located inside the boundaries of the deletion. This principle can be used to assign genes to specific bands in the chromosome.

present in all the other chromosomes consists of band 3A3. In crosses, only deletions I and II uncover the mutation *zeste (z)*, so the *z* gene must be in band 3A3, as indicated at the top. Similarly, the recessive-lethal mutation *zw2* is uncovered by all deletions except VI; therefore, the *zw2* gene must be in band 3A9. As a final example, the *w* mutation is uncovered only by deletions II, III, and IV; thus the *w* gene must be in band 3C2. The *rst* (rough eye texture) and *N* (notched wing margin) genes are not uncovered by

any of the deletions. These genes were localized by a similar analysis of overlapping deletions in regions 3C5 to 3C10.

A chromosome with a duplication has extra genes.

Some abnormal chromosomes have a region that is present twice. These chromosomes are said to have a **duplication**. A **tandem duplication** is one in which the duplicated segment is present in the same orientation immediately adjacent to the normal region in the chromosome. Tandem duplications are able to produce even more copies of the duplicated region by means of a process called **unequal crossing-over**, which is actually a type of ectopic recombination. Part A of **FIGURE 5.17** illustrates the chromosomes in meiosis of an organism that is homozygous for a tandem duplication (brown region). When they undergo synapsis, these chromosomes can mispair with each other, as illustrated in part B. A crossover within the mispaired part of the duplication (part C) will thereby produce a chromatid carrying three copies of the region, as well as a reciprocal product containing a single copy (part D).

⚛ STOP & THINK 5.2

Recessive alleles of four genes (*a*, *b*, *c*, and *d*) are clustered along a small region of chromosome, but their order is unknown. The mutants are analyzed by being made heterozygous with each of two deletions. Deletion 1 uncovers the alleles *a*, *b*, and *d*. Deletion 2 uncovers alleles *b*, *c*, and *d*. Which genes are at the ends of the cluster? What gene orders are compatible with the deletion data?

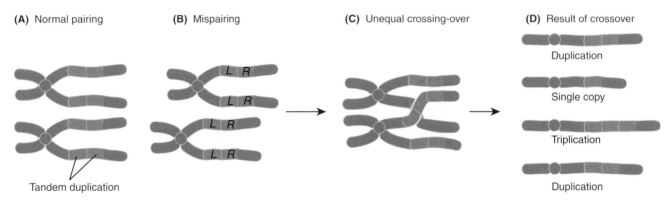

(A) Normal pairing **(B)** Mispairing **(C)** Unequal crossing-over **(D)** Result of crossover

Duplication

Single copy

Triplication

Duplication

Tandem duplication

FIGURE 5.17 An increase in the number of copies of a chromosome segment resulting from unequal crossing-over of tandem duplications (brown). (A) Normal synapsis of chromosomes with a tandem duplication. (B) Mispairing. The right-hand element of the lower chromosome is paired with the left-hand element of the upper chromosome. (C) Crossing-over within the mispaired duplication, which is called unequal crossing-over. (D) The outcome of unequal crossing-over.

Human color-blindness mutations result from unequal crossing-over.

Human color vision is mediated by three light-sensitive protein pigments present in the cone cells of the retina. Each of the pigments is related to *rhodopsin*, the pigment found in the rod cells that mediates vision in dim light. The light sensitivities of the cone pigments are toward blue, red, and green. These are our primary colors. We perceive all other colors as mixtures of these primaries. The gene for the blue-sensitive pigment is in chromosome 7, whereas the genes for the red and green pigments are in the X chromosome near the tip of the long arm, separated by less than 5 cM (roughly 5 Mb of DNA). Because the red and green pigments arose from the duplication of a single ancestral pigment gene and are still 96 percent identical in amino acid sequence, the genes are similar enough that they can pair and undergo unequal crossing-over. The process of unequal crossing-over is the genetic basis of red–green color blindness.

Almost everyone is familiar with **red–green color blindness**; it is one of the most common inherited conditions in human beings (**FIGURE 5.18**). Approximately 5 percent of males have some form of red–green color blindness. The preponderance of affected males immediately suggests X-linked inheritance, which is confirmed by pedigree studies. Affected males have normal sons and carrier daughters, and the carrier daughters have 50 percent affected sons and 50 percent carrier daughters.

Several distinct varieties of red–green color blindness can be distinguished. Defects in red vision go by the names of *protanopia*, an inability to perceive red, and *protanomaly*, an impaired ability to perceive red. The comparable defects in green perception are called *deuteranopia* and *deuteranomaly*, respectively. Isolation of the red-pigment and green-pigment genes and study of their organization in people with normal and defective color vision have indicated quite clearly

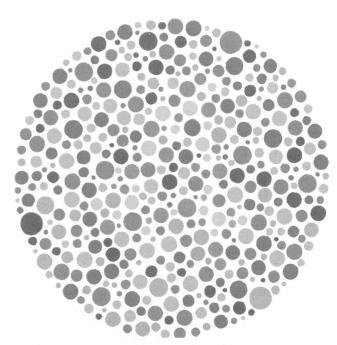

FIGURE 5.18 A standard color chart used in initial testing for color blindness. The pattern tests for an inability to distinguish red from green. Those with red–green color blindness will not be able to distinguish the red dots from the green and therefore will not see the red number.

© LuckyBall/Shutterstock.

how the "-opias" and "-omalies" differ; they have also explained why the frequency of color blindness is so relatively high.

The organization of the red-pigment and green-pigment genes in men with normal vision is illustrated in part A of **FIGURE 5.19**. Unexpectedly, a significant proportion of normal X chromosomes contain two or three green-pigment genes. How these arise by unequal crossing-over is shown in part B. The red-pigment and green-pigment genes pair, and the crossover takes place in the region of homology *between* the genes. The result is a duplication of the green-pigment gene in one chromosome and a deletion of the green-pigment gene in the other.

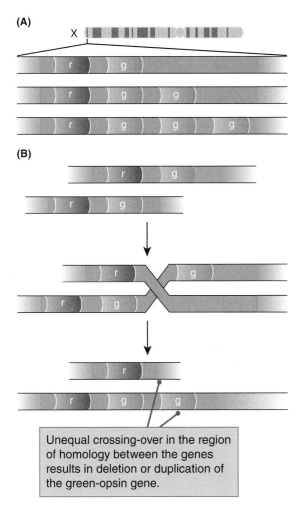

(A)

X

(B)

Unequal crossing-over in the region of homology between the genes results in deletion or duplication of the green-opsin gene.

FIGURE 5.19 (A) Organization of red-pigment and green-pigment genes in three wildtype X chromosomes. (B) Origin of multiple green-pigment genes by unequal crossing-over.

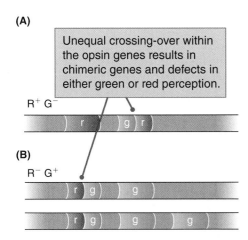

(A)

Unequal crossing-over within the opsin genes results in chimeric genes and defects in either green or red perception.

R⁺ G⁻

(B)

R⁻ G⁺

FIGURE 5.20 Genetic basis of absent or impaired red–green color vision. (A) Defects in green vision. (B) Defects in red vision.

The recombinational origin of the defects in color vision are illustrated in **FIGURE 5.20**. The top chromosome in part A is the result of deletion of the green-pigment gene shown in part B of Figure 5.19. Males with such an X chromosome have deuteranopia,

or "green blindness." Other types of abnormal pigments result when crossing-over takes place within mispaired red-pigment and green-pigment genes. Crossing-over between the genes yields a **chimeric gene**, which is a composite gene, part of one joined with part of the other. The chimeric gene in part A of Figure 5.20 joins the 5′ end of the green-pigment gene with the 3′ end of the red-pigment gene. If the crossover point is toward the 5′ end of the gene, the resulting chimeric gene is mostly "red" in sequence, and hence the chromosome will cause deuteranopia or "green blindness." However, if the crossover point is near the 3′ end of the gene, most of the green-pigment gene remains intact, and the chromosome will cause deuteranomaly.

Chromosomes associated with defects in red vision are illustrated in part B of Figure 5.20. The chimeric genes are the reciprocal products of the unequal crossovers that yield defects in green vision. In this case, the chimeric gene consists of the red-pigment gene at the 5′ end and the green-pigment gene at the 3′ end. If the crossover point is near the 5′ end, most of the red-pigment gene is replaced with the green-pigment gene. The result is protanopia, or "red blindness." The same is true of the other chromosome indicated in Figure 5.20, part B. However, if the crossover point is near the 3′ end, then most of the red-pigment gene remains intact and the result is protanomaly.

Some reciprocal deletions and duplications are associated with reciprocal risks of autism and schizophrenia.

The human genome contains repetitive DNA sequences at thousands of different locations that are similar enough to undergo recombination if they become paired in meiosis. Fortunately, chromosome pairing in meiosis is usually precise enough to keep mispairing at a very low level. But mispairing does sometimes take place, and when ectopic recombination occurs in the mispaired repeats, chromosome abnormalities result.

FIGURE 5.21, Part A shows a situation with two repeats along a chromosome flanking a region of unique sequence containing genes called *X, Y,* and *Z.* Normally they pair correctly, but occasionally they may pair as shown in Figure 5.21, part B, allowing a crossover to take place. (For simplicity, only the chromatids involved in the crossover are shown.) The result is a gamete that carries a duplication of *X, Y,* and *Z,* as well as a gamete that carries a deletion of these genes (Figure 5.21, part C). When these gametes undergo fertilization with a gamete carrying the normal homologous chromosome, one class of zygotes has three copies of genes *X, Y,* and *Z* and the other class has only one copy of genes *X, Y,* and *Z* (Figure 5.21, part D).

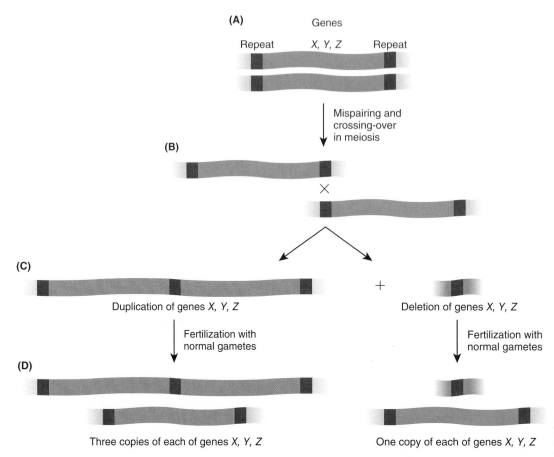

FIGURE 5.21 Origin of chromosomes bearing a duplication or deletion of a genomic region by means of unequal crossing-over between repeated DNA sequences that flank the region. The repeated sequences are indicated in red.

Unequal crossing-over of this sort is one mechanism for producing gene copy-number variation (CNV).

Many CNVs have no obvious phenotypic manifestation, but some of them do. One particularly interesting group consists of those associated with two seemingly distinct phenotypes, one phenotype connected with the CNV duplication and the other with the CNV deletion. Among these are CNVs that are associated with autism spectrum disorder and schizophrenia.

Autism comes from the Greek word for "self," and, in fact, autistic children seem focused inward. The average age at which the disease is diagnosed is 3–4 years, and about 1 in 250 to 1 in 100 children are affected, according to how strictly the condition is defined. The term *autism spectrum disorder* is often used because there is a wide spectrum of breadth and severity of the symptoms. Autism is a disorder of brain function and development with primary effects on communication skills, social skills, repetitive behavior, and fear of change (**FIGURE 5.22**). Affected children may not speak or have delayed speech. They often avoid eye contact, have a reduced sense of self, and may refer to themselves in the third person. They have difficulty in understanding and reacting to the thoughts and intentions expressed by others (impaired

mentalistic skills), and sometimes laugh and giggle at inappropriate times. Their primary emotions are basic ones like fear and anger. They may struggle with logical connections, but some are brilliant in mathematics, music, or other highly specialized skills. Many have difficulty setting goals for themselves and then following through on their pursuit. Bear in mind that these symptoms vary widely in their presence and severity and may change with age. Autism spectrum disorder is a complex disease expressed in many different ways.

Schizophrenia is also a complex disease that shows wide variation in symptoms and severity. The condition is usually diagnosed between ages 15–30, and it affects about 1 in 250 people. Many of the symptoms of schizophrenia contrast sharply with those of autism spectrum disorder (Figure 5.22). Patients may hear voices owing to auditory hallucinations, have an elevated sense of self expressed as an exaggerated feeling of power or influence, imagine conspiracies against them, have emotional swings between depression and elation, make strange and illogical connections or speak gibberish, and pursue certain goals obsessively. As with autism spectrum disorder, there is great variation in the number and severity of these symptoms among individuals.

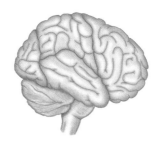

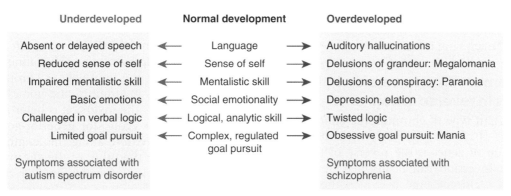

Underdeveloped	Normal development	Overdeveloped
Absent or delayed speech ←	Language →	Auditory hallucinations
Reduced sense of self ←	Sense of self →	Delusions of grandeur: Megalomania
Impaired mentalistic skill ←	Mentalistic skill →	Delusions of conspiracy: Paranoia
Basic emotions ←	Social emotionality →	Depression, elation
Challenged in verbal logic ←	Logical, analytic skill →	Twisted logic
Limited goal pursuit ←	Complex, regulated goal pursuit →	Obsessive goal pursuit: Mania
Symptoms associated with autism spectrum disorder		Symptoms associated with schizophrenia

FIGURE 5.22 Comparisons of some of the major symptoms of autism spectrum disorder and schizophrenia highlighting those that resemble polar opposites.

Based on material provided by Bernard Crespi.

TABLE 5.2 Reciprocal deletions and duplications showing reciprocal risks for autism spectrum disorder and schizophrenia

	Autism spectrum		Schizophrenia		
CNV region	Deletion	Duplication	Deletion	Duplication	Odds[‡]
1q21.1	2[†]	10	15	4	1/1000
16p11.2	14	5	5	24	1/10,000
22q11.21	1	8	16	1	1/20,000
22q13.3	5	0	0	4	1/125

[†]Entries are number of individuals in each category for each CNV.
[‡]Odds are the likelihood that chance alone could account for an association between duplication/deletion and schizophrenia/autism as strong or stronger than that actually observed.
Data from B. Crespi, P. Stead and M. Elliot. 2010. *Proc. Natl. Acad. Sci. USA* 107: 1736.

The contrasts in Figure 5.22 could be dismissed as mere word play, were it not for observations like those in **TABLE 5.2**. The table lists four regions of the genome in which CNVs occur repeatedly because they are flanked by repeats, as in Figure 5.21. Each is associated with autism spectrum disorder or schizophrenia in a reciprocal fashion (red numerals): If the deletion associates with autism spectrum, then the duplication associates with schizophrenia, and the other way around. The odds in the final column are the probabilities that an association as strong or stronger would be observed by chance alone, and in each case the results are highly unlikely to be due to chance.

What does this kind of reciprocal association imply? It implies that there is a certain stage or stages in the development of the disorders in which the same gene product has opposite effects. Hence, enhanced expression in the duplication predisposes to one disorder, and reduced expression in the deletion predisposes to the other. A similar reciprocal association to that observed for CNV is found among a number of individual genes implicated in the disorders. Importantly, not all CNVs or genes that show an association with autism spectrum disorder show a reciprocal association with schizophrenia, and not even the majority show such an association. Nevertheless, the finding that some CNVs and genes do show such opposite effects indicate the presence of shared causal links in the development of these conditions.

The results in Table 5.2 do *not* mean that autism spectrum disorder or schizophrenia are determined genetically by just a handful of genes. Both disorders are extremely complex, and there may well be different types of both diseases that masquerade with nearly the same symptoms. Environmental affects are also important, including stressors in early development, inadequate diet and nutrition, hormonal and chemical imbalance, and even viral infections.

A chromosome with an inversion has some genes in reverse order.

Another important type of chromosome abnormality is an **inversion**, a chromosome in which the linear order of a group of genes is the reverse of the normal order. An inversion can be formed by a two-break event in a chromosome in which the middle segment is reversed in orientation before the breaks are healed. An inversion can also be created by ectopic recombination between DNA sequences that are inverted repeats, as illustrated for the DNA duplexes in **FIGURE 5.23**. In this diagram the differently colored gradients represent the inverted repeats, and the letters represent the order of genes in the region between the inverted repeats. Ectopic recombination between the repeats results in a chromosome with an inversion in the order of the genes between the repeats.

In an organism that is heterozygous for an inversion, one chromosome is structurally normal (wildtype) and the other carries an inversion.

These chromosomes pass through mitosis without difficulty, because each chromosome duplicates and its chromatids are separated into the daughter cells without regard to the other chromosome.

Although a heterozygous inversion causes no problems in mitosis, there can be problems in meiosis. These result from homologous recombination in the region that is inverted. The reason is that in prophase I of meiosis the homologous chromosomes are attracted gene for gene in the process of *synapsis*. The pairing of homologous chromosomes is shown for a heterozygous inversion in **FIGURE 5.24**. In this diagram the gradients represent the orientation of the DNA sequences along the homologous chromosomes. The region in blue is inverted in one homolog but not in the other, and so the inversion is heterozygous. In an inversion heterozygote, for gene-for-gene pairing to take place everywhere along the length of the chromosome, one or the other of the chromosomes must twist into a loop in the region in which the gene order is inverted. In Figure 5.24, it is the structurally normal chromosome that is shown as looped, but in other cells it may be the inverted chromosome that is looped. In either case, the loop is called an **inversion loop**.

The inversion loop itself does not create a problem. The looping apparently takes place without difficulty and can be observed through the microscope. As long as there is no crossing-over within the inversion, the homologous chromosomes can separate normally

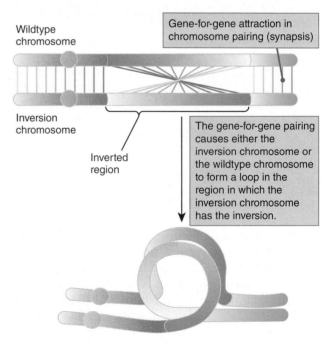

FIGURE 5.24 In an organism that carries a chromosome that is structurally normal along with a homologous chromosome with an inversion, the gene-for-gene attraction between the chromosomes during synapsis causes one of the chromosomes to form into a loop in the region in which the gene order is inverted. In this example, the structurally normal chromosome forms the loop. Only two of the four chromatids are shown.

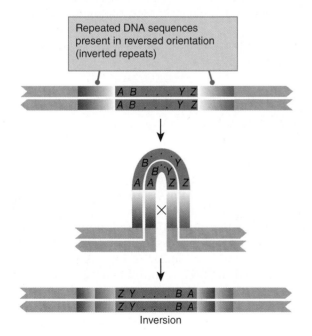

FIGURE 5.23 Ectopic recombination between inverted repeats results in an inversion.

(A) Paracentric inversion

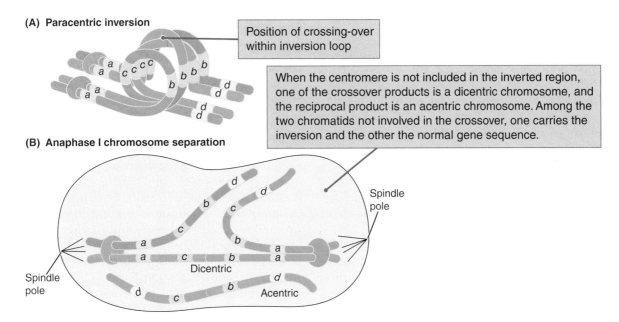

Position of crossing-over within inversion loop

When the centromere is not included in the inverted region, one of the crossover products is a dicentric chromosome, and the reciprocal product is an acentric chromosome. Among the two chromatids not involved in the crossover, one carries the inversion and the other the normal gene sequence.

(B) Anaphase I chromosome separation

Spindle pole

Spindle pole

Dicentric

Acentric

FIGURE 5.25 (A) Synapsis between homologous chromosomes, one of which contains an inversion that does not include the centromere. There is a crossover within the inversion loop. (B) Anaphase I configuration resulting from the crossover.

at anaphase I. When there is crossing-over within the inversion loop, the chromatids involved in the crossover become physically joined, and the result is the formation of chromosomes containing large duplications and deletions. **FIGURE 5.25** shows an example of pairing in a heterozygous **paracentric inversion**, which means that the centromere is not included within the inverted region. (The prefix *para-* means "beside" the inverted region.) The products of the crossover can be deduced from Figure 5.25 by tracing along the chromatids in part A. The outer chromatids are the ones that do not participate in the crossover. One of these contains the inverted sequence and the other the normal sequence, as shown in part B. Because of the crossover, the inner chromatids, which did participate in the crossover, are connected. If the centromere is not included

in the inversion loop, as is the case here, the result is a dicentric chromosome. The reciprocal product of the crossover is an acentric chromosome. Neither the dicentric chromosome nor the acentric chromosome can be included in a normal gamete. The acentric chromosome is usually lost because it lacks a centromere and, in any case, has a deletion of the *a* region and a duplication of the *d* region. The dicentric chromosome is also often lost because it is held on the meiotic spindle by the chromatid bridging between the centromeres; in any case, this chromosome is deleted for the *d* region and duplicated for the *a* region. Hence, when there is a crossover in the inversion loop, the only chromatids that can be recovered in the gametes are the chromatids that did not participate in the crossover. One of these carries the inversion and the other does not.

STOP & THINK 5.3

The accompanying illustration is a simplified version of pairing between an inversion-bearing chromosome and its normal homolog, in which only the inverted region is shown as paired. In this kind of diagram, it is straightforward to work out the consequences of double crossovers within the inverted region. Suppose that a double crossover occurs with one crossover between genes *B* and *C* and the other between genes *C* and *D*. What are the consequences if the double crossover is a two-strand double crossover? What are the consequences if the double crossover is a four-strand double crossover?

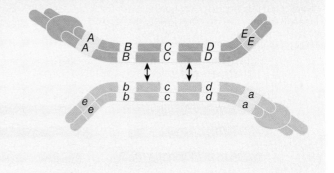

When the inversion does include the centromere, it is called a **pericentric inversion**, which means "around" (*peri-*) the centromere. Chromatids with duplications and deficiencies are also created by crossing-over within the inversion loop of a pericentric inversion, but in this case the crossover products are monocentric. The situation is illustrated in **FIGURE 5.26**, part A. The diagram is identical to that in Figure 5.25 except for the position of the centromere. The products of crossing-over can again be deduced by tracing the chromatids. In this case, both products of the crossover are monocentric, but one chromatid carries a duplication of *a* and a deletion of *d*, and the other carries a duplication of *d* and a deletion of *a* (Figure 5.26, part B). Although either of these chromosomes could be included in a gamete, the duplication and deficiency usually cause inviability. Thus, as with the paracentric inversion, the products of recombination are not recovered, but for a different reason. Among the chromatids that do not participate in the crossing-over in part A of Figure 5.26, one carries the pericentric inversion and the other has the normal sequence.

Reciprocal translocations interchange parts between nonhomologous chromosomes.

A chromosomal aberration resulting from the interchange of parts between nonhomologous chromosomes is called a **translocation**. Translocations can be formed by chromosome breakage and reunion. They can also be formed by ectopic recombination between copies of repeated sequences present in nonhomologous chromosomes. In **FIGURE 5.27**, organism A is homozygous for two pairs of structurally normal chromosomes. Organism B contains one structurally normal pair of chromosomes and another pair of chromosomes that have undergone an interchange of terminal parts. This organism is said to be *heterozygous* for the translocation. The translocation is properly called a **reciprocal translocation** because it consists of two reciprocally interchanged parts. As indicated in part C, an organism can also be homozygous for a translocation if both pairs of homologous chromosomes undergo an interchange of parts.

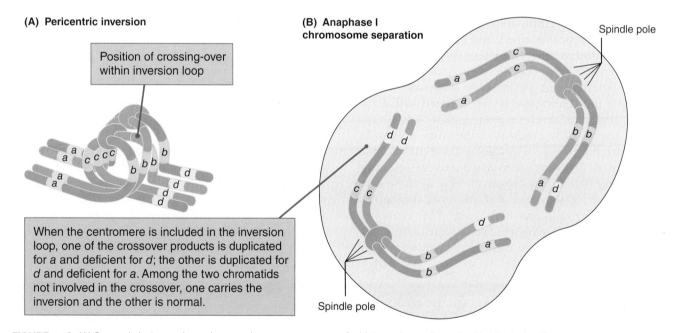

(A) Pericentric inversion

Position of crossing-over within inversion loop

When the centromere is included in the inversion loop, one of the crossover products is duplicated for *a* and deficient for *d*; the other is duplicated for *d* and deficient for *a*. Among the two chromatids not involved in the crossover, one carries the inversion and the other is normal.

(B) Anaphase I chromosome separation

Spindle pole

Spindle pole

FIGURE 5.26 (A) Synapsis between homologous chromosomes, one of which carries an inversion that includes the centromere. A crossover within the inversion loop is shown. (B) Anaphase I configuration resulting from the crossover.

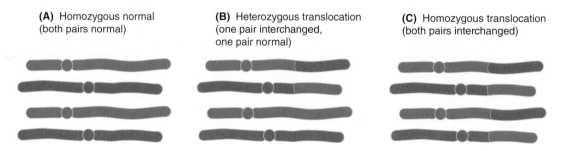

(A) Homozygous normal (both pairs normal)

(B) Heterozygous translocation (one pair interchanged, one pair normal)

(C) Homozygous translocation (both pairs interchanged)

FIGURE 5.27 (A) Two pairs of nonhomologous chromosomes in a diploid organism. (B) Heterozygous reciprocal translocation, in which two nonhomologous chromosomes (the two at the top) have interchanged terminal segments. (C) Homozygous reciprocal translocation.

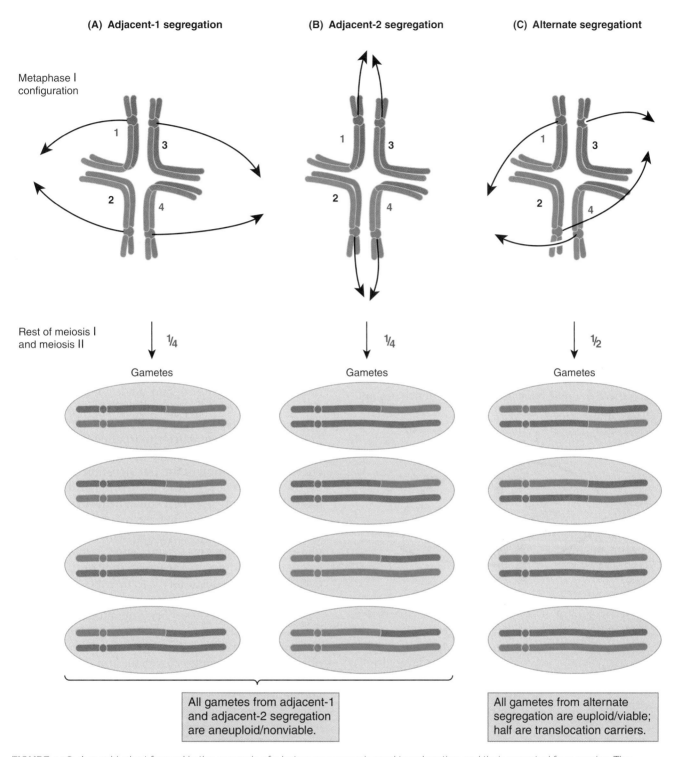

FIGURE 5.28 A quadrivalent formed in the synapsis of a heterozygous reciprocal translocation and their expected frequencies. The translocated chromosomes are numbered in red, their normal homologs in black. No chiasmata are shown. (A) Adjacent-1 segregation: homologous centromeres separate at anaphase I; all of the resulting gametes have a duplication of one terminal segment and a deficiency of the other. (B) Adjacent-2 segregation: homologous centromeres go together at anaphase I; all of the resulting gametes have a duplication of one basal segment and a deficiency of the other. (C) Alternate segregation: half of the gametes receive both parts of the reciprocal translocation and the other half receive both normal chromosomes.

An organism that is heterozygous for a reciprocal translocation usually produces only about half as many offspring as normal, which is called **semisterility**. The reason for the semisterility is difficulty in chromosome segregation in meiosis. When meiosis takes place in a translocation heterozygote, the normal and translocated chromosomes must undergo synapsis as shown in **FIGURE 5.28**. Ordinarily, there would also be chiasmata between nonsister chromatids in the arms of the homologous chromosomes, but these are not shown, as if the translocation were present in an organism with no crossing-over, such as a male *Drosophila*. Segregation from this configuration can take place in any of three ways. In the list that follows, the notation 1 + 2 ↔ 3 + 4 means that at the first meiotic anaphase, the chromosomes in Figure 5.28 labeled 1 and 2 go to one pole and those labeled 3 and 4 go to the opposite pole. The red numbers 1 and 4 indicate the two parts of the reciprocal translocation. The three types of segregation are

- **1 + 2 ↔ 3 + 4** This mode is called **adjacent-1 segregation**. Homologous centromeres go to opposite poles, but each normal chromosome goes with one part of the reciprocal translocation. All gametes formed from adjacent-1 segregation have a large duplication and deficiency for the distal part of the translocated chromosomes. (The *distal* part of a chromosome is the part farthest from the centromere.) The pair of gametes that originate from the 1 + 2 pole are duplicated for the distal part of the blue chromosome and deficient for the distal part of the red chromosome; the pair of gametes from the 3 + 4 pole have the reciprocal deficiency and duplication.

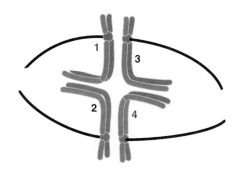

- **1 + 3 ↔ 2 + 4** This mode is **adjacent-2 segregation**, in which homologous centromeres go to the same pole at anaphase I. In this case, all gametes have a large duplication and deficiency of the proximal part of the translocated chromosome. (The *proximal* part of a chromosome is the part closest to the centromere.) The pair of gametes from the 1 + 3 pole have a duplication of the proximal part of the red chromosome and a deficiency of the proximal part of the blue chromosome; the pair of gametes from the 2 + 4 have the complementary deficiency and duplication.

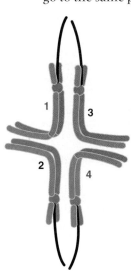

- **1 + 4 ↔ 2 + 3** In this type of segregation, which is called **alternate segregation**, the gametes are all balanced (euploid), which means that none has a duplication or a deficiency. The gametes from the 1 + 4 pole have both parts of the reciprocal translocation; those from the 2 + 3 have both normal chromosomes.

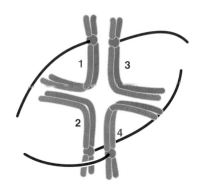

The semisterility of genotypes that are heterozygous for a reciprocal translocation results from lethality due to the duplication and deficiency gametes produced by adjacent-1 and adjacent-2 segregation. Although the expected frequencies of adjacent-1 : adjacent-2 : alternate segregation are approximately 1/4 : 1/4 : 1/2, in practice the frequency with which these types of segregation take place is strongly influenced by the position of the translocation breakpoints, by the number and distribution of chiasmata in the interstitial region between the centromere and each breakpoint, and by whether the quadrivalent tends to open out into a ring-shaped structure on the metaphase plate. Adjacent-1 segregation is usually quite frequent and adjacent-2 segregation is rare, but whatever the ratios, semisterility is to be expected from virtually all translocation heterozygotes.

Translocation semisterility is manifested in different life-history stages in plants and animals. Plants have an elaborate gametophyte phase of the life cycle—a haploid phase in which complex metabolic and developmental processes are necessary. In plants, large duplications and deficiencies are usually lethal in the gametophyte stage. Because the gametophyte produces the gametes, in higher plants the semisterility is manifested as pollen or seed lethality. In animals, by contrast, minimal gene activity is necessary in the gametes, which function in spite of very large duplications and deficiencies. In animals, therefore, the semisterility is usually manifested as zygotic lethality.

A special type of *non*reciprocal translocation is a **Robertsonian translocation**, in which two nonhomologous chromosomes undergo fusion of their short arms yielding a chromosome with a single functional centromere (**FIGURE 5.29**). Robertsonian translocations are an important risk factor to be considered in Down syndrome. When chromosome 21 is one of the

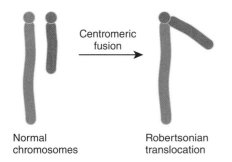

Centromeric fusion

Normal chromosomes

Robertsonian translocation

FIGURE 5.29 Formation of a Robertsonian translocation by fusion of two acrocentric chromosomes in the centromeric region.

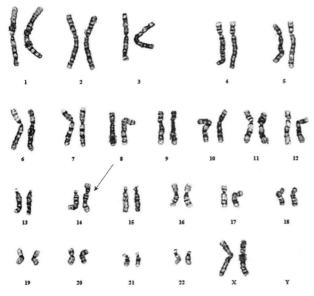

FIGURE 5.30 A karyotype of a child with Down syndrome, carrying a Robertsonian translocation of chromosomes 14 and 21 (arrow).

Courtesy of Viola Freeman, Associate Professor, Faculty of Health Sciences, Department of Pathology and Molecular Medicine, McMaster University.

acrocentrics in a Robertsonian translocation, the rearrangement leads to a familial type of Down syndrome. An example in which chromosome 21 is joined with chromosome 14 is shown in **FIGURE 5.30** (arrow). The heterozygous carrier is phenotypically normal, but a high risk of Down syndrome results from aberrant segregation in meiosis. Approximately 3 percent of children with Down syndrome are found to have one parent with such a translocation.

🧠 STOP & THINK 5.4

For an individual who is heterozygous for a reciprocal translocation, what is the ratio of unbalanced gametes to balanced gametes if the ratio of adjacent-1 : adjacent-2 : alternate segregation is 1 : 1 : 2? What is the ratio of unbalanced : balanced gametes if the ratio of the three types of segregation is 1 : 0 : 1?

5.4 Polyploid species have multiple sets of chromosomes.

The genus *Chrysanthemum* illustrates **polyploidy**, an important phenomenon found frequently in higher plants. In polyploidy, a species has a genome composed of multiple complete sets of chromosomes. One *Chrysanthemum* species, a diploid species, has 18 chromosomes. A closely related species has 36 chromosomes. However, comparison of chromosome morphology indicates that the 36-chromosome species has two complete sets of the chromosomes found in the 18-chromosome species (**FIGURE 5.31**). The basic chromosome set in the group, from which all the other genomes are formed, is called the **monoploid** chromosome set. In *Chrysanthemum*, the monoploid chromosome number is 9. The diploid species has two complete copies of the monoploid set, or 18 chromosomes altogether. The 36-chromosome species has four copies of the monoploid set ($4 \times 9 = 36$) and is a **tetraploid**. The horticulturalist's *Chrysanthemum* has 54 chromosomes (6×9, constituting the **hexaploid**). Other species have 72 chromosomes (8×9, the *octoploid*), and 90 chromosomes (10×9, the *decaploid*).

In meiosis, the chromosomes of all *Chrysanthemum* species synapse normally in pairs to form bivalents. The 18-chromosome species forms 9 bivalents, the 36-chromosome species forms 18 bivalents, the 54-chromosome species forms 27 bivalents, and so forth. Gametes receive one chromosome from each bivalent, so the number of chromosomes in the gametes of any species is exactly half the number of chromosomes in its somatic cells. The chromosomes present in the gametes of a species constitute the **haploid** set of chromosomes. In the species of *Chrysanthemum* with 54 chromosomes, for example, the haploid chromosome number is 27; in meiosis, 27 bivalents are formed, and so each gamete contains 27 chromosomes. When two such gametes come together in fertilization, the complete set of 54 chromosomes in the species is restored. Thus the gametes of a polyploid organism are not always monoploid, as they are in a diploid organism; for example, a tetraploid organism has diploid gametes.

The distinction between the term *monoploid* and the term *haploid* is subtle:

- The *monoploid* chromosome set is the basic set of chromosomes that is multiplied in a polyploid series of species, such as *Chrysanthemum*.

- The *haploid* chromosome set is the set of chromosomes present in a gamete, irrespective of the chromosome number in the species.

Polyploidy is widespread in certain plant groups. Among various groups of flowering plants, 30 to 80 percent of existing species are thought to have originated as some form of polyploid. Valuable agricultural

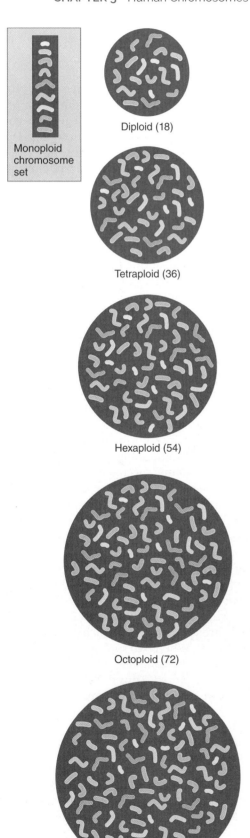

Monoploid chromosome set

Diploid (18)

Tetraploid (36)

Hexaploid (54)

Octoploid (72)

Decaploid (90)

FIGURE 5.31 Chromosome numbers in diploid and polyploid species of *Chrysanthemum*. Each set of homologous chromosomes is depicted in a different color.

crops that are polyploid include wheat, oats, cotton, potatoes, bananas, coffee, sugar cane, peanuts, and apples. Polyploidy often leads to an increase in the size of individual cells, and polyploid plants are often larger and more vigorous than their diploid ancestors; however, there are many exceptions to these generalizations. Polyploidy is rare in vertebrate animals, but it is found in a few groups of invertebrates. One reason why polyploidy is rare in animals is the difficulty in regular segregation of the sex chromosomes. For example, a tetraploid animal with XXXX females and XXYY males would produce XX eggs and XY sperm (if all chromosomes paired to form bivalents), so the progeny would be exclusively XXXY and unlike either of the parents.

Polyploid plants found in nature nearly always have an even number of sets of chromosomes, because organisms with an odd number have low fertility. Organisms with three monoploid sets of chromosomes are known as **triploids**. As far as growth is concerned, a triploid is quite normal, because the triploid condition does not interfere with mitosis; in mitosis in triploids (or any other type of polyploid), each chromosome replicates and divides just as in a diploid. However, because each chromosome has more than one pairing partner, chromosome segregation is severely upset in meiosis, and most gametes are defective. Unless the organism can perpetuate itself by means of asexual reproduction, it will eventually become extinct.

The infertility of triploids is sometimes of commercial benefit. For example, the seeds of "seedless" watermelons are small and edible because the plant is triploid and most of the seeds fail to develop to full size. In oysters, triploids are produced by treating fertilized diploid eggs with a chemical that causes the second polar body of the egg to be retained. The triploid oysters are sterile and do not spawn. In Florida and in certain other states, weed control in waterways is aided by the release of weed-eating fish (the grass carp) that do not become overpopulated, and hence a problem themselves, because the released fish are sterile triploids.

Polyploids can arise from genome duplications occurring before or after fertilization.

Polyploid organisms can be produced in two principal ways, which are illustrated in **FIGURE 5.32** for the example of tetraploidy. In the mechanism known as **sexual polyploidization**, the increase in chromosome number takes place in *meiosis* through the formation of *unreduced gametes* that have double the normal complement of chromosomes. Unreduced gametes are formed in many species at frequencies of 1 to 40 percent, and the frequency can be under genetic control. For example, in the potato, a single

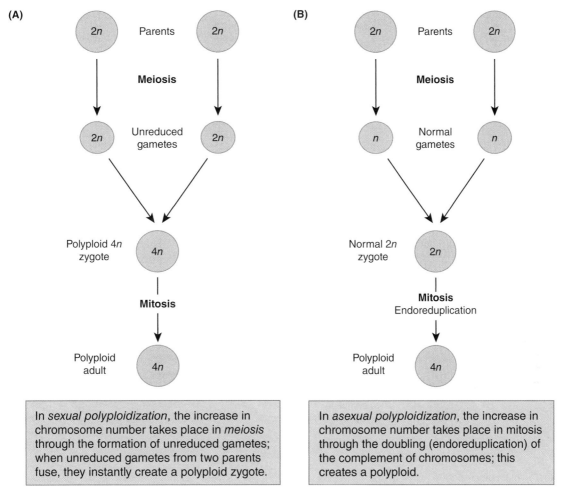

FIGURE 5.32 Formation of a tetraploid organism by (A) sexual polyploidization and (B) asexual polyploidization. The symbol *n* stands for the monoploid chromosome number.

recessive mutation that acts during pollen formation causes the first-division and second-division meiotic spindles to be oriented in the same direction (rather than being at right angles to each other as in nonmutant cells), with the result that a pollen nucleus forms around each of the two adjacent groups of telophase II chromosomes, yielding unreduced gametes. Also in the potato, a different recessive mutation acts to eliminate the second meiotic division during the formation of female gametes, again resulting in unreduced gametes. Part A of Figure 5.32 shows two unreduced *2n* gametes yielding a *4n* tetraploid, but there are many other possibilities. For example, union of an unreduced *2n* gamete with a normal *n* gamete yields a *3n* triploid.

The other principal mechanism of polyploid formation is **asexual polyploidization** (Figure 5.32, part B), in which the doubling of the chromosome number takes place in *mitosis*. Chromosome doubling through an abortive mitotic division is called **endoreduplication**. In a plant species that can undergo self-fertilization, endoreduplication creates a

new, genetically stable species, because if the chromosomes in the tetraploid can pair two by two in meiosis, they can segregate regularly and yield gametes with a full complement of chromosomes. Self-fertilization of such a tetraploid restores the chromosome number, so the tetraploid condition can be perpetuated.

The genetics of tetraploid species and that of other polyploids is more complex than that of diploid species, because the organism carries more than two alleles of any gene. With two alleles in a diploid, only three genotypes are possible: *AA*, *Aa*, and *aa*. In a tetraploid, by contrast, five genotypes are possible: *AAAA*, *AAAa*, *AAaa*, *Aaaa*, and *aaaa*. Among these genotypes, the middle three represent different types of tetraploid heterozygotes.

An octoploid species (eight sets of chromosomes) can be generated by sexual or asexual polyploidization of a tetraploid. Again, if only bivalents form in meiosis, then an octoploid organism can be perpetuated sexually by self-fertilization or through crosses with other octoploids. Furthermore, cross-fertilization between an octoploid and a tetraploid results in a

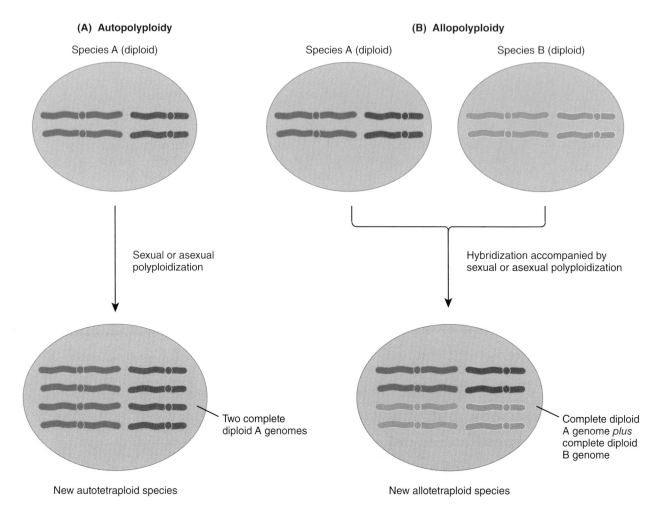

(A) Autopolyploidy

Species A (diploid)

Sexual or asexual
polyploidization

Two complete
diploid A genomes

New autotetraploid species

(B) Allopolyploidy

Species A (diploid)   Species B (diploid)

Hybridization accompanied by
sexual or asexual polyploidization

Complete diploid
A genome *plus*
complete diploid
B genome

New allotetraploid species

FIGURE 5.33 (A) Autopoly ploids have chromosome sets from a single species; (B) allopolyploids have chromosome sets from different species.

hexaploid (six sets of chromosomes). Repeated episodes of polyploidization and cross-fertilization may ultimately produce an entire polyploid series of closely related organisms that differ in chromosome number, as exemplified in *Chrysanthemum*.

Polyploids can include genomes from different species.

Chrysanthemum represents a type of polyploidy, known as **autopolyploidy**, in which all chromosomes in the polyploid species derive from a single diploid ancestral species (**FIGURE 5.33**, part A). In many cases of polyploidy, the polyploid species have complete sets of chromosomes from two or more *different* ancestral species. Such polyploids are known as **allopolyploids** (Figure 5.33, part B). They derive from occasional hybridization between different diploid species when pollen from one species germinates on the stigma of another species and sexually fertilizes the ovule, followed by endoreduplication

in the zygote to yield a hybrid plant in which each chromosome has a pairing partner in meiosis. The pollen may be carried to the wrong flower by wind, insects, or other pollinators. Part B of Figure 5.33 illustrates hybridization between species A and B in which endoreduplication leads to the formation of an allopolyploid (in this case, an *allotetraploid*), which carries a complete diploid genome from each of its two ancestral species. The formation of allopolyploids through hybridization and endoreduplication is an extremely important process in plant evolution and plant breeding. At least half of all naturally occurring polyploids are allopolyploids. Cultivated wheat provides a classic example of allopolyploidy. Cultivated bread wheat is a hexaploid with 42 chromosomes constituting a complete diploid genome of 14 chromosomes from each of three ancestral species. The 42-chromosome allopolyploid is thought to have originated by the series of hybridizations and endoreduplications outlined in **FIGURE 5.34**.

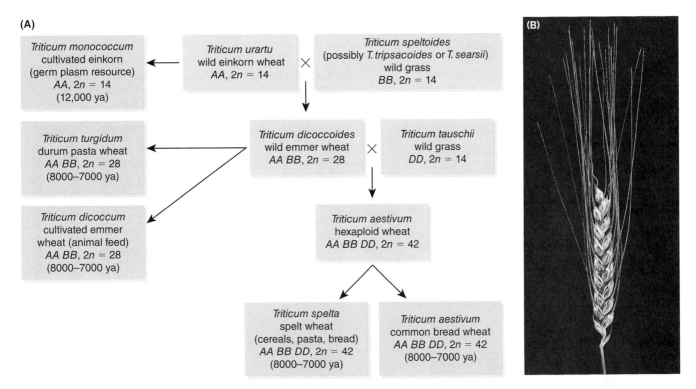

FIGURE 5.34 Repeated hybridization and polyploidization in the origin of wheat. (A) Each of the A, B, and D genomes has 7 chromosomes, and 2n is the total chromosome number for each species. Wild species are in green boxes, and domesticated species are in yellow boxes along with the approximate time of domestication (ya = years ago). (B) The spike of *T. turgidum*, one of the earliest cultivated wheats.

Photo courtesy of Gordon Kimber, Department of Agronomy, University of Missouri.

5.5 The grass family illustrates the importance of polyploidy and chromosome rearrangements in genome evolution.

The cereal grasses are our most important crop plants. They include rice, wheat, maize, millet, sugar cane, sorghum, and other cereals. The genomes of grass species vary enormously in size. The smallest, at 400 Mb, is found in rice; the largest, at 16,000 Mb, is found in wheat. Although some of the difference in genome size results from the fact that wheat is an allohexaploid whereas rice is a diploid, a far more important factor is the large variation from one species to the next in types and amount of repetitive DNA sequences present. Each chromosome in wheat contains approximately 25 times as much DNA as each chromosome in rice. For comparison, maize has a genome size of 2500 Mb; it is intermediate in size among the grasses and approximately the same size as the human genome.

In spite of the large variation in chromosome number and genome size in the grass family, there are a number of genetic and physical linkages between single-copy genes that are remarkably conserved amid a background of very rapidly evolving repetitive DNA sequences. In particular, each of the conserved regions can be identified in all the grasses and referred to a similar region in the rice genome. The situation is as depicted in **FIGURE 5.35**. The rice chromosome pairs are numbered R1 through R12, and the conserved regions within each chromosome are indicated by lowercase letters—for example, R1a and R1b. In each of the other species, each chromosome pair is diagrammed according to the arrangement of segments of the rice genome that contain single-copy DNA sequences homologous to those in the corresponding region of the chromosome of the species in question. For example, the wheat monoploid chromosome set is designated W1 through W7. One region of W1 contains single-copy sequences that are homologous to those in rice segment R5a, another contains single-copy sequences that are homologous to those in rice segment R10, and still another contains single-copy sequences that are homologous to those in rice segment R5b. The genomes of the other grass species can be aligned with those of rice as shown. Each of such conserved genetic and physical linkages is called a **synteny group**.

Synteny groups are found in other species comparisons as well. For example, the human and mouse genomes share about 180 synteny groups owing to about an equal number of chromosome rearrangements that took place in the approximately 80 million years since the species last shared a common ancestor. These synteny groups are often useful in identifying the mouse homolog of a human gene.

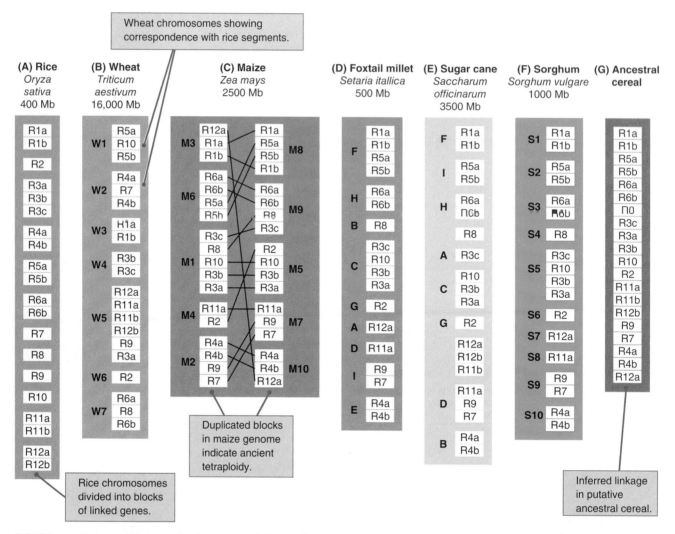

FIGURE 5.35 Conserved linkages (synteny groups) between the rice genome (A) and that of other grass species: wheat (B), maize (C), foxtail millet (D), sugar cane (E), and sorghum (F). Genome sizes are given in millions of base pairs (Mb).

Data from G. Moore, *Curr. Opin. Genet. Dev.* 5 (1995): 717–724.

CHAPTER SUMMARY

- The standard human karyotype consists of 22 pairs of autosomes and two sex chromosomes.

- In mammals, the difference in dosage of X-linked genes resulting from the presence of two X chromosomes in females but only one in males is compensated by a mechanism of X-chromosome inactivation.

- X-chromosome inactivation occurs at an early stage in embryonic development, and it results in the genetic inactivation (silencing) of most genes in all but one of the X chromosomes present in each cell.

- Chromosome abnormalities are a major factor in human spontaneous abortions and an important cause of genetic disorders such as trisomy 21 (Down syndrome).

- Aneuploid (unbalanced) chromosome rearrangements usually have greater phenotypic effects than euploid (balanced) chromosome rearrangements.

- Duplications and deficiencies refer to chromosomes that have extra or missing copies of genetic material. The effects of duplications and deficiencies depend on the size of the extra or missing region and of the particular genes present within the region.

- Inversions have a segment of chromosome present in the reverse order from wildtype. During meiosis, the inverted regions form a loop and abnormal chromosomes result from crossing-over within the loop. The specific types of abnormalities depend on whether the centromere is or is not included within the inverted region.

- Reciprocal translocations result in abnormal gametes because they upset segregation, resulting in a high risk of aneuploid gametes.
- The genetic imbalance caused by a single chromosome that is extra or missing may have a more serious phenotypic effect than an entire extra set of chromosomes.
- Duplication of the entire chromosome complement that is present in a species, or in a hybrid between species, is a major factor in the evolution of higher plants.

ISSUES AND IDEAS

- Although autosomal trisomy is common among human fetuses that undergo spontaneous abortion, autosomal monosomies are almost unknown. How can this observation be explained?
- Why do most chromosome rearrangements pass through mitosis without upsetting the process?
- What are the four major classes of abnormality in chromosome structure?
- What types of chromosomal abnormalities in meiosis are associated with an inversion? How are these related to the position of the centromere? To crossing-over?
- What types of chromosomally abnormal gametes are associated with a translocation?

- Why do inversions and translocations cause reproductive abnormalities only when they are heterozygous?
- How does the ability to form bivalents in meiosis contribute to the production of gametes that have the same number and types of chromosomes?
- Why would the presence of sex chromosomes be a hindrance to the evolution of a series of related species with different levels of polyploidy?
- Why do most naturally occurring polyploid species have an even-number multiple of the monoploid chromosome set?
- Distinguish between sexual and asexual polyploidization. In which type does the chromosome number double in meiosis? In mitosis?

SOLUTIONS: STEP BY STEP

PROBLEM 1 The genetic disorder *Charcot-Marie-Tooth disease type I* is a progressive disorder of the peripheral nerves resulting in difficulties in movement and sensation, especially in the feet. The related genetic disorder *hereditary neuropathy with pressure palsies* is also a disorder of the peripheral nerves resulting in recurrent episodes of numbness, tingling, or loss of muscle function especially in the wrists, elbows, and knees. Charcot-Marie-Tooth disease type I results from duplication of chromosomal region 17p12, whereas hereditary neuropathy with pressure palsies results from deletion of the same region. Neither condition "runs in families," yet exactly the same duplication and deletion occurs repeatedly in unrelated families. How might you explain this observation?

SOLUTION. The most straightforward hypothesis is ectopic crossing-over as diagrammed here. The repeated sequence in indicated in brown, and the region between the repeats (which may contain many genes) is indicated in green. If the repeated regions undergo mispairing and crossing-over as shown, one of the recombinant products contains a deletion of the green region and the other contains a duplication of this same region. Hence, unequal crossing-over between the repeated sequences can explain the reoccurrence of the duplications and deletions.

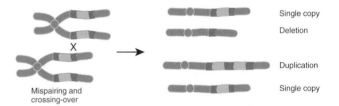

Mispairing and crossing-over

Single copy
Deletion
Duplication
Single copy

PROBLEM 2 A metacentric chromosome in maize shows an unusual property. When heterozygous with a standard chromosome carrying multiple genetic markers, it fails to yield recombinants between genetic markers in one of the arms. Cytological investigation reveals many meiotic anaphases like that shown here, in which there is a chromosome bridge connecting the centromeres of a dicentric chromosome as the centromeres are pulled to opposite poles, and an acentric chromosome remains behind on the metaphase plate.

(a) What chromosomal abnormality can account for the suppression of recombination and the anaphase bridge?
(b) Are the anaphase bridges seen at anaphase I or at anaphase II?

SOLUTION. **(a)** Suppression of recombination in one segment of a chromosome is typically associated with a heterozygous inversion. This is a reasonable hypothesis in the present case, because the suppression occurs in heterozygotes with a standard, structurally normal chromosome. The appearance of chromosome bridges indicates that the inversion is a paracentric inversion, which in this case must include most of one of the chromosome arms. (Pericentric inversions also suppress recombination but do not result in chromosome bridges.) **(b)** If there is a paracentric inversion, the bridges would be observed in anaphase I.

PROBLEM 3 When three copies of a chromosome are present during meiosis, either two of the chromosomes will form a pair and the other remain unpaired, or else all three chromosomes will come together, each chromosome pairing along part of its length with both of the others. In either case, the end result is two gametes containing two copies of the chromosome and two gametes containing one copy. In a mating between two triploid organisms containing three copies of each of five chromosomes, what is the probability that a zygote will carry a triploid complement of chromosomes?

SOLUTION. This problem can be solved with the aid of a Punnett Square, shown here for one of the chromosomes. With 1 : 1 segregation yielding gametes with a ratio of 2 copies : 1 copy in each parent, a total of $1/4 + 1/4 = 1/2$ of the zygotes will contain three copies of the chromosome. If each of the five chromosomes segregates independently, then the overall probability of three copies of each chromosome is $(1/2)^5 = 1/32$, or about 3 percent.

	Two copies $1/2$	One copy $1/2$
Two copies $1/2$	$1/4$ Four copies	$1/4$ Three copies
One copy $1/2$	$1/4$ Three copies	$1/4$ Two copies

CONCEPTS IN ACTION: PROBLEMS FOR SOLUTION

5.1 Why is it not completely correct to state that one of the X chromosomes in somatic cells of a normal female is an "inactive" X chromosome?

5.2 What is the genetic consequence of the obligatory crossover that occurs in the pseudoautosomal region between the X and Y chromosome during meiosis?

5.3 In the absence of chromosomal rearrangements, what are the most likely karyotypes of a new-born baby with 47 chromosomes? With 45 chromosomes?

5.4 A recessive mutation in the human genome results in a condition called anhidrotic ectodermal dysplasia, which is associated with an absence of sweat glands. The condition can be detected by studies of the electrical conductivity of the skin, because skin without sweat glands has a lower electrical conductivity (higher resistance) than normal skin. In kinships in which the recessive allele is segregating, affected males are found to show low conductance uniformly across their skin surface, as do affected females. However, many females show a mosaic pattern with normal conductance in some patches of skin and low conductance in others. The pattern of tissue lacking sweat glands is different for each mosaic female examined. How could this pattern of gene expression be explained?

5.5 Human chromosome 2 resulted from a Robertsonian translocation in which the short arms of two nonhomologous acrocentric chromosomes underwent fusion in a recent primate ancestor to yield a new monocentric chromosome. Imagine a primate ancestor who was heterozygous for this fusion mating with an individual who was homozygous for the unfused, ancestral chromosomes. If the ratio of adjacent-1 : adjacent-2 : alternate segregation were 1 : 1 : 2, what proportion of the zygotes would be viable?

5.6 The vast majority of progeny from either 47,XXX females or 47,XYY males are karyotypically normal 46,XX or 46,XY. Is this observation to be expected?

5.7 Inversions are often called "suppressors" of crossing-over. Is this term literally true? If not, what does it really mean?

5.8 A female cat with orange fur mates with a male with black fur. The resulting litter includes a male calico kitten which, when mature, proves to be sterile. Suggest a likely explanation.

5.9 Recessive genes *a*, *b*, *c*, *d*, *e*, and *f* are closely linked in a chromosome, but their order is unknown. Three deletions in the region are examined. One deletion uncovers *a*, *c*, and *d*; another uncovers *a*, *b*, and *f*; and the third uncovers *b* and *e*. What is the order of the genes? (*Hint*: There is enough information to order most, but not all, of the genes.)

5.10 Recessive genes *u*, *v*, *w*, *x*, *y*, and *z* are closely linked in a chromosome, but their order is unknown. Three deletions in the region are found to uncover recessive alleles of the genes as follows: deletion 1 uncovers *w*, *x*, and *z*; deletion 2 uncovers *v*, *w*, *x*, and *y*; and deletion 3 uncovers *u* and *v*. What is the order of the genes? Suggest what experiments you might carry out to complete the ordering.

5.11 Six bands in a salivary gland chromosome of *Drosophila* are shown in the accompanying figure, along with the extent of five deletions (Del1–Del5).

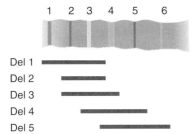

Recessive alleles *a*, *b*, *c*, *d*, *e*, and *f* are known to be in the region, but their order is unknown. When the deletions are heterozygous with each allele, the following results are obtained:

	a	*b*	*c*	*d*	*e*	*f*
Del 1	−	−	−	+	+	+
Del 2	−	+	−	+	+	+
Del 3	−	+	−	+	−	+
Del 4	+	+	−	−	−	+
Del 5	+	+	+	−	−	−

In this table, the minus sign means that the deletion is missing the corresponding wildtype allele (that is, the deletion uncovers the recessive allele), and the plus sign means that the corresponding wildtype allele is still present. Use these data to infer the position of each gene relative to the salivary gland chromosome bands.

5.12 A phenotypically normal woman has a child with Down syndrome. The woman is found to have 45 chromosomes. What kind of chromosome abnormality can account for these observations? How many chromosomes does the affected child have? How does this differ from the usual chromosome number and karyotype of a child with Down syndrome?

5.13 The most common form of color blindness in humans is due to an X-linked recessive allele. A man who is color blind has a 45,X (Turner syndrome) daughter who is also color blind. Did the nondisjunction that led to the 45,X child occur in the mother or the father? Explain the evidence supporting your answer.

5.14 The long arm of the normal human Y chromosome contains a duplicated region of about 100 kb flanking a set of genes necessary for male fertility. In some mutant Y chromosomes the male-fertility

genes are missing, and only one copy of the duplication is present. How can this observation be explained? Does the observation tell you anything about how the duplications are oriented in the chromosome?

5.15 *Drosophila virilis* has six pairs of chromosomes in somatic cells, consisting of five acrocentric chromosome pairs and one tiny chromosome pair referred to as the "dot" chromosome. The closely related species *D. texana* has five pairs of chromosomes, consisting of four acrocentric pairs, one metacentric pair, and one tiny "dot" chromosome pair. Hybrids between these species have meiotic cells in which there are four bivalents (including the "dot") and one trivalent. The trivalent consists of the metacentric chromosome of *D. texana* paired with two of the acrocentrics of *D. virilis*, oriented such that the three centromeres are close together. Suggest an explanation.

5.16 In *Drosophila melanogaster*, the genes for brown eyes (*bw*) and humpy thorax (*hy*) are about 12 map units distant on the same arm of chromosome 2. A paracentric inversion spans about one-third of this region but does not include the genes mentioned. Explain what recombinant frequency between *bw* and *hy* you would expect in females that are:

(a) Homozygous for the inversion.

(b) Heterozygous for the inversion.

5.17 Semisterile tomato plants heterozygous for a reciprocal translocation between chromosomes 5 and 11 were crossed with chromosomally normal plants homozygous for the recessive mutation *broad leaf* on chromosome 11. When semisterile F₁ plants were crossed with the plants of *broad-leaf* parental type, the following phenotypes were found in the backcross progeny:

semisterile broad-leaf	38
fertile broad-leaf	242
semisterile normal-leaf	282
fertile normal-leaf	33

(a) What is the recombination frequency between the *broad-leaf* gene and the translocation breakpoint in chromosome 11?

(b) What ratio of phenotypes in the backcross progeny would have been expected if the *broad-leaf* gene had not been on the chromosome involved in the translocation?

5.18 The herb genus *Tragopogon*, commonly known as goat's beard, shows a great deal of interspecific hybridization resulting in new species that are allopolyploids. Explain how three different species of *Tragopogon*, each with $2n = 12$ chromosomes, could hybridize and produce a new species with a chromosome number of $2n = 36$.

5.19 A genetically wildtype natural isolate of *Neurospora crassa* was crossed with a laboratory strain carrying a recessive allele *ad5* known to be 10 map units from the centromere. The resulting asci showed only 2 percent second-division segregation, and many of the asci contained inviable ascospores. One of the *ad5*-bearing ascospores from a second-division segregation ascus was germinated and mated with the original wildtype isolate. In this case, the resulting asci showed the expected 20 percent second-division segregation. How can you account for these results?

5.20 Recessive mutations *a–h* are closely linked along a chromosome, and the deletions designated Δ1–Δ5 occur in a wildtype chromosome. Each deletion eliminates the wildtype allele of one or more of the genes. A recessive allele of any gene will be "uncovered," and, therefore, expressed, in any individual in which the homologous chromosome carries a deletion that eliminates the wildtype allele. In matrix below, a minus sign indicates that a recessive allele is uncovered by the deletion, and a plus sign indicates that the wildtype allele is still present within the deleted chromosome. Deduce the order of the genes along the chromosome.

	a	b	c	d	e	f	g	h
Δ1	−	+	−	+	−	+	+	+
Δ2	+	−	−	+	−	+	−	+
Δ3	+	−	+	−	+	+	+	−
Δ4	+	−	+	−	+	−	−	−
Δ5	+	+	+	−	+	−	+	+

5.21 Shown in the accompanying gel are seven DNA fragments, designated A−G, present along the X chromosome of the laboratory rat *Rattus norvegicus*. Each fragment was amplified by means of the polymerase chain reaction (PCR), as described in Chapter 6, using a unique pair of primers. The seven fragments are present in close proximity to each other along the X chromosome. Five small deletions in the X chromosome, designated Δ1−Δ5, present in different strains, delete at least one of the templates for amplification. The gel shows the amplification products obtained from genomic DNA from males carrying each of the five deletions. Deduce the linear order of the amplified fragments A−G present in the wildtype (undeleted) X chromosome.

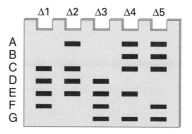

·※· STOP & THINK ANSWERS

ANSWER TO STOP & THINK 5.1

The females who are heterozygous are I-1, II-2, II-3, and II-6 because all have plasma levels of factor VIII of roughly half the normal level. (Female I-1 is certainly heterozygous because she has an affected son.) The reason for the reduced level of clotting factor in heterozygous females is that, because of the single-active-*X* principle, only about half of the cells that produce clotting factor VIII have the nonmutant X chromosome active; in other cells the mutant chromosome is the active *X*.

ANSWER TO STOP & THINK 5.2

Both deletions uncover *b* and *d*. Gene *a* cannot be between *b* and *d*, otherwise it would be uncovered by deletion 2, and gene *c* cannot be between *b* and *d*, otherwise it would be uncovered by deletion 1. Hence the genes that are at the ends of the cluster are *a* and *c*. Two possible gene orders—*a b d c* and *a d b c*—are compatible with the deletion data.

ANSWER TO STOP & THINK 5.3

The two-strand double crossover yields four monocentric products with the following gene orders (*CEN* represents the centromere): *CEN A B C D E* and *CEN a d c b e* (these come from the chromatids not involved in either crossover), and *CEN A B c D E* and *CEN a d C b e* (these come from the chromatids involved in the double crossover). The four-strand double crossover yields two dicentric chromosomes: *CEN A B c d a CEN* and *CEN A B C d a CEN*; it also yields two acentric products: *e b C D E* and *e b c D E*.

ANSWER TO STOP & THINK 5.4

Both kinds of adjacent segregation yield unbalanced gametes, whereas alternate segregation yields balanced gametes. Hence in both cases, the ratio of unbalanced to balanced gametes is 1 : 1.

CHAPTER 6

Paris japonica has the largest genome yet recorded—149 Gb, or 50 times the size of the human genome.
© H.Tanaka/Shutterstock

DNA Structure, Replication, and Manipulation

- To diagram a replication bubble that shows the leading and lagging DNA strands at each replication fork and that indicates the function and site of action of the primosome, the DNA polymerase complex, helicase, single-strand DNA binding protein, and gyrase (topoisomerase II).

- To describe the proofreading function of DNA polymerase Pol III and to predict how a mutation in the Pol III gene leading to loss of the proofreading function would affect the accuracy of DNA replication.

- For a double-stranded DNA molecule with known positions of cleavage sites for one or more restriction enzymes, to deduce the number and lengths of the fragments produced by cleaving the DNA with one or more of the enzymes, and to draw a diagram showing where these fragments would appear in a gel after electrophoresis.

- For a given sequence of double-stranded DNA, to select the sequences of primer oligonucleotides that would allow any specific fragment of the molecule to be amplified in the polymerase chain reaction.

- Given the positions and fluorescent colors of bands in a dideoxy DNA sequencing gel, to deduce the sequence of the template DNA strand and indicate its 5′ and 3′ ends.

Double-stranded DNA is a right-handed helix of paired, complementary, antiparallel strands, each composed of an ordered string of nucleotides bearing A (adenine), T (thymine), G (guanine), or C (cytosine). Watson–Crick base pairing between A and T and between G and C in the complementary strands holds the strands together. The complementarity is also the key to replication, because each strand can serve as a template for the synthesis of a new, complementary strand. In this chapter, we take a close look at the molecular biology of DNA and its replication. We also consider how knowledge of DNA structure and replication has been used in the development of laboratory techniques for isolating fragments that contain genes or parts of genes of particular interest and for determining the sequence of bases in DNA fragments.

6.1 Genome size can differ tremendously, even among closely related organisms.

The genetic complement of a cell or virus constitutes its *genome*. In eukaryotes, this term is commonly used to refer to one complete haploid set of chromosomes, such as that found in a sperm or egg. A summary of a small sample of genome sizes is shown in **TABLE 6.1**. Bacteriophage MS2 is one of the smallest viruses; it has only four genes in a single-stranded RNA molecule of about 4000 nucleotides (4 kb). SV40 virus, which infects monkey and human cells, has a genetic complement of five genes in a circular double-stranded DNA molecule of about 5 kb (5000 nucleotide pairs). The more complex phages and animal viruses have as many as 250 genes and DNA molecules ranging from 50 to 300 kb. Prokaryotic genomes are substantially larger. Archaeal genomes (for example, *Methanococcus jannaschi*) are generally similar in size to bacterial genomes. Some prokaryotic genomes consist of linear DNA, others of circular DNA. For example, the chromosome of the spirochete *Borrelia burgdorferi*, the agent of Lyme disease, is a linear DNA molecule of about 910 kb, and that of *Escherichia coli* strain K12 is a circular DNA molecule of 4600 kb. The genomes of unicellular eukaryotes are even larger. The genome size of budding yeast, *Saccharomyces cerevisiae*, is 13 Mb. The units of length of nucleic acids

TABLE 6.1 Genome Size of Some Representative Viral, Bacterial, and Eukaryotic Genomes

Genome	Approximate genome size in thousands of nucleotides	Form
Viruses		
MS2	4	Single-stranded RNA
Human immunodeficiency virus (HIV)	9	
Colorado tick fever virus	29	Linear double-stranded RNA
SV40	5	Circular double-stranded DNA
φX174	5	Circular single-stranded DNA; double-stranded replicative form
λ	50	
Herpes simplex	152	
T2,T4,T6	165	Linear double-stranded DNA
Smallpox	267	
Prokaryotes		
Methanococcus jannaschii	1,600	
Escherichia coli	4,600	Circular double-stranded DNA
Borrelia burgdorferi	910	Linear double-stranded DNA
Eukaryotes		Haploid chromosome number
Saccharomyces cerevisiae (yeast)	13,000	16
Caenorhabditis elegans (nematode)	97,000	6
Arabidopsis thaliana (mustard cress)	100,000	5
Drosophila melanogaster (fruit fly)	180,000	4
Takifugu rubripes (fish)	400,000	22
Homo sapiens (human being)	3,000,000	23
Zea mays (corn, maize)	4,500,000	10
Amphiuma means (salamander)	90,000,000	14

in which genome sizes are typically expressed are as follows:

- **kilobase (kb)** 10^3 nucleotide subunits

- **megabase (Mb)** 10^6 nucleotide subunits

- **gigabase (Gb)** 10^9 nucleotide subunits

In these terms, viral genomes are typically in the range 100–1000 kb, bacterial genomes are typically in the range 1–10 Mb, and eukaryotic genomes are typically in the range 100–1000 Mb. (The smallest eukaryotic genomes are about 10 Mb.)

Among eukaryotes, however, genome size often differs tremendously, even among closely related species. This lack of correlation is known as the **C-value paradox.**

KEY CONCEPT

The C-value paradox: Among eukaryotes, there is no consistent relationship between the C-value (that is, the DNA content of the haploid genome) and the metabolic, developmental, or behavioral complexity of the organism.

The differences are often hard to believe. Genome size among species of protozoa differs by 5800-fold: among arthropods by 250-fold, fish by 350-fold, algae by 5000-fold, and angiosperms by 1000-fold. The term *paradox* is amply justified by observing that the genome size of the Japanese pufferfish *Takifugu rubripes* is 400 Mb, whereas that of the salamander *Amphiuma means* is 90,000 Mb. The C-values differ by a factor of 225, yet both organisms are vertebrates, and there is no reason to suppose that either has more or fewer genes than the other. The main difference is that the protein-coding portion of the *Takifugu* genome makes up a much larger proportion of the total than does that in *Amphiuma*. Turning to plants, rice and maize have about the same number of genes (transcripts and proteins), but the maize genome at 2500 Mb is about six times larger than that of rice at 400 Mb. In nearly all higher animals and plants, the actual number of genes has little relationship to genome size. The reason for the discrepancy is that in higher organisms, much of the DNA has functions other than coding for the amino acid sequence of proteins.

6.2 DNA is a linear polymer of four deoxyribonucleotides.

DNA is a polymer—that is, a large molecule that contains repeating units—and is composed of 2′-deoxyribose (a five-carbon sugar), phosphoric acid, and the four nitrogen-containing bases denoted A, T, G, and C. The chemical structures of the bases are shown in **FIGURE 6.1**. Note that two of the bases have a double-ring structure; these are called **purines**. The other two bases have a single-ring structure; these are called **pyrimidines.**

- The purine bases are adenine (A) and guanine (G).

- The pyrimidine bases are thymine (T) and cytosine (C).

In DNA, each base is chemically linked to one molecule of the sugar deoxyribose, forming a compound called a **nucleoside**. When a phosphate group is also attached to the sugar, the nucleoside becomes a **nucleotide** (**FIGURE 6.2**). Thus a nucleotide is a nucleoside plus a phosphate. In the conventional numbering of the carbon atoms in the sugar in Figure 6.2, the carbon atom to which the base is

An example of the C-value paradox. (A) The Japanese pufferfish *Takifugu rubripes* has a genome size of 400 Mb. (B) The two-toed salamander *Amphiuma means* has a genome size of 90,000 Mb. The latter is no more "complex" than the former.

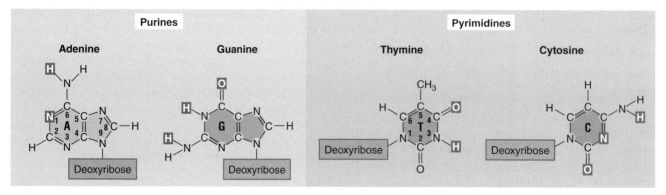

FIGURE 6.1 Chemical structures of adenine, thymine, guanine, and cytosine, the four nitrogen-containing bases in DNA. In each base, the nitrogen atom linked to the deoxyribose sugar is indicated. The atoms shown in red participate in hydrogen bonding between the DNA base pairs.

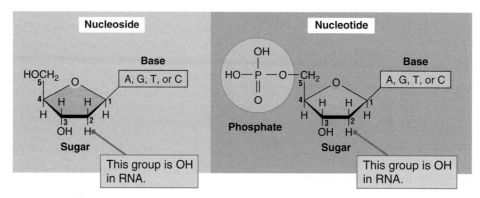

FIGURE 6.2 A typical nucleotide showing the three major components (phosphate, sugar, and base), the difference between DNA and RNA, and the distinction between a nucleoside (no phosphate group) and a nucleotide (with phosphate). Nucleotides may contain one phosphate unit (monophosphate), two units (diphosphated), or three units (triphosphated).

attached is the 1′ carbon. (The atoms in the sugar are given primed numbers to distinguish them from atoms in the bases.)

In nucleic acids, such as DNA and RNA, the nucleotides are joined to form a **polynucleotide chain** in which the phosphate attached to the 5′ carbon of one sugar is linked to the hydroxyl group attached to the 3′ carbon of the next sugar in line (**FIGURE 6.3**). The chemical bonds by which the sugar components of adjacent nucleotides are linked through the phosphate groups are called **phosphodiester bonds.** The 5′−3′−5′−3′ orientation of these linkages continues throughout the chain, which typically consists of millions of nucleotides. Note that the terminal groups of each polynucleotide chain are a 5′-phosphate (5′-P) group at one end (depicted as the "tail" of the broad arrow) and a 3′-hydroxyl (3′-OH) group at the other (depicted as the "head" of the arrow). The asymmetry of the ends of a DNA strand is the chemical basis of its polarity: One end of the strand is the **5′ end** (which terminates in a phosphate), whereas the other end is the **3′ end** (which terminates in a hydroxyl).

6.3 Duplex DNA is a double helix in which the bases form hydrogen bonds.

FIGURE 6.4 shows several representations of double-stranded DNA. The duplex molecule of DNA consists of two polynucleotide chains twisted around one another to form a right-handed helix in which adenine and thymine are paired, as are guanine and cytosine (Figure 6.4). Each chain makes one complete turn every 34 Å. The bases are spaced at 3.4 Å, so there are ten bases per helical turn in each strand, or 10 base pairs per turn of the double helix. Each base is paired to its partner base in the other strand by a hydrogen bond. A **hydrogen bond** is a weak bond in which two negatively charged atoms share a hydrogen atom. Hydrogen bonds contribute to holding the strands together, as does the stacking of the base pairs on top of one another, so as to exclude water molecules. The paired bases are planar, parallel to one another, and perpendicular to the long axis of the double helix.

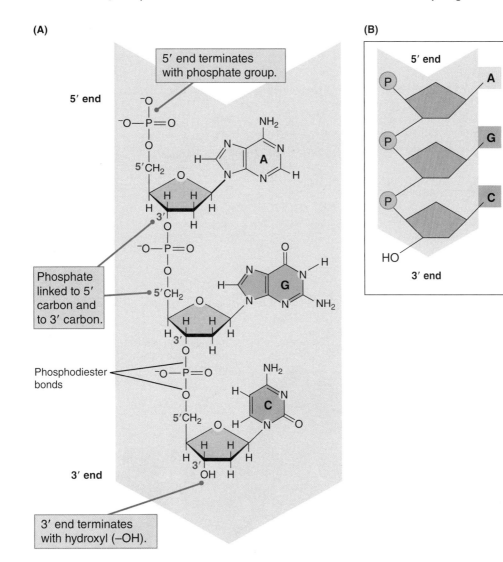

(A)

5' end

5' end terminates with phosphate group.

Phosphate linked to 5' carbon and to 3' carbon.

Phosphodiester bonds

3' end

3' end terminates with hydroxyl (–OH).

(B)

5' end

P — A

P — G

P — C

HO

3' end

FIGURE 6.3 Three nucleotides at the 5' end of a single polynucleotide strand. (A) The chemical structure of the sugar–phosphate linkages, showing the 5'-to-3' orientation of the strand (the red numbers are those assigned to the carbon atoms). (B) A common schematic way to depict a polynucleotide strand.

For encoding genetic information, the central feature of DNA structure is the A−T and G−C pairing between the bases:

The principles of A−T and G−C base pairing explain two generalizations about the relative amounts of the bases found in all double-stranded DNA:

- Number of adenine bases [A] equals number of thymine bases [T], so [A] = [T].

- Number of guanine bases [G] equals number of cytosine bases [C], so [G] = [C].

Although [A] = [T] and [G] = [C] in double-stranded DNA, the proportion of bases that are either G or C (called the *percent G + C*) varies among species but is constant in all cells of an organism. For example, human DNA has 39 percent G + C on the average, but there can be large variations in base composition along the chromosomes. The regional variation can be observed microscopically because regions relatively poor in G + C content give rise to dark bands when the chromosomes are stained appropriately.

The adenine–thymine base pair and the guanine–cytosine base pair are illustrated in **FIGURE 6.5**. Note that an A−T pair has two hydrogen bonds and a G−C pair has three hydrogen bonds. This means that the hydrogen bonding between G and C is stronger in the sense

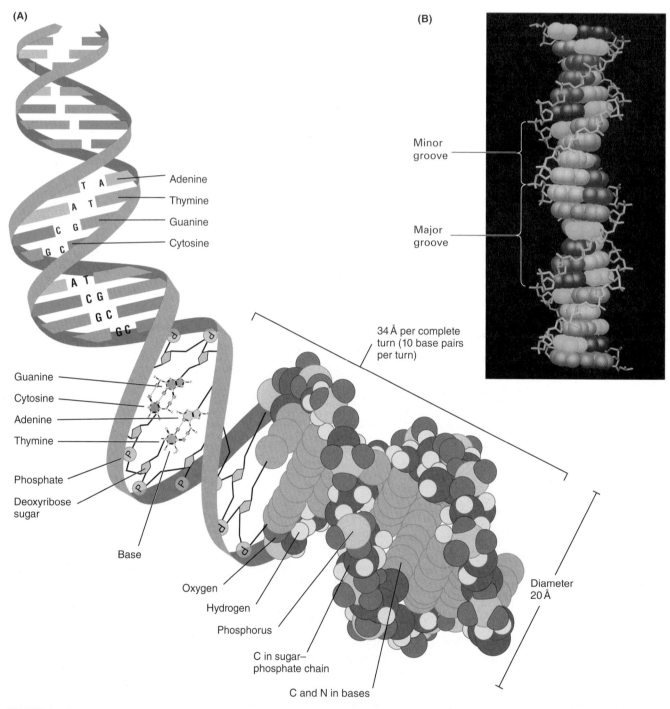

FIGURE 6.4 Two representations of DNA illustrating the three-dimensional structure of the double helix. (A) In a "ribbon diagram," the sugar–phosphate backbones are depicted as bands, with horizontal lines used to represent the base pairs. (B) A computer model of the standard form of DNA. The stick figures are the sugar–phosphate chains winding around outside the stacked base pairs, forming a major groove and a minor groove. The color coding for the base pairs is A, red or pink; T, dark green or light green; G, dark brown or beige; C, dark blue or light blue. The bases depicted in dark colors are those attached to the blue sugar–phosphate backbone; the bases depicted in light colors are attached to the beige backbone.

(B) Courtesy of Antony M. Dean, University of Minnesota.

that it requires more energy to break. The specificity of base pairing means that the sequence of bases along one polynucleotide strand of the DNA is matched (complementary) with the base sequence in the other strand. However, the base pairs along a DNA duplex can be arranged in any order, and the sequence of bases differs from one part of the molecule to another and from species to species. Because there is no restriction on the base sequence, DNA has a virtually unlimited capability to code for a variety of different protein molecules.

The backbone of each polynucleotide strand in the double helix in Figure 6.4 consists of deoxyribose

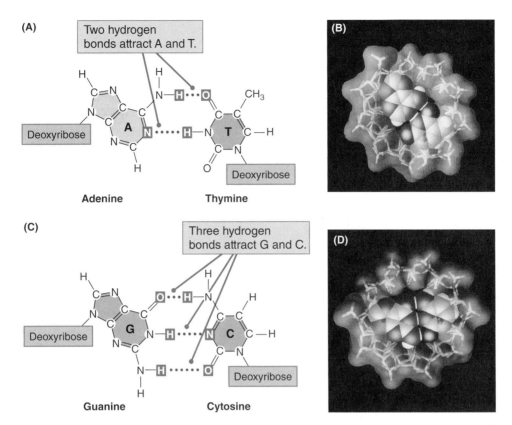

FIGURE 6.5 Normal base pairs in DNA. On the left, the hydrogen bonds (dotted lines) and the joined atoms are shown in red. (A, B) An A—T base pair. (C, D) A G—C base pair. In the space-filling models (B and D), the colors are C, gray; N, blue; O, red; and H (shown in the bases only), white. Each hydrogen bond is depicted as a white disk squeezed between the atoms that share the hydrogen. The stick figures on the outside represent the backbones winding around the stacked base pairs.

(B) and (D) Space-filling models courtesy of Antony M. Dean, University of Minnesota.

sugars alternating with phosphate groups that link the 3′ carbon atom of one sugar to the 5′ carbon of the next in line. The two polynucleotide strands of the double helix run in opposite directions, as can be seen from the orientation of the deoxyribose sugars in **FIGURE 6.6**. The paired strands are said to be **antiparallel**. Figure 6.4 also shows that there are two grooves spiraling along outside of the double helix. These grooves are not symmetrical in size. The large one is called the *major groove*, the smaller one the *minor groove*. Proteins that interact with double-stranded DNA often have regions that make contact with the base pairs by fitting into the major groove, the minor groove, or both.

The diagrams of the DNA duplexes in parts A and B of Figure 6.4 are static and so somewhat misleading. DNA is in fact a very dynamic molecule that is constantly in motion. In some regions, the strands can separate briefly and then come together again. Furthermore, although the right-handed double helix in Figure 6.4 is the standard helix, DNA can form more than 20 slightly different variants of right-handed helices, and some regions can even form helices in which the strands twist to the left. If there are complementary stretches of nucleotides in the same strand, a

single strand, separated from its partner, can fold back upon itself like a hairpin. Even triple helices, consisting of three strands, can form in regions of DNA that contain suitable base sequences.

6.4   Replication uses each DNA strand as a template for a new one.

The process of replication, in which each strand of the double helix serves as a **template** for the synthesis of a new strand, is simple in principle (**FIGURE 6.7**). It requires only that the hydrogen bonds joining the bases break to allow separation of the chains and that appropriate free nucleotides of the four types pair with the newly accessible bases in each strand. In practice, however, replication is a complex of geometric processes that require a variety of enzymes and other proteins. These processes are examined in this section.

Nucleotides are added one at a time to the growing end of a DNA strand.

The primary function of any mode of DNA replication is to reproduce the base sequence of the parent

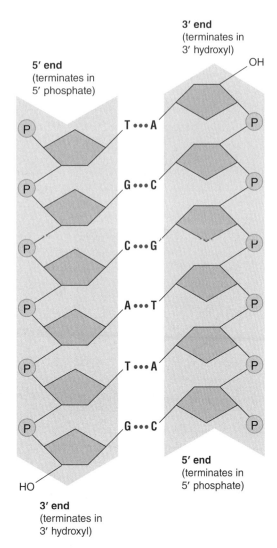

FIGURE 6.6 A segment of a DNA molecule showing the antiparallel orientation of the complementary strands. The arrows indicate the 5'-to-3' direction of each strand. The phosphate groups (P) join the 3' carbon atom of one deoxyribose to the 5' carbon atom of the adjacent deoxyribose.

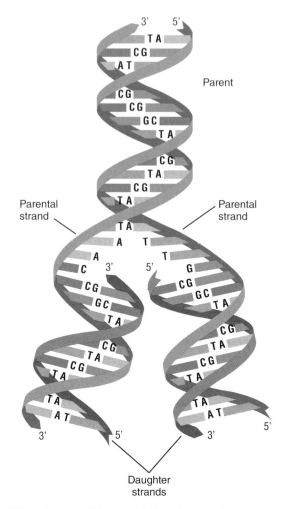

FIGURE 6.7 Watson–Crick model of DNA replication. The newly synthesized strands are in red. Each of the new strands is elongated *only at the 3' end*.

molecule. The specificity of base pairing—adenine with thymine and guanine with cytosine—provides the mechanism used by all genetic replication systems. Furthermore,

- Nucleotide monomers are added one by one to the end of a growing strand by an enzyme called a *DNA polymerase*.

- New nucleotides are added only to the 3' end of the growing strand.

- The sequence of bases in each newly replicated strand, or **daughter strand**, is complementary to the base sequence in the old strand, or **parental strand**, being replicated. For example, wherever an adenine nucleotide is present in the parental strand, a thymine nucleotide will be added to the growing end of the daughter strand.

STOP & THINK 6.1

The illustration shows the template DNA strand at the top, and the growing daughter DNA strand at the bottom. Which nucleotide base is the next one added to the daughter strand?

5' [====arrow====] 3'
    T ‖‖‖‖‖‖‖ G
3' [====arrow====] 5'

The following section explains how the two strands of a daughter molecule are physically related to the two strands of the parental molecule.

DNA replication is semiconservative: The parental strands remain intact.

The mode of replication diagrammed in Figure 6.7 is called **semiconservative replication** because each parental DNA strand serves as a template for a new strand. In the semiconservative mode of replication, each parental DNA strand serves as a template for one new strand, and as each new strand is formed, it is hydrogen-bonded to its parental template. As replication proceeds, the parental double helix unwinds and then rewinds again into two new double helices, each of which contains one originally parental strand and one newly formed daughter strand.

In theory, DNA could be replicated by a number of mechanisms other than the semiconservative mode. However, the reality of semiconservative replication was demonstrated experimentally by Matthew Meselson and Franklin Stahl in 1958. Their experiment made use of a newly developed high-speed centrifuge (an *ultracentrifuge*) that could spin a solution so fast that molecules differing only slightly in density could be separated. In their experiment, the heavy ^{15}N isotope of nitrogen was used for physical separation of parental and daughter DNA molecules. DNA isolated from the bacterium *E. coli* grown in a medium containing ^{15}N as the only available source of nitrogen is denser than DNA from bacteria grown in media with the normal ^{14}N isotope. These DNA molecules can be separated in an ultracentrifuge because they have about the same density as a very concentrated solution of cesium chloride (CsCl).

When a CsCl solution containing DNA is centrifuged at high speed, the Cs^+ ions gradually settle toward the bottom of the centrifuge tube. This movement is counteracted by diffusion (the random movement of molecules), which prevents complete sedimentation. At equilibrium, a linear gradient of increasing CsCl concentration—and of density—is present from the top to the bottom of the centrifuge tube. The DNA also moves upward or downward in the tube to a position in the gradient at which the density of the solution is equal to its own density. At equilibrium, a mixture of ^{14}N-containing ("light") and ^{15}N-containing ("heavy") *E. coli* DNA will separate into two distinct zones in a density gradient, even though they differ only slightly in density. It is for this reason that the separation technique is called *equilibrium density-gradient centrifugation*.

The Meselson–Stahl experiment is a textbook example of hypothesis-driven science. In other words, they had a hypothesis for the mechanisms of DNA replication (the Watson–Crick model), derived predictions of this model that would distinguish it from other alternatives, and then carried out an experiment to learn whether the predictions would be verified or falsified. The predictions they derived are illustrated in **FIGURE 6.8**. They imagined a situation

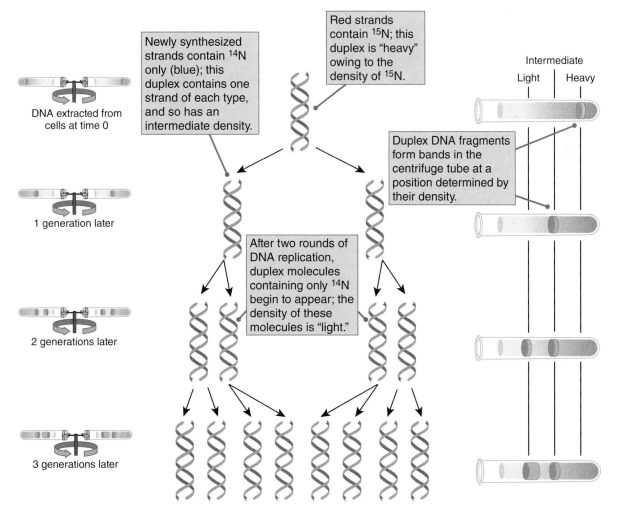

FIGURE 6.8 Predictions of semiconservative DNA replication.

in which bacteria were grown for many generations in a ^{15}N-containing medium so that all parental DNA strands would be "heavy." At one point, the cells are transferred to a ^{14}N-containing medium so that newly synthesized DNA strands will be "light." What would happen if duplex DNA were isolated from samples of cells taken from the culture at intervals, and equilibrium density-gradient centrifugation carried out to determine the density of the molecules? With semiconservative replication, the expected result of the experiment is as shown in Figure 6.8. After one round of replication, each duplex should consist of one heavy and one light strand, so all daughter molecules have intermediate density. After two rounds of replication, the duplexes containing an original parental strand would again be intermediate in density, but now there is an equal number of duplexes consisting of two light strands, so two bands differing in density are expected. After a third round of replication, DNA duplexes of light and intermediate density would again be expected, but in this generation their expected ratio of abundances are 3 : 1, as shown by the ribbon diagrams.

The actual result of the Meselson–Stahl experiment is shown in **FIGURE 6.9**. Each photograph shows the image of a centrifuge tube taken in ultraviolet light of wavelength 260 nm (nanometers), which is absorbed by DNA in solution. The positions of the DNA molecules in the density gradient are therefore indicated by the dark bands that absorb the light. Each photograph is oriented such that the bottom of the tube is at the right and the top is at the left. To the right of each photograph is a graph showing the absorbance of the ultraviolet light from the top of the centrifuge tube to the bottom. In each trace, the peaks correspond to the positions of the bands in the photographs, but the height and width of each peak allow the amount of DNA in each band to be quantified.

At the start of the experiment (generation 0), all of the DNA is heavy (^{15}N). After the transfer to ^{14}N medium, a band of lighter density begins to appear and gradually becomes more prominent as the cells replicate their DNA and divide. After 1.0 generations of growth (one round of replication of the DNA molecules and a doubling of the number of cells), all of the DNA had a "hybrid" density exactly intermediate between the densities of ^{15}N-DNA and ^{14}N-DNA. The finding of molecules with a hybrid density indicates that the replicated molecules contain equal amounts of the two nitrogen isotopes. After 1.9 generations of replication in the ^{14}N medium, approximately half of the DNA had the density of DNA with ^{14}N in both strands ("light" DNA), and the other half had the hybrid density. After 3.0 generations, the ratio of light to hybrid DNA was approximately 3 : 1, and after 4.1 generations, it was approximately 7 : 1. This distribution of ^{15}N atoms is precisely the result predicted from semiconservative replication of

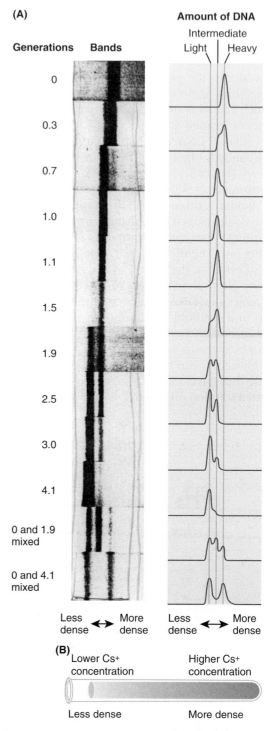

FIGURE 6.9 The Meselson–Stahl experiment on DNA replication. (A) Photographs of the centrifuge tubes taken with ultraviolet light, with the centrifuge tubes oriented as shown in part B. The smooth curves in part B show quantitatively the amount of absorption of the ultraviolet light across each tube. [Photograph reproduced from M. Meselson and F. W. Stahl, *Proc. Natl. Acad. Sci. USA* 44 (1958): 671–682.

the Watson–Crick structure, as illustrated in Figure 6.8. Similar experiments with replicating DNA from numerous viruses, bacteria, and higher organisms have also demonstrated semiconservative replication.

In the Meselson–Stahl experiment, the DNA was extensively fragmented when isolated, so the form of the molecule was unknown. Later, the isolation of unbroken molecules and their examination by other techniques showed that the DNA in *E. coli* cells is actually circular.

One alternative to semiconservative replication is *conservative replication*, in which the parental strands come apart only temporarily to serve as templates for synthesis of the daughter strands, but then come back together again as they were originally, and the two new daughter strands also form a duplex molecule. How would the DNA molecules and centrifuge tubes in Figure 6.8 have to be changed to represent the result of conservative replication after one generation of DNA replication? What single feature of Meselson and Stahl's result allowed the hypothesis of conservative replication to be rejected?

DNA strands must unwind to be replicated.

The first proof that *E. coli* DNA replicates as a circle came from an experiment in which cells were grown in a medium containing radioactive thymine (^{3}H-thymine) so that all DNA synthesized would be radioactive. The DNA was isolated without fragmentation and placed on photographic film. Each radioactive decay caused a tiny black spot to appear in the film, and after several months there were enough spots to visualize the DNA with a microscope, and the shape

of the molecule proved to be a circle with a length of 1.6 mm (4.6 Mb).

The position along a molecule at which DNA replication begins is called a **replication origin**, and the region in which parental strands are separating and new strands are being synthesized is called a **replication fork**. The process of generating a new replication fork is **initiation**. In most bacteria, bacteriophage, and viruses, DNA replication is initiated at a unique origin of replication. Furthermore, with only a few exceptions, two replication forks move in opposite directions from the origin (**FIGURE 6.10**), which means that DNA nearly always replicates bidirectionally. A replicating circle is schematically like the Greek letter θ (theta), so this mode of replication is usually called θ **replication**.

Some circular DNA molecules, including those of a number of bacterial and eukaryotic viruses, replicate by a process that does not include a θ-shaped intermediate. This replication mode is called **rolling-circle replication**. In this process, replication starts with a single-strand cleavage at a specific sugar–phosphate bond in a double-stranded circle (**FIGURE 6.11**). This cleavage produces two chemically distinct ends: a 3′ end (at which the nucleotide has a free 3′-OH group) and a 5′ end (at which the nucleotide has a free 5′-P group). The DNA is synthesized by the addition of successive deoxynucleotides to the 3′ end with simultaneous displacement of the 5′ end from the circle. As replication proceeds around the circle, the 5′ end rolls out as a tail of increasing length.

In most cases, as the tail is extended, a complementary chain is synthesized, which results in a double-stranded DNA tail. Because the displaced strand is chemically linked to the newly synthesized DNA in the circle, replication does not terminate, and extension proceeds without interruption, forming a tail that may be many times longer than the circumference of the circle. Rolling-circle replication is a common

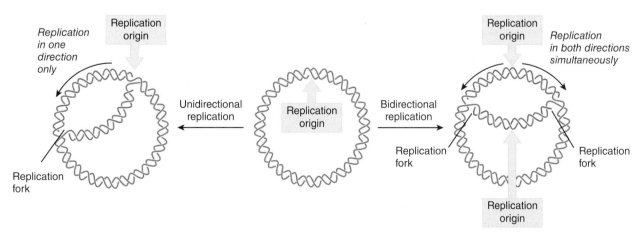

FIGURE 6.10   The distinction between unidirectional and bidirectional DNA replication. In unidirectional replication, there is only one replication fork; bidirectional replication requires two replication forks. The curved arrows indicate the direction of movement of the forks. Most DNA replicates bidirectionally.

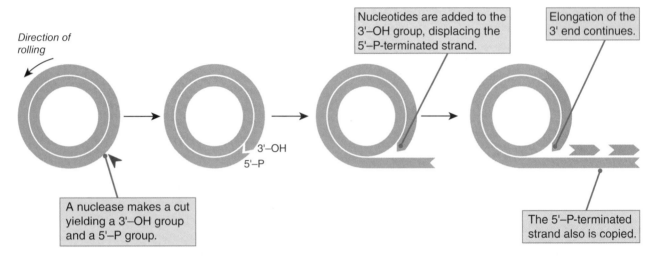

Direction of rolling

A nuclease makes a cut yielding a 3'–OH group and a 5'–P group.

3'–OH
5'–P

Nucleotides are added to the 3'–OH group, displacing the 5'–P-terminated strand.

Elongation of the 3' end continues.

The 5'–P-terminated strand also is copied.

FIGURE 6.11 Rolling-circle replication. Newly synthesized DNA is in red. The displaced strand forming the "tail" is replicated in short fragments.

feature in late stages of replication of double-stranded DNA phages that have circular intermediates.

Eukaryotic DNA molecules contain multiple origins of replication.

Although the DNA duplex in a eukaryotic chromosome is linear, it also replicates bidirectionally. Replication is initiated almost simultaneously at many sites along the DNA. The structures resulting from the numerous origins are seen in electron micrographs as multiple loops along the DNA molecule (part A of **FIGURE 6.12**). Multiple initiation is a means of reducing the total replication time of a large molecule. In eukaryotic cells, movement of each replication fork proceeds at a rate of approximately 10 to 100 nucleotide pairs per second. For example, in *D. melanogaster,* the rate of replication is about 50 nucleotide pairs per second at 25°C. Because the DNA molecule in the largest chromosome in *Drosophila* contains about 7×10^7 nucleotide pairs, replication from a single bidirectional origin of replication would take about 8 days. Developing *Drosophila* embryos actually use about 8500 replication origins per chromosome, which reduces the replication time to a few minutes. In a typical eukaryotic cell, origins are spaced about 40,000 nucleotide pairs apart, which allows each chromosome to be replicated in 15 to 30 minutes. Because not all chromosomes replicate simultaneously, complete replication of all chromosomes in eukaryotes usually takes from 5 to 10 hours.

So far, we have considered only certain geometrical features of DNA replication. In the next section, the enzymes and other proteins used in DNA replication are described.

6.5 Many proteins participate in DNA replication.

Some of the main molecular players in DNA replication are illustrated in **FIGURE 6.13**. Each player and its role will be discussed in more detail in the sections that follow, after which we will examine how they all act together. Note first that the two strands of parental DNA are replicated somewhat differently. One parental strand serves as the template for synthesis of what is called the **leading strand**, which is elongated in the direction of the replication fork in one continuous piece. The other parental strand is the template for synthesis of the **lagging strand**, which is synthesized in short **precursor fragments** that are joined together where they meet. The reason for the different modes of replication is that DNA polymerase can add nucleotides only to the 3' end of a growing chain; hence the parental strand whose 5' end is near the replication fork can be synthesized continuously, whereas the parental strand whose 3' end is near the replication fork has to be synthesized in shorter segments.

As DNA replication takes place, the parental double-stranded DNA molecule must gradually be unwound as the replication fork moves along. Unwinding the double helix to separate the parental strands requires a **helicase** protein that hydrolyzes ATP to drive the unwinding reaction. Most cells have several helicases specialized for different roles, such as replication, recombination, or repair. Once unwound, the strands of the double helix would tend to come together again spontaneously, so they must be stabilized as single strands to serve as templates for DNA synthesis. This stabilization is a function of a **single-stranded**

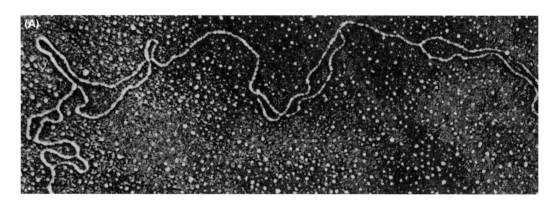

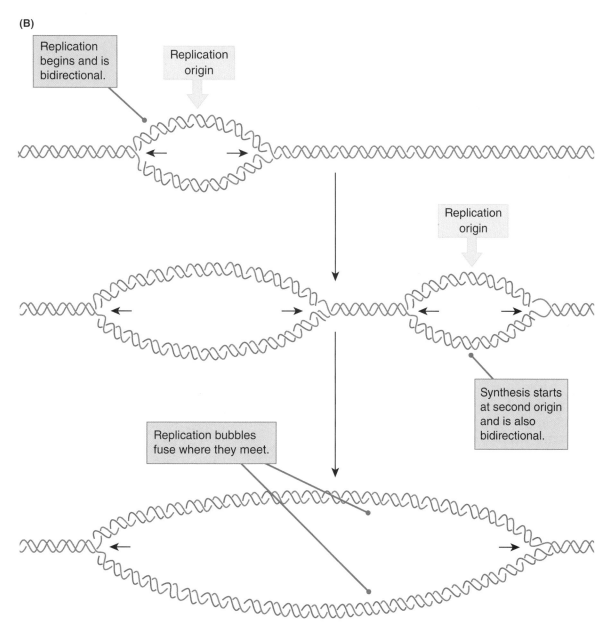

FIGURE 6.12 Replicating DNA of *Drosophila melanogaster*. (A) An electron micrograph of a 30-kb segment showing five replication loops. (B) An interpretive drawing showing how loops merge. Two replication origins are shown in the drawing. The arrows indicate the direction of movement of the replica tion forks.

Micrograph courtesy of David S. Hogness, Department of Biochemistry, Stanford School of Medicine.

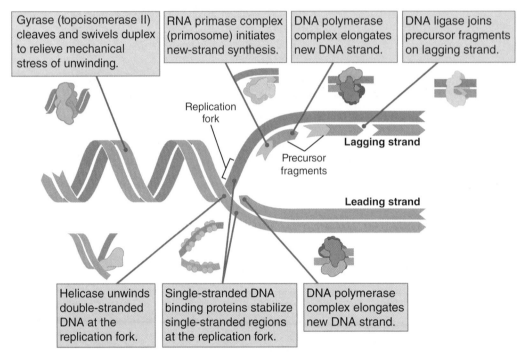

FIGURE 6.13 An overview of DNA replication highlighting the roles of some of the key proteins that are required. The DNA polymerase complex and the RNA primase complex are both composed of multiple different polypeptide subunits. The DNA polymerase that joins precursor fragments where they meet is not shown.

DNA binding protein (SSB) (Figure 6.13). The SSB binds single-stranded DNA tightly and cooperatively, and it has an affinity for single-stranded DNA at least 1000-fold greater than that for double-stranded DNA. It is this strong tendency for SSB to bind with single strands that stabilizes the templates for replication. In *E. coli*, the same SSB is used in DNA replication, recombination, and repair.

Because the two strands of a replicating helix must make a full rotation to unwind each of the turns, some kind of swivel mechanism must exist to avoid the buildup of so much stress farther along the helix that strand separation would be brought to a halt. In *E. coli*, for example, only about 10 percent of the genome could be replicated before the torsional stress caused by unwinding became too great to continue. The swivel that relieves this stress is an enzyme called **gyrase** (Figure 6.13). This enzyme cleaves both strands of a DNA duplex, swivels the ends of the broken strands to relieve the torsional stress, and then rejoins the strands (**FIGURE 6.14**). Enzymes capable of catalyzing breakage and rejoining of DNA strands are known as *topoisomerases*. Gyrase is called a topoisomerase II because it makes a double-stranded break.

As the helix is being unwound by the helicase, the template strands stabilized by SSB, and the torsional stress relieved by the gyrase, the first few nucleotides are synthesized to serve as a *primer* for elongation of the new daughter strands. Primer synthesis is considered next.

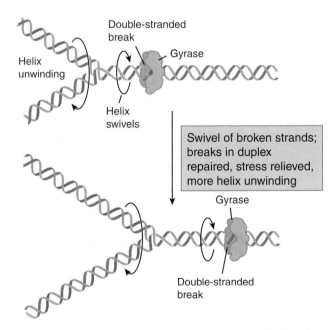

FIGURE 6.14 DNA gyrase introduces a double-stranded break ahead of the replication fork and swivels the cleaved ends around the central axis to relieve the stress of helix unwinding.

Each new DNA strand or fragment is initiated by a short RNA primer.

The major DNA polymerase is unable to initiate the synthesis of a new strand; it can only elongate an existing strand at the 3′ end. In most organisms, strand initiation is accomplished by a special type of enzyme called an **RNA polymerase**. RNA polymerases differ from DNA polymerases in that they can initiate the synthesis of RNA chains without needing a primer.

DNA synthesis is initiated by a short stretch of **primer** RNA (Figure 6.13). Bacterial primers are very short, usually two to five nucleotides. This short stretch of RNA provides a free 3′-OH onto which the DNA polymerase can add deoxynucleotides (**FIGURE 6.15**). In eukaryotic cells, the primer is synthesized by a multienzyme complex composed of 15 to 20 polypeptide chains called a **primosome**. The primer consists of an initial stretch of about 12 nucleotides of RNA to which is attached a stretch of DNA about twice as long.

🧠 STOP & THINK 6.3

A DNA template strand has the nucleotide sequence 5′–TCAAGAGT–3′. What is the nucleotide sequence of an RNA primer synthesized across this region?

DNA polymerase has a proofreading function that corrects errors in replication.

The enzyme **DNA polymerase** forms the sugar–phosphate bond (the phosphodiester bond) between adjacent nucleotides in a new DNA acid chain. The reaction catalyzed by a DNA polymerase is the formation of a phosphodiester bond between the free 3′-OH group of the chain being extended and the innermost phosphorus atom of the nucleoside triphosphate being incorporated at the 3′ end (**FIGURE 6.16**). What happens is that the 3′ hydroxyl group at the 3′ terminus of the growing strand attacks the innermost phosphate of the incoming nucleotide and forms a phosphodiester bond, releasing the two outermost phosphates. The result is as follows:

KEY CONCEPT

DNA synthesis proceeds by the elongation of primer chains, *always in the 5′ → 3′ direction*.

Recognition of the appropriate incoming nucleoside triphosphate in replication depends on base pairing with the opposite nucleotide in the template chain. DNA polymerase usually catalyzes the polymerization reaction that incorporates the new nucleotide at the primer terminus only when the correct base pair is present. The same DNA polymerase is used to add each of the four deoxynucleoside phosphates to the 3′–OH terminus of the growing strand.

Two DNA polymerases are needed for DNA replication in *E. coli*—DNA polymerase I (abbreviated Pol I) and DNA polymerase III (Pol III). Polymerase III is the major replication enzyme. Pol III exists in the cell as a large protein complex that is responsible not only for the elongation of DNA molecules but also for the initiation of the replication fork at origins of replication and the addition of deoxynucleotides to the RNA primers. Polymerase I plays an essential, but secondary, role in replication that will be described in a later section. Eukaryotic cells also contain several DNA polymerases. The key enzyme responsible for the replication of chromosomal DNA is called polymerase delta (δ). Mitochondria have their own DNA polymerase to replicate the mitochondrial DNA.

In addition to their ability to polymerize nucleotides, the major DNA polymerases also have an *exonuclease* activity that can break phosphodiester bonds

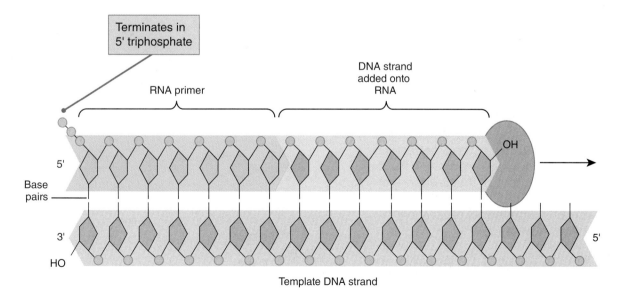

FIGURE 6.15 Each newly forming DNA strand has the structure shown here. The short stretch of RNA (green) is later removed.

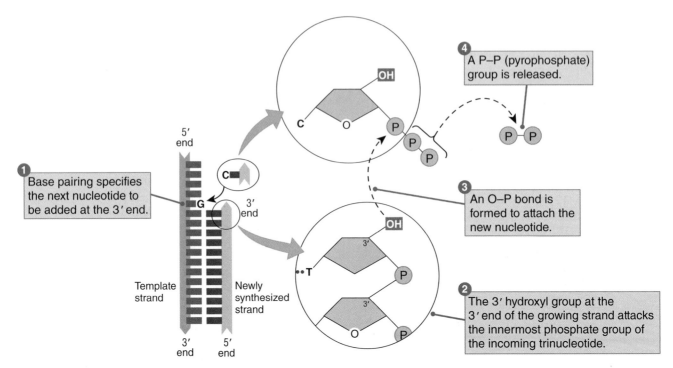

FIGURE 6.16 Addition of nucleotides to the 3′-OH terminus of a growing strand. The recognition step is shown as the formation of hydrogen bonds between the A and the T. The chemical reaction occurs when the 3′-OH group of the 3′ end of the growing chain attacks the innermost phosphate group of the incoming trinucleotide.

in the sugar–phosphate backbones of nucleic acid chains. DNA polymerases I and III of *E. coli* have an exonuclease activity that acts only at the 3′ terminus and removes the nucleotide added most recently. This exonuclease activity provides a built-in mechanism for correcting rare errors in polymerization. Occasionally, a polymerase adds to the growing chain a nucleotide with an incorrectly paired base. The presence of an unpaired base activates the exonuclease activity, which cleaves the unpaired nucleotide from the 3′-OH end of the growing chain (**FIGURE 6.17**). Because it cleaves off an incorrect nucleotide and gives the polymerase another chance to get it right, the exonuclease activity of DNA polymerase is also called the **editing function** or **proofreading function**. The proofreading function can "look back" only one base (the one added last). Nevertheless,

> ## KEY CONCEPT
>
> The genetic significance of the proofreading function is that it is an error-correcting mechanism that serves to reduce the frequency of mutation resulting from the incorporation of incorrect nucleotides in DNA replication.

One strand of replicating DNA is synthesized in pieces.

Because DNA polymerase can elongate a newly synthesized DNA strand only at its 3′ end, within a single

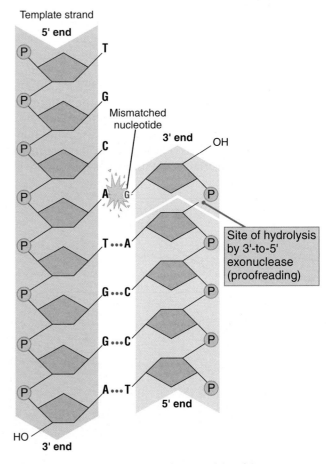

FIGURE 6.17 The 3′-to-5′ exonuclease activity of the proofreading function. The growing strand is cleaved to release a nucleotide containing the base G, which does not pair with the base A in the template strand.

⬣ THE HUMAN CONNECTION

Sickle-Cell Anemia: The First "Molecular Disease"

Vernon M. Ingram 1957

Cavendish Laboratory, University of Cambridge, England

Gene Mutations in Human Hemoglobin: The Chemical Difference Between Normal and Sickle-Cell Hemoglobin

The mutation in sickle-cell anemia results in a change in the molecular structure of hemoglobin, but what is the nature of this change? Ingram studied peptide fragments of normal and sickle-cell hemoglobin obtained by digestion with the protease enzyme trypsin (tryptic digests). He found that the only difference resided in a peptide fragment of eight amino acids. To study this fragment further, he used a method of "fingerprinting," in which digests of the peptide containing still smaller fragments were resolved into spots on a sheet of filter paper, first by separating the fragments on the basis of charge along one edge of the paper (electrophoresis) and then by separating on the basis of solubility (chromatography) in the other direction. The complete sequence of the peptide that differed between normal and sickle-cell hemoglobin was deduced after determining the amino acid sequence of each of the short peptides in the fingerprints. In this case, the normal peptide has the amino acid sequence

Val–His–Leu–Thr–Pro–<u>Glu</u>–Glu–Lys

(V–H–L–T–P–<u>E</u>–E–K in the single-letter codes), whereas that from sickle-cell hemoglobin has the sequence

Val–His–Leu–Thr–Pro–<u>Val</u> –Glu–Lys

> **❝ The difference consists in a replacement of only one of nearly 300 amino acids—a very small change indeed. ❞**

(V–H–L–T–P–<u>V</u> –E–K). The only difference is in the underlined amino acid. This was the first evidence that genes may code for polypeptides in a relatively simple manner, in which successive bits of DNA sequence encode successive amino acids in the polypeptide chain. (There were a few minor errors in Ingram's peptide sequences, which have been corrected here.)

[Among the many] amino acids in the two proteins, only one is different. Tryptic digests of the two proteins . . . were separated on paper using electrophoresis in one direction and chromatography in the other . . . All peptides had identical chromatographic properties, except for one spot . . . Partial hydrolysis of this peptide . . . followed by "fingerprinting" gave the products in the accompanying Figure. The difference consists in a replacement of only one of nearly 300 amino acids—a very small change indeed.

Nature 180: 326–328.

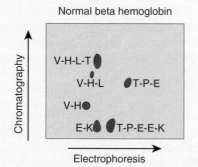
Normal beta hemoglobin

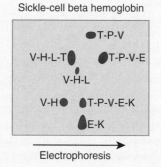
Sickle-cell beta hemoglobin

replication fork both strands grow in the 5′ → 3′ orientation, which means that they grow in opposite directions along the parental strands (Figure 6.13). One strand of the newly made DNA is synthesized continuously, while the other strand is made in small precursor fragments. The size of the precursor fragments is 1000–2000 base pairs in prokaryotic cells and 100–200 base pairs in eukaryotic cells.

Precursor fragments are joined together when they meet.

The precursor fragments are ultimately joined to yield a continuous strand of DNA. This strand contains no RNA sequences, so the final stitching together of the precursor fragments must require:

- Removal of the RNA primer
- Replacement with a DNA sequence
- Joining where adjacent DNA fragments come into contact

Primer removal and replacement in *E. coli* is accomplished by a special DNA polymerase (Pol I), which removes one ribonucleotide at a time through its exonuclease activity and replaces it with a deoxyribonucleotide through its polymerase activity. In eukaryotes, the primer RNA is removed as an intact unit (**FIGURE 6.18**). When the polymerase complex meets the RNA of the next precursor fragment in line (part A), a protein called replication protein A (RPA) joins the complex. RPA is a single-stranded DNA binding protein that unwinds the RNA and a short segment of DNA from the double helix and stabilizes the unwound single strand by binding to it (Figure 6.18, part B). RPA also recruits endonucleases that cleave the unwound single strand from the double helix, and these also cleave the bond connecting the RNA and DNA stretches in the excised segment. The polymerase complex then replaces the excised segment with DNA nucleotides, and the enzyme DNA ligase catalyzes the formation of the final bond connecting the two precursor fragments (part C). As this is happening, the RNA and DNA components of the excised segment are broken down by enzymes, and the nucleotides are recycled.

Synthesis of the leading strand and the lagging strand are coordinated.

FIGURE 6.19 is a diagram of DNA replication with all of the major components of the process in place, but it still features one major oversimplification. As it stands, Figure 6.19 makes it look like replication of each template strand takes place without regard to the other, while in reality the replication of both template strands is carefully coordinated. How this happens is shown in **FIGURE 6.20**, where for simplicity most of the participating proteins have again been ignored. Coordination of leading-strand and lagging-strand synthesis is achieved by twisting the template of the lagging strand into a loop. This brings the polymerase complex of the lagging strand into proximity with that of the leading strand, where the two are joined together by a protein clamp. Thus joined, if for any reason one or the other polymerase complex slows or stalls (for example, to allow time for damaged DNA to be repaired), the other slows or stalls also. The polymerase clamp is temporary. It is released when the polymerase complex of the lagging strand butts into the RNA primer of the

precursor fragment ahead. By this time the replication fork has moved forward, a new RNA primer is produced, the DNA polymerase complex joins the lagging strand, and it is again clamped to the polymerase complex of the leading strand. Because of the looped DNA in the template of the lagging strand, this model of DNA replication is known as the **trombone model**.

6.6 Knowledge of DNA structure makes possible the manipulation of DNA molecules.

This and the following sections show how our knowledge of DNA structure and replication has been put to practical use in the development of procedures for the isolation and manipulation of DNA.

Single strands of DNA or RNA with complementary sequences can hybridize.

One of the most important features of DNA is that the two strands of a duplex can be separated by heat without breaking any of the phosphodiester bonds that join successive nucleotides in each strand. If the temperature is maintained sufficiently high, random molecular motion will keep the strands apart. If the temperature is lowered so that hydrogen bonding between complementary base sequences is stable, then under the proper conditions, two single strands that are complementary or nearly complementary in sequence can come together to form a different double helix. The separation of DNA strands is called **denaturation**, and the coming together **renaturation**. The practical applications of denaturation and renaturation are many:

- A small part of a DNA fragment can be renatured with a much larger DNA fragment. This principle is used in identifying specific DNA fragments in a complex mixture.

- A DNA fragment from one gene can be renatured with similar fragments from other genes in the same genome; this principle is used to identify genes that are similar, but not identical, in sequence and that have related functions.

The process of renaturating DNA strands from two different sources is called **nucleic acid hybridization** because the double-stranded molecules are "hybrid." The initial phase of hybridization is a slow process because the rate is limited by the random chance that a region of two complementary strands will come together to form a short sequence of correct base pairs. This initial pairing step is followed by a rapid pairing of the remaining complementary bases and rewinding of the helix. Rewinding is accomplished in a matter of seconds, and its rate is independent of DNA concentration because the complementary strands have already found each other.

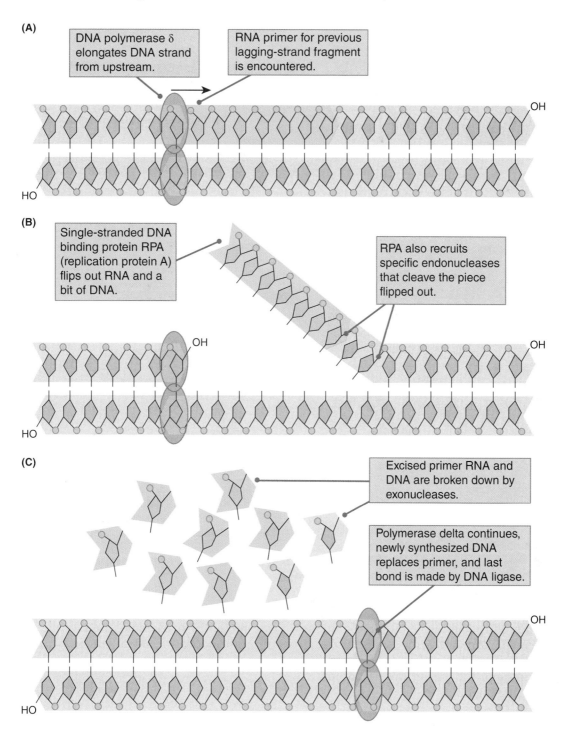

(A)

DNA polymerase δ elongates DNA strand from upstream.

RNA primer for previous lagging-strand fragment is encountered.

OH

HO

(B)

Single-stranded DNA binding protein RPA (replication protein A) flips out RNA and a bit of DNA.

RPA also recruits specific endonucleases that cleave the piece flipped out.

OH

OH

HO

(C)

Excised primer RNA and DNA are broken down by exonucleases.

Polymerase delta continues, newly synthesized DNA replaces primer, and last bond is made by DNA ligase.

OH

HO

FIGURE 6.18 Sequence of events in the joining of adjacent precursor fragments in eukaryotes.

The example of nucleic acid hybridization in **FIGURE 6.21** will enable us to understand some of the molecular details and also to see how hybridization is used to "tag" and identify a particular DNA fragment. Shown in part A is a solution of denatured DNA, called the **probe**, in which each molecule has been labeled with either radioactive atoms or light-emitting molecules. Probe DNA usually contains denatured forms of both strands present in the original duplex molecule. Part B in Figure 6.21 is a diagram of genomic DNA fragments that have been immobilized on a nitrocellulose filter. When the probe is mixed with the genomic fragments (part C), random collisions bring short, complementary stretches together. If the region of complementary sequence is short (part D), then random collision cannot initiate renaturation because the flanking sequences cannot pair; in this case the probe falls off almost immediately. If, however, a collision brings short sequences together in the correct register (part E), then this initiates renaturation, because the pairing proceeds zipper-like from the initial contact. The main point is that DNA fragments are able to hybridize only if the length of the region in which they can pair is sufficiently long. Some mismatches in

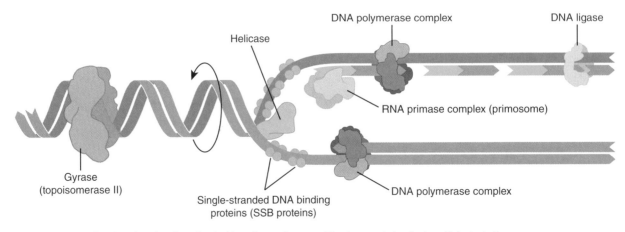

FIGURE 6.19 DNA replication showing the physical locations of some of the key proteins that participate in the process.

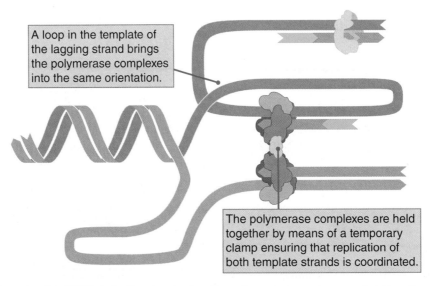

A loop in the template of the lagging strand brings the polymerase complexes into the same orientation.

The polymerase complexes are held together by means of a temporary clamp ensuring that replication of both template strands is coordinated.

FIGURE 6.20 The trombone model of DNA replication showing how the polymerase complexes are held together by a protein clamp, which allows replication of both template strands to be coordinated.

the paired region can be tolerated. How many mismatches are allowed is determined by the conditions of the experiment: The lower the temperature at which the hybridization is carried out, and the higher the salt concentration, the greater the proportion of mismatches that are tolerated.

Restriction enzymes cleave duplex DNA at particular nucleotide sequences.

One of the problems with breaking large DNA molecules into smaller fragments by random shearing is that the fragments containing a particular gene, or part of a gene, will all be of different sizes. With random shearing, because of the random length of each fragment, it is not possible to isolate and identify a *particular* DNA fragment. However, there is an important enzymatic technique, described in this section, that can be used for cleaving DNA molecules at specific sites.

Members of a class of enzymes known as **restriction enzymes** or, more specifically, as *restriction endonucleases*, are able to cleave DNA molecules at the

positions at which particular, short sequences of bases are present. For example, the enzyme *Bam*HI recognizes the double-stranded sequence

```
5'-GGATCC-3'
3'-CCTAGG-5'
```

and cleaves each strand between the G-bearing nucleotides shown in red. **FIGURE 6.22** shows the recognition sequence for *Bam*HI and the cleavage reaction that takes place.

TABLE 6.2 lists six of the several hundred restriction enzymes that are known. Most restriction enzymes are isolated from bacteria, and they are named after the species in which they were found. *Bam*HI, for example, was isolated from *Bacillus amyloliquefaciens* strain H, and it is the first (I) restriction enzyme isolated from this organism. Most restriction enzymes recognize only one short base sequence, usually four or six nucleotide pairs. The enzyme binds with the DNA at these sites and makes a break in each strand of the DNA molecule, producing 3'–OH and 5'–P groups at each position

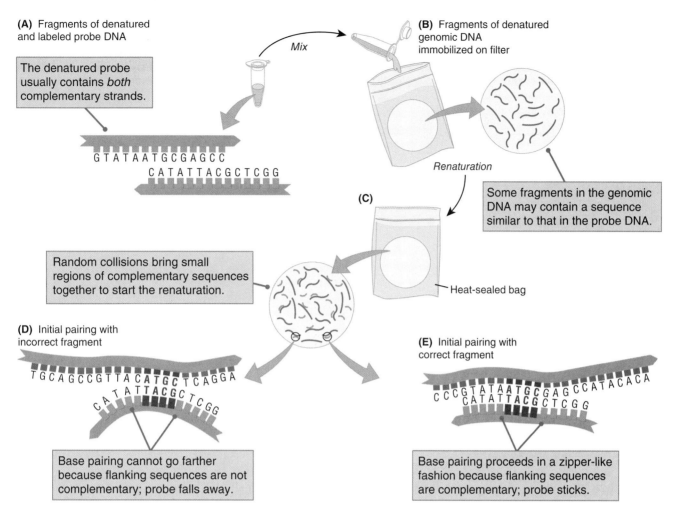

FIGURE 6.21 Nucleic acid hybridization. (A) Duplex molecules of probe DNA (obtained from a clone) are denatured and (B) placed in contact with a filter to which is attached denatured strands of genomic DNA. (C) Under the proper conditions of salt concentration and temperature, short complementary stretches come together by random collision. (D) If the sequences flanking the paired region are not complementary, then the pairing is unstable and the strands come apart again. (E) If the sequences flanking the paired region are complementary, then further base pairing stabilizes the renatured duplex.

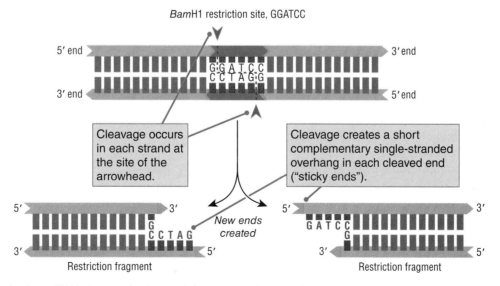

FIGURE 6.22 Mechanism of DNA cleavage by the restriction enzyme *Bam*HI. Wherever the duplex contains a *Bam*HI restriction site, the enzyme makes a single cut in the backbone of each DNA strand. Each cut creates a new 3′ end and a new 5′ end, separating the duplex into two fragments. In the case of *Bam*HI the cuts are staggered cuts, so the resulting ends terminate in single-stranded regions, each four nucleotides in length.

TABLE 6.2 Some Restriction Endonucleases, Their Sources, and Their Cleavage Sites

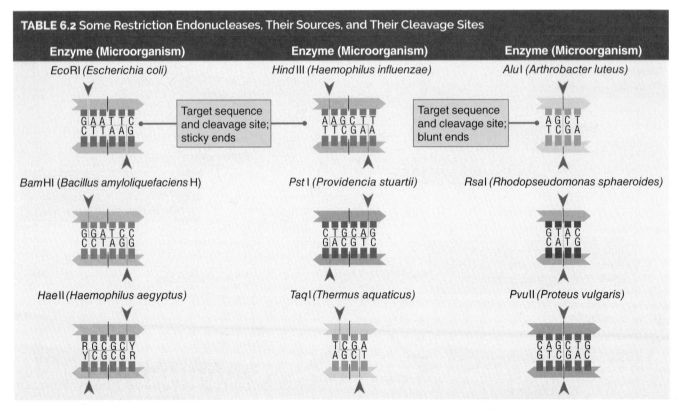

Enzyme (Microorganism)	Enzyme (Microorganism)	Enzyme (Microorganism)

Note: The vertical dashed line indicates the axis of symmetry in each sequence. Red arrows indicate the sites of cutting. The enzyme *Taq*I yields cohesive ends consisting of two nucleotides, whereas the cohesive ends produced by the other enzymes contain four nucleotides. R and Y refer respectively to any complementary purines and pyrimidines.

(Figure 6.22). The nucleotide sequence recognized for cleavage by a restriction enzyme is called the **restriction site** of the enzyme. Six restriction enzymes in Table 6.2 cleave their restriction site asymmetrically (at a different site on the two DNA strands), but three restriction enzymes cleave symmetrically (at the same site in both strands). The former leave *sticky ends* because each end of the cleaved site has a small, single-stranded overhang that is complementary in base sequence to the other end. In contrast, enzymes that have symmetrical cleavage sites yield DNA fragments that have *blunt ends*.

In virtually all cases, the restriction site of a restriction enzyme reads the same on both strands, provided that the opposite polarity of the strands is taken into account; for example, each strand in the restriction site of *Bam*HI reads 5′-GGATCC-3′. A DNA sequence with this type of symmetry is called a *palindrome*. (In ordinary English, a palindrome is a word or phrase that reads the same forward and backward; for example "madam.")

Restriction enzymes have the following important characteristics:

- Most restriction enzymes recognize a single restriction site.

- The restriction site is recognized without regard to the source of the DNA.

- Because most restriction enzymes recognize a unique restriction-site sequence, the number of

cuts in the DNA from a particular organism is determined by the number of restriction sites that are present.

The DNA fragment produced by a pair of adjacent cuts in a DNA molecule is called a **restriction fragment**. A large DNA molecule will typically be cut into many restriction fragments of different sizes. For example, an *E. coli* DNA molecule, which contains 4.6×10^6 base pairs, is cut into several hundred to several thousand fragments, and mammalian nuclear DNA is cut into more than a million fragments.

Because of the sequence specificity of cleavage, *a particular restriction enzyme produces a unique set of fragments for a particular DNA molecule*. Another enzyme will produce a different set of fragments from the same DNA molecule. In **FIGURE 6.23**, this principle is illustrated for the digestion of a circular molecule of double-stranded DNA with a length of 10 kb. When digested with the restriction enzyme *Eco*RI (Figure 6.23A), the circular molecule yields bands of 4 kb and 6 kb. This pattern would result from *Eco*RI restriction sites located in the circle at the relative positions shown beneath the gel. The circle is oriented arbitrarily with one of the *Eco*RI (*E*) sites at the top. Similarly, digestion of the circle with the enzyme *Bam*HI (Figure 6.23B) results in bands of 3 kb and 7 kb, which implies that the circle contains *Bam*HI sites at the positions indicated in the diagram beneath the gel. Again the circle is oriented arbitrarily, this time with one of the *Bam*HI (B) sites located at

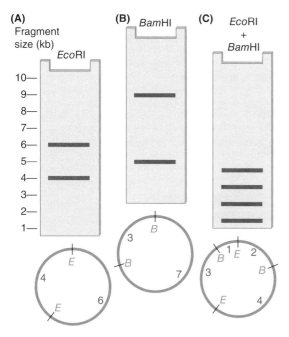

FIGURE 6.23 Gel diagrams showing the sizes of restriction fragments produced by digestion of a 10-kb circular molecule of double-stranded DNA with (A) *Eco*RI, (B) *Bam*HI, and (C) both enzymes together. Beneath each diagram is a restriction map of the circular DNA showing the locations of the restriction sites. The restriction map in (C) takes into account those in (A) and (B) as well as the fragment sizes produced by digestion with both enzymes.

the top. A diagram showing sites of cleavage of one or more restriction sites along a DNA molecule is called a **restriction map.**

When both *Eco*RI and *Bam*HI are used together, the resulting DNA fragments reveal where the *Eco*RI sites and the *Bam*HI sites are located relative to each other. In this case digestion with both enzymes yields bands of 1 kb, 2 kb, 3 kb, and 4 kb (Figure 6.23C). The restriction map shown beneath the gel indicates where the two types of restriction sites must be located in order to yield these band sizes. This restriction map can be obtained by superimposing that in part B over that in part A and rotating it until the distances between adjacent pairs of restriction sites equal 1, 2, 3, and 4 kb (not necessarily in that order). In this case one need only rotate the restriction map in part B a distance of 2 kb to the right. Note that, in the restriction-enzyme digest in Figure 6.23C, the 4-kb fragment is not the same 4-kb fragment as observed in part A, and the 3-kb fragment is not the same 3-kb fragment as observed in part B. This discordance arises because each restriction enzyme cleaves the fragments produced by the other. The orientation of the restriction map in Figure 6.23C is arbitrary. It can be flipped over or rotated by any amount in any direction, and it is still the same restriction map.

Once a restriction map of a DNA molecule has been determined, particular DNA fragments can be isolated by cutting out the small region of the gel that contains the fragment and removing the DNA from the gel. One important use of isolated restriction fragments employs the enzyme DNA ligase to insert them

into self-replicating molecules such as bacteriophage, plasmids, or even small artificial chromosomes. Using such procedures to transfer DNA from the genome of one organism into the genome of another organism constitutes one form of *genetic engineering*.

6.7 The polymerase chain reaction makes possible the amplification of a particular DNA fragment.

It is also possible to obtain large quantities of a particular DNA sequence merely by selective replication. The method for selective replication is called the **polymerase chain reaction (PCR)**, and it uses DNA polymerase and a pair of short, synthetic oligonucleotides, usually about 20 to 30 nucleotides in length, that are complementary in sequence to the ends of the DNA sequence to be amplified and so can serve as primers for strand elongation. Starting with a mixture containing as little as one molecule of the fragment of interest, repeated rounds of DNA replication increase the number of molecules exponentially. For example, starting with a single molecule, 25 rounds of DNA replication will result in $2^{25} = 3.4 \times 10^7$ molecules. This number of molecules of the amplified fragment is so much greater than that of the other unamplified molecules in the original mixture that the amplified DNA can often be used without further purification. For example, a single fragment of 3000 base pairs in *E. coli* accounts for only 0.06 percent of the total DNA in this organism. However, if this single fragment were replicated through 25 rounds of replication, 99.995 percent of the resulting mixture would consist of the amplified sequence.

An outline of the polymerase chain reaction is shown in **FIGURE 6.24**. The DNA sequence to be amplified and the oligonucleotide sequences are shown in contrasting colors. The oligonucleotides act as primers for DNA replication because they anneal to the ends of the sequence to be amplified and become the substrates for chain elongation by DNA polymerase. In the first cycle of PCR amplification, the DNA is denatured to separate the strands. The denaturation temperature is usually around 95°C. Then the temperature is decreased to allow annealing in the presence of a vast excess of the primer oligonucleotides. The annealing temperature is typically in the range of 50°C to 60°C, depending largely on the G + C content of the oligonucleotide primers. The temperature is raised slightly, to about 70°C, for the elongation of each primer. The first cycle in PCR produces two copies of each molecule containing sequences complementary to the primers. The second cycle of PCR is similar to the first. The DNA is denatured and then renatured in the presence of an excess of primer oligonucleotides, whereupon the primers are elongated by DNA polymerase; after this cycle there are four copies of each molecule present

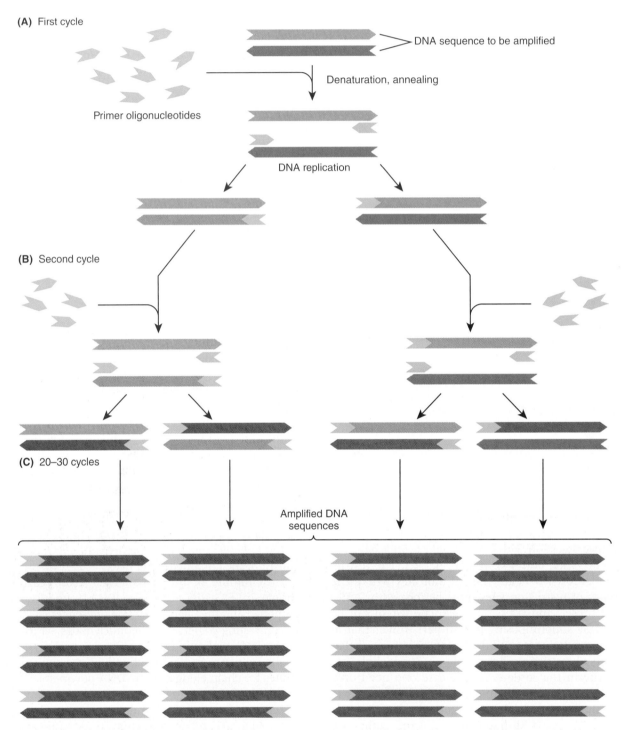

(A) First cycle

Primer oligonucleotides

DNA sequence to be amplified

Denaturation, annealing

DNA replication

(B) Second cycle

(C) 20–30 cycles

Amplified DNA sequences

FIGURE 6.24 Polymerase chain reaction (PCR) for amplification of particular DNA sequences. Only the region to be amplified is shown. Oligonucleotide primers (green) that are complementary to the ends of the target sequence (blue) are used in repeated rounds of denaturation, annealing, and DNA replication. Newly replicated DNA is shown in pink. The number of copies of the target sequence doubles in each round of replication, eventually overwhelming any other sequences that may be present.

in the original mixture. The steps of denaturation, renaturation, and replication are repeated from 20 to 30 times, and in each cycle, the number of molecules of the amplified sequence is doubled. The theoretical result of 25 rounds of amplification is 2^{25} copies of each template molecule present in the original mixture.

Implementation of PCR with conventional DNA polymerases is not practical, because at the high

temperature necessary for denaturation, the polymerase is itself irreversibly unfolded and becomes inactive. However, DNA polymerase isolated from certain bacteria is heat stable because the organisms normally live in hot springs at temperatures well above 90°C, such as are found in Yellowstone National Park. These organisms are said to be *thermophiles*. The most widely used heat-stable DNA polymerase is called *Taq* polymerase because it was originally isolated from the thermophilic bacterium *Thermus aquaticus*.

PCR amplification is very useful for generating large quantities of a specific DNA sequence. The principal limitation of the technique is that the DNA sequences at the ends of the region to be amplified must be known so that primer oligonucleotides can be synthesized. In addition, sequences longer than about 5000 base pairs cannot be replicated efficiently by conventional PCR procedures. On the other hand, there are many applications in which PCR amplification is useful. PCR can be employed to study many different mutant alleles of a gene whose wildtype sequence is known in order to identify the molecular basis of the mutations. Similarly, variation in DNA sequence among alleles present in natural populations can easily be determined using PCR. The PCR procedure has also come into widespread use in clinical laboratories for diagnosis. To take just one very important example, the presence of the human immunodeficiency virus (HIV), which causes acquired immune deficiency syndrome (AIDS), can be detected in trace quantities in blood banks by means of PCR using primers complementary to sequences in the viral genetic material. These and other applications of PCR are facilitated by the fact that the procedure lends itself to automation—for example, the use of mechanical robots to set up the reactions.

6.8   Chemical terminators and other methods are used to determine the base sequence.

A great deal of information about gene structure and gene expression can be obtained by direct determination of the sequence of bases in a DNA molecule. No technique can determine the sequence of bases in an entire chromosome in a single experiment, and so chromosomes are first cut into fragments of a size that can be sequenced easily. To obtain the sequence of a long stretch of DNA, a set of overlapping fragments must be prepared, the sequence of each is determined, and all sequences are then combined.

The **dideoxy sequencing method** employs DNA synthesis in the presence of small amounts of nucleotides that contain the sugar **dideoxyribose** instead of deoxyribose (**FIGURE 6.25**). Dideoxyribose lacks the 3'-OH group, which is essential for attachment of the next nucleotide in a growing DNA strand, so incorporation of a dideoxynucleotide instead of a deoxynucleotide immediately terminates further synthesis of the strand. To sequence a DNA strand, a DNA synthesis reaction is carried out with all four normal deoxynucleotide triphosphate precursors. The reaction mixture also contains a small amount of each of the dideoxynucleotide triphosphate analogs, each labeled with a chemical group that emits a different fluorescent

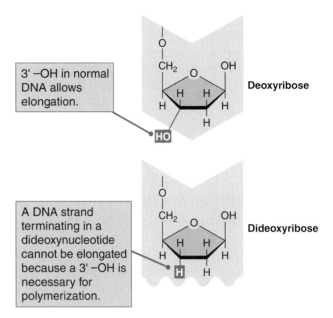

FIGURE 6.25   Structures of normal deoxyribose and the dideoxyribose sugar used in DNA sequencing. The dideoxyribose has a hydrogen atom (red) attached to the 3' carbon, in contrast with the hydroxyl group (red) at this position in deoxyribose. Because the 3' hydroxyl group is essential for the attachment of the next nucleotide in line in a growing DNA strand, the incorporation of a dideoxynucleotide immediately terminates synthesis.

wavelength. The concentrations of the normal nucleotides and the dideoxynucleotides are adjusted so that, at each step of synthesis, the daughter strand is much more likely to incorporate the normal nucleotide than the dideoxynucleotide. At each elongation step, however, a small fraction of the daughter molecules incorporates a dideoxynucleotide that prevents synthesis from continuing. The result is a set of DNA fragments of differing lengths, each of which terminates with one of the fluorescence-labeled dideoxynucleotides at its 3' end. The length of each fragment is determined by the position in the daughter strand at which the dideoxynucleotide was incorporated. The sizes of the fragments produced by chain termination are determined by gel electrophoresis, and the base sequence is then determined by the following rule:

KEY CONCEPT

If a fragment containing *n* nucleotides is generated in the reaction containing a particular dideoxynucleotide (determined by the color of the fluorescent band), then position *n* in the *daughter strand* is occupied by the base present in the dideoxynucleotide. The numbering is from the 5' nucleotide of the primer.

For example, if a 400-base fragment was terminated by the dideoxy form of dATP, then the 400th base in the daughter strand produced by DNA synthesis must be an adenine (A). Because most native duplex DNA molecules consist of complementary strands, it does not matter whether the sequence of the template strand or the daughter strand is determined. The sequence of the template strand can be deduced from the daughter strand because their nucleotide sequences are complementary. In practice, however, both strands of a molecule are usually sequenced independently and compared in order to minimize errors.

The incorporation of a dideoxynucleotide terminates strand elongation.

The procedure for sequencing a DNA fragment is diagrammed in **FIGURE 6.26**. The sequencing reaction is carried out in the presence of a small amount of fluorescently labeled dideoxynucleotides (G, black; A, green; T, red; C, purple). The products of DNA synthesis are then separated by electrophoresis in a capillary tube. In principle, the sequence can be read directly from the gel. Starting at the bottom, the sequence of the newly synthesized strand reads

5'-GACGCTGCGA-3'

However, a substantial improvement in efficiency is accomplished by continuing the electrophoresis until each band, in turn, drops off the bottom of the gel. As each band comes off the bottom of the gel, the fluorescent dye that it contains is excited by laser light, and the color of the fluorescence is read automatically by a photocell and recorded in a computer.

Frederick Sanger invented dideoxy sequencing at Cambridge University in 1977, a feat for which he was later awarded a Nobel Prize (his second—the first was for protein sequencing). Initial costs were high—upwards of $30 per base pair—but came down steadily with improvements such as fluorescent terminators. By the mid-1990s, the cost was down to about $3 per base pair, and by 2000, to about 3 cents per base pair. Currently, the cost is about 0.00003 cents per base pair. To put this in concrete terms, since the year 2000 the cost of sequencing one human genome has dropped from about $100 million to about $1000. As things stand now, it often costs more to prepare DNA for sequencing and to store the sequence in a database than it costs to determine the sequence. Since 1986, when the first automated DNA-sequencing machine was marketed, the main reason for the decrease in cost is the development of new methods of sequencing and successive generations of ever-more-powerful machines that can determine and electronically record millions of DNA sequences in parallel. These methods of high-throughput sequencing and their implications for your own personal genome are discussed in Chapter 10.

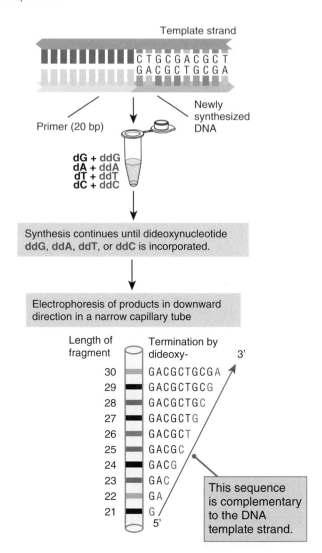

FIGURE 6.26 Dideoxy method of DNA sequencing. The terminated DNA fragments are separated by size by means of electrophoresis in a capillary tube. The sequence of the daughter strand can be read manually from the bottom to the top of the tube according to the color of each band as 5'-GACGCTGCGA-3'.

⚡ STOP & THINK 6.4

Shown here is a small part of a dideoxy sequencing gel with the DNA fragments coded by color according to the dideoxy nucleotide with which they terminate at the 3' end. The color coding is A = green, T = red, G = black, and C = purple. What is the sequence of the region of the daughter strand indicated in the gel? What is the sequence of the template strand?

Top (longer DNA fragments)   Bottom (shorter DNA fragments)

CHAPTER SUMMARY

- Prokaryotes generally have smaller genomes (less DNA) than higher eukaryotes.

- Among eukaryotes, there is no consistent relationship between genome size and organismic complexity.

- A DNA strand is a polymer of A, T, G, and C deoxyribonucleotides joined 3′ to 5′ by phosphodiester bonds.

- Hydrogen bonding between the A–T and G–C base pairs helps hold the two DNA strands in a duplex together, as does the stacking of the base pairs in the duplex molecule.

- In DNA replication, each parental strand serves as a template for a daughter strand that is synthesized in the 5′-to-3′ direction (successive nucleotides are added only at the 3′ end).

- Each type of restriction endonuclease enzyme cleaves double-stranded DNA at a particular sequence of bases, usually 4 or 6 nucleotides in length. In the polymerase chain reaction, short oligonucleotide primers are used in successive cycles of DNA replication to amplify selectively a particular region of a DNA duplex.

- The DNA fragments produced by a restriction enzyme can be separated by electrophoresis, isolated, sequenced, and manipulated in other ways.

- Dideoxynucleotides, which terminate strand elongation whenever they are incorporated into replicating DNA, are widely used in automated DNA sequencing.

- Newer methods of DNA sequencing can analyze millions of DNA strands in parallel and yield billions of base pairs of sequence per day.

ISSUES AND IDEAS

- What are the four bases commonly found in DNA? Which of these form hydrogen-bonded base pairs in duplex DNA?

- What is the relationship between the amount of DNA in a somatic cell and the amount in a gamete?

- What chemical feature at the 3′ end of a DNA strand in the process of being synthesized is essential for elongation? Can the strand also be elongated at the 5′ end?

- What does it mean to say that the two strands in duplex DNA are antiparallel?

- If the paired strands in duplex DNA were parallel rather than antiparallel, would replication still involve a leading strand and a lagging strand? Explain.

- Why is the polymerase chain reaction so extremely specific in amplifying a single region of DNA? Why is the technique so extremely powerful in multiplying the sequence?

- What feature of DNA replication guarantees that the incorporation of a dideoxynucleotide will terminate strand elongation?

SOLUTIONS: STEP BY STEP

PROBLEM 1 A researcher wishes to amplify the DNA fragment shown here by means of the polymerase chain reaction, using the 8 nucleotides at each end as binding sites for the primer oligonucleotides. The unspecified sequence in the center is 3.0 kb in length. (Primers of 8 nucleotides are chosen for simplicity, even though they would normally be too short to ensure specificity of amplification.) What sequences should be synthesized to form the primer oligonucleotides? Be sure to specify the 3′ and 5′ ends of each.

```
5′-ATGGTGCANN • • • NNAGAAGTCT-3′
3′-TACCACGTNN • • • NNTCTTCAGA-5′
```

SOLUTION. The principle to be used in this problem is that, when the primer forms a duplex with its complementary sequence in the template strand, the 3′ end of the primer should be facing the region to be amplified. This will ensure amplification of the region between the primers. In this case, the primer sequences should be 5′-ATGGTGCA-3′, which uses the bottom strand as its template and binds to the left-hand side and 3′-TCTTCAGA-5′, which uses the top strand as its template and binds to the right-hand side.

PROBLEM 2 The polymerase chain reaction is used to amplify a region of human DNA of length 3.0 kb from a DNA solution prepared from nuclei of human cells. The human genome has a size of approximately 3.0×10^9 base pairs per haploid genome.
(a) Prior to amplification, what proportion of the DNA in the solution consists of the 3.0-kb target sequence? Assume that the target sequence is present in one copy per haploid genome.
(b) Each round of amplification doubles the number of target molecules. How many rounds of replication would be required to reach a stage in which the amplified sequence constitutes more than 99.9 percent of the DNA in the solution?

SOLUTION. **(a)** The original DNA solution contains one 3.0-kb target sequence per 3000 Mb haploid genome. The proportion of DNA consisting of the target sequence is, therefore,

$$\frac{3.0 \times 10^3}{3.0 \times 10^9} = 1 \times 10^{-6} = 0.0001 \text{ percent}$$

(b) Because each round of amplification doubles the number of target molecules, after n rounds of replication there will be 2^n target molecules for each haploid human genome present in the original solution. Each of these has a length of 3000 bp, so the total amount of amplified target DNA will be $2^n \times 3000$ bp. This DNA is newly created and, therefore, increases the total amount of DNA in the solution. After n rounds of replication, the amount of DNA present per haploid genome is $2^n \times 3000$ bp (the newly created material) $+ 3.0 \times 10^9$ bp (the original material). The question asks for the value of n for which the fraction of newly created DNA constitutes 99.9 percent of the total DNA in solution. The inequality to be solved is

$$\frac{2^n \times 3000}{2^n \times 3000 + 3.0 \times 10^9} \geq 0.999$$

from which we obtain

$$n \geq \frac{1}{\log(2)} \times \log\left[\frac{(3.0 \times 10^9)(0.999)}{(3000)(1-0.999)}\right] = 29.9$$

This result implies that 30 rounds of amplification increase the percentage of target DNA in the solution by a factor of almost 10^6.

PROBLEM 3 A solution containing single-stranded DNA with the sequence

5'-ATGGTGCACCTGACTCCTGAGGAGAAGTCTNNNNNNNNN-3'

undergoes DNA replication *in vitro* in the presence of all four nucleoside triphosphates plus an amount of dideoxyadenosine triphosphate sufficient to compete for incorporation with deoxyadenosine triphosphate. The run of N's represents the nucleotides that bind with the oligonucleotide primer. What DNA fragments are expected?

SOLUTION. Replication will proceed normally for all A, G, and C nucleotides in the template strand, but it will terminate at a T wherever a dideoxyadenosine was incorporated instead of deoxyadenosine. The resulting fragments will be as shown, where XXXXXXXX represents the nucleotides in the oligonucleotide primer.

5'-XXXXXXXXA-3'
5'-XXXXXXXXAGA-3'
5'-XXXXXXXXAGACTTCTCCTCA-3'
5'-XXXXXXXXAGACTTCTCCTCAGGA-3'
5'-XXXXXXXXAGACTTCTCCTCAGGAGTCA-3'
5'-XXXXXXXXAGACTTCTCCTCAGGAGTCATTCA-3'
5'-XXXXXXXXAGACTTCTCCTCAGGAGTCATTCACA-3'
5'-XXXXXXXXAGACTTCTCCTCAGGAGTCATTCACACCA-3'
5'-XXXXXXXXAGACTTCTCCTCAGGAGTCATTCACACCAT-3'

CONCEPTS IN ACTION: PROBLEMS FOR SOLUTION

6.1 Many restriction enzymes produce restriction fragments that have "sticky ends." What does this mean?

6.2 The list below gives half of each of a set of palindromic restriction sites. Replace the N's to complete sequence of each restriction site.

  (a) 5'-ACGNNN-3'
  (b) 5'-ATCNNN-3'
  (c) 5'-AGNN-3'
  (d) 5'-NNNATC-3'

6.3 Apart from nucleotide sequence, what is different about the ends of restriction fragments produced by the following restriction enzymes? (The downward arrow represents the site of cleavage in each strand.)

  (a) *Nhe*I (5'-G $\downarrow$ CTAG-3')
  (b) *Cfo*I (5'-GCG $\downarrow$ C-3')
  (c) *Sca*I (5'-AGT $\downarrow$ ACT-3')

6.4 What is the function of the 3'-to-5' exonuclease activity associated with DNA polymerase, and what are the consequences for the cell if this function is inactivated by mutation?

6.5 What chemical groups are joined by DNA ligase? By DNA polymerase?

6.6 What is meant by the statement that the DNA replication fork is asymmetrical?

6.7 The average human chromosome contains about 130 Mb of DNA. What is the approximate length of such a DNA molecule in micrometers? (There are 10^{-4} micrometers per angstrom unit.)

6.8 Consider a duplex molecule of length 10 Mb that has only one origin of replication located exactly in the middle. If replication proceeds bidirectionally, estimate the time needed for replication of this molecule, assuming that the rate of DNA synthesis is:

  (a) 1500 nucleotide pairs per second (typical of bacterial cells)
  (b) 50 nucleotide pairs per second (typical of eukaryotic cells)

6.9 The double-stranded DNA molecule of a newly discovered virus was found by electron microscopy to have a length of 102 micrometers (102×10^4 Å).

(a) How many nucleotide pairs are present in one of these molecules?

(b) In the duplex DNA of the virus, how many complete turns of the double helix are present?

6.10 An asteroid probe brings back a bacterial species that has DNA as its genetic material. You perform a Meselson–Stahl experiment and show that, after one round of replication in ^{14}N medium, half of the daughter DNA duplexes have ^{15}N in both strands whereas the other half have ^{14}N in both strands. Interpret these data.

6.11 A DNA duplex with the sequence shown below is cleaved with *Kas*I (cleavage site 5′-G ↓ GCGCC-3′), where the arrow denotes the site of cleavage in each strand. If the resulting fragments were brought together in the same order as in the original duplex and the breaks in the backbones sealed, what possible DNA duplexes would be expected?

```
5′-CTGGGGCGCCCTCGTCAGCGAGGGGGCGCCGAT-3′
3′-GACCCCGCGGGAGCAGTCGCTCCCCCGCGGCTA-5′
```

6.12 A friend brings you three samples of nucleic acid and asks you to determine each sample's chemical identity (whether DNA or RNA) and whether the molecules are double stranded or single stranded. You use powerful nucleases to degrade each sample completely to its constituent nucleotides and then determine the approximate relative proportions of nucleotides. The results of your assay are shown here. What can you tell your friend about the nature of these samples?

Sample 1: dGMP 13% dCMP 14% dAMP 36% dTMP 37%
Sample 2: dGMP 12% dCMP 36% dAMP 47% dTMP 5%
Sample 3: GMP 22% CMP 47% AMP 17% UMP 14%

6.13 In a random sequence consisting of equal proportions of all four nucleotides, what is probability that a particular short sequence of nucleotides matches a restriction site for:

(a) A restriction enzyme with a four-base cleavage site?

(b) A restriction enzyme with a six-base cleavage site?

(c) A restriction enzyme with an eight-base cleavage site?

6.14 A 3.1-kilobase linear fragment of DNA was digested with *Pst*I and produced a 2.0-kb fragment and a 1.1-kb fragment. When the same 3.1-kb fragment was cut with *Hind*III, it yielded a 1.5-kb fragment, a 1.3-kb fragment, and a 0.3-kb fragment. When the 3.1-kb molecule was cut with a mixture of the two enzymes, fragments of 1.5, 0.8, 0.5, and 0.3 kb resulted. Draw a map of the original 3.1-kb fragment, and label the restriction sites and the distances between these sites.

6.15 The sequence of bases shown here is present along one strand of a DNA duplex that has been opened to create a replication fork. Synthesis of an RNA primer on this template begins by copying the base shown in red.

```
5′-TCTGATATCAGTACG-3′
```

If the RNA primer consists of eight nucleotides, what is its base sequence?

6.16 In *Drosophila*, the *dusky* mutation is an X-linked recessive that causes small, dark wings. In a stock of wildtype flies, you discover a single male that has the dusky phenotype. You look up *dusky* in FlyBase, the online database of *Drosophila* genetics and genomics, and learn that, in wildtype flies, the *dusky* gene is contained within an 8-kb *Xho*I restriction fragment. When you digest genomic DNA from the mutant male with *Xho*I, you find that the size of the labeled fragment is 10 kb. You clone the 10-kb fragment and use it as a probe for *in situ* hybridization with the polytene chromosomes of a number of different wildtype strains. You notice that this fragment hybridizes to multiple locations along the polytene chromosomes. Each wildtype strain has a different pattern of hybridization. What do these data suggest about the origin of the *dusky* mutation that you discovered?

6.17 For the replication bubble illustrated here, indicate the leading strand and the lagging strand at each replication fork and identify the ends as 3′ or 5′.

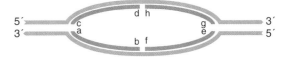

6.18 For the fluorescent color coding A = green, T = red, G = black, and C = purple, deduce the DNA sequence indicated in the accompanying gel diagram.

Top   Bottom

6.19 For the fluorescent color coding A = green, T = red, G = black, and C = purple, deduce the DNA sequence in each of the accompanying gel diagrams. How does the sequence in gel A differ from that in gel B?

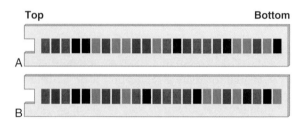

Top   Bottom

A

B

6.20 The first evidence for semiconservative replication of DNA in eukaryotic chromosomes made use of a synthetic analog of thymidine called bromodeoxyuridine (BUdR) in which the methyl group in thymine is replaced with an atom of bromine. When chromatids whose DNA contains BUdR are stained with certain fluorescent dyes, the chromatids with one strand labeled and one unlabeled fluoresce very brightly (light green in the accompanying illustration), whereas those with both strands labeled fluoresce dully (dark green). The illustration depicts the fluorescence patterns of chromosomes in mitotic metaphase after one and two rounds of DNA replication in the presence of BUdR, and the dotted lines represent the DNA strands in the DNA duplex present in each chromatid. Depict the BUdR labeling of each chromatid by (1) making the line solid if the strand is fully labeled with BUdR or (2) leaving it dashed if it is half labeled with BUdR.

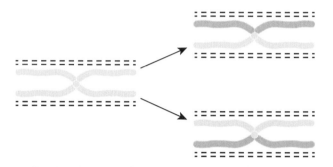

🧠 STOP & THINK ANSWERS

ANSWER TO STOP & THINK 6.1

A newly synthesized DNA strand can only be elongated at the 3′ end, hence the next nucleotide added to the bottom strand will be an A (the complement of T).

ANSWER TO STOP & THINK 6.2

The accompanying diagram shows the predicted result of conservative replication after one round of replication. Note that a hybrid molecule of one light and one heavy strand is not observed. The finding of molecules of hybrid density allowed the hypothesis of conservative replication to be rejected.

ANSWER TO STOP & THINK 6.3

The primer sequence is 3′–AGUUCUCA–5′ because RNA uses the nucleotide U to pair with A.

ANSWER TO STOP & THINK 6.4

Each DNA fragment terminates at the 3′ end, and the shortest fragments are at the bottom of the gel (on the right). The sequence of this region of the daughter strand reads 5′–CTGGAGAT–3′. The sequence of the template strand is therefore 3′–GACCTCTA–5′.

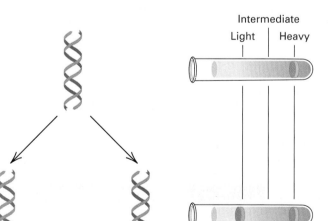

Intermediate

Light   Heavy

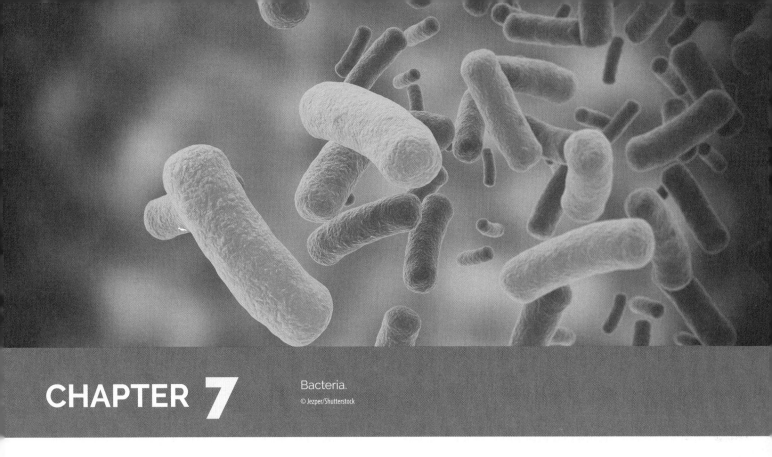

Bacteria.
© Jezper/Shutterstock

CHAPTER 7

The Genetics of Bacteria and Their Viruses

LEARNING OBJECTIVES

- To explain the role of plasmids, transposable elements, and mobile DNA in the evolution of pathogenic bacteria that are resistant to multiple, unrelated types of antibiotics, making infections extremely difficult to treat.

- To describe the methods and principal types of mutants commonly used in the study of bacterial and bacteriophage genetics.

- To construct a genetic map of mutant bacterial alleles based on frequencies of cotransduction of the wildtype alleles by a transducing phage.

- Given the times of entry of genetic markers in an interrupted mating between Hfr and F⁻ bacterial cells, to deduce the genetic map of the genes and the origin and direction of transfer of the Hfr strain.

- To diagram the processes of site-specific integration and excision of a lysogenic bacteriophage such as lambda.

Bacteria and their viruses (bacteriophage) have unique and diverse reproductive systems with multiple and novel mechanisms of genetic exchange. Some bacterial DNA sequences can become mobile by any of a variety of mechanisms. This feature enables them to become widely disseminated within a bacterial population and even to spread between species. In this chapter we discuss the genetic systems of bacteria and bacteriophage. We begin by examining **mobile DNA**: sequences that can be transferred between DNA molecules and from one cell to another. The ability to share genes in this manner, even among different bacterial species, is a unique feature of bacterial genetic systems.

7.1 Many DNA sequences in bacteria are mobile and can be transferred between individuals and among species.

A high percentage of bacteria isolated from clinical infections are resistant to one or more antibiotics. Most of them are resistant to multiple antibiotics. Some are resistant to all antibiotics in routine use. The problem has become so severe that many of the antibiotics that were at one time most effective and had the fewest side effects are now virtually useless. The widespread antibiotic-resistance genes almost never originate from new mutations in the bacterial genome. They are acquired, usually several at a time, in various forms of mobile DNA.

A plasmid is an accessory DNA molecule, often a circle.

Plasmids are nonessential DNA molecules that exist inside bacterial cells. They replicate independently of the bacterial genome and segregate to the progeny when a bacterial cell divides, so they can be maintained indefinitely in a bacterial lineage. Many plasmids are circular DNA molecules, but others are linear. The number of copies of a particular plasmid in a cell varies depending on the mechanism by which replication is regulated. High-copy-number plasmids are found in as many as 50 copies per host cell, whereas low-copy-number plasmids are present in 1 to 2 copies per cell. Plasmids range in size from a few kilobases to a few hundred kilobases (**FIGURE 7.1**) and are found in most bacterial species that have been studied. In *Escerichia coli*, most plasmids are either quite small (up to about 10 kb) or quite large (greater than 40 kb). A typical *E. coli* isolate contains three different small plasmids, each present in multiple copies per cell, and one large plasmid present in a single copy per cell. The presence of plasmids can be detected physically by electron microscopy, as in Figure 7.1, or by gel electrophoresis of DNA samples. Some plasmids can be detected because of phenotypic characteristics that they confer on the host cell. The phenotype most commonly studied is antibiotic resistance. For example, a plasmid containing a tetracycline-resistance gene (*tet-r*) will enable the host bacterial cell to form colonies on medium containing tetracycline.

Plasmids rely on the DNA-replication enzymes of the host cell for their reproduction, but the initiation

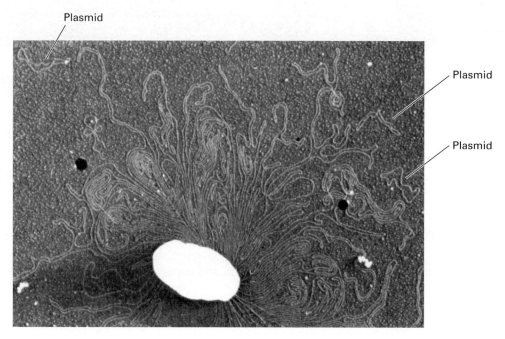

FIGURE 7.1 Electron micrograph of a ruptured *E. coli* cell, showing released chromosomal DNA and several plasmid molecules.

of replication is controlled by plasmid genes. In high-copy-number plasmids, replication is initiated multiple times during replication of the host genome, but in low-copy-number plasmids, replication is initiated only once per round of replication of the host genome. All types of plasmids contain sequences that promote their segregation into both daughter cells produced by fission of the host cell, so spontaneous loss of plasmids is uncommon.

The F plasmid is a conjugative plasmid.

Many large plasmids contain genes that enable the plasmid DNA to be transferred between cells. The transfer is mediated by a tube-like structure called a pilus (plural pili), formed between the cells, through which the plasmid DNA passes (**FIGURE 7.2**). The joining of bacterial cells in the transfer process is called **conjugation**, and the plasmids that can be transferred in this manner are called **conjugative plasmids**. Not all plasmids are conjugative plasmids. Most small plasmids are nonconjugative: They can be maintained in a bacterial lineage as the cells divide, but they do

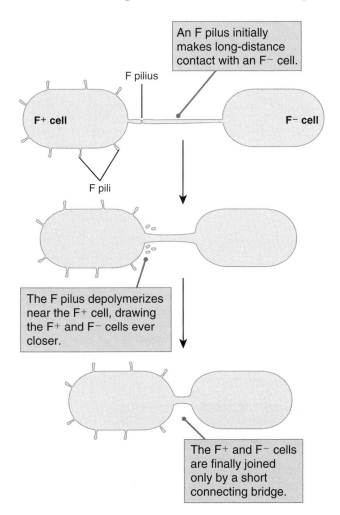

An F pilus initially makes long-distance contact with an F− cell.

F pilius

F+ cell

F− cell

F pili

The F pilus depolymerizes near the F+ cell, drawing the F+ and F− cells ever closer.

The F+ and F− cells are finally joined only by a short connecting bridge.

FIGURE 7.2 The F plasmid contains genes for producing F pili, which help the cells adhere to surfaces but can also form a bridge between two cells.

not contain the approximately 20 genes necessary for pilus assembly or those for DNA transfer. Hence they are unable to be transferred on their own. As we shall see later, however, they are able to employ the genetic trickery of recombination in order to tag along with conjugative plasmids, and in this way nonconjugative plasmids can be mobilized for cell-to-cell transfer.

The pilus between the *E. coli* cells in Figure 7.2 is an *F pilus* whose synthesis results from the presence of a conjugative plasmid called the **F factor** (the F stands for *fertility*). Cells that contain the F plasmid are donors and are designated the **F+ cells** ("F plus"); those lacking F are recipients and are designated the **F− cells** ("F minus"). The F plasmid is a low-copy-number plasmid, present in 1 to 2 copies per cell. It replicates once per cell cycle and segregates to both daughter cells in cell division. The F factor is approximately 100 kb in length and contains many genes that govern its maintenance in the cell and its transmission between cells.

Conjugation begins with physical contact between a donor cell and a recipient cell, as in Figure 7.2. The F plasmid DNA moves through a pore in the membrane from the donor to the recipient. The transfer is always accompanied by replication of the plasmid. Contact between an F+ and an F− cell initiates rolling-circle replication that results in the transfer of a single-stranded linear branch of the rolling circle to the recipient cell. During transfer, DNA is synthesized in both donor and recipient (**FIGURE 7.3**). Leading-strand synthesis in the donor replaces the transferred single strand, and lagging-strand synthesis in the recipient converts the transferred single strand into double-stranded DNA. When transfer is complete, the linear F strand becomes circular again in the recipient cell. Note that because one replica remains in the donor while the other is transferred to the recipient, after transfer both cells contain F and can function as donors. The F− cell has been converted into an F+ cell.

The transfer of the F plasmid requires only a few minutes. In laboratory cultures, if a small number of donor cells are mixed with an excess of recipient cells, F spreads throughout the population in a few hours, and all cells ultimately become F+. Transfer is not so efficient under natural conditions, and only about 10 percent of naturally occurring *E. coli* cells contain the F factor. Conjugation normally takes place only between F+ and F− cells, because the F plasmid contains two genes for *surface exclusion*, which prevents an F+ cell from conjugating with any other cell containing the same or a closely related plasmid. Most conjugative plasmids have analogous exclusion mechanisms.

Insertion sequences and transposons play a key role in bacterial populations.

Transposable elements are DNA sequences that can jump from one position to another or from one DNA

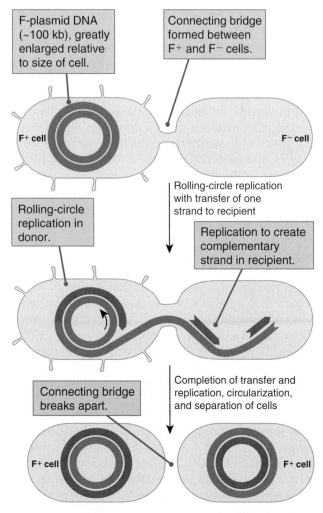

F-plasmid DNA (~100 kb), greatly enlarged relative to size of cell.

Connecting bridge formed between F+ and F− cells.

F+ cell

F− cell

Rolling-circle replication with transfer of one strand to recipient

Rolling-circle replication in donor.

Replication to create complementary strand in recipient.

Connecting bridge breaks apart.

Completion of transfer and replication, circularization, and separation of cells

F+ cell

F+ cell

FIGURE 7.3 Transfer of F from an F⁺ to an F⁻ cell. Pairing of the cells triggers rolling-circle replication. Red represents DNA synthesized during pairing. For clarity, the bacterial chromosome is not shown, and the plasmid is drawn overly large; the plasmid is in fact much smaller than a bacterial chromosome.

molecule to another. Bacteria contain a wide variety of transposable elements. The smallest and simplest are **insertion sequences**, or *IS elements*, which are typically 1–3 kb in length and usually encode only the *transposase* protein required for transposition and one or more additional proteins that regulate the rate of transposition. Like many transposable elements in eukaryotes, they possess inverted-repeat sequences at their termini, which are used by the transposase for recognizing and mobilizing the IS element. Upon insertion, they create a short, direct duplication of the target sequence at each end of the inserted element. The DNA organization of the insertion sequence IS*50* is diagrammed in part A of **FIGURE 7.4**.

Other transposable elements in bacteria contain one or more genes unrelated to transposition that can be mobilized along with the transposable element; this type of element is called a **transposon**. The length of a typical transposon is several kilobases, but a few are much longer. Much of the widespread antibiotic resistance among bacteria is due to the spread of transposons that include one or more (usually multiple) antibiotic-resistance genes. When a transposon mobilizes and inserts into a conjugative plasmid, it can be widely disseminated among different bacterial hosts by means of conjugation.

Some transposons have composite structures with antibiotic resistance sandwiched between insertion sequences, as is the case with the Tn*5* element illustrated in part B of Figure 7.4, which terminates in two IS*50* elements in inverted orientation. Transposons are usually designated by the abbreviation Tn followed by an italicized number (for example, Tn*5*). When it is necessary to refer to genes carried in such an element, the usual designations for the genes are used. For example, Tn*5* (*neo-r ble-r str-r*) contains genes

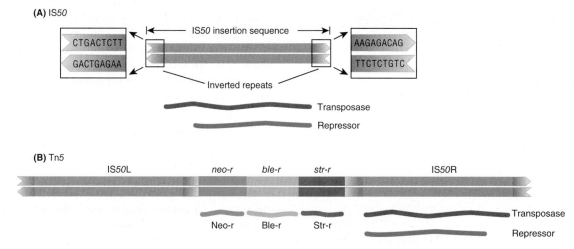

(A) IS*50*

CTGACTCTT

GACTGAGAA

IS*50* insertion sequence

AAGAGACAG

TTCTCTGTC

Inverted repeats

Transposase

Repressor

(B) Tn*5*

IS*50*L

neo-r

ble-r

str-r

IS*50*R

Neo-r

Ble-r

Str-r

Transposase

Repressor

FIGURE 7.4 Transposable elements in bacteria. (A) Insertion sequence IS*50*. The element is terminated by short, nearly perfect inverted-repeat sequences, the terminal nine base pairs of which are indicated. IS*50* contains a region that codes for the transposase and for a repressor of transposition. The coding regions are identical in the region of overlap, but the repressor is somewhat shorter because it begins at a different place. (B) Composite transposon Tn*5*. The central sequence contains genes for resistance to neomycin, *neo-r*; bleomycin, *ble-r*; and streptomycin, *str-r*. It is flanked by two copies of IS*50* in inverted orientation. The left-hand element (IS*50*L) contains mutations and is nonfunctional, so the transposase and repressor are made by the right-hand element (IS*50*R).

for resistance to three different antibiotics: neomycin, bleomycin, and streptomycin.

Nonconjugative plasmids can be mobilized by cointegration into conjugative plasmids.

Nonconjugative and conjugative plasmids typically coexist in the same cell along with host genomic DNA, and when a transposable element is mobilized, all of the DNA molecules present are potential targets for insertion. In time, many of the plasmids in a bacterial lineage can acquire copies of transposable elements present in the host DNA, and the host DNA can acquire copies of transposable elements present in the plasmids. In this manner, transposable elements become disseminated among independently replicating DNA molecules. The result is that most bacteria contain multiple copies of different types of transposable elements, some in the host genome, some in plasmids, and some in both. In *E. coli*, for example, natural isolates contain an average of 1 to 6 genomic copies of each of six naturally occurring IS elements, and among the cells that contain a particular IS element, 20 to 60 percent also contain copies in one or more plasmids.

Thus it happens that many nonconjugative and conjugative plasmids present in a bacterial cell come to carry one or more copies of the same transposable element. Because these copies are homologous DNA sequences, they can serve as substrates for recombination. When two plasmids undergo recombination in a region of homology, the result is as shown in **FIGURE 7.5**. The recombination forms a composite plasmid called a **cointegrate**. If one of the participating plasmids is nonconjugative and the other is conjugative, then the cointegrate is also a conjugative plasmid and so can be transferred in conjugation. After conjugation, the nonconjugative plasmid can become free of the cointegrate by recombination between the same sequences that created it. By the mechanism of cointegrate formation, therefore, nonconjugative plasmids can temporarily hitchhike with conjugative plasmids and be transferred from cell to cell.

Integrons have special site-specific recombinases for acquiring antibiotic-resistance cassettes.

In the evolution of multiple antibiotic resistance, bacteria have also made liberal use of a set of enzymes known as *site-specific recombinases*, which were present in bacterial populations and functioned in the evolution of other traits long before the antibiotic era.

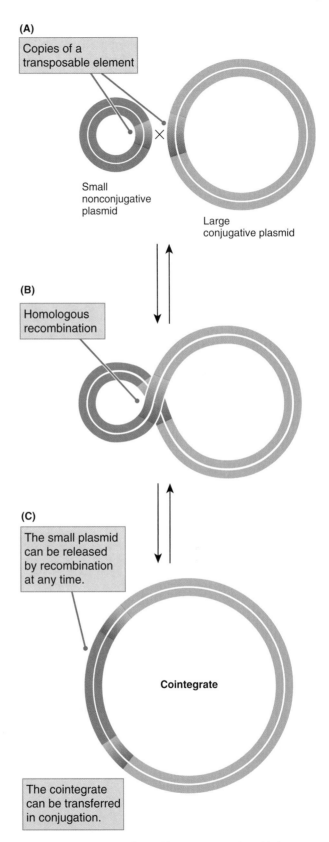

(A) Copies of a transposable element

Small nonconjugative plasmid

Large conjugative plasmid

(B) Homologous recombination

(C) The small plasmid can be released by recombination at any time.

Cointegrate

The cointegrate can be transferred in conjugation.

FIGURE 7.5 Cointegrate formed between two plasmids by recombination between homologous sequences (for example, copies of a transposable element) present in both plasmids.

Each type of **site-specific recombinase** binds with a specific nucleotide sequence in duplex DNA. When the site is present in each of two duplex DNA molecules, the recombinase brings the sites together and catalyzes a reciprocal exchange between the duplexes. An example is shown in **FIGURE 7.6**, where the site-specific recombinase joins a circular DNA molecule with a linear DNA molecule. Note that the reaction can proceed in the reverse direction, too, and free the circle from the cointegrate.

An example of a site-specific recombinase is an enzyme called the *Cre recombinase*, which is encoded in a gene in the *E. coli* bacteriophage P1. The Cre recognition sequence is called *loxP*; it is 34 bp in length and contains the central asymmetrical core sequence shown in Figure 7.6. Recombination between two *loxP* sites preserves the *loxP* sequences because the participating sites are identical, and hence the recombination reaction is reversible. Some site-specific recombinases favor the reaction that brings two molecules together into a cointegrate. Others (including Cre) favor the reaction that splits a cointegrate into two separate molecules. Some site-specific recombinases bring together and recombine sites that are similar but not identical; in these cases the recombination does not preserve the recognition sites, and so the reaction is not necessarily reversible.

Site-specific recombinases are used in the assembly of multiple-antibiotic-resistance units called *integrons*. An **integron** is a DNA element that encodes a site-specific recombinase as well as a recognition region that allows other sequences with similar recognition regions to be incorporated into the integron by recombination. The elements that integrons acquire are known as *cassettes*. In the context of integrons, a **cassette** is a circular antibiotic-resistance-coding region flanked by a recognition region for an integron. Because the site-specific recombinase integrates cassettes, the integron recombinase is usually called an **integrase**.

Several different types of integrons have been characterized. The best known of these are the Class 1 integrons, which include a site-specific recombinase denoted Int1 and, invariably, a coding region (*sul1*) that confers resistance to sulphonamide antibiotics. The molecular structure of a Class 1 integron is shown in part A of **FIGURE 7.7**. Also shown is the mechanism by which antibiotic-resistance cassettes are sequentially acquired. The Int1 integrase catalyzes a site-specific recombination between a sequence denoted *attI* present in the integron and a similar sequence denoted *attC* in the cassette. All *attC* regions are similar, but no two are identical.

Figure 7.7, part A shows how a cassette is captured by site-specific recombination between *attI* and *attC*. In general, antibiotic-resistance cassettes contain protein-coding regions but lack the promoter sequences needed to initiate transcription. They can be transcribed only by read-through transcription from an adjacent promoter. The integron provides the

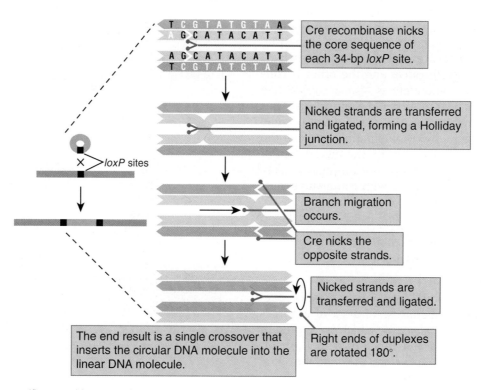

FIGURE 7.6 A site-specific recombinase catalyzes a reciprocal exchange between two specific sequences. No other sequences can serve as substrates. The recognition site for the Cre recombinase is *loxP*.

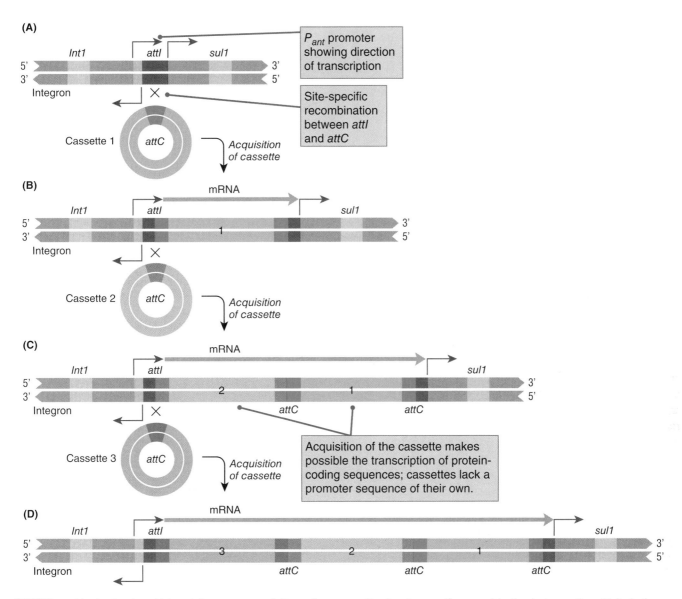

FIGURE 7.7 Mechanism by which an integron sequentially captures cassettes by site-specific recombination between the *attI* site in the integron and the *attC* site in the cassette.

needed promoter, called P_{ant}, at a position upstream from the *attI* site, so that when a cassette is captured, the coding sequence can be expressed. More than 40 different promoterless cassettes have been identified that encode proteins for resistance to antibiotics including β-lactams, aminoglycosides, chloramphenicol, trimethoprim, and streptothricin.

Once one cassette is in place, as shown in part B of Figure 7.7, a second can be captured using the same *attI* site and the *attC* present in the new cassette. Note that the new cassette is integrated immediately adjacent to the *attI* site and that the mRNA produced from the P_{ant} promoter includes the coding sequences for both cassettes. In part C of Figure 7.7, a third cassette is added to the integron, and the mRNA from P_{ant} becomes even longer. When there are multiple cassettes, as shown here, all of them are cotranscribed

from P_{ant}, but the downstream coding sequences are transcribed less frequently because there is a greater chance that transcription will terminate before reaching them. This constraint sets a practical limit on the number of cassettes that can be transcribed efficiently, but integrons with up to 10 antibiotic-resistance cassettes have been found.

The Int1 integrase can also catalyze the reverse of the cassette-capture reaction, though at a much lower level. This reaction generally uses two *attC* sites. An example is shown in **FIGURE 7.8**. Site-specific recombination between adjacent *attC* sites releases a circular cassette containing a promoterless coding sequence. The cassette cannot replicate because it lacks an origin of replication, but the capture reaction is efficient enough that the cassette will often be recaptured by the same integron (which repositions the cassette

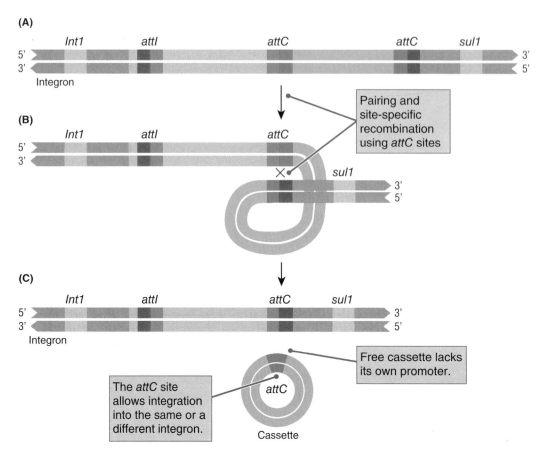

FIGURE 7.8 Mechanism of cassette excision from an integron by site-specific recombination between two *attC* sites.

immediately adjacent to *attI*) or by a different integron in the same cell (adding to the repertoire of cassettes it already contains).

Although integrons cannot mobilize themselves, they are present in transposons, conjugative plasmids, and nonconjugative plasmids, as well as in bacterial chromosomes. The integrons that are parts of mobile DNA elements are particularly important in the evolution of antibiotic resistance, because they can capture antibiotic-resistance cassettes and thereby make possible not only the transcription of the antibiotic-resistance coding sequences but also their mobilization.

Bacterial genomes can contain discrete regions of DNA from different sources.

All the comings and goings of plasmids, and genes jumping about because of transposons and integrons, may suggest that the bacterial genome is a patchwork of segments of diverse origin inserted into a core set of genes. The patchwork model first gained strong support from genome sequencing of several independent isolates (strains) of *E. coli*. The range of genome sizes among the strains was more than 10 percent—4.63 Mb in one strain versus 5.23 Mb in

another—which constitutes more than 1000 genes. All sequenced strains share a common set of about 3800 genes, which probably represents the core set of genes inherited from the original ancestor of what we now call *E. coli*.

The interspersed regions of the genome, present in some strains but not others, are due to *genomic islands* of DNA containing multiple genes that were acquired from other bacterial species. These genomic islands are said to have been acquired by means of **horizontal transmission**. When the genomic islands contain genes that cause disease, they are called **pathogenicity islands**. Examples of pathogenicity islands are found in *E. coli* strain O157:H7. This is a pathogenic strain typically spread through contaminated food or water that causes bloody diarrhea and sometimes kidney failure. The strain sickens about 100,000 people per year in the United States alone, among whom about 100 die. The O157:H7 strain contains about 1400 genes not present in the laboratory strain *E. coli* K12. These acquired genes include a pathogenicity island encoding factors that allow the cells to stick to the intestinal wall and to secrete specific proteins into the host cells. The kidney failure is promoted by a toxin encoded in an integrated bacteriophage that inhibits protein synthesis and causes vascular damage. The bloody diarrhea is promoted by genes in a plasmid whose products destroy blood coagulation factors and cause destruction of red blood cells.

Bacteria with resistance to multiple antibiotics are an increasing problem in public health.

In nature, a conjugative plasmid can, through time, accumulate different transposons containing multiple independent antibiotic-resistance genes, or transposons containing integrons that have acquired multiple antibiotic-resistance cassettes, with the result that the plasmid confers resistance to a large number of completely unrelated antibiotics. These multiple-resistance plasmids are called **R plasmids**. Some R plasmids are closely related to the F plasmid and clearly evolved from the F factor. The evolution of R plasmids is promoted by the use (and overuse) of antibiotics, which selects for resistant cells because, in the presence of antibiotics, resistant cells have a growth advantage over sensitive cells. The presence of multiple antibiotics in the environment selects for multiple-drug resistance. Serious clinical complications result when plasmids resistant to multiple drugs are transferred to bacterial pathogens, or agents of disease. Infections with some pathogens that contain R factors are extremely difficult to treat, because the pathogen may be resistant to most or all antibiotics currently in use.

7.2   Mutations that affect a cell's ability to form colonies are often used in bacterial genetics.

Bacteria can be grown both in liquid medium and on the surface of a semisolid growth medium hardened with agar. Bacteria used in genetic analysis are usually grown on an agar surface in plastic petri dishes (called *plates*). A single bacterial cell placed on a solid medium will grow and divide many times, forming a visible cluster of cells called a *colony* (**FIGURE 7.9**). The number of bacterial cells in a suspension can be determined by spreading a known volume of the suspension on an agar surface and counting the colonies that form. Typical *E. coli* cultures contain as many as 10^9 cells/ml. The appearance of colonies, or the ability or inability to form colonies, on a particular medium can sometimes be used to identify the genotype of the cell that produced the colony.

As in other organisms, genetic analysis in bacteria requires mutants. In bacteria, mutations that affect metabolic pathways or antibiotic resistance are particularly useful. There are three principal types of mutants.

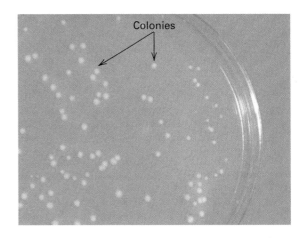

FIGURE 7.9 A petri dish with bacterial colonies that have formed on a solid medium.
Courtesy of Dr. Jim Feeley/CDC.

- **Antibiotic-resistant mutants** These mutants are able to grow in the presence of an antibiotic, such as streptomycin (Str) or tetracycline (Tet). For example, streptomycin-sensitive (Str-s) cells have the wildtype phenotype and fail to form colonies on medium containing streptomycin, but streptomycin-resistant (Str-r) mutants can form colonies.

- **Nutritional mutants** Wildtype bacteria can synthesize most of the complex nutrients they need from simple molecules present in the growth medium. The wildtype cells are said to be **prototrophs**. The ability to grow in simple medium can be lost by mutations that disable the enzymes used in synthesizing the complex nutrients. Mutant cells are unable to synthesize an essential nutrient and thus cannot grow unless the required nutrient is supplied in the medium. Such a mutant bacterium is said to be an **auxotroph** for the particular nutrient. For example, a methionine auxotroph cannot grow on a **minimal medium** containing only inorganic salts and a source of energy and carbon atoms (such as glucose), but such Met⁻ cells can grow if the minimal medium is supplemented with methionine.

- **Carbon-source mutants** These mutants cannot utilize particular substances as sources of energy or carbon atoms. For example, Lac⁻ mutants cannot utilize the sugar lactose for growth and are unable to form colonies on minimal medium containing lactose as the only carbon source.

A medium on which all wildtype cells form colonies is called a **nonselective medium**. Mutants and wildtype cells may or may not be distinguishable by growth on a nonselective medium. If the medium allows growth of only one type of cell (either wildtype

STOP & THINK 7.1

A bacterial culture consists of a mixture of the genotypes *met⁺ leu⁺ str-r*, *met⁺ leu⁻ str-r*, and *met⁻ leu⁺ str-r*, where *met⁻* cells require methionine (Met), *leu⁻* cells require leucine (Leu), and *str-r* cells are resistant to streptomycin (Str). A sample of cells is diluted appropriately, aliquots of the same volume are plated on various media, and the resulting colonies are counted. The following table shows the number of colonies observed with each type of medium:

Medium	Number of colonies
Minimal + Met + Leu + Str	400
Minimal + Met + Str	300
Minimal + Leu + Str	260
Minimal + Str	160

Based on these data, what are the estimated percentages of the genotypes *met⁺ leu⁺ str-r*, *met⁺ leu⁻ str-r*, and *met⁻ leu⁺ str-r* in the bacterial culture?

or mutant), it is said to be a **selective medium**. For example, a medium containing streptomycin is selective for the Str-r (resistant) phenotype and selective against the Str-s (sensitive) phenotype, and minimal medium containing lactose as the sole carbon source is selective for Lac⁺ cells and against Lac⁻ cells.

In bacterial genetics, phenotype and genotype are designated in the following way. A phenotype is designated by three letters, the first of which is capitalized; a superscript + or − denotes the presence or absence of the designated character; and s or r denotes sensitivity or resistance, respectively. A genotype is designated by lowercase italicized letters. Thus, a cell unable to grow without a supplement of leucine (a leucine auxotroph) has a Leu⁻ phenotype, and this would usually result from a *leu⁻* mutation in one of the genes required for leucine biosynthesis. Often the − superscript is omitted, but we will use it consistently to avoid ambiguity.

7.3 Transformation results from the uptake of DNA and recombination.

Important evidence that DNA is the genetic material came from experiments in which DNA from a heat-killed virulent strain of a pneumonia-causing bacterium was able to convert genetically cells of another strain from nonvirulent into virulent. The process of genetic alteration by pure DNA is **transformation**, and we know much more about it now than was known in 1944 when the experiments were carried out.

In transformation, recipient cells acquire genes from free DNA molecules in the surrounding medium. In laboratory experiments, DNA isolated from donor cells is added to a suspension of recipient cells. In nature, DNA can become available by spontaneous breakage (lysis) of donor cells. Either way, transformation begins with uptake of a DNA fragment from the surrounding medium by a recipient cell and terminates with *one strand* of donor DNA replacing the homologous segment in the recipient DNA. Most bacterial species are probably capable of the recombination step, but the ability of most bacteria to take up DNA efficiently is limited. Even in a species capable of transformation, DNA is able to penetrate only some of the cells in a growing population. However, appropriate chemical treatment of cells of these species yields a population of cells that are competent to take up DNA.

Transformation affords a convenient technique for gene mapping. DNA that is isolated from a donor bacterium is invariably broken into small fragments. With suitable recipient cells and excess DNA, transformation takes place at a frequency of about one transformed cell per 10^3 cells. If two genes, *a* and *b*, are so widely separated in the donor chromosome that they are always contained in two different DNA fragments, then the probability of simultaneous transformation (**cotransformation**) of an *a⁻ b⁻* recipient into wildtype *a⁺ b⁺* is the product of the probabilities of transformation of each marker separately, or roughly $10^{-3} \times 10^{-3}$, which equals one wildtype transformant per 10^6 recipient cells. However, if the two genes are so near one another that they are often present in a single donor fragment, then the frequency of cotransformation is nearly the same as the frequency of single-gene transformation, or one wildtype transformant per 10^3 recipients. The general principle is as follows:

KEY CONCEPT

Cotransformation of two genes at a frequency substantially greater than the product of the single-gene transformation frequencies implies that the two genes are close together in the bacterial chromosome.

Studies of the ability of various pairs of genes to be cotransformed also yield gene order. For example, if genes *a* and *b* can be cotransformed, and genes *b* and *c* can be cotransformed, but genes *a* and *c* cannot, then the gene order must be *a b c* (**FIGURE 7.10**).

7.4   In bacterial mating, DNA transfer is unidirectional.

Conjugation is a process in which DNA is transferred from a bacterial donor cell to a recipient cell by cell-to-cell contact. We have already examined this process in

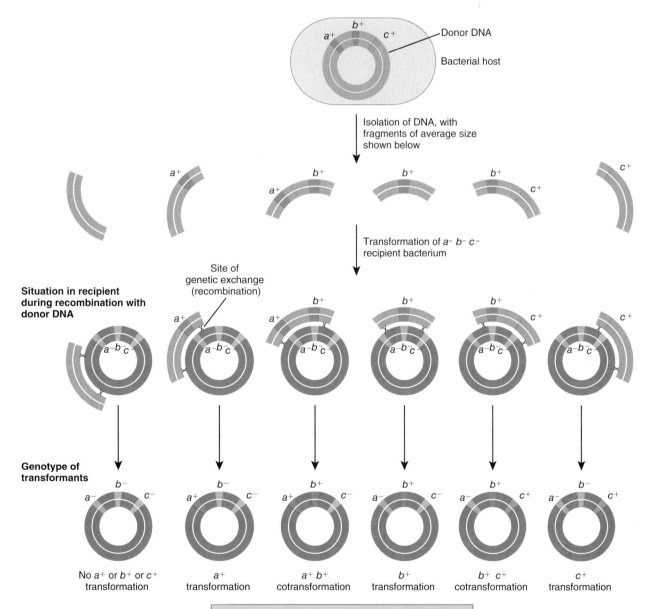

FIGURE 7.10 Cotransformation of linked markers. Markers *a* and *b* are near enough to each other that they are often present on the same donor fragment, as are markers *b* and *c*. Markers *a* and *c* are not near enough to undergo cotransformation. The gene order must therefore be *a b c*. The size of the transforming DNA, relative to that of the bacterial chromosome, is greatly exaggerated.

the context of conjugative transmission of plasmids. In this section we shall see how the same process can transfer genes present in the bacterial chromosome.

The F plasmid can integrate into the bacterial chromosome.

Transfer of chromosomal genes between *E. coli* cells was first observed by Joshua Lederberg in 1951. Although it was not known at the time, the exchange took place because the donor cells were F⁺, and in a few cells the F factor had become integrated into the bacterial chromosome (**FIGURE 7.11**). These are known as **Hfr cells**. Hfr stands for *high frequency of recombination*, which refers to the relatively high frequency with which donor genes are transferred to the recipient. The integration process is essentially the same as the formation of a cointegrate between two plasmids illustrated in Figure 7.5. Insertion sequences (Section 7.1) are key players in the origin of Hfr bacteria from F⁺ cells, because the F plasmid normally integrates through genetic exchange between an IS element present in F and a homologous copy that has transposed to an essentially random site in the bacterial chromosome. Because the F factor can

exist either separate from the chromosome or incorporated into it, it qualifies as an **episome**: a genetic element that can exist free in the cell or as a segment of DNA integrated into the chromosome.

In an Hfr cell (Figure 7.11), the bacterial chromosome remains circular, though enlarged about 2 percent by the integrated F-factor DNA. Integration of F is an infrequent event, but single cells containing integrated F can be isolated and cultured. When an Hfr cell undergoes conjugation, the process of transfer of the F factor is initiated in the same manner as in an F⁺ cell. However, because the F factor is part of the bacterial chromosome, transfer from an Hfr cell also includes DNA from the bacterial chromosome.

Chromosome transfer begins at F and proceeds in one direction.

The Hfr × F⁻ conjugation process is illustrated in **FIGURE 7.12**. The stages of transfer are much like those by which F is transferred to F⁻ cells: coming together of donor and recipient cells, rolling-circle replication in the donor cell, and conversion of the transferred single-stranded DNA into double-stranded DNA by lagging-strand synthesis in the recipient. However, in the case of Hfr matings, the transferred DNA does not become circular and is not capable of further replication in the recipient because the transferred F factor is not complete. The replication and associated transfer of the chromosomal DNA are controlled by the integrated F; they are initiated in the Hfr chromosome at the same point in F at which replication and transfer begin within an unintegrated F plasmid. A part of F is the first DNA transferred, chromosomal genes are transferred next, and the remaining part of F is the last DNA to enter the recipient. Because the conjugating cells usually break apart long before the entire bacterial chromosome is transferred, the final segment of F is almost never transferred into the recipient.

Several differences between F transfer and Hfr transfer are notable.

- It takes 100 minutes under the usual conditions for an entire bacterial chromosome to be transferred—in contrast to about 2 minutes for the transfer of F. The difference in time is a result of the relative sizes of F and the chromosome (100 kb for F versus 4600 kb for *E. coli* strain K12).

- In the transfer of Hfr DNA into a recipient cell, the mating pair usually breaks apart before the entire chromosome is transferred. Under typical experimental conditions, several hundred genes are transferred before the cells separate.

- In a mating between Hfr and F⁻ cells, the F⁻ recipient cell remains F⁻ because cell separation usually takes place before the final segment of F is transferred.

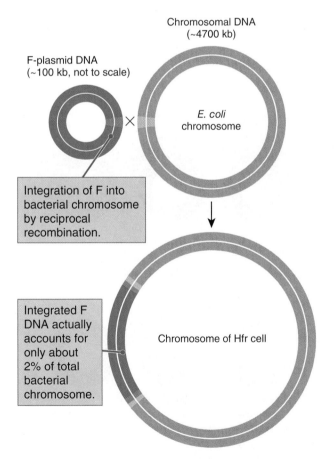

F-plasmid DNA (~100 kb, not to scale)

Chromosomal DNA (~4700 kb)

E. coli chromosome

Integration of F into bacterial chromosome by reciprocal recombination.

Integrated F DNA actually accounts for only about 2% of total bacterial chromosome.

Chromosome of Hfr cell

FIGURE 7.11 Integration of F (blue circle) by recombination between a nucleotide sequence in F and a homologous sequence (usually an insertion sequence) in the bacterial chromosome. The F-plasmid DNA is shown greatly enlarged relative to the size of the bacterial chromosome.

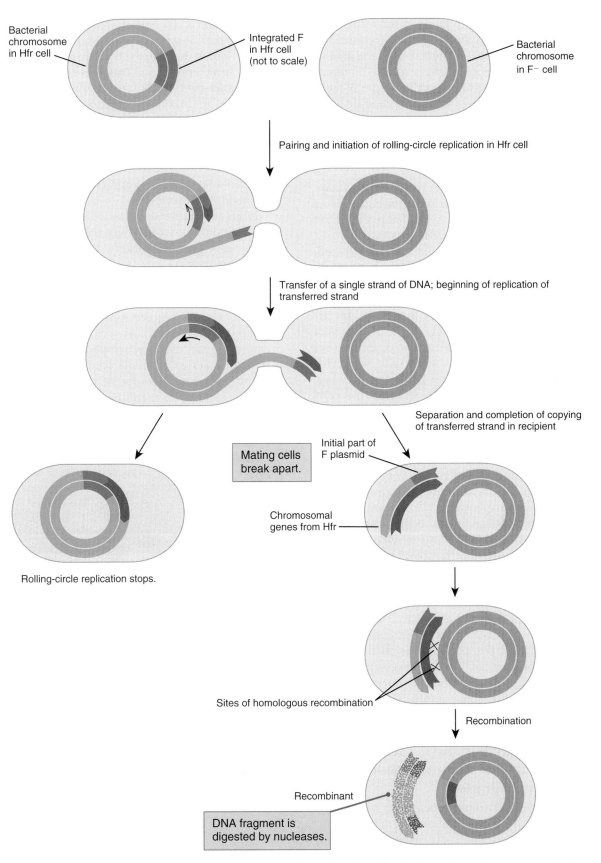

FIGURE 7.12 Stages in the transfer and production of recombinants in an Hfr × F⁻ mating. Pairing initiates rolling-circle replication within the F sequence in the Hfr cell, resulting in the transfer of a single strand of DNA. The single strand is converted into double-stranded DNA in the recipient. The mating cells usually break apart before the entire chromosome is transferred. Recombination takes place between the Hfr fragment and the F⁻ chromosome and leads to recombinants containing genes from the Hfr chromosome. Note that only a part of F is transferred. This part of F is not incorporated into the recipient chromosome. The recipient remains F⁻.

■ In Hfr transfer, some regions in the transferred DNA fragment become incorporated into the recipient chromosome. The incorporated regions replace homologous regions in the recipient chromosome. The result is that some F⁻ cells become recombinants containing one or more genes from the Hfr donor cell. For example, in a mating between Hfr *leu⁺* and F⁻ *leu⁻*, some cells arise that are F⁻ *leu⁺*. However, *the genotype of the donor Hfr cell remains unchanged.*

Genetic analysis requires that recombinant recipients be identified. Because the recombinants derive from recipient cells, a method is needed to eliminate the donor cells. The usual procedure is to use an F⁻ recipient containing an allele that can be selected. The selected allele should be located at such a place in the chromosome that most mating cells will have broken apart before the selected gene is transferred, and the selected allele must not be present in the Hfr cell. The selective agent can then be used to select the F⁻ cells and eliminate the Hfr donors. Genes that confer antibiotic resistance are especially useful for this purpose. For instance, after a mating between Hfr *leu⁺ str-s* and F⁻ *leu⁻ str-r* cells, the Hfr Str-s cells can be selectively killed by plating the mating mixture on medium containing streptomycin. A selective medium that lacks leucine can then be used to distinguish between the nonrecombinant and the recombinant recipients. The F⁻ *leu⁻* parent cannot grow in medium lacking leucine, but recombinant F⁻ *leu⁺* cells can grow because they possess a *leu⁺* gene. Thus, only recombinant recipients—that is, cells having the genotype *leu⁺ str-r*—form colonies on a selective medium that contains streptomycin and lacks leucine.

When a mating is done in this way, the transferred marker that is selected by the growth conditions (*leu⁺* in this case) is called a **selected marker**, and the marker used to prevent growth of the donor (*str-s* in this case) is called the **counterselected marker**. Selection and counterselection are necessary in bacterial matings because recombinants constitute only a small proportion of the entire population of cells (in spite of the name *high frequency of recombination*).

The unit of distance in the *E. coli* genetic map is the length of chromosomal DNA transferred in 1 minute.

Genes in the bacterial chromosome can be mapped by Hfr × F⁻ matings. However, the genetic map is quite different from linkage maps in eukaryotes in that it is based not on meiotic recombination but on transfer order. It is obtained by deliberate interruption of DNA transfer in the course of mating—for example, by violent agitation of the suspension of mating cells in a kitchen blender. The time at which a particular gene

is transferred can be determined by breaking the mating cells apart at various times and noting the earliest time at which breakage no longer prevents recombinants from appearing. This procedure is called the **interrupted-mating technique**. When it is performed with Hfr × F⁻ matings, the number of recombinants of any particular allele increases with the time during which the cells are in contact. This phenomenon is illustrated in **TABLE 7.1**. The reason for the increase in the number of recombinants is that different Hfr × F⁻ pairs initiate conjugation and chromosome transfer at slightly different times.

A greater understanding of the transfer process can be obtained by observing the results of a mating with several genetic markers. For example, consider the mating

$$\text{Hfr } a⁺ \, b⁺ \, c⁺ \, d⁺ \, e⁺ \, \text{str-s} \times \text{F}⁻ \, a⁻ \, b⁻ \, c⁻ \, d⁻ \, e⁻ \, \text{str-r}$$

in which *a⁻* cells require nutrient A, *b⁻* cells require nutrient B, and so forth. At various times after mixing of the cells, samples are agitated violently and then plated on a series of media containing streptomycin and different combinations of the five substances A through E (in each medium, one of the five is left out). Colonies that form on the medium lacking

TABLE 7.1 Data showing the production of Leu⁺ Str-r recombinants in a cross between Hfr *leu⁺ str-s* and *F⁻ leu⁻ str-r* cells when mating is interrupted at various times

Minutes after mating	Number of Leu⁺ Str-r recombinants per 100 Hfr cells
0	0
3	0
6	6
9	15
12	24
15	33
18	42
21	43
24	43
27	43

Note: Minutes after mating means minutes after the Hfr and F2 cell suspensions are mixed. Extrapolation of the recombination data to a value of zero recombinants indicates that the earliest time of entry of the leu1 marker is 4 minutes.

A are *a+ str-r*, those growing without B are *b+ str-r*, and so forth. All of these data can be plotted on a single graph to give a set of curves, as shown in part A of FIGURE 7.13. Four features of this set of curves are notable.

1. The number of recombinants in each curve increases with length of time of mating.

2. For each marker, there is a time (the **time of entry**) before which no recombinants are detected.

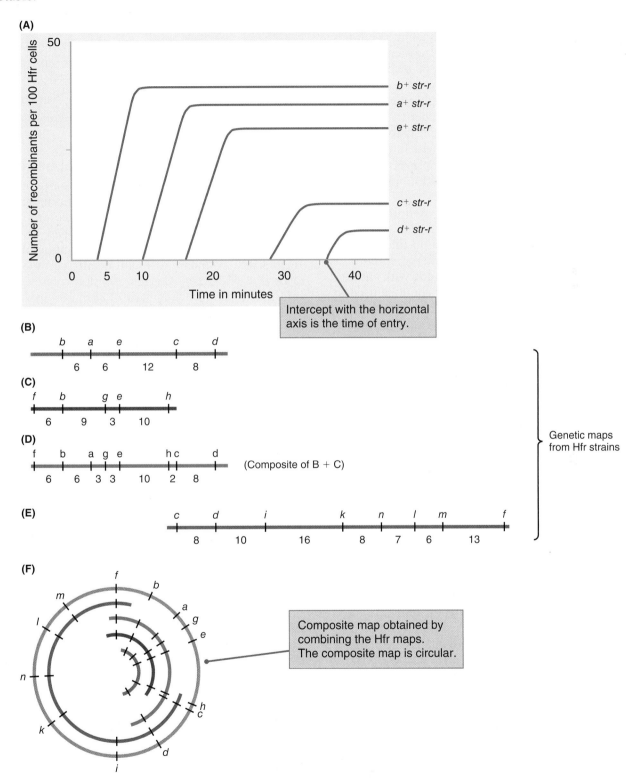

FIGURE 7.13 Time-of-entry mapping. (A) Time-of-entry curves for one Hfr strain. (B) The linear map derived from the data in part A. (C) A linear map obtained with the same Hfr but with a different F⁻ strain containing the alleles *b⁻ e⁻ f⁻ g⁻ h⁻*. (D) A composite map formed from the maps in parts B and C. (E) A linear map from another Hfr strain. (F) The circular map (gold) obtained by combining the two (green and blue) maps of parts D and E.

3. Each curve has a linear region that can be extrapolated back to the time axis, defining the time of entry of each gene a^+, b^+, . . . , e^+.

4. The number of recombinants of each type reaches a maximum, the value of which decreases with successive times of entry.

The explanation for the time-of-entry phenomenon is as follows. Not all donor cells start transferring DNA at the same time, so the number of recombinants increases with time. Transfer begins at a particular point in the Hfr chromosome (the replication origin of F). Genes are transferred in linear order to the recipient, and the time of entry of a gene is the time at which that gene first enters a recipient in the population. Separation of a mating pair prevents further transfer and limits the number of recombinants seen at a particular time.

The times of entry of the genes used in the mating just described can be placed on a map, as shown in Figure 7.13, part B. The numbers on this map are genetic distances between the markers, *measured as minutes between their times of entry*. Mating with another F$^-$ with genotype b^- e^- f^- g^- h^- str-r could be used to locate the three genes f, g, and h. Data for the second recipient would yield a map such as that shown in Figure 7.13, part C. Because genes b and e are common to both maps, the two maps can be combined to form a more complete map, as shown in Figure 7.13, part D.

Studies with different Hfr strains (Figure 7.13, part E) also are informative. It is usually found that different Hfr strains are distinguishable by their origins and directions of transfer, indicating that F can integrate at numerous sites in the chromosome and in both possible orientations. Combining the maps obtained with different Hfr strains yields a composite map that is *circular*, as illustrated in Figure 7.13, part F. The circularity of the map is a result of the circularity of the E. coli chromosome in F$^-$ cells and the multiple points of integration of the F plasmid; if F could integrate at only one site and in one orientation, the map would be linear.

A great many such mapping experiments have been carried out, and the data have been combined to provide an accurate map of approximately 2000 genes throughout the E. coli chromosome. **FIGURE 7.14** is a map of the chromosome of E. coli containing a sample of the mapped genes. Both the DNA molecule and the genetic map are circular. The entire chromosome requires 100 minutes to be transferred (it usually breaks first), so the total map length is 100 minutes. In the outer circle, the arrows indicate the direction of transcription and the coding region included in each transcript. The purple arrowheads show the origin and

direction of transfer of a number of Hfr strains. Transfer from HfrC, for example, goes counterclockwise starting with *purE acrA lac*.

⚙ STOP & THINK 7.2

Hfr KL25 has the F factor inserted at position 85 minutes in the E. coli genetic map (Figure 7.14). For Hfr KL25, is the gene *pyr* transferred early or late? How many minutes after mating is the gene *thr* transferred? How many minutes after mating is the gene *acrA* transferred?

Some F plasmids carry bacterial genes.

Occasionally, F is excised from Hfr DNA by an exchange between the same sequences used in the integration event. In some cases, however, the excision process is not a precise reversal of integration. Instead, breakage and reunion take place between the nonhomologous regions at a boundary of F and nearby chromosomal DNA (**FIGURE 7.15**). Aberrant excision creates a plasmid that contains a fragment of chromosomal DNA, which is called an **F$'$ plasmid** ("F prime"). By the use of Hfr strains that have different origins of transfer, F$'$ plasmids carrying chromosomal segments from many regions of the chromosome have been isolated. These elements are extremely useful because they render any recipient diploid for the region of the chromosome carried by the plasmid. These diploid regions allow dominance tests and gene-dosage tests (studies of the effects on gene expression of increasing the number of copies of a gene). Because only a part of the genome is diploid, cells that contain an F$'$ plasmid are **partial diploids**, also called *merodiploids*.

7.5 Some phages can transfer small pieces of bacterial DNA.

In the process of **transduction**, bacterial DNA is transferred from one bacterial cell to another by a phage particle containing the DNA. Such a particle is called a **transducing phage**. Two types of transducing phages are known. A **generalized transducing phage** produces some particles that contain only DNA obtained from the host bacterium, rather than phage DNA; the bacterial DNA fragment can be derived from *any* part of the bacterial chromosome. A **specialized transducing phage** produces particles that contain both phage and bacterial genes linked in a single DNA molecule, but the bacterial genes are obtained from a *particular* region of the bacterial chromosome. In this section, we consider E. coli phage P1, a well-studied

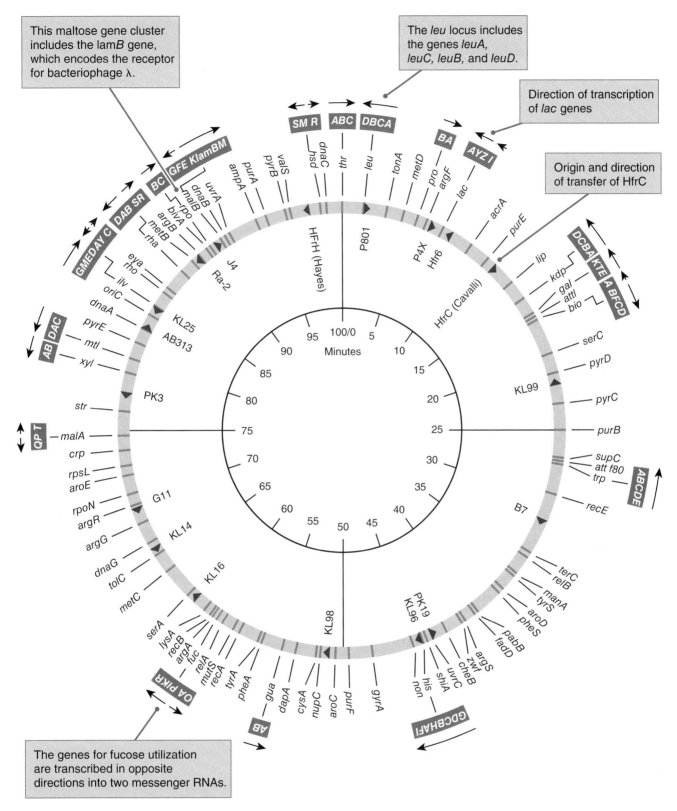

FIGURE 7.14 Circular genetic map of *E. coli*. Map distances are given in minutes; the total map length is 100 minutes. For some of the loci that encode functionally related gene products, the map order of the clustered genes is shown, along with the direction of transcription and length of transcript (black arrows). The purple arrowheads show the origin and direction of transfer of a number of Hfr strains. For example, HfrH transfers *thr* very early, followed by *leu* and other genes in a clockwise direction.

THE HUMAN CONNECTION

The Sex Life of Bacteria

Joshua Lederberg and Edward L. Tatum (1946)

Yale University, New Haven, Connecticut

Gene Recombination in Escherichia coli

After their discovery in the nineteenth century, bacteria were considered "things apart"—unlike other organisms in fundamental ways. Lederberg and Tatum's discovery of what at first appeared to be a conventional sexual cycle was a sensation, completely unexpected. It meant that bacteria could be considered "genetic organisms" along with yeast, *Neurospora, Drosophila*, and other genetic favorites. For this and related discoveries, Lederberg and Tatum were awarded the 1958 Nobel Prize, along with George W. Beadle. In this excerpt, you will note that the authors discuss bacterial recombination as requiring a cell fusion that would bring both parental genomes together. This interpretation shows that it is possible to make exactly the right observation and realize its significance, but not quite grasp what is really going on. The conclusion that bacterial recombination involved unidirectional transfer was reached much later, after the discovery of Hfr strains and the development of the interrupted-mating technique.

Lederberg and Tatum started by painstakingly generating multiple auxotrophic strains of *E. coli*, performing multiple rounds of mutagenesis with x-rays and subsequent screening by replica plating. In their studies . . . .

. . . two triple mutants have been used, one requiring threonine, leucine, and thiamin, the other requiring biotin, phenylalanine, and cystine. The strains were grown in mixed culture in complete medium. The cells were washed with

> ❝ *These types can most reasonably be interpreted as instances of the assortment of genes in new combinations.* ❞

sterile water and inoculated heavily into synthetic agar medium, ... The only new types found in "pure" cultures of the individual mutants were occasional forms that had reverted for a single factor, giving strains that required only two of the original three substances. In mixed cultures, however, a variety of types has been found. These include wildtype strains with no growth-factor deficiencies and single mutant types requiring only thiamin or phenylalanine. . . . These types can most reasonably be interpreted as instances of the assortment of genes in new combinations. In order that various genes may have the opportunity to recombine, a cell fusion would be required. . . . These experiments imply the occurrence of a sexual process in the bacterium *Escherichia coli.*

Lederberg and Tatum, like Mendel, Morgan, Beadle, and others, used analysis of phenotypic differences to make inferences about fundamental genetic processes. They did so at a time when, although DNA was increasingly recognized as the genetic material, neither its structure nor its organization in bacteria was known. Their fundamental hypothesis about recombination in bacteria has been borne out by subsequent molecular genetic analyses.

J. Lederberg and E. L. Tatum, Gene recombination in *Escherichia coli. Nature* 158 (1946): 558.

generalized transducing phage. Specialized transducing particles are discussed in Section 7.7.

During infection by P1, the phage makes a nuclease that cuts the bacterial DNA into fragments. Single fragments of bacterial DNA comparable in size to P1 DNA are occasionally packaged into phage particles in place of P1 DNA. The positions of the nuclease cuts in the host chromosome are random, so a transducing particle may contain a fragment derived from any region of the host

DNA. A large population of P1 phages will contain a few particles carrying any bacterial gene. On the average, any particular gene is present in roughly one transducing particle per 10^6 phages. When a transducing particle adsorbs to a bacterium, the bacterial DNA contained in the phage head is injected into the cell and becomes available for recombination with the homologous region of the host chromosome. A typical P1 transducing particle contains from 100 to 115 kb of bacterial DNA.

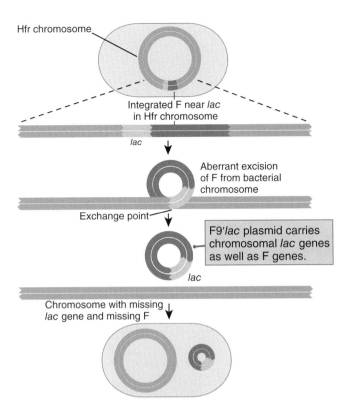

Hfr chromosome

Integrated F near *lac* in Hfr chromosome

lac

Aberrant excision of F from bacterial chromosome

Exchange point

F9'*lac* plasmid carries chromosomal *lac* genes as well as F genes.

lac

Chromosome with missing *lac* gene and missing F

FIGURE 7.15 Formation of an F' *lac* plasmid by aberrant excision of F from an Hfr chromosome. Breakage and reunion are between nonhomologous regions.

Let us now examine the events that follow infection of a bacterium by a generalized transducing particle obtained, for example, by growth of P1 on wildtype *E. coli* containing a *leu*⁺ gene (**FIGURE 7.16**). If such a particle adsorbs to a bacterial cell of *leu*⁻ genotype and injects the DNA that it contains into the cell, the cell survives because the phage head contained only bacterial genes and no phage genes. A recombination event exchanging the *leu*⁺ allele carried by the phage for the *leu*⁻ allele carried by the host converts the genotype of the host cell from *leu*⁻ into *leu*⁺. In such an experiment, typically about one *leu*⁻ cell in 10⁶ becomes *leu*⁺. Such frequencies are easily detected on selective growth medium. For example, if the infected cell is placed on solid medium lacking leucine, it is able to multiply and a *leu*⁺ colony forms. A colony does not form unless recombination inserted the *leu*⁺ allele.

The small fragment of bacterial DNA contained in a transducing particle includes about 50 genes, so transduction provides a valuable tool for genetic linkage studies of short regions of the bacterial genome. Consider a population of P1 prepared from a *leu*⁺ *gal*⁺ *bio*⁺ bacterium. This population contains particles able to transfer any of these alleles to another cell; that is, a *leu*⁺ particle can transduce a *leu*⁻ cell to *leu*⁺, or a *gal*⁺ particle can transduce a *gal*⁻ cell to *gal*⁺. Furthermore, if a *leu*⁻ *gal*⁻ culture is infected by phage, both *leu*⁺ *gal*⁻ and *leu*⁻ *gal*⁺ bacteria are produced. However, *leu*⁺ *gal*⁺ colonies do not arise, because the *leu* and *gal* genes are

too far apart to be included in the same DNA fragment (part A of **FIGURE 7.17**).

The situation is quite different for a recipient cell with genotype *gal*⁻ *bio*⁻, because the *gal* and *bio* genes are so closely linked that both genes are sometimes present in a single DNA fragment carried in a transducing particle—namely, a *gal-bio* particle (Figure 7.17, part B). However, not all *gal*⁺ transducing particles also include *bio*⁺, nor do all *bio*⁺ particles include *gal*⁺. The probability of both markers being in a single particle, and hence the probability of simultaneous transduction of both markers (**cotransduction**), depends on how close to each other the genes are. The closer they are, the greater the frequency of cotransduction. Cotransduction of the *gal*⁺-*bio*⁺ pair can be detected by plating infected cells on the appropriate growth medium. If *bio*⁺ transductants are selected by spreading the infected cells on a glucose-containing medium that lacks biotin, both *gal*⁺ *bio*⁺ and *gal*⁻ *bio*⁺ colonies grow. If these colonies are tested for the *gal* marker, 42 percent are found to be *gal*⁺ *bio*⁺ and the rest *gal*⁻ *bio*⁺; similarly, if *gal*⁺ transductants are selected, 42 percent are found to be *gal*⁺ *bio*⁺. In other words, the **frequency of cotransduction** of *gal* and *bio* is 42 percent, which means that 42 percent of all transducing particles that contain one gene also include the other.

Studies of cotransduction can be used to map closely linked genetic markers by means of three-factor crosses. Suppose, for example, that P1 is grown on wildtype bacteria and used to transduce cells carrying a mutation of each of three closely linked genes. Cotransductants that contain various pairs of wildtype alleles are examined. The gene located in the middle can be identified because its wildtype allele is nearly always cotransduced with the wildtype alleles of the genes that flank it. For example, in part B of Figure 7.17, a genetic marker located between *gal*⁺ and *bio*⁺ would almost always be present in *gal*⁺ *bio*⁺ transductants.

STOP & THINK 7.3

For generalized transduction with bacteriophage P1, the relation of the frequency of cotransduction between two genes (*c*) and the map distance between them in minutes (*m*) is given by

$$c = (1/8) \times (2-m)^3$$

What frequency of cotransduction corresponds to 0.5 minutes? 1.0 minutes? 1.5 minutes? What is the maximum distance between two genes that allows cotransduction?

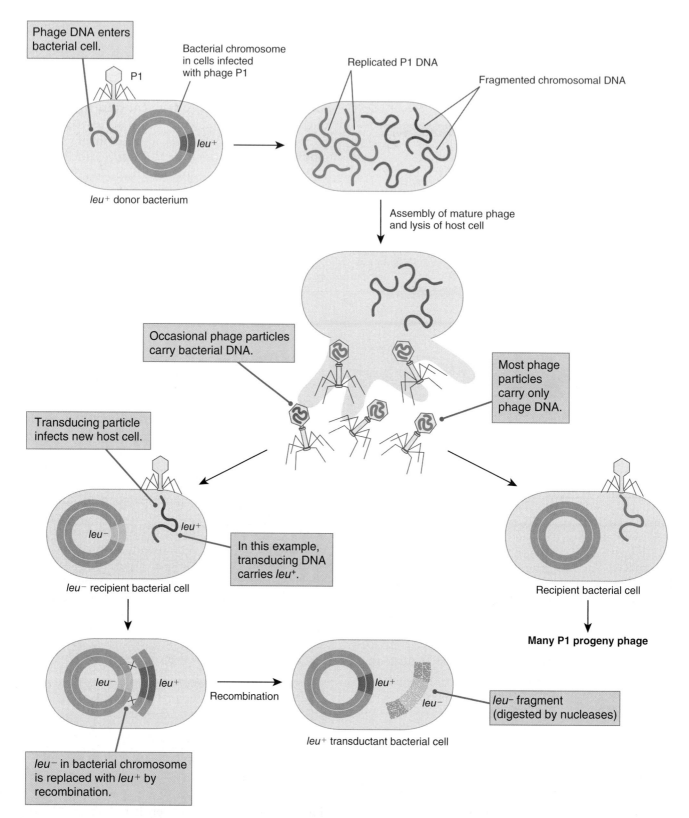

FIGURE 7.16 Transduction. Phage P1 infects a *leu⁺* donor, yielding predominantly normal P1 phages with an occasional one carrying bacterial DNA instead of phage DNA. If the phage population infects a bacterial culture, then the normal phages produce progeny phages, whereas the transducing particle yields a transductant. Note that the recombination step requires two crossovers. For clarity, double-stranded phage DNA is drawn as a single line.

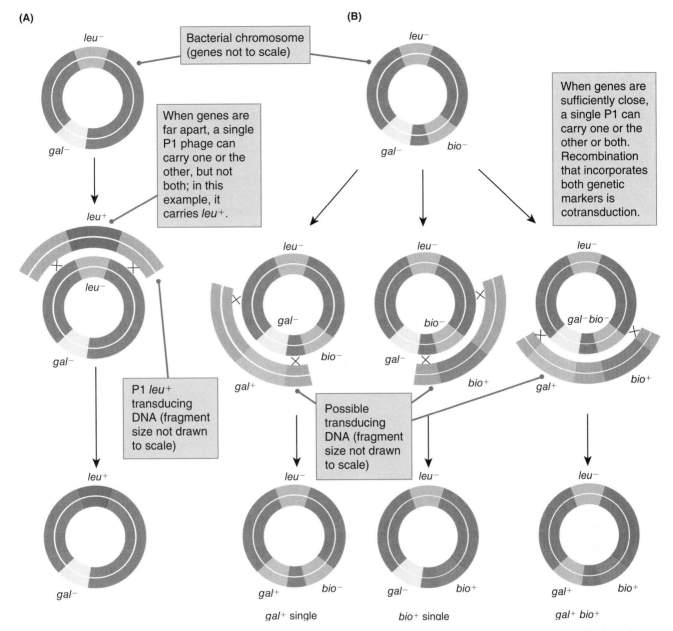

FIGURE 7.17 Demonstration of linkage of the *gal* and *bio* genes by cotransduction. (A) A P1 transducing particle carrying the *leu*+ allele can convert a *leu*– *gal*– cell into a *leu*+ *gal*– genotype (but cannot produce a *leu*+ *gal*+ genotype). (B) The transductants that could be formed by three possible types of transducing particles—one carrying *gal*+, one carrying *bio*+, and one carrying the linked alleles *gal*+ *bio*+. The third type results in cotransduction. For clarity, the distance between *gal* and *bio*, relative to that between *leu* and *gal*, is greatly exaggerated, and the size of the DNA fragment in a transducing particle, relative to the size of the bacterial chromosome, is not drawn to scale.

7.6 Bacteriophage DNA molecules in the same cell can recombine.

The reproductive cycle of a phage is called the **lytic cycle**. In the lytic cycle, phage DNA enters a cell and replicates repeatedly, bacterial ribosomes are used to produce phage protein components, the newly synthesized phage DNA molecules are packaged into protein shells to form progeny phages, and the bacterium is split open (**lysis**), releasing the progeny phages from the cell. Phage progeny from a bacterium infected by one phage have the parental genotype, except for new mutations. However, if two phage particles that have *different* genotypes infect a single bacterial cell, new genotypes can arise by genetic recombination. This process differs significantly from genetic recombination in eukaryotes in two ways: (1) The number of participating DNA molecules varies from one cell to the next, and (2) reciprocal recombinants are not always recovered in equal frequencies from a single infected cell. Recombination in bacteriophage is the subject of this section.

Bacteriophages form plaques on a lawn of bacteria.

Phages are easily detected because in a lytic cycle, an infected cell breaks open and releases phage particles into the growth medium. The test is performed as outlined in part A of **FIGURE 7.18**. A large number of bacteria (about 10^8) are placed on a solid medium. After a period of growth, a continuous turbid layer

of bacteria results. If phages are present at the time the bacteria are placed on the medium, each phage adsorbs to a cell, and shortly afterward, the infected cell lyses and releases many progeny phages. Each of the progeny phages adsorbs to a nearby bacterium, and after another lytic cycle, these bacteria in turn release progeny phages that can infect still other bacteria in the vicinity. These cycles of infection continue, and after several hours, the descendants of each phage that was originally present destroy all of the bacteria in a localized area, giving rise to a clear, transparent region—a **plaque**—in the otherwise turbid layer of confluent bacterial growth. Phages can multiply only in growing bacterial cells, so exhaustion of nutrients in the growth medium limits phage multiplication and the size of the plaque. Because a plaque is a result of

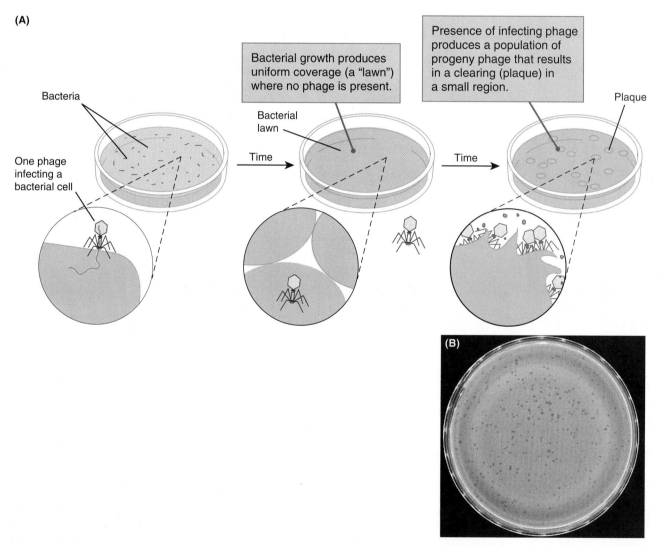

FIGURE 7.18 Plaque formation. (A) In the absence of a phage, bacterial cells grow and form a translucent lawn. Bacterial cells deposited in the vicinity of a phage are infected and lyse. Progeny phages, diffusing outward from the original site, infect other cells and cause their lysis. Because of phage infection and lysis, no bacteria can grow in a small region around the site of each phage particle that was originally present in the medium. The area devoid of bacteria remains transparent and is called a plaque. (B) Large plaques in a lawn of *E. coli* formed by infection with a mutant of bacteriophage λ. Each plaque results from an initial infection by a single bacteriophage.

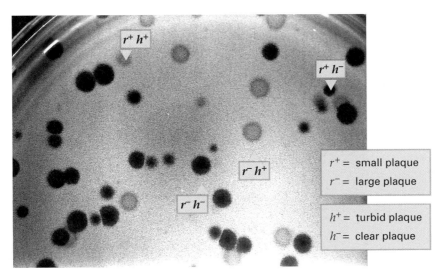

FIGURE 7.19 A phage cross is performed by infecting host cells with both parental types of phage simultaneously. This example shows the progeny of a cross between T4 phage of genotypes $r^- h^+$ and $r^+ h^-$ when both parental phage infect cells of *E. coli*. The $r^+ h^+$ and $r^- h^-$ genotypes are recombinants.

Courtesy of Leslie Smith and John W. Drake, National Institutes of Health.

an initial infection by one phage particle, the number of phage particles originally present on the medium can be counted.

The genotypes of phage mutants can be determined by studying the plaques. In some cases, the appearance of the plaque is sufficient. For example, phage mutations that decrease the number of phage progeny from infected cells often yield smaller plaques. Large plaques can be produced by mutants that cause premature lysis of infected cells, so that each round of infection proceeds more quickly (Figure 7.18, part B).

Infection with two mutant bacteriophages yields recombinant progeny.

If two phage particles with different genotypes infect a single bacterium, some phage progeny are genetically recombinant. **FIGURE 7.19** shows plaques resulting from the progeny of a mixed infection with *E. coli* phage T4 mutants. The r^- allele results in large plaques, and the h^- allele results in clear plaques. The cross is

$$r^- h^+ \qquad \times \qquad r^+ h^-$$
(large turbid plaque) (small clear plaque)

Four plaque types can be seen in Figure 7.19. Two— the large turbid plaque and the small clear plaque— correspond to the phenotypes of the parental phages. The other two phenotypes—the large clear plaque and the small turbid plaque—are recombinants that correspond to the genotypes $r^- h^-$ and $r^+ h^+$, respectively. When many bacteria are infected, approximately equal numbers of reciprocal recombinant types are usually found among the progeny phage. In an experiment like

that shown in Figure 7.19, in which each of the four genotypes yields a different phenotype of plaque morphology, the numbers of the genotypes can be counted by examining each of the plaques that is formed. The recombination frequency is the proportion of progeny phage that have recombinant genotypes.

7.7 Lysogenic bacteriophages do not necessarily kill the host.

The lytic cycle is one of two alternative phage life cycles. The alternative to the lytic cycle is called the **lysogenic cycle**, in which no progeny particles are produced, the infected bacterium survives, and a phage DNA molecule is transmitted to each bacterial progeny cell when the cell divides. All phage species can undergo a lytic cycle. Those phages that are also capable of the lysogenic cycle are called *temperate phage*, and those capable of only the lytic cycle are called *virulent phage*. In the lysogenic cycle, a replica of the infecting phage DNA becomes inserted, or *integrated*, into the bacterial chromosome (**FIGURE 7.20**). The inserted DNA is called a **prophage**, and the surviving bacterial cell is called a **lysogen**. A lysogen is denoted by the designation of the bacterial strain followed by the name of the lysogenic phage in parentheses; for example, a clone of *E. coli* strain K12 that has become lysogenic for phage λ is denoted K12(λ). Many bacterial generations after a strain has become lysogenic, the prophage can be activated and excised from the chromosome, and the lytic cycle can begin. Prophage activation results in a normal lytic cycle in which the host cell is killed and progeny phage are released.

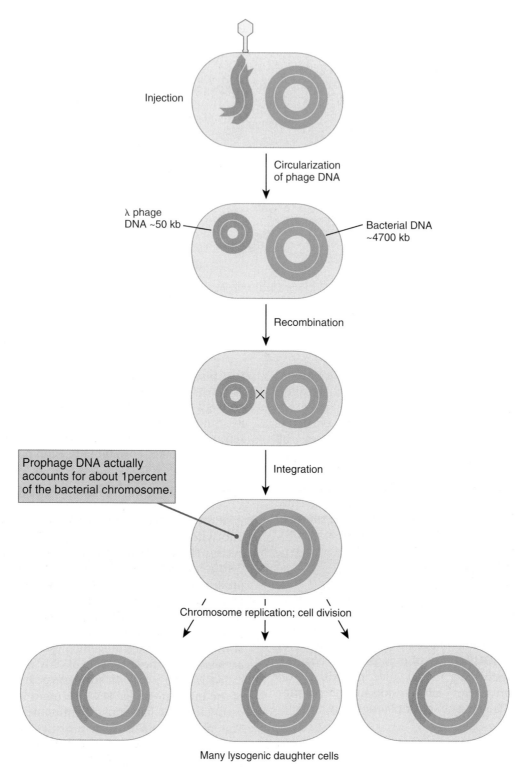

FIGURE 7.20 The general mode of lysogenization by integration of phage DNA into the bacterial chromosome. Some genes (those needed to establish lysogeny) are expressed shortly after infection and then are turned off. The inserted brown DNA is the prophage. For clarity, the phage DNA is drawn much larger than to scale; the size of phage λ DNA is actually about 1 percent of the size of the *E. coli* genome.

A temperate phage, such as *E. coli* phage λ, when reproducing in its lytic cycle, undergoes general recombination, much as phage T4 does. Physically, the DNA of phage λ is a linear, double-stranded molecule about 50 kb in length; however, at each end the molecule terminates in a single-stranded region of 12 base pairs (**FIGURE 7.21**). These single-stranded overhangs are complementary in sequence so that they can pair, forming a circular molecule. The single-stranded ends are called **cohesive ends** (*cos*) to indicate their ability

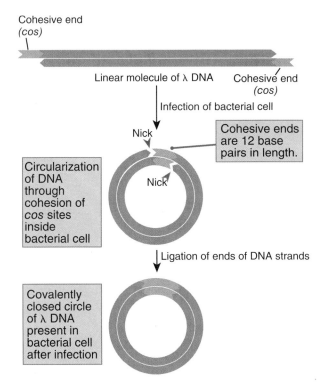

Cohesive end
(cos)

Linear molecule of λ DNA   Cohesive end
                                  (cos)

Infection of bacterial cell

Nick

Cohesive ends
are 12 base
pairs in length.

Nick

Circularization
of DNA
through
cohesion of
cos sites
inside
bacterial cell

Ligation of ends of DNA strands

Covalently
closed circle
of λ DNA
present in
bacterial cell
after infection

FIGURE 7.21 A diagram of a linear λ DNA molecule showing the cohesive ends (complementary single-stranded ends). Circularization by means of base pairing between the cohesive ends forms an open (nicked) circle, which is converted into a covalently closed (uninterrupted) circle by sealing (ligation) of the single-strand breaks. The length of the cohesive ends is 12 base pairs in a total molecule of approximately 50 kb.

to undergo base pairing. Pairing of the cohesive ends yields a circular molecule with two nicks that are joined by DNA ligase to create the closed circular molecule shown in Figure 7.21. Circularization takes place early in both the lytic and the lysogenic cycle and is necessary for both processes—for DNA replication in the lytic mode and for prophage integration in the lysogenic cycle.

The sites of breakage and rejoining in the bacterial and phage DNA are called the *bacterial attachment site* and the *phage attachment site*. Each attachment site consists of three segments. The central segment has the same nucleotide sequence in both attachment sites and is the region in which the DNA molecules are broken and rejoined. The phage attachment site is denoted *POP'* (*P* for *phage*), and the bacterial attachment site is denoted *BOB'* (*B* for *bacteria*). A phage-specific *integrase* catalyzes a site-specific recombination event that results in integration of the λ DNA molecule into the bacterial DNA (**FIGURE 7.22**). The site-specific recombination between *POP'* and *BOB'* is very similar to the site-specific recombination between *loxP* sites illustrated in Figure 7.6. The result is that the circular DNA of the phage becomes inserted into the circular DNA of the bacterial cell at the site of *BOB'*. The difference between the phage genetic map

and the prophage genetic map is a consequence of the circularization of the phage DNA and the central location of *POP'*.

A lysogenic cell can replicate nearly indefinitely without the release of phage progeny. However, the prophage can sometimes become activated to undergo a lytic cycle in which the usual number of phage progeny are produced. This phenomenon is called **prophage induction**, and it is initiated by damage to the bacterial DNA. Prophage induction is often caused by some environmental agent that damages DNA, such as chemicals or radiation. The ability to be induced allows the phage to escape from a damaged cell. The biochemical mechanism of induction is complex and will not be described in this book, but the excision of the phage is straightforward.

Excision is another site-specific recombination event that reverses the integration process. Excision requires the phage enzyme integrase and a phage protein called **excisionase**. Genetic evidence and studies of physical binding of purified excisionase, integrase, and λ DNA indicate that excisionase binds to integrase and thereby enables the latter to recognize the prophage attachment sites *BOP'* and *POB'*. Once bound to these sites, integrase makes cuts in the O sequence and recreates the *BOB'* and *POP'* sites. This reverses the integration reaction, causing excision of the prophage (Figure 7.22).

When a cell is lysogenized, a block of phage genes becomes part of the bacterial chromosome, so the phenotype of the bacterium might be expected to change—and indeed it does. Most phage genes in a prophage are kept in an inactive state by a **phage repressor** protein, the product of one of the phage genes. The repressor protein is synthesized initially by the infecting phage and then continually by the prophage. The gene that codes for the repressor is frequently the only prophage gene that is expressed in lysogens. If a lysogen is infected with a phage of the same type as the prophage—for example, λ infecting a λ lysogen—then the repressor present within the cell from the prophage prevents expression of the genes of the infecting phage. The resistance to infection by a phage identical with the prophage is called *immunity*. Thus λ does not form plaques on bacteria containing a λ prophage, because the infected cells are immune.

Specialized transducing phages carry a restricted set of bacterial genes.

When a bacterium lysogenic for phage λ is subjected to DNA damage that leads to induction, the prophage is usually excised from the chromosome precisely. However, once in every 10^6 or 10^7 cells, an excision error is made in which the sites of breakage and rejoining are displaced (**FIGURE 7.23**). The displaced sites of breakage and rejoining are not always located so as

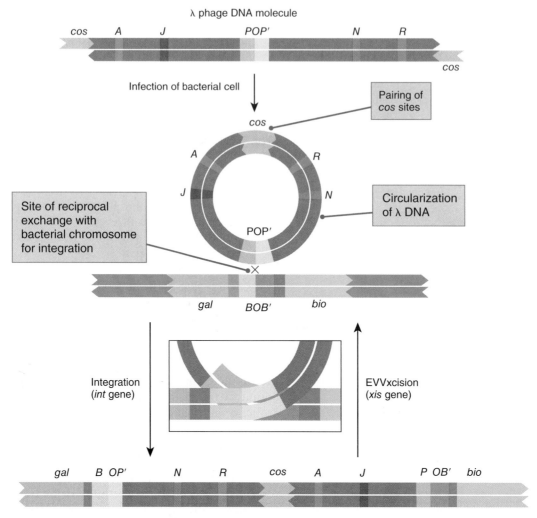

FIGURE 7.22 The geometry of integration and excision of phage λ. The phage attachment site is *POP'*. The bacterial attachment site is *BOB'*. The prophage is flanked by two hybrid attachment sites denoted *BOP'* and *POB'*.

to produce a length of DNA that can fit in a λ phage head—the DNA may be too large or too small—but sometimes a molecule forms that can replicate and be packaged. In λ lysogens, the prophage lies between the *gal* and *bio* genes, and because the aberrant cut in the bacterial DNA can be either to the right or to the left of the prophage, aberrant phage particles can carry either the *bio* genes (cut at the right) or the *gal* genes (cut at the left). They are called λ*dbio* and λ*dgal* transducing particles (Figure 7.23). These are *specialized transducing phages* because they can transduce only certain bacterial genes, in contrast to the P1-type generalized transducing particles, which can transduce any gene.

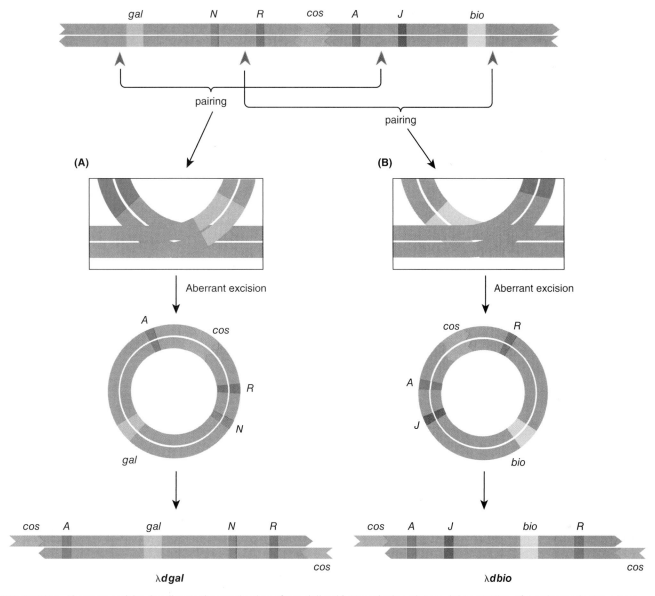

FIGURE 7.23 Aberrant excision leading to the production of specialized λ transducing phages. (A) Formation of a *gal* transducing phage (λ*dgal*). (B) Formation of a *bio* transducing phage (λ*dbio*).

CHAPTER SUMMARY

- Bacteria take advantage of several mechanisms by which DNA sequences can move from one DNA molecule to another and from one cell to another; these mechanisms have led to the evolution of bacteria resistant to multiple antibiotics.

- Bacteria also have mechanisms by which DNA can be transferred from one species to another, a process called horizontal transmission. Owing to horizontal transmission, some species of bacteria have mosaic genomes in which particular regions have been acquired from other bacterial species.

- Some bacteria are capable of DNA transfer and genetic recombination in which genes in the recipient cell are replaced with those from the donor cell.

- Transformation is a process of recombination mediated by the uptake of DNA molecules from the environment.

- In the intestinal bacterium *Escherichia coli*, the F (fertility) plasmid can mobilize the chromosome for transfer to another cell in the process of conjugation.

- Some types of bacterial viruses (bacteriophages) can incorporate bacterial genes from the host cell and transfer them into other cells, a process called transduction.

- When they are present in the same host cell, DNA molecules from related viruses, including bacterial viruses (bacteriophages), can undergo genetic recombination.

- Some bacteriophages are able to integrate their DNA into that of the host cell, where it replicates along with the host DNA and is transmitted to progeny cells.

ISSUES AND IDEAS

- How are antibiotic-resistance cassettes acquired by integrons? Once acquired, can they be lost?

- What role do antibiotic-resistance genes, transposable elements, and transmissible plasmids play in relation to certain pathogenic bacteria that are simultaneously resistant to multiple, chemically unrelated antibiotics?

- How could you distinguish between a bacterial strain that has the phenotype Lac$^+$ and one that has the phenotype Lac$^-$?

- How could you distinguish a Lac$^+$ Amp-r bacterial strain from one that is Lac$^-$ Amp-s? (Amp-r denotes resistance to the antibiotic ampicillin, and Amp-s denotes sensitivity.)

- What is the physical basis of cotransformation? If two genes can be cotransformed, what does this

observation imply about the ability of each gene to be cotransformed with a genetic marker located between them?

- How does the physical state of the F factor differ between an F$^+$ bacterial cell and an Hfr bacterial cell?

- Is the F$^+$ state of a bacterial cell infectious? Is the Hfr state infectious? Explain why or why not.

- When an Hfr bacterial cell transfers its chromosomal DNA into an F$^-$ recipient cell, where in the chromosome does the transfer process begin? How long does it take to transfer the entire chromosome? Why does complete chromosome transfer happen only rarely?

- How does the process of transduction differ from that of transformation?

SOLUTIONS: STEP BY STEP

PROBLEM 1 Four independent integrations of the F factor into the chromosome of a bacterial species closely related to *E. coli* yielded four different Hfr derivatives of the strain (HfrW, HfrX, HfrY, and HfrZ), each with a different origin and possibly a different direction of transfer of markers. These Hfr strains were examined in interrupted-mating experiments with F$^-$ recipient *E. coli* cells and were found to transfer chromosomal genes at the times shown in the accompanying table.

	Genetic marker									
Hfr	*his*	*lac*	*leu*	*lip*	*pheS*	*pyrD*	*recE*	*terC*	*tonA*	*trp*
W	—	—	—	6	—	22	—	—	—	34
X	—	30	18	42	—	—	—	—	22	—
Y	—	—	—	—	4	—	22	12	—	26
Z	38	—	—	—	26	—	8	18	—	—

(a) Draw a circular genetic map, with position 0 minutes at the top, showing the order of the chromosomal genes and the distance (in minutes) between adjacent genes. (The marker *leu* is near 2 minutes on the standard *E. coli* map.) Annotate the genetic map with arrows indicating the origin and direction of transfer of each Hfr and the distance (in minutes) from the origin of transfer to the first marker transferred.

(b) How does the genetic map of the related bacterial species compare with that of the standard *E. coli* strain in Figure 7.14? Suggest an explanation of any discrepancy between the genetic maps.

SOLUTION. (a) Consider each Hfr in turn and, starting with the earliest-entering gene, write down the name of each gene as it enters. The difference in time of entry between adjacent genes is the distance in minutes between the genes. When a partial genetic map of gene transfer from each Hfr has been made, arrange the maps so that any shared markers between two or more Hfr strains coincide. This process yields the genetic maps shown here, where the arrows indicate direction of transfer and the numbers are times of transfer in minutes.

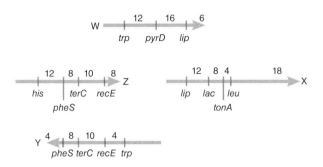

Now arrange the composite genetic map in the manner required, in the form of a circle with 0 minutes at the top and *leu* somewhat to the right. The resulting map is as shown here.

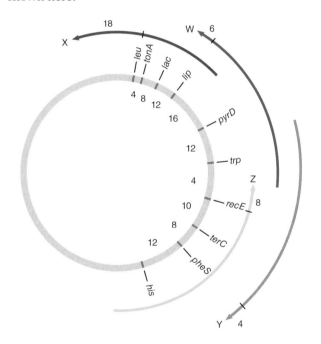

(b) Comparison of this map with that in Figure 7.14 indicates that the genetic markers are all in the same order, but the distances between the markers are about twice

as large as expected. The simplest explanation is that, for the genes studied in this experiment, the related bacterial species has the same gene order as that in *E. coli*, but that Hfr strains transfer the bacterial chromosome only half as fast as those in *E. coli*.

PROBLEM 2 Cotransduction experiments were carried out to determine the order of the closely linked genes *tolC*, *metC*, and *ebg* in the chromosome of *E. coli*. P1 phage grown on a strain of genotype

$$tolC^+ \ metC^+ \ ebg^+$$

were used to transduce a recipient strain of genotype

$$tolC^- \ metC^- \ ebg^-$$

The results were as shown in the accompanying table. What order of the genes is consistent with these results?

Selected marker	Genotypes of unselected markers		Observed percent
tolC⁺	*metC⁺*	*ebg⁺*	2
	metC⁺	*ebg⁻*	12
	metC⁻	*ebg⁺*	30
	metC⁻	*ebg⁻*	56
metC⁺	*tolC⁺*	*ebg⁺*	1
	tolC⁻	*ebg⁺*	0
	tolC⁺	*ebg⁻*	34
	tolC⁻	*ebg⁻*	65

SOLUTION. The gene order is specified if we can deduce the gene in the middle. One approach is to note that the wildtype allele of the gene in the middle will be cotransduced at high frequency (usually greater than 90 percent) when the flanking markers are both transduced. Consider the results when *tolC⁺* is the selected gene: Among the *tolC⁺ metC⁺* transductants, 2/14 are *ebg⁺*, which suggests that *ebg* is not in the middle; among the *tolC⁺ ebg⁺* transductants, 2/32 are *metC⁺*, which suggests that *metC* is not in the middle. These comparisons would suggest that *tolC* is the gene in the middle, but it is important to see whether this hypothesis is consistent with the other results. Among the *metC⁺ ebg⁺* transductants, 1/1 are *tolC*, but this is too little information for us to make an inference. However, among the *metC⁺ tolC⁺* transductants, only 1/35 are *ebg⁺*, which again implies that *ebg* is not in the middle. The data are therefore all consistent with the hypothesis that the gene order is *metC–tolC–ebg* (or *ebg–tolC–metC*) and in any case are inconsistent with either of the other genes being in the middle.

CONCEPTS IN ACTION: PROBLEMS FOR SOLUTION

7.1 Numbers of phage or bacteria in a suspension are usually so large that single colonies or plaques would be impossible to observe without suitable dilution. The usual dilutions are 100-fold, in which 0.1 milliliter (ml) of the suspension is mixed with 9.9 ml of dilution buffer, or 10-fold, in which 1 ml of suspension is mixed with 9 ml of dilution buffer. Usually serial dilutions are necessary, in which the suspension is diluted once into dilution buffer, the resulting suspension mixed thoroughly and diluted a second time, the resulting suspension mixed thoroughly and diluted a third time, and so forth. The serial dilution factors multiply, so, for example, three 100-fold and two 10-fold dilutions yield an overall dilution factor of $100 \times 100 \times 100 \times 10 \times 10 = 10^8$. A volume of 1 ml of a 10^8 dilution is, therefore, the equivalent of a volume of 10^{-8} ml of the original suspension. Suppose that a geneticist plates 1/10 ml of a 10^8 dilution of a suspension of bacteriophage T4 onto a lawn of *E. coli* and observes 16 plaques. What is the estimated number of T4 bacteriophage per milliliter in the original (undiluted) suspension?

7.2 If 1×10^6 phage are mixed with 1×10^6 bacteria and all phage adsorb, what fraction of the bacteria remain uninfected?

7.3 Naturally occurring *E. coli* strains range in genome size from about 4 Mb to about 6 Mb, owing to the presence of genomic islands in some lineages but not in others. If a strain with a genome size of 5 Mb has a genetic map length of 100 minutes, what is the map length of a strain with a genome size of 4 Mb? Of a strain with a genome size of 6 Mb?

7.4 In a cross between Hfr *met⁺ kan-s* × F⁻ *met⁻ kan-r*, where *kan-s* and *kan-r* denote sensitivity or resistance to the antibiotic kanamycin, on what medium should the mating pairs be plated to select for *met⁺ kan-r* recombinants? Which are the selected and counterselected markers? Which strain is a prototrophic for methionine production and which is auxotrophic?

7.5 Estimate the number of viable λ phage per milliliter (the phage titer) in a suspension from the following data. The original suspension was serially diluted through four dilutions of 100-fold each and one dilution of 10-fold. From the final dilution, a volume of 0.1 ml was mixed with a great excess of growing bacteria and then spread over nutrient agar in a petri dish and incubated overnight. The next day, 19 plaques were visible.

7.6 You are given a suspension of bacteria and told that it contains 5×10^7 viable cells per milliliter. What combination of 100-fold and 10-fold serial dilutions would you carry out so that 0.1 ml of the final dilution would contain approximately 50 viable cells?

7.7 A suspension of *E. coli* was serially diluted through two dilutions of 100-fold each and two dilutions of 10-fold each. From the final dilution, a volume of 0.1 ml was spread over nutrient agar in a petri dish and incubated overnight. The next day, 42 colonies were visible. Estimate the number of viable bacteria per milliliter in the original undiluted suspension.

7.8 A gene *x* undergoes transformation at a frequency of 1×10^{-3}, and another gene *y*, located more than a megabase away from *x*, undergoes transformation at a frequency of 6×10^{-4}. What is the expected frequency of cotransformation?

7.9 What is the difference between a selected marker and a counterselected marker? Why are both necessary in an Hfr × F⁻ mating?

7.10 A cross of an Hfr strain of genotype h^+ and an F⁻ strain of genotype h^- is carried out. When tetracycline sensitivity is used for counterselection, the number of recombinant colonies is 1000-fold lower than when streptomycin sensitivity is used for counterselection. In the former case the recombinants are h^+ *tet-r*, and in the latter case they are h^+ *str-r*. Suggest an explanation for the difference.

7.11 An Hfr strain transfers genes in alphabetical order, *a b c*. In an Hfr $a^+ b^+ c^+$ *str-s* × F⁻ $a^- b^- c^-$ *str-r* mating, do all b^+ *str-r* recombinants receive the a^+ allele? Are all b^+ *str-r* recombinants also a^+? Why or why not?

7.12 A temperate bacteriophage related to phage lambda has the gene order *a b c d e f g h*, whereas the prophage of the same phage has the gene order *g h a b c d e f*. What information does this permutation give you about the location of the phage attachment site?

7.13 A cotransduction experiment is carried out with three tightly linked genes, $a^+ b^+$ and c^+, using a recipient strain of genotype $a^- b^- c^-$. Among $a^+ b^+$ transductants, 14% are also c^+. Among $a^+ c^+$ transductants, 83% are also b^+. And among $b^+ c^+$ transductants, 18% are also a^+. Which of the three genes is in the middle?

7.14 Bacterial cells of genotype *pur⁻ pro⁺ his⁺* were transduced with P1 bacteriophage grown on bacteria of genotype *pur⁺ pro⁻ his⁻*. Transductants containing *pur⁺* were selected and tested for the unselected markers *pro* and *his*. The numbers of *pur⁺* colonies with each of four genotypes were as follows:

pro⁺	*his⁺*	102
pro⁻	*his⁺*	25
pro⁺	*his⁻*	160
pro⁻	*his⁻*	1

What is the gene order?

7.15 You are studying a biochemical pathway in *E. coli* that leads to the production of substance A. You isolate a set of mutants, each of which is unable to grow on minimal medium unless it is supplemented with A. By performing appropriate matings, you group all the mutants into four complementation groups (genes) designated *a1*, *a2*, *a3*, and *a4*. You know beforehand that the biochemical pathway for the production of A includes four intermediates: B, C, D, and E. You test the nutritional requirements of the mutants by growing them on minimal medium supplemented with each of these intermediates in turn. The results are summarized in the following table, where the plus signs indicate growth and the minus signs indicate failure to grow.

	A	B	C	D	E
a1	+	+	+	−	+
a2	+	−	+	−	+
a3	+	−	−	−	+
a4	+	−	−	−	−

Determine the order in which the substances A, B, C, D, and E are most likely to participate in the biochemical pathway, and indicate the enzymatic steps by arrows. Label each arrow with the name of the gene that codes for the corresponding enzyme.

7.16 The genes *A*, *B*, *G*, *H*, *I*, and *T* were tested in all possible pairs for cotransduction with bacteriophage P1. Only the following pairs were found able to be cotransduced: *G* and *H*, *G* and *I*, *T* and *A*, *I* and *B*, and *A* and *H*. What is the order of the genes along the chromosome? Explain your logic.

7.17 The order of the genes in the bacteriophage λ is *A B C D E att int xis N CI O P Q S R*.

(a) Given that the bacterial attachment site, *att*, is between *gal* and *bio* in the bacterial chromosome, what is the prophage gene order?

(b) A mutant phage is discovered that has the reverse gene order in the prophage as in the wildtype prophage. What does this say about the orientation of the *att* site in regard to the termini of the phage chromosome?

(c) A wildtype λ lysogen is infected with another λ phage carrying a genetic marker, *Z*, located between *E* and *att*. The superinfection gives rise to a rare, doubly lysogenic *E. coli* strain that carries both λ and λ*Z* prophage. Assuming that the second phage also entered the chromosome at an *att* site, diagram two possible arrangements of the prophages in the bacterial chromosome and indicate the locations of the bacterial genes *gal* and *bio*.

7.18 A bacterial geneticist hopes to map genes *a* through *g* by means of interrupted-mating experiments using three Hfr strains designated X, Y, and Z. From the data in the accompanying table, showing times of entry in minutes, deduce the genetic map of the markers *a* through *g*. Position the genes in correct order on a circle that represents the entire *E. coli* chromosome, 100 minutes in circumference, and show the distance in minutes between adjacent pairs of genes. Show the insertion point and orientation of the F plasmid in each Hfr strain.

Time of entry							
Hfr	*a*	*b*	*c*	*d*	*e*	*f*	*g*
X	11		5				17
Y		9	23		2	17	
Z	27			13			21

7.19 *Salmonella enterica* is closely related to *E. coli*. It can be infected with the F plasmid, which can integrate into the chromosome to produce Hfr strains. These can be mated with F⁻ *E. coli* to study the order and time of entry of genetic markers. The following data pertain to times of entry of four genetic markers in crosses of *E. coli* Hfr × *E. coli* F⁻ and *S. enterica* Hfr × *E. coli* F⁻.

	ile	*met*	*pro*	*arg*
E. coli Hfr × *E. coli* F⁻	28	20	6	22
S. enterica Hfr × *E. coli* F⁻	4	22	47	18

(a) How do the genetic maps of *E. coli* and *S. enterica* compare with respect to these genes?

(b) What are the origin and direction of transfer in each Hfr?

(c) Approximately how fast does the *S. enterica* Hfr transfer chromosomal DNA relative to the *E. coli* Hfr?

7.20 A time-of-entry experiment was carried out with the mating

Hfr a^+ b^+ c^+ d^+ str-s × F⁻ a^- b^- c^- d^- str-r

The data in the accompanying table were obtained. Make a graph showing the number of recombinants per 100 Hfr (*y*-axis) against time of mating (*x*-axis) for each gene, and from this graph determine the time of entry of each gene.

Time of mating in minutes	Number of recombinants of indicated genotype per 100 Hfr			
	a^+ str-r	b^+ str-r	c^+ str-r	d^+ str-r
0	0.01	0.006	0.008	0.0001
10	5	0.1	0.01	0.0001
15	50	3	0.1	0.0005
20	95	35	2	0.001
25	97	80	20	0.001
30	98	82	43	0.01
40	98	80	40	8
50	99	80	40	12
60	98	81	42	16
70	99	80	41	16

 STOP & THINK ANSWERS

ANSWER TO STOP & THINK **7.1**

The medium containing Met + Leu + Str allows all cells in the culture to grow, hence the total number of cells in the aliquot is 400. The medium containing only Str allows only the *met⁺ leu⁺ str-r* cells to grow, which means that the aliquot contains 160 cells of genotype *met⁺ leu⁺ str-r*. The medium containing Met + Str allows both *met⁻ leu⁺ str-r* and *met⁺ leu⁺ str-r* cells to grow, and therefore the number of *met⁻ leu⁺ str-r* cells in the aliquot is estimated as 300 − 160 = 140. Likewise, the medium containing Leu + Str allows both *met⁺ leu⁻ str-r* and *met⁺ leu⁺ str-r* cells to grow, and so the estimated number of *met⁺ leu⁻ str-r* cells in the aliquot is estimated as 260 − 160 = 100. The estimated percentages of the genotypes are as follows: *met⁺ leu⁺ str-r* equals 160/400 = 40 percent, *met⁺ leu⁻ str-r* equals 100/400 = 25 percent, and *met⁻ leu⁺ str-r* equals 140/400 = 35 percent.

ANSWER TO STOP & THINK **7.2**

KL25 transfers genes in a clockwise direction around the chromosome, and so *pyr* is transferred very late. The gene *thr* is at 0 minutes and hence would be transferred at 100 − 85 = 15 minutes after mating. The gene *acrA* is at 10 minutes and therefore would be transferred at 100 − 85 + 10 = 25 minutes after mating.

ANSWER TO STOP & THINK **7.3**

Recall that *c* is the frequency of cotransduction and *m* the the distance in minutes. For *m* = 0.5, 1.0, and 1.5, the values of *c* are 0.42, 0.125, and 0.016, respectively. (Note how rapidly they decrease with increasing map distance.) The maximum distance is 2 minutes, because *m* > 2.0 implies that *c* is negative.

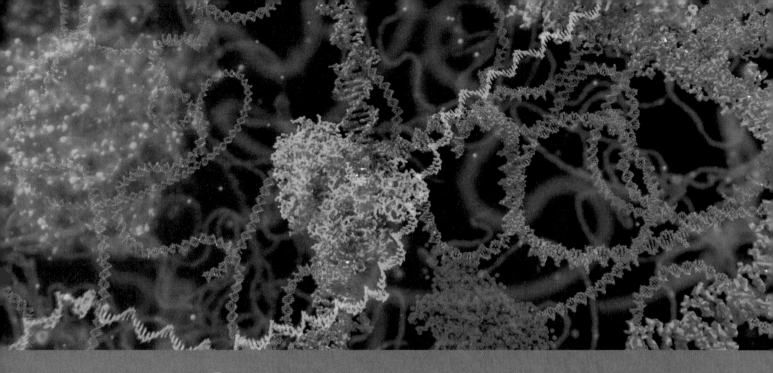

A cell nucleus in the interphase.
© Juan Gaertner/Shutterstock

CHAPTER 8

The Molecular Genetics of Gene Expression

LEARNING OBJECTIVES

- To label the R groups and the peptide bonds in a polypeptide composed of three amino acids.

- To diagram the process of transcription and describe how cells determine where transcription starts and stops.

- To draw the structure of a hypothetical primary RNA transcript in eukaryotes and its processed form, labeling the cap structure, the introns, and the poly-A tail.

- To distinguish how translation is initiated in eukaryotes and prokaryotes, and to explain how initiation in prokaryotes allows the translation of a polycistronic messenger RNA.

- Given a sequence in DNA that codes for amino acids and the direction of transcription, to deduce the sequence of the RNA transcript and the corresponding sequence of amino acids in the polypeptide chain.

- To describe the general structure of the genetic code with special reference to purines and pyrimidines in the third nucleotide position of the codon.

- To explain how the genetic code not only corresponds to codons in the messenger RNA but also to anticodons in transfer RNA and the specificity of the aminoacyl tRNA synthetase enzymes.

The term **gene expression** refers to the process by which information contained in genes is decoded to produce other molecules that determine the phenotypic traits of organisms. The process is initiated when the information contained in the base sequence of DNA is copied into a molecule of RNA, and the process culminates when the molecule of RNA is used to determine the linear order of amino acids in a polypeptide chain. This chapter will increase your understanding of these events. The principal steps in gene expression are as follows:

1. RNA molecules are synthesized by an enzyme, *RNA polymerase*, which uses a segment of a single strand of DNA as a **template strand** to produce a strand of RNA complementary in base sequence to the template DNA. The overall process by which the segment corresponding to a particular gene is selected and an RNA molecule is made is called **transcription**.

2. In the nucleus of eukaryotic cells, the RNA usually undergoes chemical modification called **RNA processing**.

3. The processed RNA molecule is used to specify the order in which amino acids are joined together to form a polypeptide chain. In this manner, the amino acid sequence in a polypeptide is a direct consequence of the base sequence in the DNA. The production of an amino acid sequence from an RNA base sequence is called **translation**, and the protein made is called the **gene product**.

8.1 Polypeptide chains are linear polymers of amino acids.

Proteins are the molecules responsible for catalyzing most intracellular chemical reactions (enzymes), for regulating gene expression (regulatory proteins), and for determining many features of the structures of cells, tissues, and viruses (structural proteins). A protein is composed of one or more chains of linked amino acids called **polypeptide chains**. Twenty different amino acids are commonly found in polypeptides, and they can be joined in any number and in any order. Because the number of amino acids in a polypeptide usually ranges from 100 to 1000, an enormous number of polypeptide chains differing in amino acid sequence can be formed.

Each amino acid contains a carbon atom (the α carbon) to which is attached one carboxyl group (−COOH), one amino group (−NH₂), and a side chain commonly called an **R group**. In **FIGURE 8.1**, the α carbon is shown in gray, the carboxyl group in red, the amino group in blue, and the R group in gold.

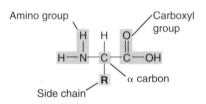

FIGURE 8.1 The general structure of an amino acid.

The R groups are generally chains or rings of carbon atoms bearing various chemical groups. The simplest R groups are those of glycine (−H) and of alanine (−CH₃). The chemical structures of all 20 amino acids are shown in **FIGURE 8.2**.

Polypeptide chains are formed when the carboxyl group of one amino acid joins with the amino group of a second amino acid to form a **peptide bond** (**FIGURE 8.3**, part A). In a polypeptide chain, the α-carbon atoms alternate with peptide groups to form a backbone that has an ordered array of side chains (Figure 8.3, part B). The opposite ends of a polypeptide molecule are chemically different. One end has a free −NH₂ group and is called the **amino terminus**; the other end has a free −COOH group and is the **carboxyl terminus**. Polypeptides are synthesized by the addition of successive amino acids to the carboxyl end of the growing chain. Conventionally, the amino acids of a polypeptide chain are numbered starting at the amino terminus. Therefore, the amino acids are numbered in the order in which they are added to the chain during synthesis.

Owing to interactions between amino acids in the polypeptide chain, most polypeptide chains fold back on themselves in a convoluted manner into a unique three-dimensional shape, in some cases assisted by interactions with other proteins in the cell. About 70 to 75 percent of polypeptide chains fold correctly within milliseconds after release from the ribosome. Exceptionally long polypeptide chains, or ones with a slow or very complex folding pathway, are assisted in their folding by specialized proteins discussed in Section 8.5.

Many protein molecules consist of more than one polypeptide chain. When this is the case, the protein is said to contain *subunits*. The subunits may be identical or different. For example, hemoglobin, the oxygen carrier of blood, consists of four subunits—two of the α polypeptide chain and two of the β polypeptide chain.

The proteins of humans and other vertebrates have a more complex domain structure than do the proteins of invertebrates.

Most polypeptide chains include regions that can fold in upon themselves to acquire well-defined structures of their own, which interact with other structures

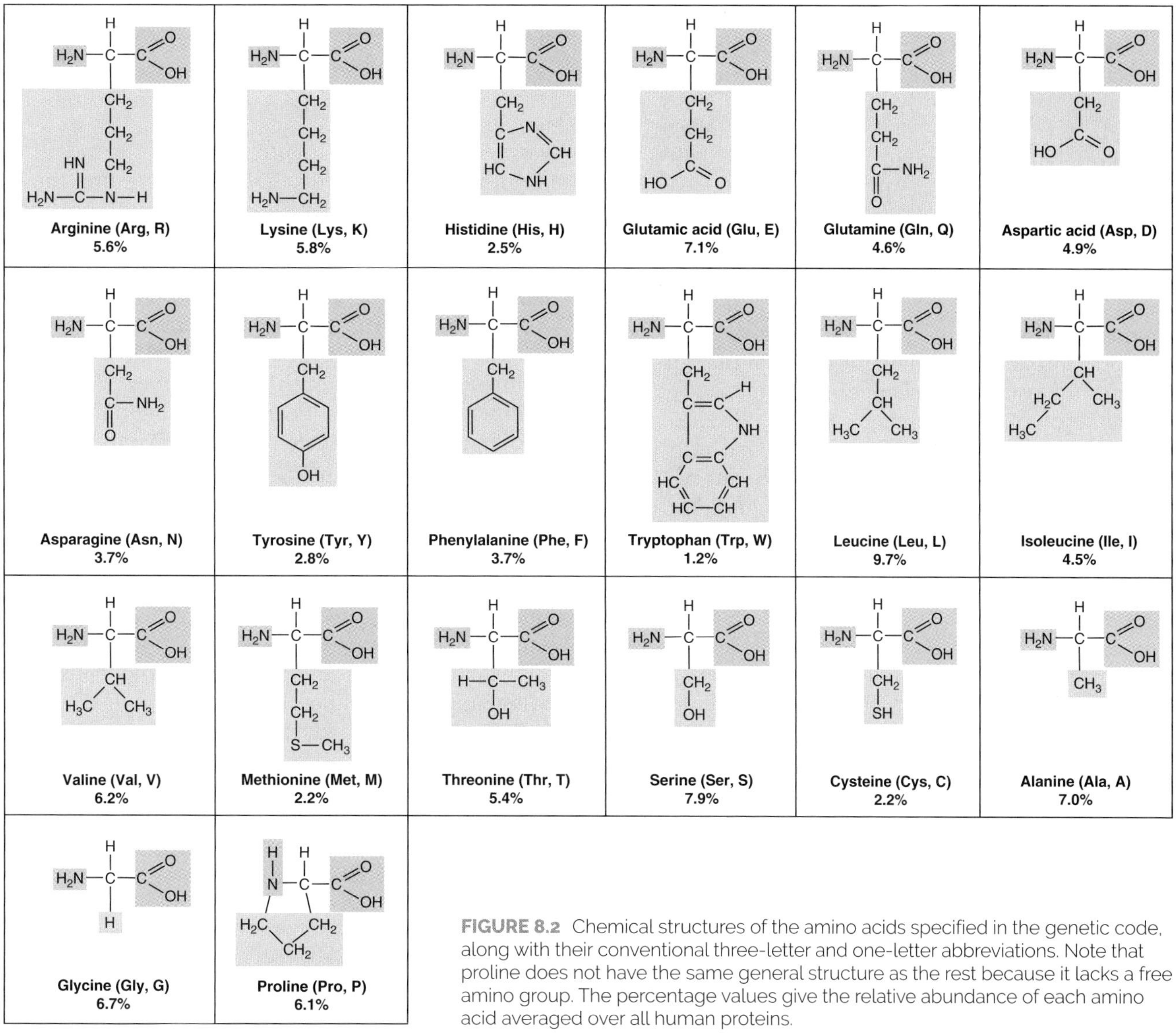

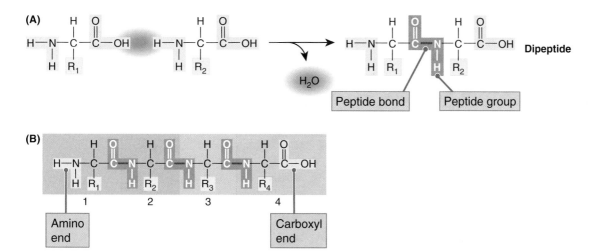

FIGURE 8.2 Chemical structures of the amino acids specified in the genetic code, along with their conventional three-letter and one-letter abbreviations. Note that proline does not have the same general structure as the rest because it lacks a free amino group. The percentage values give the relative abundance of each amino acid averaged over all human proteins.

Arginine (Arg, R) 5.6%
Lysine (Lys, K) 5.8%
Histidine (His, H) 2.5%
Glutamic acid (Glu, E) 7.1%
Glutamine (Gln, Q) 4.6%
Aspartic acid (Asp, D) 4.9%

Asparagine (Asn, N) 3.7%
Tyrosine (Tyr, Y) 2.8%
Phenylalanine (Phe, F) 3.7%
Tryptophan (Trp, W) 1.2%
Leucine (Leu, L) 9.7%
Isoleucine (Ile, I) 4.5%

Valine (Val, V) 6.2%
Methionine (Met, M) 2.2%
Threonine (Thr, T) 5.4%
Serine (Ser, S) 7.9%
Cysteine (Cys, C) 2.2%
Alanine (Ala, A) 7.0%

Glycine (Gly, G) 6.7%
Proline (Pro, P) 6.1%

(A) Dipeptide

H_2O

Peptide bond Peptide group

(B) 1 2 3 4

Amino end Carboxyl end

FIGURE 8.3 Properties of a polypeptide chain. (A) Formation of a dipeptide by reaction of the carboxyl group of one amino acid (left) with the amino group of a second amino acid (right). A molecule of water (H_2O) is eliminated to form a peptide bond (red line). (B) A tetrapeptide showing the alternation of α-carbon atoms (black) and peptide groups (blue). The four amino acids are numbered beneath the tetrapeptide.

formed in other regions of the molecule. Each of these relatively independent folding units is known as a **domain**. The domains in a protein molecule often have specialized functions, such as the binding of substrate molecules, cofactors needed for enzyme activity, or regulatory molecules that modulate activity. The individual domains in a protein usually have independent evolutionary origins, but through duplication of their coding regions and genomic rearrangements, they can come together in various combinations to create genes with novel functions of benefit to the organism. Just as the use of interchangeable parts facilitates airplane development and manufacture, so too does the use of interchangeable domains facilitate the evolution of new proteins.

Protein domains can be identified through computer analysis of the amino acid sequence. When these methods are applied to the human genome sequence, two interesting conclusions emerge:

- Only a minority (about 7 percent) of human proteins and protein domains are specific to vertebrates.

- Human proteins tend to have a more complex domain architecture (linear arrangement of domains) than proteins found in invertebrates. On average, human proteins contain about 1.8 times as many domain architectures as those of the worm or fly, and 5.8 times as many domain architectures as those of yeast.

These comparisons support the following principle:

> **KEY CONCEPT**
>
> Vertebrate genomes, including the human genome, have relatively few proteins or protein domains not found in other organisms. Their complexity arises in part from innovations in bringing together preexisting domains to create novel proteins that have more complex domain architectures than those found in other organisms.

8.2 The linear order of amino acids is encoded in a DNA base sequence.

Most genes contain the information for the synthesis of only one polypeptide chain. Furthermore, the linear order of nucleotides in a gene determines the linear order of amino acids in a polypeptide. This point was first proved by studies of the tryptophan synthase gene *trpA* in *Escherichia coli*, a gene in which many mutations had been obtained and accurately mapped genetically. The effects of numerous mutations on the amino acid sequence of the enzyme were determined by directly analyzing the amino acid sequences of the wildtype and mutant enzymes. Each mutation was found to result in a single amino acid substituting for the wildtype amino acid in the enzyme. More important, the order of the mutations in the genetic map was the same as the order of the affected amino acids in the polypeptide chain (**FIGURE 8.4**). This attribute of genes and polypeptides is called **colinearity**, which means that the sequence of base pairs (bp) in DNA determines the sequence of amino acids in the polypeptide in a colinear, or point-to-point, manner. Colinearity is universally found in prokaryotes. However, we will see later that in eukaryotes, noninformational DNA sequences interrupt the continuity of most genes; in these genes, the order but not the spacing between the mutations correlates with amino acid substitution.

8.3 The base sequence in DNA specifies the base sequence in an RNA transcript.

The first step in gene expression is the synthesis of an RNA molecule copied from the segment of DNA that constitutes the gene. The basic features of the production of RNA are described in this section.

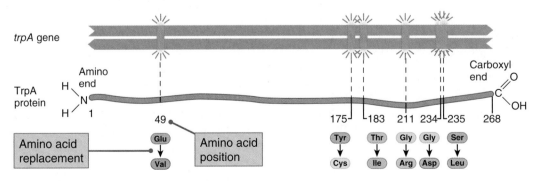

FIGURE 8.4 Colinearity of DNA and protein in the *trpA* gene of *E. coli*.

The chemical synthesis of RNA is similar to that of DNA.

Although the essential chemical characteristics of the enzymatic synthesis of RNA are generally similar to those of DNA, there are also some important differences.

- Each RNA molecule produced in transcription derives from a single strand of DNA, because in any particular region of the DNA, usually only one strand serves as a template for RNA synthesis.

- The precursors in the synthesis of RNA are the four ribonucleoside 5'-triphosphates: adenosine triphosphate (ATP), guanosine triphosphate (GTP), cytidine triphosphate (CTP), and uridine triphosphate (UTP). They differ from the DNA precursors only in that the sugar is ribose rather than deoxyribose and the base uracil (U) replaces thymine (T) (**FIGURE 8.5**).

- The sequence of bases in an RNA molecule is determined by the sequence of bases in the DNA template. Each base added to the growing end of the RNA chain is chosen for its ability to base pair with the DNA template strand. Thus the bases C, T, G, and A in the DNA template cause G, A, C, and U, respectively, to be added to the growing end of the RNA molecule.

- In the synthesis of RNA, a sugar–phosphate bond is formed between the 3'-hydroxyl group of one nucleotide and the 5'-triphosphate of the next nucleotide in line (**FIGURE 8.6**, parts A and B). The chemical bond formed is the same as that in the synthesis of DNA, but the enzyme is different. The enzyme used in transcription is RNA polymerase rather than DNA polymerase. The RNA polymerase binds to a DNA sequence of 20–200 nucleotides called a **promoter**. Transcription begins at a nucleotide in or near the promoter called the *transcription start site*.

- Nucleotides are added only to the 3'-OH end of the growing chain; as a result, the 5' end of a growing RNA molecule bears a triphosphate group. The 5' → 3' direction of chain growth is the same as that in DNA synthesis (Figure 8.6, part C).

- RNA polymerase (unlike DNA polymerase) is able to initiate chain growth without a primer (Figure 8.6, part C).

STOP & THINK 8.1

Part of a double-stranded DNA molecule has the sequence

```
5'-ATGCCGTTA-3'
3'-TACGGCAAT-5'
```

If this region is transcribed from left to right, what is the sequence of the RNA transcript?

Eukaryotes have several types of RNA polymerase.

RNA polymerases are large, multisubunit complexes whose active form is called the **RNA polymerase holoenzyme**. Bacterial cells have only one RNA polymerase, and the holoenzyme is composed of six polypeptide subunits. In transcriptional initiation, the holoenzyme contacts 70–90 bp in the promoter region, but once transcription begins, the region of contact is reduced to about 35 nucleotides centered on the nucleotide being added. The *processivity* of RNA polymerase (the number of nucleotides transcribed without dissociating from the template) is impressive: more than 10^4 nucleotides in prokaryotes and more than 10^6 nucleotides in eukaryotes. Processivity is important, because once the RNA polymerase separates from the template, it cannot resume synthesis. The rate of transcription is also impressive, approximately 70 nucleotides per second in prokaryotes and 40 nucleotides per second in eukaryotes.

Eukaryotic RNA polymerases are even larger than those in prokaryotes, and they include more subunits in the holoenzyme. There are also several different types:

1. **RNA polymerase I** is used exclusively in producing the transcript that becomes processed into

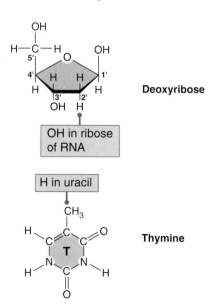

FIGURE 8.5 Differences between the structures of ribose and deoxyribose and between those of uracil and thymine.

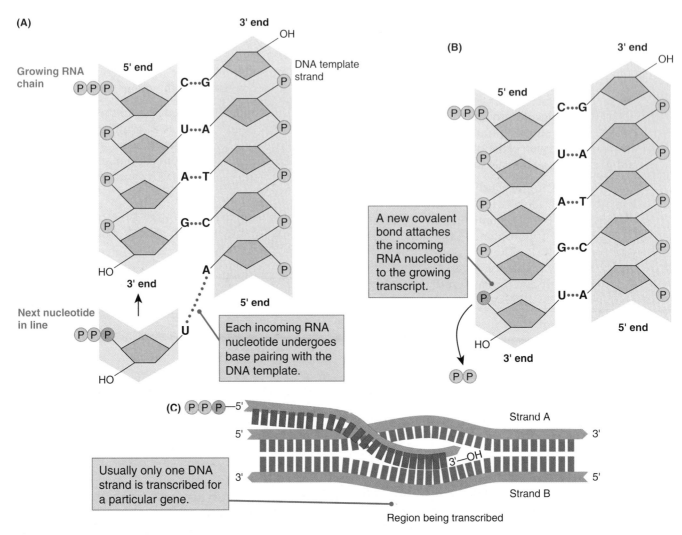

FIGURE 8.6 RNA synthesis. (A) Base pairing with the template strand. (B) The polymerization step. (C) Transcription in progress.

ribosomal RNA. The promoter region includes the transcription start site.

2. **RNA polymerase II** is the workhorse eukaryotic polymerase responsible for transcribing all protein-coding genes as well as the genes for a number of small nuclear RNAs used in RNA processing (discussed in Section 8.4). The RNA polymerase II promoter is located near the transcription start site but upstream (on the 5′ side) of it. The mechanism of Pol II is the best understood of the eukaryotic polymerases. The holoenzyme contains 12 polypeptide subunits and has a molecular mass of about 500 kD. Its structure features a groove (**FIGURE 8.7**) that helps guide the DNA template into the active site marked by the magnesium atom colored in pink.

3. **RNA polymerase III** is used in transcribing all transfer RNA genes as well as the 5S component of ribosomal RNA. The promoter for RNA polymerase III transcription is located near the transcription start site but downstream (on the 3′ side) of it.

Promoter recognition typically requires multiple DNA-binding proteins.

Many promoter sequences have been isolated and their nucleotide sequences determined. The promoters for polymerases I, II, and III show no commonality, and within each class there is substantial sequence variation. The situation in bacteria is considerably simpler. Although bacterial promoters also differ in sequence, in part because they differ in their polymerase binding affinity, certain sequence patterns or *motifs* are quite frequent. Two such patterns that are often found in promoter regions in *E. coli* are illustrated in **FIGURE 8.8**. Each pattern is defined by a **consensus sequence** of nucleotides determined from the actual sequences by majority rule: Each nucleotide in the consensus sequence is the nucleotide most often observed at that position in actual sequences. Any particular sequence may resemble the consensus sequence very well or very poorly.

In the consensus sequences shown in Figure 8.8, the transcription start site is numbered as +1. One consensus motif is TTGACA, which is called the −35 motif because it is usually located approximately 35 base pairs upstream from the transcription start site. The other consensus motif is TATAAT, which is usually located near position −10. The −10 sequence is called the **TATA box**. The positions of the promoter sequences determine where the RNA polymerase begins synthesis.

The strength of the binding of RNA polymerase to different promoters varies greatly, which causes differences in the extent of expression from one gene to

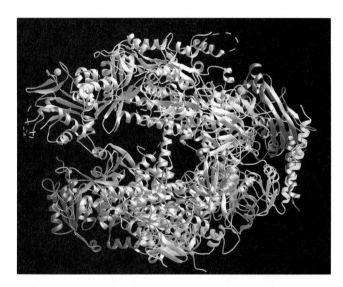

FIGURE 8.7 Structure of the Pol II holoenzyme showing the groove on the left that terminates in the active site that contains a magnesium ion (pink sphere). The positively charged magnesium ion helps orient the negatively charged phosphate group of the next nucleotide to be added to the growing chain. Different polypeptide components in the holoenzyme are shown in different colors.

Reproduced from Cramer, P., Bushnell, D. A., and Kornberg, R. D. 2001. Science 292:1863–1876. Reprinted with permission from AAAS. Photos courtesy of Roger D. Kornberg, Stanford University School of Medicine.

another. Most of the differences in promoter strength result from variations in the −35 and −10 promoter elements and in the spacing between them. Promoter strength among *E. coli* genes differs by a factor of 10^4, and most of the variation can be attributed to the promoter sequences themselves. In general, the more closely the promoter elements resemble the consensus sequence, the stronger the promoter. Mutations that change the nucleotide sequence in a promoter can alter the strength of the promoter. Changes that result in less resemblance to the consensus sequence lower the promoter's strength, whereas those with greater resemblance to the consensus increase it. Furthermore, some promoters differ greatly from the consensus sequence in the −35 region.

All promoters typically require accessory proteins to activate transcription by RNA polymerase. In bacteria, among the most important accessory proteins for transcription are sigma factors. A *sigma factor* is a protein that combines transiently with RNA polymerase to allow it to bind properly to a promoter region. All bacteria produce sigma factors that allow transcription of genes needed for their normal growth and metabolism. There are also specialized sigma factors produced only under certain conditions that enable normally untranscribed genes to be transcribed. Various kinds of stress, including heat shock or starvation, induce the production of such specialized sigma factors.

Promoter sequences in eukaryotes are generally much longer and more complex than those in prokaryotes. Many promoters recognized by Pol II include a core region containing a TATA-box motif, which is analogous to that in prokaryotes but differs in its spacing relative to the transcriptional start site (**FIGURE 8.9**). Proper binding of Pol II to the promoter also requires a set of at least 26 *general transcription factors*, but even these proteins are not sufficient. They need to be recruited to the promoter by still other proteins that bind with other sequence motifs that are often located far upstream or sometimes even downstream from the core region containing the TATA box.

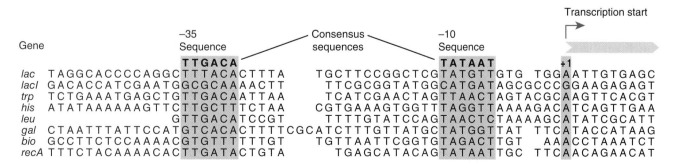

Transcription start

Gene	−35 Sequence	Consensus sequences	−10 Sequence	+1
lac	TAGGCACCCCAGGC**TTTACA**CTTTA	TGCTTCCGGCTCG**TATGTT**GTG	TGG**A**ATTGTGAGC	
lacI	GACACCATCGAATG**GCGCA**AAACTT	TTCGCGGTATGG**CATGAT**AGCGCCC**G**GAAGAGAGT		
trp	TCTGAAATGAGCTG**TTGACA**ATTAA	TCATCGAACTAG**TTAACT**AGTACGC**A**AGTTCACGT		
his	ATATAAAAAAGTTC**TTGCT**TTCTAA	CGTGAAAGTGGT**TTAGGT**TAAAAGAC**A**TCAGTTGAA		
leu	G**TTGACA**TCCGT	TTTTGTATCCAG**TAACTC**TAAAAGC**A**TATCGCATT		
gal	CTAATTTATTCCAT**GTCACA**CTTTTCGCATCTTTGTTATGC**TATGGT**TAT	TTC**A**TACCATAAG		
bio	GCCTTCTCCAAAAC**GTGTT**TTTTGT	TGTTAATTCGGTG**TAGACT**TGT	AAA**C**CTAAATCT	
recA	TTTCTACAAAACAC**TTGATA**CTGTA	TGAGCATACAG**TATAAT**TGC	TTC**A**ACAGAACAT	

FIGURE 8.8 Base sequences in promoter regions of several genes in *E. coli*. The consensus sequences located 10 and 35 nucleotides upstream from the transcription start site (+1) are indicated. Promoters vary tremendously in their ability to promote transcription. Much of the variation in promoter strength results from differences between the promoter elements and the consensus sequences at −10 and −35.

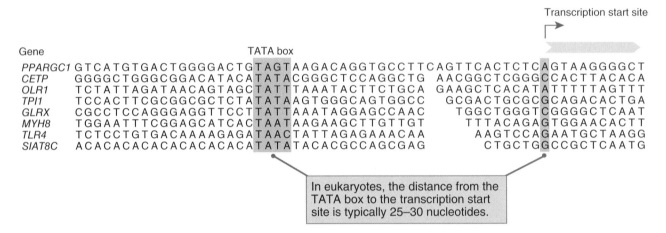

FIGURE 8.9 Some human TATA-box promoters showing sequences in the core promoter region near the transcription start file.

These sequence motifs, some of which act as *enhancers* and others as *silencers*, bind proteins that interact with the transcriptional machinery to regulate the level of transcription.

RNA polymerase is a molecular machine for transcription.

Once promoter recognition takes place, the mechanism of transcription can be described in terms of three discrete stages. These will be examined with regard to the mechanism of action of the eukaryotic Pol II polymerase.

The Pol II complex does not act alone but in combination with five general transcription factors denoted TFIIB, TFIIE, TFIID, TFIIF, and TFIIH. Although these components can correctly transcribe naked DNA, they cannot transcribe native chromatin organized into nucleosomes. In the cell, access to naked promoter DNA occurs by transient displacement of the nucleosomes through the action of *chromatin remodeling complexes*. The specificity that causes transcription to begin only at the correct start site near a promoter is brought about by the general transcription factors associated with Pol II, each of which plays an essential role.

1. **Chain initiation** In the first step in transcription, the **TATA-box binding protein** (TBP) binds the promoter DNA and bends it at almost a 90-degree angle. Physically, the TATA-box binding protein is closely associated with the polymerase, and the bend brings the promoter DNA into contact with the TFIIB component of the polymerase. The promoter DNA then follows a straight path across the top of the polymerase until, at a point 25–30 bp distant from the TATA box, the transcription start site is brought into position near the polymerase active site. At this point TFIIE joins the complex and recruits TFIIH, whose helicase activity destabilizes the DNA duplex. This

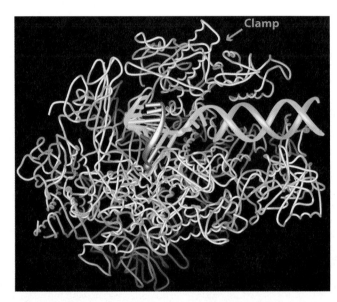

FIGURE 8.10 Pol II in action. The template strand is shown in blue and the RNA transcript in red. The magnesium ion in the active site is indicated by the pink sphere. Part of the nontemplate strand is shown in green, paired with the template strand. The replication bubble is held in place by a mobile domain of Pol II (the clamp, orange), and the two largest subunits of Pol II are held together by a helix bridging between them (green helix). The other polypeptide chains in Pol II are shown in white.

Courtesy of David A. Bushnell and Roger Kornberg, Stanford University Medical School.

makes the promoter duplex susceptible to thermal unwinding, which produces a transient unwound region or bubble. The unwound region is stabilized by TFIIF binding to the nontemplate strand, while the unbound template strand descends deeper into the groove in the polymerase where the active site is situated. At a point three nucleotides deep into the groove, the template strand undergoes a sharp bend that flips the first base to be transcribed into the active site (**FIGURE 8.10**). At the same time, a large domain of the

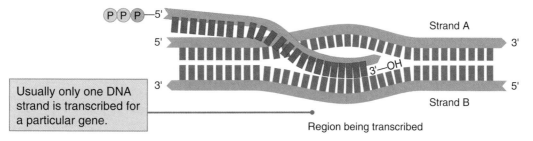

Strand A

3'

3'—OH

Strand B

Usually only one DNA strand is transcribed for a particular gene.

Region being transcribed

FIGURE 8.11 Geometry of RNA synthesis. RNA is copied from only one strand of a segment of a DNA molecule—in this example, strand B—without the need for a primer.

polymerase flips into position over the promoter DNA, forming a clamp that holds the transcription bubble in place.

Initiation of transcription now begins, a process stabilized by TFIIB. After the first nucleoside triphosphate of the RNA transcript is put in place, synthesis proceeds in the 5'-to-3' direction. After about six nucleotides have been transcribed, the TFIIB is displaced and transcription continues.

2. **Chain elongation**   As the RNA polymerase encounters new nucleotides along the template DNA strand, successive RNA nucleotides are added to the growing transcript. Only one DNA strand, the template strand, is transcribed. At steady state, the transcription bubble consists of 15 nucleotides of unwound DNA duplex, of which 8–9 are paired with the 3' end of the RNA transcript (**FIGURE 8.11**). Each incoming nucleotide is added to the 3' end of the transcript at a site three nucleotides from the point at which the DNA template strand unwinds from the nontemplate DNA.

The weak energy released by hydrogen bonding between complementary nucleotides is not adequate to account for the base-pairing specificity of transcription. The specificity is actually brought about by structural changes that take place around the active site of Pol II. The key movement is that of a domain near the active site, called the *trigger loop*, which moves into position only when correct base pairing occurs. The repositioning of the trigger loop brings a critical histidine residue into play, which promotes the flow of electrons that triggers polymerization. In this way, correct base pairing is physically coupled to formation of the phosphodiester bond in the growing RNA chain.

As each new bond is formed in the transcript, a helical segment of Pol II that is in contact with the single-stranded template lurches forward about 3 Å, which brings the next template nucleotide into the active site. About 8–9 nucleotides

behind the active site, two segments of Pol II invade the RNA–DNA hybrid, breaking the hydrogen bonds, and a few nucleotides farther behind that point, the DNA template and nontemplate DNA strands are rejoined (Figure 8.11).

3. **Chain termination**   When the RNA polymerase complex reaches a chain-termination sequence, both the newly synthesized RNA molecule and the polymerase complex are released. The requirements for transcription termination are understood best in bacteria, in which two kinds of termination events are known. In *intrinsic termination*, which is the most common case in bacteria, the signal for termination of transcription depends only on the nucleotide sequence in the DNA template. Transcription stops when the polymerase encounters a particular sequence of nucleotides in the transcribed DNA strand that is able to fold back upon itself to form a hairpin loop. An example of such a terminator found in *E. coli* is shown in **FIGURE 8.12**. The hairpin loop alone is not enough for termination of transcription; the run of U's at the end of the hairpin is also necessary. The hairpin loop terminates transcription by invading the main channel of the RNA polymerase complex near the active site, which disrupts the RNA/DNA hybrid and prevents further elongation of the chain.

A second type of termination is called *rho-dependent termination* because it requires presence of a termination protein associated with the polymerase complex. In *E. coli*, the predominant termination protein is known as *rho*. The terminator sequences recognized by the rho protein appear to be long and complex; however, the mechanism of termination is similar to that of intrinsic termination. In particular, when a termination signal is encountered, a polypeptide loop of the rho protein is formed that inserts into the main channel of the polymerase complex and stops elongation by separating the RNA/DNA hybrid.

(A) DNA

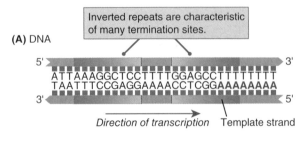

(B) Terminus of mRNA

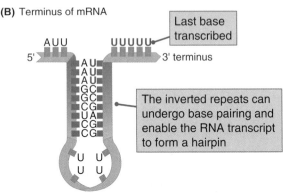

FIGURE 8.12 (A) Base sequence of the transcription termination region for the set of tryptophan-synthesizing genes in *E. coli*. (B) The 3' terminus of the RNA transcript, folded to form a stem-and-loop structure. The sequence of U's found at the end of the transcript in this and many other prokaryotic genes is shown in red. The RNA polymerase, not shown here, terminates transcription when the loop forms in the transcript.

Initiation of a second round of transcription need not await completion of the first. By the time an RNA transcript reaches a size of 50 to 60 nucleotides, the RNA polymerase has moved along the DNA far enough from the promoter that the promoter becomes available for another RNA polymerase to initiate a new transcript. Such reinitiation can take place repeatedly, and a gene can be cloaked with numerous RNA molecules in various degrees of completion.

Genetic experiments in *E. coli* yielded the first demonstration of the existence of promoters. A class of Lac⁻ mutations, denoted p^-, was isolated that was unusual in two respects:

- All p^- mutations were closely linked to the *lacZ* gene.

- Any p^- mutation eliminated activity of a wild-type *lacZ* gene *present in the same DNA molecule*.

The need for an adjacent genetic configuration to eliminate *lacZ* activity can be seen by examining a cell with two copies of the *lacZ* gene. Such cells can be produced through the use of F′*lacZ* plasmids, which contain a copy of *lacZ* in an F plasmid. Infection with an F′*lacZ* plasmid yields a cell with two copies of *lacZ*—one in the chromosome and another in the F′. Transcription of the *lacZ* gene enables the cell to synthesize the enzyme β-galactosidase. **TABLE 8.1** shows that a wildtype *lacZ* gene (*lacZ*⁺) is inactive

TABLE 8.1 Effect of Promoter Mutations on Transcription of the *lacZ* Gene

Genotype	Transcription of *lacZ*⁺ gene?
1. p^+lacZ^+	Yes
2. p^-lacZ^+	No
3. p^+lacZ^+/p^+lacZ^-	Yes
4. p^-lacZ^+/p^+lacZ^-	No
5. p^+lacZ^+/p^-lacZ^-	Yes

Note: lacZ⁺ is the wildtype gene; the lacZ⁻ mutant produces a nonfunctional enzyme.

when a p^- mutation is present in the same DNA molecule (either in the bacterial chromosome or in an F′ plasmid). This result can be seen by comparing entries 4 and 5. Analysis of the RNA shows that in a cell with the genotype p^- *lacZ*⁺, the *lacZ*⁺ gene is not transcribed. On the other hand, cells of genotype p^+lacZ^- produce a mutant RNA. The p^- mutations are called *promoter mutations*.

Mutations have also been instrumental in defining the transcription termination region. For example, mutations have been isolated that create a new termination sequence upstream from the normal one. When such a mutation is present, an RNA molecule is made that is shorter than the wildtype RNA. Other mutations eliminate the terminator, resulting in a longer transcript.

Messenger RNA directs the synthesis of a polypeptide chain.

The RNA molecule produced from a DNA template is the **primary transcript**. Each gene has only one DNA strand that serves as the template strand, but which strand is the template strand can differ from gene to gene along a DNA molecule. Therefore, in an extended segment of a DNA molecule, primary transcripts would be seen growing in either of two directions (**FIGURE 8.13**), depending on which DNA strand functions as a template in a particular gene. In prokaryotes, the primary transcript serves directly as the **messenger RNA (mRNA)** used in polypeptide synthesis. In eukaryotes, the primary transcript is generally processed before it becomes mRNA.

Not all base sequences in an mRNA molecule are translated into the amino acid sequences of polypeptides. For example, translation of an mRNA molecule rarely starts exactly at one end and proceeds to the other end; instead, initiation of polypeptide synthesis may begin many nucleotides downstream from the 5' end of the RNA. The untranslated 5' segment

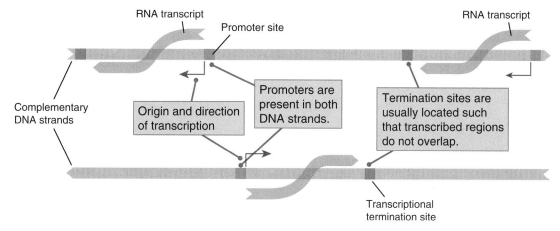

FIGURE 8.13 A typical arrangement of promoters (green) and termination sites (red) in a segment of a DNA molecule. Promoters are present in both DNA strands. Termination sites are usually located such that transcribed regions do not overlap.

of RNA is called the **5' untranslated region**. This is followed by an **open reading frame (ORF)**, which specifies the polypeptide chain. A typical ORF in an mRNA molecule is between 500 and 3000 bases long (depending on the number of amino acids in the protein), but it may be much longer. The 3' end of an mRNA molecule following the ORF also is not translated; it is called the **3' untranslated region**.

In prokaryotes, most mRNA molecules are degraded within a few minutes after synthesis. In eukaryotes, a typical lifetime is several hours, although some last only minutes whereas others persist for days. In both kinds of organisms, the degradation enables cells to dispose of molecules that are no longer needed and to recycle the nucleotides in synthesizing new RNAs. The short lifetime of prokaryotic mRNA is an important factor in regulating gene activity.

⊛ STOP & THINK 8.2

Shown here is a promoter region (light blue) located between two potentially transcribed regions (A and B) in double-stranded DNA. As indicated by the arrow, the promoter is oriented such that the template strand is on top and transcription proceeds from right to left into region A.

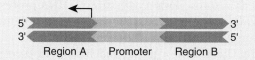

Draw a similar diagram indicating which strand would be transcribed, and in which direction, following a mutation in which the promoter region was inverted.

8.4 RNA processing converts the original RNA transcript into messenger RNA.

Although the process of transcription is very similar in prokaryotes and eukaryotes, there are major differences in the relationship between the transcript and the mRNA used for polypeptide synthesis. In prokaryotes, the immediate product of transcription (the primary transcript) is mRNA; in contrast, *the primary transcript in eukaryotes must be converted into mRNA.* The conversion of the original transcript into mRNA is called *RNA processing*. It usually consists of three types of events:

1. The 5' end is altered by the addition of a modified guanosine in an uncommon 5'−5' linkage (instead of the typical 3'−5' linkage); this terminal group is called the **cap**; the 5' cap is necessary for the ribosome to bind with the mRNA to begin protein synthesis.

2. The 3' end is usually modified by the addition of a sequence called the **poly-A tail**, which can consist of as many as 200 consecutive A-bearing nucleotides; the poly-A tail is thought to help regulate mRNA stability.

3. Certain regions internal to the transcript (*introns*) are removed by splicing. The segments that are excised from the primary transcript are called **introns** or *intervening sequences*. Accompanying the excision of introns is a rejoining of the coding segments (**exons**) to form the mRNA molecule. The excision of the introns and the joining of the exons to form the final mRNA molecule is called **RNA splicing**.

The events that constitute RNA processing begin even while transcription is still in progress, and

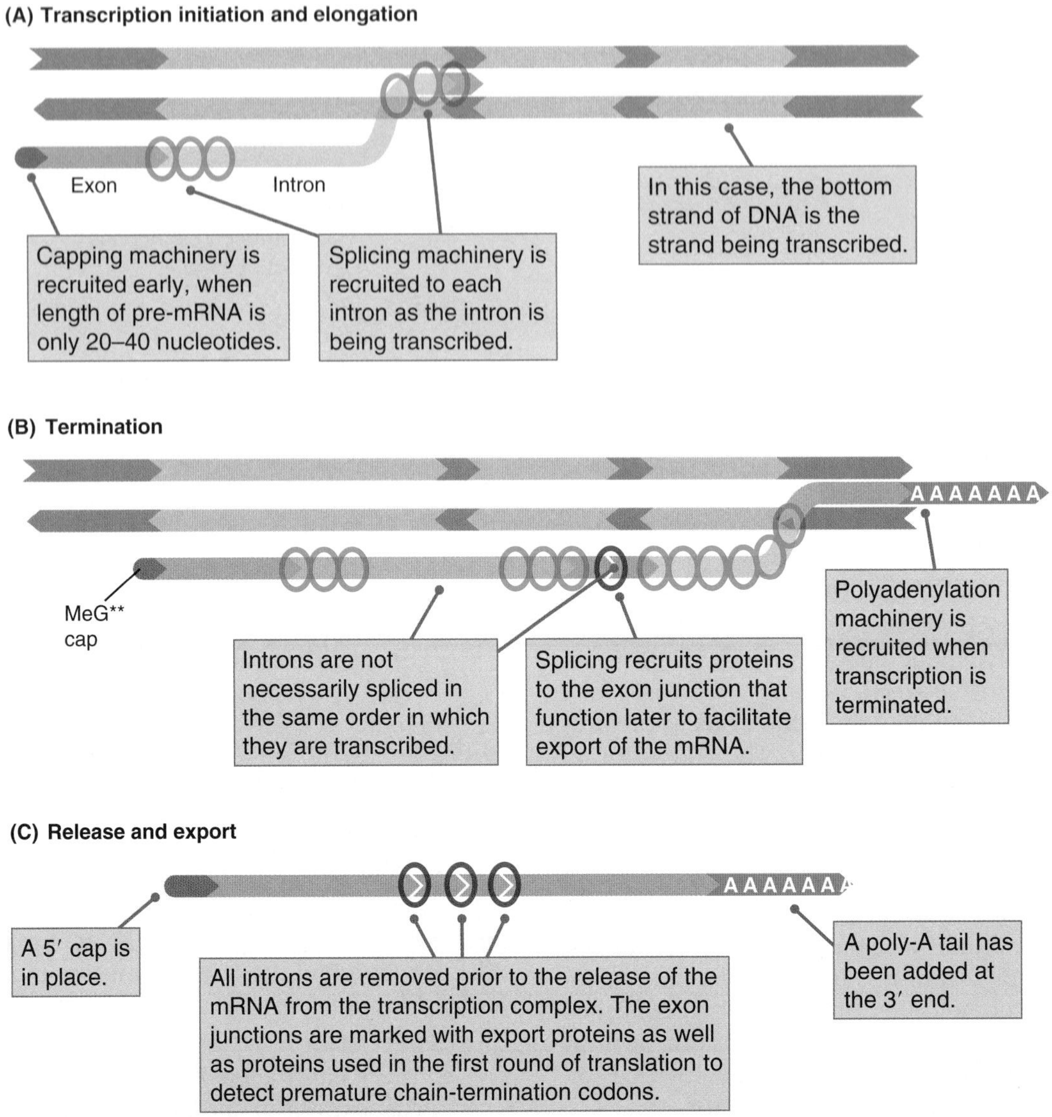

(A) Transcription initiation and elongation

Exon   Intron

Capping machinery is recruited early, when length of pre-mRNA is only 20–40 nucleotides.

Splicing machinery is recruited to each intron as the intron is being transcribed.

In this case, the bottom strand of DNA is the strand being transcribed.

(B) Termination

MeG** cap

AAAAAAA

Introns are not necessarily spliced in the same order in which they are transcribed.

Splicing recruits proteins to the exon junction that function later to facilitate export of the mRNA.

Polyadenylation machinery is recruited when transcription is terminated.

(C) Release and export

AAAAAAA

A 5′ cap is in place.

All introns are removed prior to the release of the mRNA from the transcription complex. The exon junctions are marked with export proteins as well as proteins used in the first round of translation to detect premature chain-termination codons.

A poly-A tail has been added at the 3′ end.

FIGURE 8.14 In eukaryotes, transcription and RNA processing are coupled. Each step (A, B, and C) triggers the next in line. MeG denotes 7-methylguanosine (a modified form of guanosine), and the two asterisks indicate two nucleotides whose riboses are methylated.

the events are *coupled processes*, which means that occurrence of one event initiates the next. Some of the interconnections are shown in **FIGURE 8.14**. A key player in the coupling is the carboxyl terminal domain of the large subunit of RNA polymerase II, which contains a series of nearly identical repeats of a sequence of seven amino acids. When key amino acids in this domain become phosphorylated, the RNA polymerase recruits the capping machinery, and when they are dephosphorylated the capping machinery is released. Phosphorylation of other amino acids in the domain helps recruit the machinery for splicing and polyadenylation.

The effect of the coupled processes is to greatly increase the speed and specificity of RNA processing. Without such coupling, RNA processing would be dependent on diffusion, and many mistakes would be made, especially in splicing the often large introns that

separate relatively small exons. The recruitment of the splicing machinery while transcription is still taking place greatly facilitates correct splicing. Introns are not necessarily spliced in exactly the same order in which they are transcribed, however. The order in which splicing takes place depends on the size and nucleotide composition of the introns, as well as on the overall rate of transcription.

Numerous interconnecting links couple the various steps in transcription with those in RNA processing. For example, proteins that bind with RNA polymerase to promote elongation also help recruit the splicing machinery, and the splicing machinery in turn stimulates elongation so that genes containing introns are more efficiently transcribed. The splicing machinery also helps recruit the polyadenylation machinery. The principal steps in RNA processing are all completed prior to release of the mRNA from the transcription complex. As each intron is spliced, proteins bind to the junction between the exons (Figure 8.14, parts B and C). Some of these function after release of the mRNA to facilitate its export from the nucleus to the cytoplasm. Others of these proteins function in the first round of translation to identify defective mRNA molecules that are subsequently destroyed.

Splicing removes introns from the RNA transcript.

RNA splicing takes place in nuclear particles known as **spliceosomes**. These abundant particles are composed of protein and several specialized small RNA molecules that are present in the cell as small nuclear ribonucleoprotein particles; the underlined letters give

the acronym for these particles: **snRNPs**. The specificity of splicing comes from the five small snRNP RNAs denoted U1, U2, U4, U5, and U6, which contain sequences complementary to the splice junctions, to the branchpoint region of the intron, and/or to one another; as many as 100 spliceosome proteins may also be required for splicing. The ends of the intron are brought together by U1 RNA, which forms base pairs with nucleotides in the intron at both the 5′ and the 3′ ends. U2 RNA binds to the branchpoint region. U2 RNA interacts with a paired complex of U4/U6 RNAs, resulting in a complex in which U2 RNA ends up paired with U6 RNA and the intron of the transcript (**FIGURE 8.15**). All of these dynamic interactions bring the branchpoint region near to the donor splice site and allow the A in the branchpoint to attack the G of

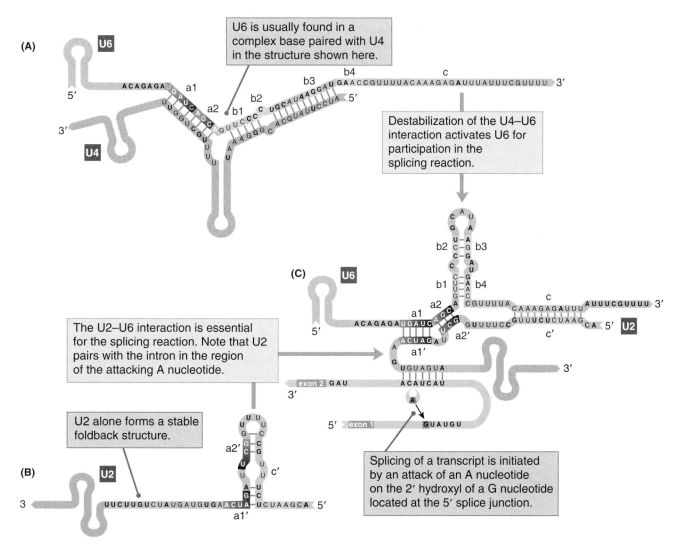

FIGURE 8.15 Dynamic interactions between some small nuclear RNAs present in snRNPs that are involved in splicing. (A) U6 snRNA is usually found complexed with U4 snRNA. (B) U2 snRNA forms a stable foldback structure on its own. (C) Essential to the splicing reaction is destabilization of the U4−U6 structure and formation of a U2−U6 structure in which U2 is base paired with part of the intron. An A in the paired region attacks the G at the 5′ splice junction, initiating the splicing reaction. The nucleotides in bold are critical to the structures, judging by their having been conserved in very diverse species. Note that G−U base pairs are allowed in double-stranded RNA.

Data from H. D. Madhani and C. Guthrie, *Annu. Rev. Genet.* 1 (1994): 1–26.

the donor splice site, freeing the upstream exon and forming the looped intermediate (Figure 8.15). U5 RNA helps line up the two exons and facilitates the final step in splicing, which results in scission of the intron from the downstream exon and in ligation of the upstream and downstream exons.

Introns are also present in some genes in organelles, such as mitochondria, but the mechanisms of their excision differ from those of introns in nuclear genes because organelles do not contain spliceosomes. In one class of organelle introns, the intron contains a sequence coding for a protein that participates in removing the intron that codes for it. The situation is even more remarkable in the splicing of a ribosomal RNA precursor in the ciliate *Tetrahymena*. In this case, the splicing reaction is intrinsic to the folding of the precursor; that is, the RNA precursor is *self-splicing* because the folded precursor RNA creates its own RNA-splicing activity. The self-splicing *Tetrahymena* RNA was the first example found of an RNA molecule that could function as an enzyme in catalyzing a chemical reaction; such enzymatic RNA molecules are usually called **ribozymes**.

The existence and the positions of introns in a particular primary transcript are readily demonstrated by renaturing the transcribed DNA with the fully processed mRNA molecule. The DNA–RNA hybrid can then be examined by electron microscopy. An example of adenovirus mRNA (fully processed) and the corresponding DNA are shown in **FIGURE 8.16**. The DNA copies of the introns appear as single-stranded loops in the hybrid molecule, because no corresponding RNA sequence is available for hybridization.

The number of introns per RNA molecule varies considerably from one gene to the next. One of the major genes for inherited breast cancer in women (*BRCA1*) contains 21 introns spread across more than 100,000 bases. More than 90 percent of the primary transcript is excised in processing, yielding a processed mRNA of about 7800 bases, which codes for a polypeptide chain of 1863 amino acids. Among human genes with a simpler intron–exon structure is that for α-globin, which contains two introns. Introns vary greatly in size as well as in number. In human beings and other mammals, most introns range in size from 100 to 10,000 bases, and in the processing of a typical primary transcript, the amount of discarded RNA ranges from about 50 percent to more than 90 percent. In lower eukaryotes, such as yeast, nematodes, and fruit flies, genes generally have fewer introns than do genes in mammals, and the introns tend to be much smaller.

Human genes tend to be very long even though they encode proteins of modest size.

TABLE 8.2 summarizes features of the "typical" human gene. Both the median and the mean values are given because many of the size distributions have a very long tail, rendering the mean potentially misleading. For example, whereas the mean number of exons is 8.8, this average is unduly influenced by some genes that have a very large number of exons, such as the gene for the muscle protein titin, which includes 178 exons (the largest number for any human gene). Similarly, the distribution of intron sizes is strongly skewed.

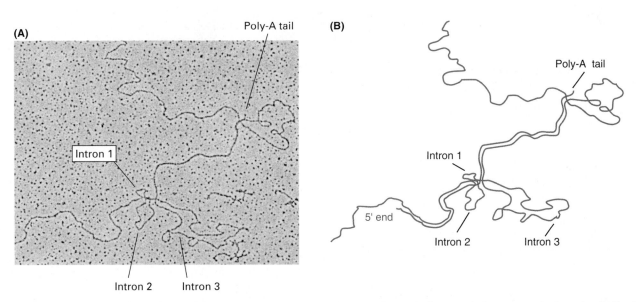

(A) Poly-A tail

Intron 1

Intron 2   Intron 3

(B) Poly-A tail

Intron 1

5' end

Intron 2   Intron 3

FIGURE 8.16 (A) An electron micrograph of a DNA–RNA hybrid obtained by annealing a single-stranded segment of adenovirus DNA with one of its mRNA molecules. The loops are single-stranded DNA. (B) An interpretive drawing. RNA and DNA strands are shown in red and blue, respectively. Four regions do not anneal, creating three single-stranded DNA segments that correspond to the introns and the poly-A tail of the mRNA molecule.

(A) courtesy of Thomas R. Broker and Louise T. Chow, University of Alabama at Birmingham. Original research completed in 1977 at the Cold Spring Harbor Laboratory, New York.

TABLE 8.2 Characteristics of Human Genes

Gene feature	Median	Mean
Size of internal exon	122 bp	145 bp
Number of exons	7	8.8
Size of introns	1023 bp	3356 bp
5′ untranslated region	240 bp	300 bp
3′ untranslated region	400 bp	770 bp
Length of coding sequence	1101 bp	1341 bp
Number of amino acids (aa)	367	447
Extent of genome occupied	14 kb	27 kb

Data from E. S. Lander et al. 2001. *Nature* 409: 860.

The most common intron length peaks at 87 nucleotides, but the tail of the distribution is so stretched out that the mean is 3365 nucleotides. The median is the value that splits the distribution in the middle: Half the values are above the median and half below.

One noteworthy feature of Table 8.2 is that human genes tend to be spread over a larger region of the genome than those in worms or flies. Most human genes consist of small exons separated by long introns, and many genes are over 100 kb in length. The average human gene occupies 27 kb of genomic DNA, yet only 1.3 kb (about 5 percent) is used to encode amino acids. The picture is not much different for the medians. The median gene length is 14 kb, of which only 1.1 kb (about 8 percent) is used to encode amino acids. Most of the added length is due to the long introns in human genes. The longest human gene is that for the muscle protein dystrophin, which is 2.4 Mb in length.

Many exons code for distinct protein-folding domains.

The existence of an elaborate splicing mechanism shared among all eukaryotes implies that introns must be very ancient. Introns may play a role in gene evolution by serving as the boundaries of exons encoding amino acid sequences that are more or less independent in their folding characteristics. For example, the central exon of the β-globin gene codes for a domain that folds around an iron-containing molecule of heme. The correlation between exons and domains found in some genes suggests that the genes were originally assembled from smaller pieces. In some cases, the ancestry of the exons can be traced. For example, the human gene for the low-density lipoprotein receptor that participates in cholesterol regulation shares exons with certain blood-clotting factors and epidermal growth factors. The model of protein evolution through the combination of different exons is called the **exon shuffle** model. The mechanism for combining exons from different genes is not known, but we have already seen that the proteins of human beings and other vertebrates tend to have more complex domain architectures than do proteins found in other organisms.

8.5 Translation into a polypeptide chain takes place on a ribosome.

The synthesis of every protein molecule in a cell is directed by an mRNA originally copied from DNA. Protein production includes two kinds of processes: (1) information-transfer processes, in which the RNA base sequence determines an amino acid sequence, and (2) chemical processes, in which the amino acids are linked together. The complete series of events is called translation.

The translation system consists of five major components:

1. **Messenger RNA, or mRNA**   Messenger RNA is needed to bring the ribosomal subunits together (described below) and to provide the coding sequence of bases that determines the amino acid sequence in the resulting polypeptide chain.

2. **Ribosomes**   These components are particles on which protein synthesis takes place. They move along an mRNA molecule and align successive transfer RNA molecules; the amino acids are attached one by one to the growing polypeptide chain by means of peptide bonds. Ribosomes consist of two separate RNA–protein particles (the small subunit and the large subunit), which come together in polypeptide synthesis to form a mature ribosome.

3. **Transfer RNA, or tRNA**   The sequence of amino acids in a polypeptide is determined by the base sequence in the mRNA by means of a set of adaptor molecules, the tRNA molecules, each of which is attached to a particular amino acid. Each successive group of three adjacent bases in the mRNA forms a **codon** that binds to a particular group of three adjacent bases in the tRNA (an anticodon), bringing the attached amino acid into line for elongation of the growing polypeptide chain.

4. **Aminoacyl-tRNA synthetases**   Each enzyme in this set of molecules catalyzes the attachment of a particular amino acid to its corresponding

tRNA molecule. A tRNA attached to its amino acid is called an **aminoacylated tRNA** or a **charged tRNA**.

5. **Initiation, elongation, and termination factors** Polypeptide synthesis can be divided into three stages: initiation, elongation, and termination. Each stage requires specialized molecules.

In prokaryotes, all of the components for translation are present throughout the cell; in eukaryotes, they are located in the cytoplasm, with specialized translation machinery in mitochondria and chloroplasts.

In eukaryotes, initiation takes place by scanning the mRNA for an initiation codon.

In overview, the process of translation begins with an mRNA molecule binding to a ribosome. The aminoacylated tRNAs are brought along sequentially, one by one, to the ribosome that is translating the mRNA molecule. Peptide bonds are made between successive amino acids. At each step, the carboxyl end of the growing chain is attached to the amino group of the amino acid on the incoming tRNA. The growing chain is thereby handed off from tRNA to tRNA until translation is completed and the finished polypeptide chain is released from the ribosome.

We will examine the processes of translation as they occur in eukaryotes, pointing out differences in the prokaryotic mechanism that are significant. In the predominant mode of translation **initiation** in eukaryotes, the 5′ cap on the mRNA is instrumental (**FIGURE 8.17**). The elongation factor eIF4F first binds to the cap and then recruits eIF4A and eIF4B (part A). This creates a binding site for the other components of the initiation complex, which consist of a charged tRNA^{Met} (that serves as an initiator tRNA), bound with elongation factor eIF2, and a small 40S ribosomal subunit together with elongation factors eIF3 and eIF5. These components all come together at the 5′ cap and form the 48S initiation complex (part B).

Once the initiation complex has formed, it moves along the mRNA in the 3′ direction, scanning for the first occurrence of the nucleotide sequence AUG, the **start codon** that signals the start of polypeptide synthesis. When this motif is encountered, the AUG is recognized as the initial methionine codon, and polypeptide synthesis begins. At this point eIF5 causes the release of all the initiation factors and the recruitment of a large 60S ribosomal subunit (part C). This subunit includes three binding sites for tRNA molecules. These sites are called the **E (exit) site**, the **P (peptidyl) site**, and the **A (aminoacyl) site**. Note that at the beginning of polypeptide synthesis, the initiator methionine tRNA is located in the P site and that the A site is the next site in line to be occupied. The tRNA binding is accomplished by hydrogen bonding between bases in the AUG codon in the mRNA and the three-base **anticodon** in the tRNA.

Elongation takes place codon by codon.

Recruitment of other elongation factors into the initiation complex begins the **elongation** phase of polypeptide synthesis. Elongation consists of three processes executed iteratively:

1. Bringing each new aminoacylated tRNA into line

2. Forming the new peptide bond to elongate the polypeptide

3. Moving the ribosome to the next codon along the mRNA

The process of elongation is illustrated in **FIGURE 8.18**. The key players in providing the energy for translation are the elongation factors EF-2 and EF-1α, which alternately occupy the same ribosomal binding site. In their active forms (EF-2-GTP and EF-1α-GTP) the molecules are bound with guanosine triphosphate (GTP). Hydrolysis of the GTP to GDP releases the energy to move the ribosomal subunits along the messenger RNA as well as to carry out the reactions needed to grow the polypeptide chain. Conversion of either elongation factor from its GTP-bound form into its GDP-bound form lowers its affinity for the ribosome, and the GDP-bound form diffuses away and is replaced by the GTP-bound form of the alternate elongation factor. GTP hydrolysis is also the source of energy for elongation in prokaryotes. In *E. coli*, the counterparts of EF-1α and EF-2 are called EF-Tu and EF-G, respectively.

In the first step of elongation, the 40S ribosomal subunit moves one codon farther along the messenger RNA, and the charged tRNA corresponding to the new codon (in this case, tRNA^{Phe}) is brought into the A site on the 60S subunit (Figure 8.18, part A). The charged tRNA comes to the ribosome in a complex that contains EF-1α-GTP, and if the codon–anticodon interaction is correct, then the EF-1α-GTP is very rapidly hydrolyzed to EF-1α-GDP. The EF-1α-GDP has a reduced affinity for the charged tRNA, and the resulting change in conformation allows the charged tRNA to fit tightly within the active site of the 60S subunit. Once the tRNA has snuggled into the active site of the 60S subunit, the peptide bond is formed by a **peptidyl transferase** activity. Peptide bond formation is a coupled reaction in which, in the example in Figure 8.18, part A, breakage of the bond connecting the methionine to the tRNA^{Met} is coupled to formation of the peptide bond connecting the methionine to phenylalanine. Peptidyl transferase activity is not due to a single molecule but requires multiple components of the 60S subunit, including

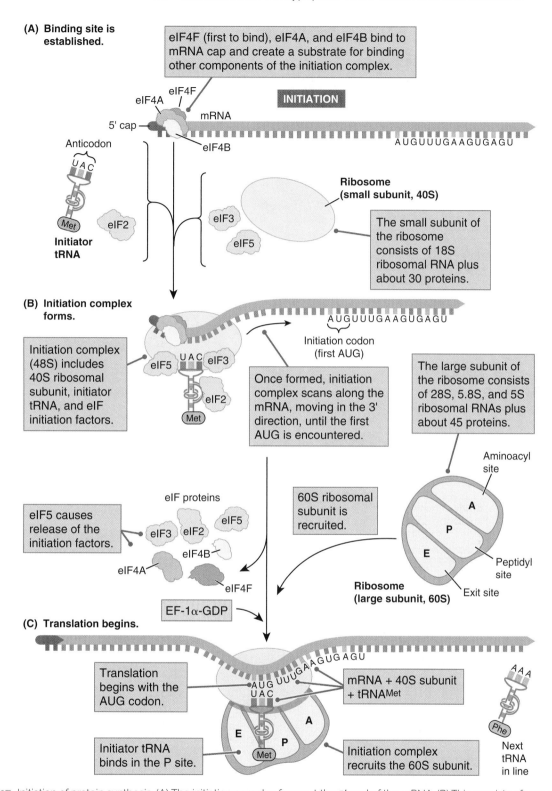

FIGURE 8.17 Initiation of protein synthesis. (A) The initiation complex forms at the 5′ end of the mRNA. (B) This consists of one 40S ribosomal subunit, the initiator tRNA^Met, and the eIF initiation factors. (C) The initiation complex recruits a 60S ribosomal subunit in which the tRNA^Met occupies the P (peptidyl) site of the ribosome. This complex travels along the mRNA until the first AUG is encountered, at which codon translation begins.

several proteins and the 28S ribosomal RNA in the 60S subunit. Some evidence indicates that the actual catalysis is carried out by the 28S RNA, which would suggest that 28S is an example of a ribozyme at work.

In the next step in chain elongation (part B), the 60S subunit swings forward to catch up with the 40S subunit, and at the same time the tRNAs in the P and A sites of the large subunit are shifted to the E and P sites, respectively.

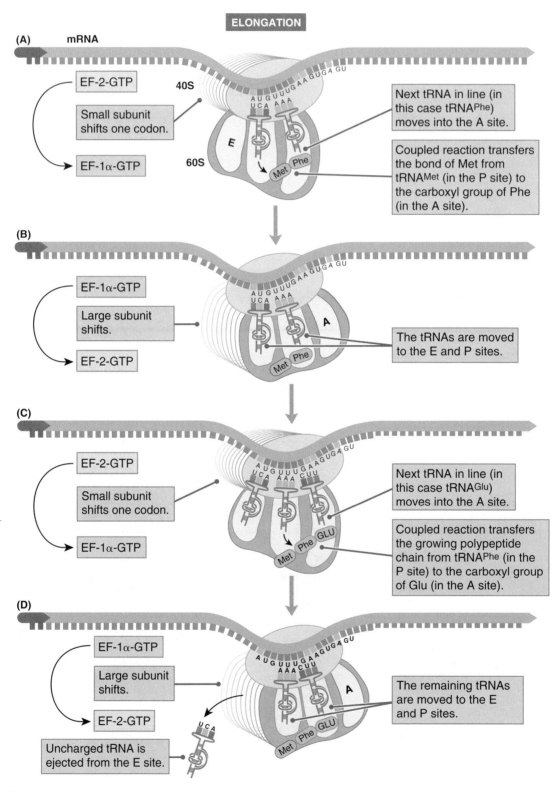

ELONGATION

(A) mRNA

EF-2-GTP

Small subunit shifts one codon.

EF-1α-GTP

40S

60S

E

Met   Phe

Next tRNA in line (in this case tRNA^Phe) moves into the A site.

Coupled reaction transfers the bond of Met from tRNA^Met (in the P site) to the carboxyl group of Phe (in the A site).

(B)

EF-1α-GTP

Large subunit shifts.

EF-2-GTP

A

Met   Phe

The tRNAs are moved to the E and P sites.

(C)

EF-2-GTP

Small subunit shifts one codon.

EF-1α-GTP

Met   Phe   GLU

Next tRNA in line (in this case tRNA^Glu) moves into the A site.

Coupled reaction transfers the growing polypeptide chain from tRNA^Phe (in the P site) to the carboxyl group of Glu (in the A site).

(D)

EF-1α-GTP

Large subunit shifts.

EF-2-GTP

Uncharged tRNA is ejected from the E site.

A

Met   Phe   GLU

The remaining tRNAs are moved to the E and P sites.

FIGURE 8.18 Elongation cycle in protein synthesis.

One cycle of elongation is now completed, and the entire procedure is repeated for the next codon (part C). The 40S subunit shifts one codon to the right, the next aminoacylated tRNA (in this case, tRNA^Glu) is brought into the A site, and a new peptide bond is formed between the carboxyl group of Phe and the amino group of Glu. As shown in part D, the large subunit swings forward while at the same time the tRNAs in the P and A sites are shifted into the E and P sites. At this point, the tRNA that formerly occupied the E site is ejected from the ribosome.

Polypeptide elongation consists of the steps C → D → C → D carried out repeatedly until a termination codon is encountered. The elongation cycle happens relatively rapidly. Under optimal conditions, eukaryotes synthesize a polypeptide chain at the rate of about 15 amino acids per second. Elongation in prokaryotes is a little faster (about 40 amino acids per second), but the essential processes are very similar.

A termination codon signals release of the finished polypeptide chain.

Compared to initiation and elongation, the termination of polypeptide synthesis—the **release phase**—is simple (FIGURE 8.19). When a stop codon is encountered, the tRNA holding the polypeptide remains in the P site, and a *release factor (RF)* binds with

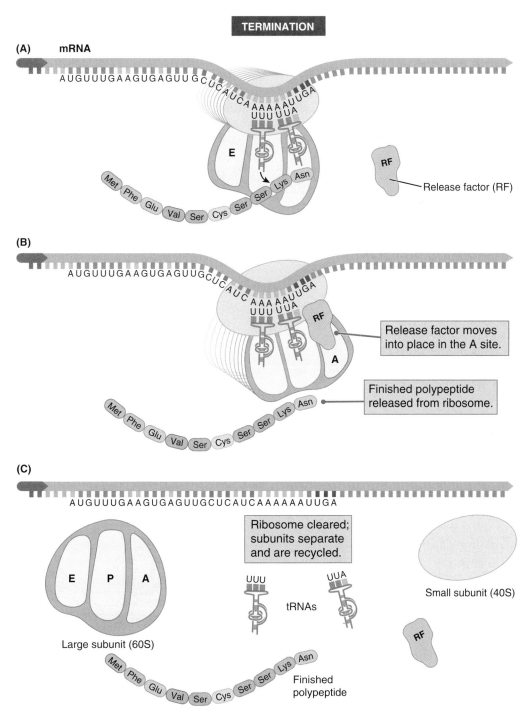

FIGURE 8.19 Termination of protein synthesis. When a stop codon is reached (A), no tRNA can bind to that site (B), which causes the release of the newly formed polypeptide and the remaining bound tRNA (C).

the ribosome. GTP hydrolysis provides the energy to cleave the polypeptide from the tRNA to which it is attached, as well as to eject the release factor and dissociate the 80S ribosome from the mRNA. At this point the 40S and 60S subunits are recycled to initiate translation of another mRNA. Eukaryotes have only one release factor that recognizes all three **stop codons**: UAA, UAG, and UGA. The situation differs in prokaryotes. In *E. coli*, the release factor RF 1 recognizes the stop codons UAA and UAG, whereas release factor RF 2 recognizes UAA and UGA. A third release factor, RF 3, plays an important role in ensuring the accuracy of translation. This topic is examined next.

Proofreading and premature termination help ensure translational accuracy.

The error rate of translation is approximately one incorrect amino acid inserted per 2000 residues. This rate would be at least 10 times greater were it not that the ribosome supervises two distinct mechanisms of quality control.

The first mechanism of quality control over translation is a type of proofreading that takes place at the A site. In the elongation process, each charged tRNA comes to the A site bound with a molecule of EF-1α-GTP. When the anticodon of the charged tRNA correctly matches the codon, hydrolysis of the GTP to GDP occurs very rapidly, and this allows the charged tRNA to fit tightly into the active site to promote formation of the peptide bond. When the codon and the anticodon do not match, however, the rate of hydrolysis of the GTP is delayed, which allows enough time for the incorrect charged tRNA to diffuse away and be replaced with the correct one. This type of proofreading at the A site is usually called *kinetic proofreading*.

The second mechanism of quality control takes place at the P site. Recall that, during elongation, after each peptide bond has been formed, the tRNA carrying the growing polypeptide chain is transferred to the P site. At this point there is a second check whether the anticodon of the tRNA is a correct match to the codon in the mRNA. If the match is correct, elongation proceeds normally. If there is a mismatch, however, it means that that last amino acid incorporated into the polypeptide chain is incorrect. When there is a mismatch at the P site, a slight change in configuration of the ribosome perturbs the fidelity of tRNA selection at the A site. One possible outcome of the perturbation is that release factor 2 (RF2) can gain access to the A site and cause premature termination, even though no termination codon is present. This type of translational termination, due to a misincorporation error, is greatly enhanced by the presence of release factor 3 (RF3). Another possible outcome is that translation continues; when this happens the perturbed A site makes it much more likely that another incorporation

error will occur. When there are two adjacent incorporation errors, the release factors are afforded even easier access to the ribosome, and the probability of chain termination can be as high as 50 percent.

Most polypeptide chains fold correctly as they exit the ribosome.

Each polypeptide chain tends to fold into a unique three-dimensional shape determined primarily by its sequence of amino acids. Generally speaking, polypeptide molecules fold so that amino acids with charged, hydrophilic side chains tend to be on the surface of the protein (in contact with water) and those with uncharged, hydrophobic side chains tend to be internal (hidden from water). Specific folded configurations also result from hydrogen bonding between peptide groups. Two fundamental polypeptide structures are the alpha (α) helix and the beta (β) sheet (**FIGURE 8.20**). An α helix is formed by hydrogen bonded between peptide groups that are close together in the polypeptide backbone. In an α helix, often represented as a coiled ribbon, the backbone is twisted so that the N−H in each peptide group is hydrogen bonded with the C=O in the peptide group located four amino acids farther along the helix. The helical twist may be right-handed or left-handed, but right-handed

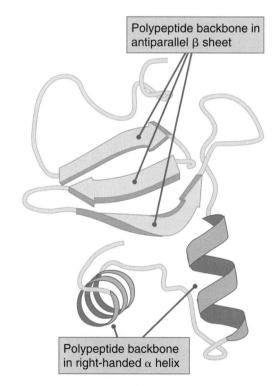

Polypeptide backbone in antiparallel β sheet

Polypeptide backbone in right-handed α helix

FIGURE 8.20 A "ribbon" diagram of the path of the backbone of a typical polypeptide showing the α-helix and β-sheet folding motifs. The α-helical regions are shown as coiled ribbons. The flat arrows represent β sheets, each of which is held to its neighboring sheet by hydrogen bonds.

Modified from W. I. Weiss, K. Drickamer, and W. A. Hendrickson, *Nature* 360 (1992): 127–134.

α helices are more common. Both α helices in Figure 8.20 are right-handed.

In contrast, a β sheet is formed by hydrogen bonding between peptide groups in distant parts of the polypeptide chain, or even in different polypeptide chains. In a β sheet, often represented as parallel "flat" ribbons, the backbones of the interacting polypeptide chains are held nearly flat and relatively rigid (forming a "sheet"), because alternate N−H groups in one backbone are hydrogen bonded with alternate C=O groups in the backbone of the adjacent chain. In each polypeptide chain, alternate C=O and N−H groups are free to form hydrogen bonds with their counterparts in a different chain on the opposite side, so a β sheet can consist of multiple aligned segments in the same or different polypeptides. The orientation of the backbones in a β sheet may be antiparallel (adjacent backbones reversed in orientation relative to their amino and carboxyl ends) or parallel, but antiparallel

is more common. In Figure 8.20, the β sheet is antiparallel. The rules of folding are so complex that, except for the simplest proteins, the final shape of a protein cannot usually be predicted from the amino acid sequence alone.

As polypeptides are being synthesized, they pass through a tunnel in the large ribosomal subunit that is long enough to include about 35 amino acids. The diameter of this tunnel is wide enough to accommodate an α helix but not so wide as to allow more complex structures to form. As the polypeptide emerges from the tunnel, it enters into a sort of cradle formed by a protein associated with the ribosome, which in prokaryotes is known as *trigger factor*. This cradle provides a protected space where the emerging polypeptide is able to undergo its folding process. About 70 to 75 percent of polypeptide chains fold properly as they emerge from the ribosomal tunnel into this protected space (**FIGURE 8.21**, part A).

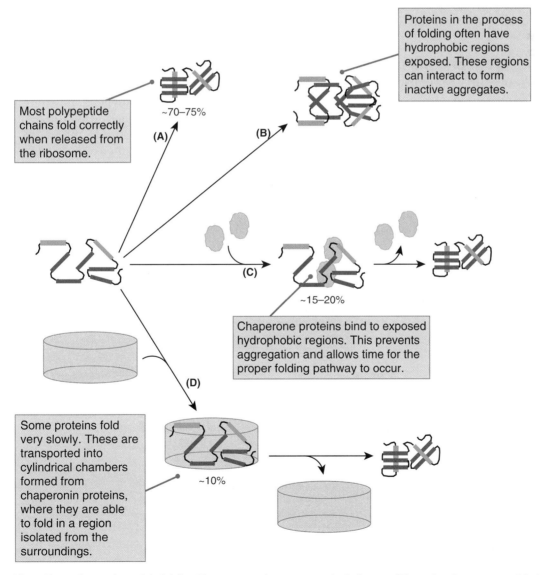

Proteins in the process of folding often have hydrophobic regions exposed. These regions can interact to form inactive aggregates.

Most polypeptide chains fold correctly when released from the ribosome.

~70–75%

(A)

(B)

(C)

~15–20%

Chaperone proteins bind to exposed hydrophobic regions. This prevents aggregation and allows time for the proper folding pathway to occur.

(D)

Some proteins fold very slowly. These are transported into cylindrical chambers formed from chaperonin proteins, where they are able to fold in a region isolated from the surroundings.

~10%

FIGURE 8.21 Alternative pathways in protein folding. The green regions represent α helices and the red regions represent β sheets.

But some polypeptide chains need additional help to fold properly. These tend to be large polypeptide chains composed of multiple folding domains that fold slowly, so that hydrophobic residues are exposed to the high concentration of macromolecules in the cytoplasm. Under such crowded conditions the exposed hydrophobic groups often attract each other and bind together, forming inactive protein aggregates (part B). The proper folding of more complex polypeptides is aided by proteins called **chaperones** (part C). These proteins bind to hydrophobic groups and unstructured regions to shield them from aggregation, and by repeated cycles of binding and release they give the polypeptide time to find its proper folding pathway. The most complex proteins with very slow and inefficient folding pathways are shielded by a special class of proteins known as *chaperonins*. These form large, hollow cylindrical structures that trap the unstable intermediates inside and allow them to fold in a protected environment (part D). In eukaryotes, the most abundant polypeptides that make use of the chaperonin cylinders for folding are the cytoskeletal proteins actin and tubulin.

Prokaryotes often encode multiple polypeptide chains in a single mRNA.

In prokaryotes, mRNA molecules have no cap, and there is no scanning mechanism to locate the first AUG. In *E. coli*, for example, translation is initiated when two initiation factors (IF-1 and IF-3) interact with the 30S subunit at the same time that another initiation factor (IF-2) binds with a special initiator tRNA charged with formylmethionine, symbolized tRNA^{fMet}. These components come together and combine with an mRNA, but not at the end. The attachment occurs by hydrogen bonding between the 3' end of the 16S RNA present in the 30S subunit and a special sequence, the **ribosome-binding site**, in the mRNA (also called the *Shine–Dalgarno sequence*). Together, the 30S + tRNA^{fMet} + mRNA complex recruits a 50S subunit, in which the tRNA^{fMet} is positioned in the P site and aligned with the AUG initiation codon, just as in part C of Figure 8.17. In the assembly of the completed ribosome, the initiation factors dissociate from the complex.

The major difference between translational initiation in prokaryotes and that in eukaryotes has an important implication. In eukaryotes, because of the scanning mechanism of initiation, a single mRNA can usually encode only one polypeptide chain. In prokaryotic mRNA, by contrast, the ribosome-binding site can be present anywhere near an AUG, so polypeptide synthesis can begin at any AUG that is closely preceded by a ribosome-binding site. Prokaryotes put this feature to good use. In prokaryotes, mRNA molecules commonly contain information for the amino acid sequences of several different polypeptide chains; such a molecule is called a **polycistronic mRNA**. (*Cistron* is a term often used to mean a base sequence that encodes a single polypeptide chain.) In a polycistronic mRNA, each polypeptide coding region is preceded by its own ribosome-binding site and AUG initiation codon. After the synthesis of one polypeptide is finished, the next along the way is translated (**FIGURE 8.22**). The genes contained in a polycistronic mRNA molecule often encode the different proteins of a metabolic pathway. For example, in *E. coli*, the ten enzymes needed to synthesize histidine are encoded by one polycistronic mRNA molecule. The use of polycistronic mRNA is an economical way for a cell to regulate the synthesis of related proteins in a coordinated manner.

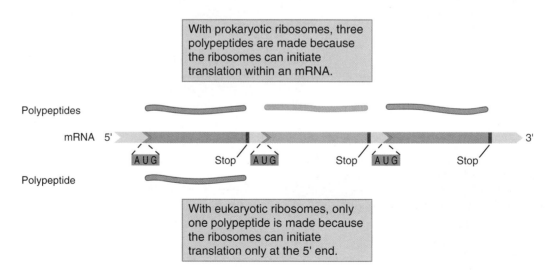

FIGURE 8.22 Different products are translated from a three-cistron mRNA molecule by the ribosomes of prokaryotes and eukaryotes. The prokaryotic ribosome translates all of the genes, but the eukaryotic ribosome translates only the gene nearest the 5' terminus of the mRNA. Translated sequences are shown in purple, yellow, and orange, stop codons in red, the ribosome binding sites in dark green, and the spacer sequences in light green.

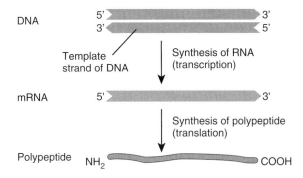

FIGURE 8.23 Direction of synthesis of RNA with respect to the coding strand of DNA, and of synthesis of protein with respect to mRNA.

In all organisms, an important feature of translation is that it proceeds in a particular direction along the mRNA and the polypeptide.

KEY CONCEPT

The mRNA is translated from an initiation codon to a stop codon in the 5′-to-3′ direction. The polypeptide is synthesized from the amino end toward the carboxyl end by the addition of amino acids, one by one, to the carboxyl end.

For example, a polypeptide with the sequence NH_2–Met–Pro–• • •–Gly–Ser–COOH would start with methionine as the first amino acid in the chain and end with serine as the last amino acid added to the chain. The directions of synthesis are illustrated schematically in FIGURE 8.23.

By convention, in writing nucleotide sequences, we place the 5′ end at the left, and in writing amino acid sequences, we place the amino end at the left. Polynucleotides are generally written so that both synthesis and translation proceed from left to right, and polypeptides are written so that synthesis proceeds from left to right. This convention is used in all of our subsequent discussions of the genetic code.

8.6 The genetic code for amino acids is a triplet code.

Only four bases in DNA are needed to specify the 20 amino acids in proteins because a combination of three adjacent bases is used for each amino acid, as well as for the signals that start and stop protein synthesis. Each sequence of three adjacent bases in mRNA is a codon that specifies a particular amino acid (or chain termination). The **genetic code** is the list of all codons and the amino acids that they encode. Before the genetic code was determined experimentally, it was assumed that if all codons had the same number of bases, then each codon would have to contain at least three bases. Codons consisting of pairs of bases would be

STOP & THINK 8.3

A double-stranded DNA molecule has the sequence

5′–ATGCCCTTTGGGCATCAT–3′
3′–TACGGGAAACCCGTAGTA-5′

If this part of the DNA molecule is transcribed from left to right, the resulting RNA has the sequence 5′–AUGCCCUUUGGGCAUCAU–3′, which is translated as Met–Pro–Phe–Gly–His–His.

What amino acid sequence would result from a single-frameshift mutation in which a T was added immediately preceding the CCC in the top strand? What amino acid sequence would result from a triple-frameshift mutation in which a T was added immediately preceding the CCC, as well as immediately following both the TTT and the GGG, in the top strand?

insufficient, because four bases can form only $4^2 = 16$ pairs; triplets of bases would suffice because four bases can form $4^3 = 64$ triplets. In fact, the genetic code is a **triplet code**, and all 64 possible codons carry information of some sort. Most amino acids are encoded by more than one codon. Furthermore, in the translation of mRNA molecules, the codons do not overlap but are used sequentially (FIGURE 8.24).

Genetic evidence for a triplet code came from three-base insertions and deletions.

Although theoretical considerations suggested that each codon must contain at least three letters, codons having more than three letters could not be ruled out. The first widely accepted proof for a triplet code came from genetic experiments using mutants of the *rII* gene in bacteriophage T4 that had been induced by replication in the presence of the chemical *proflavin*. These experiments were carried out in 1961 by Francis Crick and collaborators. Proflavin-induced mutations typically resulted in total loss of function, which the investigators suspected were due to single-base insertions or deletions. Analysis of the properties of these mutations led directly to the deduction that the code is read three nucleotides at a time from a fixed point; in other words, there is a **reading frame** to each mRNA. Mutations that delete or add a base pair shift the reading frame and are called **frameshift mutations**. FIGURE 8.25 illustrates the profound effect of a frameshift mutation on the amino acid sequence of the polypeptide produced from the mRNA of the mutant gene.

The genetic analysis of the structure of the code began with an *rII* mutation called *FC0*, which was arbitrarily designated (+), as though it had an inserted base pair. (This was a lucky guess; when *FC0* was sequenced, it did turn out to have a single-base insertion.) If *FC0* has a (+) insertion, then it should be possible to revert the *FC0* allele to "wildtype" by deletion of a nearby base. Selection for r^+ revertants was carried out by isolating plaques formed on a lawn of an *E. coli* strain K12 that was lysogenic for phage λ. The basis of the selection is that *rII* mutants are unable to propagate in K12(λ). Analysis of the revertants revealed that

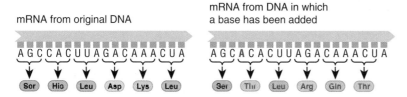

Start — Stop

5' — 3'

N N N A U G A G U C A G U G G G U C A G U C A G U C A G U C U A A N N N N

Direction of reading of codons in translation

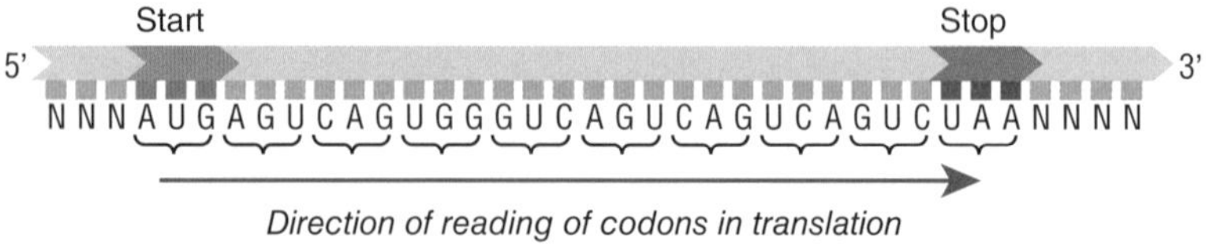

FIGURE 8.24 Bases in an RNA molecule are read sequentially in the 5' → 3' direction, in groups of three.

mRNA from original DNA

mRNA from DNA in which a base has been added

A G C C A C U U A G A C A A A C U A

A G C A C A C U U A G A C A A A C U A

Sor · Hio · Leu · Asp · Lys · Leu

Ser · Thr · Leu · Arg · Gln · Thr

FIGURE 8.25 The change in the amino acid sequence of a protein caused by the addition of an extra base, which shifts the reading frame. A deleted base also shifts the reading frame.

each still carried the original *FC0* mutation, along with a second (suppressor) mutation that reversed the effects of the *FC0* mutation. The suppressor mutations could be separated by recombination from the original mutation by crossing each revertant to wildtype; each suppressor mutation proved to be an *rII* mutation that, by itself, would cause the *r* (rapid lysis) phenotype. If *FC0* had an inserted base, then the suppressors should all result in deletion of a base pair; hence each suppressor of *FC0* was designated (−). The consequences of three such revertants for the translational reading frame are illustrated using ordinary three-letter words in **FIGURE 8.26**. The (−) mutations are designated $(-)_1$, $(-)_2$, and $(-)_3$, and those parts of the mRNA translated in the correct reading frame are indicated in green.

In the *rII* experiments, all of the individual (−) suppressor mutations were used, in turn, to select other "wildtype" revertants, with the expectation that these revertants would carry new suppressor mutations of the (+) variety, because the (−)(+) combination should yield a phage able to form plaques on K12(λ).

Various double-mutant combinations were made by recombination. Usually any (+)(−) combination, or any (−)(+) combination, resulted in a wildtype phenotype,

Phage type	Insertion/deletion	Translational reading frame of mRNA
Wildtype sequence		THE BIG BOY SAW THE NEW CAT EAT THE HOT DOG · ·
+1 insertion	(+)	THE BIG BOY SAW TTH ENE WCA TEA TTH EHO TDO G
Revertant 1	$(-)_1$ (+)	THE BIG OYS AWT THE NEW CAT EAT THE HOT DOG · ·
Revertant 2	(+) $(-)_2$	THE BIG BOY SAW TTH ENE WCA TEA THE HOT DOG · ·
Revertant 3	(+) $(-)_3$	THE BIG BOY SAW TTH ENE WAT EAT THE HOT DOG · ·
(−) deletion number 1	$(-)_1$	THE BIG OYS AWT HEN EWC ATE ATT HEH OTD OG · · ·
(−) deletion number 2	$(-)_2$	THE BIG BOY SAW THE NEW CAT EAT HEH OTD OG · · ·
(−) deletion number 3	$(-)_3$	THE BIG BOY SAW THE NEW ATE ATT HEH OTD OG · · ·
Double (−) mutant	$(-)_1$ $(-)_2$	THE BIG OYS AWT HEN EWC ATE ATH EHO TDO G · · · ·
Triple (−) mutant	$(-)_1$ $(-)_2$ $(-)_3$	THE BIG OYS AWT HEN EWA TEA THE HOT DOG · · · · ·

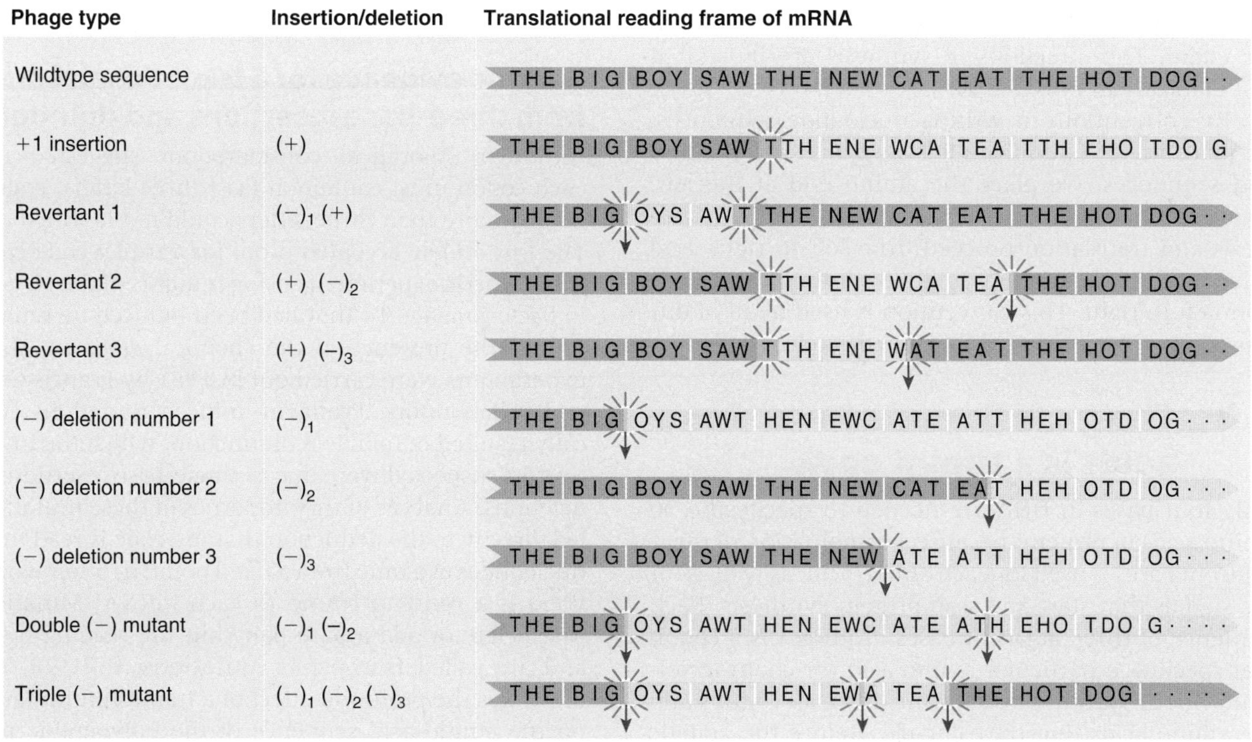

FIGURE 8.26 Interpretation of the *rII* frameshift mutations showing that combinations of appropriately positioned single-base insertions (+) and single-base deletions (−) can restore the correct reading frame (green). The key finding was that a combination of three single-base deletions, as shown in the bottom line, also restores the correct reading frame. Two single-base deletions do not restore the reading frame. These classic experiments gave strong genetic evidence that the genetic code is a triplet code.

whereas (+)(+) and (−)(−) double-mutant combinations always resulted in the mutant phenotype. The most revealing result came when triple mutants were made. Usually, the (+)(+)(+) and (−)(−)(−) triple mutants yielded the wildtype phenotype!

The phenotypes of the various (+) and (−) combinations were interpreted in terms of a reading frame. The initial *FC0* mutation, a +1 insertion, shifts the reading frame, resulting in incorrect amino acid sequence from that point on and thus a nonfunctional protein (Figure 8.26). Deletion of a base pair nearby will restore the reading frame, although the amino acid sequence encoded between the two mutations will be different and incorrect. In (+)(+) and (−)(−) double mutants, the reading frame is shifted by two bases; the protein made is still nonfunctional. However, in the (+)(+)(+) and (−)(−)(−) triple mutants, the reading frame is restored, even though all amino acids encoded within the region bracketed by the outside mutations are incorrect; the protein made is one amino acid longer for (+)(+)(+) and one amino acid shorter for (−)(−)(−) (Figure 8.26).

The genetic analysis of the (+) and (−) mutations strongly supported the following conclusions:

- Translation of an mRNA starts from a fixed point.

- There is a single reading frame maintained throughout the process of translation.

- Each codon consists of three nucleotides.

Crick and his colleagues also drew other inferences from these experiments. First, in the genetic code, most codons must function in the specification of an amino acid. Second, each amino acid must be specified by more than one codon. They reasoned that if each amino acid had only one codon, then only 20 of the 64 possible codons could be used for coding amino acids. In this case, most frameshift mutations should have affected one of the remaining 44 "noncoding" codons in the reading frame, and hence a nearby frameshift of the opposite polarity mutation should not have suppressed the original mutation. Consequently, the code was deduced to be one in which more than one codon can specify a particular amino acid.

Most of the codons were determined from *in vitro* polypeptide synthesis.

Polypeptide synthesis can be carried out in cell extracts containing ribosomes, tRNA molecules, mRNA molecules, and the various protein factors needed for translation. If radioactive amino acids are added to the extract, radioactive polypeptides are made. Synthesis continues for only a few minutes because mRNA is gradually degraded by various nucleases in the mixture. The elucidation of the genetic code began with the observation that when the degradation of mRNA was allowed

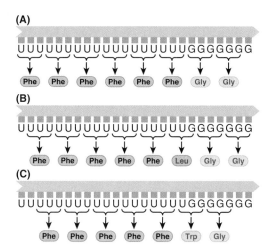

FIGURE 8.27 Polypeptide synthesis using 5′-UUUU . . . UUGGGGGGG-3′ as an mRNA in three different reading frames, showing the reasons for the incorporation of glycine, leucine, and tryptophan.

to go to completion and the synthetic polynucleotide polyuridylic acid (poly-U) was added to the mixture as an mRNA molecule, a polypeptide consisting only of phenylalanine (Phe−Phe−Phe−• • •) was synthesized. From this simple result and knowledge that the code is a triplet code, it was concluded that UUU must be a codon for the amino acid phenylalanine. Variations on this basic experiment identified other codons. For example, when a long sequence of guanines was added at the terminus of the poly-U, the polyphenylalanine was terminated by a sequence of glycines, indicating that GGG is a glycine codon (**FIGURE 8.27**). A trace of leucine or tryptophan was also present in the glycine-terminated polyphenylalanine. Incorporation of these amino acids was directed by the codons UUG and UGG at the transition point between U and G. When a single guanine was added to the terminus of a poly-U chain, the polyphenylalanine was terminated by leucine. Thus UUG is a leucine codon, and UGG must be a codon for tryptophan. Similar experiments were carried out with poly-A, which yielded polylysine, and with poly-C, which produced polyproline.

Other experiments led to a complete elucidation of the code. Three codons,

$$ \text{UAA} \qquad \text{UAG} \qquad \text{UGA} $$

were found to be stop signals for translation.

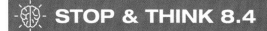

STOP & THINK 8.4

A synthetic RNA is produced with a sequence composed of tandem repeats of the sequence 5′−CAG−3′. This RNA is translated *in vitro* under conditions in which translation can begin with any codon. What are the polypeptide products of the synthetic RNA?

THE HUMAN CONNECTION

Poly-U

Marshall W. Nirenberg and J. Heinrich Matthaei (1961)
National Institutes of Health, Bethesda, Maryland

The Dependence of Cell-Free Protein Synthesis in E. coli *upon Naturally Occurring or Synthetic Polyribonucleotides*

In the years following the discovery of DNA structure by Watson and Crick in 1953, the biological implications of the discovery were largely ignored. A principal reason was that most biochemists still held strongly to the conviction that DNA had nothing to do with protein synthesis. The prevailing view was that proteins were made from small preexisting peptides by enzymes that joined the peptides together step by step in a specific order. It had been suggested that proteins might be made by amino acids being laid down in sequence upon an RNA template, but this hypothesis also was largely ignored. Not until this important paper appeared in 1961 was it shown that proteins are made by stepwise joining of individual amino acids in a sequence specified by a molecule of template RNA. The key finding was that in a cell-free mixture capable of supporting protein synthesis, the artificial polynucleotide polyuridylic acid (poly-U) resulted in the synthesis of a protein consisting of only the amino acid phenylalanine. The requirements for protein synthesis also included ribosomes (necessary for translation) and small RNA molecules (which include the charged transfer RNAs).

> **66** *The results indicate the polyuridylic acid contains the information for the synthesis of a protein having the characteristics of polyphenylalanine.* **99**

A stable cell-free system has been obtained from *E. coli* that incorporates [radioactive] valine into protein at a rapid rate. . . .

[Specifically,] the addition of polyuridylic acid resulted in a remarkable stimulation of [radioactive] phenylalanine incorporation. Phenylalanine incorporation was almost completely dependent upon the addition of polyuridylic acid, and incorporation proceeded at a linear rate for approximately 30 minutes. No other polynucleotide tested could replace polyuridylic acid. . . . The product of the reaction had the same apparent solubility as authentic polyphenylalanine . . . [and contained] phenylalanine and no other amino acids. . . .

One or more uridylic acid residues therefore appears to be the code for phenylalanine. Whether the code is of the singlet, triplet, etc., type has not yet been determined. Polyuridylic acid seemingly functions as a synthetic template or messenger RNA, and this stable, cell-free system may well synthesize any protein corresponding to meaningful information contained in added RNA.

After this paper appeared, the race was on to decipher the genetic code by which RNA specifies the amino acids in a protein.

M. W. Nirenberg and J. H. Matthaei, *Proc. Natl. Acad. Sci. USA* 47 (1961): 1588–1602.

Redundancy and near-universality are principal features of the genetic code.

The *in vitro* translation experiments with components isolated from the bacterium *E. coli* have been repeated with components obtained from many species of bacteria, yeast, plants, and animals. The standard genetic code deduced from these experiments is considered to be almost universal, because the same codon assignments can be made for nuclear genes in nearly all organisms that have been examined. However, some minor differences in codon assignments are found in certain protozoa and in the genetic codes of organelles.

The standard genetic code is shown in **TABLE 8.3**. Note that four codons—the three stop codons and the start codon—are signals. Altogether, 61 codons specify amino acids. In many cases several codons direct

TABLE 8.3 The Standard Genetic Code

First position (5′ end)	Second position				Third position (3′ end)
	U	C	A	G	
U	UUU Phe } F UUC Phe UUA Leu } L UUG Leu	UCU Ser } UCC Ser } S UCA Ser } UCG Ser	UAU Tyr } Y UAC Tyr UAA Stop UAG Stop	UGU Cys } C UGC Cys UGA Stop UGG Trp W	U C A G
C	CUU Leu } CUC Leu } L CUA Leu } CUG Leu	CCU Pro } CCC Pro } P CCA Pro } CCG Pro	CAU His } H CAC His CAA Gln } Q CAG Gln	CGU Arg } CGC Arg } R CGA Arg } CGG Arg	U C A G
A	AUU Ile } AUC Ile } I AUA Ile } AUG Met M	ACU Thr } ACC Thr } T ACA Thr } ACG Thr	AAU Asn } N AAC Asn AAA Lys } K AAG Lys	AGU Ser } S AGC Ser AGA Arg } R AGG Arg	U C A G
G	GUU Val } GUC Val } V GUA Val } GUG Val	GCU Ala } GCC Ala } A GCA Ala } GCG Ala	GAU Asp } D GAC Asp GAA Glu } E GAG Glu	GGU Gly } GGC Gly } G GGA Gly } GGG Gly	U C A G

Note: Each amino acid is given its conventional abbreviation in both the single-letter and three-letter format. The codon AUG, which codes for methionine (green), is generally used for initiation. The codons are conventionally written with the 5′ base on the left and the 3′ base on the right.

the insertion of the same amino acid into a polypeptide chain. This feature confirms the inference from the *rII* frameshift mutations that the genetic code is *redundant* (also called *degenerate*). In a redundant genetic code, some amino acids are encoded by two or more different codons. In the actual genetic code, all amino acids except tryptophan and methionine are specified by more than one codon. This **redundancy** is not random. For example, with the exception of serine, leucine, and arginine, all codons that correspond to the same amino acid are in the same box of Table 8.3; that is, *synonymous codons usually differ only in the third base*. For example, GGU, GGC, GGA, and GGG all code for glycine. Moreover, in all cases in which two codons code for the same amino acid, the third base is either A or G (both purines) or T or C (both pyrimidines).

STOP & THINK 8.5

How many different RNA sequences could code for a region of a polypeptide with the sequence Ser–Met–Ala–Arg–Thr (or, in the single-letter abbreviations, S–M–A–R–T)?

An aminoacyl-tRNA synthetase attaches an amino acid to its tRNA.

The decoding operation by which the base sequence within an mRNA molecule becomes translated into the amino acid sequence of a protein is accomplished by charged tRNA molecules, each of which is linked to the correct amino acid by an aminoacyl-tRNA synthetase.

The tRNA molecules are small single-stranded nucleic acids ranging in size from about 70 to 90 nucleotides. Like all RNA molecules, they have a 3′-OH terminus, but the opposite end terminates with a 5′-monophosphate, rather than a 5′-triphosphate, because tRNA molecules are cleaved from a larger primary transcript. Internal complementary base sequences form short double-stranded regions, causing the molecule to fold into a structure in which open loops are connected to one another by double-stranded stems (**FIGURE 8.28**). In two dimensions, a tRNA molecule is drawn as a planar cloverleaf. Its three-dimensional structure is more complex, as is shown in **FIGURE 8.29**, where part A shows a skeletal model of a yeast tRNA molecule for phenylalanine and part B is an interpretive drawing. All tRNA molecules have similar structures.

Particular regions of each tRNA molecule are used in the decoding operation. One region is the *anticodon* sequence, which consists of three bases that can form base pairs with a codon sequence in the mRNA. No normal tRNA molecule has an anticodon complementary to any of the stop codons UAG, UAA, and UGA. A second critical site, which all tRNAs share, is the CCA terminus at the 3' end where the amino acid is attached. A specific aminoacyl-tRNA synthetase transfers the amino acid onto the A residue. At least one (and usually only one) aminoacyl-tRNA synthetase exists for each amino acid. To make the correct attachment, the synthetase must be able to distinguish one tRNA molecule from another. The necessary distinction is provided by recognition regions that encompass many parts of the tRNA molecule.

The different tRNA molecules and synthetases are designated by stating the name of the amino acid that is linked to a particular tRNA molecule by a specific synthetase; for example, seryl-tRNA synthetase attaches serine to tRNA^Ser. When an amino acid has become attached to a tRNA molecule, the tRNA is said to be *charged*. An uncharged tRNA lacks an amino acid.

Much of the code's redundancy comes from wobble in codon–anticodon pairing.

Several features of the genetic code and of the decoding system suggest that base pairing between the codon and the anticodon has special features. First, the code is highly redundant. Second, the identity of the third base of a codon is often unimportant. In some cases, any nucleotide will do; in others, any purine or any pyrimidine serves the same function. Third, the number of distinct tRNA molecules present in an organism is less than the number of codons; because all codons are used, the anticodons of some tRNA

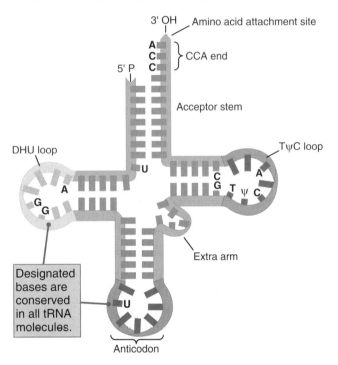

FIGURE 8.28 A tRNA cloverleaf configuration. The heavy black letters indicate bases that are conserved in the sequence of all tRNA molecules. The labeled loop regions are those found in all tRNA molecules. DHU refers to a base, dihydrouracil, found in one loop; the Greek letter ψ is a symbol for the unusual base pseudouridine.

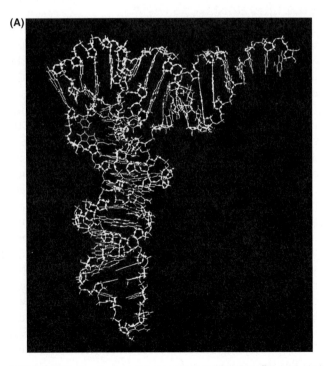

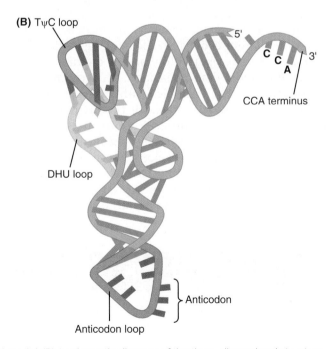

FIGURE 8.29 Yeast phenylalanine tRNA (called tRNA^Phe). (A) A skeletal model. (B) A schematic diagram of the three-dimensional structure of yeast tRNA^Phe.

molecules must be able to pair with more than one codon. Experiments with several purified tRNA molecules showed this to be the case.

To account for these observations, the **wobble** concept was advanced in 1966 by Francis Crick. He proposed that the first two bases in a codon form base pairs with the tRNA anticodon according to the usual rules (A−U and G−C) but that the base at the 5′ end of the anticodon is less spatially constrained than the first two and can form hydrogen bonds with more than one base at the 3′ end of the codon. His suggestion was essentially correct, but the allowed base pairs differ somewhat among organisms (**TABLE 8.4**).

TABLE 8.4 Wobble Rules for tRNAs of *E. coli* and *Saccharomyces cerevisiae*

First base in anticodon (5′ position)	Allowed base in third codon position (3′ position)	
	E. coli	*S. cerevisiae*
A	U	—
C	G	G
U	A or G	A
G	C or U	C or U
I	A, C, or U	C or U

Notes: In *S. cerevisiae*, an A at the 5′ position in the anticodon is always modified to I, which indicates inosine; inosine is structurally similar to adenosine except that the −NH$_2$ is replaced with −OH. Likewise, a U at the first anticodon position is often modified in this organism.

8.7   Several ribosomes can move in tandem along a messenger RNA.

In most prokaryotes and eukaryotes, the unit of translation is almost never simply one ribosome traversing an mRNA molecule. After about 25 amino acids have been joined together in a polypeptide chain, an AUG initiation codon is completely free of the ribosome, and a second initiation complex can form. The overall configuration is that of two ribosomes moving along the mRNA at the same speed. When the second ribosome has moved along a distance similar to that traversed by the first, a third ribosome can attach to the initiation site. The process of movement and reinitiation continues until the mRNA is covered with ribosomes at a density of about 1 ribosome per 80 nucleotides. This large translation unit is called a **polysome**, and this is the usual form of the translation unit in both prokaryotes and eukaryotes.

Because prokaryotes lack a nuclear envelope separating the location of DNA from that of the ribosomes, transcription and translation of mRNA can take place in rapid succession. The 5′ end of a mRNA molecule is synthesized first. This end includes the ribosome-binding site, followed, in order, by the initiating AUG codon and the rest of the coding sequence. Because translation takes place in the 5′ → 3′ direction, the first part of the mRNA becomes available for translation even before the rest of the transcript is finished. The absence of a nucleus therefore makes possible the simultaneous execution, or **coupling**, of transcription and translation. Coupled transcription and translation cannot take place in eukaryotes because the mRNA is synthesized and processed in the nucleus and is only later transported through the nuclear envelope into the cytoplasm, where the ribosomes are located.

CHAPTER SUMMARY

- In gene expression, information in the nucleotide sequence of DNA is used to dictate the linear order of amino acids in a polypeptide chain by means of an RNA intermediate called messenger RNA.

- Transcription of an RNA from one strand of the DNA is the first step in gene expression.

- The primary transcript is produced by an RNA polymerase. Eukaryotic cells have several types of RNA polymerases, only one of which is used for transcribing protein-coding genes.

- The base sequence of the primary RNA transcript is complementary to that in the template DNA strand, except that RNA contains the base uracil (U) instead of thymine (T).

- In eukaryotes, the RNA transcript is chemically modified by the addition of special nucleotides at the 5' end (the 5' cap), and the addition of a string of A's at the 3' end (the poly-A tail). Many primary transcripts also processed by RNA splicing, in which noncoding regions (introns) are removed and the coding region (exons) retained.

- The fully processed and spliced transcript constitutes the messenger RNA.

- The messenger RNA is translated on ribosomes in groups of three bases (codons), each specifying an amino acid through an interaction with molecules of transfer RNA.

- Transfer RNAs are relatively small RNA molecules, each of which has a sequence of bases (the anticodon) that pairs with the corresponding codon in the messenger RNA. For each tRNA, an enzyme attaches the correct amino acid to the 3' end of the molecule. Each codon in the messenger RNA, therefore, specifies its amino acid by undergoing codon–anticodon pairing with the tRNA. In translation, each amino acid in turn is added to the end of the growing polypeptide chain.

- The genetic code was first identified as a triplet code by means of experiments with single-base addition or deletion mutants, and later each individual codon was identified by means of the translation of chemically synthesized RNAs with known base sequence.

ISSUES AND IDEAS

- What is meant by the term *gene expression*? Would you make a distinction between gene expression and gene *regulation*? Why or why not?

- Would you regard an original text and its translation into another language as "colinear"? Explain your answer.

- In a eukaryotic cell, four general types of RNA molecules are used in gene expression. What are these types of RNA called? Which is not involved in gene expression in prokaryotic cells, and why not?

- Give an example of a genetic system that does not use the standard genetic code.

- What does it mean to say that the standard genetic code is redundant? Which (if any) amino acids are encoded by one codon? By two? By three? By four? By five? By six?

- What is a *frameshift* mutation? Explain how *rII* recombinants containing multiple, single-nucleotide frameshift mutations were used to show that the messenger RNA is translated in consecutive groups of three nucleotides.

- Suppose that a duplex DNA molecule undergoes two double-stranded breaks that tightly flank the promoter of a gene and that the promoter region is inverted before the backbones are rejoined by repair enzymes. Would you expect the inverted promoter to be able to recruit the transcription complex? What, if anything, would be wrong with the transcript of the gene?

SOLUTIONS: STEP BY STEP

PROBLEM 1 The International Union of Biochemistry and Molecular Biology (IUBMB) has designated a single-letter code for abbreviating the nucleotide bases that allows for ambiguous assignments. The code is shown in the accompanying diagram. The same code is used for DNA as for RNA. For ambiguous nucleotides, T and U are regarded as equivalent. Assuming standard Watson–Crick pairing between the two nucleotide strands shown, complete the sequence of the bottom strand, using the appropriate symbol from the standard ambiguity code.

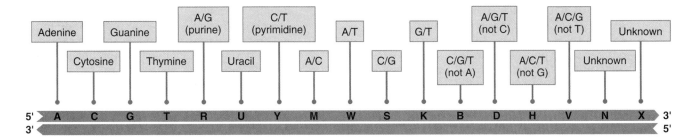

SOLUTION. The ambiguity codes are very useful not only for designating uncertain nucleotides in DNA sequences but also for summarizing the redundancies in the genetic code (see Step-by-Step Problem 2). The pairing relationships are straightforward for A, T, G, C, and U, but for ambiguous nucleotides one has to enumerate the possibilities and then select the symbol that expresses these ambiguities. One peculiar feature is that some symbols pair with themselves. For example, W (A or T) in one strand must also have a W (T or A) in the other strand, where the convention is that the paired nucleotides, though ambiguous, must obey the Watson–Crick pairing rules. All of the pairings can be worked out in this way, and the results are shown in the accompanying diagram. There are two symbols—namely N and X—in use for "any nucleotide," so these can be paired however it is convenient.

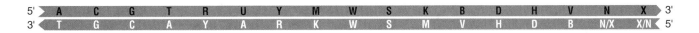

PROBLEM 2 Rewrite the genetic code table using as many as possible of the single-letter codes for ambiguous bases established by the IUBMB, as shown in Step-by-Step Problem 1.

SOLUTION. This problem requires that you examine the standard genetic code and select the proper symbol for ambiguous nucleotides. The version of the genetic code that results is shown here. It has considerably fewer entries than the standard format, and it shows the general structure of the code at a glance.

		Second nucleotide in codon			
		U	C	A	G
First nucleotide in codon	U	UUY Phe **F** UUR Leu **L**	UCN Ser **S**	UAY Tyr **Y** UAR Stop	UGY Cys **C** UGA Stop UGG Trp **W**
	C	CUN Leu **L**	CCN Pro **P**	CAY His **H** CAR Gln **Q**	CGN Arg **R**
	A	AUH Ile **I** AUG Met **M**	ACN Thr **T**	AAY Asn **N** AAR Lys **K**	AGY Ser **S** AGR Arg **R**
	G	GUN Val **V**	GCN Ala **A**	GAY Asp **D** GAR Glu **E**	GGN Gly **G**

CONCEPTS IN ACTION: PROBLEMS FOR SOLUTION

8.1 What possible amino acids are specified by a codon that consists of

(a) All pyrimidines?

(b) All purines?

8.2 A single codon in a double-stranded DNA molecule undergoes an inversion.

(a) If the original codon in the mRNA is 5′-UGG-3′, what is the codon in the transcript from the inversion?

(b) Would the inversion of any codon conserve the amino acid that is specified?

(c) What codons would, upon inversion, yield chain-termination codons?

(d) Which amino acids would inversion of each of the chain-termination codons specify?

8.3 The concept that a strand of DNA serves as a template for transcription of an RNA, which is translated into a polypeptide, is known as the "central dogma" of gene expression. All three types of molecules have a polarity. In the DNA template and the RNA transcript, the polarity is determined by the free 3′ or 5′ group at opposite ends of the polynucleotide chains; in a polypeptide chain, the polarity is determined by the free amino group (N terminal) or carboxyl group (C terminal) at opposite ends. Each of the following statements describes one possible polarity of the DNA template, the RNA transcript, and the polypeptide chain, respectively, in temporal order of use as a template or in synthesis. Which statement is correct?

(a)   5' to 3' DNA; 3' to 5' RNA; N terminal to C terminal

(b)   3' to 5' DNA; 3' to 5' RNA; N terminal to C terminal

(c)   3' to 5' DNA; 3' to 5' RNA; C terminal to N terminal

(d)   3' to 5' DNA; 5' to 3' RNA; N terminal to C terminal

(e)   5' to 3' DNA; 5' to 3' RNA; C terminal to N terminal

8.4   A part of the template strand of a DNA molecule that codes for the 5' end of an mRNA has the sequence 3'-TTTTACCGGAATTAGAGTCGCAGGATG-5'. What is the amino acid sequence of the polypeptide encoded by this region, assuming that the normal start codon is needed for initiation of polypeptide synthesis?

8.5   What codons could pair with the anticodon 5'-ICU-3', given that I (inosine) can pair with "H" (A or U or C). What amino acid would be incorporated?

8.6   How many different sequences of nine ribonucleotides would code for each of the following amino acids?

(a)   Met–His–Pro

(b)   Met–Arg–Ser

Write the sequences using the symbols Y for any pyrimidine, R for any purine, and N for any nucleotide.

8.7   If DNA consisted of only two nucleotides (say, G and C) in any sequence, what is the minimum number of adjacent nucleotides that would be needed to specify uniquely each of the 20 amino acids?

8.8   A synthetic mRNA molecule consists of the repeating base sequence

5'-GAUGAUGAUGAUGAUGA . . . -3'

This molecule is translated *in vitro* using components from cells but in the absence of living cells. The polypeptide chains result from translation of randomly broken mRNA molecules, so that translation can initiate in any of the three possible reading frames. Once translation of an mRNA begins, however, the original reading frame is preserved throughout. What polypeptide products are made from this RNA? (Note: In an *in vitro* translation system, an initial AUG is not necessary for translation; the ribosome can start at the 5' end of the mRNA no matter what initial codon is present.)

8.9   Poly–U RNA codes for polyphenylalanine. If an A is added to the 5' end of the molecule, the polyphenylalanine has a different amino acid at the amino terminus, and if an A is added to the 3' end, there is a different amino acid at the carboxyl terminus. What are the amino acids?

8.10   What polypeptide products are made when the alternating polymer UGUG . . . is used in an *in vitro* protein-synthesizing system that does not need a start codon?

8.11   Some codons in the genetic code were determined experimentally by the translation of random polymers. If a ribonucleotide polymer is synthesized that contains 3/4 A and 1/4 C in random order, which amino acids will the resulting polypeptide contain, and in what frequencies?

8.12   Part of the human gene for fibrillin 1 (a gene associated with Marfan syndrome) has the sequence shown below. If this molecule is transcribed into RNA in the direction *from right to left*, deduce the sequence of the RNA, along with its polarity (5' and 3' ends).

```
5'-CCGACTGGCTCTGGTTTCCTTCACGTT-3'
3'-GGCTGACCGAGACCAAAGGAAGTGCAA-5'
```

8.13   Suppose a primitive living organism is discovered on Mars. It has a genetic system similar to our own in that the sequence of subunits in the genetic material (nucleic acid) is used as a code to specify the linear sequence of subunits of a different type of molecule (protein). In this organism, the nucleic acids are made up of four kinds of nucleotides, but the proteins contain only five kinds of amino acids. The organism produces 100 different proteins, each 10 amino acids in length.

(a)   What is the minimum allowable number of bases in a codon in this organism?

(b)   In order to make the 10 proteins, what is the minimum number of nucleotides in the genome of the organism?

8.14   The accompanying diagram shows a fully processed eukaryotic messenger RNA molecule hybridized to the transcribed strand of DNA of a gene that contains two introns. The diagram is oriented with the promoter region of the DNA at the left.

mRNA

DNA

Use the letters that follow to label the location and/or boundaries of each segment. Some letters may be used several times, as appropriate; some, which are not applicable, may not be used at all.

(a)	5′ end	(h)	Leader region
(b)	3′ end	(i)	Ribosome-binding site
(c)	Promoter region	(j)	Translation start codon
(d)	Attenuator	(k)	Translation stop codon
(e)	Intron	(l)	5′ cap
(f)	Exon	(m)	Poly-A tail
(g)	Polyadenylation signal		

8.15 Two *E. coli* genes, *A* and *B*, are known from mapping experiments to be very close to each other. A deletion mutation is isolated that eliminates the activity of both *A* and *B*. Neither the A nor the B protein can be found in the mutant, but a novel protein is isolated in which the amino-terminal 30 amino acids are identical to those of the B gene product and the carboxyl-terminal 30 amino acids are identical to those of the *A* gene product.

    (a) With regard to the 5′ → 3′ orientation of the nontranscribed DNA strand, is the order of the genes *A B* or *B A*?

    (b) Can you make any inference about the number of bases deleted?

8.16 The accompanying table shows matching regions of the DNA, mRNA, tRNA, and amino acids encoded in a particular gene. The mRNA is shown with its 5′ end at the left, and the tRNA anticodon is shown with its 3′ end at the left. The vertical lines define the reading frame.

    (a) Complete the nucleic acid sequences, assuming normal Watson–Crick pairing between each codon and anticodon.

    (b) Is the DNA strand that is transcribed the top strand or the bottom strand?

    (c) Translate the mRNA in all three reading frames.

    (d) Specify the nucleic acid strand(s) whose sequence could be used as a probe in a Southern blot hybridization, in which the hybridization is carried out against genomic DNA.

    (e) Specify the nucleic acid strand(s) whose sequence could be used as a probe in a

---

Northern blot hybridization, in which the hybridization is carried out against mRNA.

```
DNA double helix  -C---AT-C---------AT--GT
                  T------T---------CA------
mRNA              --AAC----------GC---GC--
tRNA anticodon    ---------GCA------------
Amino acids       |   |   |   |  |Trp|  |   |   |
```

8.17 A double mutant produced by recombination contains two single nucleotide frameshifts separated by about 20 base pairs. The first is an insertion, the second a deletion. The amino acid sequences of the wildtype and mutant polypeptide in this part of the protein are as follows:

Wildtype:

    `Lys-Lys-Tyr-His-Gln-Trp-Thr-Cys-Asn`

Double mutant:

    `Lys-Gln-Ile-Pro-Pro-Val-Asp-Met-Asn`

    What are the original and the double-mutant mRNA sequences? Which nucleotide in the wildtype sequence is the frameshift addition? Which nucleotide in the double-mutant sequence is the frameshift deletion? (In working this problem, you will find it convenient to use the conventional symbols Y for unknown pyrimidine, R for unknown purine, N for unknown nucleotide, and H for A or C or T.)

8.18 What polypeptide products are made when the alternating polymer GUCGUC ... is used in an *in vitro* protein-synthesizing system that does not need a start codon?

8.19 Prior to the demonstration that messenger RNA is translated in consecutive, nonoverlapping groups of three nucleotides (codons), the possibility of an *overlapping code* had to be considered. Such overlapping codes could be rejected because they impose constraints on which consecutive pairs of amino acids could be found in proteins. To understand why, suppose that the standard triplet codons were translated with an overlap of two. (In other words, the last two nucleotides of any codon are also the first two nucleotides of the next codon in line.) Which amino acids could follow:

    (a) Lys?

    (c) Met?

    (c) Tyr?

    (d) Trp?

8.20 Protein synthesis occurs with high fidelity. In pro-karyotes, incorrect amino acids are inserted at the rate of approximately 10^{-3} (that is, one incorrect amino acid per 1000 translated).

(a) What is the probability that a polypeptide of 300 amino acids has *exactly* the amino acid sequence specified in the mRNA?

(b) For an active enzyme consisting of four subunits (a tetramer), each 1000 amino acids in length, what is the probability that every amino acid in every subunit is translated without error?

 STOP & THINK ANSWERS

ANSWER TO STOP & THINK 8.1

The RNA transcript is synthesized from 5′ to 3′. Therefore, in order to be transcribed from left to right, the template strand of the double-stranded DNA must be the bottom strand. The RNA sequence is therefore 5′–AUGCCGUUA–3′.

ANSWER TO STOP & THINK 8.2

When inverted, the promoter region is not merely reversed in orientation but is rotated through 180 degrees so that the strand originally on top is now on the bottom and in the opposite orientation. The rotation is necessary to preserve the 5′-to-3′ polarity of the antiparallel DNA strands. In the mutant, transcription begins in the opposite strand and proceeds in the reverse direction into region B, as shown here:

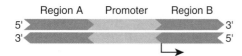

ANSWER TO STOP & THINK 8.3

The single-frameshift mutation would result in an RNA with the sequence 5′–AUGUCCCUUUGGGCAUCAU–3′, which is translated as Met–Ser–Leu–Trp–Ala–Ser with all following codons also frameshifted. The triple-frameshift mutation would yield the RNA sequence 5′–AUGUCCCUUUUGGGUCAUCAU–3′, which is translated as Met–Ser–Leu–Leu–Gly–His–His. Note that the amino acids between Met and His are incorrect, and there is an extra amino acid owing to the three added bases; however, the codons for His–His and those following are all in the correct reading frame.

ANSWER TO STOP & THINK 8.4

The synthetic RNA has the sequence 5′–CAGCAGCAG...–3′. If translation begins with CAG, the resulting polypeptide consists of repeating Gln (called polyglutamine). If translation begins with AGC, the resulting polypeptide is polyserine. And if translation begins with GCA, the resulting polypeptide is polyalanine. Hence three different polypeptides result from *in vitro* translation depending on where translation begins.

ANSWER TO STOP & THINK 8.5

The genetic code has six codons for serine, one for methionine, four for alanine, six for arginine, and four for threonine; hence the total number of RNA sequences encoding S–M–A–R–T equals $6 \times 1 \times 4 \times 6 \times 4 = 576$.

CHAPTER 9

Molecular Mechanisms of Gene Regulation

LEARNING OBJECTIVES

- To distinguish between negative regulation, positive regulation, and stochastic noise in transcription.

- To describe the mechanisms of negative and positive regulation of the lactose operon and explain the regulatory phenotype of various types of mutants of the repressor, operator, promoter, and structural genes.

- To state the role of RNA secondary structure in transcriptional termination by an attenuator or a riboswitch.

- To explain the role of topologically associating domains and nuclear compartments in restricting the regulatory activities of enhancers and silencers.

- To describe the role of siRNA and microRNA molecules in the regulation of gene expression.

- To define "epigenetic" in regard to transcriptional regulation and the main types of DNA and chromatin modification associated with epigenetic changes.

Humans and other vertebrate animals contain approximately 200 different cell types with specialized functions. Yet with very few exceptions, all cells in an organism have the same genome. The cell types differ in which genes are active. For example, the genes for hemoglobin are expressed at a high level only in the precursors of red blood cells. The subject of **gene regulation** encompasses the mechanisms that determine the types of cells in which a gene will be transcribed, when it will be transcribed, where the transcript will start along the DNA, where it will terminate, how the transcript will be spliced, when the mRNA will be exported to the cytoplasm, when and how often the mRNA will be translated, and the duration of time before the mRNA is degraded.

9.1 Regulation of transcription is a common mechanism in prokaryotes.

In bacteria and bacteriophage, on–off gene activity is often controlled through transcription. Under conditions when a gene product is needed, transcription of the gene is turned "on"; under other conditions, transcription is turned "off." The term *off* should not be taken literally. In bacteria, few examples are known of a system being switched off completely. When transcription is in the "off" state, a basal level of gene expression nearly always remains, often averaging one transcriptional event or fewer per cell generation; hence "off" really means that there is very little synthesis of the gene product. Extremely low levels of expression are also found in certain classes of genes in eukaryotes. Regulatory mechanisms other than the on–off type also are known in both prokaryotes and eukaryotes; in these examples, the level of expression of a gene may be modulated in gradations from low to high according to conditions in the cell.

In bacterial systems, when several enzymes act in sequence in a single metabolic pathway, usually either all or none of the enzymes are produced. This **coordinate regulation** results from control of the synthesis of one or more mRNA molecules that are *polycistronic*; these mRNAs encode all of the gene products that function in the same metabolic pathway. This type of regulation is not found in eukaryotic cells.

In negative regulation, the default state of transcription is "on."

The molecular mechanisms of regulation usually fall into either of two broad categories: *negative regulation* and *positive regulation*. In a system subject to **negative regulation** (FIGURE 9.1 part A), the default state is "on," and transcription takes place until it is turned "off" by a **repressor** protein that binds to the DNA upstream

from the transcriptional start site. A negatively regulated system may be either *inducible* (part B) or *repressible* (part C), depending on how the active repressor is formed. In **inducible transcription**, a repressor DNA-binding protein normally keeps transcription in the "off" state. In the presence of a small molecule called the **inducer**, the repressor binds preferentially with the inducer and loses its DNA-binding capability, allowing transcription to occur. Many degradative (catabolic) pathways are inducible and use the initial substrate of the degradative pathway as the inducer. In this way, the enzymes used for degradation are not synthesized unless the substrate is present in the cell.

In **repressible transcription** (part C), the default state is "on" until an active repressor is formed to turn it "off." In this case the regulatory protein is called an **aporepressor**, and it has no DNA-binding activity on its own. The active repressor that can bind to the DNA is formed by the combination of the aporepressor and a small molecule known as the **co-repressor**. Presence of the co-repressor thereby results in the cessation of transcription. Repressible regulation is often found in the control of the synthesis of enzymes that participate in biosynthetic (anabolic) pathways; in these cases the final product of the pathway is frequently the co-repressor. In this way, the enzymes of the biosynthetic pathway are not synthesized until the concentration of the final product becomes too low to cause repression.

In positive regulation, the default state of transcription is "off."

In a positively regulated system (FIGURE 9.2), the default state of transcription is "off," and binding with a regulatory protein is necessary to turn it "on." The protein that turns transcription on is a **transcriptional activator protein**. Negative and positive regulation are not mutually exclusive, and many systems are both positively and negatively regulated, utilizing two regulators to respond to different conditions in the cell. Negative regulation is more common in prokaryotes, positive regulation in eukaryotes.

Some genes exhibit **autoregulation**, which means that the protein product of a gene regulates its own transcription. In negative autoregulation, the protein inhibits transcription, and high concentrations of the protein result in less transcription of the mRNA. This mechanism automatically adjusts the steady-state level of the protein in the cell. In positive autoregulation, the protein stimulates transcription: As more protein is made, transcription increases to the maximum rate. Positive autoregulation is a common way for weak induction to be amplified. Only a weak signal is necessary to get production of the protein started, but then the positive autoregulation takes over and stimulates further production to the maximum level.

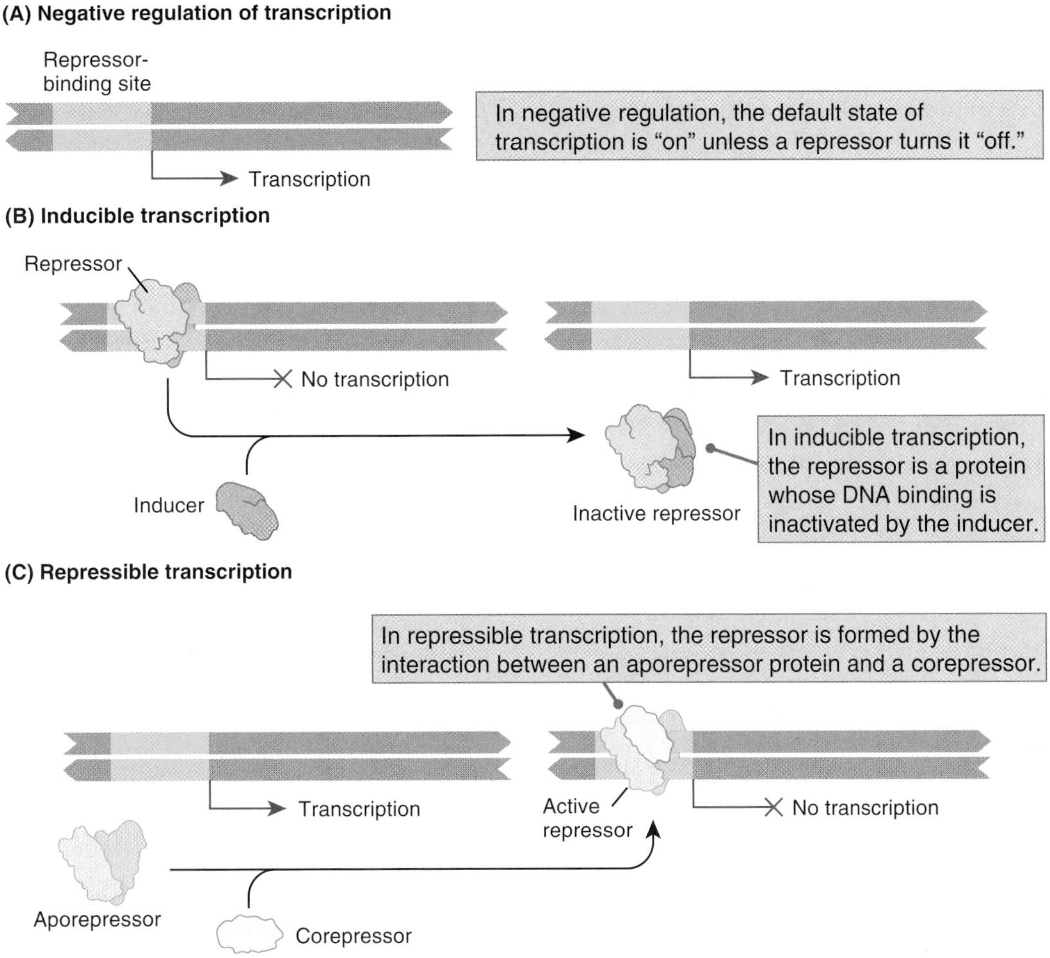

(A) Negative regulation of transcription

Repressor-
binding site

In negative regulation, the default state of
transcription is "on" unless a repressor turns it "off."

→ Transcription

(B) Inducible transcription

Repressor

✕ No transcription

Inducer

→ Transcription

Inactive repressor

In inducible transcription,
the repressor is a protein
whose DNA binding is
inactivated by the inducer.

(C) Repressible transcription

In repressible transcription, the repressor is formed by the
interaction between an aporepressor protein and a corepressor.

→ Transcription

Active
repressor

✕ No transcription

Aporepressor

Corepressor

FIGURE 9.1 Negative regulation (A) includes both inducible (B) and repressible (C) mechanisms of transcriptional control.

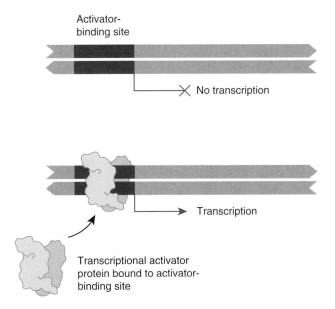

Positive regulation

Activator-
binding site

✕ No transcription

→ Transcription

Transcriptional activator
protein bound to activator-
binding site

FIGURE 9.2 In positive regulation, the default state of transcription is "off." Transcription is stimulated by the binding of a transcriptional activator protein.

Transcription sometimes occurs accidentally.

Most mechanisms of gene regulation are multilayered, with one level of control overlaid on another. Perfectly reliable control, however, is unattainable. The main reason is intrinsic unpredictability or noise in molecular processes, especially when the number of molecules is small, as is typical of regulatory molecules. Every gene is liable to a burst of RNA transcription now and again, which is followed by a burst of protein as the mRNA is translated. Such random bursts of gene expression are known as **stochastic noise**, and they occur in prokaryotic cells and eukaryotic cells at all stages of development.

Stochastic noise has been studied extensively in cells of *Escherichia coli* as well as budding yeast, using strains in which individual proteins are fused with a fluorescent polypeptide to allow even low levels of expression to be detected. In *E. coli*, about half of all proteins are produced at the level of 10–1000 molecules per cell; for these proteins the variation in expression from one cell to the next is about 30 percent. Stochastic noise is more significant for proteins expressed at lower levels. For proteins produced at an average of 0.01–1 molecules per cell, the variation in gene expression is 1–10 times the average.

Although it may seem odd that cells might depend on stochastic noise to explore different physiological states or to coordinate the expression of groups of genes, such mechanisms have evolved. In the next section we shall see an example of how *E. coli* cells use stochastic noise to respond to the presence of nutrients that appear in its environment.

STOP & THINK 9.1

Gene *A* in a bacterial species is negatively regulated by means of a repressor protein, whereas gene *B* is positively regulated by means of a transcriptional activator protein. How would gene *A* be regulated in a strain with a mutant repressor unable to bind its binding site? How would gene *A* be regulated in a strain with a mutant repressor that binds so tightly to its binding site that it cannot be displaced? How would gene *B* be regulated in strains with analogous mutations affecting the transcriptional activator protein?

9.2   In prokaryotes, groups of adjacent genes are often transcribed as a single unit.

Analysis of gene regulation was first carried out in detail for the genes responsible for degradation of the sugar lactose in *E. coli*. Much of the terminology used to describe regulation came from this genetic analysis.

The first regulatory mutations that were discovered affected lactose metabolism.

In *E. coli*, two proteins are necessary for the metabolism of lactose: the enzyme *β-galactosidase*, which cleaves lactose (a *β*-galactoside sugar) to yield galactose and glucose, and a transporter molecule, **lactose permease**, which is required for the entry of lactose into the cell. The existence of two different proteins in the lactose-utilization system was first shown by a combination of genetic experiments and biochemical analysis.

First, hundreds of mutants unable to use lactose as a carbon source, designated Lac⁻ mutants, were isolated. Some of the mutations were in the *E. coli* chromosome, and others were in an F′ *lac*, a plasmid carrying the genes for lactose utilization. By performing F′ × F⁻ matings, investigators constructed partial diploids with the genotypes F′ *lac⁻* / *lac⁺* or F′ *lac⁺* / *lac⁻*. (The genotype of the plasmid is given to the left of

the slash and that of the chromosome to the right.) It was observed that all of these partial diploids always had a Lac⁺ phenotype (that is, they made both *β*-galactosidase and permease). Other partial diploids were then constructed in which both the F′ *lac* plasmid and the chromosome carried a *lac⁻* allele. When these were tested for the Lac⁺ phenotype, it was found that all of the mutants initially isolated could be placed into two complementation groups, called *lacZ* and *lacY*, a result that implies that the *lac* system consists of at least two genes. Complementation is indicated by the observation that the partial diploids F′ *lacY⁻ lacZ⁺* / *lacY⁺ lacZ⁻* and F′ *lacY⁺ lacZ⁻* / *lacY⁻ lacZ⁺* had a Lac⁺ phenotype, producing both *β*-galactosidase and permease. However, the genotypes F′ *lacY⁻ lacZ⁺* / *lacY⁻ lacZ⁺* and F′ *lacY⁺ lacZ⁻* / *lacY⁺ lacZ⁻* had the Lac⁻ phenotype; they were unable to synthesize permease and *β*-galactosidase, respectively. Hence the *lacZ* gene codes for *β*-galactosidase and the *lacY* gene for permease. (A third gene that participates in lactose metabolism was discovered later; it was not included among the early mutants because it is not essential for growth on lactose.) Close physical proximity of the *lacZ* and *lacY* genes was deduced from a high frequency of cotransduction observed in genetic mapping experiments. In fact, *lacZ* and *lacY* are adjacent in the chromosome.

Lactose-utilizing enzymes can be inducible (regulated) or constitutive.

The on–off nature of the genes responsible for lactose utilization is evident in the following observations:

- If a culture of Lac⁺ *E. coli* is grown in a medium lacking lactose or any other *β*-galactoside, the intracellular concentrations of *β*-galactosidase and permease are exceedingly low—roughly one or two molecules per bacterial cell. However, if lactose is present in the growth medium, the number of each of these molecules is about 1000-fold higher.

- If lactose is added to a Lac⁺ culture growing in a lactose-free medium (also lacking glucose, a point we will discuss shortly), both *β*-galactosidase and permease are synthesized nearly simultaneously, as shown in **FIGURE 9.3**. Analysis of the total mRNA present in the cells before and after the addition of lactose shows that almost no *lac* mRNA (the polycistronic mRNA that codes for *β*-galactosidase and permease) is present before lactose is added and that the addition of lactose triggers synthesis of the *lac* mRNA.

These two observations led to the view that transcription of the lactose genes is *inducible transcription* and that lactose is an *inducer* of transcription. Some analogs of lactose are also inducers, such as a sulfur-containing analog denoted IPTG (isopropylthiogalactoside), which

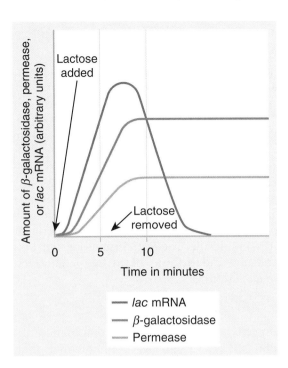

FIGURE 9.3 The "on–off" nature of the *lac* system. The *lac* mRNA appears soon after lactose or another inducer is added; β-galactosidase and permease appear at nearly the same time but are delayed with respect to mRNA synthesis because of the time required for translation. When lactose is removed, no more *lac* mRNA is made, and the amount of *lac* mRNA decreases because of the degradation of mRNA already present. Both β-galactosidase and permease are stable proteins: Their amounts remain constant even when synthesis ceases. However, their concentration per cell gradually decreases as a result of repeated cell divisions.

TABLE 9.1 Characteristics of Partial Diploids Containing Several Combinations of *lacI*, *lacO*, and *lacP* Alleles

Genotype	Synthesis of *lac* mRNA	Lac phenotype
1. F' *lacO^c lacZ^+/lacO^+ lacZ^+*	Constitutive	+
2. F' *lacO^+ lacZ^+/lacO^c lacZ^+*	Constitutive	+
3. F' *lacI^- lacZ^+/lacI^+ lacZ^+*	Inducible	+
4. F' *lacI^+ lacZ^+/lacI^- lacZ^+*	Inducible	+
5. F' *lacO^c lacZ^-/lacO^+ lacZ^+*	Inducible	+
6. F' *lacO^c lacZ^+/lacO^+ lacZ^-*	Constitutive	+
7. F' *lacI^s lacZ^+/lacI^+ lacZ^+*	Uninducible	−
8. F' *lacI^+ lacZ^+/lacI^s lacZ^+*	Uninducible	−
9. F' *lacP^- lacZ^+/lacP^+ lacZ^+*	Inducible	+
10. F' *lacP^+ lacZ^+/lacP^- lacZ^+*	Inducible	+
11. F' *lacP^+ lacZ^-/lacP^- lacZ^+*	Uninducible	−
12. F' *lacP^+ lacZ^+/lacP^- lacZ^-*	Inducible	+

is convenient for experiments because it induces, but is not cleaved by, β-galactosidase. The inducer IPTG is taken up by the cells and maintained at a constant level, whether or not the β-galactosidase enzyme is present.

Mutants were also isolated in which *lac* mRNA was synthesized, and the enzymes produced, in the *absence* of an inducer as well as in its presence. Because of their constant synthesis, with or without inducer, the mutants were called **constitutive**. They provided the key to understanding induction. Mutants were also obtained that failed to produce *lac* mRNA and the enzymes even when the inducer was present. These uninducible mutants fell into two classes, *lacI^s* and *lacP^-*. The characteristics of the mutants are shown in **TABLE 9.1** and discussed in the following sections.

Repressor shuts off messenger RNA synthesis.

In Table 9.1, genotypes 3 and 4 show that *lacI^-* mutations are recessive. In the absence of inducer, a *lacI^+* cell does not make *lac* mRNA, whereas the mRNA is made in a *lacI^-* mutant. These results suggest that

KEY CONCEPT

The *lacI* gene is a regulatory gene whose product is the repressor protein that keeps the system turned off. Because the repressor is necessary to shut off mRNA synthesis, regulation by the repressor is negative regulation.

A *lacI^-* mutant lacks the repressor and, hence, transcription is constitutive. Wildtype copies of the repressor are present in a *lacI^+* / *lacI^-* partial diploid, so transcription is repressed. It is important to note that the single *lacI^+* gene prevents synthesis of *lac* mRNA from both the F' plasmid and the chromosome. Therefore, the repressor protein must be diffusible within the cell to be able to shut off mRNA synthesis from both DNA molecules present in a partial diploid.

On the other hand, genotypes 7 and 8 indicate that the *lacI^s* mutations are dominant and act to shut off mRNA synthesis from both the F' plasmid and the chromosome, whether or not the inducer is present (the superscript in *lacI^s* signifies *super-repressor*). The *lacI^s* mutations result in repressor molecules that fail to recognize and bind the inducer and thus permanently shut off *lac* mRNA synthesis. Genetic mapping experiments placed the *lacI* gene nearly adjacent to the *lacZ* gene and established the gene order *lacI lacZ lacY*. How the *lacI* repressor prevents synthesis of *lac* mRNA will be explained shortly.

The lactose operator is an essential site for repression.

Entries 1 and 2 in Table 9.1 show that *lacOᶜ* mutants are dominant. However, the dominance is evident only in certain combinations of *lac* mutations, as can be seen by examining the partial diploids shown in entries 5 and 6. Both combinations are Lac⁺ because a functional *lacZ* gene is present. However, in the combination shown in entry 5, synthesis of β-galactosidase is inducible even though a *lacOᶜ* mutation is present. The difference between the two combinations in entries 5 and 6 is that in entry 5, the *lacOᶜ* mutation is present in the same DNA molecule as the *lacZ⁻* mutation, whereas in entry 6, *lacOᶜ* is contained in the same DNA molecule as *lacZ⁺*. The key feature of these results is that

KEY CONCEPT

A *lacOᶜ* mutation causes constitutive synthesis of β-galactosidase only when the *lacOᶜ* and *lacZ⁺* alleles are contained in the same DNA molecule.

The *lacOᶜ* mutation is said to be **cis-dominant**, because only genes in the *cis* configuration (in the same DNA molecule as that containing the mutation) are expressed in dominant fashion. Confirmation of this conclusion comes from an important biochemical observation: The mutant enzyme from the *lacZ⁻* allele is synthesized constitutively in a *lacOᶜ lacZ⁻ / lacO⁺ lacZ⁺* partial diploid (entry 5), whereas the wildtype enzyme from the *lacZ⁺* allele is synthesized only if an inducer is added. All *lacOᶜ* mutations are located between the *lacI* and *lacZ* genes; hence the gene order of the four genetic elements of the *lac* system is

lacI lacO lacZ lacY

An important feature of all *lacOᶜ* mutations is that they cannot be complemented (a characteristic feature of all *cis*-dominant mutations); that is, a *lacO⁺* allele cannot alter the constitutive activity of a *lacOᶜ* mutation. This observation implies that the *lacO* region does not encode a diffusible product and must instead define a site in the DNA that determines whether synthesis of the product of the adjacent *lacZ* gene is inducible or constitutive. The *lacO* region is called the **operator**. In a subsequent section, we will see that the operator is in fact a *binding site* in the DNA for the repressor protein.

The lactose promoter is an essential site for transcription.

Entries 11 and 12 in Table 9.1 show that *lacP⁻* mutations, like *lacOᶜ* mutations, are *cis*-dominant. The *cis*-dominance can be seen in the partial diploid in entry 11. The genotype in entry 11 is uninducible, in contrast to the partial diploid of entry 12, which is inducible. The difference between the two genotypes is that in entry 11, the *lacP⁻* mutation is in the same DNA molecule with *lacZ⁺*, whereas in entry 12, the *lacP⁻* mutation is combined with *lacZ⁻*. This observation means that a wildtype *lacZ⁺* remains inexpressible in the presence of *lacP⁻*; no *lac* mRNA is transcribed from that DNA molecule. The *lacP⁻* mutations map between *lacI* and *lacO*, and the order of the five genetic elements of the *lac* system is

lacI lacP lacO lacZ lacY

As expected because of the *cis*-dominance of *lacP⁻* mutations, they cannot be complemented; that is, a *lacP⁺* allele on another DNA molecule cannot supply the missing function to a DNA molecule carrying a *lacP⁻* mutation. Thus *lacP*, like *lacO*, must define a site that determines whether synthesis of *lac* mRNA will take place. Because synthesis does not occur if the site is defective or missing, *lacP* defines an essential site for mRNA synthesis. The *lacP* region is called the **promoter**. It is a site at which RNA polymerase binding takes place to allow initiation of transcription.

⊹ STOP & THINK 9.2

Consider a bacterial strain of genotype

F′ *lacOᶜ lacZ⁺ lacY⁻ / lacP⁻ lacZ⁻ lacY⁺*

For each of the alleles *lacZ⁺* and *lacY⁺* in this strain, classify the regulatory state of the allele as inducible, uninducible, or constitutive.

The lactose operon contains linked structural genes and regulatory sequences.

The genetic regulatory mechanism of the *lac* system was first explained by the *operon model* of François Jacob and Jacques Monod, which is illustrated in **FIGURE 9.4**. (The figure uses the alternative abbreviations *i, o, p, z, y,* and *a* for *lacI, lacO, lacP, lacZ, lacY,* and *lacA*.) The **operon model** of gene regulation has the following features:

1. The lactose-utilization system consists of two kinds of components: *structural genes* (*lacZ* and *lacY*), which encode proteins needed for the

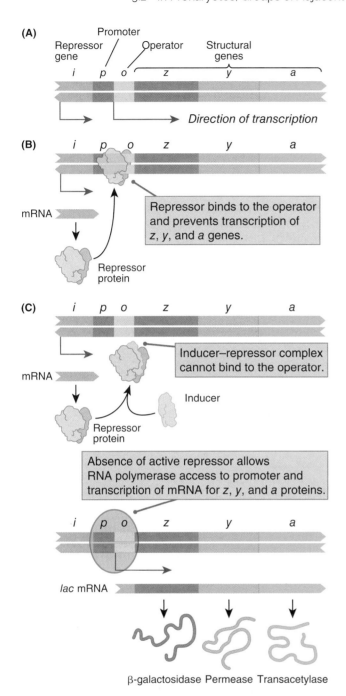

FIGURE 9.4 (A) Organization of the *lac* operon, not drawn to scale; the *p* and *o* sites are actually much smaller than the other elements and together comprise only 83 base pairs. (B) A diagram of the *lac* operon in the repressed state. (C) A diagram of the *lac* operon in the induced state. The inducer alters the shape of the repressor so that the repressor can no longer bind to the operator.

Note: The common abbreviations *i, p, o, z, y,* and *a* are used instead of *lacI, lacO,* and so forth. The *lacA* gene is not essential for lactose utilization.

transport and cleavage of lactose, and *regulatory elements* (the repressor gene *lacI*, the promoter *lacP*, and the operator *lacO*).

2. The products of the *lacZ* and *lacY* genes are coded by a single polycistronic mRNA molecule. The linked structural genes, together with *lacP* and

lacO, constitute the *lac* **operon**. (The third protein, encoded by *lacA*, is also translated from the polycistronic mRNA. This protein is the enzyme *β*-galactoside transacetylase; it is used in the metabolism of certain *β*-galactosides other than lactose and will not concern us here.)

3. The promoter mutations (*lacP⁻*) eliminate the ability to synthesize *lac* mRNA.

4. The product of the *lacI* gene is a repressor, which binds to a unique sequence of DNA bases constituting the operator.

5. When the repressor is bound to the operator, initiation of transcription of *lac* mRNA by RNA polymerase is prevented.

6. Inducers stimulate mRNA synthesis by binding to and inactivating the repressor. In the presence of an inducer, the operator is not bound with the repressor, and the promoter is available for the initiation of mRNA synthesis.

Note that regulation of the operon requires that the *lacO* operator either overlap or be adjacent to the promoter of the structural genes, because binding with the repressor prevents transcription. Proximity of *lacI* to *lacO* is not strictly necessary, because the *lacI* repressor is a soluble protein and is therefore diffusible throughout the cell. The presence of inducer has a profound effect on the DNA-binding properties of the repressor; the inducer–repressor complex has an affinity for the operator that is approximately 10^3 less than that of the repressor alone.

When the operon is induced, the numbers of protein molecules of *β*-galactosidase, permease, and transacetylase are in the ratio 1.0 : 0.5 : 0.2. These differences are partly due to the order of the genes in the mRNA. Downstream cistrons are less likely to be translated because of failure of reinitiation when an upstream cistron has finished translation.

Stochastic noise aids induction of the lactose operon.

How does lactose get into a cell in which the Lac operon is repressed when transport requires the Lac permease? The answer is that the organism depends on stochastic noise in gene expression to allow some cells to take advantage of any lactose in the growth medium. For the Lac operon, stochastic bursts of mRNA and protein expression take place once every 5–10 cell generations. The average burst results in about 40 molecules of Lac permease. These are diluted by cell division to an average of 20, 10, 5, ... molecules per daughter cell in each successive division. At steady state, under fully repressed conditions, about 35 percent of the cells have one or more molecules of Lac permease owing to stochastic noise. These cells can respond instantly to an increase in lactose in the environment by transport of lactose into the cell, where the sugar induces the operon through positive feedback. The cellular response is, therefore, "all or none." Cells that lack permease need to undergo a stochastic burst of expression before they can begin lactose utilization.

The lactose operon is also subject to positive regulation.

Mechanisms of gene regulation are often multilayered, and regulation of the Lac operon is no exception. In addition to negative regulation by lactose, the Lac operon is subject to positive regulation by glucose. Glucose is the preferred source of carbon and energy for *E. coli*; if both glucose and lactose are present in the growth medium, transcription of the *lac* operon is shut down until virtually all of the glucose in the medium has been consumed. The observation that no *lac* mRNA is made in the presence of glucose implies that another element, in addition to an inducer, is needed for initiating *lac* mRNA synthesis.

The inhibitory effect of glucose on expression of the *lac* operon is indirect. The small molecule *cyclic adenosine monophosphate* (cAMP), shown in **FIGURE 9.5**, is widely distributed in animal tissues and in multicellular eukaryotic organisms, in which it is important in mediating the action of many hormones. It is also present in *E. coli* and many other bacteria, where it has a different function. Cyclic AMP is synthesized by the enzyme *adenylate cyclase*, and the concentration of cAMP is regulated indirectly by glucose metabolism. When bacteria are growing in a medium that contains glucose, the cAMP concentration in the cells is quite low. In a medium containing glycerol or any carbon source that requires aerobic metabolism for degradation, or when the bacteria are otherwise starved of an energy source, the cAMP concentration is high (**TABLE 9.2**). Glucose levels help regulate the cAMP concentration in the cell, and *cAMP regulates the activity of the lac operon* (as well as that of several other operons that control degradative metabolic pathways).

E. coli and many other bacterial species contain a protein called the *cyclic AMP receptor protein* (CRP), which

is encoded by a gene called *crp*. Mutations of either the *crp* or the adenylate cyclase gene drastically reduce synthesis of *lac* mRNA, which indicates that both CRP function and cAMP are required for *lac* mRNA synthesis. CRP and cAMP bind to one another, forming a complex denoted **cAMP–CRP**. The presence of cAMP–CRP is necessary for full induction, because *crp⁻* and adenylate cyclase mutants are unable to make normal levels of *lac* mRNA even when a *lacI⁻* or a *lacOᶜ* mutation is present. The reason for the requirement is that transcription is impeded unless the cAMP–CRP complex is bound to a specific DNA sequence in the promoter region (**FIGURE 9.6**). Unlike the repressor, which is a *negative* regulator, the cAMP–CRP complex is a *positive* regulator. The positive and negative regulatory systems of the *lac* operon are independent of each other.

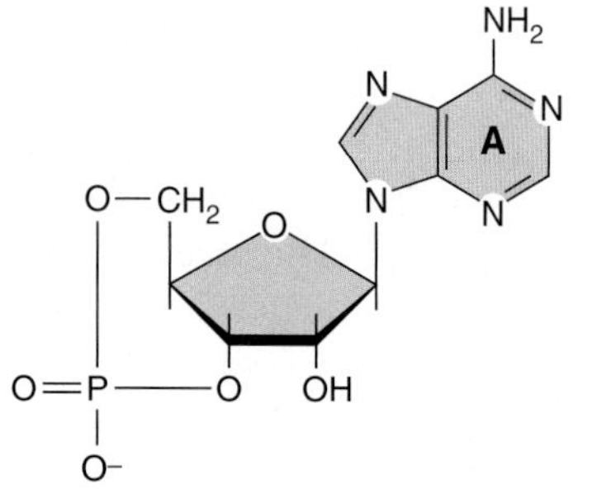

| TABLE 9.2 Concentration of Cyclic AMP in Cells Growing in Media with the Indicated Carbon Sources ||
Carbon source	cAMP concentration
Glucose	Low
Glycerol	High
Lactose	High
Lactose + glucose	Low
Lactose + glycerol	High

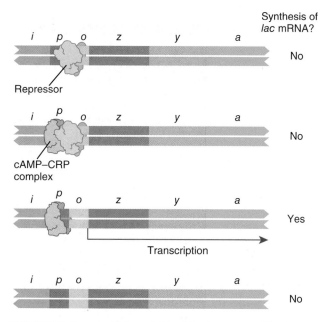

FIGURE 9.6 Four regulatory states of the *lac* operon: The *lac* mRNA is synthesized only when cAMP–CRP is present and repressor is absent.

FIGURE 9.5 Structure of cyclic AMP (cAMP).

Experiments carried out *in vitro* with purified *lac* DNA, *lac* repressor, cAMP–CRP, and RNA polymerase have established two further points:

1. In the absence of the cAMP–CRP complex, RNA polymerase binds only weakly to the promoter, but its binding is stimulated when cAMP–CRP is also bound to the DNA. The weak binding leads to reduced levels of initiation of transcription, because the correct interaction between RNA polymerase and the promoter does not take place.

2. If the repressor is bound to the operator, then RNA polymerase cannot stably bind to the promoter.

These results explain how lactose and glucose function together to regulate transcription of the *lac* operon. The relationship of these elements to one another, to the start of transcription, and to the base sequence in the region is depicted in **FIGURE 9.7**.

A great deal is also known about the three-dimensional structure of the regulatory states of the *lac* operon. **FIGURE 9.8** shows how the repressor protein (violet) binds with two operator regions to form a loop that includes the site at which the CRP protein (dark blue) binds. The region of DNA shown in Figure 9.8 corresponds to the region in Figure 9.7 that extends from Operator 3 through Operator 1, and the tabs representing the bases in Figure 9.7 are color coded to match the regions in Figure 9.8. In Figure 9.8, the DNA region in red corresponds, on the right-hand side, to Operator 1, and, on the left-hand side, to Operator 3. The *lac* repressor tetramer (violet) binds to these sites. The DNA loop is formed by the region between the repressor-binding sites and includes, in medium blue, the CRP-binding site to which the CRP protein (dark blue) is bound. The DNA regions in green are the −10 and −35 sites in the *lacP* promoter. In the configuration in Figure 9.8, the *lac* operon is not transcribed. Removal of the repressor opens up the loop and allows transcription to occur.

DNA sequencing of the *lac* operon revealed the presence of three operator sequences. Operator sequences 1 and 3 are shown in Figure 9.7, whereas Operator 2 is located about 400 nucleotides into the *lacZ* coding region. The three operators have very different efficiencies of repressor binding. Operator 1 and Operator 2 (the one within the *lacZ* gene) bind the repressor with high affinity, whereas Operator 3 binds with only about half the affinity of the other two. The most common repressed state of the operon is, therefore, one in which Operators 1 and 2 are bound with the repressor in a conformation similar to that in Figure 9.8 but with a much longer upper loop. Full repression of transcription requires all three operators. The classic genetic experiments of Jacob and Monod identified only Operator 1.

STOP & THINK 9.3

A mutant strain of *E. coli* always produces high levels of cyclic AMP. In this strain, how would the *lac* operon respond to the addition of lactose to a growth medium containing glucose? How does this response differ from that in a nonmutant strain?

Tryptophan biosynthesis is regulated by the tryptophan operon.

The tryptophan (*trp*) operon of *E. coli* contains structural genes for enzymes that synthesize the amino acid tryptophan. This operon is regulated in such a way that when adequate tryptophan is present in the growth medium, transcription of the operon is repressed. However, when the supply of tryptophan is insufficient, transcription takes place. Regulation in the *trp* operon is similar to that in the *lac* operon because mRNA synthesis is regulated negatively by a repressor. However, it differs from regulation of *lac* in that tryptophan acts as a co-repressor, which stimulates binding of the repressor to the *trp* operator to shut off synthesis. The *trp* operon is a *repressible* rather than an *inducible* operon. Furthermore, because the *trp* operon codes for a set of biosynthetic enzymes rather than degradative enzymes, neither glucose nor cAMP–CRP functions in regulation of the *trp* operon.

A simple on–off system, as in the *lac* operon, is not optimal for a biosynthetic pathway. For example, a situation may arise in which some tryptophan is present in the growth medium, but the amount is not enough to sustain optimal growth. Under these conditions, it is advantageous to synthesize tryptophan, but at less than the maximum possible rate. Cells adjust to this situation by means of a regulatory mechanism in which the amount of transcription in the derepressed state is determined by the concentration of tryptophan in the cell. This regulatory mechanism is found in many operons responsible for amino acid biosynthesis.

Tryptophan is synthesized in five steps, each requiring a particular enzyme. The genes encoding these enzymes are adjacent in the *E. coli* chromosome and are in the same linear order as the order in which the enzymes function in the biosynthetic pathway. The genes are called *trpE*, *trpD*, *trpC*, *trpB*, and *trpA*, and the enzymes are translated from a single polycistronic mRNA molecule. The *trpE* coding region is the first one translated. Upstream (on the 5′ side) of *trpE* are the promoter, the operator, and two regions called the *leader* and the *attenuator*, which are designated *trpL* and *trpa* (not *trpA*), respectively (**FIGURE 9.9**). The repressor gene, *trpR*, is located quite far from this operon.

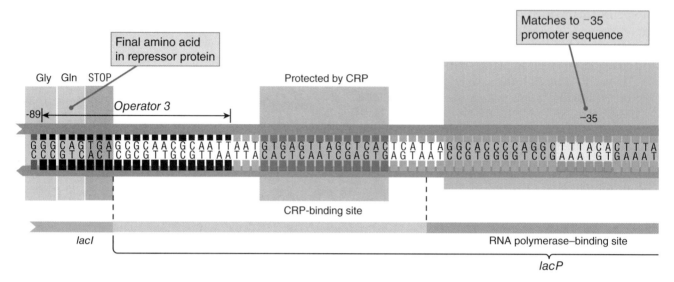

Gly Gln STOP

Final amino acid
in repressor protein

Operator 3

Protected by CRP

Matches to −35
promoter sequence

−89

−35

CRP-binding site

RNA polymerase–binding site

lacI

lacP

FIGURE 9.7 (Above and facing page) The nucleotide sequence of the regulatory region of the *lac* operon, showing regions protected from DNase digestion by the binding of various proteins. The end of the *lacI* gene is shown at the extreme left; the ribosome-binding site is the site at which the ribosome binds to the *lac* mRNA. The sites for CRP binding and for RNA polymerase binding are indicated along the bottom. The tabs representing the bases are color coded to match those in Figure 9.8.

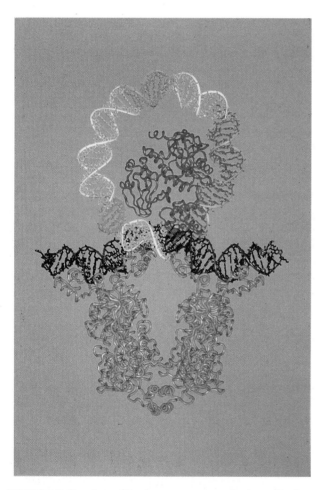

FIGURE 9.8 Structure of the *lac* operon repression loop. The *lac* repressor, shown in violet, binds to two DNA regions (red) consisting of the symmetrical operator region indicated in Figure 9.7 and a second region immediately upstream from the CRP-binding site. Within the loop is the CRP-binding site (medium blue), shown bound with CRP protein (dark blue). The −10 and −35 promoter regions are in green.

Reproduced from M. Lewis, et al., *Science* 271 (1996): 1247–1254. Reprinted with permission from AAAS. [www.sciencemag.org].

The regulatory protein of the *trp* operon is the product of the *trpR* gene, an aporepressor protein that requires tryptophan as a co-repressor in order to form the active repressor protein. Mutations in either *trpR* or the operator cause constitutive initiation of transcription of *trp* mRNA. The *trpR* gene product is the *trp* aporepressor. It does not bind to the operator unless it is first bound to tryptophan; that is, the aporepressor and the tryptophan molecule join together to form the active *trp* repressor, which binds to the operator. The reaction scheme is outlined in **FIGURE 9.10**. When there is insufficient tryptophan, the aporepressor adopts a conformation unable to bind with the *trp* operator, and the operon is transcribed (Figure 9.10, part A). When tryptophan is present at a sufficiently high concentration, some molecules bind with the aporepressor and cause it to change conformation into the active repressor. The active repressor binds with the *trp* operator and prevents transcription (Figure 9.10, part B). This is the basic on–off regulatory mechanism.

9.3 Gene activity can be regulated through transcriptional termination.

The lactose and tryptophan operons show how gene activity can be regulated through the initiation of transcription. There are also mechanisms for gene regulation through the termination of transcription. In this section we consider two examples.

Attenuation allows for fine-tuning of transcriptional regulation.

When the level of free tryptophan is low enough that transcription of the tryptophan operon is initiated, a

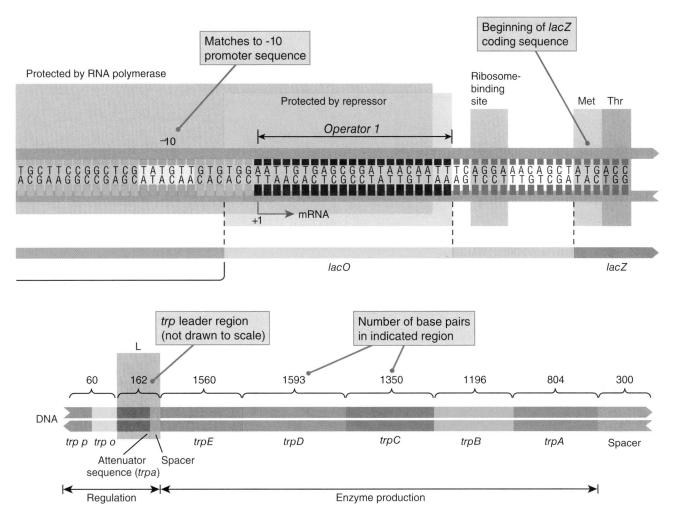

FIGURE 9.9 The *trp* operon in *E. coli*. For clarity, the regulatory region is enlarged with respect to the coding region. The actual size of each region is indicated by the number of base pairs. Region L is the leader.

still more sensitive level of transcriptional regulation is exerted based on the concentration of charged tryptophan tRNA. This type of regulation is called **attenuation**, and it uses translation to control transcription. If translation of the leader region of the mRNA takes place, it causes termination of transcription even before the first structural gene of the operon is transcribed.

Attenuation results from interactions between DNA sequences present in the leader region of the *trp* transcript. In wildtype cells, transcription of the *trp* operon is often initiated. However, in the presence of even small amounts of tryptophan, most of the mRNA molecules terminate in a specific 28-base region within the leader sequence. The result of termination is an RNA molecule containing only 140 nucleotides that stops short of the genes coding for the *trp* enzymes. The 28-base region in which termination takes place is called the **attenuator**. The base sequence of this region (**FIGURE 9.11**) contains the usual features of a termination site, including a potential stem-and-loop configuration in the mRNA followed by a sequence of eight uridylates.

In the tryptophan operon in *E. coli*, termination of transcription is determined by whether a small peptide encoded in the leader sequence can be translated. This

coding sequence, shown in **FIGURE 9.12**, specifies a **leader polypeptide** 14 amino acids in length, and it includes two adjacent tryptophan codons at positions 10 and 11. When there is sufficient charged tryptophan tRNA to allow translation of these codons, the nascent transcript adopts a conformation in which the attenuator is exposed, and transcription is terminated. On the other hand, when there is insufficient charged tryptophan tRNA to allow translation of the leader polypeptide, the ribosome stalls at the tryptophan codons; in this case the attenuator is hidden, and transcription continues through the entire operon.

The mechanism of attenuation is diagrammed in **FIGURE 9.13**. Part A shows the leader RNA molecule, including the two tryptophan codons in the leader polypeptide. Region 2 has a nucleotide sequence that enables it to pair either with region 1 or with region 3. In the purified RNA, region 1 pairs with region 2, and region 3 pairs with region 4. Part B shows the configuration in a cell in which there is sufficient tRNA^Trp to allow translation of the leader polypeptide. The ribosome moves beyond the Trp codons and blocks region 2, so the pairing that forms is between region 3 and region 4; this creates the transcriptional terminator,

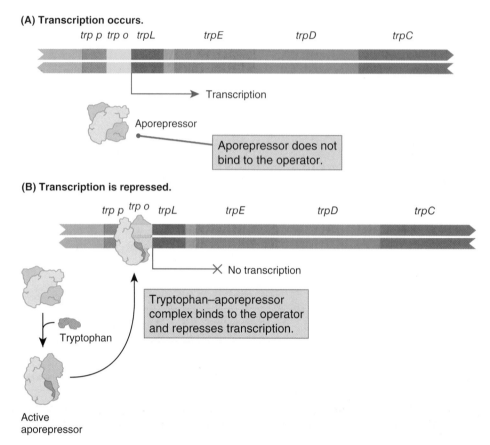

(A) Transcription occurs.

Transcription

Aporepressor

Aporepressor does not bind to the operator.

(B) Transcription is repressed.

No transcription

Tryptophan

Tryptophan–aporepressor complex binds to the operator and represses transcription.

Active aporepressor

FIGURE 9.10 Regulation of the *E. coli trp* operon. (A) By itself, the *trp* aporepressor protein does not bind to the operator, and transcription occurs. (B) In the presence of sufficient tryptophan, the combination of aporepressor and tryptophan forms the active repressor that binds to the operator, and transcription is repressed.

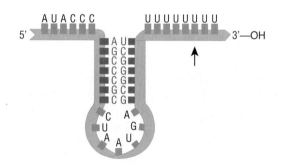

FIGURE 9.11 The terminal region of the *trp* attenuator sequence. The arrow indicates the final uridylate in attenuated RNA. Nonattenuated RNA continues past this point. The nucleotides in red letters form a stem by base-pairing within the RNA.

with termination occurring at the run of uridylates that follows region 4. Part C shows what happens when the ribosome stalls at the Trp codons as a result of insufficient tRNA^{Trp}. In this case, region 2 preferentially pairs with region 3, which disrupts the conformation of the terminator, allowing transcription to continue through the rest of the operon. The fine-tuning of this system takes place at intermediate concentrations of tryptophan, when the fraction of nascent transcripts that are completed depends on how frequently translation is stalled, which in turn depends on the intracellular concentration of charged tryptophan tRNA.

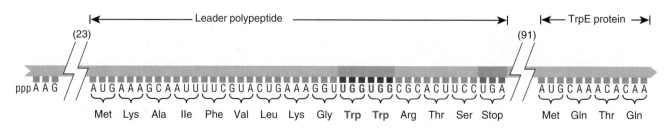

FIGURE 9.12 The sequence of bases in the *trp* leader mRNA, showing the leader polypeptide, the two tryptophan colors (red), and the beginning of the TrpE protein. The numbers 23 and 91 are the numbers of bases in the sequence that, for clarity, are not shown.

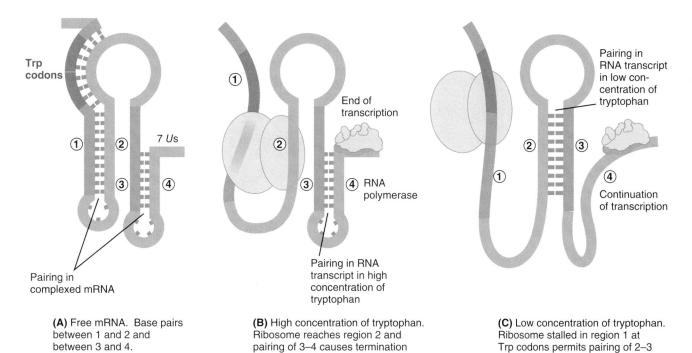

(A) Free mRNA. Base pairs between 1 and 2 and between 3 and 4.

(B) High concentration of tryptophan. Ribosome reaches region 2 and pairing of 3–4 causes termination of transcription.

(C) Low concentration of tryptophan. Ribosome stalled in region 1 at Trp codons permits pairing of 2–3 and transcription is not terminated.

FIGURE 9.13 The mechanism of attenuation in the *E. coli trp* operon. The tryptophan codons are highlighted in red.

In summary, attenuation is a fine-tuning mechanism of regulation superimposed on the basic negative control of the *trp* operon:

KEY CONCEPT

When charged tryptophan tRNA is present in amounts that support translation of the leader polypeptide, transcription is terminated, and the *trp* enzymes are not synthesized. When the level of charged tryptophan tRNA is too low, transcription is not terminated, and the *trp* enzymes are made. At intermediate concentrations, the fraction of transcription initiation events that result in completion of *trp* mRNA depends on how frequently translation is stalled, which in turn depends on the intracellular concentration of charged tryptophan tRNA.

Many operons responsible for amino acid biosynthesis (for example, the leucine, isoleucine, phenylalanine, and histidine operons) are regulated by attenuators that function by forming alternative paired regions in the transcript. In the histidine operon, the coding region for the leader polypeptide contains seven adjacent histidine codons, and in the phenylalanine operon, the coding region for the leader polypeptide contains seven phenylalanine codons. This pattern is characteristic of operons in which attenuation is coupled with translation. Through translation of these leader polypeptides, the cell monitors the level of amino-acylated tRNA charged with the amino acid

that is the end product of each amino acid biosynthetic pathway. Note that:

KEY CONCEPT

Attenuation cannot take place in eukaryotes because transcription and translation are uncoupled; transcription takes place in the nucleus and translation in the cytoplasm.

STOP & THINK 9.4

How would attenuation be affected by a mutation in the tryptophan operon in which the two codons for tryptophan in the *trp* leader RNA were deleted?

Riboswitches combine with small molecules to control transcriptional termination.

Transcription termination can also be triggered by direct binding of a small molecule to a 5′ untranslated leader mRNA. The mechanism is that the 5′ leader is able to adopt either of two conformations according to whether it binds with the small molecule. In the *antiterminator* conformation, transcription of the gene continues past the leader and through the remaining part of the gene. In the *terminator* conformation, which is triggered by binding with the small molecule,

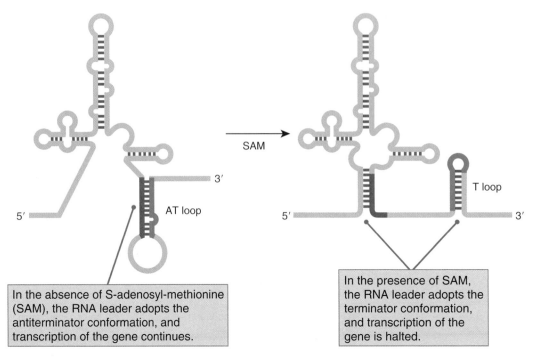

FIGURE 9.14 Riboswitch regulation of transcription termination by the *yitJ* leader RNA in *Bacillus subtilis*. The presence of S-adenosylmethionine (SAM) results in conversion from the read-through antiterminator form to the T-loop terminator form.

Pairing data from B. A. M. McDaniel, et al., *Proc. Natl. Acad. Sci. USA* 100 (2003): 3083.

transcription is terminated. An RNA leader sequence able to switch between an antiterminator conformation and a terminator conformation is known as a **riboswitch**. Comparison of genomic sequences indicates that riboswitches are present in archaea, eubacteria, and eukarya.

Riboswitches have been described that regulate synthesis or transport of many small molecules. As a specific example, **FIGURE 9.14** depicts the leader mRNA of the *yitJ* gene, which is involved in methionine biosynthesis, in *Bacillus subtilis*. The RNA regions shown in red and blue can undergo two pairing configurations. In the absence of S-adenosylmethionine (SAM), a modified form of methionine, the red region pairs with the blue, and this so-called antiterminator conformation allows transcription to continue. In the presence of SAM, the blue segment pairs with sequences nearby forming a hairpin that terminates in a string of uridylate residues, and this conformation is a transcriptional terminator.

9.4 Eukaryotes regulate transcription through transcriptional activator proteins, enhancers, and silencers.

Many eukaryotic genes are **housekeeping genes** that encode essential metabolic enzymes or cellular components and are expressed constitutively at relatively low levels in all cells. Other genes differ in their expression according to cell type or stage of the cell cycle. These genes are often regulated at the level of transcription. Typically, levels of expression of eukaryotic genes may differ 2- to 10-fold between the uninduced and induced levels. This contrasts with the more dramatic differences seen in prokaryotes, in which the ratio between the uninduced and induced levels may be as great as 1000-fold.

Galactose metabolism in yeast illustrates transcriptional regulation.

To introduce transcriptional regulation in eukaryotes, we first examine the control of galactose metabolism in yeast and compare it with the *lac* operon in *E. coli*. The first steps in the biochemical pathway for galactose degradation are illustrated in **FIGURE 9.15**. Three enzymes, encoded by the genes *GAL1*, *GAL7*, and *GAL10*, are required for conversion of galactose into glucose-1-phosphate. These three structural genes are tightly linked genetically, as shown in **FIGURE 9.16**. Despite the tight linkage of the three genes, the genes are not part of an operon. The mRNAs are monocistronic. The *GAL1* and *GAL10* mRNAs are synthesized from divergent promoters lying between the genes, and *GAL7* mRNA is synthesized from its own promoter. On the other hand, the genes are inducible because the mRNAs are synthesized only when galactose is present.

In *GAL* gene regulation, the key players are the products of the genes *GAL3*, *GAL80*, and *GAL4*, which in

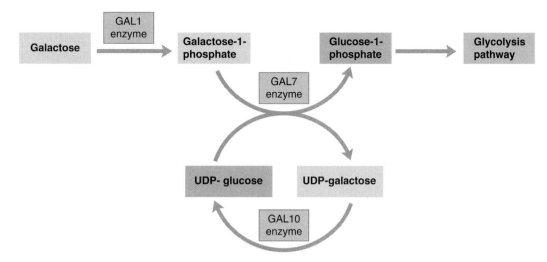

FIGURE 9.15   Metabolic pathway by which galactose is converted to glucose-1-phosphate in the yeast *Saccharomyces cerevisiae*.

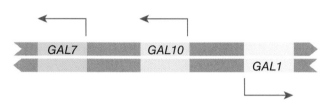

FIGURE 9.16   The linked *GAL* genes of *S. cerevisiae*. Arrows indicate the transcripts produced. The *GAL1* and *GAL10* transcripts come from divergent promoters, *GAL7* from its own promoter.

the conventions of yeast genetics are denoted GAL3p, GAL80p, and GAL4p, respectively (**FIGURE 9.17**). GAL4p is required for transcription of all three *GAL* genes. It is a positive regulatory protein that activates transcription of the three *GAL* genes individually, starting at a different site upstream from each gene. Although *GAL80* mutants, which in the terminology of yeast genetics are denoted *gal80*, superficially resemble *lacI* "repressor" mutants because *gal80/gal80* homozygous mutants produce the *GAL* enzymes constitutively, the GAL80p protein is not a repressor that binds with DNA. Rather, GAL80p has two binding sites, one for GAL4p (the transcriptional activator) and one for GAL3p.

In the mechanism of *GAL* regulation outlined in Figure 9.17, part A, GAL3p binds with galactose and ATP, and, in this state, GAL3p can bind with GAL80p and hold it in the cytoplasm. Inside the nucleus, the GAL4p protein attaches through one of its binding sites with an upstream activator sequence (UAS) located near each of the *GAL* genes. Another binding site on GAL4p recruits the transcriptional machinery, and the *GAL* genes are transcribed.

In the absence of galactose (Figure 9.17, part B), the GAL3p protein cannot bind with GAL80p. The GAL80p protein is, therefore, free to enter the nucleus. Inside the nucleus, GAL80p binds with the transcriptional activator site in GAL4p and in doing so prevents

recruitment of the transcription complex. The binding of GAL4p by GAL80p thereby prevents the *GAL* genes from being transcribed in the absence of galactose.

In the presence of galactose (Figure 9.17A), GAL4p binds with UAS sequences in the DNA. The structural nature of the GAL4p–UAS binding is shown in **FIGURE 9.18**. The GAL4p, which binds as a dimer, is shown in blue and the DNA molecule in red. The small yellow spheres represent ions of zinc, which are essential components in the DNA binding.

Transcription is stimulated by transcriptional activator proteins.

The GAL4 protein is an example of a *transcriptional activator protein*, which must bind with an upstream DNA sequence in order to prepare a gene for transcription. Some transcriptional activator proteins work by direct interaction with one or more components in the transcription complex, and in this way they recruit the transcription complex to the promoter of the gene to be activated. Other transcriptional activator proteins may initiate transcription by an already assembled transcription complex. In either case, the activator proteins are essential for the transcription of genes that are positively regulated.

Many transcriptional activator proteins can be grouped into categories on the basis of characteristics shared by their amino acid sequences. For example, one category has a *helix–turn–helix* motif, which consists of a sequence of amino acids forming a pair of α-helices separated by a bend; the helices are so situated that they can fit neatly into the grooves of a double-stranded DNA molecule. The helix–turn–helix motif is the basis of the DNA-binding ability, although the sequence specificity of the binding results from other parts of the protein.

A second large category of transcriptional activator proteins includes a DNA-binding motif called a

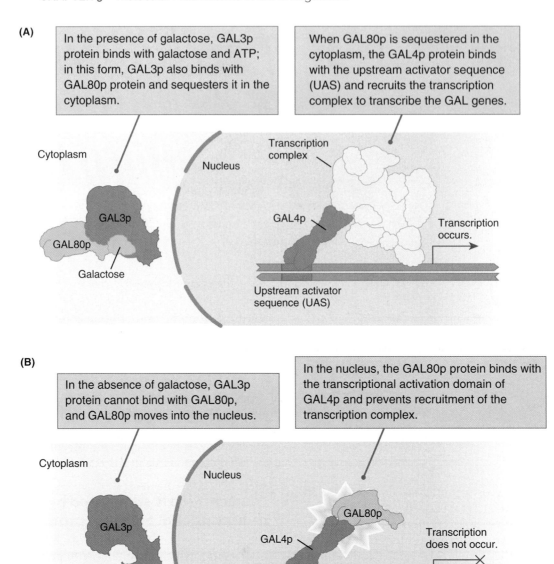

(A)

In the presence of galactose, GAL3p protein binds with galactose and ATP; in this form, GAL3p also binds with GAL80p protein and sequesters it in the cytoplasm.

When GAL80p is sequestered in the cytoplasm, the GAL4p protein binds with the upstream activator sequence (UAS) and recruits the transcription complex to transcribe the GAL genes.

Cytoplasm

Nucleus

Transcription complex

GAL3p

GAL80p

GAL4p

Galactose

Transcription occurs.

Upstream activator sequence (UAS)

(B)

In the absence of galactose, GAL3p protein cannot bind with GAL80p, and GAL80p moves into the nucleus.

In the nucleus, the GAL80p protein binds with the transcriptional activation domain of GAL4p and prevents recruitment of the transcription complex.

Cytoplasm

Nucleus

GAL3p

GAL80p

GAL4p

Transcription does not occur.

Upstream activator sequence (UAS)

FIGURE 9.17 Regulation of transcription of the *GAL* genes by the proteins encoded in *GAL3* (GAL3p), *GAL80* (GAL80p), and *GAL4* (GAL4p).

zinc finger because the folded structure incorporates a zinc ion. An already familiar example is the GAL4 transcriptional activator protein in yeast (Figure 9.18), in which the zinc ions at the extreme ends are shown in yellow. The DNA sequence recognized by the GAL4 protein is a symmetrical sequence, 17 base pairs in length, which includes a CCG triplet at each end that makes direct contact with the zinc-containing domains.

Enhancers increase transcription; silencers decrease transcription.

Some transcriptional activator proteins bind with particular DNA sequences known as **enhancers**. Enhancer sequences are typically rather short (usually fewer than 20 base pairs) and are found at a variety of locations around the gene they regulate. Most enhancers are upstream of the transcriptional start site (sometimes many kilobases away), others are in introns within the coding region, and a few are even located at the 3' end of the gene. They are able to function as enhancers irrespective of their orientation; hence, an enhancer sequence can be in either the transcribed strand or the nontranscribed strand. One of the most thoroughly studied enhancers is in the mouse mammary tumor virus and determines transcriptional activation by the glucocorticoid steroid hormone. The enhancer binds to a specific sequence of eight base pairs that is present at five different sites in the viral genome (**FIGURE 9.19**), providing five binding sites for the hormone–receptor complex that activates transcription.

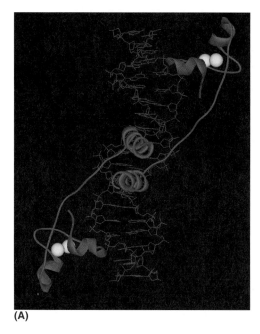
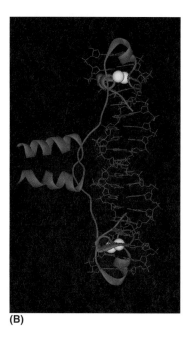

(A)   (B)

FIGURE 9.18   Three-dimensional structure of the GAL4 protein (blue) bound to DNA (red). The protein is composed of two polypeptide subunits held together by the coiled regions in the middle. The DNA-binding domains are at the extreme ends, and each physically contacts three base pairs in the major groove of the DNA. The zinc ions in the DNA-binding domains are shown in yellow. The views in (A) and (B) are at right angles.

(A) and (B) Protein Data Bank 1D66.

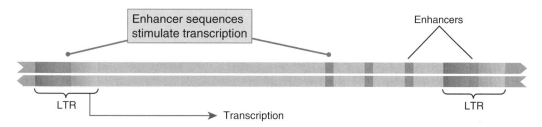

Enhancer sequences stimulate transcription

Enhancers

LTR

Transcription

LTR

FIGURE 9.19   Positions, in the mouse mammary tumor virus, of enhancers (orange) that allow transcription of the viral sequence to be induced by glucocorticoid steroid hormone. LTR stands for the long terminal repeated sequences found at the extreme ends of the virus.

Enhancers are essential components of gene organization in eukaryotes because they enable genes to be transcribed only when proper transcriptional activators are present. Some enhancers respond to molecules outside the cell—for example, steroid hormones that form receptor–hormone complexes. Other enhancers respond to molecules that are produced inside the cell (for example, during development); these enhancers enable the genes under their control to participate in cellular differentiation or to be expressed in a tissue-specific manner. Many genes are under the control of several different enhancers, so they can respond to a variety of different molecular signals, both external and internal.

Some genes are also subjected to regulation by transcriptional **silencers**, which are short nucleotide sequences that are targets for DNA-binding proteins that, once recruited to the site, promote the assembly of large protein complexes that prevent transcription of the silenced genes. Examples of such silencing

complexes include the set of *Drosophila* proteins called the PcG (Polycomb group) proteins, which silence certain genes during development.

Genome architecture consists of compact domains of associating DNA molecules.

Enhancers and silencers have the unusual properties of acting irrespective of their orientation and often at great distances from their target genes. If the genome were organized as a randomly convoluted jumble of chromatin, then nothing would prevent enhancers or silencers that evolved to regulate a given gene from affecting unrelated genes that just happen to be nearby. But genomes are not randomly convoluted jumbles of chromatin. Certain regions of the genome are physically brought into contact according to the type of tissue the cells are in. Regions of a DNA molecule may be associated for any number of reasons (**FIGURE 9.20**).

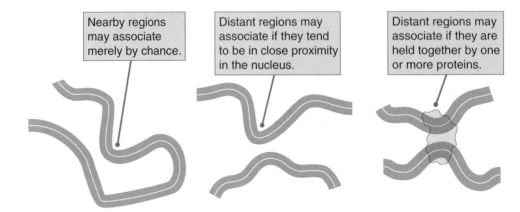

FIGURE 9.20 Different regions of a DNA molecule may be associated for any of several reasons.

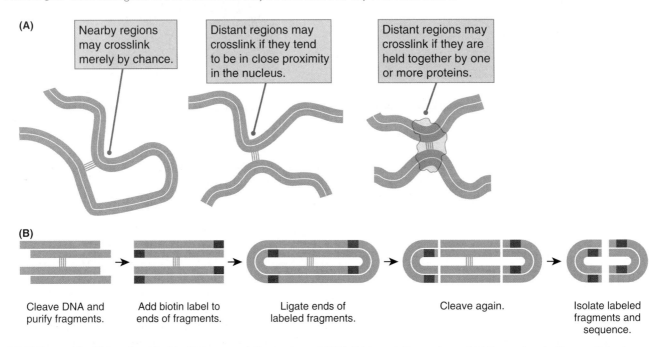

(A)

(B) Cleave DNA and purify fragments. → Add biotin label to ends of fragments. → Ligate ends of labeled fragments. → Cleave again. → Isolate labeled fragments and sequence.

FIGURE 9.21 The HiC method for identifying associating regions of DNA. (A) Associating regions of DNA are chemically crosslinked. (B) Steps in processing the crosslinked DNA to identify which regions of DNA are associating.

If the regions are physically close, they may bump into each other purely by chance. More distant regions may come into contact if they are regularly included as part of a discrete, reproducible, folded configuration. Or they might be part of a convoluted loop formed by two distant sequences being physically held together by one or more proteins.

An understanding of such higher-order genome architecture has only recently become clear owing to the development of methods to detect associated regions of DNA molecules in the nucleus by means of their sequences. The identification of associated DNA regions can be carried out in several ways that differ in technical detail and level of resolution, but here we will focus on two methods currently in use. One method, identified with the acronym *GAM* (genome architecture mapping), is conceptually simple but technically difficult. The idea is to cut very thin, randomly oriented

slices through a nucleus and to isolate and sequence the DNA molecules present in each slice. The rationale is that any two regions of DNA that consistently are closely associated in the nucleus will be found together in a thin slice significantly more often than would two regions of DNA that are not so associated.

Another method, known as *HiC* (**FIGURE 9.21**), relies on chemical crosslinking and DNA sequencing to identify DNA regions that are associated. In the first step of HiC, the DNA is crosslinked by means of a small, reactive molecule like formaldehyde (Figure 9.21, part A). The crosslinked molecules are then processed as indicated in Figure 9.21, part B. When the DNA is cleaved with one or more restriction enzymes, the associated regions remain together because of the crosslinks. After cleavage, the overhanging, single-stranded ends of the cleaved fragments are elongated with nucleotides in which the vitamin biotin is attached to the bases. The now double-stranded ends of each crosslinked fragment are ligated, and the DNA molecules are cleaved again. The fragments containing biotin are isolated using the bacterial protein streptavidin, which has a high affinity for biotin. Sequencing these biotin-labeled molecules reveals which regions of DNA were associated with one another at the time of the original crosslinking.

In presenting the results of a HiC experiment, the DNA molecule can be aligned along the *x* and *y* axes of a graph as shown in **FIGURE 9.22**. This example shows 2 Mb of a DNA molecule with the nucleotides numbered arbitrarily from 0 Mb to 2 Mb. Each little box corresponds to a pair of regions along the DNA molecule, and the intensity of color of each box depicts the frequency with which the DNA regions are found to be associated. The boxes along the diagonal are strongly colored, which means that each region of DNA is always associated with itself. The intensity of color in the off-diagonal boxes shows the strength of association between two different regions of the DNA molecule. The matrix of colors is symmetrical around the diagonal because the association of regions *i* and *j* must be the same as that between regions *j* and *i*. (Because of the symmetry, and to save space, HiC results in scientific papers are often depicted as a triangle by showing only those blocks constituting the diagonal and above.)

The key finding from HiC experiments consists of the large, square blocks of color along the diagonal. Each such block indicates a region of DNA in which there are physical associations between relatively distant sites. Each block of associating DNA regions is called a **topologically associating domain** or **TAD**. The size of TADs differs from one to the next and also differs among organisms. In mammals, TADs range in size from 0.1–2 Mb, whereas in *Drosophila* they range in size from 10–100 kb. TADs are also tissue specific, which means that the regions of DNA that are in close association differ from one type of tissue to the next.

Figure 9.22 shows two major types of topologically associated domain. One type is a *loop domain*, which has a distinctive boundary that in mammals is characterized by the presence of two proteins, namely CTCF (stands for CCCTC-binding factor) and cohesion (the same protein that holds sister chromatids together during cell division). The other type of TAD is an *ordinary domain*, the boundaries of which lack CTCF and cohesion.

TADs are important because their boundaries serve as **insulators** that prevent enhancers and silencers in one domain from affecting the expression of genes in a different domain. The chromatin in some TADs is transcriptionally active and includes all the markers of gene activity, such as sparse nucleosomes and modified histone tails associated with gene expression. In other TADs, the chromatin is repressed with more dense nucleosomes and histone modifications associated with gene silencing.

TADs are organized into nuclear **compartments** according to their transcriptional activity. TADs with active genes tend to be located toward the center of the nucleus, and those with inactive genes tend to be located toward the periphery. One compartment of inactive TADs consists of aggregates of heterochromatin. Another compartment of inactive TADs is associated with the nuclear lamina, a dense network of fibers near the inner nuclear membrane.

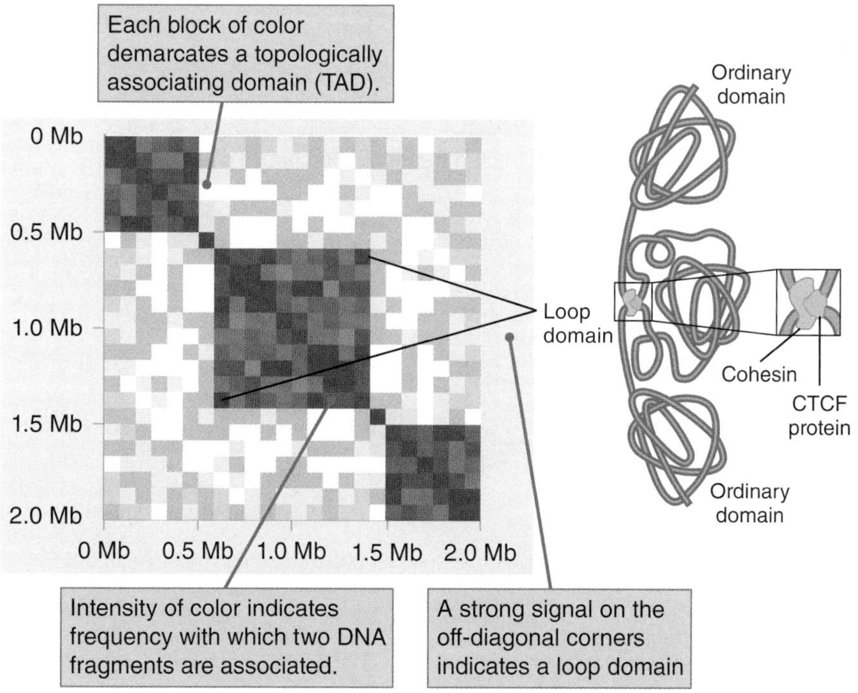

The eukaryotic transcription complex includes numerous protein factors.

The eukaryotic **transcription complex** is an aggregate of protein factors that combines with the promoter region of a gene to initiate transcription. The factors necessary for transcription include a transcriptional activator protein that interacts with at least one protein subunit of the transcription complex to recruit the transcription complex to the gene. Many enhancers activate transcription by means of **DNA looping**, which refers to physical interactions between relatively distant regions along the DNA. The mechanism is illustrated in **FIGURE 9.23**.

Each block of color demarcates a topologically associating domain (TAD).

Intensity of color indicates frequency with which two DNA fragments are associated.

A strong signal on the off-diagonal corners indicates a loop domain

Loop domain

Ordinary domain

Cohesin

CTCF protein

Ordinary domain

FIGURE 9.22 Result of a HiC experiment showing associating regions across 2 Mb of a DNA molecule. More intense colors indicate higher frequencies of association.

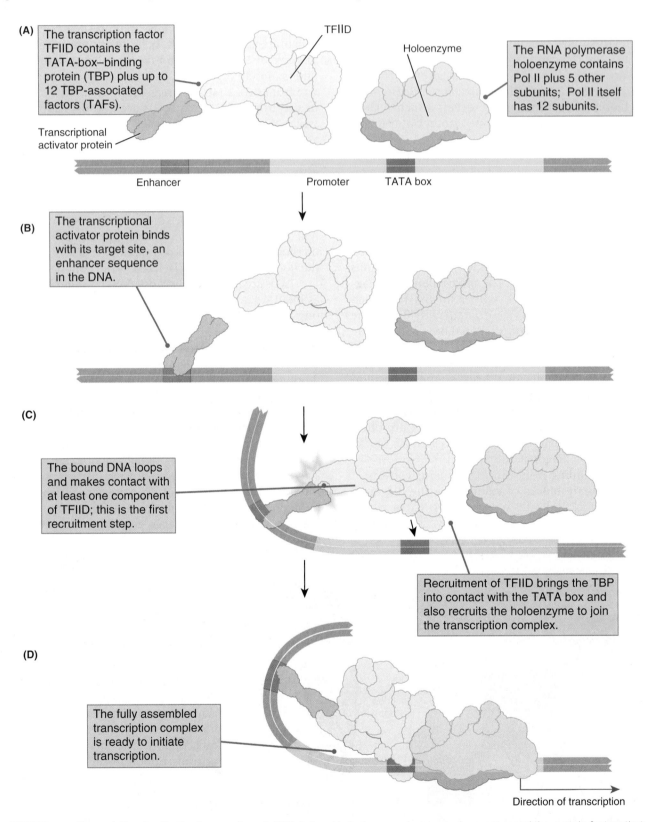

(A) The transcription factor TFIID contains the TATA-box–binding protein (TBP) plus up to 12 TBP-associated factors (TAFs).

TFIID

Holoenzyme

The RNA polymerase holoenzyme contains Pol II plus 5 other subunits; Pol II itself has 12 subunits.

Transcriptional activator protein

Enhancer   Promoter   TATA box

(B) The transcriptional activator protein binds with its target site, an enhancer sequence in the DNA.

(C) The bound DNA loops and makes contact with at least one component of TFIID; this is the first recruitment step.

Recruitment of TFIID brings the TBP into contact with the TATA box and also recruits the holoenzyme to join the transcription complex.

(D) The fully assembled transcription complex is ready to initiate transcription.

Direction of transcription

FIGURE 9.23 Transcriptional activation by recruitment. (A) Relationship between enhancer and promoter and the protein factors that bind to them. (B) Binding of the transcriptional activator protein to the enhancer. (C) Bound transcriptional activator protein makes physical contact with a subunit in the TFIID complex, which contains the TATA-box–binding protein, and attracts ("recruits") the complex to the promoter region. (D) The Pol II holoenzyme and any remaining general transcription factors are recruited by TFIID, and the transcription complex is fully assembled and ready for transcription. In the cell, not all of the Pol II is found in the holoenzyme, and not all of the TBP is found in TFIID. In this illustration, transcription factors other than those associated with TFIID and the holoenzyme are not shown.

The **basal transcription factors**, or **general transcription factors**, are proteins in the transcription complex that are used widely in the transcription of many different genes. The basal transcription factors in eukaryotes have been highly conserved in evolution. A minimal set necessary for accurate transcription *in vitro* includes TFIIB, TFIID, TFIIE, TFIIF, TFIIH, and Pol II. (TF in these designations stands for *transcription factor*.) These components can assemble *in vitro* in stepwise fashion on a promoter. The first step is recruitment of TFIID, itself a complex of proteins that includes a **TATA-box–binding protein (TBP)**, which binds with the promoter in the region of the TATA box, and about 10 other proteins, called **TBP-associated factors (TAFs)**, which are the components that respond specifically to activator proteins. The TBP binds to the DNA in the minor groove and then bends the DNA by about 80°. The Pol II RNA polymerase is also found in a complex with multiple protein subunits called the **Pol II holoenzyme.**

It is not yet clear whether the transcription complex is recruited to the promoter and assembled stepwise, as *in vitro* studies suggest, or recruited in the form of one or more large, preassembled complexes, which the composition of the Pol II holoenzyme suggests is the case. For simplicity, Figure 9.23 shows recruitment of one preassembled complex that includes TFIID, which in turn recruits the preassembled Pol II holoenzyme. To activate transcription (part B), the transcriptional activator protein binds to an enhancer in the DNA and to one of the TAF subunits in the TFIID complex. This interaction attracts ("recruits") the TFIID complex to the region of the promoter (part C). Attraction of the TFIID to the promoter also recruits the Pol II holoenzyme (part D), as well as any remaining general transcription factors. Once these components are brought together, the transcriptional complex is ready for transcription to begin.

As Figure 9.23 suggests, the fully assembled transcription complex in eukaryotes is a very large structure. A real example, taken from early development in *Drosophila*, is shown in **FIGURE 9.24**. In this case, the enhancers, located a considerable distance upstream from the gene to be activated, are bound by the transcriptional activator proteins BCD and HB, which are products of the genes *bicoid (bcd)* and *hunchback (hb)*, respectively; these transcriptional activators function in establishing the anterior–posterior axis in the embryo. Note the position of the TATA box in the promoter of the gene. The TATA box binding is the function of the TBP. The functions of a number of other components of the transcription complex have also been identified. For example, the TFIIH contains both helicase and kinase activity to separate the DNA strands and to phosphorylate RNA polymerase II.

Phosphorylation allows the polymerase to leave the promoter and elongate mRNA. The looping of the DNA effected by the transcriptional activators is an essential feature of the activation process. Transcriptional activation in eukaryotes is a complex process, especially when compared to the prokaryotic RNA polymerase, which consists of only six polypeptide chains.

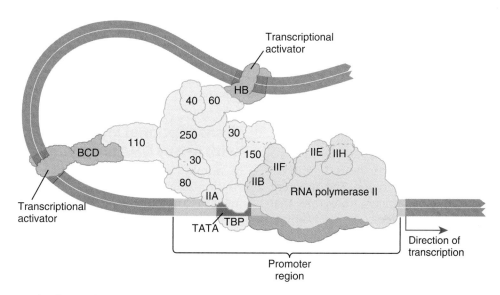

FIGURE 9.24 An example of transcriptional activation during *Drosophila* development. The transcriptional activators in this example are bicoid protein (BCD) and hunchback protein (HB). The numbered subunits are TAFs (TBP-associated factors) that, together with TBP (TATA-box–binding protein), correspond to TFIID. BCD acts through a 110-kilodalton TAF, and HB through a 60-kilodalton TAF. The transcriptional activators act via enhancers to cause recruitment of the transcriptional apparatus. The fully assembled transcription complex includes TBP and TAFs, RNA polymerase II, and general transcription factors TFIIA, TFIIB, TFIIE, TFIIF, and TFIIH.

Chromatin-remodeling complexes prepare chromatin for transcription.

Eukaryotic DNA is typically found in the form of chromatin packaged with nucleosomes. Special mechanisms are required for transcriptional activator proteins and the transcription complex to acquire access to the DNA, including the chemical modification of histone tails. The existence of such mechanisms is implied by the observation that the components of transcription sufficient to transcribe purified DNA *in vitro* are unable to initiate transcription of purified chromatin. The nucleosomes in chromatin must prevent the transcription complex from either binding to DNA or using it as a template.

Several different multiprotein complexes have been identified that can restructure chromatin and enable it to be transcribed. These are known as **chromatin-remodeling complexes (CRCs)**. All of these complexes use energy derived from ATP to restructure chromatin. The molecular mechanism of chromatin remodeling is unknown; because there are several distinct types of CRCs, there may be several mechanisms. In one general class of models, the CRC disrupts nucleosome structure without displacing the nucleosomes, rendering the DNA accessible to transcriptional activator proteins, the TATA-box–binding protein, and other components of the transcription complex. In another general class of models, the CRC repositions the nucleosomes along the DNA, making key DNA-binding sites accessible. An example is illustrated in **FIGURE 9.25**. Part A shows a transcriptionally inactive chromatin conformation, with the DNA-binding sites for a transcriptional activator protein (TAP) and TATA-box–binding protein (TBP) sequestered in nucleosomes and unavailable. Recruitment of a CRC to the site results in repositioning of the nucleosomes (part B), which renders the binding sites accessible (part C). In this chromatin configuration, TAP and TBP can bind with the DNA and recruit the rest of the transcription complex.

Some eukaryotic genes have alternative promoters.

Some eukaryotic genes have two or more promoters that are active in different cell types. The different promoters result in different primary transcripts that contain the same protein-coding regions. An example from *Drosophila* is shown in **FIGURE 9.26**. The gene code for alcohol dehydrogenase, and its organization in the genome, shown in part A, includes three protein-coding regions interrupted by two introns. Transcription in larvae (part B) uses a different promoter from that used in transcription in adults (part C). The adult transcript has a longer 5′ leader sequence, but most of this sequence is eliminated in RNA splicing. Alternative promoters make possible the independent regulation of transcription in larvae and adults.

9.5 Gene expression can be affected by heritable chemical modifications in the DNA.

In this section we discuss some examples of *epigenetic* regulation of gene activity. The prefix *epi* means "besides" or "in addition to"; **epigenetic** therefore refers to heritable changes in gene expression that are due not to changes in the DNA sequence itself, but to something "in addition to" the DNA sequence, usually either chemical modification of the bases, or protein factors bound with the DNA. We shall see that there is a great deal yet to be learned about the molecular mechanisms by which epigenetic modifications are established and maintained.

Cytosine   5-methylcytosine

In most higher eukaryotes, a proportion of the cytosine bases are modified by the addition of a methyl (CH_3) group to the number-5 carbon atom. The cytosines are incorporated in their normal, unmodified form in the course of DNA replication, and then the methyl group is added by an enzyme called **DNA methylase**. In mammals, cytosines are modified preferentially in 5′-CG-3′ dinucleotides. Many mammalian genes have CG-rich regions upstream of the coding region that provide multiple sites for methylation; these are called *CpG islands*, where the "p" represents the phosphate group in the polynucleotide backbone.

Transcriptional inactivation is associated with heavy DNA methylation.

A number of observations suggest that heavy methylation is associated with genes for which the rate of transcription is low. One example is the inactive X chromosome in mammalian cells, which is extensively methylated. In fact, in adult mammals, the majority of CpG dinucleotides in all chromosomes are methylated in somatic cells. The unmethylated CpGs are usually associated with the promoters of active housekeeping genes. The widespread methylation of inactive genes in adult somatic cells is thought to minimize accidental, low-level transcription from them.

Although there is a very strong correlation between heavy methylation and transcriptional silencing, heavy methylation may result from an earlier epigenetic

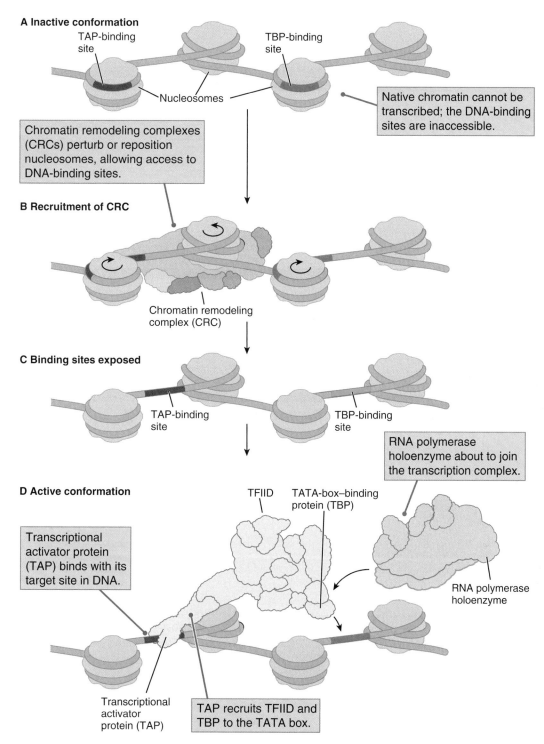

A Inactive conformation

TAP-binding site

TBP-binding site

Nucleosomes

Native chromatin cannot be transcribed; the DNA-binding sites are inaccessible.

Chromatin remodeling complexes (CRCs) perturb or reposition nucleosomes, allowing access to DNA-binding sites.

B Recruitment of CRC

Chromatin remodeling complex (CRC)

C Binding sites exposed

TAP-binding site

TBP-binding site

RNA polymerase holoenzyme about to join the transcription complex.

D Active conformation

TFIID

TATA-box–binding protein (TBP)

Transcriptional activator protein (TAP) binds with its target site in DNA.

RNA polymerase holoenzyme

Transcriptional activator protein (TAP)

TAP recruits TFIID and TBP to the TATA box.

FIGURE 9.25 Function of chromatin-remodeling complexes. (A) Native chromatin may conceal key DNA-binding sites. (B) A chromatin-remodeling complex either repositions the nucleosomes along the DNA or chemically modifies the histones. (C) DNA-binding sites become accessible. (D) The transcription complex is recruited to the site.

signal that marks a gene for silencing and that recruits the methylase. If there is such an earlier signal, then it implies that methylation is the result of gene inactivity as well as its mechanism. In any case, treatment of cells with the cytosine analog *azacytidine* reverses methylation and can restore transcriptional activity.

For example, in cell culture, some lineages of rat pituitary tumor cells express the gene for prolactin, whereas other related lineages do not. The gene is methylated in the nonproducing cells but is not methylated in the producers. Reversal of methylation in the nonproducing cells via azacytidine results in prolactin expression.

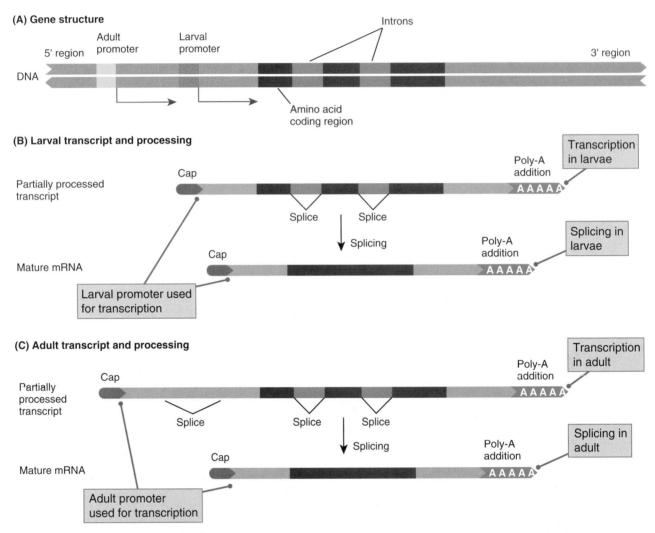

FIGURE 9.26 Use of alternative promoters in the gene for alcohol dehydrogenase in *Drosophila*. (A) The overall gene organization includes two introns within the amino acid coding region. (B) Transcription in larvae uses the promoter nearest the 5′ end of the coding region. (C) Transcription in adults uses a promoter farther upstream, and much of the larval leader sequence is removed by splicing.

In mammals, some genes are imprinted by methylation in the germ line.

Mammals feature an unusual type of epigenetic silencing known as **genomic imprinting**, a process with the following characteristics:

- Imprinting occurs in the germ line.
- It affects at most a few hundred genes (many of them located in clusters).
- It is accompanied by heavy methylation (though the primary signal for imprinting is unknown).
- Imprinted genes are differentially methylated in the female and male germ lines.
- Once imprinted and methylated, a silenced gene remains transcriptionally inactive during embryogenesis.
- Imprints are erased early in germ-line development, then later reestablished according to sex-specific patterns.

Although mammalian gametes are extensively methylated, most of the DNA is demethylated in pre-implantation development, except for imprinted genes that retain their sex-specific patterns of methylation. The embryonic DNA is remethylated beginning after implantation, gradually attaining the heavy methylation levels found in adult somatic cells. In the germ line, the original imprints are erased when the DNA is globally demethylated, and remethylation takes place later in germ-line development. All remethylated genes acquire identical patterns of methylation in the germ line of both sexes, except for those few that have sex-specific patterns of imprinting and differential methylation. The imprinted genes undergo methylation during oocyte growth prior to ovulation in females, and probably around the time of birth in males. Because the methylation associated with imprinting is retained throughout embryonic development, any gene that is imprinted in either the female or the male germ line has, effectively, only one active copy in the embryo.

The epigenetic, sex-specific gene silencing associated with imprinting is dramatically evident in a pair of syndromes characterized by neuromuscular defects, mental retardation, and other abnormalities. These are *Prader–Willi syndrome* and *Angelman syndrome*. Both conditions are associated with rare, spontaneous deletions that include chromosomal region *15q11*. If the deletion takes place in the father, the result is Prader–Willi syndrome, whereas if it takes place in the mother, the result is Angelman syndrome. The reason is that *15q11* includes at least three genes (*SNRPN*, *necdin*, and *UBE3A*) that are imprinted and differentially methylated in the gametes. Part A of **FIGURE 9.27** shows the pattern of imprinting of these three genes in a normal embryo. *SNRPN* and *necdin* are imprinted in the egg, *UBE3A* in the sperm. In the embryo, therefore, *UBE3A* is transcriptionally active in the maternal chromosome, and *SNRPN* and *necdin* in the paternal chromosome. In the germ line of female and male embryos, shown in Figure 9.27, part B, the imprints are erased and reset according to sex: In the female both homologs have *SNRPN* and *necdin* imprinted, whereas in the male both homologs have *UBE3A* imprinted. If a normal, imprinted female gamete is fertilized by a sperm with a *15q11*

deletion, the embryo has no transcriptionally active copy of either *SNRPN* or *necdin* and develops Prader–Willi syndrome. On the other hand, if a normal, imprinted male gamete fertilizes an egg with a *15q11* deletion, the embryo has no transcriptionally active copy of *UBE3A* and develops Angelman syndrome. These syndromes demonstrate not only the epigenetic control of gene expression by imprinting but also differential imprinting in the sexes and the clustering of imprinted genes in the genome.

Why is there imprinting? One suggestion is that it evolved in early mammals with polyandry (each female mating with a series of males). In such a situation, it is to a male's benefit to silence genes that conserve maternal resources at the expense of the fetus, because this strategy maximizes the father's immediate reproduction. But it is to a female's benefit to silence genes that allocate resources to the fetus at the expense of the mother, because this strategy maximizes the female's long-term reproduction. This hypothesis is supported by the fact that some imprinted genes do affect the allocation of resources between mother and fetus in the direction that would be predicted. On the other hand, many genes that are imprinted have no obvious connection to maternal–fetal conflict.

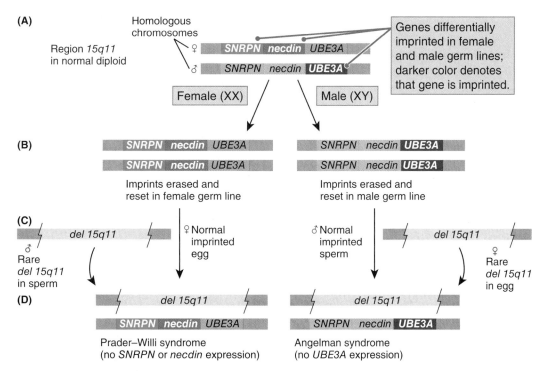

FIGURE 9.27 Imprinting of genes in chromosomal region *15q11* results in different neuromuscular syndromes, depending on which parent contributes a *15q11* deletion and which parent contributes an imprinted chromosome. (A) Pattern of imprinting in a normal diploid. The maternal chromosome is at the top, the paternal chromosome at the bottom. Imprinted and transcriptionally inactive genes are indicated. (B) In the germ line, the imprints are erased and reset in either female-specific or male-specific patterns. (C) An individual who inherits a maternally imprinted chromosome along with a *15q11* deletion has Prader–Willi syndrome, whereas one who inherits a paternally imprinted chromosome along with a *15q11* deletion has Angelman syndrome. Other genes in the region, not shown, may also be imprinted.

9.6 Regulation also takes place at the levels of RNA processing and decay.

Although transcriptional control of gene expression is of major importance, transcription is by no means the only level at which gene activity can be regulated. In this section we consider some mechanisms that act at the level of primary-transcript splicing or at the level of mRNA stability.

The primary transcripts of many genes are alternatively spliced to yield different products.

Even when the same promoter is used to transcribe a gene, different cell types can produce different quantities of the protein (or even different proteins) because of differences in the mRNA produced in processing. The reason is that the same transcript can be spliced differently from one cell type to the next. The different splicing patterns may include exactly the same protein-coding exons, in which case the protein is identical, but the rate of synthesis differs because the mRNA molecules are not translated with the same efficiency. In other cases, the protein-coding part of the transcript has a different splicing pattern in each cell type, and the resulting mRNA molecules code for proteins that are not identical even though they share certain exons. Transcripts in the human genome are frequently spliced in alternative ways; because of this, the approximately 25,000 human genes may encode 50,000 to 100,000 different proteins. Alternative RNA processing is one of the principal sources of human genetic complexity.

The insulin receptor gene in humans and other mammals provides an example of alternative splicing that results in the inclusion or exclusion of exon 11 in the messenger RNA. The resulting forms of the polypeptide chain differ in length by 12 amino acids. The relevant part of the primary transcript is shown in **FIGURE 9.28**. In the liver, all 20 exons are found in the mRNA for the long form of the receptor protein (part A), whereas in skeletal muscle exon 11 is eliminated along with the flanking introns and excluded from the mRNA for the short form (part B). The long form of the receptor shows low affinity for insulin and is expressed in tissues such as the liver that are exposed to relatively high concentrations of insulin. The short form of the

protein has a high affinity for insulin and is expressed preferentially in tissues such as skeletal muscle that are normally exposed to lower levels of insulin. Alternative splicing thus offers the possibility of generating proteins with different properties from the same gene.

The coding capacity of the human genome is enlarged by extensive alternative splicing.

Compared with genes in the worm or fly, human genes are spread over a larger region of the genome, and the primary transcripts are longer. Many human genes are alternatively spliced to yield multiple protein products. At least one-third of all human genes, and perhaps as many as two-thirds, are alternatively spliced. Among those that are alternatively spliced, the average number of distinct mRNAs produced from the primary transcript is in the range 2 to 7. The average number of different mRNAs per gene across the genome is in the range 2 to 3, which includes genes that produce a single mRNA as well as those that produce multiple different mRNAs. The alternative splicing greatly expands the number of protein products that can be encoded in a relatively small number of genes:

(A) Primary transcript

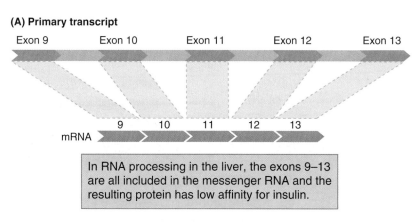

In RNA processing in the liver, the exons 9–13 are all included in the messenger RNA and the resulting protein has low affinity for insulin.

(B) Primary transcript

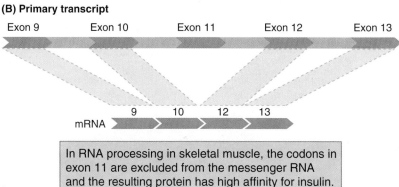

In RNA processing in skeletal muscle, the codons in exon 11 are excluded from the messenger RNA and the resulting protein has high affinity for insulin.

FIGURE 9.28 Alternative splicing of the primary transcript of the gene encoding the chain of the insulin receptor in humans and other mammals. (A) Splicing in the liver results in the low-affinity long form. (B) Splicing in skeletal muscle results in the high-affinity long form.

Different messenger RNAs can differ in their persistence in the cell.

A short-lived mRNA produces fewer protein molecules than a long-lived mRNA, so features that affect the rate of mRNA stability affect the level of gene expression. One route of degradation is the *deadenylation-dependent pathway*, which begins with enzymatic trimming of length of the poly-A tail on the mRNA. When the poly-A tail is trimmed to a length of 25 to 60 nucleotides, the mRNA becomes susceptible to a decapping enzyme that removes the 5′ cap and renders the molecule unable to initiate translation; from this state the mRNA is rapidly degraded by exonucleases. An alternative pathway is the *deadenylation-independent pathway*, which is initiated either with decapping or with endonuclease cleavage of the mRNA, after which digestion goes to completion by exonuclease activity. The deadenylation-independent pathway is particularly active for mRNAs that contain early chain termination codons or unspliced introns, and it prevents the accumulation of truncated polypeptides in the cell.

RNA interference results in the silencing of RNA transcripts.

In 1990 a group of plant geneticists reported experiments in which they manipulated genes for flower color in petunia (*Petunia hybrida*). The normal red or purple flower color in this plant results from a flavonoid pigment known as *anthocyanin*, which is synthesized via a metabolic pathway in which the rate-limiting step is catalyzed by the enzyme *chalcone synthase*. The investigators reasoned that an extra copy of this gene would increase the level of the enzyme and thus the amount of pigment, thereby yielding a darker flower color. In the actual experiment (FIGURE 9.29), the flower color of the genetically engineered plants was white! The total level of chalcone-synthase mRNA was about 50-fold lower in the engineered plants than that in control plants, and in crosses the reduced pigmentation segregated along with extra copy of the gene. Not only was the extra copy of the gene itself silent, but also its presence caused the silencing of the wild-type copies of the gene in the same plant. The mechanism of this unexpected gene silencing remained a

FIGURE 9.29   A bed of petunias with plants having wildtype flowers and those having white flowers. The white flowers are on plants with an extra copy of a gene for synthesis of the pigment anthocyanin, which was expected to make the flowers darker. Unexpectedly, the extra copy eliminated pigment altogether!

© Ivaschenko Roman/Shuttrerstock.

mystery until researchers discovered that the presence of double-stranded RNA (dsRNA) produced such silencing in the nematode worm *Caenorhabditis elegans*.

Gene silencing by dsRNA is an example of **RNA interference (RNAi)**. The ability to mount an RNAi response is widespread among eukaryotes and was probably present in the common ancestor, although this ability was lost in certain lineages including some fungi and parasitic protozoa. The molecular machinery of RNAi probably evolved originally as a defense against viruses and transposable elements that pass through a stage in which their genetic information is in the form of dsRNA. The silencing effect is highly specific and very potent, requiring only a few molecules of dsRNA per cell to be effective.

As might be expected of an RNAi response that evolved prior to the evolutionary diversification of eukaryotes, the mechanisms have been elaborated into several pathways that act somewhat differently, and organisms have made use of their components in multiple ways. Much is yet to be discovered and understood about the RNAi response, but some of the main outlines are clear. Two of the major pathways mediating an RNAi response are illustrated in FIGURE 9.30. Part A shows two major sources of dsRNA. That on the left is derived from transcription of the same duplex DNA molecule from both strands, as would happen if each strand had an upstream promoter. This is the type of dsDNA that would be produced by RNA viruses or certain transposable elements and that produced by the introduced chalcone-synthase gene in the petunias in Figure 9.29, in which the gene was inserted near a promoter that produced an *antisense transcript*, that is, a transcript from DNA strand that is not normally transcribed. When the antisense transcript formed an RNA duplex with the sense transcript, the resulting dsRNA

THE HUMAN CONNECTION

Double Trouble

Andrew Fire,[1] SiQun Xu,[1] Mary K. Montgomery,[1] Steven A. Kostas,[1] Samuel E. Driver,[2] and Craig C. Mello[2] (1998)

[1]*Carnegie Institution of Washington, Baltimore, Maryland; [2]University of Massachusetts Medical School, Worcester, Massachusetts.*

Potent and Specific Genetic Interference by Double-Stranded RNA in Caenorhabditis elegans

Weird and unexpected results began to be reported as soon as it became possible to introduce engineered RNA molecules into organisms. In extreme cases, the engineered RNA prevented the expression of endogenous host genes with sequence homology. At first, it seemed possible that the engineered RNA acted as an antisense inhibitor, in which the introduced RNA undergoes base pairing with the endogenous transcripts and interferes with their function. If this were true, the inhibitory effect of the introduced RNA should be strongly concentration dependent. In this path-breaking paper, the authors show that introduced double-stranded RNA (dsRNA) mediates the inhibitory effects, and that only a few molecules per cell are required. The nematode worm *C. elegans* proved to be ideal for these experiments because, in contrast to some other organisms, dsRNA can be transported from cell to cell and from parent to offspring.

> *66 To our surprise, we found that double-stranded RNA was substantially more effective at producing interference than was either strand individually. 99*

Experimental introduction of RNA into cells can be used in certain biological systems to interfere with the function of an endogenous gene. . . . Here we investigate the requirements for structure and delivery of the interfering RNA. To our surprise, we found that double-stranded RNA was substantially more effective at producing interference than was either strand individually. . . . Only a few molecules of injected double-stranded RNA were required per affected cell, . . . suggesting that there could be a catalytic or amplification component of the interference process. . . .

Fire and his colleagues looked more closely at this phenomenon, concentrating on the *unc-22* (uncoordinated-22) gene, loss-of-function mutations of which cause severe twitching in the worms. When they injected single-stranded RNA either identical or complementary to *unc-22* mRNA, only minimal interference was observed.

In contrast, a sense–antisense mixture produced highly effective interference with endogenous gene activity. The mixture was at least two orders of magnitude more effective than either strand alone. . . . The potent interfering activity of the sense–antisense mixture could reflect the formation of double-stranded RNA (dsRNA) or, conceivably, some other synergy between the strands. . . .

The phenotype induced by the introduced RNA was identical to that of conventional loss-of-function mutations of *unc-22*. They concluded by suggesting that RNA interference might be a more general phenomenon.

Double-stranded RNA could conceivably mediate interference more generally in other nematodes, in other invertebrates, and, potentially, in vertebrates. RNA interference might also operate in plants. . . . Genetic interference by dsRNA could be used by the organism for physiological gene silencing.

A. Fire, et al., *Nature* 391(1998): 806–810.

set off the petunia RNAi response. In these types of dsRNA the two strands are exactly (or almost exactly) matching, and the dsRNA is the source of **small interfering RNA (siRNA).**

On the right in Figure 9.30, part A is another source of dsRNA, in this case the stem of a stem-loop secondary structure formed in the transcript of a DNA duplex containing a duplicated sequence present in inverted

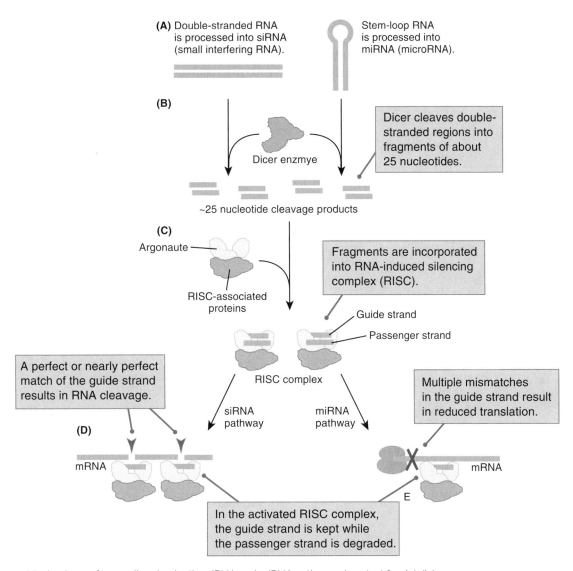

(A) Double-stranded RNA is processed into siRNA (small interfering RNA).

Stem-loop RNA is processed into miRNA (microRNA).

(B)

Dicer enzmye

Dicer cleaves double-stranded regions into fragments of about 25 nucleotides.

~25 nucleotide cleavage products

(C)

Argonaute

RISC-associated proteins

Fragments are incorporated into RNA-induced silencing complex (RISC).

Guide strand

Passenger strand

A perfect or nearly perfect match of the guide strand results in RNA cleavage.

RISC complex

Multiple mismatches in the guide strand result in reduced translation.

siRNA pathway

miRNA pathway

(D)

mRNA

mRNA

E

In the activated RISC complex, the guide strand is kept while the passenger strand is degraded.

FIGURE 9.30 Mechanisms of gene silencing by the siRNA and miRNA pathways (see text for details).

orientation. Structures like that shown are normally produced from longer transcripts by specialized enzymes present in the nucleus. The paired stem usually contains one or more base-pair mismatches, which accumulate in the inverted repeats as the genome evolves. These mismatches are the hallmark of another RNAi pathway mediated by **microRNA (miRNA)**.

Both siRNA and miRNA are produced in the cytoplasm, and their pathways make use of similar components. One of these is an enzyme known as *dicer* (Figure 9.30, part B), which does what its name implies: It cleaves dsRNA into small double-stranded pieces about 25 base pairs in length with a short, single-stranded overhang at each end. Although the figure shows a single type of dicer enzyme, in many organisms the siRNA pathway and the miRNA pathway each uses its own version of dicer encoded in a different gene.

The short pieces of dsRNA produced by dicer are then incorporated into an *RNA-induced silencing complex (RISC)*. Both strands are incorporated, but only one

serves as the *guide RNA* that identifies the target RNA by means of complementary base pairing; the other strand is a *passenger strand*, which in the activated RISC is degraded. As with dicer, in many organisms the components of the RISC differ between the siRNA pathway and the miRNA pathway, especially a key component known as *argonaute*.

After RISC formation, the siRNA and miRNA pathways function quite differently. In the siRNA pathway, the guide RNA matches the target RNA strand perfectly or almost perfectly, because the guide and the target are transcribed from opposite strands of the same duplex DNA. In this case the RISC complex cleaves the target RNA through the action of its version of argonaute (Figure 9.30, part D). It is the perfect or near-perfect match of the guide RNA that gives the siRNA pathway its great specificity. Normally only one target RNA is destroyed.

In the miRNA pathway (Figure 9.30, part E), the guide RNA and the target RNA generally include several mismatches because the guide RNA and the target RNA are transcripts from different regions in the genome. In this case, the activated RISC complex attaches to the target RNA and, through its own versions of argonaute, reduces the efficiency of translation by cleaving the mRNA, destabilizing the mRNA by shortening its poly-A tail, or inhibiting the initiation of mRNA translation. Because some mismatches are tolerated, a miRNA typically targets multiple transcripts, each from a different gene. The effects on translation are typically rather mild, however, usually reducing protein production by much less than 50 percent. Nevertheless, a single miRNA can regulate the expression of entire networks of genes. In humans, the miRNA pathway has been implicated in regulatory abnormalities in the cell cycle and the formation of tumors.

The RNAi response functions not only in regulating targets recognized through dsRNA, but components of the response are also implicated in genome structure and organization. In fission yeast, for example, argonaute protein is required for the induction and spread of heterochromatin, and dicer and other components of the siRNA pathway are used in maintaining the transcriptionally silent state of genes in heterochromatin.

The discovery of RNAi generated a great deal of excitement in genetics because of its potential in research as well as practical applications. The ability of introduced dsRNA to reduce the level of expression of genes having homologous transcripts is the basis of *genetic knockdowns* of activity. (They are called *knockdowns* rather than *knockouts* because the silencing is often incomplete.) In effect, the RNAi response affords a method of producing the equivalent of mutations that drastically reduce gene expression in organisms that do not have well-developed systems of mutagenesis and genetic manipulation.

RNAi also has important applications in biotechnology and medicine. In plant biotechnology, for example, RNAi has been used to reduce the production of toxins, much as chalcone synthase RNAi reduces the amount of anthocyanin pigment in petunia (Figure 9.29). In medicine, applications include novel therapies for controlling human immunodeficiency virus (HIV), hepatitis. influenza, measles, and other viruses, as well as new approaches to treat cancer and neurodegenerative diseases.

Some long noncoding RNA transcripts function in gene regulation.

The term long noncoding RNA (lncRNA) applies to RNA transcripts longer than 200 nucleotides that are not translated into proteins. The human genome includes at least as many long noncoding RNA transcripts as protein-coding transcripts. While some lncRNA transcripts come from intergenic regions, many originate in or near protein-coding genes and include multiple, often overlapping transcripts from either or both DNA strands using templates that may include 5' noncoding regions, introns, exons, or 3' noncoding regions. Collectively, lncRNAs are abundant, but individually they are usually transcribed at a much lower rate than protein-coding genes. Most lncRNAs are also degraded soon after synthesis. Many of these transcripts may result from stochastic noise in transcription that occurs randomly across regions of open chromatin.

On the other hand, many examples of lncRNAs that function in gene regulation are well documented. Some include self-complementary regions that are processed into microRNAs, for example. Because many of the lncRNAs are transcribed in or near protein-coding genes, they can form RNA/DNA hybrids that affect gene expression. Examples are known in which the RNA/DNA hybrid prevents assembly of the transcription complex, which silences the protein-coding gene. Other examples are known in which the RNA/DNA hybrid helps recruit the transcription complex, which activates transcription of the protein-coding gene. Some lncRNAs that are transcribed from the complementary strand of a protein-coding gene can undergo base pairing with sequences in the mRNA and affect translation.

A unique regulatory function of some lncRNAs is in epigenetic modification of gene expression through effects on DNA methylation, histone modification, chromatin remodeling, and the formation of heterochromatin. In mammals, an abundant lncRNA known as Xist is transcribed from a site in any X chromosome that is destined to become inactivated. The Xist transcript is spliced and polyadenylated, but it is not translated. Rather, it coats the inactive X throughout its length and recruits proteins for DNA methylation, histone modification, and other changes in chromatin that result in transcriptional silencing. Analogously, a lncRNA known as HOTAIR binds to its target region and serves as a scaffold on which chromatin-modifying and other proteins assemble to silence the gene.

9.7 Regulation can also take place at the level of translation.

Because transcription and translation are uncoupled in eukaryotes, gene expression can be regulated at the level of translation separately from transcription. The principal types of translational control are

- Inability of an mRNA molecule to be translated except under certain conditions

- Regulation of the overall rate of protein synthesis

- Inhibition or activation of translation by microRNAs that undergo base pairing with the mRNA (Figure 9.30, part E)

An important example of translational regulation is that of activating previously untranslated cytoplasmic mRNAs. This mechanism is prominent in early development, when newly fertilized eggs synthesize many new proteins at a rapid rate, virtually all of which derive from preexisting cytoplasmic mRNAs. In a few cases the molecular mechanism of mRNA activation is known. For example, in *Drosophila*, the mRNAs for the genes *bicoid*, *Toll*, and *torso* become activated because of the cytoplasmic elongation of their poly-A tail.

A dramatic example of translational control is the extension of the lifetime of silk fibroin mRNA in the silk-worm. During cocoon formation, the silk gland synthesizes a single type of protein, silk fibroin, in very large amounts. The amount of fibroin is increased by three different mechanisms. First, the silk-gland cells become highly polyploid, accumulating thousands of copies of each chromosome. Second, transcription of the fibroin gene is initiated at a strong promoter, which results in the creation of about 10^4 fibroin mRNA molecules per gene copy in a period of a few days. Third, the fibroin mRNA molecule has a very long lifetime. In contrast to a typical eukaryotic mRNA molecule, which has a lifetime of about 3 hours, fibroin mRNA survives for several days, during which each mRNA molecule is translated repeatedly to yield 10^5 fibroin molecules. Thus each fibroin gene copy yields about 10^9 protein molecules in the few days during which the cocoon is being created.

Small regulatory RNAs can control translation by base-pairing with the messenger RNA.

Small regulatory RNAs that control translation have been described in both prokaryotes and eukaryotes, and analyses of genome sequences suggest that there will be many more examples. The mechanism usually involves regulatory RNAs that are complementary in sequence to part of the mRNA whose translation they control. We have already seen one example of translational inhibition by microRNA (Figure 9.30, part E). An RNA sequence complementary to an mRNA is called an **antisense RNA**. The antisense regulatory RNAs act by pairing with the mRNA to either inhibit or activate translation (**FIGURE 9.31**). Bacterial regulatory RNAs

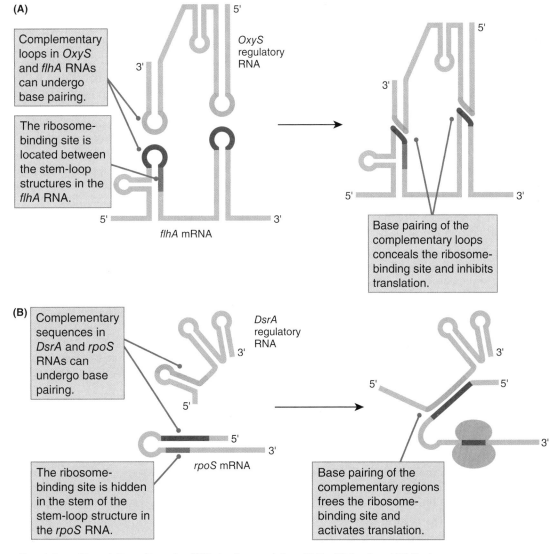

FIGURE 9.31 Regulation of translation of target mRNAs by the regulatory RNAs (A) OxyS and (B) DsrA.

Data from S. Altuvia and E. G. H. Wagner, *Proc. Natl. Acad. Sci. USA* 97 (2000): 9824–9826.

often control translation of several mRNAs and serve as global regulators of cellular processes.

Figure 9.31 shows an example of a small regulatory RNA that relieves oxidative stress in *E. coli*. One of the genes derepressed in the presence of hydrogen peroxide is *oxyS*, which encodes a regulatory RNA called OxyS. This RNA binds to several mRNAs. For any given interaction, only short stretches of OxyS are complementary to the target RNA and able to pair with it. For example, two separate regions of OxyS bind to the mRNA of the gene *flhA*, which encodes a transcriptional activator protein (part A). The complementary regions are very short, in this example only seven nucleotides. One of the complementary regions is near the AUG translational start, and the other is more than 40 nucleotides upstream. Base pairing between OxyS and the *flhA* mRNA conceals the ribosome-binding site and prevents translation. (Such a bipartite complex composed of a small regulatory RNA and an mRNA is called a *kissing complex*.)

Small regulatory RNAs can also activate translation. An example is the DsrA regulatory RNA from *E. coli* shown in part B. In this case the mRNA whose translation is controlled is from a gene *rpoS*, which encodes a sigma factor for RNA polymerase that allows transcription of a new set of RNAs from a special set of promoters at stationary phase in cell cultures when the cell density is high and the cells begin to slow their growth and division. The 5′ end of the *rpoS* mRNA is self-complementary and can form a hairpin that hides the ribosome-binding site and the translational start site. Two virtually contiguous regions of DsrA bind to the *rpoS* mRNA, and when binding occurs, the ribosome-binding site becomes free and translation can occur.

CHAPTER SUMMARY

- Genes can be regulated at any level, including transcription, RNA processing, translation, and post-translation.

- Control of transcription is an important mechanism of gene regulation.

- Transcriptional control can be negative ("on unless turned off") or positive ("off unless turned on"); many genes include regulatory regions for both types of regulation. Occasional transcripts may also be produced at random times owing to stochastic noise.

- Most genes have multiple, overlapping regulatory mechanisms that operate at more than one level, from transcription through post-translation.

- In prokaryotes, the genes coding for related functions are often clustered in the genome and controlled jointly by a regulatory protein that binds with an operator region at one end of the cluster. This type of gene organization is known as an operon.

- In eukaryotes, genes are not organized into operons. Genes at dispersed locations in the genome are coordinately controlled by one or more enhancer DNA sequences located near each gene

that interact with transcriptional activator proteins to allow gene expression.

- In eukaryotes, chromatin is organized into topologically associating domains (TADs) that are tissue specific, relatively insulated from other such domains, and spatially organized within the nucleus. In some TADs, the genes are transcriptionally active, whereas in others, the genes are repressed.

- The transcription complex in eukaryotes consists of numerous protein components that are recruited to the promoter of a gene whose chromatin has been suitably reconfigured.

- Epigenetic mechanisms of transcriptional control are hereditary changes in gene expression mediated by modification of the DNA bases (usually cytosine methylation) or by binding with regulatory proteins.

- Gene expression can also be regulated at the level of RNA processing, alternative patterns of splicing, transcript stability, or mRNA degradation.

- Double-stranded RNA molecules can be cleaved into short fragments that are used to target homologous RNA transcripts for cleavage or for blocking translation.

ISSUES AND IDEAS

- What is *positive regulation* of transcription? What is *negative regulation* of transcription? What is the role of the repressor in each case? Give an example of each type of regulation.

- What is autoregulation? Distinguish between positive and negative autoregulation. Which would be used to amplify a weak induction

signal? Which would be used to prevent overproduction?

- What class of *lac* mutants demonstrated that the presence of lactose in the growth medium was not necessary for expression of the genes for lactose utilization?

- How does an operon result in coordinated control of the genes included? Are operons usually found in eukaryotic organisms?

- In what sense does attentuation provide a "fine-tuning" mechanism for operons that control amino acid biosynthesis?

- Explain how a small molecule can regulate transcription by means of a riboswitch.

- What is a transcriptional activator protein? A transcriptional enhancer? A

- chromatin-remodeling complex? What role do these elements play in eukaryotic gene regulation?

- How does the possibility of alternative splicing affect the generality of the statement that one gene encodes one polypeptide chain?

- What is meant by the term *epigenetic regulation*? Explain how epigenetic regulation can be mediated through cytosine methylation.

- What is the phenomenon of RNA interference (RNAi)? How is RNAi used in genetic analysis?

SOLUTIONS: STEP BY STEP

PROBLEM 1 The two genotypes of *E. coli* indicated in the accompanying table were grown in the absence of lactose (uninduced) or in the presence of lactose (induced) and assayed for levels of the enzymes of the *lac* operon. Using the information provided in the table, predict the enzyme levels for the other genotypes listed in parts (a) through (d). The levels of activity are expressed in arbitrary units relative to those observed under the induced conditions.

Genotype	Absence of lactose		Presence of lactose	
	Z protein	Y protein	Z protein	Y protein
(1) $I^+ O^+ Z^+ Y^+$	0.1	0.1	100	100
(2) $I^+ O^c Z^+ Y^+$	25	25	100	100

(a) $I^- O^+ Z^+ Y^+$
(b) $F'\ I^+ O^+ Z^- Y^- / I^- O^+ Z^+ Y^+$
(c) $F'\ I^+ O^+ Z^- Y^- / I^+ O^c Z^+ Y^+$
(d) $F'\ I^+ O^+ Z^- Y^- / I^- O^c Z^+ Y^+$

SOLUTION. The data given for the wildtype operon **(1)** in the table indicate a basal level of enzyme activity of 0.1 without induction. The data for the O^c mutant **(2)** indicate that constitutive production of the enzymes occurs only at a level of 25, which means that the I^+ gene product can still bind with O^c and repress the operon to some extent. These inferences allow the phenotypes of the genotypes **(a)–(d)** to be deduced. Genotype **(a)** has no repressor and would show fully induced levels of activity (100) of both enzymes in either the absence of the presence of lactose. Genotype **(b)** is fully regulated by the product of the I^+ gene present in the F′ plasmid; its phenotype would be expected to the same as that of genotype **(1)** in the table. Genotype **(c)** has a normal repressor owing to the I^+ gene present in the F′ plasmid, but the chromosome carries O^c; this genotype would be expected to have the same phenotype as genotype **(2)** in the table. Genotype **(d)** again has a normal repressor owing to the I^+ gene present in the F′ plasmid, and the chromosome carries O^c; this genotype would be expected to have the same phenotype as genotype **(2)** in the table. The expected activities are summarized in the table shown here.

Genotype	Absence of lactose		Presence of lactose	
	Z protein	Y protein	Z protein	Y protein
(a)	100	100	100	100
(b)	0.1	0.1	100	100
(c)	25	25	100	100
(d)	25	25	100	100

PROBLEM 2 In transcriptional regulation of the genes for galactose utilization in budding yeast, deduce whether the phenotype would be constitutive, inducible by galactose, or uninducible under each of the following conditions:
(a) A mutant GAL4p unable to bind with the UAS
(b) A mutant GAL80p unable to be transported into the nucleus
(c) A mutant GAL80p unable to bind with GAL3p
(d) A mutant GAL3p unable to bind with GAL80p
(e) A mutant GAL3p unable to bind with galactose
(f) Growth of cells in the presence of a galactose analog that binds with GAL3p irreversibly
(g) Growth of cells in the presence if a galactose analog that cannot bind with GAL3p

SOLUTION. You should start by refreshing your memory about the normal transcriptional control of the genes for galactose utilization, because the phenotype of cells under each of the conditions specified can be deduced by comparison with the normal situation. In normal cells, GAL4p is transported into the nucleus, binds with the UAS, and recruits the transcription complex to transcribe the genes needed for galactose utilization. In the absence of galactose, GAP80p is also transported into the nucleus, where it binds with GAL4p and prevents the galactose-utilization genes from being transcribed. In the presence of galactose, the galactose binds with GAL3p in the cytoplasm. This complex binds with GAL80p and prevents its transport to the nucleus, and the GAL4p in the nucleus activates transcription of the galactose-utilization genes. (a) The phenotype of a mutant GAL4p unable to bind with the UAS is uninducible because the transcription complex cannot be recruited. (b) The phenotype of a mutant GAL80p unable to be transported into the nucleus is constitutive because GAL4p always activates transcription of the galactose-utilization genes. (c) The phenotype of a mutant GAL80p unable to bind with GAL3p is uninducible because GAL80p is always transported into the nucleus and prevents transcription. (d) The phenotype of a mutant GAL3p unable to bind with GAL80p is uninducible for the same reason as in part (c). (e) The phenotype of a mutant GAL3p unable to bind with galactose is uninducible because GAL80p cannot bind with GAL3p and be retained in the cytoplasm. (f) The phenotype of cells growing in the presence of a galactose analog that binds with GAL3p irreversibly is constitutive because GAL3p never releases GAL80p. (g) The phenotype of cells growing in the presence of a galactose analog that cannot bind with GAL3p is inducible by galactose (but not by the analog) because galactose can bind with GAL3p in the normal way and result in GAL80p being retained in the cytoplasm.

CONCEPTS IN ACTION: PROBLEMS FOR SOLUTION

9.1 Why are mutations of the *lac* operator often called *cis*-dominant? Why are some constitutive mutations of the *lac* repressor (*lacI*) called *trans*-recessive? Can you think of a way in which a noninducible mutation in the *lacI* gene might be *trans*-dominant?

9.2 Why is the *lac* operon of *E. coli* not inducible in the presence of glucose?

9.3 A mutation imparting constitutive synthesis of an enzyme of arginine biosynthesis in *Citrobacter* was found. The enzyme is normally repressible by arginine.

(a) What two kinds of regulatory mutations might cause this phenotype?

(b) What kind of mutation in an enzyme of arginine biosynthesis might cause this phenotype?

9.4 Among mammals, the reticulocyte cells in the bone marrow lose their nuclei in the process of differentiation into red blood cells. Yet the reticulocytes and red blood cells continue to synthesize hemoglobin. Suggest a mechanism by which hemoglobin synthesis can continue for a long period of time in the absence of the hemoglobin genes.

9.5 For each *E. coli* genotype below, indicate whether high levels of β-galactosidase (LacZ) activity will be produced when grown in lactose or glycerol. *crp* represents the gene for cyclic AMP receptor protein. Assume all other alleles are wildtype.

Genotype	β-gal activity	
	Lactose	Glycerol
lacI⁺ O⁺ Z⁺	+	−
crp⁻ lacI⁻ O⁺ Z⁺		
lacI⁻ O⁺ Z⁺		
lacIˢ O⁺ Z⁺		
F′ *lacI⁺ O⁺ Z⁺ / lacI⁺ O⁺ Z⁻*		
F′ *lacI⁺ Oᶜ Z⁻ / lacI⁺ O⁺ Z⁺*		
F′ *lacIˢ O⁺ Z⁺ / lacI⁺ Oᶜ Z⁺*		
F′ *lacIˢ O⁺ Z⁺ / lacI⁺ Oᶜ Z⁻*		

9.6 Consider a eukaryotic transcriptional activator protein that binds to an enhancer sequence and promotes transcription. What change in regulation would you expect from a duplication in which several copies of the enhancer were present instead of just one?

9.7 What change in gene regulation would you expect if a transposable element containing an insulator were to insert between an enhancer and its promoter?

9.8 Cells of genotype *lacI⁻ lacO⁺ lacZ⁺ lacY⁺* Hfr are mated with F⁻ cells of genotype *lacI⁺ lacO⁺ lacZ⁺ lacY⁺*. In the absence of any inducer in the medium, no β-galactosidase is made. However, when a *lacI⁺ lacO⁺ lacZ⁺ lacY⁺* Hfr strain is mated with an F⁻ strain of genotype *lacI⁻ lacO⁺ lacZ⁻ lacY⁻* under the same conditions, β-galactosidase is synthesized for a brief period after the *lac* operon has been transferred. Explain this observation.

9.9 When glucose is present in an *E. coli* cell, is the concentration of cyclic AMP high or low? Can a mutant with either an inactive adenyl cyclase gene or an inactive *crp* gene synthesize normal levels of β-galactosidase? Does the binding of cAMP–CRP to DNA affect the binding of the repressor?

9.10 In yeast, transcription of the *GAL* (galactose-utilization) genes is regulated by the GAL4 protein (designated GAL4p), which promotes transcription by recruiting the transcription complex to the promoter. In the absence of galactose, GAL3p binds to GAL80p and sequesters it in the cytoplasm. In the presence of galactose, the galactose binds with GAL3p and triggers the release of GAL80p.

(a) Would a mutant GAL4p that prevents it from interacting with the transcription complex (but does not interfere with its other functions) be inducible, noninducible, or constitutive? Would the mutation be dominant or recessive to the wildtype allele?

(b) Would a mutation in GAL80p that prevents it from interacting with GAL4p (but does not interfere with its other functions) be inducible, noninducible, or constitutive? Would the mutation be dominant or recessive to the wildtype allele?

9.11 A mutant strain of *E. coli* makes β-galactosidase in the absence of lactose as well as in the presence of lactose.

(a) What are two likely (haploid) genotypes of this mutant?

(b) When the mutant strain is converted into a partial diploid by introducing an F′ plasmid carrying a wildtype *lac* operon, synthesis of the LacZ and LacY enzymes becomes inducible. Does this information allow you to specify which of the possibilities in **(a)** is true?

9.12 A mutant strain of *E. coli* fails to make β-galactosidase even when lactose is present. Genetic tests show that this mutant is in the *lacI* gene and is dominant. The strain was exposed to UV light to induce additional mutations, and mutant cells were selected that could grow on lactose. Two of these new mutants are characterized and found to be constitutive for β-galactosidase expression. Mutation A maps to the *lac* operator region and mutation B maps to the *lacI* gene.

(a) Would mutation A be dominant or recessive to the wildtype allele?

(b) Briefly explain why mutation A suppresses the original dominant, noninducible mutation.

(c) Would mutation B be dominant or recessive to the wildtype allele?

(d) Briefly explain why mutation B suppresses the original dominant, noninducible mutation.

9.13 The leader sequence of the *metI* transcript involved in methionine biosynthesis in *Bacillus subtilis* acts as a riboswitch that binds S-adenosylmethionine (SAM) in a manner analogous to the *yitJ* leader RNA. What phenotype would be expected of cells with a mutation in *metI* in which the leader was unable to bind with SAM?

9.14 Imagine a bacterial species in which the methionine operon is regulated only by an attenuator and there is no repressor. In its mode of operation, the methionine attenuator is exactly analogous to the *trp* attenuator of *E. coli*. The relevant portion of the attenuator sequence in the RNA is

5′-AAA<u>A</u>UGAUGAUGAUGAUGAUGAUGGACUAA-3′

The translation start site is located upstream from this sequence, and the region shown is in the correct reading frame. What phenotype (constitutive, wildtype, or Met⁻) would you expect for each of the types of mutant RNA below? Explain your reasoning.

(a) The red A is deleted.

(b) Both the red A and the underlined A are deleted.

(c) The first three As in the sequence are deleted.

9.15 A frameshift mutation occurs near the end of an exon. Does it affect the reading frame of the next exon in the processed mRNA? Explain your answer.

9.16 You wish to create an artificial operon designated *Pb* whose protein products would allow a bacterium to survive exposure to high levels of lead. You want the *Pb* operon to be transcribed only in the presence of high levels of lead.

(a) Design and describe a simple regulatory system for the *Pb* operon that employs negative regulation of transcription.

(b) Describe a mutation in the regulatory system in **(a)** that would result in a recessive phenotype. Would the *Pb* operon be inducible or constitutive?

(c) Describe a mutation in the regulatory system in **(a)** that would result in a dominant phenotype. Would the *Pb* operon be inducible or constitutive?

9.17 In the artificial *Pb* operon described in the previous problem, you decide also to design a different regulatory system, which in this particular case may work better than negative regulation.

(a) Design and describe a simple regulatory system for the *Pb* operon that employs positive regulation of transcription.

(b) Describe a mutation in the regulatory system in **(a)** that would result in a recessive phenotype. Would the *Pb* operon be inducible or constitutive?

(c) Describe a mutation in the system in **(a)** that would result in a dominant phenotype. Would the *Pb* operon be inducible or constitutive?

9.18 Temperature-sensitive mutations in the *lacI* gene of *E. coli* render the repressor nonfunctional (unable to bind the operator) at 42°C but leave it fully functional at 30°C. In such a mutant, would β-galactosidase be expected to be produced:

(a) In the presence of lactose at 30°C?

(b) In the presence of lactose at 42°C?

(c)   In the absence of lactose at 30°C?

(d)   In the absence of lactose at 42°C?

9.19   The accompanying illustration shows a primary RNA transcript containing six exons, indicated by the rectangles labeled A–F. How many different protein products could result from alternative splicing to produce mRNAs that contain four or more of the exons?

9.20   Shown here is a primary RNA transcript with three exons (A, B, and C), which is alternatively

processed in two ways to yield either an mRNA of sequence A–B or an mRNA of sequence B–C. An organism is mutant unless it has a functional product from both mRNAs. The black lines represent loss-of-function mutations in the exons.

(a)   Draw the complementation matrix for the six mutations, using + to indicate complementation (wildtype phenotype) and − to represent lack of complementation (mutant phenotype).

(b)   Explain what is unusual about this complementation matrix.

STOP & THINK ANSWERS

ANSWER TO STOP & THINK 9.1

With negative regulation, transcription occurs unless it is prevented. Hence, for gene *A*, if the repressor cannot bind its binding site, gene *A* will always be expressed; and if the repressor binds so tightly that it cannot be displaced, gene *A* will never be expressed. With positive regulation, transcription does not occur in the absence of a transcriptional activator. Therefore, in regard to gene *B*, with a mutant transcriptional activator protein that cannot bind, gene *B* cannot be expressed; and with a mutant that binds too tightly to be displaced, gene *B* will always be expressed.

ANSWER TO STOP & THINK 9.2

The *lacZ*⁺ allele is expressed constitutively because *lacOᶜ* is *cis*-dominant. The *lacY*⁺ allele is uninducible because *lacP*⁻ is *cis*-dominant.

ANSWER TO STOP & THINK 9.3

The high level of cAMP in the mutant strain implies that the cAMP–CRP complex is always available. The addition of lactose to medium containing glucose will relieve repression by the *lacI* protein, and the presence of cAMP–CRP will allow transcription of the operon to occur. In a nonmutant strain with low levels of cAMP in the presence of glucose, a low level of cAMP–CRP will prevent *lac* operon transcription, even in the presence of lactose.

ANSWER TO STOP & THINK 9.4

Attenuation would not occur; expression of the tryptophan operon would become insensitive to the intracellular concentration of tryptophan.

CHAPTER 10

Genomics, Proteomics, and Genetic Engineering

LEARNING OBJECTIVES

- To describe at least one method for high-throughput DNA sequencing and discuss some of the applications of high-throughput sequencing in human genetics.

- In regions of the genome that code for proteins, to state the evolutionary signatures that characterize differences in DNA sequence among related species.

- To explain who the Neanderthals and Denisovans were and what methods were used to infer that interbreeding with them contributed a small proportion to the modern human genome.

- To identify some of the benefits and potential risks of sequencing the genomes of hundreds of thousands (or millions) of human genomes.

- To interpret the fluorescent colors in spots on a DNA microarray to specify which of two samples being compared has the greater level of transcription.

- To explain how ChIP-chip and ChIP-seq reveal the genes or DNA sequences that are bound with specific proteins.

- To define two-hybrid analysis and explain how the procedure helps identify proteins that come into physical contact.

- To identify restriction sites in a particular DNA fragment and the cloning site in a vector that would enable the restriction fragment to be inserted into the vector in a prescribed orientation.

- To describe the components of CRISPR-Cas9 and explain how they are used to create gene deletions or to change gene sequences.

Genome sequencing of hundreds of species has provided a colossal amount of data for analysis and comparison. In addition to the genome sequences, methods are also available for identifying which genes in the genome are transcribed in particular tissue types, at specific times in development, or at different stages of the cell cycle. These are the raw data of genomics, which deals with the DNA sequence, organization, function, and evolution of genomes. The counterpart at the level of proteins is proteomics, which aims to identify all the proteins in a cell or organism (including any posttranslationally modified forms), as well as their cellular localization, functions, and interactions. Proteomics makes use of methods discussed later in this chapter that identify which proteins in the cell undergo physical contact, thereby revealing *networks* of interacting proteins.

10.1 Genome sequencing has become rapid and inexpensive as a result of new technologies.

In 1985, the idea surfaced that it would be useful to know the complete sequence of all three billion nucleotides in the human genome. This seemed an outlandish idea at the time, because the cost of DNA sequencing was about $1 per base pair. But proponents of the idea argued that launching such a program would provide incentives for technology development, and sequencing costs would fall. The Human Genome Project was formally inaugurated in 1990; by the time the human genome sequence was completed in 2003, sequencing costs had indeed fallen and continued to fall. Today, the cost of sequencing a single human genome is about $1000.

The goals of the Human Genome Project also included sequencing the genomes of a number of model organisms used in genetic research because of their demonstrated utility in the discovery of gene function. Vast amounts of data would be generated, which would have to be stored and made accessible. A new interdisciplinary field called bioinformatics came into being, which combines computer science, engineering,

statistics, and mathematics to analyze genome sequences and other biological data. Genome information would be made available for drug development and other purposes, and ethical issues such as the privacy of one's genetic information had to be considered.

The Human Genome Project was a great success, but it will require many years, probably decades, before genome function and regulation are understood in detail. Tools for understanding the human genome include methods for annotating its content, comparative genomics, transcriptional profiling, and studying protein expression, function, and interaction. These are some of the key approaches to genomics and proteomics, and they are discussed in the following sections.

High-throughput DNA sequencing empowers personalized genomics.

You may recall from Chapter 6 that Frederick Sanger developed a method for sequencing DNA based on the termination of strand elongation during synthesis by means of dideoxynucleotides—hence the method is often called *Sanger sequencing*. Originally developed in 1977, the method has since undergone many modifications and improvements to increase speed and accuracy, and completely different approaches to DNA sequencing have also been implemented. In this section, we briefly examine four current methods for high-throughput sequencing, each having its own advantages and limitations. Then we will discuss one of main reasons motivating the push for large-scale, cost-effective genomic sequencing.

All of the sequencing methods start with purified, double-stranded DNA. In some methods, the DNA is mechanically sheared into fragments, and fragments within a desired size range are isolated. In other cases, the DNA is not sheared so as to obtain long molecules. In most methods, short, double-stranded adapter molecules are added to the ends of the DNA fragments by DNA ligase. These adaptors may be complementary to oligonucleotide primers allowing polymerase chain reaction (PCR) amplification of the fragments, and they may also be chemically modified to adhere to special surfaces.

One high-throughput method is **sequencing by synthesis**, which is outlined in **FIGURE 10.1**. The initial step is to shear genomic DNA into small pieces and to attach PCR adaptors to the 3′ and 5′ ends. The PCR adaptors are short sequences that can anneal with complementary PCR primers to allow amplification. The template DNA strands are then separated and spread out onto a flat surface densely covered with primers that are attached and stick up like blades of grass. Each template strand adheres to a primer, and PCR amplification begins. In every cycle, the DNA strand elongated from each primer is anchored in place at one end, but it is free at the other end to loop over and pair with a nearly complementary primer (the method is called *bridge PCR*). The result of multiple rounds of PCR is a tiny cluster of identical PCR products. These products now serve as sequencing templates for terminator sequencing using an ingenious chemistry of reversible terminators, in which each trinucleotide precursor has its 3′ end blocked by a chemical group and each base has a different fluorescent dye attached. The nucleotide precursors are applied in turn, and a fluorescent signal appears at any position on the plate where a particular nucleotide is incorporated. Then reagents are added that cleave the dye and remove the 3′ blocking group, and the process is repeated (Figure 10.1).

Another high-throughput method is **ion torrent sequencing** (**FIGURE 10.2**). In this case, the PCR adaptors terminate in chemical groups that adhere to microscopic beads; when the DNA sample is diluted and mixed with an excess of beads, each DNA fragment attaches to one bead. PCR reagents are added and the mixture emulsified in oil. The emulsion keeps each bead in its own little water compartment, where a local PCR reaction takes place. When each bead is studded with millions of copies of its DNA fragment, the beads are dispersed into tiny chambers of a semiconductor chip analogous to that in your digital camera, where sequencing takes place (Figure 10.2). In this method, each of the normal, unmodified trinucleotides is added in turn to the chip, and in any chamber in which a nucleotide is incorporated, a proton is released. This changes the pH, which is detected by a sensor and converted into an electrical signal. Any chamber in which the template has two or three or more identical nucleotides in a row will produce two or three or more times as many protons and therefore a greater change in pH; however,

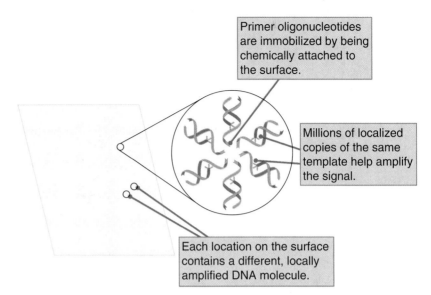

FIGURE 10.1 Sequencing by synthesis.

Primer oligonucleotides are immobilized by being chemically attached to the surface.

Millions of localized copies of the same template help amplify the signal.

Each location on the surface contains a different, locally amplified DNA molecule.

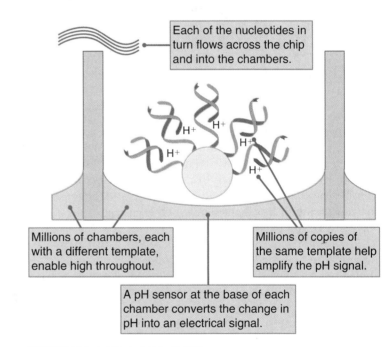

FIGURE 10.2 Ion torrent sequencing.

Each of the nucleotides in turn flows across the chip and into the chambers.

Millions of chambers, each with a different template, enable high throughout.

Millions of copies of the same template help amplify the pH signal.

A pH sensor at the base of each chamber converts the change in pH into an electrical signal.

miscounts in the number of nucleotides in longer runs are common.

FIGURE 10.3 illustrates one method of **single-molecule sequencing**, in which there is no signal amplification owing to multiple copies of the template. In this example, the adaptors are duplex molecules linked at one end. When these hairpin-shaped adaptors are ligated onto a DNA molecule and the base pairing is disrupted, the result is single-stranded, circular DNA. Such circular molecules are distributed among tiny, shallow chambers, each containing a single DNA polymerase complex tethered to the bottom (Figure 10.3). Sequencing is carried out using reversible terminators, and an optical detector converts

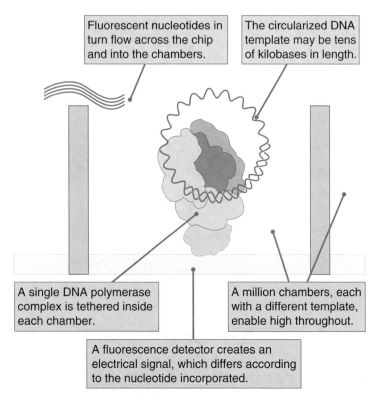

Fluorescent nucleotides in turn flow across the chip and into the chambers.

The circularized DNA template may be tens of kilobases in length.

A single DNA polymerase complex is tethered inside each chamber.

A million chambers, each with a different template, enable high throughput.

A fluorescence detector creates an electrical signal, which differs according to the nucleotide incorporated.

FIGURE 10.3 Single-molecule sequencing from a circular DNA template.

a fluorescent signal into an electronic signal. With a small circle as the template, the polymerase complex can traverse the template multiple times, which improves accuracy, but with large circles the polymerase complex may detach from the template before completing even one round. Reads of 50 kb and longer are possible with this device, but such long reads may have error rates exceeding 10 percent.

Extremely long sequencing reads (up to 1 Mb) are attainable using a method called **nanopore sequencing**, illustrated in **FIGURE 10.4**. In this method, the adaptor is a long, single strand of DNA that is recognized by a motor protein that binds the strand and propels it through a channel in a transmembrane protein. As each nucleotide passes in turn through the transmembrane protein, it changes the conductance of the membrane in accordance with the identity of the nucleotide. This tiny change in conductance is amplified and recorded. One nanopore sequencing device is smaller than a smartphone and can read almost 100 nucleotides per second. The main drawback is that nucleotides can pass through without detection, phantom nucleotides can be added, or nucleotides can be miscalled—each type of error occurring at a rate of up to 5 to 10 percent.

A genome sequence without annotation is meaningless.

A genome sequence is not self-explanatory. It is like a book printed in an alphabet of only four letters,

without spaces or punctuation, and lacking an index. To be useful, any genomic sequence must be accompanied by **genome annotation**, which refers to explanatory notes that accompany the sequence. A genome annotation specifies functional elements, notably sequences in or near coding regions that delineate protein-coding exons and introns, as well as the upstream- and downstream-binding motifs that are targets of enhancer or silencer elements. Annotations also include sequences encoding functional RNAs such as tRNAs, small nuclear RNAs involved in splicing, and microRNAs. Annotations also identify sequences corresponding to transposable elements, and so forth.

Especially for large, complex genomes in which much of the DNA does not code for proteins, and in which most protein-coding exons are relatively small and interrupted by large introns, it is a daunting challenge to parse a genomic sequence into its protein-coding exons, to identify which protein-coding exons belong to the same gene, and to recognize the upstream and downstream regulatory regions that control gene expression. The annotation of genomic

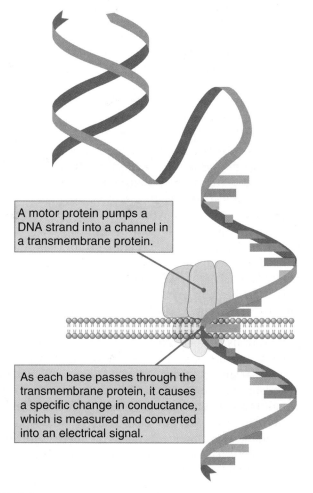

A motor protein pumps a DNA strand into a channel in a transmembrane protein.

As each base passes through the transmembrane protein, it causes a specific change in conductance, which is measured and converted into an electrical signal.

FIGURE 10.4 Nanopore sequencing.

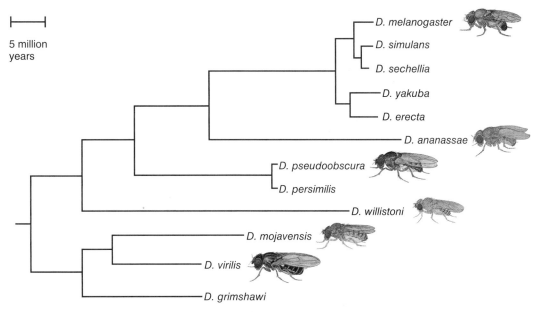

FIGURE 10.5 Evolutionary relationships among 12 *Drosophila* species whose genomes were sequenced for comparative genomics, scaled by their approximate divergence times.

Reproduced from J. T. Patterson, *Studies in the Genetics of Drosophila, Part III and Part IV*. University of Texas Publications (1943 and 1944). Used with permission of the University of Texas at Austin.

sequences at this level is one aspect of **computational genomics**, defined broadly as the use of computers in the interpretation and management of biological data.

Furthermore, especially in multicellular eukaryotes, even for genes whose functions can be assigned, it is not usually known when during the life cycle each gene is expressed, in which tissues it is expressed, or the presence, patterns, or tissue specificity of alternative splicing. Interactions among genes and gene products are also typically unknown. The greatest challenge is to understand how the genes in the genome function and are coordinately regulated to control development, metabolism, reproduction, behavior, and response to the environment.

Comparison among genomes is an aid to annotation.

In many cases, useful information can be gained by identifying genes with similar sequences in other organisms, but if the organisms diverged from a common ancestor too long ago, there is the problem of recognizing which sequences are sufficiently similar to be regarded as functionally equivalent. One way to get around this problem is to compare the genome sequences of groups of related species that have a graded series of divergence times. This approach is known as **comparative genomics**, which has become one of the most powerful strategies for identifying genetic elements in the human genome and those of model organisms.

The fruits of comparative genomics are exemplified in the genome sequences of 12 *Drosophila* species.

FIGURE 10.5 summarizes the evolutionary relationships among the species and their approximate divergence times. The species are very diverse in their geographical origins, global distribution, morphology, behavior, feeding habits, and other phenotypes, yet they share a similar cellular physiology, developmental program, and life cycle. Their genomes show substantial differences in sequence (5 million years in the scale of Figure 10.5 corresponds to about 1 nucleotide difference per 10 nucleotide sites), and they have also undergone multiple gene rearrangements primarily due to inversions. The 12-genome comparison, therefore, reveals how conserved gene functions are maintained in spite of extensive changes in genome structure and sequence.

Comparative genomics derives its power from the distinctive evolutionary patterns, called *evolutionary signatures*, that different types of functional elements exhibit. Some examples from the 12 *Drosophila* species are illustrated in **FIGURE 10.6**. Part A shows characteristic evolutionary signatures of DNA sequences that do not code for protein. The nucleotide differences between species occur virtually at random throughout the sequence (rust), changes that would correspond to chain-terminating (nonsense) codons come and go (yellow), and small insertions or deletions (gray) can consist of any number of nucleotides.

Contrast the pattern of nucleotide differences in noncoding regions with that of protein-coding regions in Figure 10.6, part B. In coding sequences, the characteristic evolutionary signatures show a pronounced triplet periodicity uninterrupted by stop codons.

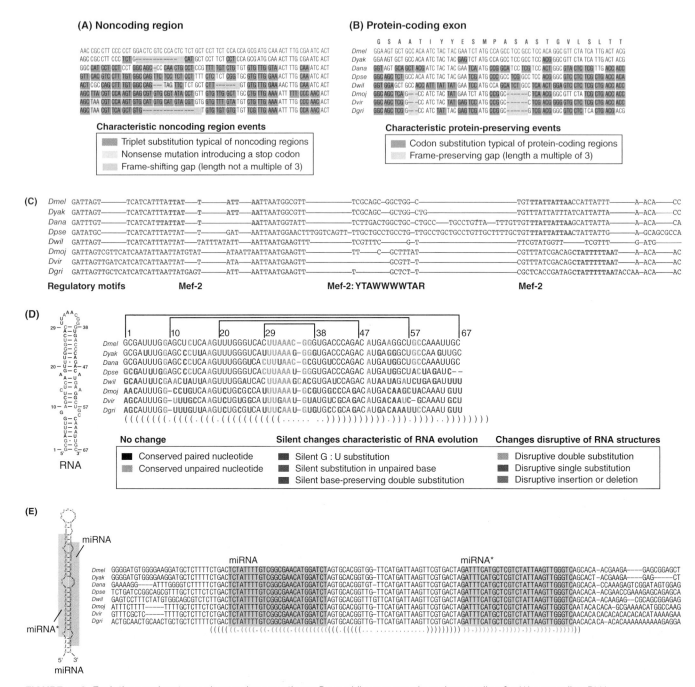

FIGURE 10.6 Evolutionary signatures observed among the 12 *Drosophila* genomes in regions coding for (A) noncoding RNA, (B) protein, (C) transcription-factor binding sites, (D) a stem-loop secondary structure in RNA, and (E) microRNA.

(A) through (E) Data from A. Stark, et al., *Nature* 450 (2007): 219–232.

Many of the nucleotide differences between species are in the third codon position, and the variant codons often encode the same amino acid (green). Deletions, when they occur, remove a number of nucleotides that is a multiple of three (beige), which conserves the proper reading frame.

Comparative genomics also helps to identify regulatory motifs that are the targets of enhancers and silencers. These are often difficult to recognize because they are relatively short, can be present on either DNA strand, and can change position within the gene promoter. The example in Figure 10.6, part C shows binding sites for the protein Mef-2. The consensus binding site has the sequence YTAWWWWTAR, where Y is any pyrimidine, R is any purine, and W means either A or T. The 12-species comparison shows the differing sequence and location of the Mef-2 biding site in one of its target genes. Some of the species have the binding site toward the 5′ end of the region shown, others have it near the 3′

end, and several species have a Mef-2 binding site at both locations.

RNA transcripts that form foldback secondary structures, such as tRNAs, rRNAs, and some snRNAs, have distinctive evolutionary signature of their own. In Figure 10.6, part D, for example, the matched parentheses show the conserved base pairs in the paired stem structures, but the identities of the paired bases often differ among species (paired nucleotides are color coded). An example involves the nucleotides at positions 29 and 38, which in *D. melanogaster* constitute a C–G pair but in *D. yakuba* constitute a U–G pair (U pairs with G as well as with A in double-stranded RNA).

MicroRNAs, important for their regulatory functions in the RNAi pathways, show yet another type of evolutionary signature (Figure 10.6, part E). In this case, changes in the stem regions, even those that are complementary, are not well tolerated, but differences in the loop and other nonpaired regions are found.

In *Drosophila*, comparisons among the 12 genomes have been instrumental in correctly annotating hundreds of protein-coding genes in the *D. melanogaster* sequence, predicting the secondary structures of many noncoding RNAs likely to be involved in gene regulation, showing that some microRNA genes have multiple functional products that increase their regulatory repertoire, and revealing a network of pretranscriptional and posttranscriptional miRNA regulatory targets. The utility of a graded series of divergence times was also validated by the observation that the optimal divergence time for identifying evolutionary signatures depends on the length of the functional element. Longer functional elements are most easily recognized in closely related species, whereas shorter ones are most efficiently identified in more distantly related species.

STOP & THINK 10.1

Explain briefly why comparative genomics is useful in identifying enhancers and silencers in genome sequences.

Ancient DNA indicates interbreeding between our ancestors and archaic human groups that became extinct.

On a hot August day in 1856, two quarrymen in the Neander Valley near Düssledorf, Germany spied bones they thought were those of a bear. They turned them over to a local teacher, an amateur naturalist who recognized them as human—but thought them

unusual. He showed them to a professional anatomist, who realized that the bones belonged to a prehistoric human-like organism from a group now called the **Neanderthals**. Modern methods date the bones to 42,000 years ago. Since that time, more than 400 Neanderthal remains have been found in Europe, the Middle East, and Western Asia, and they range in age from about 400,000 to 30,000 years ago. Although the term Neanderthal evokes the notions of "crude, brutish, and stupid," this stereotype is wrong. Yes, they were stocky, big-boned, barrel-chested, and short-limbed with a prominent brow ridge, broad nose, and small chin. But their brain size was somewhat larger than that of modern humans. They were geographically widespread but lived in small, relatively isolated social groups. They hunted game with spears, butchered it with stone tools, used fire for warmth and cooking, and buried their dead.

As modern humans migrated out of Africa into the Middle East about 60,000 years ago, they encountered Neanderthal groups that were there already (**FIGURE 10.7**). The two groups coexisted for at least 10,000 years, prompting much curiosity on the part of modern researchers regarding whether they interbred. The answer became clear when ancient DNA from the Neanderthal genome was extracted and sequenced. Comparisons of the Neanderthal sequence with that of modern human populations indicated that all human populations outside of Africa include DNA derived from Neanderthals in their genomes. Specifically, about 2.5 percent of all non-African genomes can be traced to Neanderthal ancestors. The principal observations are that

- The only populations without traces of Neanderthal DNA are native Africans.

- All non-African populations derive from anatomically modern humans who migrated to the Middle East about 65,000 years ago.

- Neanderthals and anatomically modern humans coexisted in the Middle East.

These findings argue that some interbreeding took place between Neanderthals and anatomically modern humans, probably in the Middle East, prior to the time that anatomically modern humans began their worldwide spread (Figure 10.7). Interestingly, the DNA derived from Neanderthals present in human populations today is found only in the nucleus. No lineages of Neanderthal mitochondrial DNA are found in modern human populations. Because mitochondrial DNA is inherited only through the female, the absence of Neanderthal mitochondrial lineages implies that the interbreeding that left its traces in modern genomes was between Neanderthal males and anatomically modern females.

As anatomically modern humans spread throughout the world, they encountered still other resident

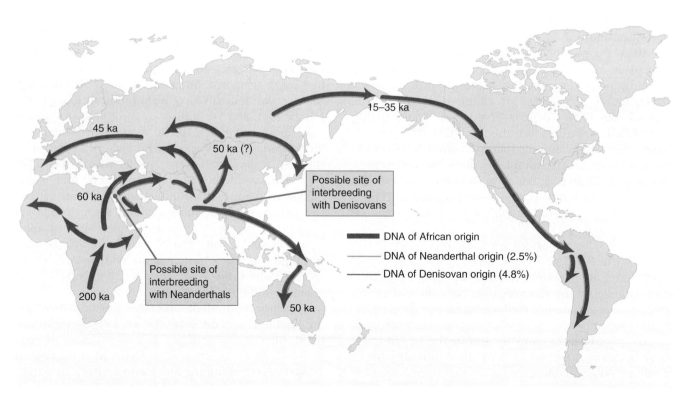

FIGURE 10.7 Evidence suggests that anatomically modern humans evolved in southern Africa, then migrated north to populate Africa, while a subgroup moved still farther into the Middle East and from there spread throughout the world. Interbreeding with Neanderthals and Denisovans is evident because of DNA sequences derived from these archaic groups in the nuclear genomes of all (Neanderthal) or some (Denisovan) modern non-African populations. The suggested sites of interbreeding are plausible but unproven. ka = kiloannum (1000 years).

archaic human groups. One such group was an off-shoot of the Neanderthals known as the **Denisovans** owing to the discovery of the first fossil bone in Denisova Cave in the Altai Mountains in Siberia. The bone was dated to about 41,000 years ago, a time when Neanderthals also lived in the same region.

Comparison of Denisovan DNA and the genomes of modern human populations indicate yet more interbreeding in our ancestry. Approximately 4.8 percent of the nuclear genomes in modern Melanesians, Australian Aborigines, and scattered groups in Southeast Asia and the Philippines are derived from Denisovans. The interbreeding must have taken place prior to about 50,000 years ago, before the ancestors of these groups left continental Asia. Once again, only nuclear DNA shows evidence of interbreeding, indicating that the matings that contributed genetic material to today's populations were between Denisovan males and anatomically modern human females. These findings suggest that at one time the Denisovans must have been quite widespread in eastern Eurasia, and a possible site of interbreeding with anatomically modern humans is indicated in Figure 10.7. One chromosomal region derived from the Denisovan genome is found at high frequency only in Tibetans. It is a region in chromosome 2 containing the gene *EPAS1*, which has played an important role in the adaptation of the Tibetan population to live with the low oxygen levels present at high altitudes.

Your genome sequence can help personalize your medical care.

Genome sequencing was initially motivated by the desire to understand how genomes are organized, what they contain, and how they function. As the cost of sequencing decreased, it became possible to imagine a time that genome sequencing could become part of routine clinical practice. This would allow medical treatments to be based on the genetic makeup of the individual patient. Experts estimate that most drugs work as advertised for only about half the people who take them, irrespective of the disease. Treatment failures cost about $150 billion per year in wasted drug costs and expose millions of people to potentially harmful side effects. Much of the variation in drug response reflects genetic variation among patients. For example, the drugs cetuximab (Erbitux) and panitumumab (Vectibix) for colon cancer fail about half the time because they work only on tumors that have a particular mutation. Genome sequencing of tumor cells would identify responders and nonresponders in advance and lead to more rapid deployment of alternative treatment for the nonresponders. As another example, the drug tamoxifen (Nolvadex) used to treat breast cancer is ineffective in about 10 percent of women who lack a particular enzyme and works poorly in another 20 to 40 percent who have reduced levels of the enzyme.

THE HUMAN CONNECTION

Skeletons in Our Closet

Richard E. Green,[1] Johannes Krause,[1] Adrian W. Briggs,[1] Tomislav Maricic,[1] Udo Stenzel,[1] Martin Kircher,[1] Nick Patterson,[2] and 49 other authors (2010)

[1]Max-Planck Institute for Evolutionary Anthropology, Leipzig, Germany; [2]The Broad Institute of MIT and Harvard, Cambridge, Massachusetts

A Draft Sequence of the Neanderthal Genome

Much of this important paper is concerned with technical issues such as how to extract DNA from bone without contamination with modern human DNA, how to deal with the 95 to 99 percent of the extracted DNA derived from microbes that invaded the bones after death, and how to correct for sequencing errors from damage to DNA as it lies in the ground for long periods. But the payoff for dealing successfully with these difficulties was a far deeper understanding of human origins.

Fossilized Neanderthal bones dating from 400,000 years ago to 30,000 years ago are found at various sites in Europe. In this paper, 5.3 Gb of Neanderthal genomic DNA was analyzed. The Neanderthals sequences were more similar to modern non-Africans than to modern Africans. This finding suggests that Neanderthals and early modern humans hybridized, probably in the Middle East some time between 100,000 years ago and 50,000 years ago,

> **66 Neanderthals share significantly more derived alleles with non-Africans than with Africans. 99**

when both populations coexisted in the same regions. The actual amount of interbreeding might have been very limited, as only 1 to 4 percent of the genome of present-day non-Africans seems to derive from Neanderthals.

We analyzed Neanderthal bones from Vindija Cave in Croatia [and] generated DNA sequence. . . . Neanderthals share significantly more derived alleles with [present-day] non-Africans than with [present-day] Africans. . . . A parsimonious explanation is that Neanderthals exchanged genes with the ancestors of non-Africans, . . . most likely before the divergence of Europeans, East Asians, and Papuans. This may be explained by mixing of early modern humans in the Middle East before their expansion into Eurasia.

R. E. Green, et al. *Science* 328 (2010): 710–722.

Treatment tailored to the individual patient is known as **personalized medicine**, sometimes called *precision medicine*, which means that medical practices, interventions, or treatments are chosen according to an individual patient's likelihood of disease or success of response. Personalized medicine has been around for a long time. Prescription eyeglasses, which seem to have been invented in Italy shortly before the year 1300, are a good example. It is only the role of genome sequencing in treatment choice that makes personalized medicine seem so contemporary.

But sequencing hundreds of thousands (or millions) of individual genomes is not without its ethical challenges and risks. This was recognized early on when the National Institutes of Health established the Ethical, Legal, and Social Implications (ELSI) program, the purpose of which was "to study the ethical, legal, and social implications of genetic and genomic research for individuals, families, and communities."

Some of the ethical, legal, and social issues are posed below as questions. We list these as questions for discussion, recognizing that opinions may differ, and reasonable answers may change with time and experience as well as with use and potential misuse.

- *Who decides on sequencing your genome?* You? Your physician? Your insurance company? And who pays for the sequencing?

- *Who performs the sequencing?* This is an issue of quality control. Can any company advertise itself as a genome sequencer? What standards, if any, should govern sample tracking so that individual identities are protected and DNA samples are not misidentified or mishandled? What standards, if any, should there be for sequencing accuracy? These issues are already prominent in current **direct-to-consumer (DTC)** genetic testing, in which consumers send off their credit card information and a cheek swab and in return get a report on their inferred ancestry and their genotype for a number of known genetic risk factors. Similar issues also apply to **over-the-counter (OTC)** genetic testing, in which consumers can purchase an at-home kit testing for a number of known genetic risk factors for diseases such as Parkinson's, late-onset Alzheimer's, celiac disease (gluten sensitivity), and alpha-1 antitrypsin deficiency (a lung condition).

- *Who interprets your genome sequence?* What information comes with your genome sequence? Does the sequencing company send you an online link and tell you that you're on your own? Most people, including most genetic researchers, are not trained to interpret genome sequences. Ideally, a qualified professional would make the interpretation, but what qualifications should be required, and who should pay for the consultation? There is also an issue of what action to take in light of genetic risk factors that you may or may not possess. Many genetic risk factors increase disease risk by only a modest amount, but someone who carries such a factor may feel that biology is destiny and fall into despair. At the other extreme, a person who learns they lack the known genetic risk factors for lung cancer may think this means it's safe to continue smoking tobacco, when in fact smoking is the single greatest risk factor for the disease.

- *Who owns your genome sequence, who keeps the record, who may access it, and for how long?* This is an issue of privacy. Your genome sequence can reveal your sex, age (by telomere length), skin color, hair color, ethnicity, and many other personal features. Who can have access to your genome sequence? Your employer? Your prospective spouse? Your insurance company? If it is maintained in a database, what protection is there against it being hacked? If third parties have access to the sequence, can it be sold to other parties for applications like microtargeting advertisements based on your genetic makeup—weight-reduction pills for those with risk factors for diabetes, gym memberships for those at risk of heart disease, quit-smoking remedies for those at risk of emphysema, etc.? How long is your genome sequence maintained? Can other parties, such as the employers or insurers of your children, grandchildren, or other relatives, access it after your death?

As we noted earlier, many of these questions are difficult and have far-reaching ethical, legal, and social implications. There are no clear-cut answers, but policy decisions will eventually become necessary.

10.2 Genomics and proteomics reveal genome-wide patterns of gene expression and networks of protein interactions.

Genomic sequencing has made possible a new approach to genetics called **functional genomics**, which focuses on genome-wide patterns of gene expression and the mechanisms by which gene expression is coordinated. As changes take place in the cellular environment—for example, through development, aging, or changes in the external conditions—the patterns of gene expression also change. But genes are usually deployed in sets, not individually. As the level of expression of one coordinated set is decreased, the level of expression of a different coordinated set may be increased. How can one study tens of thousands of genes all at the same time?

DNA microarrays and RNA-seq are used to estimate the relative level of gene expression of each gene in the genome.

The study of genome-wide patterns of gene expression became feasible with the development of the **DNA microarray** (or *chip*), a flat surface about the size of a postage stamp on which 10,000 to 100,000 distinct spots are present, each containing a different immobilized DNA sequence suitable for hybridization with DNA or RNA isolated from cells growing under different conditions, from cells exposed or not exposed to a drug or toxic chemical, from cells at different stages of development, or cells from different types or stages of a disease such as cancer. Two types of DNA chips are presently in use:

- A chip arrayed with oligonucleotides synthesized directly on the chip, one nucleotide at a time, by automated procedures; these chips typically have hundreds of thousands of spots per array.

- A chip arrayed with denatured, double-stranded DNA sequences of 500 to 5000 bp, in which the spots, each about a millionth of a drop in volume, are deposited by capillary action from miniaturized, fountain-pen–like devices mounted on the movable head of a flatbed robotic workstation; these chips typically have tens of thousands of spots per array.

FIGURE 10.8 shows one method by which DNA chips are used to assay the genome-wide levels of gene expression in an experimental sample relative to a control. At the upper right are shown six adjacent spots in the microarray, each of which contains a DNA sequence that serves as a probe for a different

(A)

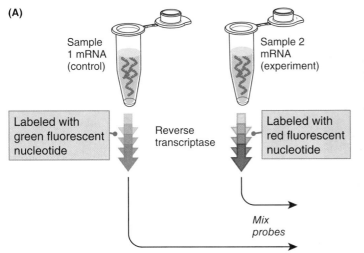

(B) DNA chip

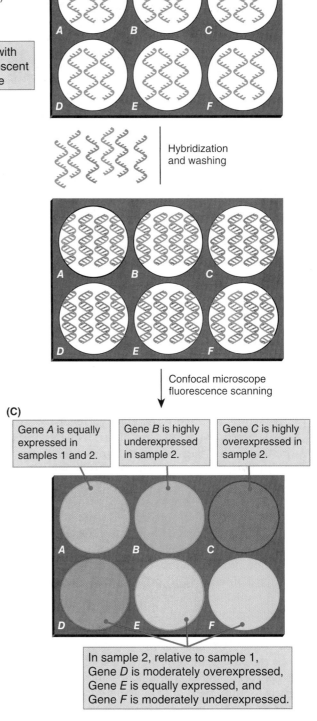

FIGURE 10.8 Principle of operation of one type of DNA microarray. (A) Dried microdrops, each of which contains immobilized DNA strands from a different gene (*A–F*). These are hybridized with a mixture of fluorescence-labeled DNA samples obtained by reverse transcription of cellular mRNA. (B) Competitive hybridization of red (experimental) and green (control) label is proportional to the relative abundance of each mRNA species in the samples. The relative levels of red and green fluorescence of each spot are assayed by microscopic scanning and displayed as a single color. (C) Red or orange indicates overexpression in the experimental sample, green or yellow-green indicates underexpression in the experimental sample, and yellow indicates equal expression.

gene, *A* through *F*. Part A shows the experimental protocol. Messenger RNA is first extracted from both the experimental and the control samples. This material is then subjected to one or more rounds of amplification into DNA copies using the enzyme *reverse transcriptase*. In the experimental material (sample 2), the primer for reverse transcription includes a red fluorescent label; in the control material (sample 1), the primer includes a green fluorescent label. When a sufficient quantity of labeled DNA strands has accumulated, the fluorescent samples are mixed and hybridized with the DNA chip.

The result of hybridization is shown in part B of Figure 10.8. Because the samples are mixed, the hybridization is competitive, and therefore the density of red or green strands bound to the DNA chip is proportional to the concentration of red or green molecules in the mixture. Genes that are overexpressed in sample 2 relative to sample 1 will have more red strands hybridized to the spot, whereas those that are underexpressed in sample 2 relative to sample 1 will have more green strands hybridized to the spot.

After hybridization, the DNA chip is placed in a confocal fluorescence scanner that scans each *pixel* (the smallest discrete element in a visual image) first to record the intensity of one fluorescent label and then again to record the intensity of the other fluorescent label. These signals are synthesized to produce

the signal value for each spot in the microarray. The signals indicate the relative levels of gene expression by color, as shown in **FIGURE 10.9**. A spot that is red or orange indicates high or moderate overexpression of the gene in the experimental sample, a spot that is green or yellow-green indicates high or moderate

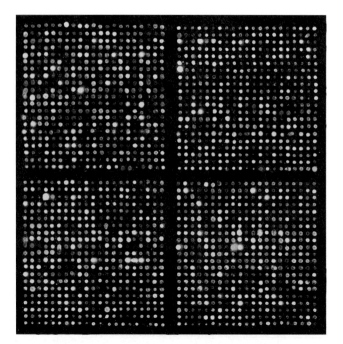

FIGURE 10.9 Small part of a yeast DNA chip showing 1764 spots, each specific for hybridization with a different mRNA sequence. The color of each spot indicates the relative level of gene expression in experimental and control samples. The complete chip for all yeast open reading frames includes over 6200 spots.

Courtesy of Jeffrey P. Townsend, Yale University and Duccio Cavalieri, University of Florence.

underexpression of the gene in the experimental sample, and a spot that is perfectly yellow indicates equal levels of gene expression in the samples. In this manner, DNA chips can assay the relative levels of any mRNA species that has an abundance in the sample of more than one molecule per 10^5, and differences in expression as small as approximately twofold can be detected.

An alternative to microarrays for transcriptional profiling makes use of the techniques of massively parallel sequencing. In this method, called **RNA-seq**, the poly-A tail of messenger RNA is targeted by a poly-T oligonucleotide primer, and reverse transcriptase is used to produce a single-stranded DNA complementary to each of the mRNA molecules. These DNA strands are then replicated to produce a double-stranded **complementary DNA (cDNA)** corresponding to the population of mRNA molecules present in the cells at the time of extraction. The collection of cDNAs is analyzed with massively parallel sequencing, and each cDNA sequence is compared with the reference genome of the organism to identify the gene to which the transcript corresponds.

RNA-seq has many advantages over microarray hybridization. For example, differences in efficiency of hybridization among transcripts can affect results from microarrays but not those from RNA-seq, because in RNA-seq, each transcript is identified according to sequence. Another advantage of RNA-seq is that

differences in the level of transcription of alternative alleles in heterozygous genotypes can be detected. Unbalanced allelic expression can result from differences in the promoter sequences of the alleles, differences in chromatin remodeling to enable transcription, and other factors. If two alleles in a heterozygous genotype differ on average at one nucleotide site across the length of each sequence obtained, then about two-thirds of all RNA-seq reads will differ at one or more sites and, therefore, allow the allele of origin to be identified. There are limitations to RNA-seq, however. Chief among these is difficulty in detecting or accurately estimating the abundance of rare transcripts present in just a few molecules per cell. For this and other reasons, DNA microarrays and RNA-seq should be considered complementary approaches to transcriptional profiling.

Transcriptional profiling reveals groups of genes that are coordinately expressed during development.

Gene-expression arrays have been used to identify groups of genes that are coordinately regulated in development. The example in **FIGURE 10.10** shows expression profiles for 20 groups of genes in the early stages of development in *Caenorhabditis elegans*. In these experiments, time in development was measured in minutes relative to the four-cell stage. Relative levels of gene expression are plotted on a logarithmic scale and, hence, the changes in relative transcript abundance are often two or three orders of magnitude. Over the time period examined, the embryo undergoes a transition from control through maternal transcripts present in the egg to those transcribed in the embryo itself, and this time period includes the times during which most of the major cell fates are specified. The microarrays used in these experiments allowed detection of transcripts from almost 9000 open reading frames, and the plots include traces for approximately 2500 genes, about 80% of all those that showed significant changes in transcript abundance over the time interval shown.

Up to the four-cell stage of development, the patterns of transcription are all quite stable, but after that they begin to change rapidly. Development in the earliest stages is supported largely by maternally derived transcripts. Many of these are cleared rapidly as development proceeds—for example, as shown in the transcripts plotted for clusters of 141, 244, and 568 genes in the lower right panels of Figure 10.10. Production of transcripts from embryonic cells is clearly induced, as evidenced by the patterns for clusters of 431 and 153 genes in the panels at the upper right. The curves showing the disappearance of the maternal transcripts and appearance of the embryonic transcripts intersect at about the time of gastrulation, indicating a somewhat

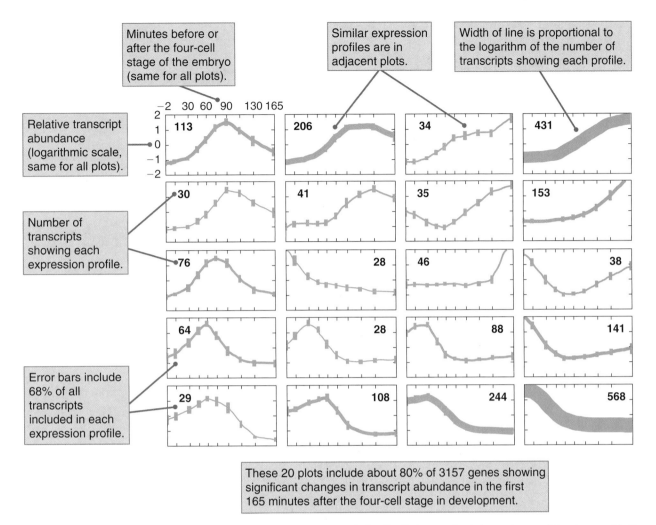

FIGURE 10.10   Patterns of transcriptional regulation of about 2500 genes during the first approximately 2.75 hours of development in *C. elegans*. Complete development requires about 14 hours.

Reproduced from L. R. Baugh, et al., *Development* 130 (2003): 889–900 [http://dev.biologists.org/cgi/content/abstract/130/5/889]. Reproduced with permission of the Company of Biologists.

earlier (mid-blastula) transition from maternal to embryonic control of development. Many of the gene transcription patterns are very complex, with a transient peak of expression suggesting that the transcript (though not necessarily the protein product) is needed for only a brief period in development. All five panels along the left-hand side of Figure 10.10 show this kind of pattern.

Although the transcriptional analysis in Figure 10.10 is a rather coarse, bird's-eye view of what takes place during development, the identification of groups of coordinately expressed genes is of considerable value in itself because it suggests that these genes may share common or overlapping *cis*-acting regulatory sequences that are controlled by common or overlapping sets of transcriptional activator proteins.

Chromatin immunoprecipitation (ChIP) reveals protein–DNA interactions.

It is important not only to measure the extent to which genes differ in their levels of transcription but

also to understand why they differ. Much of the variation in transcriptional levels results from the interaction between DNA and proteins—for example, transcription factors or chromatin proteins. A method for studying such interactions makes use of antibodies that combine with specific types of DNA-binding proteins. The antibody forms a complex with the protein attached to its binding sites in DNA, and the antibody–protein–DNA complex is precipitated and its components isolated for further study.

The technique for isolating protein–DNA complexes is known as **chromatin immunoprecipitation** or **ChIP**, and it is outlined in **FIGURE 10.11**. Part A illustrates a segment of chromatin with nucleosomes including a region of DNA bound with a transcription factor (red) and a nucleosome in which a histone has an amino acid that has been chemically modified by, for example, methylation or acetylation (green). In the first step of ChIP, the proteins and DNA are chemically crosslinked by treatment with a chemical such

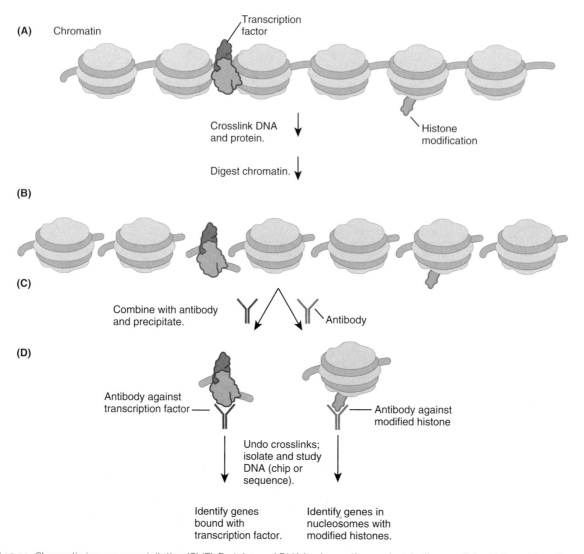

FIGURE 10.11 Chromatin immunoprecipitation (ChIP). Proteins and DNA in chromatin are chemically crosslinked (A), and then the chromatin is sheared into fragments (B). Specific antibodies are added that combine with a transcription factor (red) or a modified histone (green), and the antibody–protein–DNA complex is then precipitated (C). The crosslinks are reversed to allow further analysis of the DNA present in the complex (D).

as formaldehyde. The chromatin is then digested by enzymes or sheared into small fragments containing the crosslinked protein and DNA (Figure 10.11, part B). Then the sample is divided, and antibodies are added that combine specifically with either the transcription factor or the modified histone. In this example, the antibody for the transcription factor is indicated in red, and that for the modified histone in green. The complex of antibody, protein, and DNA that forms is precipitated and purified (Figure 10.11, part C). At this point, the chemical crosslinks are reversed, which frees the bound DNA for further analysis to identify which specific genes or DNA sequences are associated with the transcription factor or with the modified histone (Figure 10.11, part D).

The DNA fragments isolated by means of ChIP are typically analyzed by either of two methods. In one, known as *ChIP-chip*, the DNA is fluorescently labeled and used in hybridization with a microarray chip. In the other, known as *ChIP-seq*, the DNA is analyzed using massively parallel sequencing technology. Either method reveals the sequences associated with the proteins targeted by the antibody and thus indicates which proteins bind with specific DNA sequences and either enhance or repress transcription of nearby genes.

Yeast two-hybrid analysis reveals networks of protein interactions.

Protein–protein interactions are also important for understanding biological processes because proteins that participate in related cellular processes often come into physical contact. Hence, knowing which proteins contact one another can provide clues to the possible functions of otherwise anonymous proteins.

One method for identifying protein–protein interactions makes use of the GAL4 transcriptional activator protein in budding yeast. The GAL4 protein includes two separate domains or regions, both of which are necessary for transcriptional activation. One domain is the zinc-finger DNA-binding domain that binds with the target site in the promoter of the *GAL* genes that are activated, and the other domain is the transcriptional activation domain that makes contact with the transcriptional complex and actually triggers transcription. In the wildtype GAL4 protein, these domains are tethered together because they are parts of the same polypeptide chain.

The key to identifying protein–protein interactions through the use of GAL4 is that the coding regions for the separate domains can be taken apart and each fused to a coding region for a different protein. The strategy is shown in **FIGURE 10.12**, part A, where the GAL4 DNA-binding domain and the transcriptional activation domain are depicted as separate entities, each fused to a different polypeptide chain, shown in the vicinity of a *GAL* promoter. The promoter is attached to a **reporter gene** the transcription of which can be detected by means of, for example, a color change in the colony, the production of a fluorescent protein, or the ability of the cells to grow in the presence of an antibiotic. The fused DNA-binding domain and the fused transcriptional activation domain are both hybrid proteins, and for this reason the test system is called a **two-hybrid analysis**. In Figure 10.12, part A, the proteins fused to

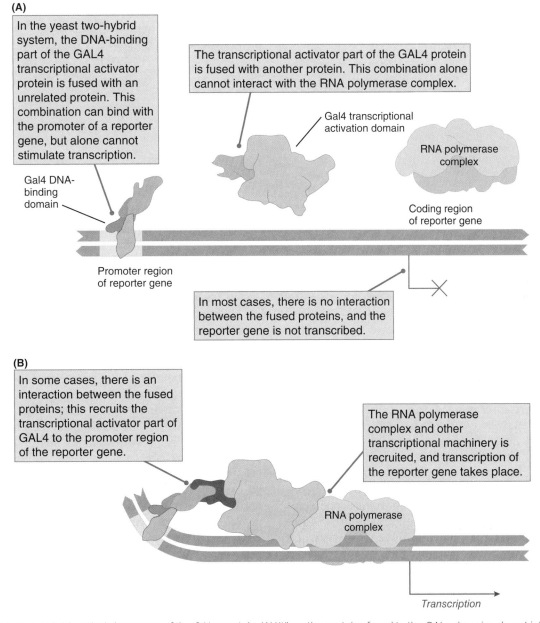

(A)

In the yeast two-hybrid system, the DNA-binding part of the GAL4 transcriptional activator protein is fused with an unrelated protein. This combination can bind with the promoter of a reporter gene, but alone cannot stimulate transcription.

The transcriptional activator part of the GAL4 protein is fused with another protein. This combination alone cannot interact with the RNA polymerase complex.

Gal4 transcriptional activation domain

RNA polymerase complex

Gal4 DNA-binding domain

Coding region of reporter gene

Promoter region of reporter gene

In most cases, there is no interaction between the fused proteins, and the reporter gene is not transcribed.

(B)

In some cases, there is an interaction between the fused proteins; this recruits the transcriptional activator part of GAL4 to the promoter region of the reporter gene.

The RNA polymerase complex and other transcriptional machinery is recruited, and transcription of the reporter gene takes place.

RNA polymerase complex

Transcription

FIGURE 10.12 Two-hybrid analysis by means of the GAL4 protein. (A) When the proteins fused to the GAL4 domains do not interact, transcription of the reporter gene does not take place. (B) When the proteins do interact, the reporter gene is transcribed.

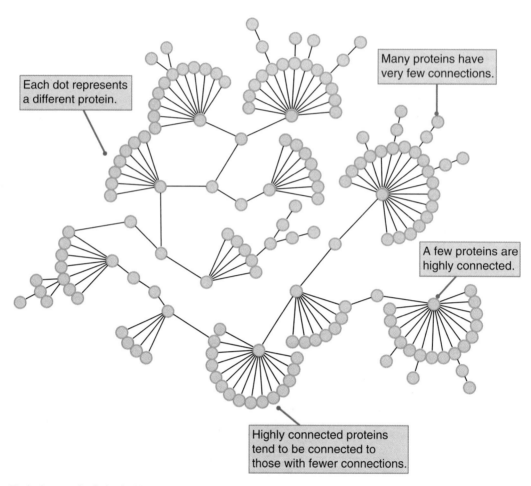

FIGURE 10.13 Typical network of physical interactions among nuclear proteins.

the GAL4 domains do not interact within the nucleus. The DNA-binding domain therefore remains separated from the transcriptional activation domain, and transcription of the reporter genes does not occur.

Figure 10.12, part B shows a case in which the protein fused to the GAL4 domains do interact. In this case, the DNA-binding domain and the transcriptional activation domain are brought into contact, and transcription of the reporter gene does take place. In this manner, transcription of the reporter gene in the two-hybrid analysis indicates that the proteins fused to the GAL4 domains undergo a physical interaction that brings the two hybrid proteins together.

A typical network of protein–protein interactions in the nucleus revealed by two-hybrid analysis is shown in **FIGURE 10.13**. An important property of this example and many other protein networks is that there are fewer than expected interactions between proteins that are already highly connected. In other words, proteins that are highly connected to other proteins through many interactions tend to be connected to proteins with fewer connections rather than to other highly connected proteins. The systematic suppression of links between highly connected proteins has the effect of minimizing the extent to which a random

environmental or genetic perturbation in one part of the network spreads to other parts of the network.

Two-hybrid analysis affords a powerful approach to discovering protein–protein interactions because it can be performed on a large scale, requires no protein purification, detects interactions that occur in living cells, and requires no information about the function of the proteins being tested. The method, however, does have some limitations. For example, the two-hybrid assay is qualitative, not quantitative, and so weak interactions cannot easily be distinguished from strong ones. The hybrid proteins are usually highly expressed to enhance the reliability of the assay, and so interactions can take place that would not take place at normal concentrations. The two-hybrid assay also requires that the protein–protein interactions take place in the nucleus, whereas some proteins may interact only in the environment of the cytoplasm. Finally, hybrid proteins may fold differently than native proteins, and the misfolded proteins may fail to interact when the native conformations do, or they may interact when the native conformations do not. The conclusion is that results from two-hybrid analyses need to be interpreted with care; nevertheless, the method has already yielded much valuable information.

STOP & THINK 10.2

State concisely why a network structure in which highly connected proteins are linked mostly to proteins with many fewer connections is a structure that minimizes the effects of random environmental or genetic perturbations.

10.3 Recombinant DNA is produced by the manipulation of DNA fragments.

Genomics was made possible by the invention of techniques originally devised for the manipulation of genes and the creation of genetically engineered organisms with novel genotypes and phenotypes. We refer to this approach as **recombinant DNA**, but it also goes by the names *gene cloning* or *genetic engineering*. The basic technique is quite simple: DNA is isolated and cut into fragments by one or more restriction enzymes; then the fragments are joined together in a new combination and introduced back into a cell or organism to change its genotype in a directed, predetermined way. Such genetically engineered organisms are called **transgenic organisms**. Transgenic organisms are often created for experimental studies, but an important application is the development of improved varieties of domesticated animals and crop plants, in which case a transgenic organism is often called a *genetically modified organism (GMO)*. Specific examples of genetically modified organisms are considered later in this chapter.

In recombinant DNA, the immediate goal of an experiment is usually to insert a particular fragment of chromosomal DNA into a plasmid or a viral DNA molecule. This is accomplished by techniques for breaking DNA molecules at specific sites and for isolating particular DNA fragments. The fragments that are manipulated are typically smaller than 100 kb because dealing with much larger fragments presents major technical challenges.

Restriction enzymes cleave DNA into fragments with defined ends.

DNA fragments are usually obtained by the treatment of DNA samples with restriction enzymes. *Restriction enzymes* are nucleases that cleave DNA wherever it contains a particular short sequence of nucleotides that matches the *restriction site* of the enzyme. Most restriction sites consist of four or six nucleotides, within which the restriction enzyme makes two single-strand breaks, one in each strand, generating 3'-OH and 5'-P groups at each position. About 1000 restriction enzymes, nearly all with different restriction site specificities, have been isolated from microorganisms.

Most restriction sites are symmetrical in the sense that the sequence is identical in both strands of the DNA duplex. For example, the restriction enzyme *Eco*RI, isolated from *Escherichia coli*, has the restriction site 5'-GAATTC-3'; the sequence of the other strand is 3'-CTTAAG-5', which is identical but written with the 3' end at the left. *Eco*RI cuts each strand between the G and the A. The term *palindrome* is used to denote this type of symmetrical sequence.

Soon after restriction enzymes were discovered, observations with the electron microscope indicated that the fragments produced by many restriction enzymes could spontaneously form circles. The circles could be made linear again by heating. On the other hand, if the circles that formed spontaneously were treated with DNA ligase, which joins 3'-OH and 5'-P groups, then they could no longer be made linear with heat because the ends were covalently linked by the DNA ligase. This observation was the first evidence for three important features of restriction enzymes:

- Restriction enzymes cleave DNA molecules in palindromic sequences.

- The breaks need not be directly opposite one another in the two DNA strands.

- Enzymes that cleave the DNA strands asymmetrically generate DNA fragments with complementary ends.

These properties are illustrated for *Eco*RI in **FIGURE 10.14**.

Most restriction enzymes are like *Eco*RI in that they make staggered cuts in the DNA strands, producing single-stranded ends called **sticky ends** that can adhere to each other because they contain complementary nucleotide sequences. Some restriction enzymes (such as *Eco*RI) leave a single-stranded overhang at the 5' end (**FIGURE 10.15**, part A); others leave a 3' overhang. A number of restriction enzymes cleave both DNA strands at the center of symmetry, forming **blunt ends**. Part B of Figure 10.15 shows the blunt ends produced by the enzyme *Bal*I. Blunt ends also can be ligated by DNA ligase. However, whereas ligation of sticky ends recreates the original restriction site, any blunt end can join with any other blunt end and not necessarily create a restriction site.

Most restriction enzymes recognize their restriction sequence without regard to the source of the DNA. Thus:

KEY CONCEPT

Restriction fragments of DNA obtained from one organism have the same sticky ends as restriction fragments from another organism if they were produced by the same restriction enzyme

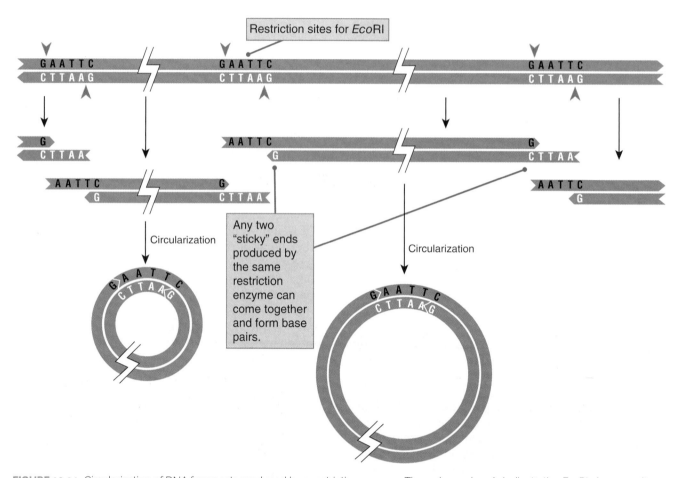

FIGURE 10.14 Circularization of DNA fragments produced by a restriction enzyme. The red arrowheads indicate the *Eco*RI cleavage sites.

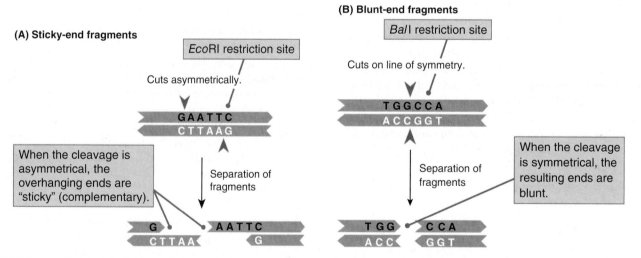

FIGURE 10.15 Two types of cuts made by restriction enzymes. The red arrowheads indicate the cleavage sites. (A) Cuts made in each strand at an equal distance from the center of symmetry of the restriction site. (B) Cuts made in each strand at the center of symmetry of the restriction site.

This principle is one of the foundations of recombinant DNA technology.

Because most restriction enzymes recognize a unique sequence, the number of cuts made in the DNA of an organism by a particular enzyme is limited. For example, an *E. coli* DNA molecule contains 4.6×10^6 base pairs, and any enzyme that cleaves a six-base restriction site will cut the molecule into about 1000 fragments. This number of fragments follows from the fact that any particular six-base sequence (including a six-base restriction site) is expected to occur in a random sequence every $4^6 = 4096$ base pairs, on average, assuming equal frequencies of the four bases. For the same reason, mammalian nuclear DNA would be cut into about 1 million fragments. These large numbers are still small compared with the number that would be produced if breakage occurred at completely random sequences. Of special interest are the smaller DNA molecules, such as viral or plasmid DNA, which may have from only 1 to 10 sites of cutting (or even none) for particular enzymes. Plasmids that contain a single site for a particular enzyme are especially valuable, as we will see shortly.

Restriction fragments are joined end to end to produce recombinant DNA.

In recombinant DNA, a particular DNA fragment of interest is joined to a *vector*, a relatively small DNA molecule that is able to replicate inside a cell and that usually contains one or more sequences able to confer antibiotic resistance (or some other detectable phenotype) on the cell. The simplest types of vectors are plasmids whose DNA is double stranded and circular (**FIGURE 10.16**). When the DNA fragment of interest has been joined to the vector, the recombinant molecule is introduced into a cell by means of DNA transformation (**FIGURE 10.17**). Inside the cell, the recombinant molecule is replicated as the cell replicates its own DNA, and as the cell divides, the recombinant molecule is transmitted to the progeny cells. When a transformant containing the recombinant molecule has been isolated, the DNA fragment linked to the vector is said to be **cloned**. A **vector** is therefore a DNA molecule into which another DNA fragment can be cloned; it is a carrier for recombinant DNA. In the following sections, several types of vectors are described.

A vector is a carrier for recombinant DNA.

The most generally useful vectors have three properties:

1. The vector DNA can be introduced into a host cell relatively easily.

2. The vector contains a replication origin and so can replicate inside the host cell.

3. Cells containing the vector can usually be selected by a straightforward assay, most conveniently by allowing growth of the host cell on a solid, selective medium.

The vectors most commonly used in *E. coli* are plasmids and derivatives of the bacteriophages λ and M13. Many other plasmids and viruses also have been developed for cloning into cells of animals, plants, and other bacteria. Recombinant DNA can be detected in host cells by means of genetic markers

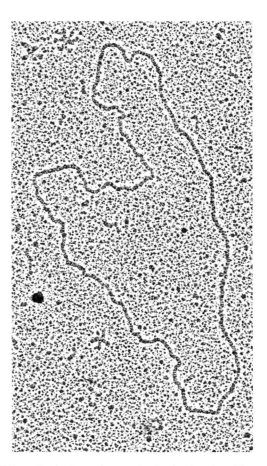

FIGURE 10.16 Electron micrograph of a circular plasmid used as a vector for cloning in *E. coli*.

or phenotypic characteristics that are evident in the appearance of colonies or plaques. Plasmid and phage DNA can be introduced into cells by transformation, in which cells gain the ability to take up free DNA by exposure to a calcium chloride solution. Recombinant DNA can also be introduced into cells by a kind of electrophoretic procedure called *electroporation*. After introduction of the DNA, the cells that contain the recombinant DNA are plated on a solid medium. If the added DNA is a plasmid, colonies consisting of bacterial cells that contain the recombinant plasmid are formed, and the transformants can usually be detected by the phenotype that the plasmid confers on the host cell. For example, plasmid vectors typically include one or more genes for resistance to antibiotics, and plating the transformed cells on a selective medium with antibiotic prevents all but the plasmid-containing cells from growing. Alternatively, if the vector is phage DNA, the infected cells are plated in the usual way to yield plaques. Variants of these procedures are used to transform animal or plant cells with suitable vectors, but the technical details may differ considerably.

Three types of vectors commonly used for cloning into *E. coli* are illustrated in **FIGURE 10.18**. Plasmids (Figure 10.18, part A) are most convenient for

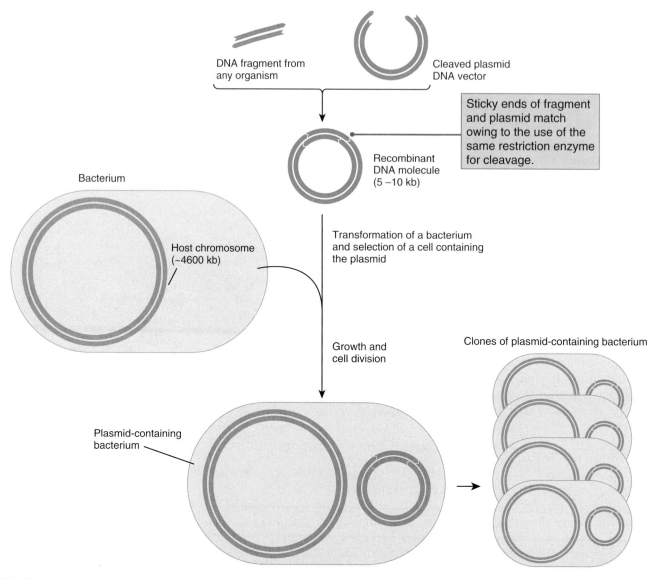

DNA fragment from any organism

Cleaved plasmid DNA vector

Sticky ends of fragment and plasmid match owing to the use of the same restriction enzyme for cleavage.

Recombinant DNA molecule (5 –10 kb)

Bacterium

Host chromosome (~4600 kb)

Transformation of a bacterium and selection of a cell containing the plasmid

Growth and cell division

Clones of plasmid-containing bacterium

Plasmid-containing bacterium

FIGURE 10.17 An example of cloning. A fragment of DNA from any organism is joined to a cleaved plasmid. The recombinant plasmid is then used to transform a bacterial cell, where the recombinant plasmid is replicated and transmitted to the progeny bacteria. The bacterial host chromosome is not drawn to scale. It is typically about 1000 times larger than the plasmid.

cloning relatively small DNA fragments (5 to 10 kb). Somewhat larger fragments can be cloned with bacteriophage λ (Figure 10.18, part B). The wildtype phage is approximately 50 kb in length, but the central portion of the genome is not essential for lytic growth and can be removed and replaced with donor DNA. After the donor DNA has been ligated in place, the recombinant DNA is packaged into mature phage *in vitro*, and the phage is used to infect bacterial cells. However, to be packaged into a phage head, the recombinant DNA must be neither too large nor too small, which means that the donor DNA must be roughly the same size as the portion of the λ genome that was removed. Most λ cloning vectors accept inserts ranging in size from 12 to 20 kb. Still larger DNA fragments can be inserted into cosmid vectors (Figure 10.18, part C).

These vectors can exist as plasmids, but they also contain the complementary overhanging single-stranded ends of phage λ, which enables them to be packaged into mature phages. The size limitation on cosmid inserts usually ranges from 40 to 45 kb.

Some vectors can accept large DNA fragments in the size range 100 to 200 kb. These vectors are called *artificial chromosomes*. Among the most widely used are **bacterial artificial chromosomes (BACs)**. The BAC vector illustrated in Figure 10.18, part D includes functions derived from the F factor of *E. coli*, which plays an important role in bacterial conjugation. The essential functions included in the 6.8-kb BAC vector are genes for replication (*repE* and *oriS*), for regulating copy number (*parA* and *parB*), and for chloramphenicol resistance. DNA fragments suitable for cloning in BAC

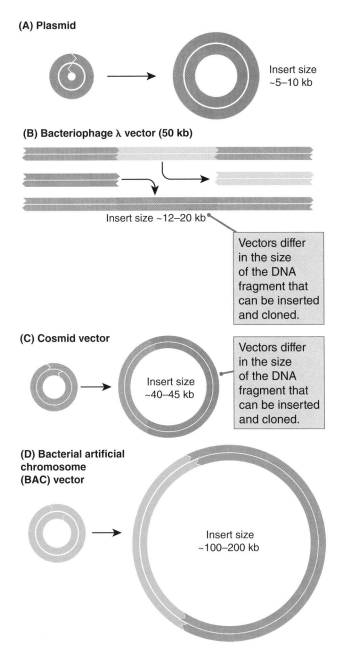

(A) Plasmid

Insert size ~5–10 kb

(B) Bacteriophage λ vector (50 kb)

Insert size ~12–20 kb

Vectors differ in the size of the DNA fragment that can be inserted and cloned.

(C) Cosmid vector

Insert size ~40–45 kb

Vectors differ in the size of the DNA fragment that can be inserted and cloned.

(D) Bacterial artificial chromosome (BAC) vector

Insert size ~100–200 kb

FIGURE 10.18 Common cloning vectors for use with *E. coli*, not drawn to scale. (A) Plasmid vectors are ideal for cloning relatively small fragments of DNA. (B) Bacteriophage λ vectors contain convenient restriction sites for removing the middle section of the phage and replacing it with the DNA of interest. (C) Cosmid vectors are useful for cloning DNA fragments up to about 40 to 45 kb; they can replicate as plasmids but contain the cohesive ends of phage λ and so can be packaged in phage particles. (D) BAC vectors are useful because of the relatively large DNA fragments that they can carry.

Vector and target DNA fragments are joined with DNA ligase.

The circularization of restriction fragments that have terminal single-stranded regions with complementary bases is illustrated in Figure 10.14. Because a particular restriction enzyme produces fragments with *identical* sticky ends, without regard for the source of the DNA, fragments from DNA molecules isolated from two different organisms can be joined, as shown in **FIGURE 10.19**. In this example, the restriction enzyme *Eco*RI is used to digest DNA from any organism of interest and to cleave a bacterial plasmid that contains only one *Eco*RI restriction site. The donor DNA is digested into many fragments (one of which is shown) and the plasmid into a single linear fragment. When the donor fragment and the linearized plasmid are mixed, recombinant molecules can form by base pairing between the complementary single-stranded ends. At this point, the DNA is treated with DNA ligase to seal the joints, and the donor fragment becomes permanently joined in a combination that may never have existed before. The ability to join a donor DNA fragment of interest to a vector is the basis of recombinant DNA technology.

Joining sticky ends does not always produce a DNA sequence that has functional genes. For example, consider a linear DNA molecule cleaved into four fragments—A, B, C, and D—in which the sequence in the original molecule was ABCD. Reassembly of the fragments can yield the original molecule, but because B and C have the same pair of sticky ends, molecules with the fragment arrangements ACBD and BADC can also form with the same probability as ABCD. Restriction fragments from the vector can also join together in the wrong order, but this potential problem can be eliminated by using a vector that has only one cleavage site for a particular restriction enzyme. Plasmids of this type are available (most have been created by genetic engineering). Many vectors contain unique sites for several different restriction enzymes, but generally only one enzyme is used at a time.

DNA molecules that lack sticky ends also can be joined. A direct method uses the DNA ligase made by *E. coli* phage T4. This enzyme differs from other DNA ligases in that it not only heals single-stranded breaks in double-stranded DNA but also can join molecules with blunt ends.

A recombinant cDNA contains the coding sequence of a eukaryotic gene.

Many genes in higher eukaryotes are very large. They can extend over hundreds of thousands of base pairs. Much of the length is made up of introns, which are excised from the mRNA in processing. With such large

vectors can be produced by breaking larger molecules into fragments of the desired size by physical means, by treatment with restriction enzymes that have infrequent cleavage sites (for example, enzymes such as *Not*I and *Sfi*I), or by treatment with ordinary restriction enzymes under conditions in which only a fraction of the restriction sites are cleaved (*partial digestion*).

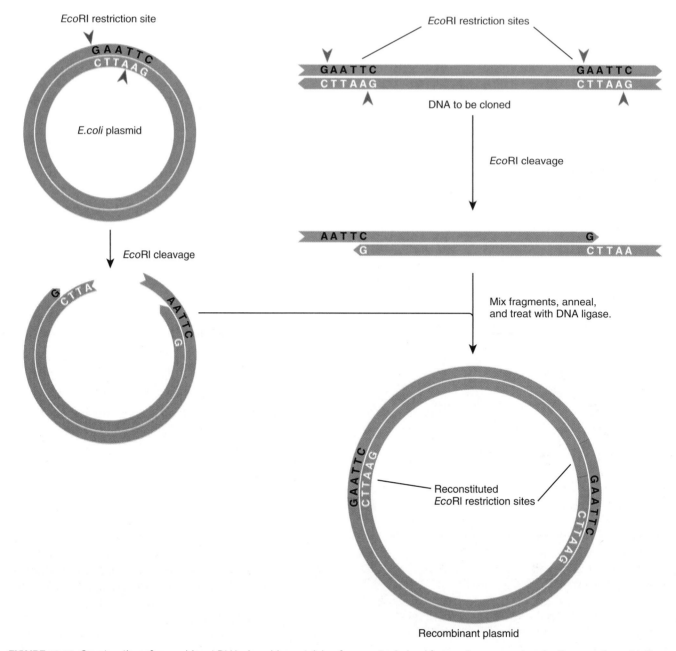

FIGURE 10.19 Construction of recombinant DNA plasmids containing fragments derived from a donor organism, by the use of a restriction enzyme (in this example *Eco*RI) and the joining of complementary (sticky) ends. Red arrowheads indicate cleavage sites.

genes, the length of the spliced mRNA is usually much less than the length of the gene. Even if the large DNA sequence were cloned, expression of the gene product in bacterial cells would be impossible because bacterial cells are not capable of RNA splicing. Therefore, when a gene is so large that it is difficult to clone and express directly, it would be desirable to clone the coding sequence present in the mRNA to determine the base sequence and study the polypeptide gene product. The method illustrated in **FIGURE 10.20** makes possible the direct cloning of any eukaryotic coding sequence from cells in which the mRNA is present.

Cloning from mRNA molecules depends on an unusual polymerase, **reverse transcriptase**, which

can use a single-stranded RNA molecule as a template and synthesize a complementary strand of DNA called **complementary DNA**, or **cDNA**. Like other DNA polymerases, reverse transcriptase requires a primer. The stretch of A nucleotides usually found at the 3′ end of eukaryotic mRNA serves as a convenient priming site, because the primer can be an oligonucleotide consisting of poly-T (Figure 10.20). Like any other single-stranded DNA molecule, the single strand of DNA produced from the RNA template can fold back upon itself at the extreme 3′ end to form a "hairpin" structure that includes a very short double-stranded region consisting of a few base pairs. The 3′ end of the hairpin serves as a primer for second-strand synthesis.

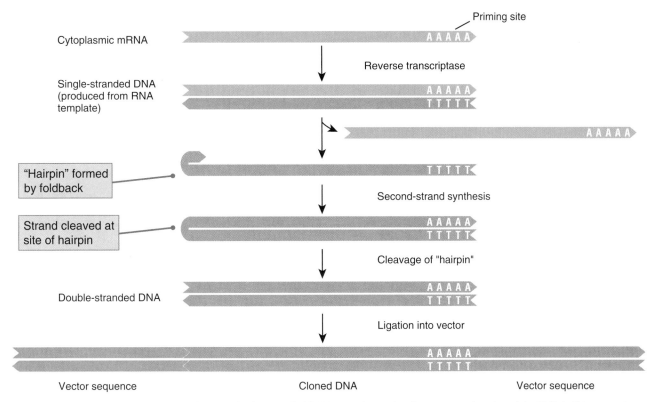

FIGURE 10.20 Reverse transcriptase produces a single-stranded DNA complementary in sequence to a template RNA. In this example, a cytoplasmic mRNA is copied. As indicated here, most eukaryotic mRNA molecules have a tract of consecutive A nucleotides at the 3′ end, which serves as a convenient priming site. After the single-stranded DNA is produced, a foldback at the 3′ end forms a hairpin that serves as a primer for second-strand synthesis. After the hairpin is cleaved, the resulting double-stranded DNA can be ligated into an appropriate vector either immediately or after PCR amplification. The resulting clone contains the entire coding region for the protein product of the gene.

The second strand can be synthesized either by DNA polymerase or by reverse transcriptase itself. Reverse transcriptase is the source of the second strand in RNA-based viruses that use reverse transcriptase, such as the human immunodeficiency virus (HIV). Conversion into a conventional double-stranded DNA molecule is achieved by cleavage of the hairpin by a nuclease.

In the reverse transcription of an mRNA molecule, the resulting full-length cDNA contains an uninterrupted coding sequence for the protein of interest. Eukaryotic genes often contain DNA sequences, called *introns*, that are initially transcribed into RNA but are removed in the production of the mature mRNA. Because the introns are absent from the mRNA, the cDNA sequence is not identical with that in the genome of the original donor organism. However, if the purpose of forming the recombinant DNA molecule is to identify the coding sequence or to synthesize the gene product in a bacterial cell, then cDNA formed from processed mRNA is the material of choice for cloning. The joining of cDNA to a vector can be accomplished by available procedures for joining blunt-ended molecules (Figure 10.20).

Some specialized animal cells make only one protein, or a very small number of proteins, in large amounts. In these cells, the cytoplasm contains a great abundance of specific mRNA molecules, which constitute a large fraction of the total mRNA synthesized.

An example is the mRNA for globin, which is highly abundant in reticulocytes while they are producing hemoglobin. The cDNA produced from purified mRNA from these cells is greatly enriched for the globin cDNA. Genes that are not highly expressed are represented by mRNA molecules whose abundance ranges from low to exceedingly low. The cDNA molecules produced from such rare RNAs will also be rare. The efficiency of cloning rare cDNA molecules can be markedly increased by PCR amplification prior to ligation into the vector. The only limitation on the procedure is the requirement that enough DNA sequence be known at both ends of the cDNA for appropriate oligonucleotide primers to be designed. PCR amplification of the cDNA produced by reverse transcriptase is called **reverse transcriptase PCR (RT-PCR)**. The resulting amplified molecules contain the coding sequence of the gene of interest with very little contaminating DNA.

Loss of β-galactosidase activity is often used to detect recombinant vectors.

When a vector is cleaved by a restriction enzyme and renatured in the presence of many different restriction fragments from a particular organism, many types of molecules result, including such examples as a self-joined circular vector that has not acquired any

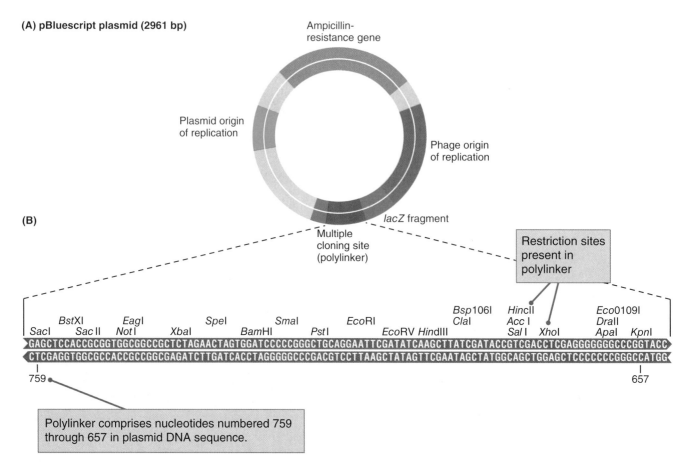

FIGURE 10.21 (A) Diagram of the cloning vector pBluescript II. It contains a plasmid origin of replication, an ampicillin-resistance gene, a multiple cloning site (polylinker) within a fragment of the *lacZ* gene from *E. coli*, and a bacteriophage origin of replication. (B) Sequence of the multiple cloning site showing the unique restriction sites at which the vector can be opened for the insertion of DNA fragments. The numbers 657 and 759 refer to the position of the base pairs in the complete sequence of pBluescript.

Courtesy of Agilent Technologies, Inc, Stratagene Products Division.

fragments, a vector containing one or more fragments, and a molecule consisting only of many joined fragments. To facilitate the isolation of a vector containing a particular gene, some means is needed to ensure (1) that the vector does indeed possess an inserted DNA fragment, and (2) that the fragment is in fact the DNA segment of interest. This section describes several useful procedures for detecting the correct products.

In the use of transformation to introduce recombinant plasmids into bacterial cells, the initial goal is to isolate bacteria that contain the plasmid from a mixture of plasmid-free and plasmid-containing cells. A common procedure is to use a plasmid that possesses an antibiotic-resistance marker and to grow the transformed bacteria on a medium that contains the antibiotic: Only cells that contain plasmid can form a colony. An example of a cloning vector is the pBluescript plasmid illustrated in **FIGURE 10.21**, part A. The entire plasmid is 2961 base pairs. Different regions contribute to its utility as a cloning vector.

- The plasmid origin of replication is derived from the *E. coli* plasmid ColE1. The ColE1 is a high-copy-number plasmid, and its origin of replication

enables pBluescript and its recombinant derivatives to exist in approximately 300 copies per cell.

- The ampicillin-resistance gene allows for selection of transformed cells in medium containing ampicillin.

- The cloning site is called a *multiple cloning site* (MCS), or *polylinker*, because it contains unique cleavage sites for many different restriction enzymes and enables many types of restriction fragments to be inserted. In pBluescript, the MCS is a 108-bp sequence that contains cloning sites for 23 different restriction enzymes (Figure 10.21, part B).

- The detection of recombinant plasmids is by means of a region containing the *lacZ* gene from *E. coli*, shown in blue in Figure 10.21, part A. The basis of the selection is illustrated in **FIGURE 10.22**. When the *lacZ* region is interrupted by a fragment of DNA inserted into the MCS, the recombinant plasmid yields Lac⁻ cells because the interruption renders the lacZ region

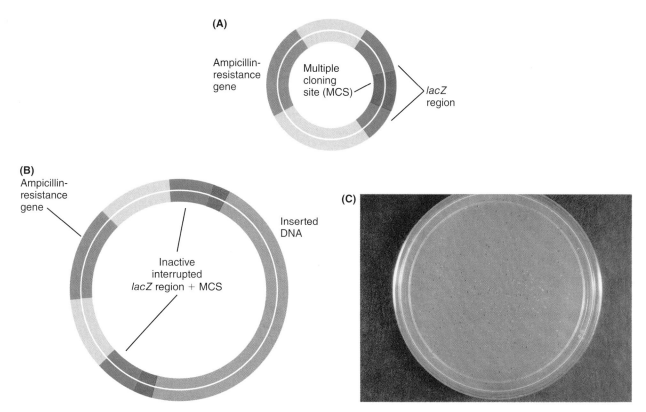

FIGURE 10.22 Detection of recombinant plasmids through insertional inactivation of a fragment of the *lacZ* gene from *E. coli*. (A) Nonrecombinant plasmid containing an uninterrupted *lacZ* region. The multiple cloning site (MCS) within the region (not drawn to scale) is sufficiently small that the plasmid still confers β-galactosidase activity. (B) Recombinant plasmid with donor DNA inserted into the multiple cloning site. This plasmid confers ampicillin resistance but not β-galactosidase activity, because the donor DNA interrupting the *lacZ* region is large enough to render the region nonfunctional. (C) Transformed bacterial colonies. Cells in the white colonies contain plasmids with inserts that disrupt the *lacZ* region; those in the blue colonies do not.

nonfunctional. Nonrecombinant plasmids do not contain a DNA fragment in the MCS and yield Lac⁺ colonies. The Lac⁺ and Lac⁻ phenotypes can be distinguished by color when the cells are grown on a special β-galactoside compound called X-gal, which releases a deep blue dye when cleaved. On medium containing X-gal, Lac⁺ colonies contain nonrecombinant plasmids and are a deep blue, whereas Lac⁻ colonies contain recombinant plasmids and are white.

- The bacteriophage origin of replication is from the single-stranded DNA phage f1. When cells that contain a recombinant plasmid are infected with an f1 helper phage, the f1 origin enables a single strand of the inserted fragment, starting with *lacZ*, to be packaged in progeny phage. This feature is very convenient because it yields single-stranded DNA for sequencing. The plasmid shown in part A of Figure 10.21 is the SK(+) variety. There is also an SK(−) variety in which the f1 origin is in the opposite orientation and packages the complementary DNA strand.

All good cloning vectors have an efficient origin of replication, at least one unique cloning site for the insertion of DNA fragments, and a second gene whose interruption by inserted DNA yields a phenotype indicative of a recombinant plasmid. Once a **library**, or large set of clones, has been obtained in a particular vector, the next problem is how to identify the particular recombinant clones that contain the DNA fragment of interest. These clones can be identified in any number of ways including hybridization with a labeled probe that is complementary to the cloned DNA (Figure 6.21 in Chapter 6). Recombinant clones can also be assayed for whether the DNA fragment they contain can be amplified by PCR primers complementary to the fragment of interest.

10.4 CRISPR-Cas9 technology for gene editing has revolutionized genetic engineering.

It rarely happens in science that a new technique is invented that revolutionizes how research is carried out. DNA sequencing was one such technique, the polymerase chain reaction another, and more recently, there is CRISPR-Cas9. The term **CRISPR-Cas9** refers

🧠 STOP & THINK 10.3

Presence of a polylinker, or multiple cloning site (MCS), in a vector makes possible *directional cloning*. In this approach, the vector and the target sequence are both cleaved with the same two restriction enzymes that are chosen so their complementary sticky ends will ensure that the fragment of interest is inserted in a particular orientation in the vector. Consider, for example, the restriction sites in the vector MCS on the left and the target sequence on the right shown in the accompanying diagram.

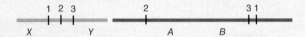

The vector sequence *X* to the left of the MCS is a promoter sequence, and the vector sequence *Y* to the right of the MCS is a transcriptional terminator. The target sequence is a protein-coding region that in order to be expressed must be oriented with *A* adjacent to, and to the right of, the promoter *X*.

Suppose you are a geneticist who wishes to create a recombinant molecule with the sequences oriented as *X—A—B—Y*. The positions of the cleavage sites of three restriction enzymes 1, 2, and 3 are shown. Each enzyme produces unique sticky ends. For directional cloning, which restriction enzymes would you use to digest the vector and target so that, after mixing and ligation of the fragments, the cloned DNA would have the sequence *X—A—B—Y*?

to a simple method for precise manipulation of the genome by altering specific sequences of DNA. The method is efficient and versatile, and it has applications in basic research, development, medicine, biotechnology, horticulture, agriculture, and other fields of biology.

CRISPR (rhymes with whisper) is an acronym for *clustered regularly interspaced short palindromic repeats.* Such clustered repeats were first discovered in the genome of *E. coli* in 1985; within a few years they were found to be widespread in bacteria and archaea. By 2005, it became clear that the repeated sequences derived from a wide variety of viruses and plasmids, and it began to dawn on researchers that CRISPR might be part of a prokaryotic immune system. This proved to be the case, although it was not until 2012 that the molecular mechanism was worked out. Please note the timeline: It was nearly 30 years—a whole human generation—between the time that CRISPR was first discovered and the time that the system was understood well enough to be applied to genetic engineering. This example is typical of the timescale between important basic-science discoveries and their practical applications. It also illustrates the utter unpredictability of which basic-science discoveries made at any given time will turn out to have practical implications.

CRISPR-Cas9 refers specifically to the bacterial immune system of *Streptococcus pyogenes*, the first to be described in detail. Other prokaryotic species have slightly different systems that work similarly; however, the term CRISPR-Cas9 is often used generically to refer to any such system.

Before discussing how CRISPR-Cas9 is used in genetic engineering, let's first examine how the components function in bacterial immunity. Part A in **FIGURE 10.23** lays out the players. First, there is a CRISPR repeat in the bacterial genome, which for concreteness we suppose is part of the genome of a bacterial virus. This repeat is transcribed into a *guide RNA (gRNA)*. The guide RNA is one key component of the system. The other two components are a *trans*-acting CRISPR RNA (*tracrRNA*) and the CRISPR-associated protein 9 nuclease (*Cas9*). The gRNA initiates a series of events that ultimately results in destruction of the genome of any invading virus that contains a complementary sequence. In the first step (Figure 10.23, part B), the gRNA invades the viral target DNA by base pairing with its complementary DNA strand. At this point (Figure 10.23, part C), the tracrRNA joins the complex and recruits the Cas9 protein. Cas9 protein is a nuclease, and it cleaves both strands of the target viral DNA (Figure 10.23, part C). The cleaved ends of the target DNA are then attacked by exonucleases in the cell, and the target viral DNA is degraded (Figure 10.23, part D).

CRISPR-Cas9 can be used to create knockout mutations of any gene.

The most straightforward application of CRISPR-Cas9 is to create targeted deletions, often called **knockout mutations** because a deletion results in loss of function. In applying CRISPR-Cas9 to other organisms, the components have been simplified because the gRNA

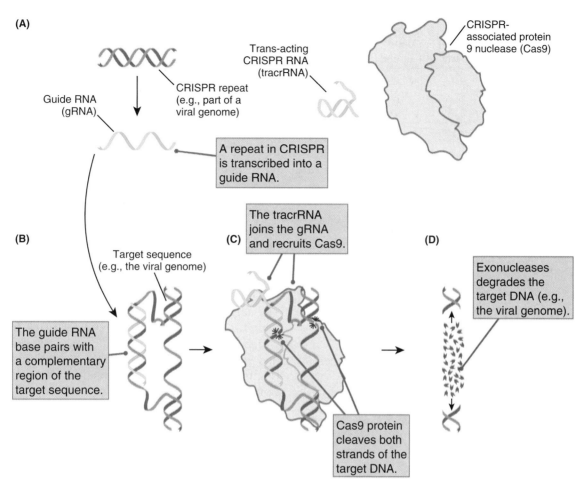

(A)

Guide RNA (gRNA)

CRISPR repeat (e.g., part of a viral genome)

Trans-acting CRISPR RNA (tracrRNA)

CRISPR-associated protein 9 nuclease (Cas9)

A repeat in CRISPR is transcribed into a guide RNA.

The tracrRNA joins the gRNA and recruits Cas9.

(B)

Target sequence (e.g., the viral genome)

(C)

(D)

Exonucleases degrades the target DNA (e.g., the viral genome).

The guide RNA base pairs with a complementary region of the target sequence.

Cas9 protein cleaves both strands of the target DNA.

FIGURE 10.23 CRISPR-Cas9 in bacterial immunity. (A) Components of the system. (B) Invasion of target DNA. (C) Recuitment of tracrRNA and Cas9 results in cleavage of the target DNA. (D) Exonucleases degrade target DNA from the cleaved ends.

still works when joined to the tracrRNA. The joined components constitute what is known as **synthetic guide RNA (sgRNA)**.

Creation of targeted deletions is outlined in **FIGURE 10.24**. The first step is to synthesize an sgRNA that is complementary to the genomic sequence to be deleted, and the second is to introduce the sgRNA and Cas9 into the nucleus of cells to be engineered. As shown in Figure 10.24, the sgRNA invades the target DNA by base pairing with its complementary strand, and then the tracrRNA component of sgRNA recruits Cas9, which cleaves both strands of the target DNA. Exonucleases then begin to degrade the target DNA from the cleaved ends, enlarging the gap.

The DNA in a eukaryotic chromosome is much longer than that in a bacterial virus, and eukaryotic cells have mechanisms to detect broken DNA and to rejoin the ends. One such mechanism is known as **nonhomologous end joining**, which, in the case of Figure 10.24, joins the ends of the gapped DNA strand. The end joining cannot recover the already degraded nucleotides, however, and the result is a deletion of part of the target sequence. The targeted-deletion

method is very efficient. In many cases, 50 percent or more of treated cells have a knockout of one or both alleles of the target sequence.

CRISPR-Cas9 can be used to edit the sequence of any gene.

Cells have an alternative to nonhomologous end joining, which is known as **template-directed gap repair**. In this process, a DNA duplex present elsewhere in the same nucleus is used as a template for replacing the missing nucleotides in the gap. All that is necessary is that the template DNA has sufficient homology to allow base pairing with the DNA strands at each end of the gap. When used with template DNA to alter a gene sequence, the CRISPR-Cas9 method is known as **DNA editing** or *gene editing*.

Use of CRISPR-Cas9 for DNA editing is outlined in **FIGURE 10.25**. As we have already seen, the sgRNA together with Cas9 nuclease produces a double-stranded break at a specific site in the target DNA, and exonucleases widen the gap. When the treated cells also contain a template DNA, the

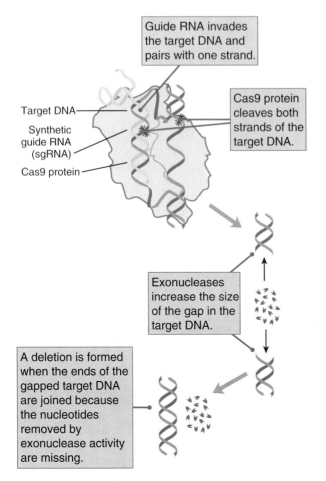

Guide RNA invades the target DNA and pairs with one strand.

Target DNA

Synthetic guide RNA (sgRNA)

Cas9 protein

Cas9 protein cleaves both strands of the target DNA.

Exonucleases increase the size of the gap in the target DNA.

A deletion is formed when the ends of the gapped target DNA are joined because the nucleotides removed by exonuclease activity are missing.

FIGURE 10.24 Use of CRISPR-Cas9 to create targeted deletions.

gapped target DNA can invade the template by base pairing; each strand of the template is used to restore the sequences that were in the gap. The use of the template to repair the gap is virtually identical to double-strand break repair as it takes place in genetic recombination.

In DNA editing, however, the template sequence need not be identical to the original target. It could have one or more nucleotide substitutions, which in a coding region would result in one or more amino acid replacements. The template DNA could be even longer than the original target DNA and contain one or more genes not present in the unedited genome. In this manner, a gene from any organism can be introduced into the genome of any other organism.

Methods for using CRISPR-Cas9 depend on the organism.

We've already seen how CRISPR-Cas9 works in producing targeted deletions or in DNA editing. How is the procedure carried out in practice? The methods depend on the situation. In cells in culture, sgRNA, Cas9 coding sequence, and template DNA can be introduced by DNA transformation using plasmids or artificial chromosomes created by the methods discussed in Section 10.3. In insects, the CRISPR-Cas9 components can be injected into early embryos, where they are taken up by the nuclei of germ-line cells and function efficiently to alter the genome. Genetic markers are usually incorporated into the target DNA to identify progeny whose genomes have been altered. In some insects, including *Drosophila*, special strains have been engineered to produce Cas9 protein, and in these strains all that is needed for genome editing is to introduce sgRNA and template DNA.

One method for using CRISPR-Cas9 in mice is shown in **FIGURE 10.26**. In this case, the components of the system are injected into single-cell embryos, where they are taken up and function in the nucleus, and groups of injected embryos are introduced in to the uterus of a foster mother (Figure 10.26, part A). The target DNA usually includes a genetic marker, which in this example is a gene for black fur color, which results in offspring that can easily be identified if their genomes have been successfully edited (Figure 10.26, part B). (In practice, a green fluorescent protein is often used as a marker and, viewed under ultraviolet light, the transformed progeny have an eerie greenish glow.)

CRISPR-Cas9 can also be used in plants.

A procedure for using CRISPR/Cas9 in plant cells makes use of a plasmid found in the soil bacterium *Agrobacterium tumefaciens* and related species. Infection of susceptible plants with this bacterium results in the growth of what are known as crown gall tumors at the entry site, which is usually a wound. Susceptible plants comprise about 160,000 species of flowering plants, known as the dicots, and include the great majority of the most common flowering plants.

The *Agrobacterium* contains a large plasmid of approximately 200 kb called the *Ti* **plasmid**, which includes a smaller region of about 25 kb known as the *T* **DNA** flanked by 25-bp direct repeats (**FIGURE 10.27**). In its natural state, the *Agrobacterium* causes a profound change in the metabolism of infected cells because the *T* DNA is transferred into the plant genome. The *T* DNA contains genes coding for proteins that stimulate division of infected cells, thereby causing the tumor, and also coding for enzymes that convert the amino acid arginine into an unusual derivative, generally nopaline or octopine, that the bacterium needs in order to grow. The transfer functions are present not in the *T* DNA itself but in another region of the plasmid called the *vir* (stands for virulence) region of about 40 kb that includes six genes necessary for transfer (Figure 10.27, part A).

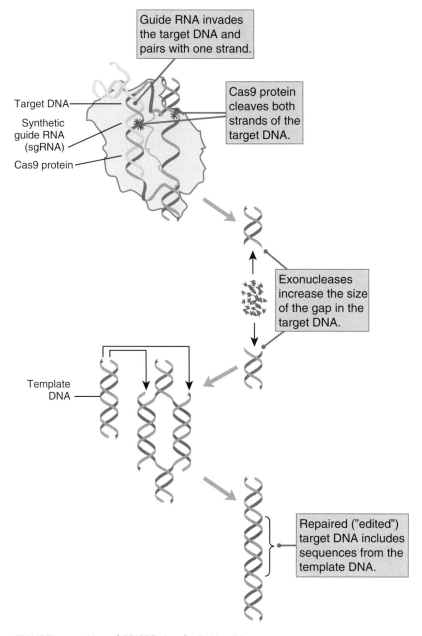

Guide RNA invades the target DNA and pairs with one strand.

Target DNA

Synthetic guide RNA (sgRNA)

Cas9 protein

Cas9 protein cleaves both strands of the target DNA.

Exonucleases increase the size of the gap in the target DNA.

Template DNA

Repaired ("edited") target DNA includes sequences from the template DNA.

FIGURE 10.25 Use of CRISPR-Cas for DNA editing.

In its use as a vector for CRISPR/Cas9, the *T* DNA is modified to include the template DNA as well as sequences encoding sgRNA and Cas9 protein (Figure 10.27). Transfer of *T* DNA into the plant genome is similar in some key respects to bacterial conjugation. As illustrated in Figure 10.27, transfer begins with the formation of a nick that frees one end of the *T* DNA, which peels off the plasmid and is replaced by rolling-circle replication. The region of the plasmid that is transferred is delimited by a second nick at the other end of the *T* DNA, but the position of this nick is variable. The resulting single-stranded *T* DNA is bound with molecules of a single-stranded binding protein and is transferred into the plant cell and incorporated

into the nucleus. There it is integrated into the chromosomal DNA and the components of the CRISPR-Cas9 system carry out their functions (part C).

10.5 Genetic engineering is applied in medicine, industry, agriculture, and research.

CRISPR-Cas9 and earlier technologies have revolutionized modern biology, not only by opening up new approaches in basic research but also by making possible the creation of organisms with novel genotypes for practical use in agriculture and industry. In this section, we examine a few of many applications of recombinant DNA.

Animal growth rate can be genetically engineered.

In many animals, the rate of growth is controlled by the amount of growth hormone produced. Transgenic animals with a growth-hormone gene under the control of a highly active promoter to drive transcription often grow larger than their normal counterparts. An example of a highly active promoter is found in the gene for *metallothionein*. The metallothioneins are proteins that bind heavy metals. They are ubiquitous in eukaryotic organisms and are encoded by members of a family of related genes. The human genome, for example, includes more than 10 metallothionein genes that can be separated into two major groups according to their sequences. The promoter region of a metallothionein gene drives transcription of any gene to which it is attached, in response to heavy metals or steroid hormones. For example, when DNA constructs consisting of a rat growth-hormone gene under metallothionein control are used to produce transgenic mice, the resulting animals grow about twice as large as normal mice.

The effect of another growth-hormone construct is shown in **FIGURE 10.28**. The fish are coho salmon at 14 months of age. Those on the left are normal, whereas those on the right are transgenic animals that contain a salmon growth-hormone gene driven

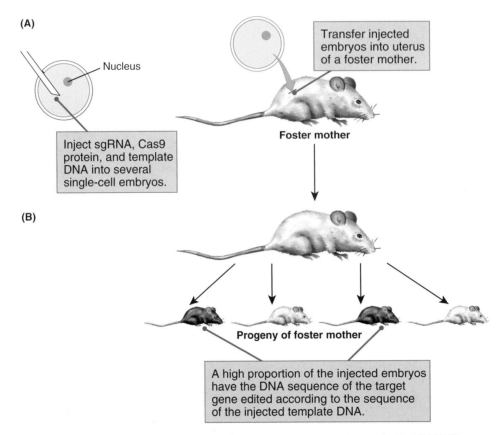

(A)

Nucleus

Inject sgRNA, Cas9 protein, and template DNA into several single-cell embryos.

Transfer injected embryos into uterus of a foster mother.

Foster mother

(B)

Progeny of foster mother

A high proportion of the injected embryos have the DNA sequence of the target gene edited according to the sequence of the injected template DNA.

FIGURE 10.26 One method for using CRISPR/Cas9 in mice. (A) Single-cell embryos are injected with CRISPR/Cas9 components, and groups of injected embryos are introduced into foster mothers. (B) A genetic marker in the template DNA (in this example, a gene for black coat color) identifies the offspring whose genome has been successfully edited.

by a metallothionein regulatory region. Both the growth-hormone gene and the metallothionein gene were cloned from the sockeye salmon. As an indicator of size, the largest transgenic fish on the right has a length of about 42cm. On average, the transgenic fish are 11 times heavier than their normal counterparts; the largest transgenic fish was 37 times the average weight of the nontransgenic animals. Not only do the transgenic salmon grow faster and become larger than normal salmon; they also mature faster.

Crop plants with improved nutritional qualities can be created.

Beyond the manipulation of single genes, it is also possible to create transgenic organisms that have entirely new metabolic pathways introduced. A remarkable example is in the creation of a genetically engineered rice that contains an introduced biochemical pathway for the synthesis of β-carotene, a precursor of vitamin A found primarily in yellow vegetables and greens. (Deficiency of vitamin A affects some 400 million people throughout the world, predisposing them to skin disorders and night blindness.) The β-carotene pathway includes four enzymes, which in the engineered rice are encoded in genes from different organisms (**FIGURE 10.29**).

Two of the genes come from the common daffodil (*Narcissus pseudonarcissus*), whereas the other two come from the bacterium *Erwinia uredovora*. Each pair of genes was cloned into *T* DNA and transformed into rice using *Agrobacterium tumefaciens* (Figure 10.27). Transgenic plants were then crossed to produce progeny containing all four enzymes. The engineered rice seeds contain enough β-carotene to provide the daily requirement of vitamin A in 300 grams of cooked rice; they even have a yellow tinge (Figure 10.29, part B).

People on high-rice diets are also prone to iron deficiency because rice contains a small phosphorus-storage molecule called *phytate*, which binds with iron and interferes with its absorption through the intestine. The transgenic β-carotene rice was also engineered to minimize this problem by introducing the fungal enzyme from *Aspergillus ficuum* that breaks down phytate, along with a gene encoding the iron-storage protein ferritin from the French bean, *Phaseolus vulgaris*, plus yet another gene from basmati rice that encodes a metallothionein-like gene that facilitates iron absorption in the human gut. Altogether, then, the transgenic rice strain rich in β-carotene and available iron contains six new genes taken from four unrelated species plus one gene from a totally different strain of rice!

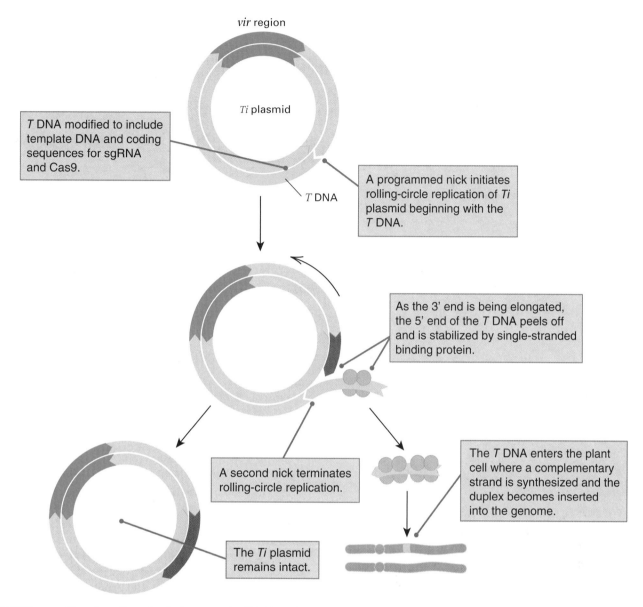

FIGURE 10.27 Transformation of a plant genome by *T* DNA engineered to contain components of the CRISPR-Cas9 system for DNA editing.

The production of useful proteins is a primary impetus for recombinant DNA.

Among the most important applications of genetic engineering is the production of large quantities of particular proteins that are otherwise difficult to obtain (for example, proteins that are present in only a few molecules per cell or that are produced in only a small number of cells or only in human cells). The method is simple in principle. A DNA sequence coding for the desired protein is cloned in a vector adjacent to an appropriate regulatory sequence. This step is usually done with cDNA, because cDNA has all the coding sequences spliced together in the right order. Using a vector with a high copy number ensures that many copies of the coding sequence will be present in each bacterial cell, which can result in synthesis of the gene product at concentrations ranging from 1 to 5 percent of the total cellular protein. In practice, the production of large quantities of a protein

FIGURE 10.28 Normal coho salmon (left) and genetically engineered coho salmon (right) containing a sockeye salmon growth-hormone gene driven by the regulatory region from a metallothionein gene. The transgenic salmon average 11 times the weight of the nontransgenic fish. The smallest fish on the left is about 4 inches long.

Courtesy of R. H. Devlin, Fisheries and Oceans Canada (after Devlin et. al. *Nature* 371 (1994): 209–210).

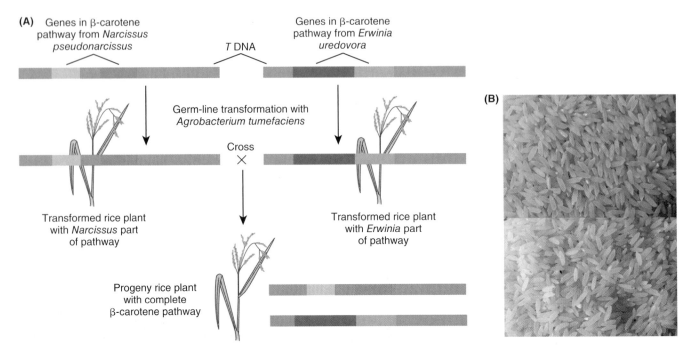

FIGURE 10.29 Genetically engineered rice containing a biosynthetic pathway for β-carotene. (A) Enzymes in the pathway derive from genes in two different species. (B) Rice plants with both parts of the pathway produce grains with a yellowish cast (top) because of the β-carotene they contain, in contrast to the pure white grains (bottom) of normal plants.

(B) Courtesy of Ingo Potrykus, Institute für Pflanzenwissenschaften, ETH Zurich.

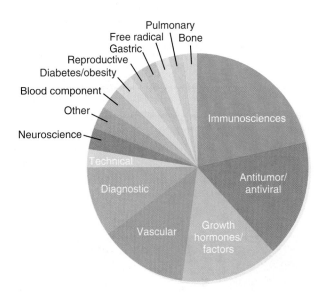

FIGURE 10.30 Relative numbers of patents issued for various clinical applications of the products of genetically engineered human genes.

Data from S. M. Thomas, et al., *Nature* 380 (1996): 387-388.

in bacterial cells is straightforward, but there are often problems that must be overcome, because in the bacterial cell, which is a prokaryotic cell, the eukaryotic protein may be unstable, may not fold properly, or may fail to undergo necessary chemical modification. Many important proteins are currently produced in bacterial cells, including human growth hormone, blood-clotting factors, and insulin. Patent offices in Europe and the United States have issued more than 50,000 patents for the clinical use of the products of genetically engineered human genes. **FIGURE 10.30** gives a breakdown of the approximate numbers of patents issued for various clinical applications.

CHAPTER SUMMARY

- High-throughput automated DNA sequencing has resulted in the complete sequence of the genomes of many species of bacteria, archaeons, and eukaryotes. It may soon become part of routine medical diagnostics and personalized medicine.

- Comparisons among genomes of related species help discover coding sequences and other functional genetic elements.

- Genomic sequences of contemporary non-African human populations indicate genomic regions inherited from interbreeding with Neanderthals or a smaller offshoot Denisovan population.

- Functional genomics using DNA microarrays enables the level of gene expression of all genes in the genome to be assayed simultaneously, which allows global patterns and coordinated regulation of gene expression to be investigated.

- Proteomics methods, such as two-hybrid analysis of proteins, allow protein–protein interaction networks to be identified.

- In recombinant DNA (gene cloning), DNA fragments are isolated, inserted into suitable vector molecules, and introduced into host cells (usually bacteria or yeast), where they are replicated.

- Recombinant DNA is widely used in research, medical diagnostics, and the manufacture of drugs and other commercial products.

- CRISPR-Cas9 is a simple, efficient, and versatile method for producing knockout mutations or for DNA editing to alter genomes or introduce new genes.

- Transgenic organisms carry DNA sequences that have been introduced by CRISPR-Cas9 or other methods.

ISSUES AND IDEAS

- What does the term *recombinant DNA* mean? What are some of the practical uses of recombinant DNA?

- What features are essential in a bacterial cloning vector? How can a vector have more than one cloning site?

- What is the reaction catalyzed by the enzyme reverse transcriptase? How is this enzyme used in recombinant DNA technology?

- What is meant by the term *genome annotation*? Explain why genome sequences need to be annotated.

- What are DNA microarrays and how are they used in functional genomics?

- Describe the two-hybrid system that makes use of the yeast GAL4 protein and explain how the two-hybrid system detects interaction between proteins.

- What is a transgenic organism? What are some of the practical uses of transgenic organisms?

- Explain how CRISPR-Cas9 is used to create knockout mutations of specific genes and explain how the system is used in DNA editing.

SOLUTIONS: STEP BY STEP

PROBLEM 1 What is the average distance between restriction sites for each of the following restriction enzymes? Assume that the DNA substrate has a random sequence with equal amounts of each base. The symbol N stands for any nucleotide, R for any purine (A or G), and Y for any pyrimidine (T or C).
(a) TCGA (*Taq*I)
(b) GGTACC (*Kpn*I)
(c) GTNAC (*Mae*III)
(d) GGNNCC (*Nla*IV)
(e) GRCGYC (*Acy*I)

SOLUTION. **(a)** The average distance between restriction sites equals the reciprocal of the probability of occurrence of the restriction site. You must, therefore, calculate the probability of occurrence of each restriction site in a random DNA sequence. The probability of the sequence TCGA is

$$1/4 \times 1/4 \times 1/4 \times 1/4 = (1/4)^4 = 1/256$$

so 256 bases is the average distance between *Taq*I sites. **(b)** By the same reasoning, the probability of a GGTACC site is $(1/4)^6 \times 1/4096$, and so 4096 bases is the average distance between *Kpn*I sites. **(c)** The probability of N (any nucleotide at a site) is 1, and, hence, the probability of the sequence GTNAC is

$$1/4 \times 1/4 \times 1 \times 1/4 \times 1/4 = (1/4)^4 = 1/256$$

Therefore, 256 is the average distance between *Mae*III sites. **(d)** The same reasoning yields the average distance between GGNNCC (*Nla*IV) sites as $1/4 \times 1/4 \times 1 \times 1 \times 1/4 \times 1/4 = 1/256$ bases. **(e)** The probability of an R (A or G) at a site is 1/2, and the probability of a Y (T or C) at a site is 1/2. Hence, the probability of the sequence GRCGYC is

$$1/4 \times 1/2 \times 1/4 \times 1/4 \times 1/2 \times 1/4 = 1/1024$$

so the average number of bases between *Acy*I sites is 1024 bases.

PROBLEM 2 How many clones are needed to establish a library of DNA from a species of lemur with a diploid genome size of 6×10^9 base pairs if (1) the clones contain fragments of average size 2×10^4 base pairs, and (2) one wants 99 percent of the genomic sequences to be present in at least one clone in the library? (*Hint*: If a genome is cloned at random into a library with x-fold coverage, the probability that a particular sequence will be missing from the library is e^{-x}.)

SOLUTION. The hint says that if the genome were represented x times in the library, the probability that a particular sequence would be missing is e^{-x}, which we want to equal 0.01. Hence, the required library should have x-fold coverage, where $e^{-x} = 0.01$ or $x = -\ln(0.01) = 4.6$. Because one haploid representation of the genome equals $(6 \times 10^9)/2 = 3 \times 10^9$ base pairs, and the average insert size is 2×10^4 base pairs, the required library should include $[(3 \times 10^9)/(2 \times 10^4)] \times 4.6 = 6.9 \times 10^5$ clones.

CONCEPTS IN ACTION: PROBLEMS FOR SOLUTION

10.1   Will the sequences 5′–GGCC–3′ and 3′–GGCC–5′ in a double-stranded DNA molecule be cut by the same restriction enzyme?

10.2   A circular plasmid has two restriction sites for the enzyme *Zsp2*I, which cleaves the site ATGCA↓T (the arrow indicates the position of the cleavage). After digestion the fragments are ligated together, and a circular product is isolated that includes one copy of each of the fragments. Does this mean that the ligated plasmid is the same as the original? Explain.

10.3   In recombinant DNA, researchers typically prefer ligating restriction fragments that have sticky ends (single-stranded overhangs) rather than those that have blunt ends. Can you propose a reason why?

10.4   A *kan-r tet-r* plasmid is treated with the restriction enzyme *Bgl*I, which cleaves the *kan* (kanamycin) gene. The DNA is annealed with a *Bgl*I digest of *Neurospora* DNA and after ligation used to transform *E. coli*.

   (a)   What antibiotic would you put into the growth medium to ensure that each colony has the plasmid?

   (b)   What antibiotic-resistance phenotypes would be found among the resulting colonies?

   (c)   Which phenotype is expected to contain *Neurospora* DNA inserts?

10.5   You want to introduce the human insulin gene into a bacterial host in hopes of producing a large amount of human insulin. Should you use the genomic DNA or the cDNA? Explain your reasoning.

10.6   You decide to clone your pet dog, which is brown with black spots. You take a few somatic cells from your dog and perform a somatic cell nuclear transfer procedure using an egg from a female dog that is black. In this procedure, the egg nucleus is removed and replaced with that from a somatic cell. What color fur will the puppy clone of your dog have?

10.7   After doing a restriction digest with the enzyme *Sse*I, which has the recognition site 5′–CCTGCA↓GG–3′ (the arrow indicates the position of the cleavage), you wish to separate the fragments in an agarose gel. In order to choose the proper concentration of agarose, you need to know the expected size of the fragments. Assuming equivalent amounts of each of the four nucleotides in the target DNA, what average fragment size would you expect?

10.8   The restriction enzymes *Acc*651 and *Kpn*I have the restriction sites

G↓GTACCT (*Acc*651)
GGTAC↓C (*Kpn*I)

where the 5′ end is written at the left and the arrow indicates the position of the cleavage. Are the sticky ends produced by these restriction enzymes compatible? Explain.

10.9   In cloning into bacterial vectors, why is it useful to insert DNA fragments to be cloned into a restriction site inside an antibiotic-resistance gene? Why is another gene for resistance to a second antibiotic also required?

10.10   A mutant allele is found to express the wildtype gene product but at only about 20 percent of the wildtype level. The mutation is traced to an intron whose size has increased by 3.1 kb because of the presence of a DNA fragment with the restriction map shown here. The symbols *A, B, C, D, E, H, K, P, S,* and *X* represent cleavage sites for the restriction enzymes *Alu*I, *Bam*HI, *Cla*I, *Dde*I, *Eco*RI, *Hin*dIII, *Kpn*I, *Pst*I, *Sac*I, and *Xho*II, respectively. Does the restriction map of the insertion give any clues to what it is?

EX S EBP   D   P C   H A   S   K PBE S XE
├┼─┼┼┼┤──────┼───┼──┼─────┼─┼────────┼────────┼──┼┼┼┼─┼┤

10.11 If the genomic and cDNA sequences of a gene are compared, what information does the cDNA sequence provide that is not obvious from the genomic sequence? What information does the genomic sequence contain that is not in the cDNA?

10.12 In studies of the operator region of an inducible operon in *E. coli*, the four constructs shown below were examined for level of transcription *in vitro*. The number associated with each construct is the relative level of transcription observed in the presence of the repressor protein. The symbols *E*, *B*, *H*, and *S* stand for the restriction sites *Eco*RI, *Bam*HI, *Hind*III, and *Sac*I. Construct **(a)** is the wildtype operator region, and in parts **(b–d)**, the open boxes indicate restriction fragments that were deleted. What hypothesis about repressor–operator interactions can explain these results? How could this hypothesis be tested?

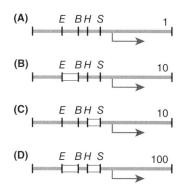

10.13 The Hessian fly *Mayetiola destructor* has among the smallest genomes in insects, with a haploid genome size of about 88 Mb. If this genome is digested with *Not*I (an eight-base cutter), approximately how many DNA fragments would be produced? Assume equal and random frequencies of the four nucleotides.

10.14 A circular plasmid of 8 kb is digested with *Eco*RI (E) and/or *Bam*HI (B); the digests are run on an agarose gel and stained. The results are shown below; molecular size standards are shown.

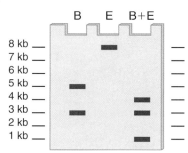

Draw the map of the plasmid.

10.15 How frequently would the restriction enzymes *Taq*I (restriction site TCGA) and *Mae*III (restriction site GTNAC, in which N is any nucleotide) cleave double-stranded DNA molecules containing each of the following random sequences?

(a) 20% A, 20% T, 30% G, and 30% C

(b) 30% A, 30% T, 20% G, and 20% C

10.16 How many clones are needed to establish a library of DNA from a species of grasshopper with a diploid genome size of 1.6×10^{10} base pairs if (1) fragments of average size 1×10^4 base pairs are used, and (2) one wants 95 percent of the genomic sequences to be in the library? (*Hint:* If the genome is cloned at random with *x*-fold coverage, the probability that a particular sequence will be missing is e^{-x}.)

10.17 Suppose that you digest the genomic DNA of a particular organism with *Sau*3A ($\downarrow$GATC), where the arrow represents the cleavage site. Then you ligate the resulting fragments into a unique *Bam*HI (G$\downarrow$GATCC) cloning site of a plasmid vector. Would it be possible to isolate the cloned fragments from the vector using *Bam*HI? From what proportion of clones would it be possible?

10.18 A DNA microarray is hybridized with fluorescently labeled reverse-transcribed DNA as described in the text, where the control mRNA (C) is labeled with a green fluorescent compound and the experimental mRNA (E) with a red fluorescent compound. Indicate what you can conclude about the relative levels of expression of a spot in the microarray that fluoresces:

(a) Red

(b) Green

(c) Yellow

(d) Orange

(e) Lime green

10.19 Shown here is a restriction map of a 12-kb linear plasmid isolated from cells of *Borrelia burgdorferi*, a spirochete bacterium transmitted by the bite of *Ixodes* ticks that causes Lyme disease. The symbols *D*, *P*, *C*, *H*, *K*, *S*, and *A* represent cleavage sites for the restriction enzymes *Dde*I, *Pst*I, *Cla*I, *Hind*III, *Kpn*I, *Sac*I, and *Alu*I, respectively. In the accompanying gel diagram, show the positions at which bands would be found after digestion of the plasmid with the indicated restriction enzyme or enzymes.

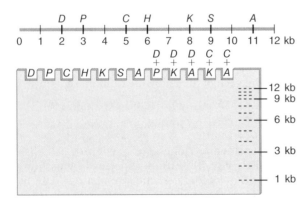

10.20 A functional genomics experiment is carried out using a DNA microarray to assay levels of gene expression in a species of bacteria. What genes would you expect to find overexpressed in cells grown in minimal medium compared to cells grown in complete medium?

10.21 A functional genomics experiment is carried out in *E. coli* to examine global levels of gene expression in various types of minimal growth medium. RNA extracted from the experimental culture is labeled with a molecule that fluoresces red, and RNA extracted from the control culture is labeled with a molecule that fluoresces green. The experimental and control samples are mixed prior to hybridization. Shown here are spots on the microarray corresponding to five genes: *trpE* (the first gene in the tryptophan bio-synthetic operon), *lacI*, *lacZ*, *lacY*, and *crp* (which encodes the cAMP receptor protein). Color the spots red, green, or yellow according to the relative levels of expression of each gene in the experimental and control cultures. (*Hint*: Before answering, think carefully about how the cAMP receptor protein co-regulates the *lac* operon.)

Experimental minimal medium	Control minimal medium	Transcript				
		trpE	*lacI*	*lacZ*	*lacY*	*crp*
Glucose	Glucose	○	○	○	○	○
Glucose	Glycerol	○	○	○	○	○
Glycerol	Glucose	○	○	○	○	○
Lactose	Glucose	○	○	○	○	○
Glucose	Lactose	○	○	○	○	○
Lactose	Glycerol	○	○	○	○	○
Glycerol	Lactose	○	○	○	○	○

STOP & THINK ANSWERS

ANSWER TO STOP & THINK 10.1

Enhancers and silencers are relatively short sequences that can differ in their orientation relative to the gene they affect and in their distance from it. Comparing the genomes of different but related species reveals which short sequences are conserved between the species irrespective of their orientation and distance from the gene they affect. These short, conserved sequences are good candidates for enhancers or silencers.

ANSWER TO STOP & THINK 10.2

If highly connected proteins were connected with many other highly connected proteins, any perturbation would spread across much or most of the network. Minimizing the connections of proteins to which highly connected proteins limits the spread of perturbations mainly to proteins connected to one hub.

ANSWER TO STOP & THINK 10.3

You should use enzymes 2 and 3. These produce the fragment 2—*A*—*B*—3, which inserts into the vector with *A* immediately to the right of *X*. Using 1 and 2 results in 2—*A*—*B*—1, which inserts in the opposite orientation.

Design Credits: Stop & Think icon made by Darius Dan from www.flaticon.com; The Human Connection icon made by Daniel Bruce from www.flaticon.com; Elephant image: © NickBiemans/GettyImages.

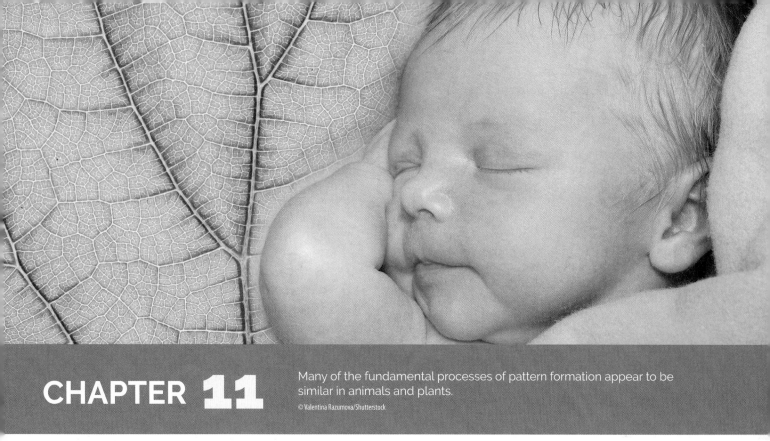

CHAPTER **11**

Many of the fundamental processes of pattern formation appear to be similar in animals and plants.

The Genetic Control of Development

LEARNING OBJECTIVES

- To distinguish between a ligand and a receptor and explain their respective roles in developmental processes.

- To explain how the principle of epistasis is applied to deduce the temporal order in which gene products function in a linear switch-regulation developmental pathway.

- To describe the role of coordinate genes, gap genes, pair-rule genes, segment-polarity genes, and homeotic genes in *Drosophila* development, and describe the characteristic phenotypes associated with mutants of each type of gene.

- To design an experiment to determine whether a *Pax6*-like gene is involved in the development of eyes in a newly discovered species of animal.

- To describe the ABC model of flower development in *Arabidopsis thaliana*, and explain how some of the transcription factors act combinatorially in the development of certain floral organs.

In the development of an organism, genes are expressed according to a prescribed program to ensure that as the fertilized egg divides repeatedly the resulting cells become specialized in an orderly way to give rise to the fully differentiated organism. Within what is usually a wide range of environments, the genotype determines not only the events that take place in development but also the temporal order in which the events unfold. The key process in development is **pattern formation**, which means the emergence of the spatially organized and specialized cells in the embryo from cell division and differentiation of the fertilized egg.

Genetic analyses of development often make use of mutations that alter developmental patterns. These mutations make it possible to identify genes that control development and to study the interactions among them. This chapter demonstrates how genetics is used in the study of development. To illustrate the principles, we focus on a specific example from each of three key model organisms: *Caenorhabditis elegans, Drosophila melanogaster,* and *Arabidopsis thaliana.*

11.1   The determination of cell fate in *C. elegans* development is largely autonomous.

The soil nematode *Caenorhabditis elegans* (**FIGURE 11.1**) is popular for genetic studies because it is small, easy to culture, and has a short generation time with a large number of offspring. The worms are grown on agar surfaces in petri dishes and feed on bacterial cells such as *Escherichia coli.* Because they are microscopic in size, as many as 10^5 animals can be contained in a single petri dish. Sexually mature adults of *C. elegans* are capable of laying more than 300 eggs within a few days. At 20°C, it requires about 60 hours for the eggs to hatch, undergo four larval molts, and become sexually mature adults.

Nematodes are diploid organisms with two sexes. In *C. elegans*, the two sexes are the hermaphrodite and the male. The hermaphrodite contains two X chromosomes (XX), produces both functional eggs and functional sperm, and is capable of self-fertilization. The male produces only sperm and fertilizes the hermaphrodites. The sex-chromosome constitution of *C. elegans* consists of a single X chromosome; there is no Y chromosome, and the male karyotype is XO.

Development in *C. elegans* exhibits a fixed pattern of cell divisions and cell lineages.

The transparent body wall of the worm has made it possible to study the division, migration, and death or differentiation of all cells present in the course of development. Nematode development is unusual in that the pattern of cell division and differentiation is virtually identical from one individual to the next. The result is that each sex shows the same geometry in the number and arrangement of somatic cells. The hermaphrodite contains exactly 959 somatic cells, and the male contains exactly 1031 somatic cells. The complete developmental history of each somatic cell is known.

The mechanisms that control early development can be studied genetically by isolating mutants with early developmental abnormalities and altered cell fates. In most organisms, it is difficult to trace the lineage of individual cells in development because the embryo is not transparent, the cells are small and numerous, and cell migrations are extensive. The *lineage* of a cell refers to the ancestor–descendant relationships among a group of cells. A cell lineage can be illustrated with a **lineage diagram**, a sort of cell pedigree that shows each cell division and indicates the

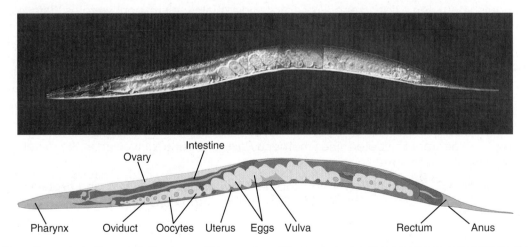

FIGURE 11.1 The soil nematode *Caenorhabditis elegans.* This organism offers several advantages for the genetic analysis of development, including the fact that each individual of each sex exhibits an identical pattern of cell lineages in the development of the somatic cells. DNA sequencing of the 100-Mb genome was the first eukaryote completed.

Photo courtesy of Tim Schedl, Washington University School of Medicine.

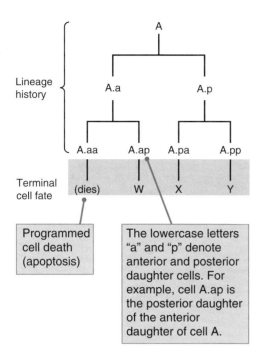

FIGURE 11.2 Hypothetical cell-lineage diagram. Different terminally differentiated cell fates are denoted W, X, and Y. One cell in the lineage (cell A.aa) undergoes programmed cell death.

terminal differentiated state of each cell. **FIGURE 11.2** is a lineage diagram of a hypothetical cell A in which the cell fate is either programmed cell death or one of the terminally differentiated cell types designated W, X, and Y. The letter symbols are the kind normally used for cells in nematodes, in which the name denotes the cell lineage according to ancestry and position in the embryo. For example, the cells A.a and A.p are, respectively, the anterior and posterior daughters of cell A, and A.aa and A.ap are the anterior and posterior daughters of cell A.a.

Cell fate is determined by autonomous development and/or intercellular signaling.

Two principal mechanisms progressively restrict the cell fate, or developmental outcome, of cells within a lineage.

- Developmental restriction may be autonomous, which means that it is determined by genetically programmed changes in the cells themselves.

- Cells may respond to positional information, which means that developmental restrictions are imposed by the position of cells within the embryo. Positional information may be mediated by signaling interactions between neighboring cells or by gradients in concentration of particular molecules.

Nematode development is largely autonomous, which means that in most cells, the developmental program unfolds automatically without the need for interactions with other cells. However, in the early embryo, some of the developmental fates are established by interactions among the cells. In later stages of development of these cells, the fates established early are reinforced by still other interactions between cells.

Worm development also provides important examples of the effects of intercellular signaling on determination. **FIGURE 11.3**, part A illustrates the first three cell divisions in the development of *C. elegans*, which result in eight embryonic cells that differ in genetic activity and developmental fate. The determination of cell fate in these early divisions is in part autonomous and in part results from interactions between cells. Figure 11.3, part B shows the lineage relationships between the cells. Cell-autonomous mechanisms are illustrated by the transmission of cytoplasmic particles called *polar granules* from the cells P0 to P1 to P2 to P3. Polar granules are ribonuclear protein complexes that function mainly in posttranscriptional regulation. Normal segregation of the polar granules is a function of microfilaments in the cytoskeleton. Cell-signaling mechanisms are illustrated by the effects of P2 on EMS and on ABp. The EMS fate is determined by the activity of the *mom-2* gene in P2. The P2 cell also produces a signaling molecule, APX-1, which determines the fate of ABp through the cell-surface receptor GLP-1. In contrast to *C. elegans*, in which many developmental decisions are cell autonomous, in *Drosophila* and *Mus* (the mouse), regulation by cell-to-cell signaling is more the rule than the exception. The use of cell signaling to regulate development provides a sort of insurance that helps to overcome the death of individual cells in development that might happen by accident.

Developmental mutations often affect cell lineages.

Many mutations that affect cell lineages have been studied in nematodes, and they reveal several general features by which genes control development.

- The division pattern and fate of a cell are generally affected by more than one gene and can be disrupted by mutations in any of them.

- Most genes that affect development are active in more than one type of cell.

- Complex cell lineages often include simpler, genetically determined lineages within them; these components are called *sublineages* because they are expressed as an integrated pattern of cell division and terminal differentiation.

- The lineage of a cell may be triggered autonomously within the cell itself or by signaling interactions with other cells.

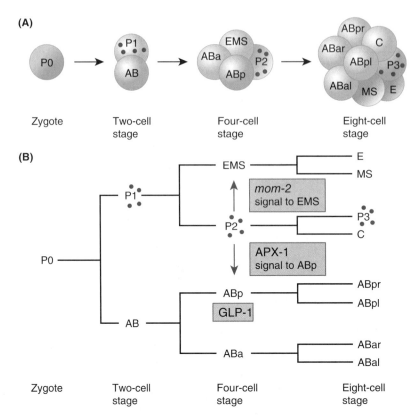

FIGURE 11.3 Early cell divisions in *C. elegans* development. (A) Spatial organization of cells. (B) Lineage relationships of the cells. The transmission of the polar granules illustrates cell-autonomous development. The arrows denote cell-to-cell signaling mechanisms that determine developmental fate.

- Regulation of development is controlled by genes that determine the different sublineages that cells can undergo and the individual steps within each sublineage.

The next section deals with some of the types of mutations that affect cell lineages and development.

Transmembrane receptors often mediate signaling between cells.

The controlling genes that cause cells to diverge in developmental fate are not always easy to recognize. For example, a mutant allele may identify a gene that is *necessary* for the expression of a particular developmental fate, but the gene may not be *sufficient* to determine the developmental fate of the cells in which it is expressed. This possibility complicates the search for genes that control major developmental decisions.

Genes that control decisions about cell fate can sometimes be identified by the unusual characteristic that dominant or recessive mutations have opposite effects. That is, if alternative alleles of a gene result in opposite cell fates, then the product of the gene must be both necessary and sufficient for expression of the fate. Recessive mutations in genes that control development often result from **loss of function** in that the mRNA is not produced or the protein is inactive. Dominant mutations in developmental-control genes often

result from **gain of function** in that the gene is overexpressed or is expressed at the wrong time.

In *C. elegans*, a relatively small number of genes have dominant and recessive alleles that affect the same cells in opposite ways. Among them is the *lin-12* gene, which controls developmental decisions in a number of cells. One example involves the cells denoted Z1.ppp and Z4.aaa in part A of **FIGURE 11.4**. These cells lie side by side in the embryo, but they have quite different lineages. Normally, one of the cells differentiates into an *anchor cell* (AC), which participates in development of the vulva, and the other one differentiates into a *ventral uterine precursor cell* (VU). Z1.ppp and Z4.aaa are equally likely to become the anchor cell.

Direct cell–cell interaction between Z1.ppp and Z4.aaa controls the AC–VU decision. If either cell is burned away (ablated) by a laser microbeam, the remaining cell differentiates into an anchor cell (part B). This result implies that the preprogrammed fate of both Z1.ppp and Z4.aaa is that of an anchor cell. When either cell becomes committed to the anchor-cell fate, its contact with the other cell elicits the ventral-uterine-precursor-cell fate. As noted, recessive and dominant mutations of *lin-12* have opposite effects. Mutations in which *lin-12* activity is lacking or greatly reduced are denoted *lin-12(0)*. These mutations are recessive, and in the mutants

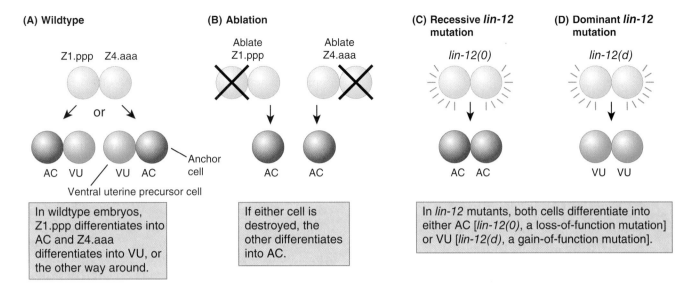

(A) Wildtype

Z1.ppp   Z4.aaa

or

AC   VU   /   VU   AC
Ventral uterine precursor cell

Anchor cell

In wildtype embryos, Z1.ppp differentiates into AC and Z4.aaa differentiates into VU, or the other way around.

(B) Ablation

Ablate Z1.ppp

Ablate Z4.aaa

AC   AC

If either cell is destroyed, the other differentiates into AC.

(C) Recessive *lin-12* mutation

lin-12(0)

AC   AC

In *lin-12* mutants, both cells differentiate into either AC [*lin-12(0)*, a loss-of-function mutation] or VU [*lin-12(d)*, a gain-of-function mutation].

(D) Dominant *lin-12* mutation

lin-12(d)

VU   VU

FIGURE 11.4   Control of the fates of Z1.ppp and Z4.aaa in vulval development and genetic control of cell fate by the *lin-12* gene. In recessive loss-of-function mutants [*lin-12(0)*], both cells become anchor cells; in dominant gain-of-function mutants [*lin-12(d)*], both cells become ventral uterine precursor cells.

both Z1.aaa and Z4.aaa become anchor cells (part C). In contrast, *lin-12(d)* mutations are those that cause *lin-12* activity to be overexpressed. These mutations are dominant or partly dominant, and in the mutants both Z1.aaa and Z4.ppp become ventral uterine precursor cells (part D).

The effects of *lin-12* mutations suggest that the wildtype gene product is a receptor of a developmental signal. The molecular structure of the *lin-12* gene product is typical of a **transmembrane receptor** protein containing regions that span the cell membrane. The LIN-12 protein shares domains with other proteins important in developmental control (**FIGURE 11.5**). The transmembrane region separates the LIN-12 protein into an extracellular part (the amino end) and an intracellular part (the carboxyl end). The extracellular part contains 13 repeats of a domain found in a mammalian peptide hormone, epidermal growth factor (EGF), as well as in the product of the *Notch* gene in *Drosophila*, which controls the decision between epidermal-cell and neural-cell fates. Nearer the transmembrane region, the amino end contains three repeats of a cysteine-rich domain also found in the *Notch* gene product. Inside the cell, the carboxyl part of the LIN-12 protein contains six repeats of a domain also found in the SWI6 proteins, which control cell division in yeast.

Cells can determine the fate of other cells through ligands that bind with their transmembrane receptors.

The anchor cell expresses a signaling gene, called *lin-3*, that controls the fate of other cells in the development of the vulva. **FIGURE 11.6** illustrates five precursor

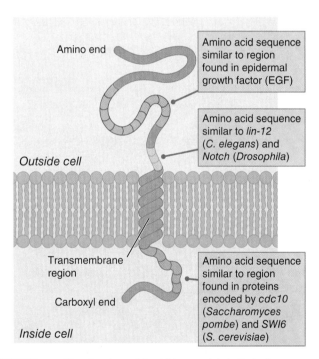

Amino end

Amino acid sequence similar to region found in epidermal growth factor (EGF)

Amino acid sequence similar to *lin-12* (*C. elegans*) and *Notch* (*Drosophila*)

Outside cell

Transmembrane region

Carboxyl end

Inside cell

Amino acid sequence similar to region found in proteins encoded by *cdc10* (*Saccharomyces pombe*) and *SWI6* (*S. cerevisiae*)

FIGURE 11.5   The structure of the LIN-12 protein is that of a receptor protein containing a transmembrane region and various types of repeated units that resemble those in epidermal growth factor (EGF) and other developmental control genes.

cells, P4.p through P8.p, that participate in vulval development. Each precursor cell has the capability of differentiating into one of three fates, called the 1°, 2°, and 3° lineages, which differ according to whether descendant cells remain in a syncytium (S) or divide longitudinally (L), transversely (T), or not at all (N). The precursor cells normally differentiate as shown in Figure 11.6, giving five lineages in the order 3°-2°-1°-2°-3°. The vulva itself is formed from the 1° and 2°

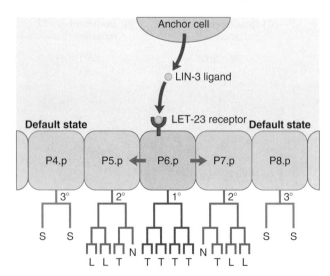

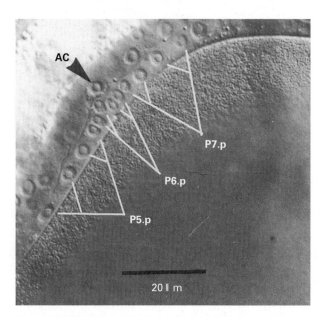

FIGURE 11.6 Determination of vulval differentiation by means of intercellular signaling. Cells P4.p through P8.p in the hermaphrodite give rise to lineages in the development of the vulva. The three types of lineages are designated 1°, 2°, and 3°. The 1° lineage is induced in P6.p by the ligand LIN-3 produced in the anchor cell (AC), which stimulates the LET-23 receptor tyrosine kinase in P6.p. The P6.p cell, in turn, produces a ligand that stimulates receptors in P5.p and P7.p to induce the 2° fate. On the other hand, the 3° fate is the default or baseline condition, which P4.p and P8.p adopt normally and all cells adopt in the absence of AC.

FIGURE 11.7 Spatial organization of cells in the vulva, including the anchor cell (black arrowhead) and the daughter cells produced by the first two divisions of P5.p through P7.p (white tree diagrams). The length of the scale bar equals 20 μm.

cell lineages. The spatial arrangement of some of the key cells is shown in **FIGURE 11.7**. The black arrow indicates the anchor cell, and the white lines show the pedigrees of 12 cells. The four cells in the middle derive from P6.p, and the four on each side derive from P5.p and P7.p.

The important role of the *lin-3* gene product (LIN-3) is suggested by the opposite phenotypes of loss-of-function and gain-of-function alleles. Loss of LIN-3 results in the complete absence of vulval development, whereas overexpression of LIN-3 results in excess vulval induction. LIN-3 is a typical example of an interacting molecule, or **ligand**, that binds with an EGF-type transmembrane receptor. In this case the receptor is located in cell P6.p and is the product of the gene *let-23*. The LET-23 protein is a tyrosine-kinase receptor that, when bound with the LIN-3 ligand, stimulates a series of intracellular signaling events that ultimately results in the synthesis of transcription factors that determine the 1° fate. Among the genes that are induced is a gene for yet another ligand, which binds with receptors on the cells P5.p and P7.p, causing these cells to adopt the 2° fate (horizontal arrows in Figure 11.6).

In vulval development, the adoption of the 3° lineages by the P4.p and P8.p cells is determined not by a positive signal but by the lack of a signal, because in the absence of the anchor cell, all of the cells P4.p through P8.p express the 3° lineage. Thus development of the 3° lineage is the uninduced or *default* state, which means that the 3° fate is preprogrammed into the cell and must be overridden by another signal if the cell's fate is to be altered.

11.2 Epistatic interactions between mutant alleles can help define signaling pathways.

Analysis of the interactions between mutant alleles can reveal the temporal order in which genes function in a developmental pathway. The logic is based on the principle that developmental pathways are *switch-regulation pathways*, in which each component in the pathway either stimulates or inhibits the activity of the next component in line. Components that stimulate the pathway are *positive regulators*, and those that inhibit the pathway are *negative regulators*.

The control of vulval induction in *C. elegans* illustrates the logic behind the genetic analysis of switch-regulation pathways. Before proceeding we need to emphasize some caveats:

- The recessive mutant alleles that are analyzed must be complete loss-of-function alleles, because in some cases even a residual activity will give misleading results.

- Each mutant gene must have a unique and nonredundant function in the pathway. Pathways with genetic redundancy resulting from duplicate genes or genes with overlapping functions are not suitable for this type of analysis.

■ The mutant alleles should affect components in the same developmental pathway.

■ The regulatory pathway should be linear, with each component interacting only with its downstream neighbor, without branching or parallel signaling.

Despite the caveats, many switch-regulation pathways do lend themselves to genetic analysis. The first step is to isolate a large number of recessive loss-of-functional alleles of different genes that encode components of the pathway. In switch-regulation pathways, some mutant alleles block the pathway whereas others activate it, leading to contrasting phenotypes. For example, in the developmental determination of the vulva in *C. elegans*, some mutant alleles result in no vulva (called vulvaless) whereas other mutant alleles result in the development of multiple vulvas (multivulva). For any given gene in the pathway, one extreme phenotype is observed for loss-of-function alleles whereas the opposite extreme phenotype is observed for gain-of-functional alleles.

The genetic analysis of a switch-regulation pathway is based on the phenotypes of double mutants of pairs of genes in which the mutant alleles show contrasting phenotypes. For vulva development, for example, we would make all possible pairs of vulvaless–multivulva double mutants. The order of the components in the switch-regulation pathway is determined by the type of epistasis observed in the double mutants. A classical definition of *epistasis* is any interaction between mutant alleles that alters the 9 : 3 : 3 : 1 ratio expected from independent assortment of two genes. In the analysis of switch-regulation pathways the term is used in a somewhat different sense. For double mutants with alleles in genes with contrasting phenotypes, a gene is called an **epistatic gene** if its mutant phenotype masks the mutant phenotype of another gene. For example, if the phenotype of *aa bb* is the same as that of *aa* b^+b^+, then the gene *a* is said to be epistatic to the gene *b*. The gene whose mutant phenotype is concealed is called a **hypostatic gene**, and in the previous example the *b* gene is hypostatic to the *a* gene. To take a specific example, in a vulvaless–multivulva double mutant, if the double mutant is vulvaless, then the vulvaless gene is epistatic to the multivulva gene. Equivalently, we could say that the multivulva gene is hypostatic to the vulvaless gene. Epistasis helps to determine the order of components in a developmental pathway for the following reason:

Principle of epistasis: In a linear switch-regulation pathway, the product of the epistatic gene acts *downstream* in the pathway relative to the product of the hypostatic gene; to say the same thing in another way, the product of the hypostatic gene acts *upstream* relative to that of the epistatic gene.

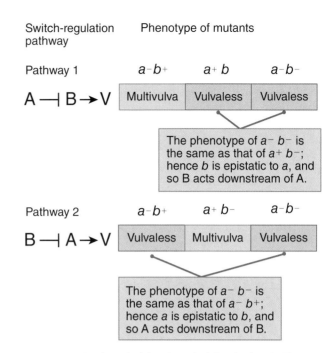

FIGURE 11.8 Logic underlying the principle of epistasis. The gene product of the epistatic gene acts downstream of the product of the hypostatic gene.

The logic behind this principle is illustrated in **FIGURE 11.8**. Two switch-regulation pathways are shown in which mutants yield opposite phenotypes, either vulvaless or multivulva. In both pathways, A and B represent the wildtype gene products of the genes *a* and *b*, respectively. By convention, the arrowhead implies positive regulation (stimulation) and the T-bar implies negative regulation (inhibition). The gene products A and B need not interact directly. If there are intervening components, then the arrows and bars represent the net effect on the intervening components.

The arrows and bars in Figure 11.8 were chosen so that a^-a^- b^+b^+ and a^+a^+ b^-b^- mutants result in different phenotypes. In pathway 1, A inhibits B and B stimulates vulva development (V). The a^-a^- b^+b^+ genotype therefore has less inhibition of B, and the resulting greater activity of B implies a phenotype of multivulva. In contrast, the a^+a^+ b^-b^- genotype lacks B and hence no vulva induction occurs, yielding a phenotype of vulvaless. Pathway 2 has the order of A and B interchanged, and, in this case, the a^-a^- b^+b^+ phenotype is vulvaless and that of a^+a^+ b^-b^- is multivulva.

The epistatic interactions of *a* and *b* are shown in the last column. In pathway 1 the a^-a^- b^-b^- double mutant is vulvaless because the animal lacks B, whereas in pathway 2 the a^-a^- b^-b^- double mutant is vulvaless because the animal lacks A. In both cases, in accord with the principle of epistasis, the gene that is downstream in the pathway is epistatic to the gene that is upstream in the pathway. The principle makes intuitive sense because the downstream mutant has the last word: If the downstream mutant blocks the pathway, the pathway is blocked; and if the downstream mutant activates the pathway, the pathway is activated.

Now we are in a position to apply the principle of epistasis to a number of mutants that affect vulval induction in *C. elegans*. Consider the mutant alleles in **FIGURE 11.9**, part A: the mutant alleles *let-23* and

lin-45 result in the vulvaless phenotype, and the mutant alleles *lin-1* and *let-60* result in the multivulva phenotype. The phenotypes of the mutants are shown color coded, brown for vulvaless and green for multivulva. To define the switch-regulation pathway based on epistasis, a researcher would examine the phenotype of each of the double mutants shown in the square, which are also color coded. The principle of epistasis says that the phenotype of the double mutant is the same as that of the single mutant whose product acts farther downstream. Each row and each column in the matrix, therefore, provides some information about the linear order in which the gene products function. This information is summarized in part B. In cases when order of action of two gene products cannot be determined, the genes are listed vertically in square brackets. The pathway symbols are dashes instead of arrows (stimulation) and T-bars (inhibition). The reason is that the linear order of components in the pathway should be deduced first. Deciding which dashes indicate stimulation and which indicate inhibition comes later. Based on the rows and columns in Figure 11.9, part A and the deductions in part B, the gene products act in the order shown in part C.

Now we are in a position to convert the dashes in Figure 11.9, part C into arrows (indicating stimulation) or T-bars (indicating inhibition). A strategy for doing this is to start with the last component in the pathway and work backwards. The result is shown in **FIGURE 11.10**. The reasoning is based on the nature of each type of mutant (loss of function or gain of function). For example, *lin-1* is a loss of function allele that results in a multivulva phenotype. Since loss of *lin-1* activity promotes vulva formation, the wildtype activity of the Lin-1 protein must be inhibitory; hence, there is a T-bar between *lin-1* and vulva. The *lin-45* mutation is also a loss-of-function allele, but it results in a vulvaless phenotype. Because the vulvaless phenotype result from overactivity of Lin-1 protein, the wildtype activity of Lin-45 protein must be to inhibit *lin-1* as shown in Figure 11.10. In this case, a *lin-45* loss-of-functional allele would result in greater activity of *lin-1* and, hence, yield the vulvaless phenotype observed.

Taking one step backward, we next examine *let-60*. This mutation is a gain-of-function allele that results in a multivulva phenotype. Since excess *lin-45* activity would result in greater inhibition of *lin-1* and, in turn, less inhibition of vulva induction, it follows that the effect of wildtype Let-60 protein is greater activity of

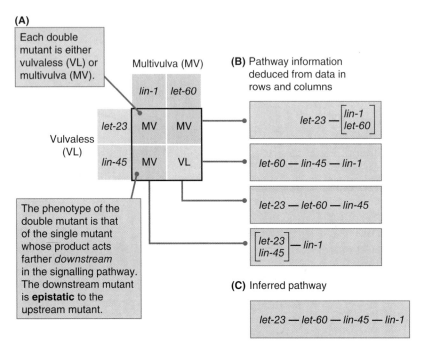

(A)

Each double mutant is either vulvaless (VL) or multivulva (MV).

Multivulva (MV)

Vulvaless (VL)

	lin-1	let-60
let-23	MV	MV
lin-45	MV	VL

The phenotype of the double mutant is that of the single mutant whose product acts farther *downstream* in the signalling pathway. The downstream mutant is **epistatic** to the upstream mutant.

(B) Pathway information deduced from data in rows and columns

$$let\text{-}23 - \begin{bmatrix} lin\text{-}1 \\ let\text{-}60 \end{bmatrix}$$

$$let\text{-}60 - lin\text{-}45 - lin\text{-}1$$

$$let\text{-}23 - let\text{-}60 - lin\text{-}45$$

$$\begin{bmatrix} let\text{-}23 \\ lin\text{-}45 \end{bmatrix} - lin\text{-}1$$

(C) Inferred pathway

$$let\text{-}23 - let\text{-}60 - lin\text{-}45 - lin\text{-}1$$

FIGURE 11.9 Application of the principle of epistasis to vulval induction. (A) Experimental results. (B) Information implied about the switch-regulation pathway from each row and column in the data matrix. (C) Inferred pathway connecting all of the genes.

⟐ STOP & THINK 11.1

In the linear switch-regulation pathway illustrated below, X and Y are gene products, the arrow denotes positive regulation, and the T-bar denotes negative regulation (inhibition).

$$X \dashv Y \rightarrow \text{Thoracic spots}$$

The pathway controls spots on the thorax of a certain insect. The wildtype phenotype has a single spot on the thorax. A homozygous *aa* mutant has no spots on the thorax, whereas a homozygous *bb* mutant has multiple spots. Both *a* and *b* are loss-of-function mutations. The double mutant *aa bb* has no spots on the thorax. Let A denote the gene product of the nonmutant *a+* allele and B denote the gene product of the nonmutant *b+* allele. Does X correspond to A or B? Which gene product does Y correspond to?

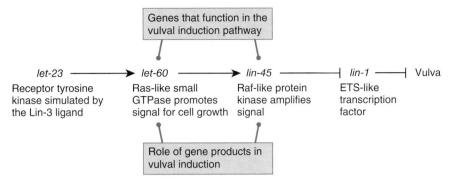

FIGURE 11.10 Some of the key genes in vulva induction and their protein products, showing the net effect at each step. Arrows indicate stimulation; T-bars indicate inhibition.

Lin-45. The *let-60* gain-of-functional allele, therefore, results in the multivulva phenotype observed. (This inference also implies that a *let-60* loss-of-function allele would yield a vulvaless phenotype, which is in fact also observed.) We leave it as an exercise to puzzle out that *let-23*, a loss-of-function allele, results in the observed vulvaless phenotype only if the wild-type function of Let-23 protein is to stimulate the Let-60 protein.

Why the emphasis on vulval induction in worms? One reason is that vulval induction demonstrates how epistasis is used in the analysis of switch-regulation pathways. Figure 11.10 includes the types of proteins known to be encoded by each of the genes. The switch-regulation pathway actually includes more components than shown here, including some components that act between *let-23* and *let-60* and between *lin-45* and *lin-1*. In spite of the missing players, the analysis of epistasis in the switch-regulation pathway yields the correct order of all components included in the analysis, and the correct net effect (stimulatory or inhibitory) or each step. This pathway is of general interest because ligands for epidermal growth-factor receptors that activate Ras and downstream protein kinases are widespread in the regulation of cell growth and development and have also been implicated in many human cancers. Many of the human cancers have mutations that are analogous to the gain-of-function mutation in *let-60*.

11.3 Development in *Drosophila* illustrates progressive regionalization and specification of cell fate.

Many important insights into developmental processes have been gained from genetic analysis in *Drosophila*. The developmental cycle of *D. melanogaster*, summarized in **FIGURE 11.11**, includes egg, larval, pupal, and adult stages. Early development includes a series of cell divisions, migrations, and infoldings that result in the *gastrula*. About 24 hours after fertilization, the first-stage larva, composed of about 10^4 cells, emerges from the egg. Each larval stage is called an *instar*. Two successive larval molts that give rise to the second- and third-instar larvae are followed by pupation and a complex metamorphosis that gives rise to the adult fly composed of more than 10^6 cells. In wildtype strains reared at 25°C, development requires from 10 to 12 days.

Early development in *Drosophila* takes place within the egg case (**FIGURE 11.12**, part A). The first nine mitotic divisions occur in rapid succession without division of the cytoplasm and produce a cluster of nuclei within the egg (part B). The nuclei migrate to the periphery, and the germ line is formed from about 10 **pole cells** set off at the posterior end (part C); the pole cells undergo two additional divisions and are reincorporated into the embryo by invagination. The nuclei within the embryo undergo four more mitotic divisions without division of the cytoplasm, forming the *syncytial blastoderm*, which contains about 6000 nuclei (part D). Cellularization of the blastoderm takes place from about 150 to 195 minutes after fertilization by the synthesis of membranes that separate the nuclei. The **blastoderm** formed by cellularization (part E) is a flattened hollow ball of cells that corresponds to the blastula in other animals.

The experimental destruction of patches of cells within a *Drosophila* blastoderm results in localized defects in the larva and adult. This finding implies that cells in the blastoderm have predetermined developmental fates, with little ability to substitute in development for other, sometimes even adjacent, cells. Further evidence for this conclusion comes from experiments in which cells from a genetically marked blastoderm are implanted into host blastoderms. Blastoderm cells implanted into the equivalent regions of the host become part of the normal adult structures. However, blastoderm cells implanted into different regions develop autonomously and are not integrated into host structures.

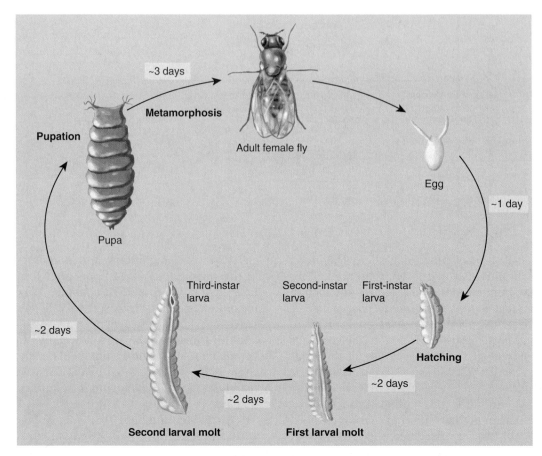

FIGURE 11.11 Developmental program of *Drosophila melanogaster*. The durations of the stages are at 25°C.

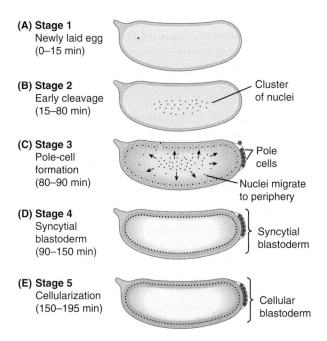

(A) Stage 1
Newly laid egg
(0–15 min)

(B) Stage 2
Early cleavage
(15–80 min)

Cluster
of nuclei

(C) Stage 3
Pole-cell
formation
(80–90 min)

Pole
cells

Nuclei migrate
to periphery

(D) Stage 4
Syncytial
blastoderm
(90–150 min)

Syncytial
blastoderm

(E) Stage 5
Cellularization
(150–195 min)

Cellular
blastoderm

FIGURE 11.12 Early development in *Drosophila*. (A) The nucleus in the fertilized egg. (B) Mitotic divisions take place synchronously within a syncytium. (C) Some nuclei migrate to the periphery of the embryo, and at the posterior end, the pole cells (which form the germ line) become cellularized. (D) Additional mitotic divisions occur within the syncytial blastoderm. (E) Membranes are formed around the nuclei, giving rise to the cellular blastoderm.

Because of the relatively high degree of determination in the blastoderm, genetic analysis of *Drosophila* development has tended to focus on the early stages of development, when the basic body plan of the embryo is established and key regulatory processes become activated. The following sections summarize the genetic control of these early events.

Mutations in a maternal-effect gene result in defective oocytes.

Early development in *Drosophila* requires translation of maternal mRNA molecules present in the oocyte. Blockage of protein synthesis during this period arrests the early cleavage divisions. Expression of the zygote genome is also required, but the timing is different. Blockage of transcription of the zygote genome at any time after the ninth cleavage division prevents formation of the blastoderm.

Because the earliest stages of *Drosophila* development are programmed in the oocyte, mutations that affect oocyte composition or structure can upset development of the embryo. Genes that function in the mother that are needed for development of the embryo are called **maternal-effect genes**, and developmental genes that function in the embryo are called zygotic genes. The interplay between the two types of genes is as follows:

Mutations in maternal-effect genes result in a phenotype in which homozygous females produce eggs unable to support normal embryonic development, whereas homozygous males produce normal sperm. Therefore, reciprocal crosses give dramatically different results. For example, a recessive maternal-effect mutation, *m*, will yield the following results in reciprocal crosses:

$$m/m\female \times +/+\male \rightarrow +/m \text{ progeny}$$
(abnormal development)

$$+/+\female \times m/m\male \rightarrow +/m \text{ progeny}$$
(normal development)

The $+/m$ progeny of the reciprocal crosses are genetically identical, but development is upset when the mother is homozygous m/m.

The reason why maternal-effect genes are needed in the mother is that the maternal-effect genes establish the polarity of the *Drosophila* oocyte even before fertilization takes place. They are active during the earliest stages of embryonic development, and they determine the basic body plan of the embryo. Maternal-effect mutations provide a valuable tool for investigating the genetic control of pattern formation and for identifying the molecules important in morphogenesis.

Embryonic pattern formation is under genetic control.

Some of the early stages in *Drosophila* development are shown in **FIGURE 11.13**. The larva that hatches from the egg features 14 superficially similar repeating units visible as a pattern of stripes along the main trunk (**FIGURE 11.14**). The stripes can be recognized externally by the bands of *denticles*, which are tiny, pigmented, tooth-like projections from the surface of the larva. The 14 stripes in the larva correspond to the segments that form from the embryo. Each **segment** is defined morphologically as the region between successive indentations formed by the sites of muscle attachment in the larval cuticle. The designations of the segments are indicated in Figure 11.14. There are three head segments (C1−C3), three thoracic segments (T1−T3), and eight abdominal segments (A1−A8). In addition to the segments, another type of repeating unit is also important in development. These repeating units are called **parasegments**; each parasegment consists of the posterior region of one segment and the anterior region of the adjacent segment. Parasegments have a transient existence in embryonic development. Although they are not visible morphologically, they are important in gene expression because the boundaries of expression of many genes coincide with the boundaries of the parasegments rather than with those of the segments.

The early stages of pattern formation are determined by genes that are often called **segmentation genes** because they determine the origin and fate of the segments and parasegments. There are four classes of segmentation genes that differ in their times and patterns of expression in the embryo.

1. The *coordinate genes* determine the principal coordinate axes of the embryo: the anterior–posterior axis, which defines the front and rear; and the dorsal–ventral axis, which defines the top and bottom.

2. The *gap genes* are expressed in contiguous groups of segments along the embryo (**FIGURE 11.15**, part A), and they establish the next level of spatial organization. Mutations in gap genes result in the absence of contiguous body segments, so gaps appear in the normal pattern of structures in the embryo.

3. The *pair-rule* genes determine the separation of the embryo into discrete segments (part B). Mutations in pair-rule genes result in missing pattern elements in alternate segments. The reason for the two-segment periodicity of pair-rule genes is that the genes are expressed in a zebra stripe pattern along the embryo.

4. The *segment-polarity* genes determine the pattern of anterior–posterior development within each segment of the embryo (part C). Mutations in segment-polarity genes affect all segments or parasegments in which the normal gene is active. Many segment-polarity mutants have the normal number of segments, but part of each segment is deleted and the remainder is duplicated in mirror-image symmetry.

Evidence for the existence of the four classes of segmentation genes—coordinate genes, gap genes, pair-rule genes, and segment-polarity genes—is presented in the following sections.

THE HUMAN CONNECTION

Distinguished Lineages

J. E. Sulston,[1] **E. Schierenberg,**[2] **J. G. White,**[1] **and J. N. Thomson**[1] **(1983)**
[1]*Medical Research Council Laboratory for Molecular Biology, Cambridge, United Kingdom;*
[2]*Max-Planck Institute for Experimental Medicine, Gottingen, Germany*

The Embryonic Cell Lineage of the Nematode Caenorhabditis elegans

The data produced in this landmark study form the basis for interpreting developmental mutants in the nematode worm. This long paper offers voluminous data and is available through the Internet. During embryogenesis, 1030 cells are generated; 131 of these, or 13 percent, undergo programmed cell death. What is the reason for such a high proportion of programmed cell deaths? The embryonic lineage is highly invariant—the same from one organism to the next. Why isn't there more developmental flexibility, as is found in most other organisms? These issues are addressed in this excerpt, in which the emphasis is on the historical background and motivation of the study, the big picture of development, and interpretation of the lineage in terms of the evolution of the nematode.

The technique of Nomarski microscopy mentioned in this excerpt is a modern invention that is also called differential interference contrast microscopy. When light passes through living material, it changes phase according to the refractive index of the material. Adjacent parts of a cell or organism that differ in refractive index cause different changes in phase. When two sets of waves combine after passing through an object, the difference in phase creates an interference pattern that yields an image of the object. The major advantage of Nomarski microscopy is that it can be used to observe living tissue.

This report marks the completion of a project begun over one hundred years ago—namely, the determination of the entire cell lineage of a nematode. . . . By the technique of Nomarski microscopy, which is nondestructive and yet produces high resolution, cells can be followed in living larvae. The use of living material . . . has permitted the origin and fate of every cell in one nematode species [*Caenorhabditis elegans*] to

be determined. Thus, not only are the broad relationships between tissues now known unambiguously, but also the detailed pattern of cell fates is clearly revealed. . . .

Both the classical analyses of fixed tissues and the more modern ones involving living tissue pointed to the conclusion that the overall lineages of embryos are invariant—all embryos develop via the same patterns of cell division, programmed cell death, and terminal differentiation. However, when sublineages—patterns of differentiation programmed into a particular cell—are examined, two features emerge:

> **66** *The nematode belongs to an ancient phylum, and its cell lineage is a piece of frozen evolution. In the course of time, new cell types were generated from precursors selected not so much for their intrinsic properties as for the accident of their position in the embryo.* **99**

Perhaps the most striking findings are firstly the complexity and secondly the cell autonomy of the lineages. . . . The nematode belongs to an ancient phylum, and its cell lineage is a piece of frozen evolution. . . . Cell–cell interactions that were initially necessary for developmental decisions may have been gradually supplanted by autonomous programs that were fast, economical, and reliable, the loss of flexibility being outweighed by the gain in efficiency. On this view, . . . all the features that could, it seems, be eliminated from a more efficient design—are so many developmental fossils.

Another way to state this hypothesis is that in the evolutionary ancestors of nematodes, cell lineages and fates were determined largely by mechanisms of positional information, but with time autonomous mechanisms evolved and supplanted them. Thus, what we see today is the result of an evolutionary process that has been constrained by the cellular organization of an evolutionary ancestor and is thus not the most "efficient" developmental pathway for a contemporary nematode.

J. E. Sulston, et al., *Dev. Biol.* 100 (1983): 64–119.

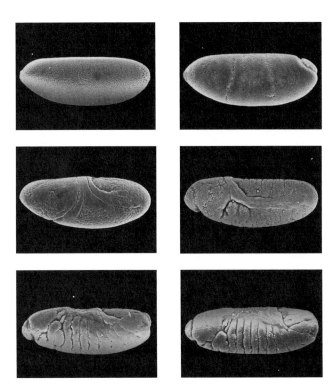

FIGURE 11.13 Representative stages of early development in *Drosophila* showing the pattern of segmentation that gives rise to the larval body plan.

Courtesy of Thomas Kaufman and F. Rudolf Turner, Indiana University at Bloomington.

Coordinate genes establish the main body axes.

The **coordinate genes** are maternal-effect genes that establish early polarity through the presence of their products at defined positions within the oocyte or through gradients of concentration of their products. The genes that determine the anterior–posterior axis can be classified into three groups according to the effects of mutations in them, as illustrated in **FIGURE 11.16**.

1. The first group of coordinate genes includes the *anterior genes*, which affect the head and thorax. The key gene in this class is *bicoid*. Mutations in *bicoid* produce embryos lacking the head and thorax that occasionally have abdominal segments in reverse polarity duplicated at the anterior end. The *bicoid* gene product is a transcription factor for genes determining anterior structures. Because the *bicoid* mRNA is localized in the anterior part of the early-cleavage embryo, these genes are activated primarily in the anterior region. The *bicoid* mRNA is produced in nurse cells (cells surrounding the oocyte) and exported to a localized region at the anterior pole of the oocyte. The protein product is less localized and, during the syncytial cleavages, forms an anterior–posterior concentration gradient with the maximum at the anterior tip of the embryo. The Bicoid protein

is a transcriptional activator containing a helix–turn–helix motif for DNA binding. Genes affected by the Bicoid protein contain multiple upstream binding domains that consist of nine nucleotides resembling the consensus sequence 5'-TCTAATCCC-3'. Binding sites that differ by as many as two base pairs from the consensus sequence bind the Bicoid protein with high affinity, and sites that contain four mismatches bind with low affinity. The combination of high- and low-affinity binding sites determines the concentration of Bicoid protein needed for gene activation; genes with many high-affinity binding sites can be activated at low concentrations, but those with many low-affinity binding sites need higher concentrations. Such differences in binding affinity mean that the level of gene expression can differ from one regulated gene to the next along the Bicoid concentration gradient. It is the local concentration of the Bicoid protein that regulates the expression of critical gap genes along the embryo—for example, *hunchback*.

2. The second group of coordinate genes includes the *posterior genes*, which affect the abdominal segments (Figure 11.16). Some of the mutants also lack pole cells. One of the posterior mutations, *nanos*, yields embryos with defective abdominal segmentation but normal pole cells. The *nanos* mRNA is localized tightly to the posterior pole of the oocyte, and the gene product is a repressor of translation. Among the genes whose mRNA is not translated in the presence of Nanos protein is the gene *hunchback*. Hence *hunchback* expression is controlled jointly by the Bicoid and Nanos proteins, Bicoid protein activating transcription in an anterior–posterior gradient, and Nanos protein repressing translation in the posterior region.

3. The third group of coordinate genes includes the *terminal genes*, which simultaneously affect the most anterior structure (the acron) and the most posterior structure (the telson) (Figure 11.16). The key gene in this class is *torso*, which codes for a transmembrane receptor that is uniformly distributed throughout the embryo in the early developmental stages. The Torso receptor is activated by a signal released only at the poles of the egg by the nurse cells in that location.

Apart from the three sets of genes that determine the anterior–posterior axis of the embryo, a fourth set of genes determines the dorsal–ventral axis. The morphogen for dorsal/ventral determination is the product of the gene *dorsal*, which is present in a pronounced ventral-to-dorsal gradient in the late syncytial blastoderm.

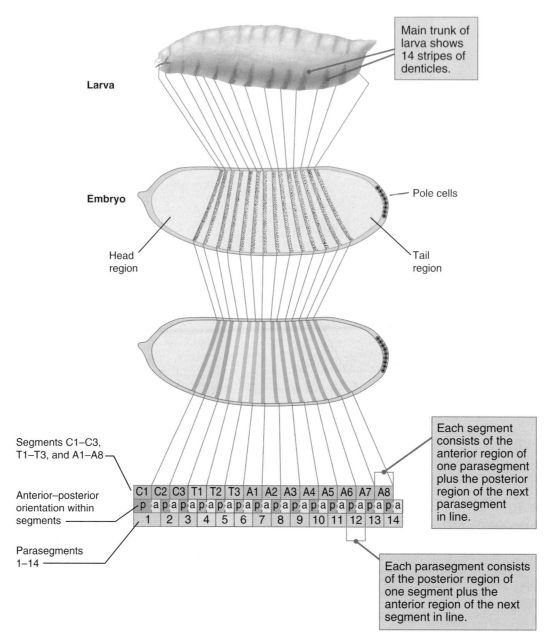

FIGURE 11.14 Segmental organization of the *Drosophila* embryo and larva. The segments are defined by successive indentations formed by the sites of muscle attachment in the larval cuticle. The parasegments are not apparent morphologically but include the anterior and posterior regions of adjacent segments.

Gap genes regulate other genes in broad anterior–posterior regions.

The main role of the coordinate genes is to regulate the expression of a small group of genes along the anterior–posterior axis. The genes are called **gap genes** because mutations in them result in the absence of pattern elements derived from a group of contiguous segments (Figure 11.15). Gap genes are zygotic genes. The gene *hunchback* serves as an example of the class because *hunchback* expression is controlled by offsetting effects of Bicoid and Nanos. Transcription of *hunchback* is stimulated in an anterior-to-posterior gradient by the Bicoid transcription factor, but posterior *hunchback*

expression is prevented by translational repression owing to the posteriorly localized Nanos protein. In the early *Drosophila* embryo in part A of **FIGURE 11.17**, the gradient of *hunchback* expression is indicated by the green fluorescence of an antibody specific to the hunchback gene product. The superimposed red fluorescence results from antibody specific to the product of *Krüppel*, another gap gene. The region of overlapping gene expression appears in yellow. The products of both *hunchback* and *Krüppel* are transcription factors of the zinc-finger type. Other gap genes also are transcription factors. Together, the gap genes have a pattern of regional specificity and partly overlapping domains of expression that enable them to act in combinatorial

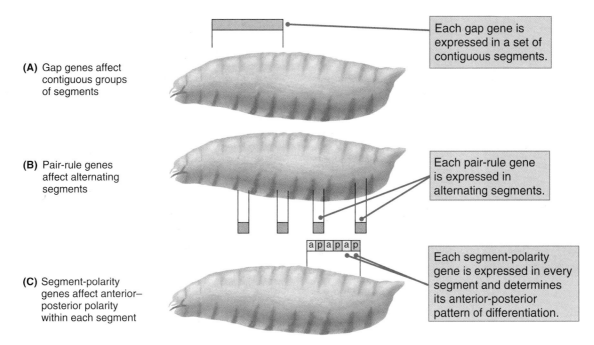

(A) Gap genes affect contiguous groups of segments

Each gap gene is expressed in a set of contiguous segments.

(B) Pair-rule genes affect alternating segments

Each pair-rule gene is expressed in alternating segments.

(C) Segment-polarity genes affect anterior–posterior polarity within each segment

Each segment-polarity gene is expressed in every segment and determines its anterior-posterior pattern of differentiation.

FIGURE 11.15 Patterns of expression of different types of segmentation genes.

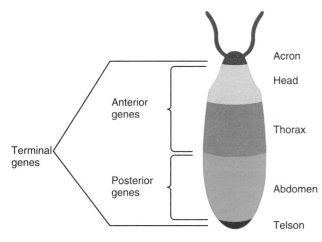

Acron
Head
Thorax
Abdomen
Telson

Terminal genes

Anterior genes

Posterior genes

FIGURE 11.16 Regional differentiation of the early *Drosophila* embryo along the anterior–posterior axis. Mutations in any of the classes of genes shown result in elimination of the corresponding region of the embryo.

fashion to control the next set of genes in the segmentation hierarchy, the pair-rule genes.

Pair-rule genes are expressed in alternating segments or parasegments.

The coordinate and gap genes determine the polarity of the embryo and establish broad regions within which subsequent development takes place. As development proceeds, the progressively more refined organization of the embryo is correlated with the patterns of expression of the segmentation genes. Among these are the **pair-rule genes**, in which the mutant phenotype has alternating segments absent or malformed (Figure 11.15). For example, mutations of

the pair-rule gene *even-skipped* affect even-numbered segments, and those of another pair-rule gene, *odd-skipped*, affect odd-numbered segments. The function of the pair-rule genes is to give the early *Drosophila* larva a segmented body pattern with both repetitiveness and individuality of segments. For example, there are eight abdominal segments that are repetitive in that they are regularly spaced and share several common features, but they differ in the details of their differentiation.

One of the earliest pair-rule genes expressed is *hairy*, whose pattern of expression is under both positive and negative regulation by the products of *hunchback*, *Krüppel*, and other gap genes. Expression of *hairy* occurs in seven stripes (Figure 11.17, part B). The striped pattern of pair-rule gene expression is typical, but the stripes of expression of one gene are usually slightly out of register with those of another. Together with the continued regional expression of the gap genes, the combinatorial patterns of gene expression in the embryo are already complex and linearly differentiated. Part C shows an embryo stained for the products of three genes—*hairy* (green), *Krüppel* (red), and *giant* (blue). The regions of overlapping expression appear as color mixtures—orange, yellow, light green, or purple. Even at this early stage in development, there is a unique combinatorial pattern of gene expression in every segment and parasegment. The complexity of combinatorial control can be appreciated by considering that the expression of the *hairy* gene in stripe 7 depends on a promoter element smaller than 1.5 kb that contains a series of binding sites for the protein products of the genes *caudal, hunchback, knirps, Krüppel, tailless, huckebein, bicoid,* and perhaps still other

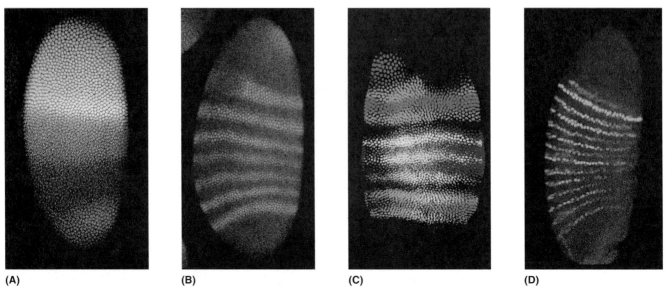

(A)   (B)   (C)   (D)

FIGURE 11.17 (A) An embryo of *Drosophila*, approximately 2.5 hours after fertilization, showing the regional localization of the *hunchback* gene product (green), the *Krüppel* gene product (red), and their overlap (yellow). (B) Characteristic seven stripes of expression of the gene *hairy* in a *Drosophila* embryo approximately 3 hours after fertilization. (C) Combined patterns of expression of *hairy* (green), *Krüppel* (red), and *giant* (blue) in a *Drosophila* embryo approximately 3 hours after fertilization. Already there is considerable linear differentiation apparent in the patterns of gene expression. (D) Expression of the segment-polarity gene *engrailed* partitions the early *Drosophila* embryo into 14 regions. These eventually differentiate into three head segments, three thoracic segments, and eight abdominal segments.

proteins yet to be identified. The combinatorial patterns of gene expression of the pair-rule genes define the boundaries of expression of the segment-polarity genes, which function next in the hierarchy.

Segment-polarity genes govern differentiation within segments.

Whereas the pair-rule genes determine the body plan at the level of segments and parasegments, the **segment-polarity genes** create a spatial differentiation within each segment. The mutant phenotype has repetitive deletions of pattern along the embryo (Figure 11.15) and usually a mirror-image duplication of the part that remains. Among the earliest segment-polarity genes expressed is *engrailed*, whose stripes of expression approximately coincide with the boundaries of the parasegments and so divide each segment into anterior and posterior domains (Figure 11.17, part D).

Expression of the segment-polarity genes finally establishes the early polarity and linear differentiation of the embryo into segments and parasegments.

Interactions among genes in the regulatory hierarchy ensure an orderly progression of developmental events.

Genes in the regulatory hierarchy are controlled by a complex set of interactions that ensure an orderly progression through the molecular events of development. Interactions among some of the coordinate genes, gap genes, pair-rule genes, and segment-polarity genes are shown in **FIGURE 11.18**. Many of these interactions were originally inferred from genetic analysis using the principle of epistasis discussed in Section 11.2. The green connectors indicate stimulatory effects, and the red connectors indicate inhibitory effects. Most of the genes are controlled by a complex set of stimulatory and inhibitory effects acting together. The coordinate genes act first to establish the polarity of the embryo, then the gap genes to differentiate large regions, after which the pair-rule genes establish the periodicity of the embryo indicated by the zebra stripes, and finally the segment-polarity genes act in the specification of the developmental identity and fate of each of the body segments. At each level in the regulatory hierarchy, the genes act to regulate other genes expressed at the same level, and also act to regulate the activity of genes that are expressed in the next downstream level in the hierarchy.

The segment-polarity genes also act to regulate downstream developmental genes that control the pathways of differentiation in each segment or parasegment, resulting ultimately in the morphology of the adult fly. The metamorphosis of the adult fly and how it emerges are discussed next.

Homeotic genes function in the specification of segment identity.

As with many other insects, the larvae and adults of *Drosophila* have a segmented body plan consisting of a head formed from segments C1–C3, a thorax formed

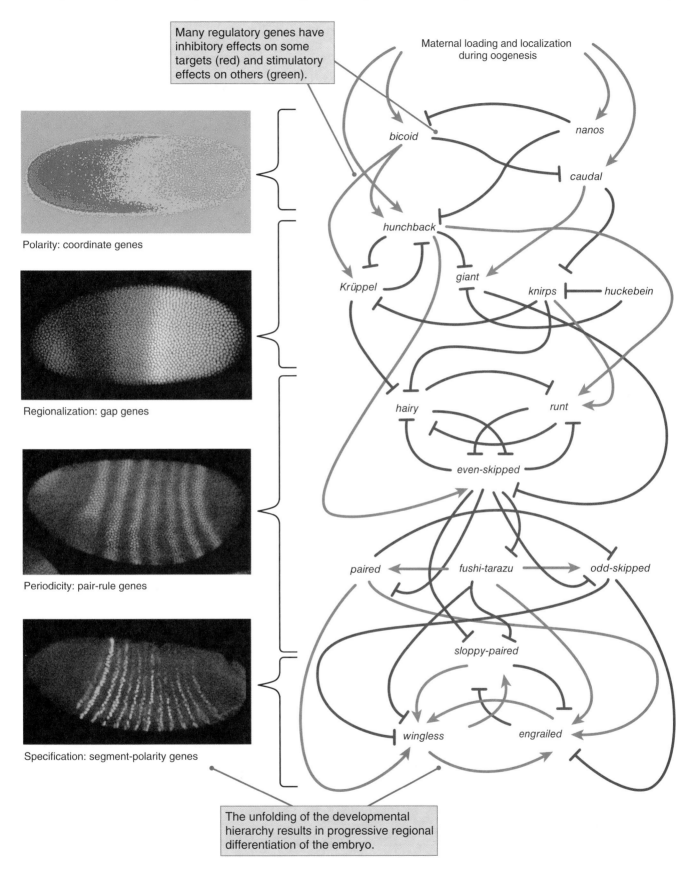

Many regulatory genes have inhibitory effects on some targets (red) and stimulatory effects on others (green).

Maternal loading and localization during oogenesis

Polarity: coordinate genes

Regionalization: gap genes

Periodicity: pair-rule genes

Specification: segment-polarity genes

The unfolding of the developmental hierarchy results in progressive regional differentiation of the embryo.

FIGURE 11.18 Hierarchy of regulatory interactions among genes controlling early development in *Drosophila*.

Photos courtesy of James Langeland, Sean Carroll, and Stephen Paddock, University of Wisconsin at Madison. Diagram adapted from an illustration by George von Dassow, Center for Cell Dynamics, University of Washington.

from segments T1–T3, and an abdomen formed from segments A1–A8 (**FIGURE 11.19**). Metamorphosis makes use of about 20 structures called **imaginal disks** present inside the larvae (**FIGURE 11.20**). Formed early in development, the imaginal disks ultimately give rise to the principal structures and tissues in the adult organism. Examples of imaginal disks include the pair of wing disks (one on each side of the body) that give rise to the wings and their attachments on thoracic segment T2, and the pair of haltere disks that give rise to the halteres (flight balancers) on thoracic segment T3 (Figure 11.19). During the pupal stage, when many larval tissues and organs break down, the imaginal disks progressively unfold and differentiate into adult structures. The morphogenic events that take place in the pupa are initiated by the hormone *ecdysone*, secreted by the larval brain.

As in the early embryo, overlapping patterns of gene expression and combinatorial control guide later events in *Drosophila* development. The pattern of expression of a key gene in wing development, *vestigial*, in a wing disk is shown in part A of **FIGURE 11.21**. The apparently uniform and approximately circular pattern of expression is actually the summation of *vestigial* response to two separate signaling pathways shown in **FIGURE 11.22**, which result in the cross-shaped and four-part patterns of expression shown at the bottom. Separate visualization of these patterns in the wing disk is shown in Figure 11.21, part B. The signaling pathway A in Figure 11.22 consists of the products of the genes *apterous, fringe, serrate*, and so forth; and pathway B consists of the products of the genes *engrailed, hedgehog*, and so forth. The Suppressor-of-Hairy protein binds to a *boundary enhancer* in the *vestigial* gene, which induces gene expression in the cross-shaped pattern. The Mad protein binds to a separate *quadrant enhancer*, which induces gene expression in the quadrant pattern. Such overlapping patterns of gene expression of *vestigial* and other genes in wing development ultimately yield the exquisitely fine level of cellular and morphological differentiation observed in the adult animal.

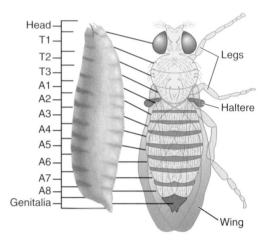

FIGURE 11.19 Relationship between larval and adult segmentation in *Drosophila*. Each of the three thoracic segments in the adult carries a pair of legs. The wings develop on the second thoracic segment (T2) and the halteres (flight balancers) on the third thoracic segment (T3).

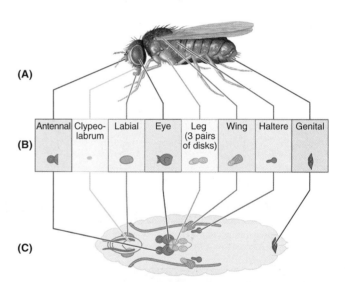

FIGURE 11.20 (A) Structures in the adult *Drosophila* correlated with the imaginal disks from which they arise. (B) General morphology of the disks late in larval development. (C) Larval locations of the imaginal disks.

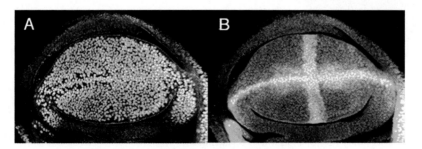

FIGURE 11.21 (A) Expression of the *vestigial* gene (green) in the developing wing imaginal disk. The approximately circular area of expression gives rise to the wing proper. (B) Visualization of the underlying boundary and quadrant patterns of *vestigial* expression in the same disk.

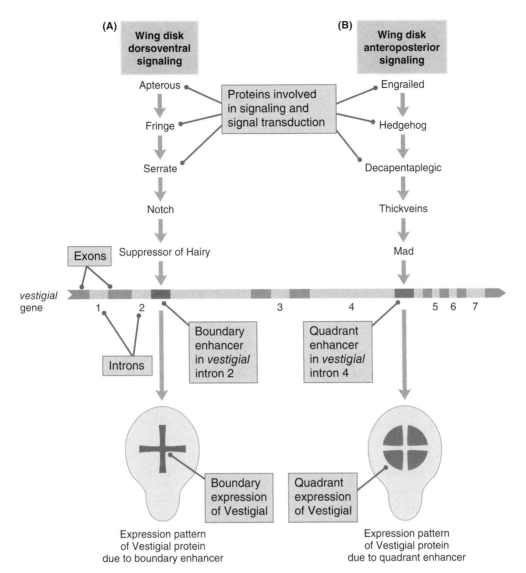

FIGURE 11.22 The uniform pattern of *vestigial* expression in the wing imaginal disk results from the superposition of two separate patterns. (A) The boundary expression pattern is determined by a dorsoventral signaling pathway. (B) The quadrant expression pattern is determined by the anteroposterior signaling pathway.

Among the genes that transform the periodicity of the *Drosophila* embryo into a body plan with linear differentiation are two small sets of **homeotic**, or *Hox*, **genes**. Homeotic mutations result in the transformation of one body segment into another, which is recognized by the misplaced development of structures that are normally present elsewhere in the embryo. One class of homeotic mutation is illustrated by *bithorax*, which causes transformation of the anterior part of the third thoracic segment into the anterior part of the second thoracic segment, with the result that the halteres normally formed from segment T3 are transformed into a pair of wings in addition to the pair normally formed from segment T2 (**FIGURE 11.23**, part B). The other class of homeotic mutation is illustrated by *Antennapedia*, which results in transformation of the antennae into legs. The *Hox* genes represented by

bithorax and *Antennapedia* are in fact gene clusters. The cluster containing *bithorax* is designated BX-C (stands for *bithorax*-complex), and that containing *Antennapedia* is called ANT-C (stands for *Antennapedia*-complex). Both gene clusters were initially discovered through their homeotic effects in adults. Later they were shown to affect the identity of larval segments. The BX-C is primarily concerned with the development of larval segments T3 through A8 (Figure 11.19), with principal effects in T3 and A1. The ANT-C is primarily concerned with the development of the head (H) and thoracic segments T1 and T2.

The homeotic genes are transcriptional activators of other genes. Most *Hox* genes contain one or more copies of a characteristic sequence of about 180 nucleotides called a **homeobox**, which is also found in key genes concerned with the development of embryonic

FIGURE 11.23 (A) Wildtype *Drosophila* showing wings and halteres (the pair of knob-like structures protruding posterior to the wings). (B) A fly with four wings produced by mutations in the *bithorax* complex. The mutations convert the third thoracic segment into the second thoracic segment, and the halteres normally present on the third thoracic segment become converted into the posterior pair of wings.

(A) and (B) Courtesy of Edward B. Lewis. Used with permission of Hugh Lewis.

segmentation in organisms as diverse as segmented worms, frogs, chickens, mice, and human beings. Homeobox sequences are present in exons and code for a protein-folding domain that includes a helix–turn–helix DNA-binding motif.

Hox genes are important master control genes in animal development.

The *Hox* genes are important in part because they function as master control genes in specifying the body plan of animals along the anterior–posterior axis. From organisms with fixed cell lineages like *Caenorhabditis*, to those with well-defined segments such as *Drosophila*, to those with variable cell lineages including humans and other mammals, the *Hox* genes control the differentiation in the anterior-posterior direction of the nervous system, musculature, skeletal elements, and so forth. The *Hox* genes are highly conserved in nucleotide sequence and in their orientation along the chromosome. Those controlling the anterior structures are located at the 3′ end of each Hox gene cluster, and those controlling posterior structures are located at the 5′ end. This orientation coincides with the temporal order in which the genes are expressed during development. The gene order is conserved in part because key regulatory elements that control gene expression are located in regions flanking each *Hox* gene cluster.

Pax6 is a master regulator of eye development.

The evolutionary conservation of master control genes in development is well illustrated by a class of homeotic genes known as *Pax*, which encode PAX proteins that contain a homeobox and, in addition, another DNA-binding domain known as the *paired box*. The best known of these genes is *Pax6*, which was originally discovered as the cause of a small-eye phenotype in mutant mice and then shown to be mutated in a hereditary form of impaired vision or blindness in humans known as *aniridia* (absence of the iris). In both cases, the affected individuals are heterozygous, as homozygous mutant genotypes do not survive owing to severe developmental abnormalities of the head and brain.

The PAX6 protein has an identical amino acid sequence in humans and mice, and the extreme sequence conservation prompted a search for similar genes in other organisms. A *Pax*-related gene was soon discovered in *Drosophila*, and it proved to be the gene *eyeless* that had been discovered nearly a century earlier. As its name implies, *eyeless* mutants have drastically reduced or absent eyes.

A key experimental observation is that the *Pax6* gene from the mouse is able to induce eye development in *Drosophila*! The resulting eyes are not mouse eyes, however, but the normal compound eyes of flies. Hence, *Pax6* merely turns on the hierarchy of genes needed for normal development of the compound eye.

How far back in evolution do the *Pax* genes function as master control genes for eye development? In all cases so far examined in detail, these genes are implicated in eye development (**FIGURE 11.24**). These cases include the simple eye spots of planarians (Figure 11.24, part A), each consisting only of an elongated light-sensitive nerve cell oriented in a cup-shaped pigment cell. They include the camera-type eyes found in jellyfish, octopus, and vertebrates (Figure 11.24, parts B–D), each consisting of a single lens projecting light onto a light-sensitive retina. As we have seen, *Pax* genes also control the development of compound eyes in insects (Figure 11.24, part E), which consist of hundreds of cylinders, each containing a set of photoreceptor cells and a lens. The same family of genes specifies the development of the mirror-type eyes around the periphery of the scallop (Figure 11.24, part F), in which light not only is projected onto a retina but also is reflected to increase sensitivity. The key role of *Pax* and related genes in the development of these diverse types of eyes over many hundreds of millions of years of evolutionary history suggests that the eyes evolved only once very

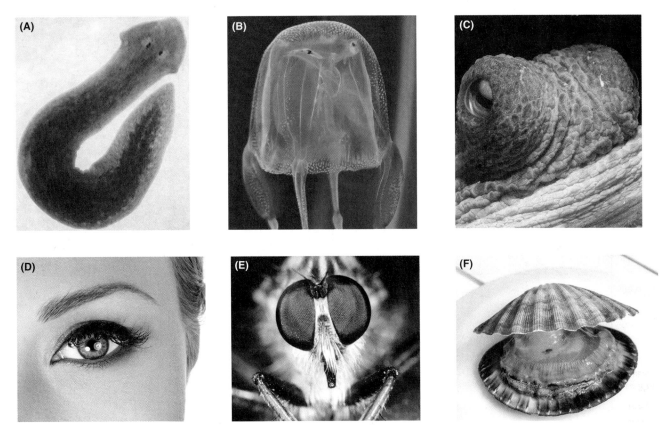

FIGURE 11.24 Major types of eyes found in animals. (A) planaria, (B) jellyfish, (C) octopus, (D) vertebrate, (E) dragonfly, and (F) scallop.

early in evolution, and that the various types of eyes found in fossil and living organisms evolved from that common ancestor using the same master control genes over and over again.

11.4 Floral development in *Arabidopsis* illustrates combinatorial control of gene expression.

As we have seen, most of the major developmental decisions in animals are made early in life, during embryogenesis. In higher plants, differentiation takes place almost continuously throughout life in regions of actively dividing cells called **meristems** in both the vegetative organs (root, stem, and leaves) and the floral organs (sepal, petal, stamen, and carpel). The shoot and root meristems are formed during embryogenesis and consist of cells that divide in distinctive geometric planes and at different rates to produce the basic morphological pattern of each organ system. The floral meristems are established by a reorganization of the shoot meristem after embryogenesis and eventually differentiate into floral structures characteristic of each particular species. One important difference between animal and plant development is that

KEY CONCEPT

In higher plants, as groups of cells leave the proliferating region of the meristem and undergo further differentiation into vegetative or floral tissue, their developmental fate is determined almost entirely by their position relative to neighboring cells.

The critical role of positional information in higher plant development stands in contrast to animal development, in which cell lineage often plays a key role in determining cell fate.

The plastic or "indeterminate" growth patterns of higher plants are the result of continuous production of both vegetative and floral organ systems. These patterns are conditioned largely by day length and the quality and intensity of light. The plasticity of plant development gives plants a remarkable ability to adjust to environmental insults. **FIGURE 11.25** shows a tree that, over time, adjusts to the presence of a nearby fence by engulfing it into the trunk. Higher plants can also adjust remarkably well to a variety of genetic aberrations.

FIGURE 11.25 The ability of plant development to adjust to perturbations is illustrated by this tree. Encountering a fence, it eventually incorporates the fence into the trunk.

Courtesy of Robert E. Pruitt, Purdue University.

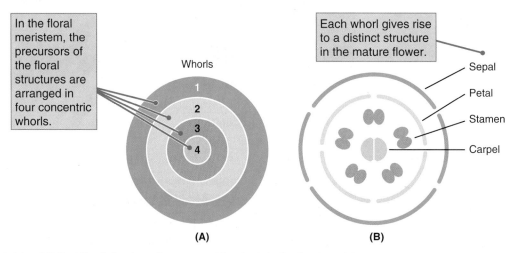

FIGURE 11.26 Origin of distinct floral structures from concentric whorls in the floral meristem.

Flower development in *Arabidopsis* is controlled by MADS box transcription factors.

Genetic analysis of *Arabidopsis thaliana*, a member of the mustard family, has revealed important principles in the genetic determination of floral structures. As is typical of flowering plants, the flowers of *Arabidopsis* are composed of four types of organs arranged in concentric rings, or whorls. **FIGURE 11.26** illustrates the geometry, looking down at a flower from the top. From outermost to innermost, the whorls are designated 1, 2, 3, and 4 (part A). In the development of the flower, each whorl gives rise to a different floral organ (part B). Whorl 1 yields the sepals (the green, outermost floral leaves), whorl 2 the petals (the white, inner floral leaves), whorl 3 the stamens (the male organs, which form pollen), and whorl 4 the carpels (which fuse to form the ovary).

Mutations that affect floral development fall into three major classes, each with a characteristic phenotype (**FIGURE 11.27**). Compared with the wildtype flower (panel A), one class lacks sepals and petals (panel B), another class lacks petals and stamens (panel C), and the third class lacks stamens and carpels (panel D). On the basis of crosses between homozygous mutant organisms, these classes of mutants can be assigned to different complementation groups, each of which defines a different gene. The key genes and their mutant phenotypes are listed in **TABLE 11.1**.

- The phenotype lacking sepals and petals is caused by mutations in the gene *ap2 (apetala-2)*.

- The phenotype lacking stamens and petals is caused by a mutation in either of two genes, *ap3 (apetala-3)* or *pi (pistillata)*.

- The phenotype lacking stamens and carpels is caused by mutations in the gene *ag (agamous)*.

(A) Wildtype

(B) *apetala-2* (*ap2*)

(C) *pistillata* (*pi*)

(D) *agamous* (*ag*)

FIGURE 11.27 Phenotypes of the major classes of floral mutations in *Arabidopsis*. (A) The wildtype floral pattern consists of concentric whorls of sepals, petals, stamens, and carpels. (B) The homozygous mutation *ap2* (*apetala-2*) results in flowers missing sepals and petals. (C) Genotypes that are homozygous for either *ap3* (*apetala-3*) or *pi* (*pistillata*) yield flowers that have sepals and carpels but lack petals and stamens. (D) The homozygous mutation *ag* (*agamous*) yields flowers that have sepals and petals but lack stamens and carpels.

(A) Courtesy of Elliot M. Meyerowitz, California Institute of Technology.; (B) Reproduced from E. M. Meyerowitz and J. L. Bowman, *Sci. Am.* 271 (1994): 56-65. Used with permission of Elliot M. Meyerowitz, California Institute of Technology.; (C) and (D) Courtesy of Elliot M. Meyerowitz, California Institute of Technology.

TABLE 11.1 Floral Development in Mutants of *Arabidopsis*

Genotype	Whorl			
	1	2	3	4
Wildtype	Sepals	Petals	Stamens	Carpels
ap2/ap2	Carpels	Stamens	Stamens	Carpels
ap3/ap3	Sepals	Sepals	Carpels	Carpels
pi/pi	Sepals	Sepals	Carpels	Carpels
ag/ag	Sepals	Petals	Petals	Sepals

These genes encode transcription factors that are members of the *MADS box* family of transcription factors. MADS box transcription factors include a common sequence motif consisting of 58 amino acids. They are involved frequently in transcriptional regulation in plants and to a lesser extent in animals.

Flower development in *Arabidopsis* is controlled by the combination of genes expressed in each concentric whorl.

The role of the *ap2, ap3, pi,* and *ag* transcription factors in the determination of floral organs can be inferred from the phenotypes of the mutations. The logic of the inference is based on the observation (see Table 11.1) that mutation in any of the genes eliminates two floral organs that arise from adjacent whorls. This pattern suggests that *ap2* is necessary for sepals and petals, *ap3* and *pi* are both necessary for petals and stamens, and *ag* is necessary for stamens and carpels. Because the mutant phenotypes are caused by loss-of-function alleles of the genes, it may be inferred that *ap2* is expressed in whorls 1 and 2, that *ap3* and *pi* are expressed in whorls 2 and 3, and that *ag* is expressed in whorls 3 and 4. The overlapping patterns of expression are shown in **TABLE 11.2**.

The model of gene expression in Table 11.2 suggests that floral development is controlled in combinatorial fashion by the four genes. Sepals develop from tissue in which only *ap2* is active; petals are evoked by a combination of *ap2, ap3*, and *pi*; stamens are determined by a combination of *ap3, pi*, and *ag*; and carpels derive from tissue in which only *ag* is expressed. This model is illustrated graphically in **FIGURE 11.28**. This model of floral determination is often called the **flower ABC model** because the wildtype activity of *ap2* was originally designated A, that of *ap3* and *pi* acting together as B, and that of *ag* as C. Therefore, the combination of activities present in each whorl would be represented as A in whorl 1, AB in whorl 2, BC in whorl 3, and C in whorl 4 (Figure 11.28).

You may have noted already that the model in Table 11.2 does not account for all of the phenotypic features of the *ap2* and *ag* mutations in Table 11.1. In particular, according to the combinatorial model in Table 11.2, the development of carpels and stamens from whorls 1 and 2 in homozygous *ap2* plants would require expression of *ag* in whorls 1 and 2. Similarly, the development of petals and sepals from whorls 3 and 4 in homozygous *ag* plants would require expression of *ap2* in whorls 3 and 4. This discrepancy can be explained if it is assumed that *ap2* expression and *ag* expression are mutually exclusive: In the presence of the AP2 transcription factor, *ag* is repressed; in the presence of the AG transcription factor, *ap2* is repressed. If this were the case, then in *ap2* mutants, *ag* expression

TABLE 11.2 Domains of Expression of Genes Determining Floral Development

Whorl	Genes expressed	Determination
1	*ap2*	Sepal
2	*ap2 + ap3* and *pi*	Petal
3	*ap3* and *pi + ag*	Stamen
4	*ag*	Carpel

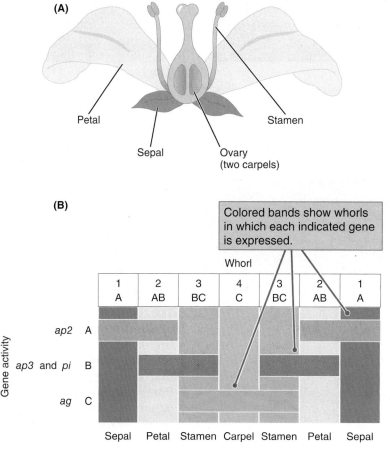

FIGURE 11.28 Control of floral development in *Arabidopsis* by the overlapping expression of four genes. (A) The sepals, petals, stamens, and carpels are floral organ systems that form in concentric rings, or whorls. The developmental identity of each concentric ring is determined by the genes *ap2, ap3* and *pi*, and *ag*, each of which is expressed in two adjacent rings. (B) Therefore, each whorl has a unique combination of active genes indicated by the combinations of A, B, and C.

would spread into whorls 1 and 2; in *ag* mutants, *ap2* expression would spread into whorls 3 and 4. This additional assumption enables us to explain the phenotypes of the single and even double mutants.

With the additional assumption we have made about *ap2* and *ag* interaction, the model in Table 11.2 fits the data. But is the model correct? For these genes, the patterns of gene expression, assayed by *in situ* hybridization of RNA in floral cells with labeled probes for each of the genes, fit the patterns in Table 11.2. In particular, *ap2* is expressed in whorls 1 and 2, *ap3* and *pi* in whorls 2 and 3, and *ag* in whorls 3 and 4. Furthermore, the seemingly arbitrary assumption about *ap2* and *ag* expression being mutually exclusive turns out to be true. In *ap2* mutants, *ag* is expressed in whorls 1 and 2; reciprocally, in *ag* mutants, *ap2* is expressed in whorls 3 and 4. It is also known how *ap3* and *pi* work together. The active transcription factor that corresponds to these genes is a dimeric protein composed of Ap3 and Pi polypeptides. Each component polypeptide, in the absence of the other, remains inactive in the cytoplasm. Together, they form an active dimeric transcription factor that migrates into the nucleus.

STOP & THINK 11.3

Given the critical role of the Ap2, Ap3/Pi, and Ag transcription factors in floral determination, it might be speculated that triple mutants lacking all three types of transcription factors would have very strange flowers. Can you predict what the floral phenotype of an *ap2 pi ag* triple mutant would be?

CHAPTER SUMMARY

- In animal cells, maternal gene products in the oocyte control the earliest stages of development, including the establishment of the main body axes.

- Developmental genes are often controlled by gradients of gene products, either within cells or across parts of the embryo.

- Regulation of developmental genes is hierarchical— genes expressed early in development regulate the activities of genes expressed later.

- Regulation of developmental genes is combinatorial— each gene is controlled by a combination of other genes.

- For genes that control development, the phenotype of loss-of-function mutants is often the opposite of that of gain-of-function mutants.

- The principle of epistasis helps to determine the order in which genes act in a linear switch– regulation developmental pathway. This principle asserts that the epistatic gene acts downstream of the hypostatic gene.

- Many of the fundamental processes of pattern formation appear to be similar in animals and plants.

ISSUES AND IDEAS

- What is meant by *positional information* in regard to development? How can positional information affect cell fate?

- If a gene is both necessary and sufficient for determining a developmental pathway, why would loss-of-function mutants be expected to have a different phenotype than gain-of-function mutants?

- What is a transmembrane receptor? What is a ligand? What role do these types of molecules play in signaling between cells?

- What is the principle of epistasis? How does this principle help identify the order of action of gene products along a linear switch–regulation pathway?

- Why was the study of maternal-effect lethal genes a key to deciphering the genetic control of early embryogenesis in *Drosophila*?

- Do plants have a germ line in the same sense as animals? What does the difference in germ-cell origin imply about the potential role of "somatic" mutations in the evolution of each type of organism?

- How does the genetic determination of floral development in *Arabidopsis* illustrate the principle of combinatorial control?

SOLUTIONS: STEP BY STEP

PROBLEM 1 In the diagram shown here, substance A inhibits the development of red pigments in a flower. The wildtype color of the flower is pink, but mutants are known that are either white or red. Assuming that substance A is the product of gene A, what flower-color phenotype would you expect of a loss-of-function mutation in gene A? What flower-color phenotype would you expect of a gain-of-function mutation in gene A? What principle in developmental genetics does this situation exemplify?

SOLUTION. Because substance A inhibits flower color, and the wildtype phenotype is pink, loss of function in gene A would eliminate A and, hence, reduce the amount of inhibition; the expected phenotype would, therefore, be red. Similarly, a gain-of-function mutation in gene A would increase the amount of A and, therefore, intensify the inhibition, and the expected flower-color phenotype would be white. This situation exemplifies the principle that, in developmental-control genes, a loss-of-function mutation often has the opposite phenotypic effect of a gain-of-function mutation.

PROBLEM 2 Shown here are three possible phenotypes observed on a distal appendage of a certain insect species. The wildtype phenotype consists of a single row of bristles. Some mutants yield a phenotype of no bristles, whereas others yield a phenotype in which the appendage has two rows of bristles. The development of these bristles is known to be due to a linear switch–regulation pathway involving the substances A, B, C, and D, which are encoded in the genes *A*, *B*, *C*, and *D*, respectively. The order of action of A, B, C, and D is unknown. The mutant alleles *a* and *b* result in missing bristles (M), whereas the mutant alleles c and d result in an extra row of bristles (E). The matrix shows the phenotype observed in all possible double mutants. What is the implied order of gene action of substances A, B, C, and D?

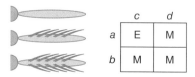

	c	d
a	E	M
b	M	M

SOLUTION. According to the principle of epistasis, the product of the epistatic gene in a double-mutant combination acts downstream of the product of the hypostatic gene. The first row, therefore, implies the order A–C and D–A, whereas the second row implies the order C–B and D–B. Putting all this information together, the order of action of the substances must be D–A–C–B.

PROBLEM 3 Consider a hypothetical mutant protease that affects floral development in *Arabidopsis thaliana*. The protease has an altered substrate specificity that enables it to cleave and inactivate both Ap2 and Ag proteins (the products of *ap2* and *ag*, respectively). In view of the fact that tissue containing the Ap3/Pi dimeric protein, but neither Ap2 nor Ag alone, develops into floral organs intermediate between petals and stamens, what floral phenotype would be expected in the protease mutant?

SOLUTION. In whorl 1, Ap2 activity is missing, so this region will develop as a whorl of leaves. Likewise, in whorl 4, Ag activity is missing, so this region will develop as a whorl of leaves. In whorls 2 and 3, only Ap3/Pi is present, so these will develop as whorls of tissue intermediate between petals and stamens. Therefore, the flower phenotype will be leaves in whorls 1 and 4 and petal/stamen intermediates in whorls 2 and 3.

CONCEPTS IN ACTION: PROBLEMS FOR SOLUTION

11.1 Why is it important for an embryo initially to have a large supply of rRNA?

11.2 Distinguish between a loss-of-function mutation and a gain-of-function mutation. Can the same gene undergo both types of mutations? Can the same allele have both types of effects?

11.3 Classify each of the following mutant alleles as a cell-lineage mutation, a homeotic mutation, or a pair-rule mutation.

(a) A mutant allele in C. *elegans* in which a cell that normally produces two daughter cells with different fates gives rise to two daughter cells with identical fates.

(b) A mutation in *Drosophila* causes an antenna to appear at the normal site of a leg.

(c) A lethal mutation in *Drosophila* results in abnormal gene expression in alternating segments of the embryo.

11.4 Programmed cell death (apoptosis) is responsible in part for shaping many organs and tissues in normal development. Consider a group of cells in the duck leg primordium that normally undergo apoptosis. If the cells are transplanted from their normal leg site to another part of the embryo just before they would normally die, they still die on schedule. But if the same operation is performed a few hours earlier, the cells do not die. How can you explain this observation?

11.5 What are the developmental consequences of mutants that cannot execute programmed cell death in C. *elegans*?

11.6 Which of the following statements is true?

(a) The *odd-skipped* mutation exemplifies a pair-rule gene. One would expect that a loss-of-function phenotype of an *odd-skipped* mutant would be deletion of alternating segments of the larva.

(b) Mutations in several different genes result in flies lacking wings, but when each of these genes is expressed alone, it does not lead to wing development. Thus, these genes are necessary but not sufficient for wing development.

(c) The phenotype of a particular *Drosophila* mutant is the development of a larva lacking all denticle belts but having the normal number of segments. This is a mutation in a gap gene.

11.7 A mutant allele in the axolotl designated o is a maternal-effect lethal because embryos from oo females die at gastrulation, irrespective of their own genotype. However, the embryos can be rescued by injecting oocytes from oo females with an extract of nuclei from either o^+o^+ or o^+o eggs. Injection of cytoplasm is not effective. Suggest an explanation for these results.

11.8 The same transmembrane receptor protein encoded by the *lin-12* gene is used in the determination of different developmental fates in different cell lineages. Suggest a mechanism by which the same receptor can determine different fates in different cell types.

11.9 Can the key genes involved in the specification of floral organ identity in *Arabidopsis* be regarded as homeotic genes? Explain your answer.

11.10 A particular gene is necessary, but not sufficient, for a certain developmental fate. What is the expected phenotype of a loss-of-function mutation in the gene? Is the allele expected to be dominant or recessive?

11.11 In *Drosophila*, the Bicoid and Nanos proteins work together to restrict expression of the gap gene *hunchback* to the anterior one-third of the embryo. What phenotype would you expect of a mutant larva with a loss-of-function mutation in *hunchback*?

11.12 What floral phenotype would be expected of a mutation resulting in a loss of function of *ap3* and *pi*? A gain-of-function mutation in which both *ap3* and *pi* were expressed in whorls 1 and 4 as well as whorls 2 and 3? What do these results imply about Ap3/Pi being necessary or necessary and sufficient for the developmental fate of the four whorls?

11.13 What phenotype would be expected of a gain-of-function mutation in *Arabidopsis* that resulted in expression of *Ap3* and *Pi* in whorl 1?

11.14 The homeotic floral-identity genes X, Y, and Z in whirligigs act either separately or in pairs to control the identity of floral organs. The floral structure consists of four whorls, the outermost being whorl 1 and the innermost whorl 4. Gene product X is expressed in whorls 1 and 2, Y in 2 and 3, and Z in 3 and 4. The presence of X without Y induces sepals, X + Y together induce petals, Y + Z together induce stamens, and Z without Y induces carpels. In the absence of X, the domain of activity of Z expands to all four whorls. In the absence of Z, the domain of activity of X expands to all four whorls. Mutant alleles of X, Y, and Z eliminate characteristic floral organs from a specific whorl, and a different floral organ appears in its place. What would you expect for the phenotype of each of the following?

(a) A loss-of-function X allele

(b) A loss-of-function Y allele

(c) A loss-of-function Z allele

11.15 Using the information presented in the foregoing whirligig problem, deduce the expected floral phenotype of a double mutant homozygous for loss-of-function alleles of both X and Y.

11.16 Two classes of genes involved in segmentation of the *Drosophila* embryo are gap genes, which are expressed in one region of the developing embryo, and pair-rule genes, which are expressed in seven stripes. Larvae that are homozygous for recessive mutations in gap genes lack a continuous block of larval segments, and those that are homozygous for recessive mutations in pair-rule genes lack alternating segments. You examine gene expression by means of mRNA hybridization *in situ* and find that (1) the embryonic expression pattern of gap genes is normal in all pair-rule mutants, and (2) the pair-rule gene expression pattern is abnormal in all gap gene mutants. What do these observations tell you about the temporal hierarchy of gap genes and pair-rule genes in the developmental pathway of segmentation?

11.17 The nuclei of brain cells in the adult frog normally do not synthesize DNA or undergo mitosis. However, when transplanted into developing oocytes, the brain cell nuclei behave as follows: (a) In rapidly growing premeiosis oocytes, they synthesize RNA. (b) In more mature oocytes, they do not synthesize DNA or RNA, but their chromosomes condense and they begin meiosis. How would you explain these results?

11.18 Edward B. Lewis, Christianne Nüsslein-Volhard, and Eric Wieschaus shared a 1995 Nobel Prize in Physiology or Medicine for their work on the developmental genetics of *Drosophila*. In their screen for developmental genes, Nüsslein-Volhard and Wieschaus initially identified 20 lines bearing maternal-effect mutations that produced embryos lacking anterior structures but having the posterior structures duplicated. When Nüsslein-Volhard mentioned this result to a colleague, the colleague was astounded to learn that mutations in 20 genes could give rise to this phenotype. Explain why his surprise was completely unfounded.

11.19 Explain why, in accounting for the phenotypes of floral mutants in *Arabidopsis*, it was necessary to postulate that:

(a) In *agamous* mutants, the domain of expression of *apetala-2* expands to whorls 3 and 4.

(b) In *apetala-2* mutants, the domain of expression of *agamous* expands to whorls 1 and 2.

11.20 The autosomal gene *rosy (ry)* in *Drosophila* is the structural gene for the enzyme xanthine dehydrogenase (XDH), which is necessary for wild-type eye pigmentation. Flies of genotype *ry/ry* lack XDH activity and have rosy eyes. The X-linked gene *maroonlike (mal)* is also necessary for XDH activity, and *mal/mal*; ry^+/ry^+ females and *mal/Y*; ry^+/ry^+ males also lack XDH activity; they have maroonlike eyes. The cross mal^+/mal; *ry/ry* females × *mal/Y*; ry^+/ry^+ males produces *mal/mal*; ry^+/ry females and *mal/Y*; ry^+/ry males that have wildtype eye color, even though their genotypes would imply that they should have rosy eyes. Suggest an explanation.

 STOP & THINK ANSWERS

ANSWER TO STOP & THINK 11.1

Because *aa bb* has the same phenotype as *aa b^+b^+*, gene *a* is epistatic to gene *b*, which means that the product A of *a^+* acts downstream of the product B of *b^+*. In other words, A corresponds to Y, and B corresponds to X.

ANSWER TO STOP & THINK 11.2

Females of genotype *m/m* can be produced by crossing heterozygous *m/+* females with either *m/m* or *m/+* males. Homozygous *m/m* females produce defective eggs, but *m/+* females produce normal eggs, and when an *m*-bearing egg is fertilized by an *m*-bearing sperm, a viable *m/m* female results.

ANSWER TO STOP & THINK 11.3

The developing flower of the *ap2 pi ag* triple mutant still consists of four concentric whorls, but each whorl lacks any of the combinations of transcription factors necessary for the development of sepals, petals, stamens, or carpels. Each whorl, therefore, develops according to its default state, which is that of a leaf. Hence, the predicted floral phenotype lacks all of the normal floral organs. The flowers consist merely of leaves arranged in concentric whorls, which is very strange indeed.

Design Credits: Stop & Think icon made by Darius Dan from www.flaticon.com; The Human Connection icon made by Daniel Bruce from www.flaticon.com; Elephant image: © NickBiemans/GettyImages.

CHAPTER 12

Molecular Mechanisms of Mutation and DNA Repair

LEARNING OBJECTIVES

- To distinguish between mutations that are spontaneous or induced, germ line or somatic, conditional or unconditional, and loss of function or gain of function.

- To explain why nucleotide substitutions in the first or second positions of codons are more likely to be nonsynonymous than synonymous, and why third-position transitions are more likely to be synonymous than third-position transversions.

- To describe the replica-plating experiments that showed that antibiotic-resistance mutations arise in the absence of the antibiotic, and hence, that mutations occur without regard to their favorable or unfavorable effects on the organism.

- To predict the types of mutations likely to occur upon treatment of cells with a base analogue, x rays, or ultraviolet light.

- To distinguish between mismatch repair and nucleotide excision repair and explain why mismatch repair is the more important for the accuracy of DNA replication.

- To describe the main mechanisms of repair of deamination of cytosine resulting in uracil and deamination of 5-methylcytosine resulting in thymine and explain why these mechanisms are not the same.

A **mutation** is any heritable change in the genetic material. In this chapter, we examine the nature of mutations at the molecular level. You will learn how mutations are created, how they are detected phenotypically, and the means by which many mutations are corrected by special DNA repair enzymes almost immediately after they occur. You will see that mutations can be induced by radiation and a variety of chemical agents that produce strand breakage and other types of damage to DNA.

12.1 Mutations are classified in a variety of ways.

The principal ways in which mutations are classified are listed in **TABLE 12.1**. The first five categories pertain to any type of mutation, whereas the last pertains only to mutations in regions of DNA that code for proteins.

Mutagens increase the chance that a gene undergoes mutation.

Most mutations are **spontaneous**, which means that they are statistically random, unpredictable events. Nevertheless, each gene has a characteristic **rate of mutation**, measured as the probability of undergoing a change in DNA sequence in the time span of a single generation. Rates of mutation can be increased by treatment with a chemical **mutagen** or radiation, in which case the mutations are said to be **induced**. However, some of the mutations that take place in the presence of a mutagen would have taken place anyway, so it is usually impossible to state positively whether a particular mutation was or was not induced by a mutagen. For example, if treatment with a mutagen increases the spontaneous mutation rate by a factor of 10, then for every mutation that would have occurred anyway, there will now be 10. This means that 1 out of 10, or 10 percent, of all the mutations that take place in the presence of the mutagen would have occurred even in its absence.

Germ-line mutations are inherited; somatic mutations are not.

In multicellular organisms, one important distinction is based on the type of cell in which a mutation first occurs. Mutations that arise in cells that ultimately form gametes are **germ-line mutations**; all others are **somatic mutations**. A somatic mutation yields

TABLE 12.1 Major Types of Mutations and Their Distinguishing Features

Basis of classification	Major types of mutations	Major features
Origin	Spontaneous	Occurs in absence of known mutagen.
	Induced	Occurs in presence of known mutagen.
Cell type	Somatic	Occurs in nonreproductive cells.
	Germ line	Occurs in reproductive cells.
Expression	Conditional	Expressed only under restrictive conditions (such as high temperature).
	Unconditional	Expressed under permissive conditions as well as restrictive conditions.
Effect on function	Loss of function (knockout, null)	Eliminates normal function.
	Hypomorphic (leaky)	Reduces normal function.
	Hypermorphic	Increases normal function.
	Gain of function (ectopic expression)	Expressed at incorrect time or in inappropriate cell types.
Molecular change	Base substitution	One base pair in duplex DNA replaced with a different base pair.
	Transition	Pyrimidine (T or C) to pyrimidine, or purine (A or G) to purine.
	Transversion	Pyrimidine (T or C) to purine, or purine (A or G) to pyrimidine.
	Insertion	One or more extra nucleotides present.
	Deletion	One or more missing nucleotides.
Effect on translation	Synonymous (silent)	No change in amino acid encoded.
	Missense (nonsynonymous)	Change in amino acid encoded.
	Nonsense (termination)	Creates translational termination codon (UAA, UAG, or UGA).
	Frameshift	Shifts triplet reading of codons out of correct phase.

an organism that is genotypically a mixture (*mosaic*) of normal and mutant tissue. Most common cancers result from somatic-cell mutations. In animals, a somatic mutation cannot be transmitted to the progeny. In higher plants, somatic mutations can often be propagated by vegetative means without going through seed production, such as by grafting or the rooting of stem cuttings. Vegetative propagation is typical of many commercially important fruits, such as the "Delicious" apple and the "Florida" navel orange.

Conditional mutations are expressed only under certain conditions.

Among the mutations that are most useful for genetic analysis are those whose effects can be turned on or off by the experimenter. These are called **conditional mutations** because they produce changes in phenotype in one set of environmental conditions (called the **restrictive conditions**) but not in another (called the **permissive conditions**). For example, a **temperature-sensitive mutation** is a conditional mutation whose expression depends on temperature. Usually, the restrictive temperature is high (in *Drosophila*, 29°C), and the organism exhibits a mutant phenotype above this critical temperature; the permissive temperature is lower (in *Drosophila*, 18°C), and under permissive conditions the phenotype is wildtype or nearly wildtype. Proteins containing amino acid replacements are often temperature sensitive: The protein folds properly and functions nearly normally under permissive conditions, but it is unstable and denatures under restrictive conditions. Temperature-sensitive amino acid replacements are frequently used to block particular biochemical pathways under restrictive conditions, in order to test the importance of the pathways in various cellular processes, such as DNA replication.

An example of temperature sensitivity is found in the Siamese cat, with its black-tipped paws, ears, and

FIGURE 12.1 A Siamese cat showing the characteristic pattern of pigment deposition.

© OrangeGroup/Shutterstock.

tail (**FIGURE 12.1**). In this breed, an enzyme in the pathway for deposition of the black pigment melanin is temperature sensitive. The pathway is blocked at normal body temperature, and pigment is not deposited over most of the body. Pigment is deposited in the tips of the legs, ears, snout, and tail because these extremities are cooler than the rest of the body.

Mutations can affect the amount or activity of the gene product, or the time or tissue specificity of expression.

Mutations can also be classified according to their effects on gene function. The major categories are described in Table 12.1.

- A mutation that results in complete gene inactivation or in a completely nonfunctional gene product is a **loss-of-function mutation**, also called a *knockout* or *null* mutation. Examples include a deletion of all or part of a gene, and an amino acid replacement that inactivates the protein.

- A mutation that reduces, but does not eliminate, the level of expression of a gene or the activity of the gene product is called a **hypomorphic mutation**. Typically resulting from a nucleotide substitution that reduces the level of transcription, or from an amino acid replacement that impairs protein function, this type of mutation is sometimes referred to as *leaky*. The basis of the term is that because the level of expression or activity differs from individual to individual by chance, a few individuals have enough enzyme activity to "leak through" to produce a quasi-normal phenotype.

- The opposite of a hypomorphic mutation is a **hypermorphic mutation**. As the prefix *hyper* implies, a hypermorphic mutant produces a greater-than-normal level of gene expression, typically because the mutation changes the regulation of the gene so that the gene product is overproduced.

■ A gain-of-function mutation is one that qualitatively alters the action of a gene. For example, a gain-of-function mutation may cause a gene to become active in a type of cell or tissue in which the gene is not normally active. Or it may result in the expression of a gene in development at a time during which the wildtype gene is not normally expressed. Whereas most loss-of-function and hypomorphic mutations are recessive, many gain-of-function mutations are dominant. Expression of a wildtype gene in an abnormal location is also called *ectopic expression*. For example, expression of the wildtype gene product of the *Drosophila* gene *eyeless* in tissues that do not normally form eyes results in the development of parts of compound eyes, complete with eye pigments, in abnormal locations such as on the legs or mouthparts, in the abdomen, or on the wings.

12.2 Mutations result from changes in DNA sequence.

All mutations result from changes in the nucleotide sequence of DNA or from deletions, insertions, or rearrangement of DNA sequences in the genome. Mutations that substitute a single base pair with a different base pair or that add or delete a single nucleotide pair in the DNA are known as point mutations.

A base substitution replaces one nucleotide pair with another.

The simplest type of mutation is a base substitution (Table 12.1), in which a nucleotide pair in a DNA duplex is replaced with a different nucleotide pair. For example, in an A → G substitution, an A is replaced with a G in one of the DNA strands. This substitution temporarily creates a mismatched G−T base pair, but at the very next replication the mismatch is resolved as a proper G−C base pair in one daughter molecule and as a proper A−T base pair in the other daughter molecule. In this case, the G−C base pair is mutant and the A−T base pair is nonmutant. Similarly, in an A → T substitution, an A is replaced with a T in one strand, creating a temporary T−T mismatch, which is also resolved by replication as T−A in one daughter molecule and A−T in the other. In this example, the T−A base pair is mutant and the A−T base pair is nonmutant. The T−A and the A−T are not equivalent, as may be seen by considering the polarity. If the original

unmutated DNA strand has the sequence 5'-GAC-3', for example, then the mutant strand has the sequence 5'-GTC-3' (which we have written as T−A), and the nonmutant strand has the sequence 5'-GAC-3' (which we have written as A−T).

Some base substitutions replace one pyrimidine base with the other or one purine base with the other. These are called transition mutations. The four possible transition mutations are

$$T \rightarrow C \quad \text{or} \quad C \rightarrow T$$
$$(\text{pyrimidine} \rightarrow \text{pyrimidine})$$
$$A \rightarrow G \quad \text{or} \quad G \rightarrow A$$
$$(\text{purine} \rightarrow \text{purine})$$

Other base substitutions replace a pyrimidine with a purine or the other way around. These are called transversion mutations. The eight possible transversion mutations are

$$T \rightarrow A \quad T \rightarrow G \quad C \rightarrow A \quad \text{or} \quad C \rightarrow G$$
$$(\text{pyrimidine} \rightarrow \text{purine})$$
$$A \rightarrow T \quad A \rightarrow C \quad G \rightarrow T \quad \text{or} \quad G \rightarrow C$$
$$(\text{purine} \rightarrow \text{pyrimidine})$$

Because there are four possible transitions and eight possible transversions, if base substitutions were strictly random, one would expect a 1 : 2 ratio of transitions to transversions. However,

KEY CONCEPT

Spontaneous base substitutions are biased in favor of transitions. Among spontaneous base substitutions, the ratio of transitions to transversions is approximately 2 : 1.

Mutations in protein-coding regions can change an amino acid, truncate the protein, or shift the reading frame.

Most base substitutions in coding regions result in one amino acid being replaced with another; these are called missense mutations or *nonsynonymous mutations* (Table 12.1). A single amino acid replacement in a protein may alter the biological properties of the protein. An example is the R408W amino acid replacement in phenylalanine hydroxylase, which results from the base-pair substitution C−G to T−A at the first position in codon 408 of the gene; codon 408 in the mRNA is thereby changed from CGG (R, arginine) to UGG (W, tryptophan). This change inactivates the enzyme and results in phenylketonuria.

Examination of the genetic code shows that not all base substitutions cause amino acid replacements, particularly if they occur in the third codon position. In all codons with a pyrimidine in the third position, the particular pyrimidine present does not matter; likewise, in most codons ending in a purine, either purine will do. This means that most transition mutations in the third codon position do not change the amino acid that is encoded. Such mutations change the nucleotide sequence without changing the amino acid sequence; they are called synonymous substitutions or silent substitutions (Table 12.1) because they are not detectable by changes in phenotype.

Occasionally a base substitution creates a new stop codon UAA, UAG, or UGA. For example, a G → A change at the third position of the normal tryptophan codon UGG converts the codon into UGA. The result is that translation is terminated at the position of the mutant codon, and the polypeptide is truncated. A base substitution that creates a new stop codon is called a nonsense mutation. Nonsense mutations almost always result in loss of gene function, in many cases because RNA processing is disrupted and the transcript is degraded before it even leaves the nucleus.

Small insertions or deletions, when they take place in coding regions, can add or delete amino acids, provided that the number of nucleotides added or deleted is an exact multiple of three (the length of a codon). Otherwise, the insertion or deletion shifts the phase in which the ribosome reads the triplet codons and, consequently, alters all of the amino acids downstream from the site of the mutation. Mutations that shift the reading frame of the codons in the mRNA are called frameshift mutations. A common type of frameshift mutation is a single-base addition or deletion. The consequences of a frameshift can be illustrated by the insertion of an adenine at the position of the arrow in the following mRNA sequence:

LeuLeuLeuLeu
. . . CUGCUGCUGCUG . . .

↓

. . . CUGCAUGCUGCUG . . .
LeuHisAlaAla

Because of the frameshift, all of the amino acids downstream from the insertion are different from the original. Any addition or deletion that is not a multiple of three nucleotides will produce a frameshift. Unless it is very near the carboxyl terminus of a protein, a frameshift mutation usually results in the synthesis of a nonfunctional protein.

Sickle-cell anemia results from a missense mutation that confers resistance to malaria.

A classic example of the sometimes profound phenotypic effects of a single amino acid replacement is the mutation responsible for the human hereditary disease sickle-cell anemia. The molecular basis of sickle-cell anemia is a mutant gene for β-globin, one component of the hemoglobin present in red blood cells (FIGURE 12.2). The sickle-cell mutation changes the sixth codon in the coding sequence from the normal GAG, which codes for glutamic acid, into the codon GUG, which codes for valine. In the DNA, the mutant has an A−T base pair (transcribed as the middle A in the codon) replaced with a T−A base pair (transcribed as the middle U in the mutant codon). One consequence of the seemingly simple Glu → Val replacement is that hemoglobin containing the defective β polypeptide chain has a tendency to form long, needle-like crystals. Red blood cells in which crystallization happens become deformed into crescent, sickle-like shapes. Some of the deformed red blood cells are destroyed immediately (reducing the oxygen-carrying capacity of the blood and causing the anemia), whereas others may clump together and clog the blood circulation in the capillaries. The impaired circulation affects the heart, lungs, brain, spleen, kidneys, bone marrow, muscles, and joints. Patients suffer bouts of severe pain. The anemia causes impaired growth, weakness, jaundice, and other symptoms. Affected people are so generally weakened that they are susceptible to bacterial infections; infections are the most common cause of death in children with the disease.

Sickle-cell anemia is a severe genetic disease that often results in premature death. Yet it is a relatively common disease in areas of Africa and the Middle East in which malaria, caused by the protozoan parasite *Plasmodium falciparum*, is widespread. The association between sickle-cell anemia and malaria is not coincidental: It results from the ability of the mutant hemoglobin to afford some protection against malarial infection. In the life cycle of the parasite, it passes from a mosquito to a human being through the mosquito's bite. The initial stages of infection take place in cells in the liver where specialized forms of the parasite are produced that are able to infect and multiply in red blood cells. Widespread infection of red blood cells impairs the ability of the blood to carry oxygen, causing the weakness, anemia, and jaundice characteristic of malaria. In people with the mutant hemoglobin, however, infection with malaria is less likely and also less severe.

There is consequently a genetic balancing act between the prevalence of the genetic disease sickle-cell anemia and the parasitic disease malaria. If the mutant hemoglobin becomes too frequent, more lives are lost from sickle-cell anemia than are saved by the protection it affords against malaria; on the other hand, if the mutant hemoglobin becomes too rare, fewer lives are lost from sickle-cell anemia but the gain is offset by more deaths from malaria. The end result is a kind of genetic balancing act.

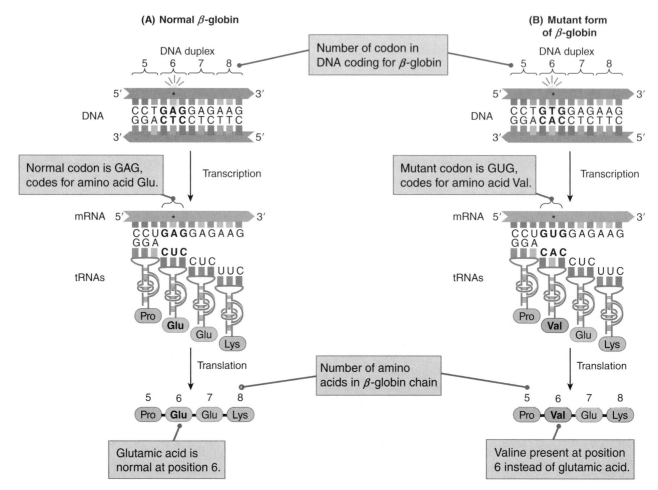

FIGURE 12.2 Molecular basis of sickle-cell anemia. (A) Part of the DNA in the normal β-globin gene. (B) Mutation of the normal A−T base pair to a T−A base pair results in the codon GUG (valine) instead of GAG (glutamic acid).

In contrast to the situation with sickle-cell anemia, an amino acid replacement does not always create a mutant phenotype. For instance, replacement of one amino acid by another with the same charge (say, lysine for arginine) may in some cases have no effect on either protein structure or phenotype. Whether the substitution of a similar amino acid for another produces an effect depends on the precise role of that particular amino acid in the structure and function of the protein. Any change in the active site of an enzyme usually decreases enzymatic activity.

In the human genome, some trinucleotide repeats have high rates of mutation.

Genetic studies of an X-linked form of mental retardation revealed an unexpected class of mutations called **dynamic mutations** because of the extraordinary genetic instability of the region of DNA involved. The X-linked condition, one of at least 12 genetic disorders associated with dynamic mutation, is associated with a class of X chromosomes that tends to fracture

in cultured cells that are starved for DNA precursors. The position of the fracture is in region Xq27.3, near the end of the long arm. The X chromosomes containing this site are called *fragile-X* chromosomes, and the associated form of mental retardation is the *fragile-X syndrome*. The fragile-X syndrome affects about 1 in 2500 children. It accounts for about one-half of all cases of X-linked mental retardation and is second only to Down syndrome as a cause of inherited mental impairment.

The fragile-X syndrome is highly variable in severity. Males are usually more severely affected than females. Developmental delays in speech and communication skills are common, as well as delays in gross motor skills such as sitting up and walking. Physical symptoms may include a long face with protruding ears, weakness in connective tissues resulting in poor muscle tone and extremely flexible joints, and enlarged testicles in males past puberty. Mental retardation is usually moderate in males and mild in females. Behavioral effects may include anxiety, poor concentration, trouble coping with sensory stimuli, avoidance of eye contact, and tantrums or emotional outbursts. These

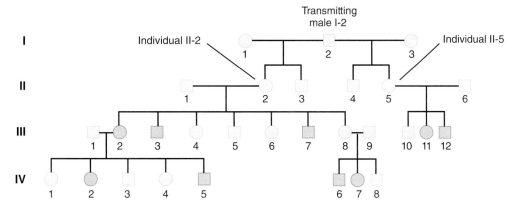

FIGURE 12.3 Pedigree showing transmission of the fragile-X syndrome. Male I-2 is not affected, but his daughters (II-2 and II-5) have affected children and grandchildren.

Data from C. D. Laird, *Genetics* 117 (1987): 587–599.

symptoms are nonspecific and overlap with such conditions as autism and attention deficit-hyperactivity disorder.

A hint of something unusual about the fragile-X syndrome was the paradoxical pattern of its inheritance, key features of which are illustrated in **FIGURE 12.3**. Approximately 1 in 5 males who carry the fragile-X chromosome are themselves phenotypically normal and also have phenotypically normal children. The oddity is that the heterozygous daughters of such a "transmitting male" often have affected children of both sexes. In Figure 12.3, the transmitting male denoted I-2 is not affected, but the X chromosome that he transmits to his daughters (II-2 and II-5) somehow becomes altered in the females in such a way that sons and daughters in the next generation (III) are affected. Both affected and normal granddaughters of the transmitting male may have affected progeny (generation IV).

The molecular basis of the fragile-X chromosome has been traced to a **trinucleotide repeat** of the form CGG (or, equivalently, CCG on the other strand) present in the DNA at the site where the breakage takes place (**FIGURE 12.4**). Normal X chromosomes have 6 to 54 tandem copies of the repeating unit, with an average of about 30, whereas affected persons have 230 to 2300 or more copies of the repeat. The trinucleotide repeat in the X chromosome in transmitting males is called the *premutation* and has an intermediate number of copies, ranging from 52 to 230. Approximately 1 in 250 females and 1 in 800 males carries an X chromosome with the premutation. The unprecedented feature of the trinucleotide premutation is that when transmitted by females (and only by females), it often increases in copy number (called *trinucleotide expansion*) to a level of 230 copies or more, at which stage the chromosome causes the fragile-X syndrome. The amplification does not take place in transmission through a male. The functional basis of the disorder

is that an excessive number of copies of the CGG repeat cause loss of function of a gene designated *FMR1* (fragile-site mental retardation-1) in which the CGG repeat is present. Most fragile-X patients exhibit no *FMR1* messenger RNA, whereas normal persons and carriers do show expression. The *FMR1* gene is expressed primarily in brain and testes, which explains the strange association between mental and testes abnormalities in affected males.

There is about an 80 percent chance that a premutation transmitted by a female will undergo amplification. Surprisingly, the amplification does not take place in the mother's germ line, but in somatic cells of the early embryo. Amplification occurs to a different extent in different somatic cells, and so individuals with the fragile-X syndrome are somatic mosaics for cells with different numbers of copies of the CGG repeat in the X chromosome. This accounts for the great variation in severity of the fragile-X syndrome from one affected individual to the next.

Other genetic diseases associated with dynamic mutation include the neurological disorders myotonic dystrophy (with an unstable repeat of CTG), Kennedy disease (AGC), Friedreich ataxia (AAG), spinocerebellar ataxia type 1 (AGC), and Huntington disease (AGC). For the fragile-X syndrome and myotonic dystrophy, the trinucleotide expansions occur primarily or exclusively when transmitted by females, but for spinocerebellar ataxia type I and Huntington disease, they occur primarily or exclusively when transmitted by males. Some trinucleotide repeats can undergo amplification when transmitted by either sex.

The molecular mechanism of trinucleotide expansion is illustrated in **FIGURE 12.5**. The process is called **replication slippage** (also called *slipped-strand mispairing*). As replication is proceeding along a template strand containing the repeats (part A), the replication complex momentarily dissociates from the template strand. In reassociating with the template, the 3' end

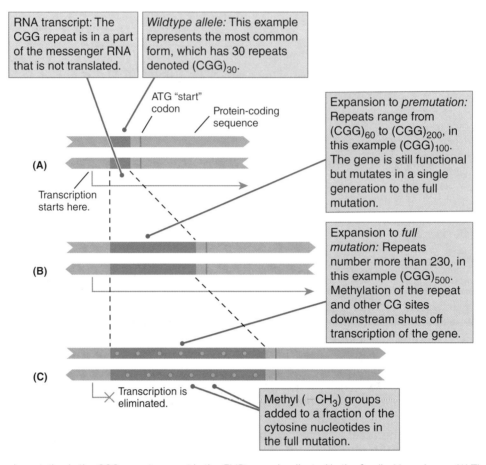

RNA transcript: The CGG repeat is in a part of the messenger RNA that is not translated.

Wildtype allele: This example represents the most common form, which has 30 repeats denoted (CGG)$_{30}$.

ATG "start" codon

Protein-coding sequence

(A)

Transcription starts here.

Expansion to *premutation:* Repeats range from (CGG)$_{60}$ to (CGG)$_{200}$, in this example (CGG)$_{100}$. The gene is still functional but mutates in a single generation to the full mutation.

(B)

Expansion to *full mutation:* Repeats number more than 230, in this example (CGG)$_{500}$. Methylation of the repeat and other CG sites downstream shuts off transcription of the gene.

(C)

Transcription is eliminated.

Methyl (CH$_3$) groups added to a fraction of the cytosine nucleotides in the full mutation.

FIGURE 12.4 Dynamic mutation in the CGG repeat present in the *FMR1* gene implicated in the fragile-X syndrome. (A) The wildtype allele typically has 30 copies of the repeat. (B) The premutation has 60 to 200 copies, which predisposes to further amplification when transmitted through a female. (C) The full mutation contains > 230 copies. In the full mutation, absence of transcription of the gene is associated with methylation of certain CG dinucleotides in the region.

of the new strand backtracks along the template and pairs with an upstream set of repeats (part B). Replication continues normally from this point (part C), but some of the repeats will be replicated twice (expanded); the level of expansion depends on how far the replication complex backtracked in reassociating. The template and the daughter strand cannot pair properly because they have a different number of repeats, but this situation is corrected by nucleotide-excision repair (Section 12.6); one outcome of repair is that the expanded region is introduced into the template strand (part D). Although the mechanism of dynamic mutation is known, it is not known why some trinucleotide repeats in the genome are genetically unstable whereas others are stable, or why the trinucleotide premutation state is uniquely prone to expansion whereas chromosomes that may have only somewhat fewer copies are genetically stable.

The molecular mechanism of *FMR1* inactivation is associated with the enzymatic addition of a methyl (–CH3) group to each of certain of the cytosine nucleotides in the 5′ region of the *FMR1* gene (Figure 12.5). Cytosine methylation occurs at a

fraction of the cytosine nucleotides in many higher eukaryotes, and in mammals it occurs preferentially at CG dinucleotides. Each CGG repeat in the amplified region of *FMR1* includes a potential methylation site. A high density of methylated CG dinucleotides is usually associated with repression of transcription of the affected gene. In the case of *FMR1*, the lack of transcription of the gene in affected individuals is associated with the methylation of the expanded CGG repeat as well as increased methylation of other CG dinucleotides nearby.

What does the *FMR1* protein do? The protein, called FMRP (for fragile-X mental retardation protein), is an RNA-binding protein that binds with the 5′ end of certain messenger RNAs and regulates either their translation into protein, their localization in the cytoplasm, or both. FMRP does not bind all mRNA molecules, but only a specific subset that encode proteins that function in the development of the facial bones and the nervous system or that function in learning and memory. Many of them are proteins that function in the communication between neurons.

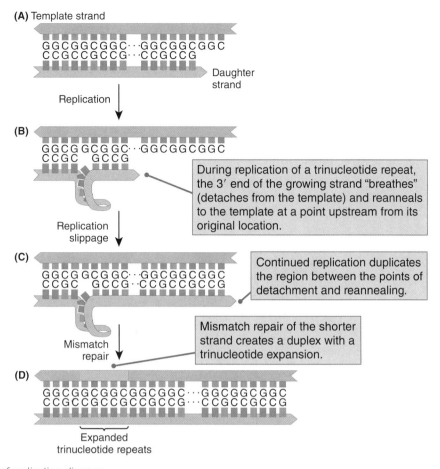

(A) Template strand

GGCGGCGGC···GGCGGCGGC
CCGCCGCCG···CCGCCG

Daughter strand

Replication

(B)

GGCGGCGGC···GGCGGCGGC
CCGC   GCCG

During replication of a trinucleotide repeat, the 3′ end of the growing strand "breathes" (detaches from the template) and reanneals to the template at a point upstream from its original location.

Replication slippage

(C)

GGCGGCGGC···GGCGGCGGC
CCGC   GCCG···CCGCCGCCG

Continued replication duplicates the region between the points of detachment and reannealing.

Mismatch repair of the shorter strand creates a duplex with a trinucleotide expansion.

Mismatch repair

(D)

GGCGGCGGCGGCGGC···GGCGGCGGC
CCGCCGCCGCCGCCG···CCGCCGCCG

Expanded trinucleotide repeats

FIGURE 12.5 Model of replication slippage.

12.3 Transposable elements are agents of mutation.

In a 1940s study of the genetics of kernel mottling in maize (**FIGURE 12.6**), Barbara McClintock discovered a genetic element that not only regulated the mottling, but also caused chromosome breakage. She called this element *Dissociation (Ds)*. Genetic mapping showed that the chromosome breakage always occurs at or very near the location of *Ds*. McClintock's critical observation was that *Ds* does not have a constant location but occasionally moves to a new position (**transposition**), causing chromosome breakage at the new site. Furthermore, *Ds* moves only if a second element, called *Activator (Ac)*, is also present in the same genome. In addition, *Ac* itself moves within the genome and can also cause modification in the expression of genes at or near its insertion site. Since McClintock's original discovery, many other **transposable elements** have been discovered. They can be grouped into "families" based on similarity in DNA sequence. The genomes of most organisms contain multiple copies of each of several distinct families of transposable elements. Once situated in the genome, transposable elements

FIGURE 12.6 Sectors of purple and yellow tissue in the endosperm of maize kernels resulting from the presence of the transposable elements *Ds* and *Ac*. The different level of sectoring in some ears results from dosage effects of *Ac*.

Courtesy of Jerry L. Kermicle, Professor Emeritus, University of Wisconsin at Madison.

can persist for long periods and undergo multiple mutational changes. Approximately 50 percent of the human genome consists of transposable elements; as we shall see later, most of these are evolutionary remnants no longer able to transpose.

Some transposable elements transpose via a DNA intermediate, others via an RNA intermediate.

The molecular mechanism of *Ds* transposition is illustrated in **FIGURE 12.7**. In this example the insertion goes into the wildtype *shrunken* gene in maize chromosome 9, causing a knockout mutation. To initiate the process, the target site for insertion is cleaved with a staggered cut, leaving a 3′ overhang of eight nucleotides on each strand (part A). The overhanging 3′ ends are ligated with the 5′ ends of the *Ds* element

to be inserted, leaving an eight-nucleotide gap in each strand (part B). When the gap is filled by repair enzymes, the result is a new insertion of *Ds* flanked by an eight-bp duplication of the target sequence (part C). The *Ds* element can insert in either orientation.

The presence of a target-site duplication is characteristic of most transposable element insertions, and it results from asymmetrical cleavage of the target sequence. For elements like *Ds*, target-site cleavage is a function of a transposase protein that catalyzes transposition. Each family of transposable elements has its own **transposase** that determines the distance

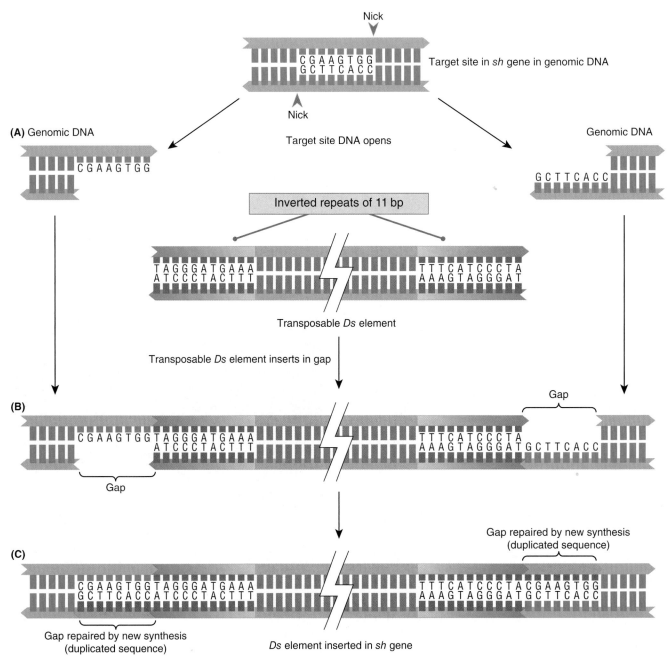

FIGURE 12.7 The sequence arrangement of a cut-and-paste transposable element (in this case the *Ds* of maize) and the changes that take place when it inserts into the genome. *Ds* is inserted into the maize *sh* gene at the position indicated. In the insertion process, a sequence of eight bp next to the site of insertion is duplicated and flanks the *Ds* element.

between the cuts made in the target DNA strands. Depending on the particular transposable element, the distance may be 1 to 12 bp, and this determines the length of the target-site duplication. Most transposable elements have many potential target sites scattered throughout the genome, and they usually show little or no sequence similarity from one site to the next.

The *Ds* element is one of a large class of elements called **DNA transposons** that transpose via a mechanism known as **cut-and-paste transposition**, in which the transposon is cleaved from one position in the genome and the same molecule is inserted somewhere else. Characteristic of DNA transposons is the presence of *terminal inverted repeats*, a sequence repeated in inverted orientation at each end of the element. In the case of *Ds*, the terminal inverted repeats are 11 bp in length (Figure 12.7), but in other families of DNA transposons, they can be up to a few hundred base pairs long. The terminal repeats are usually essential for efficient transposition, because they contain binding sites for the transposase that allow the element to be recognized and ligated into the cleaved target site. Many transposable elements encode their own transposase in sequences located in the central region between the terminal inverted repeats, so these elements are able to promote their own transposition. Elements in which the transposase gene has been deleted or inactivated by mutation are transposable

only if another member of the family, encoding a functional transposase, is present in the genome to provide this activity. The inability of the maize *Ds* element to transpose without *Ac* results from the absence of a functional transposase gene in *Ds*. The presence of an *Ac* element provides *transactivation* that enables a *Ds* element to transpose.

Another large class of transposable elements possess terminal direct repeats, typically 200 to 500 bp in length, called *long terminal repeats*, or LTRs. As the name implies, terminal direct repeats are present in the same orientation at both ends of the element (**FIGURE 12.8**, part A), whereas terminal inverted repeats are present in reverse orientation (Figure 12.8, part B). Transposable elements with long terminal repeats are called **LTR retrotransposons** because they transpose using an RNA transcript as an intermediate. A typical LTR retrotransposon is the *copia* element from *Drosophila* illustrated in **FIGURE 12.9**; in this case each LTR is itself flanked by short terminal inverted repeats. Transposition of a retrotransposon begins with transcription of the element into an RNA copy. Among the encoded proteins is an enzyme known as *reverse transcriptase*, which can "reverse-transcribe," using the RNA transcript as a template for making a complementary DNA daughter strand. A primer is needed for reverse transcription. For retrotransposons the primer is usually a cellular transfer RNA molecule whose 3' end is complementary to part of the LTR. The

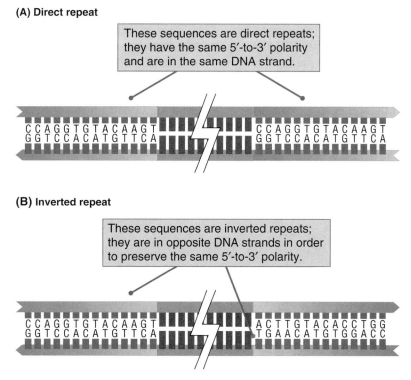

(A) Direct repeat

These sequences are direct repeats; they have the same 5'-to-3' polarity and are in the same DNA strand.

```
CCAGGTGTACAAGT          CCAGGTGTACAAGT
GGTCCACATGTTCA          GGTCCACATGTTCA
```

(B) Inverted repeat

These sequences are inverted repeats; they are in opposite DNA strands in order to preserve the same 5'-to-3' polarity.

```
CCAGGTGTACAAGT          ACTTGTACACCTGG
GGTCCACATGTTCA          TGAACATGTGGACC
```

FIGURE 12.8 (A) In a direct repeat, a DNA sequence is repeated in the same left-to-right orientation. (B) In an inverted repeat, the sequence is repeated in the reverse left-to-right orientation *in the opposite strand*. The opposite strand is necessary in order to maintain the correct 5'-to-3' polarity.

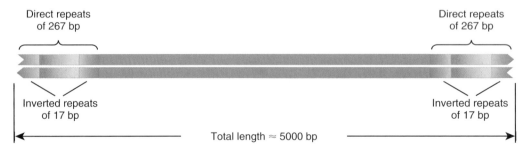

FIGURE 12.9 Sequence organization of a *copia* retrotransposable element of *Drosophila melanogaster*.

reverse transcriptase adds successive deoxyribonucleotides to the 3′ end of the tRNA, using the original RNA transcript as a template. Single-stranded cleavage of the RNA template by an element-encoded RNase provides a primer for second-strand DNA synthesis using the first DNA strand as a template. In this way a double-stranded DNA copy is made of the RNA transcript, and this is inserted into the target site.

Some retrotransposable elements have no terminal repeats and are called **non-LTR retrotransposons**. This class includes elements denoted **LINE elements** (long interspersed elements) and **SINE elements** (short interspersed elements). LINE and SINE elements are the most abundant types of transposable elements in mammalian genomes, although DNA and LTR retrotransposons are also found. An example of a SINE in the human genome is a set of related sequences called the AluI family because its members contain a characteristic restriction site for the restriction endonuclease *Alu*I. The AluI sequences are about 300 bp in length and are present in approximately one million copies in the human genome. The AluI family alone accounts for about 11 percent of human DNA. In many organisms, transposable elements of various families constitute a significant part of the total genome size.

Transposable elements can cause mutations by insertion or by recombination.

Transposable elements can cause mutations. For example, in some genes in *Drosophila*, approximately one-half of all spontaneous mutations that have visible phenotypic effects result from insertions of transposable elements. We have already seen (Figure 12.7) an example of mutation associated with the insertion of *Ds* into the *shrunken* gene in maize. The wrinkled-seed mutation in Mendel's peas is another good example. In this case, the transposable element is a DNA element related to the maize *Ac* element that also produces an eight-bp target-site duplication. The insertion site is in the gene for starch-branching enzyme I (SBEI), and the insertion creates a loss-of-function allele. Most transposable elements are present in nonessential

regions of the genome and usually cause no detectable phenotypic change. But when an element transposes, it can insert into an essential region and cause a mutant phenotype. If transposition inserts an element into a coding region of DNA, then the inserted element interrupts the coding region. Because most transposable elements contain coding regions of their own, either transcription of the transposable element interferes with transcription of the gene into which it is inserted, or transcription of the gene terminates within the transposable element. The insertion therefore causes a knockout mutation. Even if transcription proceeds through the element, the phenotype will be mutant because the coding region then contains incorrect sequences.

Genetic aberrations are occasionally caused by ectopic recombination between two copies of a transposable element present at different locations in the genome. **FIGURE 12.10** illustrates two possible outcomes of recombination between copies present in the same chromosome. In part A the copies are present in direct orientation. In this case, pairing between the copies forms a loop. Recombination between the copies results in the formation of a free circle of DNA that contains the region between the elements. Since this circle lacks a centromere, it will be lost. The reciprocal product of the recombination is a chromosome that has a *deletion* of the region originally between the copies of the transposable element. In part B of Figure 12.10, the copies are shown as being present in inverted orientation. In this case, pairing between the copies creates a hairpin structure instead of a loop, and recombination between the copies results in an *inversion* in which the order of the genes between the copies is reversed.

Recombination between two copies of a transposable element present in nonhomologous chromosomes is illustrated in **FIGURE 12.11**. When the transposable elements are present in the orientation shown here, the result is the interchange of terminal segments between the nonhomologous chromosomes. The two products of the recombination constitute the parts of a *reciprocal translocation* between nonhomologous chromosomes.

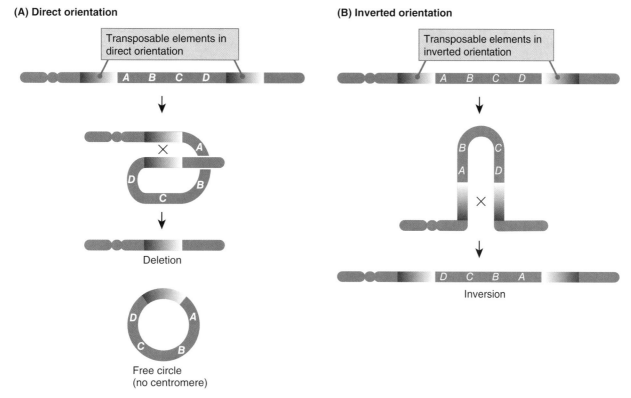

FIGURE 12.10 Recombination between transposable elements (or other repeated sequences) in the same chromosome. (A) If the repeats are in direct orientation, then recombination results in a deleted duplex and a circular molecule containing the deleted region. (B) If the repeats are in inverted orientation, then recombination results in an inversion of the region between them.

⦿ STOP & THINK 12.2

A metacentric chromosome (one with its centromere approximately in the middle) has a copy of a transposable element in each arm located very close to the telomere. The two copies are present in the same left-to-right orientation. Draw a diagram of a single chromatid in which the two copies are paired and undergo a crossover within the transposable element. What kind of chromosomes result from the reciprocal products of the crossover?

Almost 50 percent of the human genome consists of transposable elements, most of them no longer able to transpose.

Part of the reason why the human genome is so relatively large yet contains only about 25,000 genes is that it includes a high proportion of transposable elements. The principal categories and their abundances are shown in **TABLE 12.2**. The human genome consists of almost 50 percent transposable elements. The largest single category consists of SINE elements, of which the AluI family are the most abundant. Although

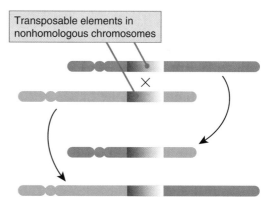

Partner chromosomes of a reciprocal translocation

FIGURE 12.11 Unequal crossing-over between copies of transposable elements that are present in the same orientation in nonhomologous chromosomes results in a reciprocal translocation, in which segments of the chromosomes are interchanged. Crossing-over takes place at the four-strand stage of meiosis, but only the two strands participating in the crossover are shown.

transposable elements were long regarded as "selfish DNA"—a sort of genomic parasite—evidence is beginning to suggest that at least the AluI family may benefit the human genome. First, AluI elements are disproportionately represented in gene-rich regions of the genome that are high in G + C content, which suggests that they play some functional role. Second,

TABLE 12.2 Transposable Elements in the Human Genome

Type	Number of copies	Percentage of total genome
SINEs	1,558,000	13.1
*Alu*I	1,090,000	10.6
LINEs	868,000	20.4
LINE1	516,000	16.9
LTR elements	443,000	8.3
DNA elements	294,000	2.8
mariner	14,000	0.1
Unclassified	3,000	0.1
Total of all types		**44.7**

Data from E. S. Landers, et al., *Nature* 409 (2001): 860.

in human beings, as in some other organisms, SINE elements are transcribed when the organism is under stress. The resulting transcripts can bind to a particular protein kinase that normally blocks translation under stress. In this way SINE elements may be able to promote translation under organismic stress.

The second major class of human transposable elements consists of LINE elements, of which LINE1 is the most abundant. Third on the list are the LTR retrotransposable elements, followed a distant fourth by DNA elements, of which the *mariner* transposon is an example. The *mariner* transposon is of some interest because it is widespread among eukaryotic genomes. About 14 percent of all insect species carry *mariner*, for example. One reason for its wide distribution is that *mariner* is relatively efficient in being transferred from one species to another, even unrelated, species, but the mechanisms by which this *horizontal transmission* takes place are largely matters of speculation.

The human genomic DNA sequence implies that most transposons in the genome are no longer capable of transposition. One type of evidence derives from comparing sequences of different copies of the same element throughout the genome. Because a transpositionally active element will give rise to new copies that are identical or nearly identical in sequence from one to the next, close sequence similarity among copies suggests active transposition. On the other hand, copies of transposons that can no longer move are free to change in sequence as successive mutations take place and are incorporated into the population, and so large sequence differences among copies suggest

a low rate of transposition. Because the average rate of nucleotide substitution per base pair in the human genome is roughly constant through time, the amount of sequence divergence between copies can be used to estimate the time since transposition.

The analysis of sequence differences among human transposable elements suggests that the overall activity of transposable elements in the human genome has decreased substantially, and quite steadily, over the past 35 to 50 million years. The ancient times mean that the decrease in transposition was taking place in the hominid lineage long before human beings existed as a species. Other mammals that have been studied show greater and more typical rates of transposition. In the mouse, for example, the rate of transposition of SINE and LINE elements, relative to that in the human genome, has increased from 1.7-fold higher in the past 100 million years to 2.6-fold higher in the past 25 million years. This comparison is consistent with the finding that about 1 in 10 new mutations in the mouse is due to transposition, whereas only about 1 in 600 new mutations in the human genome is due to transposition. LTR retrotransposons exhibit no convincing evidence of ongoing transposition in the human genome, and DNA transposons seem to have lost their ability to transpose about 50 million years ago. Hence, human beings stand in contrast to many other organisms, including other mammals, in which transposition is a major source of mutation as well as evolutionary innovation.

12.4 Mutations are statistically random events.

There is no way of predicting when, or in which cell, a mutation will take place, but because every gene mutates spontaneously at a characteristic rate, it is possible to assign probabilities to particular mutational events. In other words, there is a definite probability that a specified gene will mutate in a particular cell; likewise there is a definite probability that a mutant allele of a specified gene will appear in a population of a designated size. The various kinds of mutational alterations in DNA differ substantially in complexity, so their probabilities of occurrence are quite different. A fundamental principle concerning mutation is that:

KEY CONCEPT

The mutational process is also random in the sense that whether a particular mutation happens is unrelated to any adaptive advantage it may confer on the organism in its environment. A potentially favorable mutation does not arise *because* the organism has a need for it.

The experimental basis for this conclusion is presented in the next section.

Mutations arise without reference to the adaptive needs of the organism.

The concept that mutations are spontaneous, statistically random events unrelated to adaptation was not widely accepted until the late 1940s. Before that time, it was believed that mutations occurred in bacterial populations *in response to* particular selective conditions. The basis for this belief was the observation that when antibiotic-sensitive bacteria are spread on a solid growth medium containing the antibiotic, some colonies form that consist of cells having an inherited resistance to the drug. The initial interpretation of this observation (and similar ones) was that

these adaptive variations were *induced* by the selective agent itself.

Several types of experiments showed that adaptive mutations take place spontaneously and hence were present at low frequency in the bacterial population even *before* it was exposed to the antibiotic. One experiment utilized a technique developed by Joshua and Esther Lederberg called **replica plating** (**FIGURE 12.12**). In this procedure, a suspension of bacterial cells is spread on a solid medium. After colonies have formed, a piece of sterile velvet mounted on a solid support is pressed onto the surface of the plate. Some bacteria from each colony stick to the fibers, as shown in part A of Figure 12.12. Then the velvet is pressed onto the surface of fresh medium, transferring some of the cells from each colony, which give rise to new colonies that have positions identical to

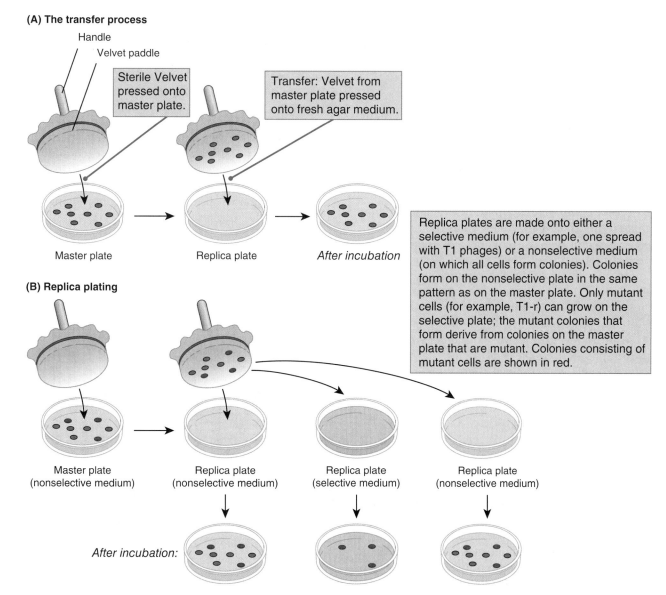

(A) The transfer process

Handle
Velvet paddle

Sterile Velvet pressed onto master plate.

Transfer: Velvet from master plate pressed onto fresh agar medium.

Master plate   Replica plate   *After incubation*

Replica plates are made onto either a selective medium (for example, one spread with T1 phages) or a nonselective medium (on which all cells form colonies). Colonies form on the nonselective plate in the same pattern as on the master plate. Only mutant cells (for example, T1-r) can grow on the selective plate; the mutant colonies that form derive from colonies on the master plate that are mutant. Colonies consisting of mutant cells are shown in red.

(B) Replica plating

Master plate (nonselective medium)   Replica plate (nonselective medium)   Replica plate (selective medium)   Replica plate (nonselective medium)

After incubation:

FIGURE 12.12 Replica plating. (A) In the transfer process, a velvet-covered disk is pressed onto the surface of a master plate in order to transfer cells from colonies on that plate to a second medium. (B) For the detection of mutants, cells are transferred onto successive plates.

those on the first plate. Part B of Figure 12.12 shows how this method was used to demonstrate the spontaneous origin of phage T1-r mutants. A master plate containing about 10^7 cells growing on nonselective medium (lacking phage) was replica-plated onto a series of plates that had been spread with about 10^9 T1 phages. After incubation for a time sufficient for colony formation, a few colonies of phage-resistant bacteria appeared in the same positions on each of the selective replica plates. This meant that the T1-r cells that formed the colonies must have been transferred from corresponding positions on the master plate. Because the colonies on the master plate had never been exposed to the phage, the mutations to resistance must have been present, by chance, in a few original cells not exposed to the phage.

The replica-plating experiment illustrates the following principle:

KEY CONCEPT

Selective techniques merely select mutants that preexist in a population.

This principle is the basis for understanding how natural populations of rodents, insects, and disease-causing bacteria become resistant to the chemical substances used to control them. A familiar example is the high level of resistance to insecticides, such as DDT, that now exists in many insect populations, the result of selection for spontaneous mutations affecting behavioral, anatomical, and enzymatic traits that enable the insect to avoid or resist the chemical. Similar problems are encountered in controlling plant pathogens. For example, the introduction of a new variety of a crop plant resistant to a particular strain of disease-causing fungus results in only temporary protection against the disease. The resistance inevitably breaks down because of the occurrence of spontaneous mutations in the fungus that enable it to attack the new plant genotype. Such mutations confer a clear selective advantage, and the mutant alleles rapidly become widespread in the fungal population.

The surprisingly large number of new mutations in human gametes increases with father's age.

In traditional genetic studies of mutation, new mutations could be detected only if they resulted in some recognizable phenotype. Mutations with no phenotype effect went undetected, and the rate at which they occurred was unknown. The advent of high-throughput, highly accurate genome sequencing enables all new mutations to be detected irrespective

of their phenotypic effects. In a study of 78 mother-father-offspring trios in Iceland, the total number of new base-substitution mutations that occur in each generation of the human genome proved to be surprising large. On average, the number of new base-substitution mutations in a newly fertilized egg is about 60, with about 15 inherited from the mother and 45 from the father. The average base-substitution rate per nucleotide is about one new substitution per 10^8 nucleotides per generation — but the rate increases according to the age of the father.

Sixty new mutations per zygote per generation may seem like a lot, but let's put it in context. We've already seen that human genomes contain millions of single-nucleotide polymorphisms (Chapter 4), each of which arose as a new mutation in some previous generation. The average zygote is heterozygous for roughly 3 million SNPs. This number implies that, for every *new* base-substitution mutation present in a zygote, there are about 50,000 *old* base-substitution mutations inherited from previous generations.

FIGURE 12.13 shows how the average number of new mutations per sperm increases as a function of paternal age. The increase is linear, from about 30 new mutations per sperm at age 15 to about 95 at age 45, which amounts to approximately two new base-substitution mutations per sperm for each additional year of the father's age. The dashed line in Figure 12.13 is the number of new mutations per egg nucleus, which is equal to about 15 and does not change with the age of the mother.

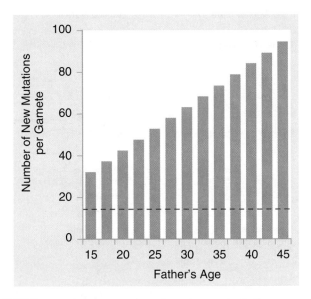

FIGURE 12.13 Average number of new base-substitution mutations per gamete as a function of the father's age. The dashed line indicates the average number of new mutations in eggs, which is about 15 irrespective of the mother's age.

Data from A. Kong, et al., *Nature* 488 (2012): 471–475.

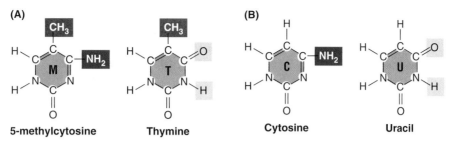

FIGURE 12.14 (A) Spontaneous loss of the amino group from 5-methylcytosine yields thymine. (B) Loss of the amino group from normal cytosine yields uracil.

The age effect in males but not in females may at first seem paradoxical, but the results are in complete accord with what is known about gamete formation in the sexes. In females, the cells destined to form eggs all undergo DNA replication and initiate meiosis early in life. The majority of oocytes are actually produced in the fetus *in utero*; however, the first meiotic division proceeds only to the diplotene stage and then undergoes a sort of dormancy until puberty, when, stimulated by a surge of luteinizing hormone, meiosis resumes and proceeds through metaphase II, where again there is a pause. When a sperm enters the egg, meiosis is completed and the egg and sperm nuclei fuse.

In males, by contrast, cells destined to form sperm continue to undergo DNA replication and mitotic cell divisions throughout life. They undergo meiosis immediately prior to spermatogenesis. Sperm from older males have therefore undergone more rounds of DNA replication than sperm from younger males. Because most base-substitution mutations occur during DNA replication, the greater number of replications in older males implies that sperm from older males will contain a greater number of new mutations than sperm from younger males.

Mutations are nonrandom with respect to position in a gene or genome.

Certain DNA sequences are called mutational **hotspots** because they are more likely to undergo mutation than others. Mutational hotspots include unstable trinucleotide repeats that can expand by replication slippage (Figure 12.5). Hotspots are found at many sites throughout the genome and within genes. For genetic studies of mutation, the existence of hotspots means that a relatively small number of sites account for a disproportionately large fraction of all mutations.

Sites of cytosine methylation are usually highly mutable, and the mutations are usually G—C → A—T transitions. In many organisms, including bacteria, maize, and mammals (but not *Drosophila*), a few percent of the cytosine bases are methylated at the carbon-5 position, yielding 5-methylcytosine instead of ordinary cytosine (**FIGURE 12.14**). A special enzyme adds the methyl group to the cytosine base in certain target sequences of DNA. In DNA replication, the 5-methylcytosine pairs with guanine and replicates normally.

Cytosine methylation is an important contributor to mutational hotspots, as illustrated in Figure 12.14. Both 5-methylcytosine and cytosine are subject to occasional loss of an amino group, a process called *deamination*. When 5-methylcytosine is deaminated, it becomes converted into normal thymine (part A). In duplex DNA this creates a temporary G—T mismatch, which has a chance of being repaired by the mismatch repair system. If it is repaired to A—T, then the duplex has undergone a transition mutation; and if it is not repaired immediately, then in the next generation the T-bearing strand pairs normally with A, yielding a mutant A—T base pair in this generation.

Whereas the deamination of 5-methylcytosine is often mutagenic, that of normal cytosine is not. The reason is that deamination of normal cytosine changes cytosine into uracil (part B). Fortunately, as we shall see in Section 12.6, uracil is one of the bases recognized by a DNA repair system called *base-excision repair*. In this repair process, the uracil is removed and replaced with normal cytosine. Hence, deamination of normal cytosine resulting in uracil is easily detected and repaired.

12.5 Spontaneous and induced mutations have similar chemistries.

Almost any kind of mutation that can be induced by a mutagen can also occur spontaneously, but mutagens bias the types of mutations that occur according to the type of damage to the DNA that they produce. For the geneticist, the use of mutagens is a means of greatly increasing the number of mutants that can be isolated in an experiment. But mutagens are also of great importance in public health because many environmental contaminants are mutagenic, as are numerous chemicals found in tobacco products.

TABLE 12.3 Major Agents of Mutation and Their Mechanisms of Action

Agent of mutation	Examples	Principal mechanism of mutagenesis
Water	Hydrolysis	Depurination (A or G detached from its deoxyribose sugar).
Oxidizing agent	Nitrous acid	Deamination ($-NH_2 \rightarrow = O$): C → U, 5-MeC → T, A → Hypoxanthine.
Base analog	5-bromodeoxyuridine	Increased rate of base mispairing.
Alkylating agent	Ethylmethane sulfonate Nitrogen mustard	Bulky attachments made to side groups on bases.
Intercalating agent	Proflavin	Causes topoisomerase II to leave a nick in DNA strand; misrepair results in the insertion or deletion of one or a few nucleotides.
Ultraviolet (UV) light	Natural sunlight UV lamps	Forms pyrimidine dimers (covalent bonds between adjacent pyrimidines, primarily T) present in the same DNA strand.
Ionizing radiation	X rays Radon gas Radioactive materials	Single- and double-stranded breaks in DNA; damage to nucleotides.

Purine bases are susceptible to spontaneous loss.

Some of the principal agents that damage DNA are listed in **TABLE 12.3**, along with the major types of damage they produce. At the head of the list is water. In purine nucleotides, the sugar–purine bonds are relatively labile and subject to hydrolysis. The loss of the purine base, called **depurination**, is illustrated in **FIGURE 12.15**. Depurination is not always mutagenic, because the site lacking the base can be corrected by the same system that repairs sites from which uracil has been removed. If, however, the replication fork reaches the apurinic site before repair has taken place, then replication almost always inserts an adenine nucleotide in the daughter strand opposite the apurinic site. After another round of replication, what was originally a G−C pair becomes a T−A pair, which is an example of a transversion mutation.

In air, the rate of spontaneous depurination is approximately 3×10^{-9} depurinations per purine nucleotide per minute. This rate is at least tenfold greater than any other single source of spontaneous DNA degradation. At this rate, the half-life of a purine nucleotide exposed to air is about 300 years. This sets a practical limit to how long DNA can persist in the environment before losing its biological activity.

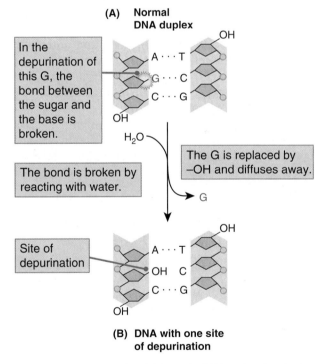

(A) Normal DNA duplex

In the depurination of this G, the bond between the sugar and the base is broken.

The bond is broken by reacting with water.

The G is replaced by −OH and diffuses away.

Site of depurination

(B) DNA with one site of depurination

FIGURE 12.15 Depuration. (A) Part of a DNA molecule prior to depurination. The bond between the labeled G and the deoxyribose to which it is attached is about to be hydrolyzed. (B) Hydrolysis of the bond releases the G purine, which diffuses away from the molecule and leaves a hydroxyl (−OH) in its place in the depurinated DNA.

Some weak acids are mutagenic.

Many mutagens are chemicals that react with DNA and change the hydrogen-bonding properties of the bases. An example is **nitrous acid**, which acts as a mutagen by deamination of the bases adenine, cytosine, and guanine.

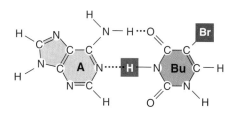

Adenine → **Hypoxanthine**

FIGURE 12.16 Deamination of adenine results in hypoxanthine.

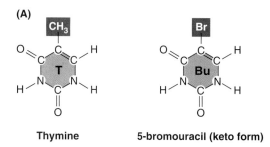

(A)

Thymine   **5-bromouracil (keto form)**

Deamination alters the hydrogen-bonding specificity of each base. As we have seen in Figure 12.14, deamination of 5-methylcytosine results in thymine, and deamination of cytosine results in uracil. The result of deamination of adenine is illustrated in **FIGURE 12.16**. The product is a base called *hypoxanthine*, which pairs with cytosine rather than thymine, so the result of deamination of A is an A−T → G−C transition.

A base analog masquerades as the real thing.

A **base analog** is a molecule sufficiently similar to one of the four DNA bases that it can be incorporated into a DNA duplex in the course of normal replication. Such a substance must be able to pair with a base in the template strand. Some base analogs are mutagenic because they are more prone to mispairing than are the normal nucleotides. The molecular basis of the mutagenesis can be illustrated with 5-bromouracil (Bu), a commonly used base analog that is efficiently incorporated into the DNA of bacteria and viruses.

The base 5-bromouracil is an analog of thymine, and the bromine atom is about the same size as the methyl group of thymine (**FIGURE 12.17**, part A). Normally, 5-bromouracil is in the *keto form*, in which it pairs with adenine (part B), but it occasionally shifts its configuration to the *enol form*, in which it pairs with guanine (part C). The shift is influenced by the bromine atom and takes place in 5-bromouracil more frequently than in thymine.

There are two pathways by which 5-bromouracil can be mutagenic. These are illustrated in **FIGURE 12.18**. In pathway A, the 5-bromouracil is incorporated in its enol form, paired with G. This mode of incorporation is rare, but the mutagenic base pair is created in the first round of replication. In the next round of replication, the Bu will usually pair with A, which leads to a G−C → A−T transition. In pathway B, the 5-bromouracil is incorporated in its keto form, paired with A. This is by far the more frequent mode of incorporation, but the mutagenic base pair is not formed

(B) Adenine paired with 5-bromouracil

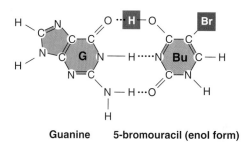

Adenine   **5-bromouracil (keto form)**

(C) Guanine paired with 5-bromouracil

Guanine   **5-bromouracil (enol form)**

FIGURE 12.17 Mispairing mutagenesis by 5-bromouracil. (A) Structures of thymine and 5-bromouracil. (B) A base pair between adenine and the keto form of 5-bromouracil. (C) A base pair between guanine and the rare enol form of 5-bromouracil. One of the hydrogen atoms (shown in red) changes position when the molecule is in the keto form.

until a later round of replication when Bu pairs with G. In this case the result is an A−T → G−C transition.

Highly reactive chemicals damage DNA.

Some mutagens react with DNA in a variety of different ways and produce a broad spectrum of effects. Among these are the **alkylating agents**, which are highly reactive chemicals that act as potent mutagens in both prokaryotes and eukaryotes. Examples of alkylating agents are ethyl methanesulfonate (EMS) and nitrogen mustard, the structures of which are shown in **FIGURE 12.19**. Nitrogen mustard is a gas causing extreme pain and extensive lung damage when inhaled, and was used for chemical warfare in Europe in the First World War (1914–1918).

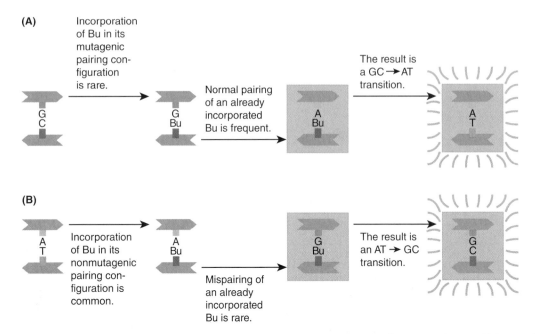

FIGURE 12.18 Shown are two pathways for mutagenesis by 5-bromouracil (Bu). The position of the arrow shows which strand of each DNA duplex is being followed through the next round of replication. (A) Incorporation paired with G is rare. In this case the mutagenic base pair is formed in the first round of replication. (B) Incorporation paired with A is frequent. In this case the mutagenic base pair is formed after the first round of replication.

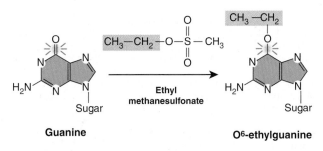

FIGURE 12.19 The chemical structures of two highly mutagenic alkylating agents; the alkyl groups are in the pink rectangles.

FIGURE 12.20 Mutagenesis of guanine by ethyl methanesulfonate (EMS).

Ethyl methanesulfonate is a soluble solid and has been used widely to induce mutations for genetic research. The alkylating agents add bulky side groups to the DNA bases that either alter their base-pairing properties or cause structural distortion of the DNA molecule. For example, the reaction of EMS with guanine results in O^6-ethylguanine (**FIGURE 12.20**). Alkylation of either guanine or thymine causes mispairing, leading to the transitions A—T to G—C or G—C to A—T. EMS reacts less readily with adenine and cytosine than with thymine and guanine.

Some agents cause base-pair additions or deletions.

The **acridine** molecules are planar three-ringed molecules whose dimensions are roughly the same as those of a purine–pyrimidine pair. Acridine orange is an example whose structure is shown below.

Once thought to insert between the base pairs in DNA, the acridines actually do their damage by interfering with topoisomerase II, which relieves torsional stress in DNA by making a double-stranded break, rotating the free ends, and then sealing the break. In the presence of acridine, the enzyme leaves the DNA nicked. Failure of prompt repair results in the addition or deletion of one or a few base pairs at the site. The result of a single-base addition or deletion in a coding region is a frameshift mutation.

Ultraviolet radiation absorbed by DNA is mutagenic.

Ultraviolet (UV) light is mutagenic in all viruses and cells. The effects are caused by chemical changes in the bases resulting from absorption of the energy of the light. The major products formed in DNA after UV irradiation are covalently joined pyrimidines (**pyrimidine dimers**), primarily thymine (**FIGURE 12.21**, part A), that are adjacent in the same polynucleotide strand.

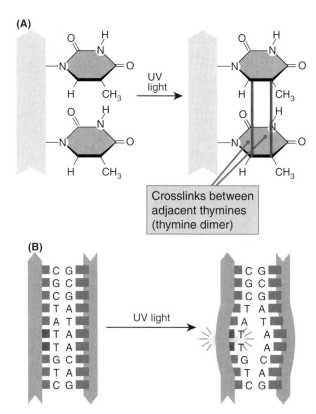

FIGURE 12.21 (A) Structural view of the formation of a thymine dimer. Adjacent thymines in a DNA strand that have been subjected to ultraviolet (UV) irradiation are joined by formation of the bonds shown in red. Other types of bonds between the thymine rings also are possible. Although they are not drawn to scale, these bonds are considerably shorter than the spacing between the planes of adjacent thymines, so that the double-stranded structure becomes distorted. The shape of each thymine ring also changes. (B) The distortion of the DNA helix caused by two thymines moving closer together when joined in a dimer.

This chemical linkage brings the bases closer together, causing a distortion of the helix (part B), which blocks transcription and transiently blocks DNA replication. Pyrimidine dimers can be repaired in ways discussed later in this chapter. Nevertheless, excessive exposure of the skin to the UV rays in sunlight increases the risk of skin cancer.

Ionizing radiation is a potent mutagen.

Ionizing radiation includes x rays and the particles and radiation released by radioactive elements (α and β particles and γ rays). When x rays were first discovered late in the nineteenth century, their power to pass through solid materials was regarded as a harmless entertainment and source of great amusement. Witness this account from one history of the period:

By 1898, personal x rays had become a popular status symbol in New York. The *New York Times* reported that "there is quite as much difference in the appearance of the hand of a washerwoman and the hand of a fine lady in an x-ray picture as in reality." The hit of the exhibition season was Dr. W. J. Morton's full-length portrait of "the *x-ray lady*," a "fashionable woman who had evidently a scientific desire to see her bones." The portrait was said to be a "fascinating and coquettish" picture, the lady having agreed to be photographed without her stays and corset, the better to satisfy the "longing to have a portrait of well-developed ribs." Dr. Morton said women were not afraid of x rays: "After being assured that there is *no danger* they take the rays without fear."

The titillating possibility of using x rays to see through clothing or to invade the privacy of locked rooms was a familiar theme in popular discussions of x rays and in cartoons and jokes. Newspapers carried advertisements for "*x ray proof underclothing*" for those seeking to protect themselves from x ray inspection.

The luminous properties of radium soon produced a full-fledged radium craze. A famous woman dancer performed *radium dances* using veils dipped in fluorescent salts containing radium. *Radium roulette* was popular at New York casinos, featuring a "roulette wheel washed with a radium solution, such that it glowed brightly in the darkness; an unseen hand cast the ball on the turning wheel and sparks marked its course as it bounded from pocket to glimmery pocket." A patent was issued for a process for making women's gowns luminous with radium, and Broadway producer Florenz Ziegfeld snapped up the rights for his stage extravaganzas.

Even while the unrestrained use of x rays and radium was growing, evidence was accumulating that the new forces might not be so benign after all. Hailed as tools for fighting cancer, they could also cause cancer. Doctors using x rays were the first to learn this bitter lesson.

Quoted from S. Hilgartner, R. C. Bell, and R. O'Connor. 1982. *Nukespeak*. Sierra Club Books, San Francisco, CA.

Doctors were indeed the first to learn the lesson. Many suffered severe x-ray burns or required amputation of overexposed hands or arms. Many others died from radiation poisoning or from radiation-induced cancer. By the mid-1930s, the number of x-ray deaths had grown so large that a monument to the "x-ray martyrs" was erected in a hospital courtyard in Hamburg, Germany. Yet the full hazards of x-ray exposure were not widely appreciated until the 1960s.

When ionizing radiation interacts with water or with living tissue, highly reactive ions called *free radicals* are formed. The free radicals react with other molecules, including DNA, which results in the carcinogenic and mutagenic effects. The intensity of a beam of ionizing radiation can be described quantitatively in several ways. There are, in fact, a bewildering variety of units in common use (**TABLE 12.4**). Some of the units (becquerel, curie) deal with the number of disintegrations emanating from a material, others (roentgen) with the number of ionizations the

TABLE 12.4 Units of Radiation	
Unit (abbreviation)	Magnitude
Becquerel (Bq)*	1 disintegration/second = 2.7 × 10⁻¹¹ Ci
Curie (Ci)	3.7 × 10¹⁰ disintegrations/second = 3.7 × 10¹⁰ Bq
Gray (Gy)*	1 joule/kilogram = 100 rad
Rad (rad)	100 ergs/gram = 0.01 Gy
Rem (rem)	Damage to living tissue done by 1 rad = 0.01 Sv
Roentgen (R)	Produces 1 electrostatic unit of charge per cubic centimeter of dry air under normal conditions of pressure and temperature. (By definition, 1 electrostatic unit repels with a force of 1 dyne at a distance of 1 centimeter.)
Sievert (Sv)	100 rem

*Units officially recognized by the International System of Units as defined by the General Conference on Weights and Measures.

radiation produces in air, still others (gray, rad) with the amount of energy imparted to material exposed to the radiation, and some (rem, sievert) with the effects of radiation on living tissue. The types of units have proliferated through the years in attempts to encompass different types of radiation, including nonionizing radiation, in a common frame of reference. The units in Table 12.4 are presented only as an aid in interpreting the multitude of units found in the literature on the health effects of radiation.

Genetic studies of ionizing radiation support the following general principle:

KEY CONCEPT

Over a wide range of x-ray doses, the frequency of mutations induced by x rays is proportional to the radiation dose.

One type of evidence supporting this principle is the frequency with which X-chromosome recessive lethals are induced in *Drosophila* (**FIGURE 12.22**). The mutation rate increases linearly with increasing x-ray dose. For example, an exposure of 10 sieverts increases the frequency from the spontaneous value of 0.15 percent to about 3 percent. The mutagenic and lethal effects of ionizing radiation at low to moderate doses result primarily from damage to DNA. Three types of damage in DNA are produced by ionizing radiation:

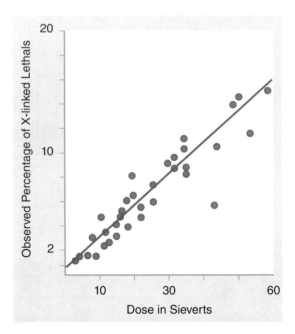

FIGURE 12.22 The relationship between the percentage of X-linked recessive lethals in *D. melanogaster* and x-ray dose. The frequency of spontaneous X-linked lethal mutations is 0.15 percent per X chromosome per generation.

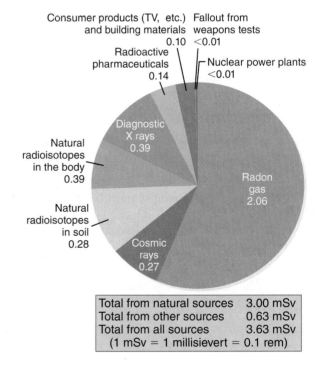

Total from natural sources	3.00 mSv
Total from other sources	0.63 mSv
Total from all sources	3.63 mSv
(1 mSv = 1 millisievert = 0.1 rem)	

FIGURE 12.23 Annual exposure of human beings in the United States to various forms of ionizing radiation.
Data from the National Research Council.

single-strand breakage (in the sugar–phosphate backbone), double-strand breakage, and alterations in nucleotide bases. The single-strand breaks are usually efficiently repaired, but the other damage is responsible for mutation. In eukaryotes, ionizing radiation also results in chromosome breaks. Although systems exist for repairing the breaks, the repair often leads to translocations, inversions, duplications, and deletions.

STOP & THINK 12.3

The graphs shown here show the increase in the number of point mutations versus the number of reciprocal translocations observed as a function of increasing x-ray dose. The point mutations increase linearly with dose (as in Figure 12.22); however, the reciprocal translocations increase more slowly at first but then more rapidly as the dose increases, very roughly in proportion to the square of the dose. Propose a hypothesis to explain why point mutations should increase linearly, but reciprocal translocations increase nonlinearly.

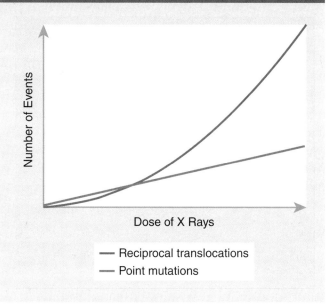

— Reciprocal translocations
— Point mutations

In human cells in culture, a dose of 0.2 sievert results in an average of one visible chromosome break per cell.

Ionizing radiation is widely used in tumor therapy. The basis for the treatment is the increased frequency of chromosomal breakage (and the consequent lethality) in cells undergoing division. Tumors usually contain many more mitotic cells than do most normal tissues, so more tumor cells than normal cells are destroyed. Because all tumor cells are not in mitosis at the same time, irradiation is carried out at intervals of several days to allow interphase tumor cells to enter mitosis. Over a period of time, most tumor cells are destroyed.

FIGURE 12.23 gives representative values of doses of ionizing radiation received by human beings in the United States in the course of a year. The unit of measure is the millisievert (mSv), which equals 0.1 rem. The exposures in Figure 12.23 are on a yearly basis, so over the course of a generation, the total exposure is approximately 100 mSv. Note that, with the exception of diagnostic x rays, which yield important compensating benefits, most of the total radiation exposure comes from natural sources, particularly radon gas. Less than 20 percent of the average radiation exposure comes from artificial sources. Nevertheless, there are dangers inherent in any exposure to ionizing radiation, particularly an increased risk of leukemia and certain other cancers in the exposed persons. In regard to increased genetic diseases in future generations resulting from the mutagenic effects of radiation, the risk of a small amount of additional radiation is low enough that most geneticists are currently more concerned about the effects of the many mutagenic (as well as carcinogenic)

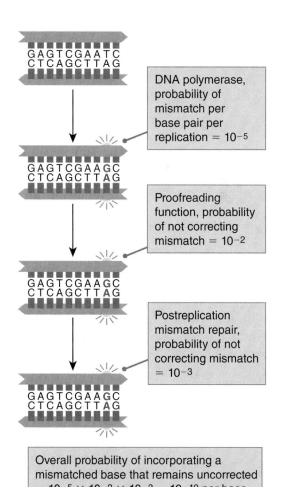

DNA polymerase, probability of mismatch per base pair per replication = 10^{-5}

Proofreading function, probability of not correcting mismatch = 10^{-2}

Postreplication mismatch repair, probability of not correcting mismatch = 10^{-3}

Overall probability of incorporating a mismatched base that remains uncorrected = $10^{-5} \times 10^{-2} \times 10^{-3} = 10^{-10}$ per base pair per replication

FIGURE 12.24 Summary of rates of error in DNA polymerization, proofreading, and postreplication mismatch repair. The overall rate of misincorporated nucleotides that are not repaired is 10^{-10} per base pair per replication.

TABLE 12.5 Estimated Genetic Effects of an Additional 10 Millisieverts per Generation

Type of disorder	Current incidence per million liveborn	Additional cases per million liveborn per 10 mSv per generation	
		First generation	At equilibrium
Autosomal dominant			
Clinically severe	2500	5–20	25
Clinically mild	7500	1–15	75
X-linked	400	<1	<5
Autosomal recessive	2500	<1	Very slow increase
Chromosomal			
Unbalanced translocations	600	<5	Very little increase
Trisomy	3800	<1	<1
Congenital abnormalities (multifactorial)	20,000–30,000	10	10–100
Heart disease	600,000	Unknown	Unknown
Cancer	300,000	Unknown	Unknown
Others	300,000	Unknown	Unknown

Data from National Research Council, *Health Effects of Exposure to Low Levels of Ionizing Radiation (BEIR V)*, 1990, Washington, D.C.

chemicals that are introduced into the environment from a variety of sources.

The National Academy of Sciences of the United States regularly updates the estimated risks of radiation exposure. The latest estimates are summarized in **TABLE 12.5**. The message is that an additional 10 millisieverts of radiation per generation (about a 10 percent increase in the annual exposure) is expected to cause a relatively modest increase in diseases that are wholly or partly due to genetic factors. The most common conditions in the table are heart disease and cancer. No estimate for the radiation-induced increase is given for either of these traits because the genetic contribution to the total is still uncertain.

12.6 Many types of DNA damage can be repaired.

Spontaneous damage to DNA in human cells takes place at a rate of approximately one event per billion nucleotide pairs per minute (or, expressed per nucleotide pair, at a rate of 1×10^{-9} per nucleotide

pair per minute). This may seem quite a small rate, but it implies that every 24 hours, in every human cell, the DNA is damaged at approximately 10,000 different sites. Fortunately for us, and for all living organisms, much of the damage done to DNA by spontaneous chemical reactions in the nucleus, by chemical mutagens, and by radiation can be repaired. **TABLE 12.6** summarizes some of the most important mechanisms of DNA repair. The first on the list is straightforward. A nick in a DNA strand is a site at which one of the phosphodiester bonds along the backbone is broken. Nicks are repaired by the enzyme DNA ligase, which restores the covalent bond. In the following sections, we examine other key molecular mechanisms for the repair of aberrant or damaged DNA.

Mismatch repair fixes incorrectly matched base pairs.

Mismatch repair functions as a "last chance" error-correcting mechanism in DNA replication. During DNA

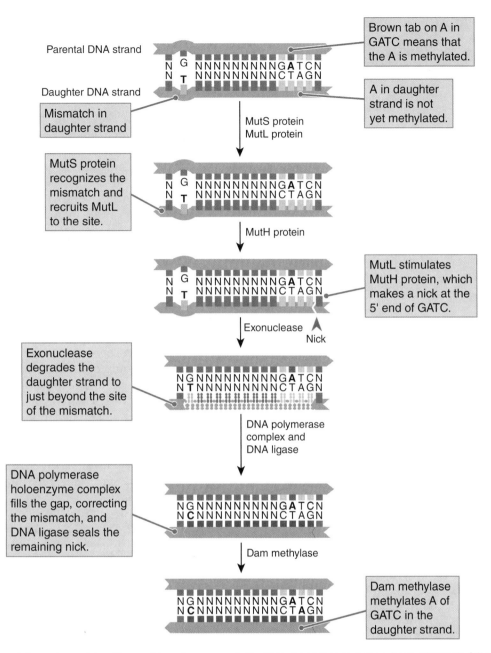

FIGURE 12.25 Mismatch repair consists of the excision of a segment of a DNA strand that contains a base mismatch, followed by repair synthesis. In *E. coli*, cleavage takes place at the nearest methylated GATC sequence in the unmethylated strand. An exonuclease removes successive nucleotides until just past the mismatch, and the resulting gap is repaired. Either strand can be excised and corrected, but in newly synthesized DNA, methylated bases in the template strand often direct the excision mechanism to the newly synthesized strand that contains the incorrect nucleotide.

replication, mismatched nucleotides are incorporated at the rate of about 10^{-5} per template nucleotide per round of replication. As shown in **FIGURE 12.24**, approximately 99 percent of these are immediately corrected by the proofreading function (3′-to-5′ exonuclease activity) of the major replication polymerase. This leaves a mismatch rate of 10^{-7} per template nucleotide per round of replication; 99.9 percent of the remaining mismatches are corrected by the mismatch repair system, yielding an overall mismatch rate of

$$10^{-5} \times 10^{-2} \times 10^{-3} = 10^{-10}$$

The operation of **mismatch repair** is illustrated in **FIGURE 12.25**. When a mismatched base is detected, one of the strands is cut in two places and a region around the mismatch is removed. The excised region is variable in size. In *Escherichia coli*, an enzyme first cleaves the DNA at the nearest GATC sequence on the unmethylated (newly synthesized) strand, provided that the site is within about 1 kb of the mismatch. Then an exonuclease degrades the cleaved DNA strand to a position on the other side of the mismatch (the excision step), creating a single-stranded gap. After excision, DNA polymerase fills in the gap by using the

TABLE 12.6 Types of DNA Damage and Mechanism of Repair

Type of damage	Major mechanism of repair
Nicks in DNA strand	Repaired by DNA ligase
Chemically modified base	Base excision (removal by base-specific DNA glycosylase)
Mismatched bases	Corrected by mismatch repair (excision and resynthesis)
Apurinic or apyrimidinic site	Fixed by AP endonuclease repair system
Damaged region of DNA	Nucleotide excision repair (excision and resynthesis across partner strand)
Pyrimidine dimers	Enzymatically reversed (from UV light)
Damaged region of DNA	DNA damage bypass (sequence in damaged region recovered via recombination)

remaining strand as a template, thereby eliminating the mismatch. The mismatch-repair system also corrects most small insertions or deletions.

If the DNA strand that is removed in mismatch repair is chosen at random, then the repair process sometimes creates a mutant molecule by cutting the strand that contains the correct base and using the mutant strand as a template. However, this is prevented from happening in newly synthesized DNA because the daughter strand is less methylated than the parental strand. The mismatch-repair system recognizes the degree of methylation of a strand and *preferentially excises nucleotides from the undermethylated strand*. This helps ensure that incorrect nucleotides incorporated into the daughter strand in replication will be removed and repaired. The daughter strand is always the undermethylated strand because its methylation lags somewhat behind the moving replication fork, whereas the parental strand was fully methylated in the preceding round of replication.

The mechanism of mismatch repair has been studied extensively in bacteria. Mutants defective in the process were identified as having high rates of spontaneous mutation. The products of two genes, *mutL* and *mutS*, recognize and bind to a mismatched base pair. This triggers the excision of a tract of nucleotides from the newly synthesized strand. Experiments carried out in yeast revealed that tandem repeats of short nucleotide sequences are a 1000-fold less stable in mutants deficient in mismatch repair than in wildtype yeast. By that time it was already known that some forms of human hereditary colorectal cancer result in decreased stability of simple repeats, and the yeast researchers suggested that these high-risk families might be segregating for an allele causing a mismatch-repair deficiency. Within less than 2 years, scientists in several laboratories had identified four human genes homologous to *mutL* or *mutS*, any one of which, when mutated, results in hereditary nonpolyposis colorectal cancer (HNPCC). Most cases of this type of cancer may be caused by mutations in one of these four mismatch-repair genes.

Base excision removes damaged bases from DNA.

We have already seen that DNA duplexes may contain uracil formed from the deamination of cytosine. When deamination of guanine takes place, the result is 8-oxoguanine. These and other incorrect bases in DNA also commonly occur at the rate of many thousands per day per cell. The vast majority of these are removed almost immediately by a process of base excision repair using any of about eight enzymes that recognize specific incorrect bases. For example, the enzyme *DNA uracil glycosylase* recognizes an incorrect G–U base pair in duplex DNA, and the enzyme *DNA 8-oxoguanine glycosylase* recognizes an incorrect 8-oxoG–C base pair.

The base excision part of the repair process is shown in **FIGURE 12.26**. Part A shows deamination of cytosine leading to the presence of a uracil-containing base. The incorrect G–U base pair is recognized by its specific glycosylase, which cleaves the uracil from the deoxyribose sugar to which it is attached. The enzyme works by scanning along duplex DNA until a uracil nucleotide is encountered, whereupon the enzyme intrudes into the DNA through the minor groove and flips the uracil out of the double helix. The exposed uracil is then accessible to cleavage by the uracil glycosylase. Removal of the uracil leaves behind a deoxyribose sugar in the duplex DNA that lacks a pyrimidine base, which is known as an *apyrimidinic site*. Figure 12.26, part B shows how 8-oxoguanine is removed by its specific glycosylase, resulting in this case in an *apurinic site*. Both apyrimidinic sites and apurinic sites are repaired by the mechanism discussed next.

AP endonuclease repairs nucleotide sites at which a base has been lost.

Restoration of apurinic and apyrimidinic sites occurs by means of **AP repair**. The process is illustrated for an apurinic site in **FIGURE 12.27**. The key enzyme is the *AP endonuclease*, which cleaves the baseless sugar

🔅 THE HUMAN CONNECTION

Damage Beyond Repair

Frederick S. Leach and 34 other investigators (1993)

Johns Hopkins University, Baltimore, MD and 10 other research institutions

Mutations of a mutS Homolog in Hereditary Nonpolyposis Colorectal Cancer

Hereditary nonpolyposis colorectal cancer (HNPCC) is one of the most common conditions predisposing to colon cancer, affecting as many as 1 in 200 individuals in the Western world. Among several forms of the disease, one form had been traced to a mutant gene in chromosome 2. Certain short, repeating nucleotide sequences (dinucleotide repeats) were known to be genetically unstable during DNA replication in cells with this form of familial colorectal cancer. The accurate replication of such sequences involves the mismatch-repair system, which had been studied extensively in bacteria. Mutants defective in the process were identified as having high rates of spontaneous mutation. In bacteria, proteins encoded in the *mutL* and *mutS* genes recognize and bind to a mismatched base pair. The binding triggers the excision of a tract of nucleotides from the newly synthesized strand. Eukaryotes have similar mismatch-repair systems, and yeast enzymes involved in mismatch repair have amino acid sequences similar to those of bacterial enzymes. The yeast gene that has a product similar to that of *mutS* is called *MSH2*, and in yeast mutants with defective *MSH2*, dinucleotide repeats in DNA are 1000-fold less stable. Recognizing

> 66 *All eleven affected individuals contained one allele with the C to T transition, while all ten unaffected members contained two normal alleles.* 99

the potential relevance of these findings to the genetic instability in HNPCC, the authors of this paper hypothesized that high-risk families might be segregating for an allele causing a mismatch-repair deficiency. Their results identified a human gene (denoted *hMSH2*) in human chromosome 2 in which the product is similar in amino acid sequence to the product of *MSH2*. After comparing affected and nonaffected members in an extensive kindred, the study strongly supported the hypothesis that mutations in *hMSH2* are responsible for HNPCC.

Twenty-one members of the kindred [with multiple affected individuals] were analyzed . . . . All eleven affected individuals contained one [*hMSH2*] allele with the C to T transition, while all ten unaffected members contained two normal alleles, thus documenting perfect segregation with the disease. Importantly, the proline [changed in the mutant *hMSH2* allele] was at a highly conserved position, the identical residue being found in all known *mutS*-related genes from prokaryotes and eukaryotes.

F. S. Leach, et al., *Cell* 75: 1215–1225.

from the DNA, leaving a single-stranded gap that is repaired by a specialized DNA polymerase (polymerase beta). The gap filling leaves one nick remaining in the repaired strand, which is closed by DNA ligase, completing the repair.

Nucleotide excision repair works on a wide variety of DNA damage.

Nucleotide excision repair is a ubiquitous multistep enzymatic process by which a stretch of a damaged DNA strand is removed from a duplex molecule and replaced by resynthesis using the undamaged strand as a template (**FIGURE 12.28**). The substrate

for repair can be any distortion in the duplex molecule. The distortion recruits repair proteins to the site, and the DNA duplex is unwound. After the unwinding, repair endonucleases make two cuts in the sugar–phosphate backbone. The excised region is usually quite precise: In prokaryotes, the cleavage sites are eight nucleotides from the 5′ end and five nucleotides from the 3′ end of the damage; and in eukaryotes, the cleavage sites are 24 nucleotides from the 5′ end and five nucleotides from the 3′ end of the damage. The free 3′ hydroxyl group at one end of the gap serves as a primer for synthesis of a new strand using special DNA polymerases. The final step of the repair

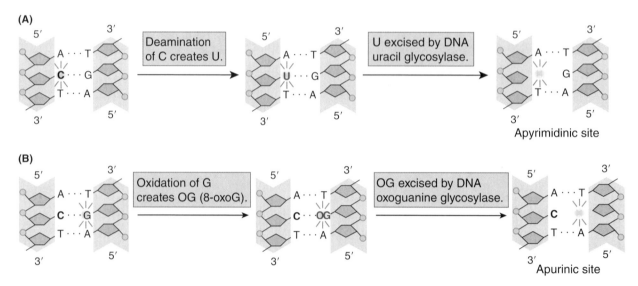

FIGURE 12.26 Base excision. (A) Excision of uracil results in an apyrimidinic site. (B) Excision of 8-oxoguanine results in an apurinic site.

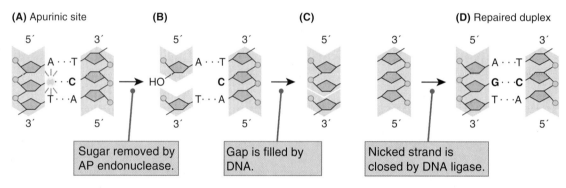

FIGURE 12.27 Action of AP endonuclease. (A) An apurinic site in a DNA duplex. (B) AP endonuclease excises the empty deoxyribose from the DNA strand. (C) A specialized DNA polymerase fills the gap using the continuous strand as a template. (D) The remaining nick is closed by DNA ligase, restoring the original sequence. The AP endonuclease acts similarly to repair apyrimidinic sites.

process is the joining of the newly synthesized segment to the contiguous strand by DNA ligase.

Several disease syndromes are associated with defects in nucleotide excision repair. These include xeroderma pigmentosum, Cockayne's syndrome, and trichothiodystrophy. Although these syndromes share such symptoms as skin abnormalities or neurological defects, they differ dramatically in light sensitivity and predisposition to cancer, stature, hair texture, and presence or absence of facial abnormalities. For example, xeroderma pigmentosum patients have severe light sensitivity and a high incidence of early-onset skin cancer, Cockayne's syndrome is associated with dwarfism, and trichothiodystrophy patients have sulfur-deficient, brittle hair.

Special enzymes repair damage to DNA caused by ultraviolet light.

Various enzymes can recognize and catalyze the direct reversal of specific DNA damage. A classic example

found in some organisms is the reversal of UV-induced pyrimidine dimers by enzymes that break the bonds that join the pyrimidines in the dimer, thereby restoring the original bases. Some enzymes of this type require light to work. They bind to the dimers in the dark but need the energy of blue light to cleave the bonds.

DNA damage bypass skips over damaged bases.

Sometimes DNA damage persists rather than being reversed or removed, but its harmful effects can be minimized. This often requires that the damaged area be skipped over during replication, so the process is called **DNA damage bypass**. For example, when DNA polymerase reaches a damaged site (such as a pyrimidine dimer), it stops synthesis of the strand (**FIGURE 12.29**, part A). However, after a brief time, synthesis is reinitiated beyond the damage, and chain growth continues, producing a gapped strand with the

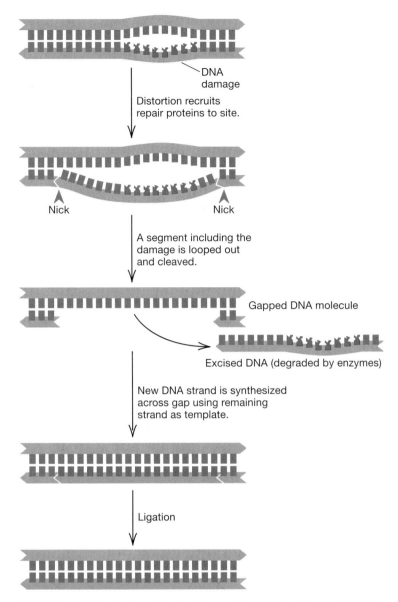

FIGURE 12.28 Mechanism of nucleotide excision repair of damage to DNA.

damaged spot in the gap (part B). The gap can be filled by strand exchange with the parental strand having the same polarity (part C), and then the secondary gap produced in the undamaged strand is filled by repair synthesis (part D). The products of this exchange and resynthesis are two intact single strands, each of which can then serve in the next round of replication as a template for the synthesis of an undamaged DNA molecule.

Double-stranded gaps can be repaired using a homologous molecule as a template.

Double-stranded gaps in DNA (**FIGURE 12.30**, part A) are common. They can be caused by replication

across an unrepaired nick in a template strand, by replication across templates that are heavily damaged or crosslinked, enzyme cleavage, excision of transposable elements, inhibition of topoisomerase, treatment with ionizing radiation, and many other factors.

In an enzymatic process known as **nonhomologous end joining**, the ends of the DNA strands flanking the gap are simply rejoined. This type of resolution results in a deletion of the DNA sequence originally present in the gap. Alternatively, when a homologous DNA molecule is present in the cell, then this molecule can be used as a template for DNA synthesis to restore the nucleotide sequence missing from the gap. The repair template may be the molecule present in the homologous chromosome or the molecule present in the sister chromatid.

The process of **template-directed gap repair** is outlined in Figure 12.30. In double-stranded gaps, the ends flanking the gaps usually have 3' overhangs because the 5' ends are trimmed back by specific nucleases present in the cell (Figure 12.30, part A). The 3' overhangs are stabilized by binding with a number of single-strand binding proteins; however, a particular protein known as RAD51 binds with a 3' overhang and forms a filament that can find and displace a homologous sequence in a template molecule (part B). The strand invasion allows the broken 3' end to be elongated by new synthesis along the template (part C). The products of the human breast cancer genes *BRCA1* and *BRCA2* participate in double-stranded gap repair.

When the 3' end of one of the broken strands has become sufficiently elongated, it can return to its original pairing partner (part D), producing a molecule that has a single-stranded gap in each strand. However, each of the single-stranded gaps has a free 3' end that can be elongated across the gap, and so the end result is a completely restored molecule (part E).

Double-stranded breaks in DNA are the normal substrate that initiates homologous recombination during meiosis. In fact, the repair process in Figure 12.30 uses the same proteins as those of homologous recombination (RecA in bacteria is the counterpart of RAD51 in eukaryotes). Although the type of restitution of the DNA molecules in Figure 12.30 does not result in recombination, there is an alterative type of restitution in which the gapped molecule and the template molecule do undergo recombination. Double-stranded gap repair is also a key process in DNA editing by means of CRISPR-Cas9 (Chapter 10).

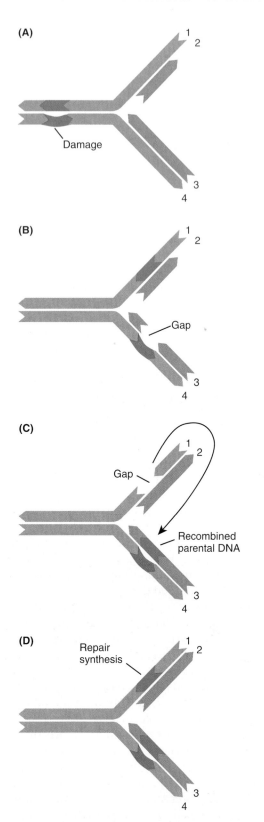

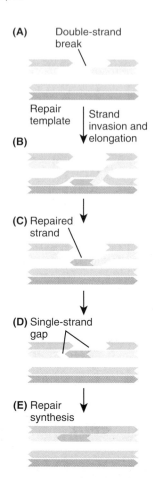

FIGURE 12.30 Template-directed gap repair. (A) Gapped DNA molecule with 3' overhangs. (B) Invasion of the template and elongation of the 3' end. (C) Release of the elongated 3' end. (D) Reassociation of the 3' end with its original partner. (E) New synthesis across the single-stranded gaps restores the original molecule, except for possible differences picked up from the template.

12.7 Genetic tests are useful for detecting agents that cause mutations and cancer.

A genetic test for mutations in bacteria is widely used for the detection of chemical mutagens. In view of the increased number of chemicals used and present as environmental contaminants, tests for the mutagenicity of these substances has become important. Furthermore, most agents that cause cancer (*carcinogens*) are also mutagens, and so mutagenicity provides an initial screening for potential hazardous agents.

In the **Ames test** for mutation, histidine-requiring (His⁻) mutants of the bacterium *Salmonella typhimurium*, containing either a base substitution or a frameshift mutation, are tested for backmutation

FIGURE 12.29 DNA damage bypass. (A) A molecule with DNA damage in strand 4 is being replicated. (B) By reinitiating synthesis beyond the damage, a gap is formed in strand 3. (C) A segment of parental strand 1 is excised and inserted in strand 3. (D) The gap in strand 1 is next filled in by repair synthesis.

(**reversion**) to His$^+$. In addition, the bacterial strains have been made more sensitive to mutagenesis by the incorporation of several mutant alleles that inactivate the excision-repair system and that make the cells more permeable to foreign molecules. Because some mutagens act only on replicating DNA, the solid medium used contains enough histidine to support a few rounds of replication but not enough to permit the formation of a visible colony. The medium also contains the potential mutagen to be tested and an extract of rat liver. The role of the liver extract is to permit identification of substances that are not directly mutagenic (or carcinogenic) but are converted into mutagens by enzymatic reactions that take place in the livers of animals. The normal function of these enzymes is to protect the organism from various naturally occurring harmful substances by converting them into soluble nontoxic substances that can be disposed of in the urine. However, when the enzymes encounter certain artificial and natural compounds, they convert these substances, which may not be harmful in themselves, into mutagens or carcinogens.

In the Ames test, if the test substance is a mutagen or is converted into a mutagen, some colonies are formed. A quantitative analysis of reversion frequency can also be carried out by incorporating various amounts of the potential mutagen in the medium. The reversion frequency generally depends on the concentration of the substance being tested and, for a known carcinogen or mutagen, correlates roughly with its carcinogenic potency in animals. The Ames test is simple, rapid, inexpensive, and exquisitely sensitive. Some chemicals can be detected to be mutagenic in amounts as small as 10^{-9} g, and a condensate of as little as 1/100 of a cigarette can be shown to be mutagenic. The test is also highly quantitative (**FIGURE 12.31**). Chemicals need not be classified simplistically as "mutagenic" or "nonmutagenic." They can be classified according to their potency as mutagens, because more than a millionfold range in potency can be detected in the *Salmonella* test.

An example of the importance of adequate testing involves a chemical known as tris-BP. This substance was used as a flame retardant in children's polyester pajamas from 1972 to 1977. Then it was discovered that tris-BP is a potent mutagen in *Salmonella* and also in *Drosophila*. Further studies showed that tris-BP interacts with human DNA and damages mammalian chromosomes. It was also found to be a carcinogen in experimental tests in rats and mice, and it was found to be capable of causing sterility in laboratory animals. Moreover, the substance was shown to be absorbed through the skin, and its breakdown products could be detected in the urine of children

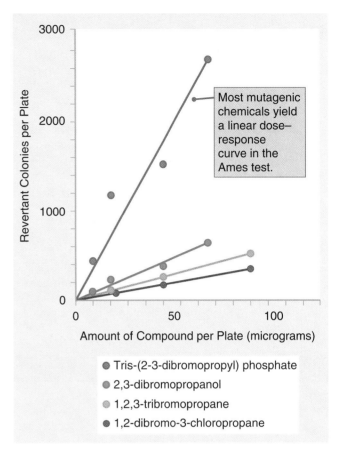

FIGURE 12.31 Linear dose–response relationships obtained with various chemical mutagens in the Ames test.

Data from B. N. Ames, *Science* 204 (1979): 587–593.

who were wearing the treated sleepwear. Before this information became known and use of the substance in clothing was discontinued, more than 50 million children were exposed to the chemical through contact with their nightclothes.

The Ames test has been used with thousands of substances and mixtures (such as industrial chemicals, food additives, pesticides, hair dyes, and cosmetics), and numerous unsuspected substances have been found to stimulate reversion in this test. A high frequency of reversion does not necessarily indicate that the substance is definitely a carcinogen but only that it has a high probability of being so. As a result of these tests, many industries have reformulated their products; for example, the cosmetics industry has changed the formulation of many hair dyes and cosmetics to render them nonmutagenic. Ultimate proof of carcinogenicity is determined by testing for tumor formation in laboratory animals. However, only a few percent of the substances known from animal experiments to be carcinogens failed to increase the reversion frequency in the Ames test.

CHAPTER SUMMARY

- Mutations can be classified by the type of cell in which they occur (for example, somatic cell versus germ-line cell), the type of change at the molecular level, conditions of expression (for example, temperature-sensitive mutations), effects on gene function, and in other ways.

- Substitution of one base for another is an important mechanism of spontaneous mutation. A single-base substitution in a codon may be silent (synonymous) and not change the amino acid specified by the codon, or it may be nonsynonymous and result in an amino acid replacement; a single-base deletion or insertion results in a shifted reading frame.

- In the human genome, some inherited diseases, including the fragile-X syndrome and Huntington disease, are associated with a sudden, dramatic increase in the number of copies of a trinucleotide repeat.

- Transposable elements are DNA sequences that are able to change their location within a chromosome or to move between chromosomes. Activities associated with these mobile elements are an important mechanism of spontaneous mutation.

- Mutations can be induced by various agents, including some classes of chemicals and various types of radiation.

- Cells contain enzymatic pathways for the repair of different types of damage to DNA. Among the most important repair mechanisms are mismatch repair of mispaired nucleotides, AP repair of apurinic or apyrimidinic sites, base excision repair, nucleotide excision repair, and repair following replication bypass of severe DNA damage.

- Most agents that cause mutation also cause cancer.

ISSUES AND IDEAS

- If a mutation is a conditional mutation, what determines whether the mutant phenotype will be expressed?

- How can an organism with a temperature-sensitive, recessive-lethal mutation survive as a homozygous genotype?

- What does it mean to say that a particular allele has a mutation rate of 10^{-6} per gene per generation?

- How does replica plating demonstrate that mutations to antibiotic resistance can arise even in cells that have never been exposed to the antibiotic?

- Mutations in genes whose products are involved in DNA repair are often associated with an increased risk of cancer. What does this observation imply about the role of spontaneous mutation in the development of cancer?

- What is "mismatched" in the process of mismatch repair?

- What is the purpose of the liver extract in the Ames test for chemical mutagens?

SOLUTIONS: STEP BY STEP

PROBLEM 1 The molecule 2-aminopurine (Ap) is an analog of adenine that pairs with thymine. It also occasionally pairs with cytosine. What pathways of mutation are possible, and what types of mutations are formed?

SOLUTION. In problems of this sort, first note the base pair that is usually formed (the nonmutagenic base pair), and then note the base pair that is rarely formed (the mutagenic base pair). In this case, the nonmutagenic base pair is Ap–T and the mutagenic base pair is Ap–C. There are two pathways of mutation, analogous to those for 5-bromouracil illustrated in Figure 12.18. If the Ap is incorporated in its mutagenic mode, it is incorporated opposite C, creating an Ap–C base pair. In the subsequent round of replication, the Ap will usually pair with T, forming an Ap–T base pair. The end result is a G–C to A–T transition. On the other hand, if the Ap is incorporated

in its nonmutagenic mode, it is incorporated opposite T, creating an Ap–T base pair. In a subsequent round of replication, the Ap may pair with C, forming an Ap–C base pair. The end result is an A–T to G–C transition.

PROBLEM 2 Mutant strains of *E. coli* are compared. What would you expect to be the phenotype of each the following mutations affecting DNA repair, with respect to (1) the overall mutation rate, and (2) whether there is a marked difference in the rates of substitution among nucleotide pairs?
(a) The mutation is in a gene whose product plays a major role in mismatch repair.
(b) The mutation is in the gene for DNA uracil glycosylase.

SOLUTION. (a) The mismatch-repair mutant cannot repair errors in incorporation in DNA synthesis. A general increase in the mutation rate is expected. There should be no marked preference for one type of nucleotide substitution over another. **(b)** Uracil (which normally pairs with A) in DNA results from deamination of cytosine (which normally pairs with G). A general increase in the mutation rate is expected. There should, however, be a marked preference of C–G to T–A mutations over any other nucleotide substitutions.

CONCEPTS IN ACTION: PROBLEMS FOR SOLUTION

12.1 Ultraviolet light primarily damages DNA by

  (a) Inactivation of mismatch-repair enzymes

  (b) Formation of crosslinked thymine dimers

  (c) Alkylation of bases in DNA

  (d) Deamination of bases in DNA

  (e) Interaction with water to create free radicals

12.2 Occasionally, a person is found who has one blue eye and one brown eye or who has a sector of one eye a different color from the rest. Can these phenotypes be explained by new mutations? If so, in what types of cells must the mutations occur?

12.3 What is the overall probability of incorporating a mismatched base that remains uncorrected by the proofreading function of DNA polymerase as well as by postreplication mismatch repair?

12.4 Torsion dystonia is an autosomal dominant disorder with a mutation rate estimated as 2×10^{-4} per generation. What is the number of gametes that contain, on average, one new mutation?

12.5 If spontaneous depurination of DNA occurs at the rate of approximately 3×10^{-9} depurinations per purine nucleotide per minute, then considering that a diploid human cell has a genome size of 6×10^9 base pairs, approximately how many spontaneous depurinations must be repaired in each cell per day?

12.6 How many different codons can result from a single-base substitution in DNA coding for the cysteine codon UGC? Classify each as synonymous (silent), nonsynonymous (missense), or chain termination.

12.7 A cytosine deamination occurs in the top strand of the following sequence:

```
5'-TTGGGCA-3'
3'-AACCCGT-5'
```

  (a) If there is no repair before the next round of DNA replication, what is the sequence of the affected strand and its newly replicated complementary strand?

  (b) If the damaged base is repaired by uracil DNA glycosylase before the next round of DNA replication, what is the sequence of the affected strand and its newly replicated complementary strand?

  (c) If the uracil DNA glycosylase repair mechanism is inactivated and the mismatch repair process repairs the damaged base before the next round of DNA replication, what is the sequence of the affected strand and its newly replicated complementary strand?

12.8 Weedy plants that are resistant to the herbicide atrazine have a single amino acid substitution in the gene *psbA* that results in the replacement of a serine with an alanine in the polypeptide. Is the base change in the *psbA* gene that results in this amino acid replacement a transition or a transversion?

12.9 What is the minimum number of single-nucleotide substitutions that would be necessary for each of the following amino acid replacements?

  (a) Trp → Asn   **(b)** Tyr → Ala

  (c) Met → Lys   **(d)** Ala → Asp

12.10 A *Drosophila* male carries an X-linked temperature-sensitive recessive allele that is viable at 18°C but lethal at 29°C. What sex ratio would be expected among the progeny if the progeny were reared at 29°C and the male were mated to:

  (a) A normal XX female?

  (b) An attached-X female having both X chromosomes joined to one centromere?

12.11 How many amino acids can replace tyrosine by a single-base substitution in the DNA? (Do not assume that you know which tyrosine codon is being used.)

12.12 Mutations caused by the insertion of a DNA transposon are often genetically unstable, reverting to wildtype (or a phenotype resembling wildtype) at a relatively high rate. Suggest a reason why this might be expected.

12.13 A population of 1×10^6 bacterial cells undergoes one round of DNA replication and cell division. The forward mutation rate of a gene is 1×10^{-6} per replication.

(a) What is the expected number of mutant cells after cell division?

(b) What is the probability that the population contains no mutant cells?

12.14 In the mouse, a dose of approximately 1 sievert (Sv) of x rays produces a rate of induced mutation equal to the rate of spontaneous mutation. Expressed as a multiple of the spontaneous mutation rate, what is the total mutation rate at 1 Sv? Assuming that the total mutation rate is proportional to the x-ray dose, what dose of x rays will increase the mutation rate by 50 percent? What dose will increase the mutation rate by 20 percent?

12.15 If every human gamete contains 25,000 genes; if the forward mutation rate is between 1×10^{-5} and 1×10^{-6} new mutations per gene per generation, what is the average number of new mutations per gamete per generation?

12.16 Human hemoglobin C is a variant in which a lysine in the beta-hemoglobin chain is substituted for a particular glutamic acid. What single-base substitution can account for the hemoglobin-C mutation?

12.17 Gene conversion results from mismatch repair in a heteroduplex DNA molecule that gives rise to fungal asci with ratios of alleles such as 3 A : 1 *a* and 1 A : 3 *a*, instead of 2 A : 2 *a*.

(a) Why is a ratio of 2 A : 2 *a* expected?

(b) Why, among a large number of gametes chosen at random, is the Mendelian segregation ratio of 1 A : 1 *a* still observed in spite of gene conversion?

12.18 This problem illustrates how conditional mutations can be used to determine the order of genetically controlled steps in a developmental pathway. A certain organ undergoes development in the sequence of stages A → B → C, and both gene *X* and gene *Y* are necessary for the

sequence to proceed. A conditional mutation X' is sensitive to heat (the gene product is inactivated at high temperatures), and a conditional mutation Y' is sensitive to cold (the gene product is inactivated at cold temperatures). The double mutant X'/X'; Y'/Y' is created and reared at either high or low temperatures. To what stage would development proceed in each of the following cases at the high temperature and at the low temperature?

(a) Both X and Y are necessary for the A → B step.

(b) Both X and Y are necessary for the B → C step.

(c) X is necessary for the A → B step, and Y is necessary for the B → C step.

(d) Y is necessary for the A → B step, and X is necessary for the B → C step.

12.19 The accompanying diagrams show two nonhomologous chromosomes, each containing a copy of a transposable element (shaded), that can be oriented either **(a)** in the same direction or **(b)** in opposite directions with respect to the centromere. For clarity, the length of the transposable element is greatly exaggerated relative to the length of the chromosome. (In reality, the average transposable element in *Drosophila* is about 0.01 percent of the length of a chromosome.) Draw diagrams illustrating the consequences of ectopic recombination between the transposable elements.

12.20 You carry out a large-scale cross of genotypes *A m B* × *a + b* of two strains of *Neurospora crassa* and observe a number of aberrant asci, some of which are shown here.

A m B	A m B	A m B	A m B	A m B
A m B	A m B	A m B	A m B	A m B
A m b	A m b	A m b	A m b	A + b
A + b	A m b	A + b	A m b	A + b
a m B	a m B	a + B	a m B	a + B
a + B	a + B	a + B	a m B	a + B
a + b	a + b	a + b	a + b	a + b
a + b	a + b	a + b	a + b	a + b

Your colleague who provided the strains insists that they are both deficient in the same gene in the DNA mismatch-repair pathway. The results of your cross seem to contradict this assertion.

(a)   Which of the asci depicted here offer evidence that your colleague is incorrect?

(b)   Which ascus (or asci) would you exhibit as definitive evidence that DNA mismatch repair in these strains is not 100 percent efficient?

 STOP & THINK ANSWERS

ANSWER TO STOP & THINK **12.1**

In this cross, all of the female offspring survive at both temperatures because they inherit a nonmutant X chromosome from their father. Half the male offspring receive the nonmutant X chromosome from their mother and half receive the mutant X chromosome with the temperature-sensitive lethal. Males with the mutant X chromosome survive at 18°C (the permissive temperature); therefore at 18°C the expected ratio of females : males is 1 : 1. Males with the mutant X chromosome die at 30°C (the restrictive temperature); therefore at 30°C the expected ratio of females : males is 2 : 1

ANSWER TO STOP & THINK **12.2**

The accompanying diagram shows the pairing configuration and the position of the crossover. Tracing along the chromatid, you can see that one product is a ring chromosome, which bears the centromere; the other product is an acentric fragment bearing a telomere at each end.

ANSWER TO STOP & THINK **12.3**

A point mutation results from one-hit event, in which the track of free radicals from one x ray causes a single break in a chromosome that is incorrectly repaired. The number of single breaks is therefore expected to increase in direct proportion to the x-ray dose. On the other hand, a reciprocal translocation results from a two-hit event, because two different chromosomes must be broken in the same cell at the same time. A two-hit event requires two tracks of free radicals, and the likelihood of such an event is expected to increase as the square of the x-ray dose. (The exponent is actually a little smaller than 2 because some free-radical tracks can produce two breaks.)

CHAPTER **13**

Cancer is a genetic disease.

Molecular Genetics of the Cell Cycle and Cancer

LEARNING OBJECTIVES

- To describe the roles of cyclins, cyclin-dependent protein kinases, and protein degradation in the progression of cells through the cell cycle.

- To identify the major checkpoints in the cell cycle and explain how checkpoint failure can result in mutations, chromosome aberrations, nondisjunction, and polyploidy.

- To explain the central role of the p53 protein in arresting the cell cycle to allow DNA-damage repair or to promote apoptosis.

- To define tumor-suppressor genes and oncogenes and describe how mutations in each type of gene can result in cancer.

- To distinguish a promoter fusion from a gene fusion in the origin of leukemia.

Cancer is a disease characterized by the uncontrolled proliferation of cells. The normal mechanisms that regulate cellular growth and division break down. Cancer is a genetic disease. It results from mutations that overcome the normal limits to the number of cell divisions that can take place in a cell lineage. These mutations usually occur in somatic cells, and full-blown malignant cancer usually requires a number of sequential mutations to get started. But occasionally a mutation affecting cell-cycle regulation is inherited through the germ line, and persons who inherit the mutation have a greatly increased risk (sometimes approaching 100 percent) of developing malignancies due to additional somatic mutations. To understand cancer, therefore, one must understand the normal mechanisms that control the cell cycle and that prevent the proliferation of genetically damaged cells. This is where we begin.

13.1 The cell cycle is under genetic control.

The cell cycle is divided into a three-part *interphase* composed of G_1 *(gap 1)*, S *(DNA synthesis)*, and G_2 *(gap 2)*, which occur in that order, followed by *M (mitosis)* proper, in which the sister chromatids are physically separated into the two daughter nuclei. The essential functions of the mitotic cell cycle are:

1. To ensure that each chromosomal DNA molecule is replicated once and only once per cycle.

2. To ensure that the identical replicas of each chromosome (the sister chromatids) are distributed equally to the two daughter cells.

Some of the key events that must take place for the proper duplication and distribution of chromosomes are highlighted in **FIGURE 13.1**. The spindle pole is organized around a small region of clear cytoplasm near the interphase nucleus called the **centrosome**; in many organisms this role is played by a pair of **centrioles**, which are more particulate in appearance. Both are microtubule-organizing centers that must be duplicated and positioned. In most cells, centrosome duplication begins late in G_1 and is completed during S phase. The duplicated poles then slowly begin to migrate to positions on opposite sides of the nucleus. Meanwhile, within the nucleus during the S phase, DNA replication takes place. Completion of DNA replication marks the beginning of G_2. Soon after the M phase commences, the centrosomes reach their final destinations and the chromosomes begin to condense. Each centrosome organizes one pole of the spindle by nucleating the formation of spindle and astral microtubules, and the condensed chromosomes become attached to spindle microtubules on both sides of the kinetochore (centromere) as the nuclear envelope breaks down. Each chromosome is thus physically attached to both spindle poles and is maneuvered to a position approximately halfway between them. As anaphase begins, the spindle elongates, the centromeres separate, and the sister chromatids migrate toward opposite poles. Once the

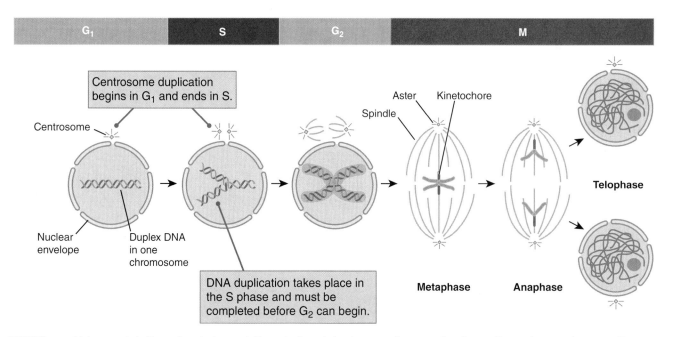

FIGURE 13.1 Major events in the cell cycle. In yeast, the spindle pole body serves the same function as the centrosome in many other organisms: Both are microtubule-organizing centers from which the spindle emerges. In most other organisms, nuclear division takes place and the pinching off of cells (cytokinesis) follows. In yeast, the "shell" of the daughter cell forms and enlarges before nuclear division takes place. The nucleus (its membrane never breaks down) moves into the bridge between the mother and daughter cell, and nuclear division occurs there. After the daughter nuclei move into the two cell bodies, a septum is laid down between the two cells.

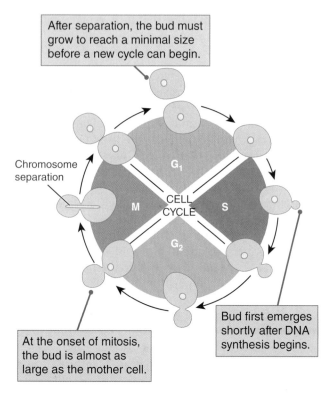

After separation, the bud must grow to reach a minimal size before a new cycle can begin.

Chromosome separation

CELL CYCLE

G₁

S

G₂

M

At the onset of mitosis, the bud is almost as large as the mother cell.

Bud first emerges shortly after DNA synthesis begins.

FIGURE 13.2 The cell cycle of budding yeast, *Saccharomyces cerevisiae*.

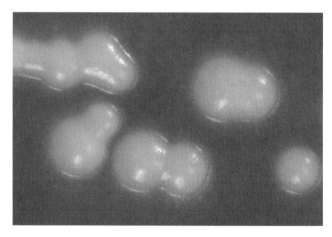

FIGURE 13.3 Photograph of cells of diploid budding yeast *S. cerevisiae* in various stages of the cell cycle. Bud size is correlated with the position of the cell within the cell cycle. Here the buds range in size from quite small to nearly as large as the mother cell.

© Rattiya Thongdumhyu/Shutterstock.

daughter chromosomes reach the poles, the spindle disassembles, the chromosomes decondense, and the nuclear membrane is formed again. These events return the cell to the G₁ phase.

The mechanisms by which cell growth is regulated and the cell cycle controlled have been approached through biochemistry, cell biology, and genetics. Some of the most extensive genetic studies have focused on budding yeast *Saccharomyces cerevisiae*. As its common name implies, *S. cerevisiae* multiplies by budding. The position of a cell within the cell cycle can be monitored visually by the size of the bud. This convenient feature of the cell cycle is depicted diagrammatically in **FIGURE 13.2**. The bud first emerges shortly after entry into S phase and grows throughout the cell cycle. Mitosis takes place *within* the nucleus without breakdown of the nuclear envelope. The spindle poles are actually embedded within the nuclear envelope, and one pole moves to a position exactly opposite the second pole across the way. Mitosis occurs when the bud has grown to a size nearly as large as the mother cell. Shortly after chromosome separation, a barrier is laid down between the mother and daughter cells. At cell separation, the daughter cell is only slightly smaller than the mother cell, but it typically must grow this extra bit to achieve the optimal size before it can start its own cell cycle. Various stages of bud growth from beginning to end can be identified in the photograph in **FIGURE 13.3**.

Many genes are transcribed during the cell cycle just before their product is needed.

Because the DNA sequence of the entire genome is known for budding yeast, it is possible to analyze the transcription patterns of all of the approximately 6000 genes in the cell in a single experiment using high-density DNA microarrays. Such experiments have shown that the transcript levels of about 800 genes vary in a periodic or cyclic pattern through the cell cycle. Typically, genes encoding proteins that are needed in one part of the cycle are transcribed in the immediately preceding period. For example, enzymes needed for synthesis of the trinucleotide precursors of DNA and for DNA replication are made in G₁ immediately prior to their use in S phase. Similarly, the histone proteins are synthesized during S phase immediately prior to their incorporation into chromatin and their use in chromosome condensation.

Mutations affecting the cell cycle have helped to identify the key regulatory pathways.

The ready assignment of a yeast cell to a position in the cell cycle on the basis of the relative sizes of the mother cell and its bud has made possible extensive genetic analyses of the cell cycle through the isolation and study of temperature-sensitive mutants. Temperature-sensitive **cell division cycle (*cdc*) mutants** are typically wildtype at 23°C (the *permissive* temperature) but unable to complete the cell cycle at 36°C (the *restrictive* temperature). At the higher temperature, mutant cells accumulate at a characteristic stage in the cell cycle. This is the stage at which their progression in the cell cycle is halted. The stage-specific stop is exceptional among mutations affecting cellular functions. For example, temperature-sensitive mutants with defects in protein synthesis do not cease growing abruptly when the temperature is raised. Each cell continues along until it runs out of functional protein, and new protein needs to be synthesized for continued progression in the cell cycle. The stopping point differs from cell to cell, so the mutant cells stop growing at different stages in the cell cycle. In the microscope one sees cells with no bud, cells with small buds, and cells with large buds—essentially the same distribution one sees in an asynchronous cell population. But for each *cdc* mutant the stop is at a specific stage, which differs from one mutant to the next and is related to the function of the mutant gene product. Once it has been established where in the cell cycle a *cdc* mutant is blocked, further analysis can be used to determine whether specific processes can occur in the mutants, such as DNA synthesis or spindle formation.

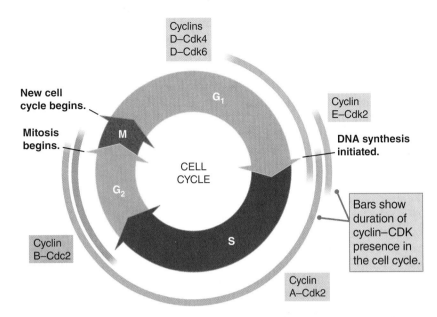

FIGURE 13.4 The temporal expression pattern of activities of the cyclin–CDKs in mammalian cells.

Cyclins and cyclin-dependent protein kinases propel the cell through the cell cycle.

In the early stages of the cell cycle, progression from one phase to the next is controlled by characteristic protein complexes that are called **cyclin–CDK complexes** because they are composed of **cyclin** subunits combined with **cyclin-dependent protein kinase (CDK)** subunits (**FIGURE 13.4**). All eukaryotes regulate progression through the cell cycle by means of cyclin–CDK complexes, although the details of their structure and their mechanisms of action may differ slightly from one organism to another.

The term *cyclin* is apt because the abundance of these proteins changes cyclically in the cell cycle, and some are present only at specific times. Most cyclins appear abruptly and disappear a short time later (Figure 13.4). More than one cyclin may be present in cells at the same time in the cell cycle, and some CDKs are able to form complexes with different cyclins. Each cyclin appears at its characteristic time in the cell cycle because its transcription is linked to the cell cycle via a previously expressed cyclin. Typically, the presence of an active cyclin–CDK complex results in the activation of a transcription factor that leads to transcription of the next cyclin needed in the cell cycle. The cyclin disappears when its gene is no longer transcribed and the previously made mRNA and protein are degraded. Active cyclin–CDK complexes also cause the transcriptional activation of genes other than cyclins.

The cyclin–CDK complexes control progression through the cell cycle through their activity as protein kinases. They attach phosphate groups to the hydroxyl groups present in the amino acids serine, threonine, and tyrosine found in certain proteins. The cyclin component of the complex binds to the protein substrate and tethers it, allowing the CDK component to phosphorylate the tethered substrate. Once the targeted protein is phosphorylated, it dissociates from the cyclin–CDK complex. The activities of the phosphorylated form of a protein often differ dramatically from those of the unphosphorylated form. Phosphorylation may activate one enzyme but inactivate another. Even in a single enzyme molecule, phosphorylation at different sites may have opposite effects.

Another route of cell-cycle regulation is mediated through *phosphatase* enzymes that dephosphorylate proteins that the cyclin-dependent kinases have phosphorylated. Reversing the effects of CDKs, the phosphatases activate enzymes that are inactive or deactivate ones that are active.

The retinoblastoma protein controls the initiation of DNA synthesis.

Our first connection between cell-cycle control and cancer came from a protein known to be involved in tumors of the retina. The type of cancer is called *retinoblastoma*, and we will examine the condition in some detail in Section 13.4. The normal role of the **retinoblastoma (RB) protein** is to maintain cycling cells at a point in G_1 called the **G_1 restriction point** or *start*, until the cell has attained proper size. The RB protein acts by binding to the transcription factor E2F, which is needed for further progression in the cell cycle (**FIGURE 13.5**, part A). If the cycling cell is progressing properly, cyclin D is produced in the middle of G_1. The RB protein then begins to be phosphorylated by both the cyclin D–Cdk4 kinase and the cyclin D–Cdk6 kinase. Late in G_1, the RB phosphorylation is completed by the cyclin E–Cdk2 kinase as cells approach the **G_1/S transition**, when they become committed to DNA synthesis. Phosphorylation of RB eliminates its ability to bind the E2F transcription factor (part A). Release of the E2F results in transcription of the genes and translation of the enzymes responsible for DNA synthesis, including DNA polymerase. E2F also activates transcription of the gene for E2F itself (an example of positive autoregulation), as well as those for the cyclins E and A and the Cdk2 kinase subunit.

Once cells have entered S phase, the cyclin A–Cdk2 phosphorylates E2F and inhibits its binding to DNA, thus inactivating its function as a transcription factor. The cyclin A–Cdk2 activity is required throughout S phase, apparently to keep the RB protein heavily

phosphorylated. The progression from G_2 to M (the **G_2/M transition**) is controlled by a cyclin B–Cdc2 complex also known as *maturation-promoting factor*. At this transition point, the chromosomes condense and assemble onto the spindle, and the chromosome segregation machinery becomes active.

Protein degradation also helps regulate the cell cycle.

A fundamental feature of the cell cycle is that it is a true cycle: It is not reversible. The cycle is propelled forward by a process of protein degradation that complements the periodic activation of cyclin–CDK complexes. Protein degradation (*proteolysis*) eliminates proteins that were used in the preceding phase as well as proteins that would inhibit progression into the next phase. In the early stages of the cell cycle, progression requires the sequential activation of cyclin–CDKs. Entry into each new phase also requires destruction of the cyclins used in the preceding phase. In the later stages of the cell cycle, progression is propelled by

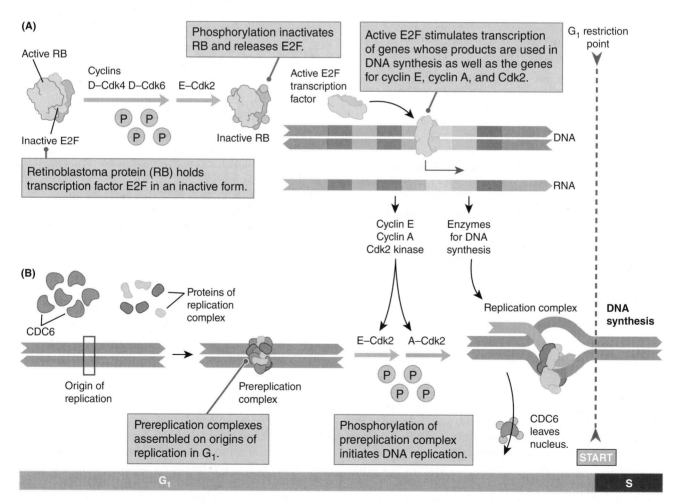

FIGURE 13.5 (A) Role of the retinoblastoma protein RB in controlling the transition from G_1 phase to S phase. The cyclin D-dependent kinases Cdk4 and Cdk6 initiate phosphorylation of RB in mid-G_1 phase, a process completed by cyclin E–Cdk2; this frees the transcription factor E2F, which activates transcription of enzymes for DNA synthesis. (B) The free E2F also activates transcription of the genes for cyclin A, cyclin E, and Cdc2, which help convert prereplication complexes to replication complexes for transition to S phase.

proteolysis alone. This process is best understood in yeast, and we describe it here using the yeast terminology. In the completion of mitosis and the return to G_1 phase, two key regulatory events must occur:

1. The sister chromatids must separate (marking the onset of anaphase).

2. The cells must exit from mitosis, which entails chromosome decondensation, spindle disassembly, inactivation of the chromosome segregation machinery, and cytokinesis.

Both of these key events are triggered by protein degradation, as indicated in **FIGURE 13.6** for yeast. Exit from mitosis requires the destruction of cyclin B. Cyclin B is most abundant when cells enter mitosis, but it disappears after chromosome disjunction has occurred at the transition from metaphase to anaphase. Cyclin B is marked for destruction by the **anaphase-promoting complex (APC/C)**, which is a ubiquitin–protein ligase responsible for adding the 76-amino-acid protein *ubiquitin* to its target proteins and marking them for destruction in the *proteasome*—a large, multifunctional, multi-subunit complex responsible for most of the cytoplasmic proteolysis in the cell beyond that which takes place in lysosomes. Other substrates for the APC/C are the protein *securin*, which inhibits a protease called *separase*, and Ase1p, a microtubule-associated protein that binds antiparallel microtubules from the spindle pole bodies in the midzone of the spindle. As the securin is degraded, the separase protein becomes free to degrade Scc1p, a component of the *cohesin complex* that condenses chromosomes and also holds sister chromatids together. Cleavage of Scc1p results in its dissociation from the chromatin, which allows the sister chromatids to come apart and be pulled toward opposite poles of the spindle.

13.2 Checkpoints in the cell cycle allow damaged cells to repair themselves or to self-destruct.

Cells monitor their external environment as well as their internal physiological state and functions. In the absence of needed nutrients or growth factors, animal cells may exit from the cell cycle and become quiescent. Upon growth stimulation, they reenter the cell cycle in a process that requires cyclin D–CDK activity. Cells also have mechanisms that respond to symptoms of stress, including DNA damage, oxygen depletion, inadequate pools of nucleoside triphosphates, and (in the case of animal cells) loss of intercellular adhesion. Inside the cell, several key events in the cell cycle are monitored. When defects are identified, progression through the cell cycle is halted at a **checkpoint**.

KEY CONCEPT

Checkpoints in the cell cycle serve to maintain the correct order of steps with respect to each other as the cycle progresses; they do this by causing the cell cycle to pause while defects are corrected or repaired.

Three principal checkpoints that function to maintain the genetic integrity of cells are

- A DNA damage checkpoint

- A centrosome duplication checkpoint

- A spindle checkpoint

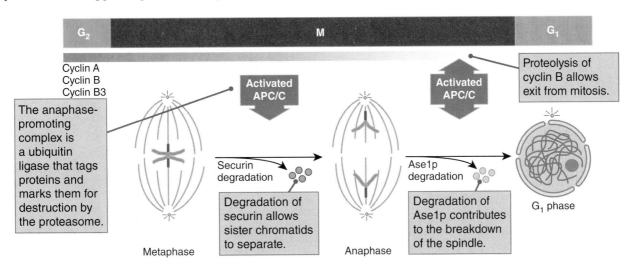

FIGURE 13.6 Role of activated anaphase-promoting complex (APC/C) in controlling the proteolysis necessary for the transition to anaphase and the exit from mitosis. APC/C is a ubiquitin ligase that marks proteins for proteasome destruction. Its substrates include securin, which unless destroyed inhibits breakdown of the protein "glue" (Scc1p) that holds the sister chromatids together, and Ase1p, a microtubule-associated protein that binds antiparallel microtubules from the spindle pole bodies in the spindle midzone. Destruction of cyclin B somewhat later allows the exit from mitosis.

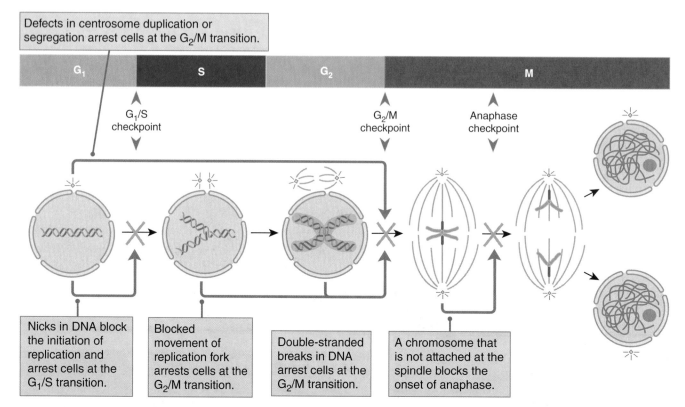

Defects in centrosome duplication or segregation arrest cells at the G₂/M transition.

G_1 — S — G_2 — M

G_1/S checkpoint

G_2/M checkpoint

Anaphase checkpoint

Nicks in DNA block the initiation of replication and arrest cells at the G_1/S transition.

Blocked movement of replication fork arrests cells at the G_2/M transition.

Double-stranded breaks in DNA arrest cells at the G_2/M transition.

A chromosome that is not attached at the spindle blocks the onset of anaphase.

FIGURE 13.7 Key cell-cycle checkpoints that act to maintain the genetic stability of cells. The text in the boxes explains the events monitored and the steps affected.

These checkpoints are summarized in **FIGURE 13.7**. All three types of checkpoints are important in maintaining the stability of the chromosome complement. When there is an error in any of the three processes monitored by the checkpoints, failure to stop at the checkpoint may lead to aneuploidy (extra or missing chromosomes), polyploidy, or an increased number of mutations. In the following sections, we shall examine how some of these checkpoints work.

The p53 transcription factor is a key player in the DNA damage checkpoint.

A **DNA damage checkpoint** arrests the cell cycle when DNA is damaged or replication is not completed. DNA damage includes either modification of nitrogenous bases or breakage of the phosphodiester backbone. When modified nucleotides are repaired by an excision repair pathway, the repair entails removal of the affected nucleotides, followed by resynthesis with a repair polymerase and finally ligation. DNA molecules broken in the backbone are repaired by homology-based recombination, nonhomologous end joining, or addition of new telomeres. In animal cells, a DNA damage checkpoint acts at three stages in the cell cycle: at the G_1/S transition, in the S period (DNA synthesis), and at the G_2/M boundary. The S-period checkpoint continuously monitors the progression of DNA synthesis as it takes place. There are three

DNA-damage checkpoints; each monitors DNA damage or incomplete replication. If either type of problem is detected, the DNA checkpoint acts to block the cell cycle at multiple points.

Key proteins in the mammalian cell's response to stress in general, and to DNA damage in particular, are several slightly different forms of a protein called the **p53 transcription factor**. In normal cells, the level of activated p53 protein is very low, even though substantial amounts of p53 mRNA and protein are present. The activity of p53 is kept low by another protein called Mdm2. To function as a transcription factor, p53 protein must be activated first by phosphorylation and then by acetylation. Mdm2 binds to p53 and prevents phosphorylation and subsequent steps in the activation of p53 as a transcription factor. In addition, Mdm2 continuously shuttles between the nucleus and the cytoplasm, and in this process it continuously exports p53 from the nucleus for degradation in the cytoplasm. The p53–Mdm2 export cycle is illustrated in **FIGURE 13.8**.

When cells are treated with agents that damage DNA, some cells arrest in G_1 and others arrest in G_2. The DNA damage signal is sensed and transmitted; this causes p53 protein to become activated by protein kinases and acetylases, overriding the inhibitory effect of Mdm2. Activation of p53 causes its release from Mdm2, which stabilizes the active p53 transcription factor and results in increased levels of the p53

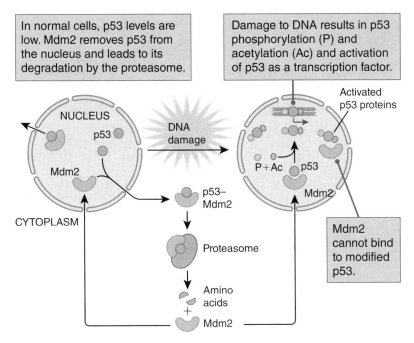

In normal cells, p53 levels are low. Mdm2 removes p53 from the nucleus and leads to its degradation by the proteasome.

Damage to DNA results in p53 phosphorylation (P) and acetylation (Ac) and activation of p53 as a transcription factor.

Mdm2 cannot bind to modified p53.

FIGURE 13.8 The effect of DNA damage on transcription factor p53. In normal cells the level of p53 is low, in part because Mdm2 shuttles it to the cytoplasm where it is destroyed by the proteasome. DNA damage results in phosphorylation and acetylation of p53, rendering it unable to bind with Mdm2. Hence, p53 levels in the nucleus increase.

protein in the nucleus. The activated p53 then triggers the transcription of a number of genes and the repression of others. Some of the key proteins whose genes are transcriptionally activated by p53 are listed in **TABLE 13.1**, and a flow chart of how these proteins affect processes in the cell cycle and other cellular properties is given in **FIGURE 13.9**.

Increased transcription of the genes for p21, GADD45, 14-3-3σ, and miRNA34, as well as decreased transcription of the gene for cyclin B, all serve to block the cell cycle at particular points, as illustrated in **FIGURE 13.10**. The G_1/S transition checkpoint is mediated by the increased level of activated p21, which results in inhibition of the G_1 cyclin–CDKs and in this way blocks the G_1/S transition. Hence if DNA damage is detected in G_1, the cell is blocked in the G_1/S transition. The S-period response to DNA damage is mediated by the p21 protein and GADD45; these form a complex with another protein, which results in a reduction in the processivity of DNA polymerase. The **processivity** of a DNA polymerase is the number of consecutive nucleotides in the template strand that are replicated before the polymerase detaches from the template. Decreasing the processivity of DNA polymerase therefore slows down DNA synthesis, in effect buying time for the cell to repair DNA damage. The G_2/M checkpoint is mediated by the 14-3-3σ protein, which hinders activation of cyclin B–Cdc2, thus blocking the G_2/M transition. At the same time, the decrease in the level of cyclin B reduces the level of the active cyclin B–Cdc2 complex, which also ensures that

the cell remains in G_2. Hence if DNA damage is detected in the S period or in G_2, the cell cycle arrests at the G_2/M checkpoint.

DNA damage also triggers activation of another pathway, a pathway for **apoptosis**, or **programmed cell death**. When the apoptotic pathway is activated, a cascade of proteolysis is initiated that culminates in cell suicide. The proteases involved are called *caspases*. Their activation ultimately results in destruction of the cellular DNA, internal organelles, and the actin cytoskeleton (**FIGURE 13.11**), and it is accompanied by nuclear condensation and usually followed by engulfment of the cellular remnants by phagocytes. The p53 transcription factor also activates the apoptotic pathway by activating transcription of *Bax* and *Apaf1* and inhibiting synthesis of Bcl2. As illustrated in **FIGURE 13.12**, Bcl2 is an inhibitor of apoptosis that normally forms a dimer with Bax protein. When p53 activates transcription of *Bax* and inhibits synthesis of Bcl2, the balance is tilted in favor of Bax homodimers and against the Bax/Bcl2 heterodimer, which promotes apoptosis and self-destruction of the cell. On the other hand, activation of **oncogenes**, which are genes associated with cancers, can increase the level of activated (phosphorylated) Bcl2, which prevents apoptosis and allows the affected cells to grow and divide indefinitely. Cellular immortality does not always follow oncogene activation, because in some cases it activates the apoptotic pathway, possibly through an oncogene-sensing function of p53. In all of its functions, p53 acts to protect the integrity of the genome with respect to nucleotide sequence and strand integrity and with respect to euploidy. Instability in the genome, unbalanced genomes, and damaged DNA pose a hazard to the organism. Apoptosis is a mechanism for killing such damaged—and thus dangerous—cells.

To gain an appreciation of the importance of the DNA checkpoint, look again at Figure 13.10 and consider what would happen if part or all of this failsafe mechanism should malfunction. Suppose, for example, that cells lacked p21 protein but retained the rest of the mechanism. If DNA damage were properly sensed, and p53 functioned as usual to increase levels of 14-3-3σ and GADD45, cells would accumulate in G_2 and be unable to undergo mitosis. However, DNA would continue to be synthesized. In addition, lacking p21 protein, the cells would have a defective G_1/S checkpoint and so could embark on additional rounds of DNA synthesis. The expected result of p21 malfunction would therefore be polyploid cells. This, in fact, is the phenotype of cells that are mutant for *p21*.

TABLE 13.1 Products of Genes Transcriptionally Activated by p53

Gene product	Function
p21	Inhibits several cyclin-dependent kinases; arrests cells at G1/S boundary.
14-3-3σ	Predicted to bind to and sequester phosphorylated Cdc25C phosphatase in the cytoplasm, which prevents Cdc25C from activating the cyclin B–Cdc2 kinase; arrests cells at the G2/M boundary.
GADD45	Binds to proliferating cell nuclear antigen (PCNA), blocking its role as a processivity factor for DNA polymerase and hence blocking DNA replication; functions directly in DNA repair.
Bax	Acts as a positive regulator of apoptosis (programmed cell death).
Apaf1	Scaffold protein that, when activated by cytochrome c, oligomerizes caspases into the complex that promotes apoptosis.
Maspin	Acts as an inhibitor of serine proteases, and is an inhibitor of angiogenesis (formation of blood vessels) and metastasis.
miRNA34a, b/c	MicroRNAs that induce senescence and apoptosis.

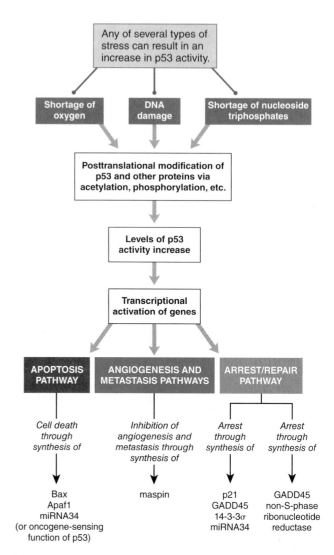

FIGURE 13.9 Downstream events triggered by p53 include transcriptional activation of the genes for p21, GADD45, 14-3-3σ, Bax, maspin, and Apaf1. Activation of the apoptosis pathway by p53 follows transcriptional activation of Bax and, possibly, direct sensing of activated oncogenes (cancer-causing genes) of either cellular or viral origin.

To take another example, consider the consequence of loss of p53 function. Even if DNA damage were detected, the cell would be unable to mount a response—it would be unable to buy time to repair the DNA damage. The cells could initiate new rounds of synthesis with damaged chromosomes. Such synthesis would not only result in an increased frequency of mutation but would also permit gene amplification. Amplification of genes encoding cyclin D or Cdk4 would allow cells to escape the normal controls on DNA synthesis and proliferation. Cells already in S or G₂ would enter mitosis with damaged chromosomes, because not enough time would have elapsed for repair of lesions. In addition, the organism would have lost its ultimate protection against such damaged cells, apoptosis. The absence of p53 means that transcription of *Bax* and *Apaf1* would not be increased and that the apoptotic pathway would not be turned on; the normal balance between Bax and Bcl2 would be maintained, ensuring the survival of these damaged cells and thus putting the organism at risk. If the description of these runaway cells reminds you of the unchecked proliferation of cancer cells, this is because cancer cells *become* cancer cells by subverting the checkpoint mechanisms.

STOP & THINK 13.1

The metallothionein promoter can be induced to increase transcription by 10 fold or more by adding copper ions to the growth medium. Suppose you use CRISPR-Cas9 to edit the genome of cultured cells. In the edited cell line, you put the *Bax* gene under the control of the metallothionein promoter. How would you expect cells of this line to react when copper ions are added to the medium?

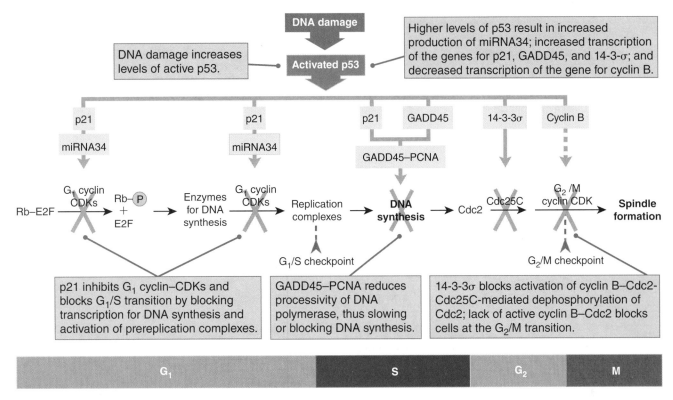

FIGURE 13.10   Role of activated p53 protein in the DNA damage checkpoint.

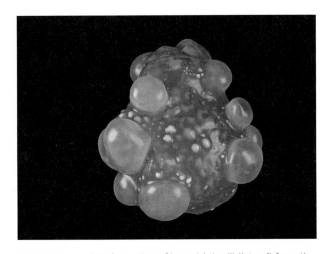

FIGURE 13.11   The formation of large blebs ("blisters") from the cell membrane is a characteristic of cells undergoing apoptosis.

© Science artwork/Science Photo Library/Getty images.

The centrosome duplication checkpoint and the spindle checkpoint function to maintain the normal complement of chromosomes.

Recall that the *centrosome* is the cellular organelle around which the bipolar spindle is organized. A **centrosome duplication checkpoint** monitors formation of the spindle. It seems to be coordinated with entry into mitosis, because activation of cyclin B–Cdc2 kinase is correlated with centrosome duplication and formation

of the spindle. In some organisms, the centrosome duplication checkpoint may also be coordinated with the spindle checkpoint and the exit from mitosis.

The **spindle assembly checkpoint** monitors assembly of the spindle and attachment of the kinetochores to the spindle. (The *kinetochore* is the spindle-fiber attachment site on the chromosome.) Improper or incomplete spindle assembly triggers a block in the separation of the sister chromatids by preventing activation of the anaphase-promoting complex needed for entry into anaphase.

Cells can detect a single unattached or misattached chromosome and delay anaphase. Studies on insect and mammalian chromosomes suggest that the absence of tension at the kinetochore is the initiating signal for cell-cycle arrest. Tension on the kinetochore is related to the level of phosphorylation at kinetochores; unattached kinetochores have relatively more of a phosphorylated protein than do attached kinetochores. If all chromosomes form stable, bipolar attachments to spindle microtubules at their kinetochores, then the anaphase-promoting complex is activated. As illustrated in **FIGURE 13.13**, when an unattached kinetochore is detected, the Bub, Mad, and Mps1 proteins act to block the onset of anaphase, apparently by inhibiting the protein Cdc20. Because Cdc20 is not activated, the anaphase-promoting complex is not activated, and the cells remain at metaphase until all chromosomes form stable, bipolar attachments to spindle microtubules.

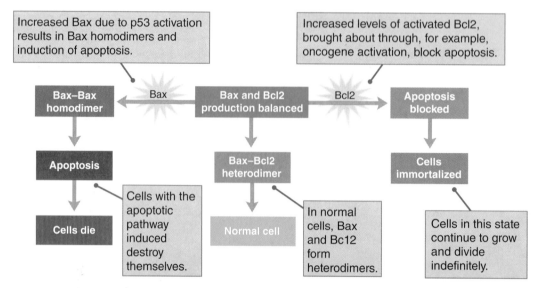

FIGURE 13.12 Mechanism by which Bax protein and Bcl2 protein interact to regulate apoptosis.

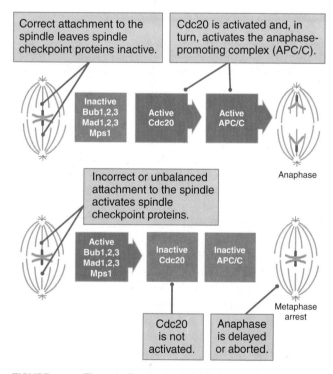

FIGURE 13.13 The spindle checkpoint. All chromosomes must form stable, bipolar attachments to spindle microtubules to activate the anaphase-promoting complex, which promotes the onset of anaphase and separation of the sister chromatids.

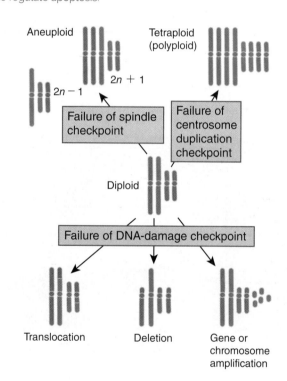

FIGURE 13.14 Contributions of checkpoint failures to genetic instability.

Failure of any checkpoint results in genetic instability. The particular kinds of genomic instability associated with defects in the three checkpoints we have discussed are summarized in **FIGURE 13.14**. Malfunctioning of the spindle itself can lead to aneuploidy, whereas a failure to duplicate a spindle pole can lead to polyploidy. Defects in the DNA damage checkpoints can result in chromosomal aberrations of various kinds, including translocations, deletions, and amplification of genes or subregions of chromosomes.

The amplified genes may be found as tandem repeats within a chromosome or as extrachromosomal circles lacking a centromere and telomeres.

13.3 Cancer cells have a small number of mutations that prevent normal checkpoint function.

Cancer is not one disease but rather many diseases that share similar cellular attributes. All cancer cells show uncontrolled growth as a result of mutations that affect a relatively small number of genes. It is a disease of somatic cells. About 1 percent of cancer cases are **familial**, which means there is clear evidence for segregation of a gene in the pedigree that predisposes cells in affected individuals to progress to the cancerous state. The other 99 percent of cases are called **sporadic**, which in this context means *not familial*, and are the result of genetic changes in somatic cells. **FIGURE 13.15** highlights six attributes of cancer cells that are not found in normal cells:

- Loss of growth-factor dependence

- Insensitivity to anti-growth signals

- Evasion of apoptosis

- Immortality (no cell senescence)

- Ability to metastasize and invade other tissues

- Sustained angiogenesis (formation of blood vessels)

Within an organism, tumor cells are *clonal*, which means that they are descendants from a single ancestral cell that became cancerous. This conclusion is based in part on the observation that tumor cells in a female express the genes in only one of the X chromosomes, as a result of the normal inactivation of the other X chromosome. If the tumor cells were not clonal, gene expression from both X chromosomes would be expected, reflecting the random inactivation of one X chromosome in each cell lineage. The finding of expression from a single X chromosome must therefore reflect clonality.

Luckily for most people, the conversion of normal somatic cells into cancer cells is a process that requires multiple mutational steps. An important contributor to cancer conversion is genetic instability in the cell population that serves as the precursor to cancer cells. This genetic instability may occur at the level of nucleotide sequences in the DNA or at the level of chromosome structure or number. The genetic instability is manifested as an increased number of mutations, gene amplification, chromosomal rearrangements, or aneuploidy. The mutations result in a cell population that is genetically heterogeneous. In such a mixed population, any cell that has a proliferative advantage will contribute a greater fraction of descendants to the future cell population than will its neighbors, and because of this advantage, its clone expands at the expense of others. Subsequent mutations in the descendants of this cell, and further clonal expansions of the derivative cells, can give rise to a clone of cells with the proliferative capacity typical of cancer cells.

Many cancers are the result of alterations in cell-cycle control, particularly in control of the G_1-to-S transition, and in the G_1/S checkpoint associated with this transition. These alterations also affect apoptosis through their interactions with p53. **FIGURE 13.16** summarizes the key elements of the regulatory circuitry that governs the G_1/S transition, including the proteins that promote cell-cycle progression, that activate the checkpoint, and that govern the apoptotic pathway. Tumor cells commonly show altered expression or inactivation of function of one or more of these genes. Some genes for which alterations have been detected in cancer cells are listed in **TABLE 13.2**.

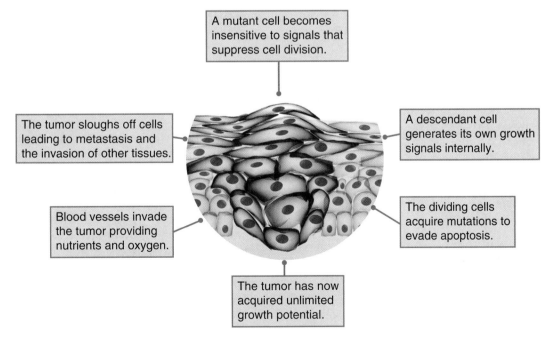

A mutant cell becomes insensitive to signals that suppress cell division.

A descendant cell generates its own growth signals internally.

The dividing cells acquire mutations to evade apoptosis.

The tumor has now acquired unlimited growth potential.

Blood vessels invade the tumor providing nutrients and oxygen.

The tumor sloughs off cells leading to metastasis and the invasion of other tissues.

FIGURE 13.15 Capabilities acquired by cancer cells.

Photo © Inbevel/Shutterstock.

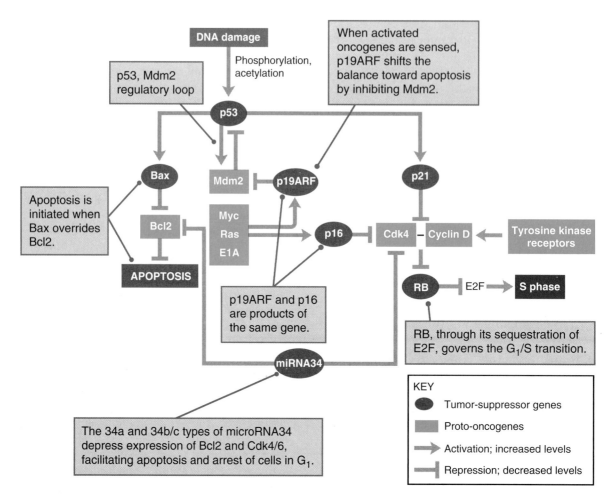

FIGURE 13.16 Interactions between the p53 pathway and the RB (retinoblastoma protein) pathway in controlling apoptosis or DNA synthesis.

The major mutational targets for the multistep cancer progression are of two types: *proto-oncogenes* and *tumor-suppressor genes*.

> ## KEY CONCEPT
>
> The normal function of proto-oncogenes is to promote cell division or to prevent apoptosis; the normal function of tumor-suppressor genes is to prevent cell division or to promote apoptosis.

These types of genes are discussed in the following sections.

Proto-oncogenes normally function to promote cell proliferation or to prevent apoptosis.

Oncogenes are gain-of-function mutations associated with cancer progression. They are derived from normal cellular genes called **proto-oncogenes**. Oncogenes are gain-of-function mutations because they

improperly enhance the expression of genes that promote cell proliferation or inhibit apoptosis. In Figure 13.21, the proteins enclosed in rectangles are the products of proto-oncogenes, which when mutated can give rise to oncogenes. Examples of oncogenes found mutated in some cancers are listed in Table 13.2. Three of these warrant a closer look.

Growth-factor receptors and Ras Cellular growth factors stimulate growth by binding to a growth-factor receptor at the membrane. The binding activates a signal transduction pathway that acts through Ras, cyclin D, and its partner CDKs. The gene that encodes the receptor for epidermal growth factor (EGFR) or the genes for fibroblast growth-factor receptor (FGFR) are amplified in many cancers.

Activated tyrosine kinase receptors such as EGFR and FGFR activate a signal transduction pathway that activates a small signaling protein called a **G protein**, as illustrated in **FIGURE 13.17** for the G protein Ras. Ras is usually bound to GDP, and in this state it is inactive. Receptor activation results in downstream activation of a protein that stimulates the exchange of GTP

TABLE 13.2 Cell-Cycle Regulatory Genes Affected in Tumors

Protein oncogenes	Alteration	Consequence
Cyclin D	Amplification or overexpression	Promotes entry into S phase.
Cdk4	Amplification	Promotes entry into S phase.
Cdk4	Mutation	Cdk4 resistant to inhibition by p16; promotes entry into S phase.
EGFR (epidermal growth-factor receptor, a tyrosine kinase receptor)	Amplification	Promotes proliferation by constitutive activation of pathway from growth-factor receptor.
FGFR (fibroblast growth-factor receptor)	Amplification	Promotes proliferation by constitutive activation of pathway from growth-factor receptor.
Ras	Amplification	Promotes proliferation by constitutively transmitting growth signal.
Ras	Mutation	Inactivates GTPase activity; constitutive activation of pathway from growth-factor receptor.
Bcl2	Overexpression by translocation next to strong enhancer	Blocks apoptosis.
Mdm2	Amplification	Mimics loss of p53 with loss of G_1/S, S, and G_2/M checkpoint functions.
Telomerase	Overexpression	Cells no longer undergo senescence.

Tumor-suppressor genes	Alteration	Consequence
p53	Mutation	Loss of G_1/S, S, and G_2/M checkpoint functions.
p21	Mutation	Loss of G_1/S and S checkpoint functions.
RB	Mutation	Promotes proliferation; E2F uninhibited.
Bax	Mutation	Failure to promote apoptosis of damaged cells.
Bub1p	Mutation	Loss of spindle assembly checkpoint function.
E-cadherin	Mutation	Loss of contact inhibition; tissue invasion and metastasis.

to replace GDP in the Ras protein, and the Ras protein is active when it is bound to GTP. The Ras–GTP, in turn, activates another downstream protein, propagating the growth signal. The activation of Ras as Ras–GTP is transient. Yet another protein (GAP) activates the intrinsic GTPase activity of Ras, so Ras hydrolyzes the GTP and returns itself to the inactive state of Ras–GDP. Mutations in the *RAS* gene occur that inactivate its GTPase function; in these mutants, the GTPase activity cannot be activated by GAP, and RAS remains in its active form of Ras–GTP, whether or not the cell is receiving signals from growth-factor receptors. The signal for cellular growth is transmitted constitutively, and therefore unrestrained growth and division take place. This loss of growth-factor dependence is one of the hallmarks of cancer cells.

EGFR and FGFR   Amplification or overexpression of EGFR or FGFR results in self-activation of the receptor and transmission of a constitutive growth signal acting

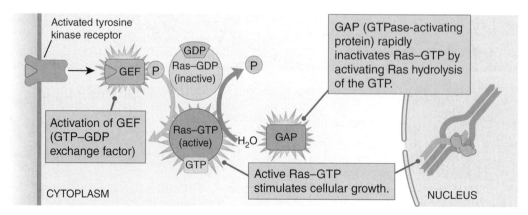

FIGURE 13.17 Function of the Ras protein, which acts as a switch in stimulating cellular growth in the presence of growth factors. Ras is the product of a proto-oncogene. Certain mutant Ras proteins, such as a G19V (valine replacement for glycine at position 19), lack GTPase activity and remain in the form of Ras–GTP; hence a growth-promoting signal is present even in the absence of growth factors.

through the Ras pathway. Overexpression of Ras also leads to enhanced signal transduction and renewed cycles of proliferation.

Telomerase Normal cells in culture cease to divide after about 50–70 doublings depending on the growth conditions, which is a process called *senescence*. Cancer cells in culture divide indefinitely, however—they are immortal. The senescent behavior of normal cells is associated with a loss of telomerase activity. The telomeres are no longer elongated, which contributes to the onset of senescence and cell death. Cancer cells have high levels of telomerase, which help to protect them from senescence, making them immortal.

Tumor-suppressor genes normally act to inhibit cell proliferation or to promote apoptosis.

Tumor-suppressor genes are genes that normally negatively control cell proliferation or that activate the apoptotic pathway. Loss-of-function mutations in tumor-suppressor genes contribute to cancer progression. Examples of tumor-suppressor genes found mutated in some cancers are listed in Table 13.2. Three of these deserve brief further discussion.

p53 Loss of function of p53 results in acquisition of two characteristics of cancer cells: insensitivity to anti-growth signals and evasion of apoptosis. Loss of function of p53 eliminates the DNA checkpoint that monitors DNA damage in G_1 and S. In the absence of functional p53, the proteins and microRNAs responsible for arresting cells in G_1 or G_2 are not synthesized in response to DNA damage. There is consequently no block to a cell's proceeding into S phase or into M phase even if it has damaged chromosomes, altered chromosome number, or amplified genes. In addition, loss of p53 function costs the organism its ultimate defense against aberrant cells, because DNA damage

is no longer able to trigger enhanced expression of Bax and Apaf1 and reduced expression of Bcl2. The result is less self-destruction of aberrant cells; the damaged cells survive and proliferate. Furthermore, their genetic instability increases the probability of additional genetic changes and thus progression toward the cancerous state. Given this scenario, and the key role of p53 in protecting the cell against the consequences of DNA damage, it is not surprising that p53 proves to be nonfunctional in more than half of all cancers.

RB The retinoblastoma protein controls the transition from G_1 to S phase by controlling the activity of the transcription factor E2F. Loss of RB function frees E2F to initiate transcription of the enzymes for DNA synthesis at all times in the cell cycle; hence excessive rounds of DNA synthesis are continuously being initiated. Tumor cells are nearly always altered for RB, p16, cyclin D, or Cdk4, but rarely for more than one of them. This finding suggests that loss of function of either p16 or RB is equivalent to overexpression of either cyclin D or Cdk4. Any one of these defects leads to the same result: The cells lose control of the G_1/S transition and embark upon unscheduled rounds of DNA synthesis, and they become insensitive to anti-growth signals.

E-cadherin In normal cells, cell-to-cell contact inhibits further growth and division, a process called **contact inhibition**. Cancer cells have lost contact inhibition. They continue to grow and divide and even pile up on one another. E-cadherin is ubiquitously expressed on epithelial cells. Interaction between molecules on adjacent cells allows transmission of anti-growth factors and other signals via the cytoplasmic contacts. E-cadherin acts to suppress invasion and metastasis by epithelial cells, and inactivation is a key step in acquisition of the capability for metastasis.

13.4   Mutations that predispose to cancer can be inherited through the germ line.

As we have noted, approximately 99 percent of cancers are sporadic: All of the genetic changes resulting in the cancer take place in somatic cells. Furthermore, the change from a normal cell into a tumor cell is progressive. It goes step by step as each new somatic mutation along the way compromises a different mechanism of cell-cycle control.

Cancer initiation and progression occur through mutations that allow affected cells to evade normal cell-cycle checkpoints.

A small minority of cancers—approximately 1 percent—are familial. In these cases, one of the mutations associated with cancer progression is inherited through the germ line. The presence of this mutation predisposes the individual to cancer because it reduces the number of additional somatic mutations necessary for a precancerous cell to progress to malignancy. FIGURE 13.18 illustrates some of the genetic mutations and changes in cell morphology that take place

in the progression of a type of familial colon cancer called *adenomatous polyposis*. In this case, the inherited mutation is in the gene *APC* in chromosome 5, which is a tumor-suppressor gene whose normal function is to transduce the signal of contact inhibition into the cell to inhibit further growth and division. Subsequent mutations in the progression to malignancy include an oncogenic mutation in a *Ras* gene and mutation or loss of the gene encoding p53. This scenario is only one possible route of progression. Different tumors may progress by different pathways, depending on what mutations occur and in what order.

Familial adenomatous polyposis of the colon and some other important familial cancer syndromes are identified in **TABLE 13.3**. The Li–Fraumeni syndrome, familial retinoblastoma, and familial melanoma are all associated with germ-line mutations in genes that are also found to be mutant in some sporadic tumors. The genes affected in these syndromes are the *p53* gene, the *RB1* gene, and the *p16* gene, respectively. A pedigree of a cancer-prone family segregating for a mutation in the *p53* gene (Li–Fraumeni syndrome) is given in FIGURE 13.19. This syndrome shows clear autosomal dominant inheritance. However, the affected individuals show a range of different tumors and often have more than one cancer, including osteosarcoma, leukemia, breast cancer, lung cancer, soft-tissue sarcoma,

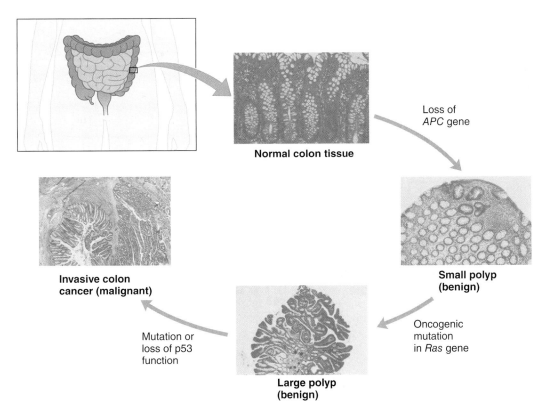

FIGURE 13.18   Somatic mutations and cell morphology in the progression of invasive colon cancer.

Photos courtesy of Kathleen R. Cho, University of Michigan Medical School.

TABLE 13.3 Inherited Cancer Syndromes

Syndrome	Primary tumors	Associated tumors	Chromosome	Gene	Proposed function
Li–Fraumeni syndrome	Sarcomas, breast cancer	Brain tumors, leukemia	17p13	p53	Transcription factor
Familial retinoblastoma	Retinoblastoma	Osteosarcoma	13q14	RB1	Cell-cycle regulator
Familial melanoma	Melanoma	Pancreatic cancer	9p21	p16	Inhibitor of Cdk4 and Cdk6
Hereditary nonpolyposis colorectal cancer (HNPCC)	Colorectal cancer	Ovarian cancer, glioblastoma	2p22 3p21 2q32 7p22	MSH2 MLH1 PMS1 PMS2	DNA-mismatch repair
Familial breast cancer	Breast cancer	Ovarian cancer	17q21	BRCA1	Repair of DNA double-strand breaks
Familial adenomatous polyposis of the colon	Colorectal cancer	Other gastrointestinal tumors	5q21	APC	Regulation of β-catenin
Xeroderma pigmentosum	Skin cancer		Several complementation groups	XPB XPD XPA	DNA-repair helicases, nucleotide excision repair

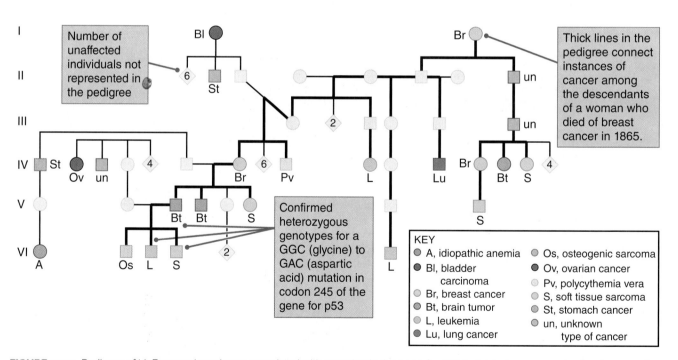

FIGURE 13.19 Pedigree of Li–Fraumeni syndrome associated with mutation in the gene for p53. Individuals are prone to develop any of a variety of cancers.

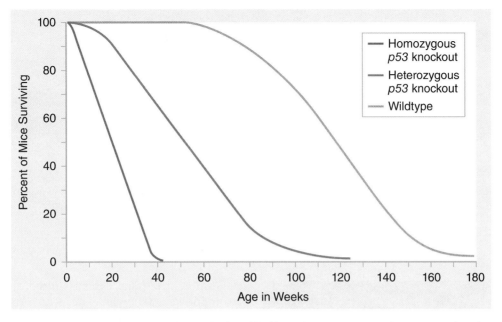

FIGURE 13.20 Survivorship curves for genetically engineered mice with either a heterozygous or a homozygous knockout (complete loss of function) mutation in the *p53* gene.

and brain tumors. A large fraction of Li–Fraumeni families show segregation for a mutation in the *p53* gene. For this family, affected members are heterozygous.

A situation analogous to the human Li–Fraumeni syndrome has been created in mice by experimental knockout (loss of function) of the *p53* gene via the germ-line transformation methods discussed in Chapter 11. Animals heterozygous or homozygous for the *p53* knockouts were compared to wildtype mice (**FIGURE 13.20**). Normal mice do not develop tumors; 50 percent of them live to the age of 120 weeks, and about 5 percent live for at least 160 weeks. In contrast, half of the heterozygous *p53* knockout mice are dead by 55 weeks, and almost none of them live as long as 120 weeks. Furthermore, 50 percent of these mice develop tumors (mostly osteosarcomas) by 18 months of age. The homozygous knockout mice fare even worse: 75 percent develop tumors by 6 months, 50 percent are dead by 20 weeks, and all are dead by 40 weeks. In the homozygous genotype, the tumors are lymphomas rather than osteosarcomas, possibly because normal development of the immune system requires massive apoptosis within the thymus gland, and in the p53 homozygous knockouts, apoptosis is abolished because Bax is not up-regulated as in normal mice and p53 knockout heterozygotes.

The fact that animals heterozygous for the *p53* knockout are severely affected for both longevity and incidence of tumors does not necessarily imply that the *p53* mutation is dominant at the level of the individual cell. In the case of the *p53* knockout mutation, the effects of the mutation are manifested only in somatic cells that have become homozygous for the knockout mutation or in those that have lost (or have undergone somatic mutation at) the wildtype allele in the homologous chromosome. In a heterozygous animal, only one copy of the wildtype *p53* gene is present to protect the cell; inactivation of the lone wildtype allele disables the checkpoints that depend on the p53 protein. Because there are so many somatic cells in which such a rare aberration can occur, it is nearly certain that inactivation or loss of *p53* will take place somewhere in the heterozygous organism, initiating the sequence of mutations that results in cancer.

Most mutations in the p53 gene are *dominant-negative mutations*, however. They have an amino acid replacement that inactivates the protein, and they show dominance at both the organismic and cellular levels. The reason for the dominance is depicted in **FIGURE 13.21**. The p53 protein functions as a tetramer. In the heterozygous genotype for a dominant-negative mutation, 15 of the 16 possible types of tetramers contain at least one mutant subunit. When there is an amino acid replacement in which any mutant subunit "poisons" the activity of the whole tetramer, the mutation is dominant because the remaining 1/16 of the tetramers with normal activity are not sufficient for normal function.

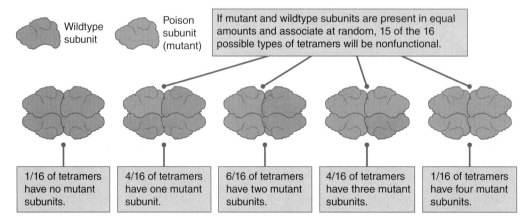

FIGURE 13.21 Consequences for p53 function of a dominant-negative mutation in the *p53* gene.

STOP & THINK 13.2

Consider a mutation that encodes a poison subunit in a protein whose functional form includes three subunits (that is, a trimeric protein). In a heterozygous genotype, assuming that the wildtype and mutant subunits are equally abundant and associate at random, what fraction of the trimeric molecules is expected to be functional?

Retinoblastoma is an inherited cancer syndrome associated with loss of heterozygosity in the tumor cells.

In the cell cycle, the cells monitor their internal and external conditions. Until the conditions are suitable to initiate a division cycle, the cells accumulate at a point in G_1 called the G_1 restriction point, or *start*. In animal cells, a protein called the retinoblastoma (RB) protein holds cells at the restriction point by binding to and sequestering the transcription factor E2F, which is needed for further progression. The RB protein was first identified in human pedigrees because mutant forms of the protein are associated with the formation of malignant tumors in the retina, which can require surgical removal of the eyes.

The idea that loss of the wildtype allele of a tumor-suppressor gene might be the triggering event at the cellular level for tumors in heterozygous genotypes was first suggested from studies on retinoblastoma by Alfred Knudson in 1971, long before the function of the *RB* gene was identified. Knudson noted that sporadic cases of retinoblastoma usually had a single tumor, whereas familial retinoblastoma cases usually had bilateral tumors and more than one tumor. On the basis of a statistical analysis of the time of tumor diagnosis, he suggested that genesis of

a tumor in familial cases required a "single hit" in a somatic cell, whereas genesis of a tumor in sporadic cases required "two hits," which would happen only rarely. The two-hit model would also explain why sporadic cases rarely have a second tumor: Each individual "hit" is a rare event.

Retinoblastoma, like the *p53* deficiency, is inherited in pedigrees as a simple Mendelian dominant. But Knudson's hypothesis implied that even in familial cases, there must be another mutational event that triggers tumor development. Once the gene itself, called *RB1*, was located, analysis of genetic markers around the gene in tumor cells revealed that the triggering event is the loss of the wildtype *RB1* allele. Any of several mechanisms can uncover the mutant allele, including chromosome loss, mitotic recombination, deletion, and inactivating nucleotide substitutions. These can be distinguished from one another on the basis of the genetic markers that are lost or retained in the tumor cells; several of the mechanisms known to occur are diagrammed in **FIGURE 13.22**. At the organismic level, expression of the mutant gene is dominant. However, at the cellular level, expression of the mutant gene is recessive. Uncovering of the recessive allele by various mechanisms is called **loss of heterozygosity**.

The inherited cancer syndromes listed in Table 13.3 result from mutations in tumor-suppressor genes. All show autosomal dominant inheritance except for xeroderma pigmentosum, which is inherited as a recessive. In all cancer syndromes that show dominant inheritance, loss of heterozygosity is required for manifestation of the tumor phenotype. Expression of the mutant allele *predisposes* the cell to become cancerous but is not in itself sufficient for generation of a cancer cell. Tumor progression still occurs only when additional somatic mutations and clonal expansions take place. The germ-line mutations do not themselves cause cancer; they merely make it much more likely that the progression will occur, because the tumor, in effect, has been given a "head start."

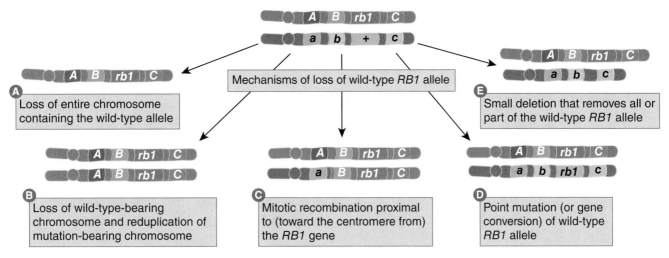

FIGURE 13.22 Genetic mechanisms for loss of heterozygosity of the wildtype *RB1* allele in patients with familial retinoblastoma.

STOP & THINK 13.3

Separation of DNA fragments by electrophoresis can be used to sort out different causes of loss of heterozygosity in retinoblastoma and other inherited cancer syndromes. The accompanying gel diagram shows size of an *Eco*RI restriction fragment from a nonmutant *RB1*⁺ allele and from a mutant *RB1*⁻ allele in which there is an internal 10-kb deletion in the *Eco*RI fragment. A scale showing the electrophoretic mobility of DNA fragments of various sizes appears at the right.

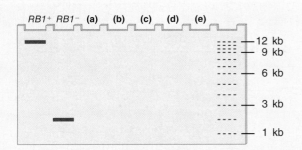

Diagram the pattern of bands that would be expected in each of the following: (a) normal retinal cells from a heterozygous *RB1*⁺/*RB1*⁻ individual, (b) cells of retinoblastoma tumors caused by loss of the wildtype *RB1*⁺− bearing chromosome; (c) cells of retinoblastoma tumors caused by mitotic recombination, (d) cells of retinoblastoma tumors caused by a new missense substitution in *RB1*⁺, and (e) cells of retinoblastoma tumors caused by a 6-kb deletion in the *Eco*RI fragment.

Some inherited cancer syndromes result from defects in processes of DNA repair.

Genetic instability clearly contributes to the origin of tumor cells. We know from studies on yeast and bacteria that cells have extensive mechanisms for repairing DNA lesions. Defects in these processes result in greatly elevated mutation rates and genetic instability. It was therefore not surprising when hereditary cancer syndromes that result from inherited defects in DNA repair were discovered.

Several of the inherited cancer syndromes listed in Table 13.3 are the consequence of defects in DNA repair. Defects in any of four genes that encode proteins involved in DNA mismatch repair cause hereditary nonpolyposis colorectal cancer, which shows autosomal dominant inheritance. Mutant cells have higher mutation rates, which promote progression toward the cancerous state. Inherited breast cancer syndromes prove to be associated with mutations in either of two genes, *BRCA1* or *BRCA2*, which are involved in repair of double-strand breaks. Inherited skin cancer syndromes are called xeroderma pigmentosum. Xeroderma pigmentosum cells are defective in nucleotide excision repair; they are unable to repair defects such as thymine dimers that are induced by ultraviolet light. Individuals with this syndrome are very sensitive to the ultraviolet light that is present in sunlight and emitted by fluorescent lights.

13.5 Acute leukemias are proliferative diseases of white blood cells and their precursors.

Acute *leukemia* is a malignant disease of the bone marrow, spleen, and lymph nodes associated with uncontrolled proliferation of white blood cells (**leukocytes**)

THE HUMAN CONNECTION

Two Hits, Two Errors

Alfred G. Knudson (1971)

M. D. Anderson Hospital, University of Texas, Houston, Texas

Mutation and Cancer: Statistical Study of Retinoblastoma

This landmark paper was a turning point in the thinking about genetics and cancer. Pedigree studies had already shown that there were hereditary predispositions to certain cancers, such as retinoblastoma. But sporadic cases also occurred, in which only one member of a kindred was affected. What is the relationship between these two forms? By an ingenious statistical analysis, Knudson showed that the sporadic cases exhibit two-hit kinetics as a function of age, indicating that two independent mutations in the same retinal cell are involved, whereas familial cases exhibit one-hit kinetics. The simplest interpretation is that in sporadic cases, the first mutation knocks out a key gene and the second knocks out its allele; in familial cases, the initial mutant allele is inherited, so only one mutation (in the remaining functional allele) is needed to cause the disease. At the level of the individual cell, therefore, a mutation in the retinoblastoma gene (*RB1*) is recessive, whereas at the familial level, the disease is inherited in a dominant fashion, because almost every person who inherits one mutant allele will undergo the second mutation in at least one retinal cell.

The hypothesis is developed that retinoblastoma is a cancer caused by two mutational events. In the dominantly inherited form, one mutation is inherited via the germinal cells and the second occurs in somatic cells. In the nonhereditary form, both mutations occur in somatic cells. . . . Several authors have concluded that retinoblastoma may be caused by either a germinal or a somatic mutation. . . . All bilateral cases [25–30 percent] should be counted as hereditary because the proportion of affected offspring closely approximates the 50 percent expected with dominant inheritance. . . .

> **Based upon observations on 48 cases of retinoblastoma and published reports, the hypothesis is developed that retinoblastoma is a cancer caused by two mutational events. In the dominantly inherited form, one mutation is inherited via the germinal cells and the second occurs in somatic cells. In the nonhereditary form, both mutations occur in somatic cells.**

Based on this hypothesis, Knudson reasoned that the timing of occurrence of inherited (bilateral) cases should be predictable—the somatic mutation referred to above will occur in cells at some rate over time, and thus the distribution of bilateral cases with time should be an exponential function, i.e., the fraction of the total cases that develop should be constant, as expressed in the relationship $dS/dt = -kS$ and $ln(S) = -kt$, where S is the fraction of survivors not yet diagnosed at time t, and dS is the change in this fraction in the interval dt. As shown in the figure, this is indeed the case. By contrast, the fractional

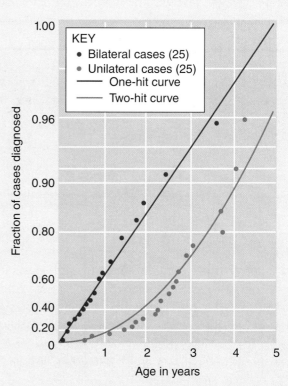

Data from Knudson, A. G. (1971). Mutation and Cancer: Statistical Study of Retinoblastoma. *Proceedings of the National Academy of Sciences of the United States of America*, 68(4), 820–823.

decrease in unilateral cases per unit time does not show this relationship.

The slope of the log-linear regression of the occurrence of bilateral cases on age at diagnosis can be used to determine the rate at which those somatic mutations occur.

. . . . The exponential decline in new hereditary cases with time reflects the occurrence of a

second event at a constant rate . . . of the order of 2×10^{-7} per year. . . . The two-mutation hypothesis is consistent with current thought . . . that the common cancers are produced by about 3–7 mutations. Interestingly, one of the lowest estimates [is] for brain tumors, which are, like retinoblastoma, derived from neural elements.

A. G. Knudson, *Proc. Natl. Acad. Sci.* USA 68 (1971): 820–823.

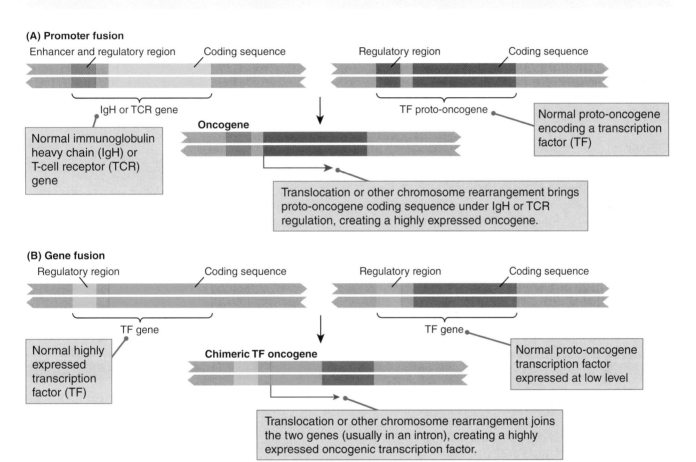

(A) Promoter fusion

Enhancer and regulatory region — Coding sequence

IgH or TCR gene

Normal immunoglobulin heavy chain (IgH) or T-cell receptor (TCR) gene

Oncogene

Regulatory region — Coding sequence

TF proto-oncogene

Normal proto-oncogene encoding a transcription factor (TF)

Translocation or other chromosome rearrangement brings proto-oncogene coding sequence under IgH or TCR regulation, creating a highly expressed oncogene.

(B) Gene fusion

Regulatory region — Coding sequence

TF gene

Normal highly expressed transcription factor (TF)

Chimeric TF oncogene

Regulatory region — Coding sequence

TF gene

Normal proto-oncogene transcription factor expressed at low level

Translocation or other chromosome rearrangement joins the two genes (usually in an intron), creating a highly expressed oncogenic transcription factor.

FIGURE 13.23 Translocations that aberrantly activate transcription factors in acute leukemias. (A) Promoter fusion. (B) Gene fusion.

and their precursors in the bone marrow. The initial genetic events that result in the acute leukemias are quite different from those indicated in the cancers examined previously. They do not arise as a consequence of alterations in cell cycle regulation or checkpoints, nor are they familial.

Some acute leukemias result from a chromosomal translocation that fuses a transcription factor with a leukocyte regulatory sequence.

Up to 65 percent of the acute leukemias arise as a consequence of chromosomal translocations involving

genes that encode transcription factors that play a role in blood-cell development (hematopoiesis). The translocations are of the two types illustrated in **FIGURE 13.23**. In a **promoter fusion** (part A), the coding region for a gene that encodes a transcription factor is translocated near an enhancer for an immunoglobulin heavy-chain gene or a T-cell receptor gene. The result is overexpression of the transcription factor, derangement of normal hematopoiesis, and overproduction of lymphocytes. One important aspect of normal hematopoiesis is the apoptotic destruction of progenitor cells that fail to rearrange their antigen-receptor genes correctly. Estimates are that 75 percent of B-cell and 95 percent of T-cell precursors self-destruct during normal development. *Bcl2* (*Bcl* stands for B-cell lymphoma) is an example of an oncogene in Table 13.3 that was originally discovered in a promoter fusion that placed *Bcl2* next to an immunoglobulin heavy-chain gene enhancer. Overexpression of Bcl2 blocks apoptosis by preventing formation of Bax homodimers (Figure 13.16). These "undeservedly alive" lymphocytes proliferate and come to dominate the population of white blood cells, but they are useless in fighting infection because they have defective immunoglobulin genes. In such cases the bone marrow becomes almost totally occupied by the cancerous leukocytes, leading to severe anemia and bleeding. Chemotherapy can sometimes be quite effective in treating acute leukemias in young children by activating the *p53* gene, provided that the aberrant leukocytes still carry the wildtype alleles encoding p53.

Other acute leukemias result from a chromosomal translocation that fuses two genes to create a novel chimeric gene.

The second type of translocation associated with acute leukemia is a **gene fusion** (part B of Figure 13.23). This type of rearrangement is found more frequently in acute leukemia than a promoter fusion. Most commonly, the translocation breakpoints occur in introns of genes that encode transcription factors in two different chromosomes. The result is a fusion gene called a **chimeric gene** composed of parts of the original genes. The chimeric gene produces a chimeric protein. In the case of acute leukemia, the chimeric protein is a transcription factor with an altered function that interferes with normal hematopoiesis. The uniqueness of these chimeric proteins, and the fact that they are present only in cancer cells and not in normal cells, make them inviting targets for drugs or chemotherapy. If one could successfully attack cells expressing the chimeric protein, one could selectively kill the cancer cells.

Chronic myeloid leukemia (CML) accounts for 15 to 20 percent of all cases of leukemia. The hallmark of CML is the presence of the *Philadelphia chromosome*, which results from a reciprocal translocation between chromosomes 9 and 22 in hematopoietic stem cells of the bone marrow. These are the cells that differentiate into various specialized types of cells that become part of the blood and immune system. The molecular result of the *t(9, 22)* translocation is replacement of the first exon of *c-abl* with sequences from the *bcr* gene, resulting in a Bcr-Abl fusion protein. Because the N-terminal region of Abl normally inhibits function of the catalytic domain, the fusion protein, having lost the inhibitory domain, is a constitutively active protein kinase.

CHAPTER SUMMARY

- Progression from one stage of the cell cycle to the next is controlled by protein complexes called cyclin-dependent kinase (CDK) complexes. These are made up of a cyclin component and a cyclin-dependent protein kinase. Protein degradation is also important, especially in later stages of the cell cycle.

- Checkpoints monitor a dividing cell for DNA damage, cellular defects, other abnormalities in the cell cycle, and cell size. Detection of abnormalities elicits a response that arrests the cell cycle, allowing time for repair of defects (or cell death—apoptosis) and ensuring that the phases remain in the correct order.

- Cancer cells show uncontrolled growth and proliferation and loss of contact inhibition.

- Progression from a normal to a cancerous state requires several genetic changes. Most cancers are sporadic (not inherited).

- A small proportion of cancers are associated with mutations transmitted through the germ line, which predispose the somatic cells of people carrying them to undergo cancer progression.

- The genetic changes that take place in cancer progression often involve defects or overexpression of genes that function in cell-cycle regulation or checkpoint control.

ISSUES AND IDEAS

- What role does the centriole (in some organisms, the centrosome) play in cell division?

- What molecular process that takes place in the nucleus defines the S period?

- In yeast, what phenotype defines a *cdc* mutant? For a temperature-sensitive *cdc* mutant grown at the restrictive temperature, why do cells arrest at a particular stage in the cell cycle?

- What is a cell-cycle checkpoint? Which checkpoints are emphasized in this chapter, and what does each "check" for?

- What is apoptosis, and what role does it play in preserving the integrity of the genome of a multicellular organism?

- Mutations in genes whose products are involved in DNA repair are often associated with an increased risk of cancer. What does this observation imply about the role of spontaneous mutations in the development of cancer?

- How can one reconcile the following statements: "Cancer is a genetic disease" and "Most cancers are sporadic (not inherited)."

- Distinguish between proto-oncogenes and tumor-suppressor genes; give one example of each. In which class of genes does a loss-of-function mutation predispose to cancer? A gain-of-function mutation? Explain your answer.

- What is "loss of heterozygosity," and how is this phenomenon related to the progression of some types of cancer?

SOLUTIONS: STEP BY STEP

PROBLEM 1 Indicate which of the checkpoints in the cell cycle discussed in this chapter would be expected to be the "first line of defense" in preventing changes in the genome due to:
(a) The effects of x rays and other forms of ionizing radiation.
(b) Endoreduplication as a cause of polyploidy.
(c) Nondisjunction as a cause of polysomy.

SOLUTION. (a) Because x rays and other forms of ionizing radiation produce single-stranded and double-stranded breaks in DNA molecules, the DNA damage checkpoint is primarily responsible for preventing mutations due to these agents, though not always successfully. (b) Endoreduplication occurs when a cell undergoes chromosome replication without chromosome separation, which represents a failure of the spindle to form properly. In this case the centrosome duplication checkpoint is the first line of defense. (c) Polysomy, in contrast to polyploidy, results from the misbehavior of a single chromosome at anaphase, rather than the whole set of chromosomes. The spindle assembly checkpoint, therefore, is expected to be the first line of defense against nondisjunction.

PROBLEM 2 Distinguish between an oncogene and a tumor-suppressor gene. Why are oncogene mutations associated with cancer gain-of-function mutations, whereas tumor-suppressor genes are associated with cancer loss-of-function mutations?

SOLUTION. Oncogenes are mutant genes whose normal counterparts (called proto-oncogenes) encode products that promote cell proliferation or inhibit apoptosis.

The products of tumor-suppressor genes inhibit cell proliferation or activate the apoptotic pathway. Oncogenes are gain-of-function mutations because their enhanced expression makes uncontrolled cell division possible or prevents the apoptosis pathway. Tumor-suppressor mutations associated with cancer are loss-of-function mutations because absence of a functional gene product fails in inhibiting cell division or in activating the apoptosis pathway.

PROBLEM 3 The accompanying pedigree includes individuals affected with retinoblastoma, and the diagram of the gel shows a restriction fragment from a number of alleles of the *RB1* gene, mutant forms of which are associated with this cancer. Three sizes of restriction fragments (a, b, and c) are observed. Individuals in generations I and II are old enough to have developed the cancer if they carry a mutant *RB1* allele, but the individuals in generation III are all too young to have developed the disease. Identify the high-risk individuals in generation III and those who are not at risk.

SOLUTION. First compare the pedigree with the gel patterns to identify which restriction fragment is associated with the mutant *RB1* allele in this pedigree. Because the father in generation I (individual 10) is affected, the mutant allele must be associated with either band a or band b. All of the affected individuals in generation II have band b, so this band must mark the mutant *RB1* allele. Therefore, the individuals at high risk in generation III are those with band b, whereas the others are not at risk. The high-risk individuals are numbers 1, 2, 8, 11, 13, and 18; those not at risk are individuals 3, 6, 7, 12, 16, and 17.

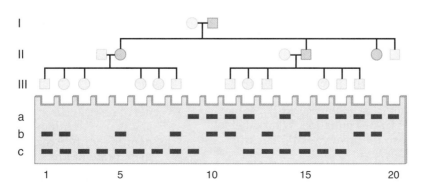

CONCEPTS IN ACTION: PROBLEMS FOR SOLUTION

13.1 Mutations in the *KRAS* gene are frequently found in colon cancer. These mutations are always missense mutations, primarily in codons 12 or 61. Mutations in the *APC* gene are also associated with colon cancer. In this case, however, nonsense and frameshift mutations are frequent and are scattered throughout most of the coding region. From this information, answer the following and explain.

    (a) Is *KRAS* an oncogene or a tumor-suppressor gene?

    (b) Is *APC* an oncogene or a tumor-suppressor gene?

13.2 If cancer is a "genetic disease," how can it be true that most cases are sporadic (that is, not familial)?

13.3 A human cell in culture is homozygous for a temperature-sensitive mutation in a gene necessary to repair double-stranded breaks in DNA. If cells were irradiated at the restrictive temperature, at what stage of the cell division cycle would you expect the mutant cells to accumulate?

13.4 What role does Ras–GTP play in intracellular signaling that makes *Ras* a proto-oncogene?

13.5 Bax is a protein that promotes apoptosis and is normally kept inactive in healthy cells by Bcl2, which forms heterodimers with Bax. Mutations in the gene *Bax* are found in cancer cells.

    (a) Is *Bax* an oncogene or a tumor-suppressor gene?

    (b) What type of mutations might be found in this gene in cancer cells?

    (c) Will these mutations be recessive or dominant?

13.6 The p53 protein is defective in more than half of all cancers. Why might this be expected?

13.7 What does it mean to say that mutations in the retinoblastoma gene *RB1* are "dominant at the organismic level but recessive at the cellular level"?

13.8 Draw a diagram showing how recombination between homologous chromosomes during mitosis can result in a cell lineage with loss of heterozygosity for a *p53* mutation. What other genes also lose heterozygosity in this process?

13.9 How does the normal retinoblastoma protein function to hold mammalian cells at the G1/S restriction point ("start")?

13.10 What cell-cycle checkpoints were highlighted in this chapter? What event or events cause each checkpoint to be activated to stop the cell division cycle?

13.11 Many types of cancer cells have defects in the G1/S checkpoint. These also tend to have abnormalities in chromosome number or structure. Why would chromosome abnormalities be expected in such cases?

13.12 DNA from cells of a patient with retinoblastoma was analyzed with regard to a particular restriction fragment in the *RB1* gene. The result from nontumor cells and the results from cells taken from a tumor in each eye are shown in the accompanying diagram.

    (a) Which band should be associated with the mutant allele and which with the nonmutant allele?

    (b) How is it possible for the bands from the tumor in the left eye to be different from those from the tumor in the right eye?

    (c) Explain how cells in the tumor in the right eye can have a "loss of heterozygosity" even though the bands are indistinguishable from those observed from nontumor cells.

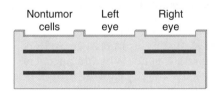

13.13 A woman has a mammogram (breast x ray) that reveals a suspicious lump of tissue. Cytological analysis of the lump reveals cells with a highly variable chromosome number and many chromosome rearrangements. What does this finding suggest about the malignant or nonmalignant nature of the suspicious growth? Explain your answer.

13.14 In familial retinoblastoma, there is an average of three retinal tumors per heterozygous carrier of the mutation. Assuming that the number of retinal cells at risk is 2×10^6 in each eye and that each tumor results from an independent loss of heterozygosity, what is the estimated rate of loss of heterozygosity per cell?

Should this be regarded as a "mutation rate"? Why or why not?

13.15 In an inherited cancer syndrome for pancreatic cancer there is an average of two independent tumors resulting from loss of heterozygosity, and 8×10^6 cells are at risk of causing this type of cancer. What is the penetrance of the familial form of the disease?

13.16 In patients with bilateral retinoblastoma, would the mechanism of loss of heterozygosity in tumors in different eyes be expected to be the same or different? Explain your answer.

13.17 The accompanying gel diagram shows the pattern of bands observed for restriction fragments located in or near the *p53* gene in which Li–Fraumeni syndrome is found. Shown are the bands observed in an affected mother, an unaffected father, and one affected son. The bands *AL* and *AS* correspond to long versus short restriction fragments from a region a few megabases on one side of *p53*, and the bands *BL* and *BS* correspond to long versus short restriction fragments from a region a few megabases on the other side of *p53*. The *p53*⁺ and *p53*⁻ bands are for the restriction fragment within the *p53* gene itself, where the wildtype *p53*⁺ allele yields the band denoted *p53*⁺ and the mutant *p53*⁻ allele yields the band denoted *p53*⁻ containing a 1-kb insertion. The order of the three regions along the chromosome arm is *centromere–A–p53–B–telomere*. Considering the mother and the father, determine which restriction fragments are linked with the *p53*⁺ allele in the son. If DNA from tumor cells in the son were assayed for these restriction fragments, what pattern of bands would be expected from each of the following cell types?

(a) Cells that had lost the *p53*⁺-bearing chromosome and had a reduplication of the *p53*⁻-bearing homolog

(b) Cells that had undergone mitotic recombination in the *centromere*–A interval

(c) Cells that had undergone mitotic recombination in the A–*p53* interval

(d) Cells that had a new nonsense mutation in the *p53*⁺ allele

(e) Cells that had undergone gene conversion of *p53*⁺ to *p53*⁻

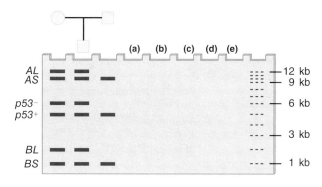

13.18 Mutagenesis of a *RAS* gene of budding yeast yields a temperature-sensitive conditional mutation.

(a) Would you expect a cell that is carrying a mutation that prevents Ras from exchanging GDP for GTP at the restrictive temperature to continue to divide at this temperature?

(b) Would the mutation be dominant or recessive?

(c) Would you expect a cell that is carrying a mutation that inactivates the GTPase activity of Ras at 36°C to continue to divide at the restrictive temperature?

(d) Would this mutation be dominant or recessive?

13.19 Human papilloma virus (HPV) is present in greater than 90 percent of cervical cancers. HPV encodes two proteins, E6 and E7, that are potent contributors to its tumorigenicity. E7 is known to disable RB; E6 binds to p53 and targets it for degradation. Discuss how these activities might contribute to the development of the cancerous state in infected cells.

13.20 The accompanying pedigree includes individuals affected with adenomatous polyposis, and the diagram of the gel shows a restriction fragment from a number of alleles of *APC*, mutant forms of which are associated with this cancer. Four sizes of restriction fragments (a through d) are observed. Individuals in generations I and II are old enough to have developed the cancer if they carry a mutant *APC* allele, but the individuals in generation III are all too young to have developed the disease. Identify the high-risk individuals in generation III and those who are not at risk. (Note that a mutant allele and a nonmutant allele can yield the same size restriction fragment.)

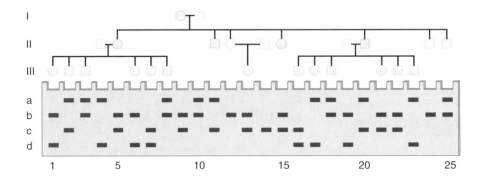

STOP & THINK ANSWERS

ANSWER TO STOP & THINK 13.1

Figure 13.12 implies that *Bax* overproduction will result in apoptosis; this is the expected result of adding copper ions to the medium.

ANSWER TO STOP & THINK 13.2

A functional trimer requires three wildtype subunits, hence the expected fraction of functional trimmers equals $1/2 \times 1/2 \times 1/2 = 1/8$.

ANSWER TO STOP & THINK 13.3

To deduce the expected pattern of bands in each case, first think through the consequences of each type of event with respect to the *RB1* alleles present in the cells with loss of heterozygosity. (a) Normal heterozygous cells will contain both DNA fragments. (b) Loss of the wildtype chromosome eliminates the *RB1*⁺ allele. (c) Mitotic recombination makes the *RB1*⁻ allele homozygous in the tumor cells. (d) A missense mutation in the *RB1*⁺ allele does not change the size of the *Eco*RI fragment (except in the extremely unlikely case that the mutation hits one of the *Eco*RI sites). (e) In the new *RB1* deletion mutation, the size of the *Eco*RI fragment will be 6 kb instead of 12 kb. Therefore, the expected gel patterns are as shown in the illustration.

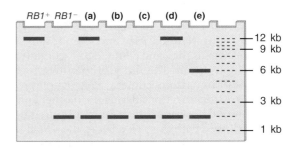

CHAPTER **14**

Population genetics is the application of genetic principles to entire populations of organisms.
© Hobbit/Shutterstock

Molecular Evolution and Population Genetics

LEARNING OBJECTIVES

- To distinguish between a gene tree and a species tree and state a reason why they may differ.

- To explain why some nucleotide sites in a genome may undergo evolutionary change much more rapidly than others.

- Given a set of genotype frequencies for two alleles of a single gene, to calculate the allele frequencies and, for any specified allele frequencies, deduce the expected genotype frequencies under random mating.

- To describe the principal effect of inbreeding for a rare recessive allele.

- To translate "mutation-selection balance" into everyday language.

- To define "random genetic drift" (also called "random failure to breed") and explain why its effect on allele frequency depends on the number of breeding individuals in the population.

- To summarize the evidence that all mitochondrial DNA sequences in modern humans derive from a common ancestor about 200,000 years ago.

Molecular genetics has brought about major changes in the way biologists study genetic differences within and among natural populations of organisms. Traditionally, the study of genetic differences among organisms required controlled matings and analysis of the progeny. This requirement made it almost impossible to study genetic differences between species, because all but the most closely related species either do not mate or they yield progeny that are inviable or sterile. But the discovery that DNA is the genetic material made it possible to compare corresponding genes even in distantly related species. As we shall see in this chapter, studies of DNA sequences from different species can reveal important information not only about how genes evolve, but also about the evolutionary relationships among the species. Comparative study of macromolecules within and among species constitutes the field of *molecular evolution*.

Molecular methods have also transformed the study of genetic variation within species. Traditionally, genetic differences among individuals of a species were undetectable unless they caused a difference in phenotype. The primacy of phenotypic differences severely limited the types of population studies that could be carried out, because most genetic differences among individuals in a population cause no detectable difference in any aspect of the phenotype. But the study of molecular markers revealed that natural populations contain abundant genetic variation at the molecular level. This type of genetic variation can be used to investigate population history, subdivision, and the genealogical relationships among individuals, as well as to identify the chromosomal locations of genetic risk factors for inherited diseases. The application of genetic principles to entire populations of organisms constitutes the subject of *population genetics*.

14.1 DNA and protein sequences contain information about the evolutionary relationships among species.

Macromolecules such as DNA, RNA, and protein are linear polymers of subunits. The specific sequence of subunits along each molecule determines its information content or function. With the vast outpouring of data from large-scale genomic sequencing, there is great interest in comparing the sequences of related molecules among species, motivated in part by the hope of correlating differences in sequence with differences in function, especially in proteins.

Although the sequences of macromolecules contain information about function, they also contain information about evolutionary history. Sequences

change through time even among macromolecules whose function remains identical. In fact, it is often difficult to distinguish which differences in sequence between species are important to the function of a molecule, and which differences have such small effects that they simply reflect changes that take place by chance over evolutionary time.

The study of how (and why) the sequences of macromolecules change through time constitutes **molecular evolution**. In this section, we consider several aspects of molecular evolution. We begin with **molecular phylogenetics**, in which genome sequences are analyzed in order to infer the evolutionary relationships among species, and we end with the origin of new genes.

The ancestral history of species is recorded in their genome sequences.

Charles Darwin was the first to realize that evolution naturally results in a treelike structure of relatedness among organisms. In an evolutionary tree, the tips of the branches represent extant species and the internal nodes represent extinct common ancestral species (**FIGURE 14.1**). In sexually reproducing eukaryotes, the criterion for assigning organisms to the same **species** is that they be able to mate and produce fully viable and fertile hybrid offspring. In asexual eukaryotes and prokaryotes, the matter is not so simple; however, a definition of species that suits our purposes is a group of organisms that are more closely related to one another than they are to members of other such groups. As it happens, the groups of organisms defined by reproductive compatibility in sexual eukaryotes are also more closely related to each other than to members of other such groups.

Relatedness among organisms is reflected most directly in the similarity of their genome sequences. Because genome sequences change through time, it follows that sequence differences accumulate through time. The accumulation of differences is the basis of sequence divergence between genomes, and the flip side of sequence divergence is sequence similarity.

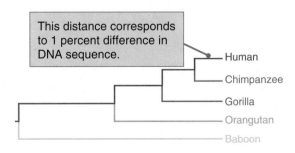

FIGURE 14.1 Evolutionary relations among higher primates based on similarity in DNA sequence. A divergence of 1 percent implies that 1 percent of the nucleotides differ between the genomic DNA sequences.

Two randomly chosen human genomes are identical at an average of about 99.9 percent of their nucleotide sites. A human genome and a chimpanzee genome are identical at about 98 percent of their nucleotide sites (Figure 13.1). Species that are more distantly related show greater sequence divergence.

Analyzing sequence divergence across the entire genome avoids three potential problems in making inferences from a single gene or small groups of genes. The first problem is that changes in sequence are a matter of chance, depending on which mutations take place and their likelihood of being fixed in the population. (A mutant allele is said to be **fixed** if it replaces all other alleles in the population.) Because new mutations occur at random times, any two molecules separated by the same length of time may differ at more or fewer sites purely by chance. In practice, this problem can be minimized by avoiding sequences that are so closely related that the expected number of differences between them is of the same magnitude as the random variation.

The second problem is that some mutations may increase the ability of the organism to survive and reproduce, and these mutations have a much better chance of being fixed in the population. This problem can be minimized by studying regions of DNA, RNA, or protein that are less likely to include favorable mutations, but it is often difficult to know where these regions are.

The third problem with reconstructing ancestral history from too few genes is the subject of the next section.

A gene tree is a diagram of the inferred ancestral history of a group of gene sequences.

A frequent goal of molecular evolution is to estimate the pattern of evolutionary relations among sequences of the same gene present in different species, which is called a **gene tree** because it is based on a single gene. It seems reasonable to suppose that the evolutionary relations among a set of genes from different species (the gene tree) must be the same as the evolutionary relations among the species themselves (the **species tree**), because the genes are present in the genomes of the species. But this is not the case. The gene tree is not necessarily the same as the species tree.

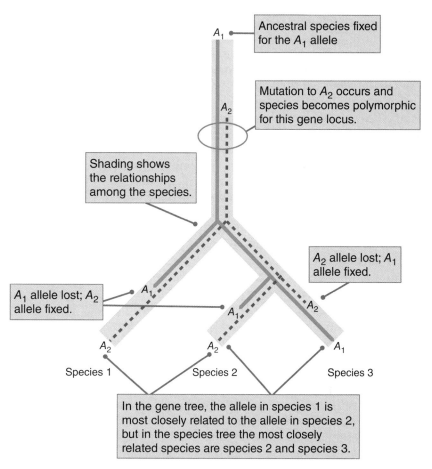

FIGURE 14.2 A gene tree may not coincide with a species tree because of the sorting of polymorphic alleles in the different lineages.

One way in which the gene tree can differ from the species tree is shown in **FIGURE 14.2**. The ancestral population is originally fixed for the A_1 allele (top), but then an A_2 mutation occurs and the population becomes polymorphic. The polymorphism is maintained even as species 1 splits off and as species 2 and 3 become separated. However, loss of one allele (and fixation of the other) eventually takes place, and in this example species 1 and 2 become fixed for allele A_2 whereas species 3 becomes fixed for allele A_1. This means that the gene tree would group the alleles from species 1 and 2 as being the most closely related, whereas the species tree shows that species 2 and 3 are actually the most closely related.

The situation becomes even more complicated because of recombination. When genes can undergo recombination, it implies that different genes can have different evolutionary histories. Because recombination within a gene can occur, it is even possible for different parts of the same gene to have different evolutionary histories. This is one reason why the noncombining part of the Y chromosome and the entirety of mitochondrial DNA are of such utility in tracing the recent history of human populations. However, these studies yield estimates of either the Y chromosome tree or the mitochondrial DNA tree, and it is important to bear in mind that the gene trees for nuclear genes will not necessarily be the same, and in fact may differ from one nuclear gene to the next.

On the other hand, for genes with polymorphisms that persist for relatively short times or for species that are sufficiently old, gene trees often do coincide with species trees. Discordance between gene trees and species trees is a potential problem primarily for genes that can maintain polymorphisms for long periods of time or for species that are closely related. The ancient polymorphism problem is exemplified by many genes that function in the immune system that are highly polymorphic and have been polymorphic for periods that are as long or longer than the time required for the formation of new species. In such cases, the allele sorting process in Figure 14.2 can result in discrepancies between the gene tree and the species tree. The young taxa problem is exemplified by the branching order between human, chimpanzee, and gorilla. The species are still so relatively young that polymorphisms in the ancestral populations have been sorted so that some genes support one branching order and other genes support another. In the human genome, for example, about two-thirds of the genes are most closely related to their chimpanzee counterpart and about one-third are most closely related to their gorilla counterpart. At one time, this discrepancy caused considerable confusion and debate; however, complete genome sequencing convincingly supports a branching order in which the gorilla lineage is the first to split off from the common ancestor.

Rates of evolution can differ dramatically from one protein to another.

By the *rate* of sequence evolution of a molecule, we mean the fraction of sites that undergo a change in some designated interval of time. For example, in the entire β globin molecule between mouse and human, the rate of sequence evolution is 1.23 amino acid replacements per amino acid site per billion years, or 1.23×10^{-9} amino acid replacements per amino acid site per year.

Different proteins (and sometimes different parts of proteins) evolve at very different rates. Molecules such as the histones H3 and H4 evolve very slowly. For example, among the 103 amino acids in histone H4, the molecules in rice and humans differ at only two sites. Other molecules evolve relatively rapidly. To illustrate, between mouse and humans the antiviral protein gamma interferon has evolved at a rate of about 5×10^{-9} amino acid replacements per amino acid site per year, a rate approximately 4 times faster than β globin and very much faster than histone H4 (there are no differences in histone H4 between mouse and human).

Although protein sequences evolve at very different rates, some proteins in some groups of organisms show a rough constancy in their rate of amino acid replacement over long periods of evolutionary time. The apparently constant rate of sequence change has been called a **molecular clock**, and affords a basis for attaching a time scale to a gene tree and therefore a time scale for the branching of species independent of the fossil record. There is an elegant theoretical argument that explains why a constant rate of sequence evolution might be expected, at least in the simplest cases. Consider a gene in a population of N diploid individuals, so that in the entire population there are $2N$ copies of the gene. Suppose that some new mutations are **selectively neutral**, which means that they have no effects on the ability of the organisms to survive and reproduce, and that the rate of mutation to selectively neutral alleles is μ per gene per generation. Then in any one generation, the expected number of new selectively neutral alleles is $2N\mu$. As time goes on, because the population is finite in size, some of the lineages of these genes will become extinct by chance and they will be replaced with other gene lineages. Eventually a time will come when all the gene lineages will have become extinct except one. The probability that any particular gene lineage replace all others is $1/(2N)$, and because there are $2N\mu$ new mutations, the expected number of new mutations (sequence changes) that become fixed in each generation is

$$\text{Rate of neutral evolution} = \frac{2N\mu}{2N} = \mu$$

The expected rate of neutral evolution is therefore equal to the rate of neutral mutation, which constitutes a sort of molecular clock whose ticks are mutations that become fixed.

In applying this model there are some important caveats. One warning is that the neutral mutation rate can differ from one gene to the next, according to what fraction of new mutations is neutral or nearly neutral. For genes that evolve extremely slow, like histone H4, most amino acid replacements are probably very deleterious, and so the neutral mutation rate will be very low. On the other hand, for genes that evolve extremely fast, such as gamma interferon, a significant fraction of the amino acid replacements that become fixed may be favorable mutations, which violate the assumption of neutrality. A second caveat is that the molecular clock is not like a timepiece that ticks at reproducible intervals. It is a random or stochastic clock, in which only the average interval between ticks is predictable. An analogy may be made with radioactive decay, which is a random but clocklike process, but even this analogy has the shortcoming that the random variation in a molecular clock is much larger, relative to the mean, than the random variation in an atomic clock. A third warning is that in many cases molecular clocklike behavior is not observed, because there are different rates of sequence evolution along different branches of the gene tree. As an example, in the lineages of humans and mice, the rate of sequence evolution along the branch leading to mice is about two times faster than that along the branch leading to humans, which is thought to be due to the shorter generation time of organisms in the mouse lineage, resulting in more generations along that lineage. The best methods for reconstructing gene trees take this source of variation into account.

Rates of evolution of nucleotide sites differ according to their function.

The general principles of the molecular evolution of protein sequences also apply to DNA sequences, but there are some important differences. Proteins consist of 20 amino acids, but DNA consists of only 4 nucleotides. The smaller number of subunits means that independent but identical changes at nucleotide sites are much more likely than at amino acid sites, and also that a DNA site that undergoes two sequential mutations is more likely to return to the original state than is an amino acid site. In analyzing DNA sequences, these kinds of events often require that multiple-hit corrections be made.

There are also different kinds of nucleotide sites depending on their position and function in the genome, and these evolve at different rates (**FIGURE 14.3**). Note that the vertical scale in Figure 14.3 is logarithmic and covers three orders of magnitude, so the differences are very large. On the right are shown the rates of synonymous substitution (red) and nonsynonymous substitution (blue) in each of the 43 genes. A **synonymous substitution** in a coding sequence does not result in an amino acid replacement. Synonymous sites are sites at which synonymous substitutions can occur, primarily at the third codon position. A **nonsynonymous substitution** does result in an amino acid replacement. Nonsynonymous nucleotide sites occur primarily at first and second codon positions. Reflecting the great variation in rate of amino acid replacement among different proteins, the rates of nonsynonymous substitution in Figure 14.3 are highly variable among genes. The rates of synonymous substitution in the same genes are much less variable and also much faster.

Plotted on the left in Figure 14.3 are the average rates of nucleotide substitution for different classes of DNA sequence. The fastest evolving DNA sequences are those of **pseudogenes**, which are duplicate genes that have lost their function because of mutation. Introns and fourfold degenerate sites also evolve

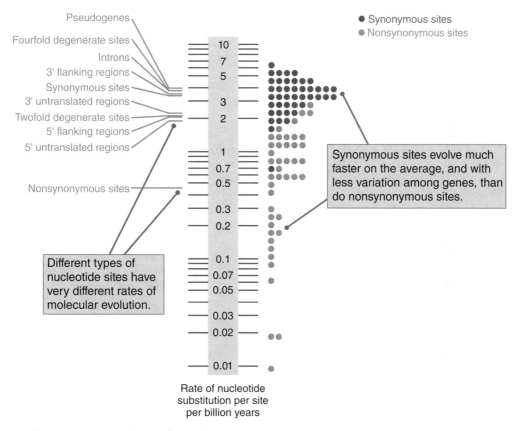

FIGURE 14.3 Rates of nucleotide substitution in mammalian genes. The left side shows the average rates for different classes of sequence; the right side shows the rates for synonymous and nonsynonymous sites in a sample of 43 genes.

very rapidly. (A *fourfold degenerate site* is a synonymous nucleotide site at which the same amino acid is specified whatever the identity of the nucleotide; a *twofold degenerate site* is one at which the encoded amino acid depends only on whether the nucleotide at the site is a pyrimidine or a purine.) The differences in the average rate of nucleotide substitution among the different classes of DNA sequence are thought to reflect differing tolerance for nucleotide substitutions. DNA sequences in which most nucleotide substitutions are deleterious are relatively intolerant to nucleotide substitutions, and the rate of nucleotide substitution is relatively low. Conversely, among sequences in which many nucleotide substitutions are equivalent or nearly equivalent in their effects on survival and reproduction, the rate of nucleotide substitution is relatively high. The high rate of nucleotide substitution in pseudogenes is understandable from this point of view, as are the high rates in fourfold degenerate sites and introns.

The K_a/K_s ratio can reveal selection acting across a protein-coding sequence.

In protein-coding regions, comparison of the relative rates at which nonsynonymous (amino-acid changing) and synonymous nucleotide substitutions accumulate over evolutionary time can reveal whether evolution of the protein has been driven by selection. Most synonymous substitutions take place in the third nucleotide position of the codon, where for some amino acids any nucleotide encodes the same amino acid, for others either pyrimidine will do, and for still others either purine is equivalent.

Because most nucleotide substitutions at the first or second codon position are nonsynonymous whereas many at the third codon position are synonymous, the number of nucleotide substitutions of each type across a protein-coding region must be compared relative to the number of possible nucleotide substitutions that could yield a change of each type. The comparison of nonsynonymous with synonymous nucleotide substitutions is often expressed as a ratio denoted K_a/K_s. The K_a/K_s ratio is a quantitative measure of the actual number of amino acid changes in a protein, relative to the number of possible amino-acid changing mutations (K_a), compared to the actual number of synonymous changes in the coding sequence, relative to the number of possible synonymous mutations (K_s).

If a protein-coding gene is evolving neutrally, which means randomly like a pseudogene, without regard to its nucleotide or amino acid sequence, then the expected value of $K_a/K_s = 1.0$. If $K_a/K_s < 1.0$, then nonsynonymous substitutions take place less frequently than with neutrality, which is the pattern expected if the protein is subject to selective constraints on its sequence so that many of the theoretically

possible amino acid replacements are deleterious and, therefore, eliminated by selection. Most protein-coding genes have $K_a/K_s < 1.0$.

Occasionally, however, proteins show $K_a/K_s > 1.0$, which implies that amino acid changes take place faster than with neutrality, a result expected if amino acid changes were favorable. Examples with $K_a/K_s > 1.0$ include the human sex-determining gene *Sry*, some homeodomain proteins, vertebrate growth hormone, some smell and taste receptors in certain species of *Drosophila*, and histone proteins specialized for spermatogenesis.

Comparison of K_a/K_s between human and chimpanzee genes has revealed a number of proteins that have apparently been under strong selection in our recent ancestry. These include skin-cell proteins, keratin proteins in hair, and proteins that function in pregnancy, the sense of smell, immunity, and regulation of cell physiology. Transcription factors are among the classes of proteins that show an excess with $K_a/K_s > 1.0$ between humans and chimpanzees, especially those that function in early development.

The interpretation of the K_a/K_s ratio is not, however, as straightforward as it may at first appear. Some of the complications are

- Mutation may be biased. For example, most organisms show a mutational bias toward transitions over transversions. Any such bias can change the value of the neutral expectation of K_a/K_s.

- Over sufficiently long periods of evolutionary time, mutations can take place at the same site, with one nucleotide changing to another and then later changing to still another or back again to the original. Proper estimates of K_a/K_s need to be corrected for such "multiple hits."

- A value of K_a/K_s sufficiently greater than 1.0 to be statistically significant requires strong selection among many amino acids spread across the molecule. Amino acid changes may, therefore, be driven by selection but fail to produce a K_a/K_s significantly greater than 1.0.

- K_a/K_s provides information only about possible selection at the amino acid level. It is completely insensitive to changes in level of gene expression; therefore, a gene may be under strong positive selection for regulatory changes but not show an elevated value of K_a/K_s.

New genes usually evolve through duplication and divergence.

In the course of evolution, new genes usually come from preexisting genes, and new gene functions evolve from previous gene functions. The raw material for new genes comes from duplications of regions of the genome, which may include one or more genes. Duplications take place relatively frequently. Analysis of the genomic sequences of a wide variety of eukaryotes suggests that a eukaryotic genome containing 30,000 genes may be expected to undergo roughly 60 to 600 duplications per million years.

From an evolutionary standpoint, two types of duplications need to be distinguished. One is gene duplication that results from speciation. Genes in related species that retain the same function are known as **orthologous genes**. It may seem odd to regard such genes as "duplications," yet they are duplicated, in the sense that the gene in each of the derived species is a copy of the gene that existed in the ancestral species. In contrast, gene duplications that take place in the genome of a single species result in **paralogous genes**. The principal distinction between orthologous genes and paralogous genes is that orthologous genes have their duplicates in the genomes of different species, whereas paralogous genes have their duplicates within the genome of a single species.

When a paralogous gene duplication takes place, the paralogs are redundant and one copy is free to evolve along any path. Probably the most common event is that one of the paralogs undergoes a mutation that destroys its function, or a deletion that eliminates it. But occasionally mutations take place that cause the functions of the paralogs to diverge. They may evolve different pH optima, for example, so that one gene product performs optimally in compartments of the cell that are relatively basic and the other in compartments that are relatively acidic. Or genetic rearrangements can fuse two unrelated genes and yield a new activity. For instance, translocations that fuse a transcription factor with an oncogene can lead to acute leukemia. These particular rearrangements are extremely deleterious, but the creation of chimeric genes affords an illustration of how new functions can be acquired.

Gene duplications also allow the paralogous copies to evolve more specialized functions. An example is shown in **FIGURE 14.4**. At the top is a gene in a multicellular eukaryote that has enhancers for expression in tissue types 1 (yellow) and 2 (green). The other panels show a gene duplication followed by sequential mutations that knock out either the type-2 enhancer elements (left) or the type-1 enhancer elements (right). The result is that the paralog on the left is expressed only in tissue type 1 and that on the right only in tissue type 2. Specialization of paralogs accompanying loss of functional capabilities is known as **subfunctionalization**. It may often be advantageous because each of the specialized genes is free to evolve toward optimal function in its own domain of expression.

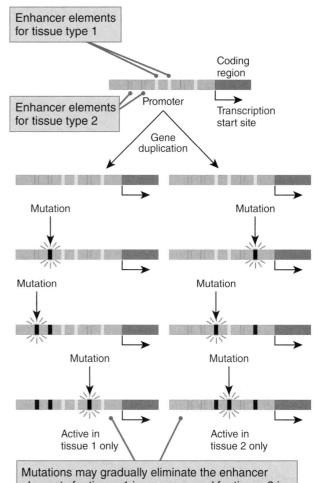

FIGURE 14.4 Specialization of paralogous genes by subfunctionalization. In this example, the paralogs become specialized for tissue-specific expression.

14.2 Genotypes may differ in frequency from one population to another.

Our discussion of molecular evolution focused on new mutations that become fixed in lineages because the rate of sequence evolution depends on how rapidly such fixations can take place. Between the time that a new mutation occurs and the time that it ultimately becomes fixed or lost are important processes that determine what the ultimate fate of the mutation will be and how rapidly the fate will be realized. The study of the processes that determine the fate of alleles in populations is part of population genetics.

The term **population** refers to a group of organisms of the same species living within a prescribed geographical area. Most widespread populations are subdivided by geographical or other features into smaller units called **subpopulations** or *local populations*. In the human population, for example, we may distinguish the subpopulation of people who live in the United States. Although matings occur primarily within each subpopulation, because of occasional migration between subpopulations they all share a common pool of genetic information called the **gene pool**. This section begins with an analysis of local populations with respect to a phenotype determined by two alleles. The phenotype is of considerable interest because it is associated with genetic resistance to AIDS (acquired immune deficiency syndrome).

Allele frequencies are estimated from genotype frequencies.

The genetic composition of a population can often be described in terms of the frequencies, or relative abundances, in which alternative alleles are found. The concepts can be illustrated with respect to an AIDS-resistance phenotype determined by a relatively common mutant allele. The gene in question is a chemokine receptor gene, *CCR5*, found in the human population. (*Chemokines* are molecules that white blood cells of the immune system use to attract one another.) The *CCR5* receptor enables the human immunodeficiency virus (HIV) to combine with the plasma membrane and infect the CD4(+) class of T cells of the immune system, which is necessary for an HIV infection to progress to full-blown AIDS. Most

human subpopulations contain a *CCR5* allele known as Δ*32* because it has a 32-bp deletion within the coding region. The molecular consequences of the deletion are shown in **FIGURE 14.5**. The relevant part of the normal *CCR5* DNA sequence is shown, grouped into codons, along with the amino acid sequence of the polypeptide, given in the single-letter abbreviations. The nucleotides missing in the Δ*32* deletion are highlighted in red. Above the DNA sequence is the normal CCR5 polypeptide; below is the Δ*32* mutant polypeptide. The deletion creates a frameshift in translation following codon 184, which results in the insertion of 31 incorrect amino acids until a termination codon is encountered after amino acid 215. The truncated protein is nonfunctional and does not support HIV entry into CD4(+) cells. The Δ*32* mutation was originally discovered among persons infected with HIV-1 who had remained free of AIDS for at least 10 years. The usual frequency of heterozygous genotypes, as we shall see, is about 20 percent, but among AIDS nonprogressors the frequency is approximately 40 percent. The homozygous Δ*32* genotypes, which are much less frequent, seem to have even greater protection.

Why is the Δ*32* homozygous genotype much less frequent than the Δ*32* heterozygous genotype? And what is the relationship between the homozygous and heterozygous frequencies? To begin to answer these questions, we should look at some data. For convenience, we will represent the normal *CCR5* allele as *A* and the Δ*32* deletion allele as *a*. In one study of 1000 French people whose DNA was genotyped for

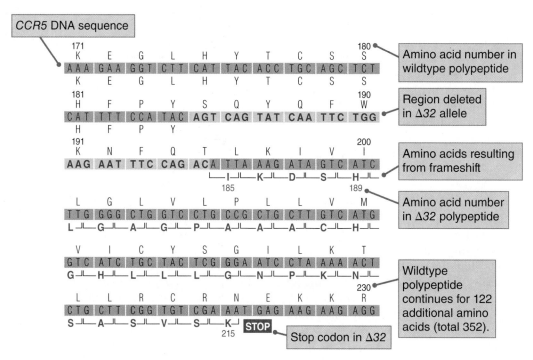

FIGURE 14.5 DNA sequence of a portion of the normal *CCR5* chemokine receptor gene and of the Δ*32* deletion allele, along with their products of translation.

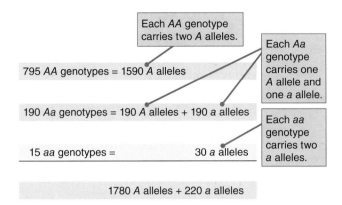

FIGURE 14.6 Analysis of the alleles present among a sample of 1000 French people genotyped for the *CCR5* receptor gene. The symbol *a* refers to the allele containing the **Δ**32 deletion.

CCR5, the numbers of homozygous normal (*AA*), heterozygous **Δ**32 (*Aa*), and homozygous **Δ**32 (*aa*) individuals were as follows:

$$795\ AA \quad 190\ Aa \quad 15\ aa$$

These numbers contain a great deal of information about the population, such as whether it is a single homogeneous population or a mixture of genetically somewhat different subpopulations. To interpret this information, first note that the sample contains two types of data: (1) the number of each of the three genotypes, and (2) the number of individual *CCR5* wildtype (*A*) and **Δ**32 (*a*) alleles. Furthermore, the 1000 genotypes represent 2000 alleles of the *CCR5* gene, because each human genome is diploid. These alleles break down as shown in **FIGURE 14.6**. Each homozygous *AA* genotype represents two *A* alleles, each homozygous *aa* genotype two *a* alleles, and each heterozygous *Aa* genotype one allele of each type. By the kind of allele counting shown in Figure 14.6, the sample of 1000 people therefore represents 1780 *CCR5* normal (*A*) alleles and 220 **Δ**32 (*a*) alleles.

Usually, it is more convenient to analyze the data in terms of the relative frequencies of genotypes and alleles than in terms of the observed numbers. For genotypes, the **genotype frequency** in a population is the proportion of organisms that have the particular genotype. For any specified allele, the **allele frequency** is the proportion of all alleles that are of the specified type. In the *CCR5* example, the genotype frequencies are obtained by dividing the observed numbers by the total sample size—in this case, 1000. Therefore, the genotype frequencies are

$$0.795\ AA \quad 0.190\ Aa \quad 0.015\ aa$$

Similarly, the allele frequencies are obtained by dividing the observed number of each allele by the total number of alleles (in this case, 2000):

$$\text{Allele frequency of } A = \frac{1780}{2000} = 0.89$$

$$\text{Allele frequency of } a = \frac{220}{2000} = 0.11$$

Note that the genotype frequencies add up to 1.0, as do the allele frequencies. This is a consequence of their definition in terms of proportions: They must add up to 1.0 when all of the possibilities are taken into account. An allele with a frequency of 1.0 is *fixed*, and an allele whose frequency has reached 0 is *lost*.

The allele frequencies among gametes equal those among reproducing adults.

One question we can ask at this stage is whether the genotypes in the sample have the frequencies we might expect if the alleles were joined in pairs at random. Two *A* alleles (*AA*) would then be paired with a frequency of 0.89 × 0.89 = 0.7921, an *A* allele would be paired with an *a* allele (*Aa*) with a frequency of 2 × 0.89 × 0.11 = 0.1958, and two *a* alleles (*aa*) would be paired with a frequency of 0.11 × 0.11 = 0.0121. These values are, in fact, quite close to the observed genotype frequencies. It is also noteworthy that the allele frequencies among adults, as we have calculated them in Figure 14.8, are the same as those among the gametes produced by the adults. This is true because Mendelian segregation ensures that each heterozygous genotype will produce equal numbers of each type of gamete (**FIGURE 14.7**). Thus when we calculate the gametic frequencies, the heterozygous genotypes contribute equally to both classes of gametes, whereas the homozygous *AA* and homozygous *aa* genotypes contribute only *A*-bearing or only *a*-bearing gametes, respectively. Consequently, when the genotype frequencies among the parents are taken into account, as well as Mendelian segregation in the heterozygous *Aa* genotypes, the gametes produced in the population have the

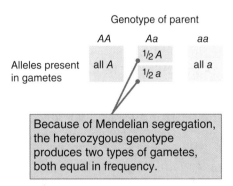

FIGURE 14.7 Mendelian considerations in population genetics. Each homozygous genotype produces gametes containing a single allele, but a heterozygous genotype, because of Mendelian segregation, produces two types of gametes in equal frequency.

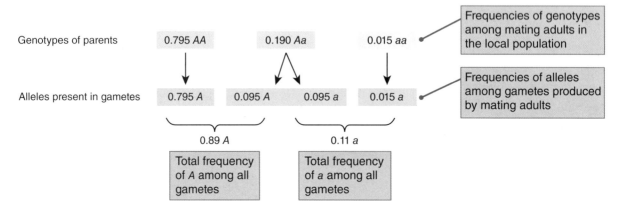

FIGURE 14.8 Calculation of the allele frequencies in gametes. In any population, the frequency of gametes containing any particular allele equals the frequency of the genotypes that are homozygous for the allele, plus one-half the frequency of all genotypes that are heterozygous for the allele.

composition deduced in **FIGURE 14.8**. The equality between the allele frequencies among the adults and those among the gametes produced by the adults—namely 0.89 *A* and 0.11 *a*—must be true whenever each adult in the population produces the same number of functional gametes. The apparent random pairing of alleles in the adult genotypes is also important, because it is a clue as to how the gametes are united in fertilization. This issue is examined in the next section.

14.3 Random mating means that mates pair without regard to genotype.

When a local population undergoes **random mating**, it means that organisms in the local population form mating pairs independently of genotype. Each type of mating pair is formed as often as would be expected by chance encounters. Random mating is by far the most prevalent mating system for most species of animals and plants, except for plants that regularly reproduce through self-fertilization. Self-fertilization is an extreme example of another important type of mating system, which is called *inbreeding*, or mating between relatives. **FIGURE 14.9** represents an example of inbreeding. In this case, the female I is the offspring of a first-cousin mating (G with H). The closed loop in the pedigree is diagnostic of inbreeding, and the individuals designated A and B are called *common ancestors* of I, because they are ancestors of both of the parents of I. Because A and B are common ancestors, a particular allele present in A (or in B) could, by chance, be transmitted in inheritance down both sides of the pedigree, to meet again in the formation of I. This possibility is the most important and characteristic consequence of inbreeding, and it will be discussed later in this chapter.

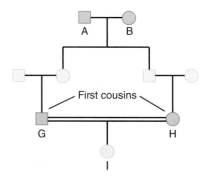

FIGURE 14.9 An inbreeding pedigree in which individual I is the result of a mating between first cousins.

Random mating implies a pleasingly simple relationship between allele frequency and genotype frequency. This is summarized in the following principle:

KEY CONCEPT

Random mating of individuals is equivalent to the random union of gametes.

On the basis of this principle, we may imagine random mating to be equivalent to drawing female and male gametes at random from a large container. Each zygote genotype is formed from the pairing of one female and one male gamete. To be specific, consider two alleles *A* and *a* with allele frequencies p and q, respectively, where $p + q = 1$. In the sample of *CCR5* receptor genotypes considered earlier in this chapter, we calculated

$$p = 0.89 \text{ for the } A \text{ allele}$$
$$q = 0.11 \text{ for the } a \text{ allele}$$

The genotype frequencies expected with random mating can be deduced from the tree diagram in **FIGURE 14.10**. The gametes at the left represent the

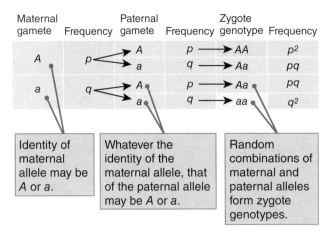

FIGURE 14.10 When gametes containing either of two alleles unite at random to form the next generation, the genotype frequencies among the zygotes are given by the ratio $p^2 : 2pq : q^2$. This constitutes the Hardy–Weinberg principle.

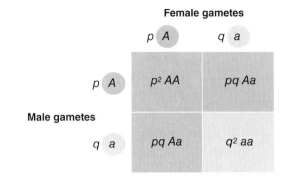

FIGURE 14.11 A Punnett square showing, in a cross-multiplication format, the ratio $p^2 : 2pq : q^2$ that is characteristic of random mating.

eggs and those in the middle the sperm. The genotypes that can be formed with two alleles are shown at the right, and with random mating, the frequency of each genotype is calculated by multiplying the allele frequencies of the corresponding gametes. However, the genotype Aa can be formed in two ways: The A allele could have come from the mother (top part of diagram) or from the father (bottom part of diagram). In each case, the frequency of the Aa genotype is pq; considering both possibilities, we find that the frequency of Aa is $pq + pq = 2pq$. Consequently, the overall genotype frequencies expected with random mating are

$$AA: p^2 \quad Aa: 2pq \quad aa: q^2 \qquad (14.1)$$

The frequencies p^2, $2pq$, and q^2 result from random mating for a gene with two alleles; they constitute the **Hardy–Weinberg principle**, named after Godfrey Hardy and Wilhelm Weinberg, who derived it independently of each other in 1908. Sometimes the Hardy–Weinberg principle is demonstrated by the type of Punnett square illustrated in **FIGURE 14.11**. Such a square is completely equivalent to the tree diagram used in Figure 14.10. Although the Hardy–Weinberg principle is exceedingly simple, it has a number of important implications that are not obvious. These are described in the following sections.

The Hardy–Weinberg principle has important implications for population genetics.

One important implication of the Hardy–Weinberg principle is that *the allele frequencies remain constant from generation to generation*. To understand why, consider again a gene with two alleles, A and a, having frequencies p and q, respectively ($p + q = 1$). With random mating, the frequencies of genotypes AA, Aa,

and aa among zygotes are p^2, $2pq$, and q^2, respectively. Assuming equal ability to survive among the genotypes, these frequencies equal those among adults. If, in addition, all of the adult genotypes are equally fertile, then the frequency of allele A among gametes that form the zygotes of the next generation can be calculated in terms of the frequency of the A allele in the previous generation. If we use a prime (') to denote the frequency of the A allele in the next generation, then the allele frequency is p'. In terms of the alleles present in the previous generation, p' includes all of the A alleles in homozygous AA genotypes (frequency p^2) plus half of the alleles in heterozygous Aa genotypes (frequency $2pq$). The Aa heterozygotes are multiplied by 1/2 because of Mendelian segregation; only one-half of the gametes from Aa genotypes carry A. Putting all this together, the frequency p' of the A allele in the next generation is

$$p' = p^2 + 2pq/2 = p(p + q) = p$$

The final equality follows because $p + q = 1$. We have therefore shown that the frequency of allele A remains constant at the value of p through the passage of one or more complete generations. This principle depends on certain assumptions, of which the most important are the following:

- Mating is random; there are no subpopulations that differ in allele frequency.

- Allele frequencies are the same in males and females.

- All the genotypes are equal in survival and fertility (*selection* does not operate).

- Mutation does not occur.

- Migration into the population is absent.

- The population is sufficiently large that the frequencies of alleles do not change from generation to generation because of chance.

As a practical application of the Hardy–Weinberg principle, consider again the *CCR5* receptor discussed in Section 14.2. The frequencies of the *CCR5* normal allele (*A*) and the Δ*32* allele (*a*) among 1000 adults were 0.89 and 0.11, respectively. Assuming random mating, the expected genotype frequencies can be calculated from Equation (14.1) as

$$AA: \quad (0.89)^2 = 0.7921$$
$$Aa: \quad 2(0.89)(0.11) = 0.1958$$
$$aa: \quad (0.11)^2 = 0.0121$$

Note that these are the same frequencies calculated earlier on the basis of the supposition that the alleles were joined in pairs at random. Now we know this means that the genotype frequencies are given by the Hardy–Weinberg principle. Because the total number of people in the sample is 1000, the expected numbers of the genotypes are 792.1 *AA*, 195.8 *Aa*, and 12.1 *aa*. The *observed* numbers were 795 *AA*, 190 *Aa*, and 15 *aa*. Goodness of fit between the observed and expected numbers can be determined by means of the χ^2 test. In this case,

$$\chi^2 = \frac{(795 - 792.1)^2}{792.1}$$
$$+ \frac{(190 - 195.8)^2}{195.8} + \frac{(15 - 12.1)^2}{12.1}$$
$$= 0.877$$

The fit is obviously satisfactory, but to estimate the *P* value we need to adjust the degrees of freedom. Normally, with 3 classes of data, we would have 2 degrees of freedom. But when any quantity, such as an allele frequency, is estimated from the data, 1 degree of freedom must be deducted for each quantity estimated. In this case we estimated the allele frequency *p* (*q* = 1 − *p* follows automatically), so we lose 1 of the 2 degrees of freedom we would otherwise have, and thus the appropriate number of degrees of freedom for the test is 1. For this we obtain a *P* value of 0.35, which means that the hypothesis of random mating can account for the data. On the other hand, the χ^2 test detects only deviations that are rather large, so a good fit to Hardy–Weinberg frequencies should not be over interpreted, because

KEY CONCEPT

It is entirely possible for one or more assumptions of the Hardy–Weinberg principle to be violated, including the assumption of random mating, and still not produce deviations from the expected genotype frequencies that are large enough to be detected by the χ^2 test.

STOP & THINK 14.1

The bands in the electrophoresis gel shown here are fast-moving (*F*) and slow-moving (*S*) restriction fragments that distinguish two alleles (*F* and *S*) of a gene in a diploid population. The numbers across the top are the numbers of individuals of each genotype (*F/F*, *F/S*, or *S/S*) observed in a sample of 100 individuals. What is the estimated allele frequency (*p*) of the *F* allele and the estimated allele frequency (*q*) of the slow (*S*) allele? Assuming random mating, what is the expected number of genotype based on the Hardy–Weinberg principle?

48 38 14

S

F

If an allele is rare, it is found mostly in heterozygous genotypes.

Another important implication of the Hardy–Weinberg principle is that *for a rare allele, the frequency of heterozygotes far exceeds the frequency of the rare homozygote.* For example, when the frequency of the rarer allele is *q* = 0.1, the ratio of heterozygotes to homozygotes equals $2pq/q^2 = 2(0.9)/(0.1)$, or approximately 20; when *q* = 0.01, this ratio is about 200; and when *q* = 0.001, it is about 2000. In other words,

KEY CONCEPT

When an allele is rare, there are many more heterozygotes than there are homozygotes for the rare allele.

The reason for this perhaps unexpected relationship is shown in **FIGURE 14.12**, which plots the frequencies of homozygous and heterozygous genotypes with random mating. Note that at allele frequencies near 0 or 1, the frequency of the heterozygous genotype goes to 0 much more slowly than does the frequency of the rarer homozygous genotype.

One practical implication of this principle is seen in the instance of cystic fibrosis, an inherited secretory disorder of the pancreas and lungs. Cystic fibrosis is one of the most common recessively inherited severe disorders among Caucasians; it affects about 1 in 1700 newborns. In this case, the heterozygotes cannot readily be identified by phenotype, so a method of calculating allele frequencies that is different from the

$q = 1 - p$
1.0 0.9 0.8 0.7 0.6 0.5 0.4 0.3 0.2 0.1 0
Genotype frequency
1.0
0.9
0.8
0.7
0.6
0.5
0.4
0.3
0.2
0.1
0
Frequency of heterozygotes is maximum when $p = q = 1/2$.
0 0.1 0.2 0.3 0.4 0.5 0.6 0.7 0.8 0.9 1.0
$p = 1 - q$
q^2 (aa)
$2pq$ (Aa)
p^2 (AA)

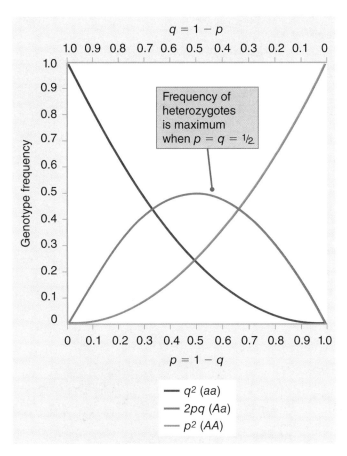

FIGURE 14.12 Graphs of p^2, $2pq$, and q^2. If the allele frequencies are between 1/3 and 2/3, the heterozygote is the genotype with the highest frequency.

gene-counting method used earlier must be applied. This method is straightforward because with random mating, the frequency of recessive homozygotes must correspond to q^2. In the case of cystic fibrosis,

$$q^2 = 1/1700 = 0.00059$$
$$\text{or}$$
$$q = (0.00059)^{1/2} = 0.024$$

and consequently,

$$p = 1 - q = 1 - 0.024 = 0.976$$

The frequency of heterozygous genotypes that carry a mutant allele for cystic fibrosis is calculated as

$$2pq = 2(0.976)(0.024) = 0.047 = 1/21$$

This calculation implies that for cystic fibrosis, although only 1 person in 1700 is affected with the disease (homozygous), about 1 person in 21 is a carrier (heterozygous). The calculation should be regarded as approximate because it is based on the assumption of Hardy–Weinberg genotype frequencies. Nevertheless, considerations like these are important in predicting the outcome of population screening for the detection of carriers of harmful recessive alleles, which is essential in evaluating the potential benefits of such programs.

Hardy–Weinberg frequencies can be extended to multiple alleles.

Extension of the Hardy–Weinberg principle to multiple alleles of a single autosomal gene can be illustrated by a three-allele case. **FIGURE 14.13** shows the results of random mating in which three alleles are considered. The alleles are designated A_1, A_2, and A_3, where the uppercase letter represents the gene and the subscript designates the particular allele. The allele frequencies are p_1, p_2, and p_3, respectively. With three alleles (as with any number of alleles), the allele frequencies of all alleles must sum to 1; in this case, $p_1 + p_2 + p_3 = 1.0$. As in Figure 14.11, the entry in each square is obtained by multiplying the frequencies of the alleles at the corresponding margins; any homozygous genotype (such as A_1A_1) has a random-mating frequency equal to the square of the corresponding allele frequency (in this case, p_1^2). Any heterozygous genotype (such as A_1A_2) has a random-mating frequency equal to twice the product of the corresponding allele frequencies (in this case, $2p_1p_2$). The extension to any number of alleles is straightforward:

Frequency of any homozygous genotype
= square of allele frequency
Frequency of any heterozygous genotype
= 2 × product of allele frequencies (14.2)

Multiple alleles determine the human ABO blood groups. The gene has three principal alleles, designated I^A, I^B, and I^O. In one study of 3977 Swiss people, the allele frequencies were found to be 0.27 I^A, 0.06 I^B, and 0.67 I^O. Applying the rules for multiple alleles, we can

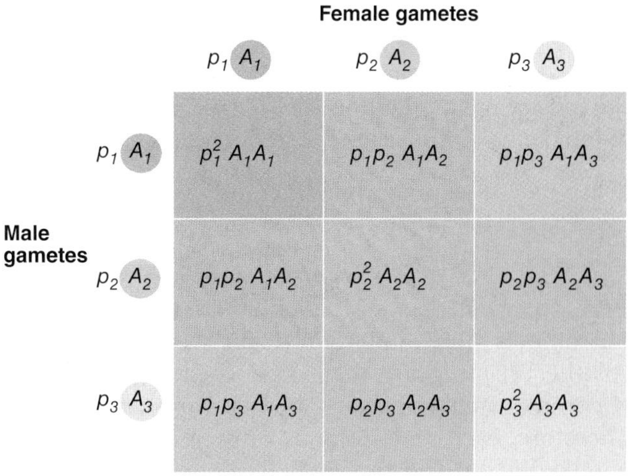

Female gametes
p_1 A_1
p_2 A_2
p_3 A_3
Male gametes
p_1 A_1
p_1^2 A_1A_1
p_1p_2 A_1A_2
p_1p_3 A_1A_3
p_2 A_2
p_1p_2 A_1A_2
p_2^2 A_2A_2
p_2p_3 A_2A_3
p_3 A_3
p_1p_3 A_1A_3
p_2p_3 A_2A_3
p_3^2 A_3A_3

FIGURE 14.13 Punnett square showing the results of random mating with three alleles.

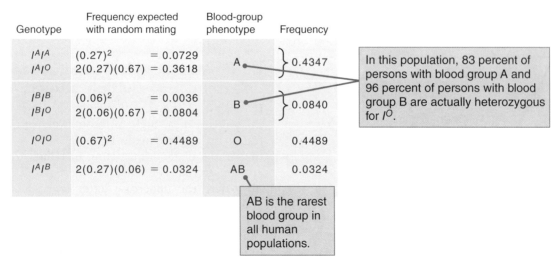

Genotype	Frequency expected with random mating		Blood-group phenotype	Frequency
$I^A I^A$	$(0.27)^2$	$= 0.0729$	A	0.4347
$I^A I^O$	$2(0.27)(0.67)$	$= 0.3618$		
$I^B I^B$	$(0.06)^2$	$= 0.0036$	B	0.0840
$I^B I^O$	$2(0.06)(0.67)$	$= 0.0804$		
$I^O I^O$	$(0.67)^2$	$= 0.4489$	O	0.4489
$I^A I^B$	$2(0.27)(0.06)$	$= 0.0324$	AB	0.0324

In this population, 83 percent of persons with blood group A and 96 percent of persons with blood group B are actually heterozygous for I^O.

AB is the rarest blood group in all human populations.

FIGURE 14.14 Random-mating frequencies for the three alleles governing the ABO blood groups.

expect the genotype frequencies that result from random mating to be as shown in **FIGURE 14.14**. Because both I^A and I^B are dominant to I^O, the expected frequency of blood-group *phenotypes* is that shown at the right in the illustration. Note that the majority of A and B phenotypes are actually heterozygous for the I^O allele; this is because the I^O allele has such a high frequency in the population.

X-linked genes are a special case because males have only one X chromosome.

The implications of random mating for two X-linked alleles (H and h) are illustrated in **FIGURE 14.15**. The principles are the same as those considered earlier, but male gametes carrying the X chromosome (part A) must be distinguished from those carrying the Y chromosome (part B). When the male gamete carries an X chromosome, the Punnett square is exactly the same as that for the two-allele autosomal gene in Figure 14.11. However, because the male gamete carries an X chromosome, all the offspring in question are female. Consequently, among females, the genotype frequencies are

Frequency of *HH* females $= p^2$
Frequency of *Hh* females $= 2pq$
Frequency of *hh* females $= q^2$

When the male gamete carries a Y chromosome, the outcome is quite different (Figure 14.15, part B). All the offspring are male, and each has only one X chromosome, which is inherited from the mother. Therefore, each male receives only one copy of each X-linked gene, and the genotype frequencies among males are the same as the allele frequencies:

Frequency of *H* males $= p$
Frequency of *h* males $= q$

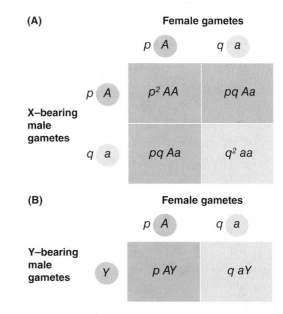

(A) Female gametes

		p A	q a
X–bearing male gametes	p A	p^2 AA	pq Aa
	q a	pq Aa	q^2 aa

(B) Female gametes

		p A	q a
Y–bearing male gametes	Y	p AY	q aY

FIGURE 14.15 The results of random mating for an X-linked gene. (A) Genotype frequencies in females. (B) Genotype frequencies in males.

An important implication of Figure 14.15 is that if h is a rare recessive allele, then there will be many more males exhibiting the trait than females because the frequency of affected females (q^2) will be much smaller than the frequency of affected males (q). This principle is illustrated in **FIGURE 14.16**. As the allele frequency of the recessive decreases toward 0, the frequencies of affected males and females both decrease, but the frequency of affected females decreases faster. The result is that the ratio of affected males to affected females increases. At an allele frequency of $q = 0.3$, for example, the ratio of affected males to affected females is 3.33; but for an allele frequency of $q = 0.1$, the ratio of affected males to affected females is 10.0. In general, the ratio of affected males to affected females is q/q^2, or $1/q$.

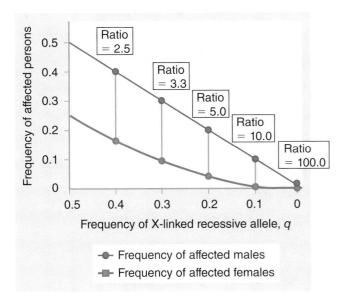

FIGURE 14.16 For a recessive X-linked allele, the upper curve gives the frequency of affected males (q), and the lower curve the frequency of affected females (q^2), for values of $q < 0.5$. Although both frequencies decrease as the recessive allele becomes rare, the frequency of affected females decreases more rapidly. The result is that the ratio of affected males to affected females increases as the allele frequency decreases.

For an X-linked recessive trait, the frequency of affected males provides an estimate of the frequency of the recessive allele. A specific example is found in the common form of X-linked color blindness in human beings. This trait affects about 1 in 20 males among Caucasians, so $q = 1/20 = 0.05$. The expected frequency of color-blind females is, therefore, estimated as $q^2 = (0.05)^2 = 0.0025$, or about 1 in 400.

14.4 Highly polymorphic sequences are used in DNA typing.

Many genes in human populations are **polymorphic**, which means that they have two or more alleles that are common in the population. (Quantitatively speaking, a gene is generally regarded as polymorphic if the frequency of heterozygous genotypes is 10 percent or greater.) Averaged across the human genome, about 1 per 1000 nucleotide sites differs from one individual to the next at a frequency high enough to be regarded as polymorphic. Owing to the high level of genetic variation, it is virtually impossible for two human beings to be genetically identical. The only exceptions are identical twins, identical triplets, and so forth. Each human genotype is unique. The theoretical principle of genetic uniqueness has become a practical reality through the study of DNA polymorphisms. Small samples of human material from an unknown person (for example, material left at the scene of a crime) often contain enough DNA that the genotype

can be determined for a number of polymorphisms and matched against those present among a group of suspects. Typical examples of crime-scene evidence include blood, semen, hair roots, and skin cells. Even a small number of cells is sufficient, because predetermined regions of DNA can be amplified by the polymerase chain reaction.

If a suspect's DNA contains sequences that are clearly not present in the crime-scene sample, then the sample must have originated from a different person. On the other hand, if a suspect's DNA *does* match that of the crime-scene sample, then the suspect could be the source. The strength of the DNA evidence depends on the number of polymorphisms that are examined and the number of alleles present in the population. The greater the number of polymorphisms that match, especially if they are highly polymorphic, the stronger the evidence linking the suspect to the sample taken from the scene of the crime. The use of polymorphisms in DNA to link suspects with samples of human material is called **DNA typing**. This method of identifying individuals is generally regarded as the most important innovation in criminal investigation since the development of fingerprinting more than a century ago.

FIGURE 14.17 illustrates one type of polymorphism used in DNA typing. The restriction fragments corresponding to each allele differ in length because they contain different numbers of units repeated in tandem. A polymorphism of this type is called an **STR**, which stands for **simple tandem repeat**. STRs are abundant in the human genome (**TABLE 14.1**). They account for about 3 percent of the total DNA and have a density of about one STR per 2 kb.

The STRs used in DNA typing usually have repeat units of 14 to 500 bases. STRs are of value in DNA typing because many alleles are possible, owing to the variable number of repeats present at the site from one chromosome to the next. Although many alleles may be present in the population as a whole, each person can have no more than two alleles for each STR polymorphism. An example of an STR used in DNA typing is shown in **FIGURE 14.18**. The lanes in the gel labeled M contain multiple DNA fragments of known size to serve as molecular-weight markers. Each lane numbered 1 through 9 contains DNA from a single person.

Two typical features of STRs are to be noted:

1. Most people are heterozygous for STR alleles that produce restriction fragments of different sizes. Heterozygosity is indicated by the presence of two distinct bands. In Figure 14.18, only the person numbered 1 appears to be homozygous for a particular allele.

2. The restriction fragments from different people cover a wide range of sizes. The variability in size indicates that the population as a whole contains many STR alleles.

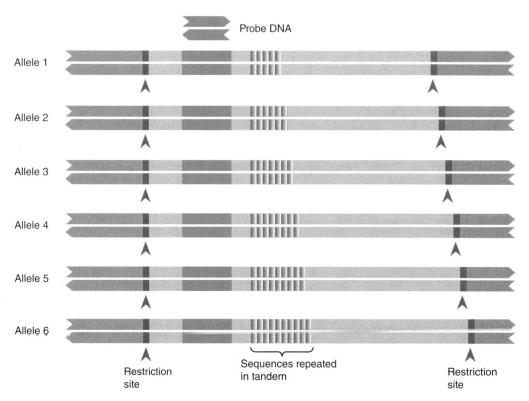

FIGURE 14.17 Allelic variation resulting from a variable number of units repeated in tandem in a nonessential region of a gene. The probe DNA detects a restriction fragment for each allele. The length of the fragment depends on the number of repeating units present.

TABLE 14.1 Simple Tandem Repeats (STRs) in the Human Genome		
Length of repeat unit	Average number of occurrences per Mb	Average number of repeats per occurrence
1	36.7	45
2	43.1	59
3	11.8	29
4	32.5	26
5	17.6	31
6	15.2	15
7	8.4	15
8	11.1	13
9	8.6	12
10	8.6	18
11	8.7	8

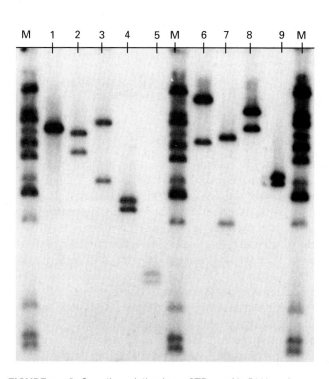

FIGURE 14.18 Genetic variation in an STR used in DNA typing. Each numbered lane contains DNA from a single person; the DNA has been cleaved with a restriction enzyme, separated by electrophoresis, and hybridized with radioactive probe DNA. The lanes labeled M contain molecular-weight markers.

Courtesy of Robert W. Allen, Human Identity Testing Laboratory, Oklahoma State University.

The reason why STRs are useful in DNA typing is also evident in Figure 14.18: Each of the nine people tested has a different pattern of bands and thus could be identified uniquely by means of this STR. On the other hand, the uniqueness of each person in Figure 14.18 is due in part to the high degree of polymorphism of the STR and in part to the small sample size. If more people were examined, then individuals that matched in their STR types by chance would certainly be found.

DNA exclusions are definitive.

DNA typing cannot only implicate the guilty, it can also exonerate the innocent whose DNA type does *not* match the evidence. For example, if the DNA type of a suspected rapist fails to match the DNA type of semen taken from the victim, then the suspect could not be the perpetrator. Another example, which illustrates the use of DNA typing in paternity testing, is given in FIGURE 14.19. The numbers 1 and 2 designate different cases, in each of which a man was alleged to have fathered a child. In each case, DNA was obtained from the mother (M), the child (C), and the man (A). The pattern of bands obtained for one STR is shown. The lanes labeled A + C contain a mixture of DNA from the man and the child. In case 1, the lower band in the child was inherited from the mother and the upper band from the father; because the upper band is the same size as one of those in the alleged father, he could have contributed this allele to the child. This finding does not prove that this individual is the father; it says only that he cannot be excluded on the basis of this particular STR. (However, if enough STRs are studied,

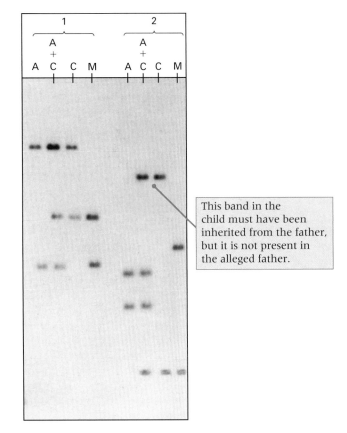

This band in the child must have been inherited from the father, but it is not present in the alleged father.

FIGURE 14.19 Use of DNA typing in paternity testing. The sets of lanes numbered 1 and 2 contain DNA samples from two different paternity cases. In each case, the lanes contain DNA fragments from the following sources: M, the mother; C, the child; A, the alleged father. The lanes labeled A + C contain a mixture of DNA fragments from the alleged father and the child.

Courtesy of Robert W. Allen, Human Identity Testing Laboratory, Oklahoma State University.

STOP & THINK 14.2

In the early-morning hours of July 17, 1918, on orders of Bolshevik leader Vladimir Lenin, the last Russian tsar, Nicholas II, was machine gunned to death in the basement of a home in Ekaterinburg (now Sverdlovsk), Russia. Executed with him were his wife, Tsarina Alexandra; their four daughters Olga, Tatiana, Marie, and Anastasia; and their hemophiliac son, the Tsarevitch Alexis. Also executed were the family's male personal physician and three female servants. In 1979, the remains of nine human beings were found in a shallow grave in the area where the bodies were

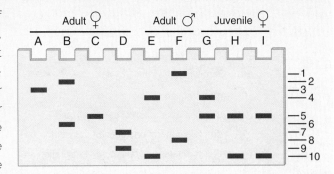

rumored to have been disposed of, and in 1991, they became available for DNA testing. The gel pattern obtained for one STR used in DNA typing is shown in the accompanying diagram.

The bands are numbered in order of increasing electrophoretic mobility. Degree of bone development and presence of Y-chromosome–specific DNA indicated that the remains included those of four adult females (A–D), two adult males (E–F), and three juvenile females (G–I). (a) Are the DNA typing results consistent with the presence of six unrelated adult persons? (b) Are they consistent with a group including a mother, father, and three of their daughters? (c) Identify the lanes containing DNA from Tsar Nicholas II and Tsarina Alexandra.

and the man cannot be excluded by any of them, it does make it more likely that he really is the father.) Case 2 in Figure 14.19 is an exclusion. The small band lowest in the gel is the band inherited by the child from the mother. The other band in the child does not match either of the bands in the accused father, so the accused man could not be the biological father. (In theory, mutation could be invoked to explain the result, but this is extremely unlikely.)

14.5 Inbreeding means mating between relatives.

Inbreeding means mating between relatives, such as first cousins. The principal consequence of inbreeding is that the frequency of heterozygous offspring is smaller than it is with random mating. This effect is seen most dramatically in repeated self-fertilization, which occurs naturally in certain plants. To understand why the frequency of heterozygous genotypes decreases with self-fertilization, consider a hypothetical initial population consisting exclusively of *Aa* heterozygotes (**FIGURE 14.20**). With self-fertilization, each plant would produce offspring in the proportions 1/4 *AA*, 1/2 *Aa*, and 1/4 *aa*. Thus one generation of self-fertilization reduces the proportion of heterozygotes from 1 to 1/2. In the second generation, only the heterozygous plants can again produce heterozygous offspring, and only half of their offspring will again be heterozygous. Heterozygosity is therefore reduced to 1/4 of what it was originally. Three generations of self-fertilization reduce the heterozygosity to $1/4 \times 1/2 = 1/8$, and so forth. The remainder of this section demonstrates how the reduction in heterozygosity because of inbreeding can be expressed quantitatively.

Repeated self-fertilization is a particularly intense form of inbreeding, but weaker forms of inbreeding are qualitatively similar in that they also lead to a reduction in heterozygosity. A convenient measure

of the effect of inbreeding is based on the reduction in heterozygosity. Suppose that H_I is the frequency of heterozygous genotypes in a population of inbred organisms. The most widely used measure of inbreeding is called the **inbreeding coefficient**, symbolized *F*, which is defined as the proportionate reduction in H_I compared with the value of $2pq$ that would be expected with random mating:

$$F = (2pq - H_I)/2pq$$

This equation can be rearranged as

$$H_I = 2pq(1 - F),$$

which says that in a population of organisms having an inbreeding coefficient of *F*, the proportion of heterozygous genotypes is reduced by the fraction *F* relative to what it would be with random mating. As the proportion of heterozygous genotypes decreases in frequency, the proportion of homozygous genotypes increases correspondingly. The overall genotype frequencies in an inbred population are given by

Frequency of *AA* genotype = $p^2(1 - F) + pF$
Frequency of *Aa* genotype = $2pq(1 - F)$ (14.3)
Frequency of *aa* genotype = $q^2(1 - F) + qF$

These frequencies are a modification of the Hardy–Weinberg principle that takes inbreeding into account. When $F = 0$ (no inbreeding), the genotype frequencies are the same as those given in the Hardy–Weinberg principle in Equation (14.1), namely p^2, $2pq$, and q^2. At the other extreme, when $F = 1$ (complete inbreeding), the inbred population consists entirely of *AA* and *aa* genotypes in the frequencies *p* and *q*, respectively. Whatever the value of *F*, however, the allele frequencies remain at the values of *p* and *q* because

$$p^2(1 - F) + pF + (1/2)[2pq(1 - F)]$$
$$= p(p + q)(1 - F) + pF = p$$

A graphical representation of the genotype frequencies in Equation (14.3) is shown in **FIGURE 14.21**. For each organism, the population is divided conceptually into two groups. In one group (amounting to a proportion *F* of the population), the gene in question has been affected by the inbreeding, which means that the two alleles present in the organism are identical by descent, owing to DNA replication in a common ancestor of the inbred organism. In the other group (amounting to a proportion $1 - F$ of the population), the gene in question has, by chance, escaped the effects of inbreeding, which means that the genotype frequencies are those expected with random mating. Taking both groups into account results in the genotype frequencies in Equation (14.3).

The reduction in heterozygosity due to inbreeding is shown in **FIGURE 14.22**. Each curve is given by an equation of the form $H_I = 2pq(1 - F)$. The case $F = 0$ corresponds to random mating. The curves show that

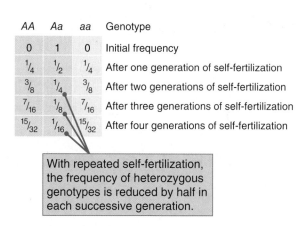

AA	Aa	aa	Genotype
0	1	0	Initial frequency
$\frac{1}{4}$	$\frac{1}{2}$	$\frac{1}{4}$	After one generation of self-fertilization
$\frac{3}{8}$	$\frac{1}{4}$	$\frac{3}{8}$	After two generations of self-fertilization
$\frac{7}{16}$	$\frac{1}{8}$	$\frac{7}{16}$	After three generations of self-fertilization
$\frac{15}{32}$	$\frac{1}{16}$	$\frac{15}{32}$	After four generations of self-fertilization

With repeated self-fertilization, the frequency of heterozygous genotypes is reduced by half in each successive generation.

FIGURE 14.20 Effects of repeated self-fertilization on the genotype frequencies. In each generation, the proportion of heterozygous genotypes decreases to half of its value in the previous generation.

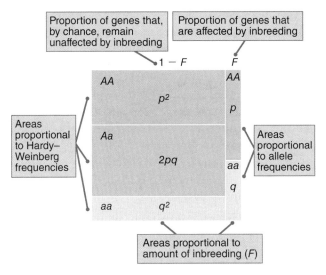

Proportion of genes that, by chance, remain unaffected by inbreeding

Proportion of genes that are affected by inbreeding

Areas proportional to Hardy–Weinberg frequencies

Areas proportional to allele frequencies

Areas proportional to amount of inbreeding (F)

FIGURE 14.21 Effect of inbreeding on genotype frequencies. The large rectangles on the left pertain to alleles whose ancestries, by chance, are not affected by inbreeding and for which the genotype frequencies remain in Hardy–Weinberg proportions. The narrow rectangles on the right pertain to alleles whose ancestries are affected by the inbreeding; in this case genotype frequencies of *AA* and aa are related as *p* : *q*. (Note that there are no heterozygous genotypes in the latter case.)

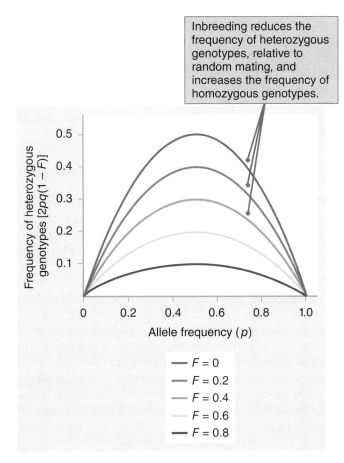

Inbreeding reduces the frequency of heterozygous genotypes, relative to random mating, and increases the frequency of homozygous genotypes.

— $F = 0$
— $F = 0.2$
— $F = 0.4$
— $F = 0.6$
— $F = 0.8$

FIGURE 14.22 Frequency of heterozygous genotypes in an inbred population (y-axis) against allele frequency (x-axis). As the inbreeding becomes more intense (greater inbreeding coefficient F), the proportion of heterozygous genotypes decreases.

as the inbreeding coefficient increases, the frequency of heterozygous genotypes decreases proportionately until, when $F = 1$, there are no heterozygotes remaining in the inbred population. In other words, in a highly inbred population with $F = 1$, the genotypes consist of *AA* and *aa* in the relative proportions *p* and *q*, respectively.

Inbreeding results in an excess of homozygotes compared with random mating.

The effects of inbreeding differ according to the normal mating system of an organism. At one extreme, in regularly self-fertilizing plants, inbreeding is already so intense and the organisms are so highly homozygous that additional inbreeding has virtually no effect. However,

KEY CONCEPT

In most species, inbreeding is harmful, and much of the effect is due to rare recessive alleles that would not otherwise become homozygous.

Among human beings, close inbreeding is uncommon in much of the world because of social conventions. The closest type of inbreeding usually found is mating between first cousins. In small, isolated populations (such as aboriginal groups and religious communities), some inbreeding is inevitable as a result of matings between remote relatives. The effect of inbreeding is always an increase in the frequency of genotypes that are homozygous for rare, usually harmful recessives. For example, among American whites, the frequency of albinism among offspring of matings between nonrelatives is approximately 1 in 20,000, but among offspring of first-cousin matings, the frequency is approximately 1 in 2000. The reason for the increased risk may be understood by comparing the genotype frequencies of homozygous recessives in Equation (14.3) (for inbreeding) and Equation (14.1) (for random mating). In the most common form of inbreeding among human beings (mating between first cousins), $F = 0.062$ ($= 1/16$) among the offspring. Therefore, the frequency of homozygous recessives produced by first-cousin mating is

$$q^2(1 - 0.062) + q(0.062),$$

whereas, among the offspring of nonrelatives, the frequency of homozygous recessives is q^2. For albinism, $q = 0.007$ (approximately), and the calculated frequencies are 5×10^{-4} for the offspring of first cousins and 5×10^{-5} for the offspring of nonrelatives.

The effect of first-cousin mating in causing an increased frequency of offspring homozygous for a rare recessive allele is shown in **FIGURE 14.23**.

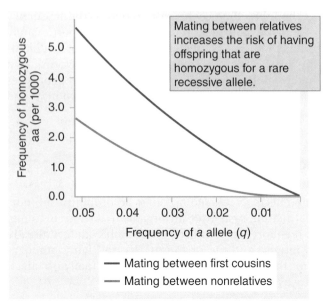

FIGURE 14.23 Effect of first-cousin mating on the frequency of offspring genotypes that are homozygous for a rare autosomal recessive allele. Although both curves decrease as the allele becomes rare, the curve for mating between nonrelatives decreases more rapidly. As a consequence, the more rare the recessive allele, the greater the proportion of all affected individuals that result from first-cousin matings.

The increase occurs for any allele frequency, but the relative increase is greater for alleles that are more rare. Considering the curves in Figure 14.23, with an allele frequency $q = 0.05$, for example, the relative risk of producing a homozygous offspring is $0.0335938/0.0025 = 13.4$; whereas with an allele frequency $q = 0.01$, the relative risk of a homozygous offspring is $0.00634375/0.0001 = 63.4$.

⚛ STOP & THINK 14.3

In the United States, about 1 percent of all marriages are between first cousins. Consider a rare mutant allele for which $q = 0.001$, so that the proportion of homozygous offspring from unrelated parents equals $q^2 = 1 \times 10^{-6}$ and that from first-cousin parents equals $q^2(1 - 0.062) + q(0.062) = 6 \times 10^{-5}$. Among newborns that are homozygous for a mutant allele, what is the probability that the parents are first cousins?

14.6   Evolution is accompanied by genetic changes in species.

In its broadest sense, the term **evolution** means any change in the gene pool of a population or in the allele frequencies present in a population. Evolution is possible because genetic variation exists in populations. Four processes account for most of the evolutionary changes. They form the basis of cumulative change in the genetic characteristics of populations, leading to the descent with modification that characterizes the process of evolution. Most evolutionary biologists believe that these same processes, when carried out continuously over a long time, also account for the formation of new species. These processes are

1. **Mutation**, the origin of new genetic capabilities in populations by means of spontaneous heritable changes in genes.

2. **Migration**, the movement of organisms among subpopulations within a larger population.

3. **Natural selection**, resulting from the different abilities of organisms to survive and reproduce in their environment. Natural selection is the primary process by which populations of organisms become progressively better adapted to their environments.

4. **Random genetic drift**, the random, undirected changes in allele frequency that happen by chance in all populations, but particularly in small ones.

Evolution is a population phenomenon, so it is conveniently discussed in terms of allele frequencies. In the following sections, we consider some of the population genetic implications of the major evolutionary processes.

14.7   Mutation and migration bring new alleles into populations.

Mutation is the ultimate source of genetic variation. It is an essential process in evolution, but it is a relatively weak force for changing allele frequencies, primarily because typical mutation rates are so low. Moreover, most newly arising mutations with phenotypic effects are harmful to the organism.

Migration is similar to mutation in that new alleles can be introduced into a local population, although the alleles derive from another subpopulation rather than from new mutations. In the absence of migration, the allele frequencies in each local population can change independently, so local populations may undergo considerable genetic differentiation. Genetic differentiation among subpopulations means that there are differing frequencies of common alleles among the local populations or that some local populations possess certain rare alleles not found in others. The accumulation of genetic differences among subpopulations can be minimized if some migration of organisms among the subpopulations is possible. In fact, only a relatively small amount of migration among subpopulations (on the order of just a few migrant organisms in each local population in each generation) is usually sufficient to prevent high levels of genetic differentiation. On the other hand, some genetic differentiation can accumulate in spite of migration if other evolutionary forces, such as natural selection for adaptation to the local environments, are sufficiently strong.

14.8   Natural selection favors genotypes that are better able to survive and reproduce.

The driving force of adaptive evolution is natural selection, which is a consequence of hereditary differences among organisms in their ability to survive and reproduce in the prevailing environment. Since it was first proposed by Charles Darwin in 1859, the concept of natural selection has been revised and extended, most notably by the incorporation of genetic concepts. In its modern formulation, the concept of natural selection rests on three premises:

- In all organisms, more offspring are produced than survive and reproduce.

- Organisms differ in their ability to survive and reproduce, and some of these differences are due to genotype.

- In every generation, genotypes that promote survival in the prevailing environment (favored genotypes) are present in excess at the reproductive age; hence they contribute disproportionately to the offspring of the next generation. In this way, the alleles that enhance survival and reproduction increase in frequency from generation to generation, and the population becomes progressively better able to survive and reproduce in its environment. This progressive genetic improvement in populations constitutes the process of evolutionary adaptation.

Fitness is the relative ability of genotypes to survive and reproduce.

Selection over many generations can be studied in bacterial populations because of the short generation time (about 30 minutes). **FIGURE 14.24** shows the result of competition between two bacterial genotypes, A and B. Genotype A is the superior competitor under the particular conditions. In the experiment, the competition was allowed to continue for 120 generations, during which time the proportion of A genotypes (p) increased from 0.60 to 0.9995 and that of B genotypes decreased from 0.40 to 0.0005. The data points give a satisfactory fit to an equation of the form

$$\frac{P_n}{q_n} = \left(\frac{p_o}{q_o}\right)\left(\frac{1}{w}\right)^n \qquad (14.4)$$

in which p_0 and q_0 are the initial frequencies of A and B (in this case 0.6 and 0.4, respectively), p_n and q_n are the frequencies after n generations of competition, n is the number of generations of competition, and w is a measure of the competitive ability of B when competing against A under the conditions of the experiment. The theoretical derivation of Equation (14.4) is based on the definition of w as the rate of survival and/or reproduction of genotype B, relative to genotype A, under the conditions of the experiment, and assuming that w is constant through all generations. What this means is that if the relative frequencies of $A : B$ in generation 0 are $p_0 : q_0$, then in the next generation they will be in the relative frequencies $p_0 : q_0 w$, which is to say that $p_1/q_1 = (p_0/q_0)(1/w)$. Likewise, in the next generation, the relative frequencies will be given by $p_2/q_2 = (p_1/q_1)(1/w)$, and by substituting for p_1/q_1, we obtain $p_2/q_2 = (p_0/q_0)(1/w)^2$. Continuing in this manner, generation after generation, we obtain Equation (14.4). Because $q = 1 - p$ in every generation, Equation (14.4) implies that

$$p_n = \frac{p_o}{p_o + q_o w^n} \qquad (14.5)$$

This is the smooth curve plotted in Figure 14.24 for the values $p_0 = 0.6$, $q_0 = 0.4$, and $w = 0.96$. The dots show the good fit with the experimental data.

The value of $w = 0.96$ is called the **relative fitness** of the B genotype relative to the A genotype under these particular conditions. As we have seen, the relative fitness measures the comparative contribution of

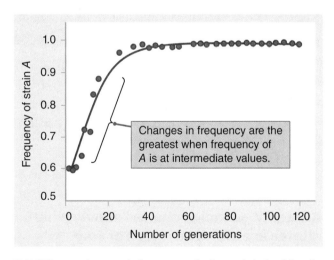

FIGURE 14.24 Increase in frequency of a favored strain of *E. coli* resulting from selection in a continuously growing population. The y value for each point is the number of A cells at any time divided by the total number of cells ($A + B$). Note that the changes in frequency are greatest when the frequency of the favored strain, A, is at intermediate values.

each parental genotype to the pool of offspring genotypes produced in each generation. A value $w = 0.96$ means that for each offspring cell produced by an A genotype, a B genotype produces, on the average, 0.96 offspring cell.

In population genetics, relative fitnesses are usually calculated with the superior genotype (A in this case) taken as the standard with a fitness of 1.0. However, the selective disadvantage of a genotype is often of greater interest than its relative fitness. The selective disadvantage of a disfavored genotype is called the **selection coefficient** associated with the genotype, and it is calculated as the difference between the fitness of the standard (taken as 1.0) and the relative fitness of the genotype in question. In the present example, the selection coefficient against B, denoted s, is

$$s = 1.000 - 0.96 = 0.04 \qquad (14.6)$$

The specific meaning of s in this example is that the selective disadvantage of strain B is 4 percent per generation. When the selection coefficient is known, Equation (14.5) also makes it possible to predict the allele frequencies in any future generation, given the original frequencies. Alternatively, it can be used to calculate the number of generations required for selection to change the allele frequencies from any specified initial values to any later ones. For example, from the relative fitnesses of A and B just estimated, what is the number of generations necessary to change the frequency of A from 0.1 to 0.8? In this particular problem, $p_0/q_0 = 0.1/0.9$, $p_n/q_n = 0.8/0.2$, and $w = 0.96$. A little manipulation of Equation (14.4) gives

$$n = [\log(0.1/0.9) - \log(0.8/0.2)]/\log(0.96)$$
$$= 87.8 \text{ generations}$$

🧠 STOP & THINK 14.4

In the example just discussed, the selection coefficient against the deleterious allele equals $s = 1 - 0.96 = 0.04$. Consider a less deleterious allele with a selection coefficient of $s = 0.02$. How long would it take for this allele to decrease from an allele frequency of 0.9 to 0.2?

Allele frequencies change slowly when alleles are either very rare or very common.

Selection in diploids is analogous to that in haploids, but dominance and recessiveness create additional complications. **FIGURE 14.25** shows the change in allele frequencies for both a favored dominant and

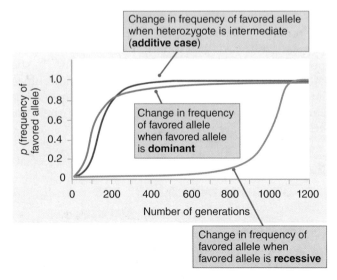

FIGURE 14.25 Theoretically expected change in frequency of an allele favored by selection in a diploid organism undergoing random mating when the favored allele is dominant or recessive or when the heterozygote is exactly intermediate in fitness (additive). In each case, the selection coefficient against the least fit homozygous genotype is 5 percent.

a favored recessive. The striking feature of the figure is that the frequency of the favored dominant allele changes very slowly when the allele is common, and the frequency of the favored recessive allele changes very slowly when the allele is rare. The reason is that rare alleles are found much more frequently in heterozygotes than in homozygotes. With a favored dominant at high frequency, most of the recessive alleles are present in heterozygotes, and the heterozygotes are not exposed to selection and hence do not contribute to change in allele frequency. Conversely, with a favored recessive at low frequency, most of the favored alleles are in heterozygotes, and again the heterozygotes are not exposed to selection and do not contribute to change in allele frequency. The principle is quite general:

KEY CONCEPT

Selection for or against recessive alleles is very inefficient when the recessive allele is rare.

One simple example of this principle is selection against a recessive lethal. In this case, it can be shown that the number of generations required to reduce the frequency of the recessive allele from q to $q/2$ equals $1/q$ generations. For example, if $q = 0.01$, then successive halvings of the frequency of the recessive allele require 100, 200, 400, 800, 1600, ... generations, so a recessive lethal allele is eliminated very slowly.

The inefficiency of selection against rare recessive alleles has an important practical implication. There is a widely held belief that medical treatment to save the lives of persons with rare recessive disorders will cause a deterioration of the human gene pool because of the reproduction of persons who carry the harmful genes. This belief is unfounded. With rare alleles, the proportion of homozygotes is so small that reproduction of the homozygotes contributes a negligible amount to change in allele frequency. Considering their low frequency, it matters little whether homozygotes reproduce or not. Similar reasoning applies to eugenic proposals to "cleanse" the human gene pool of harmful recessives by preventing the reproduction of affected persons. People with severe genetic disorders rarely reproduce anyway and, even when they do, they have essentially no effect on allele frequency. In other words,

KEY CONCEPT

The largest reservoir of harmful recessive alleles is in the genomes of heterozygous carriers, who are phenotypically normal.

Selection can be balanced by new mutations.

It is apparent from Figure 14.25 that selection tends to eliminate harmful alleles from a population. However, harmful alleles can never be eliminated totally because recurrent mutation of the normal allele continually creates new harmful alleles. These new mutations tend to replenish the harmful alleles eliminated by

 # THE HUMAN CONNECTION

Resistance in the Blood

Anthony C. Allison (1954)

Radcliffe Infirmary, Oxford, England

Protection Afforded by Sickle-Cell Trait Against Subtertian Malarial Infection

Malaria is the most prevalent infectious disease in tropical and subtropical regions of the world, infecting up to 300 million people each year and causing as many as 1 million deaths. The disease is characterized by recurrent episodes of fever with alternating shivering and sweating. Patients suffer anemia because of the destruction of red blood cells, as well as enlargement of the spleen, inflammation of the digestive system, bronchitis, and many other severe complications. The type of malaria caused by the protozoan parasite *Plasmodium falciparum* is called "subtertian" malaria because there is an interval of less than 3 days between bouts of fever. This parasite is spread through bites by *Anopheles* mosquitoes. Upon transmission, the parasites multiply in the liver for about 1 week and then begin to infect and multiply in red blood cells, which are destroyed after a few days.

In parts of Africa, the Middle East, the Mediterranean region, and India where *P. falciparum* malaria is endemic, there is also a relatively high frequency of the sickle-cell

> 66 *It became imperative to ascertain whether sickle cells can afford some degree of protection against malarial infection.* 99

mutation affecting the amino acid sequence of the beta chain of hemoglobin. Heterozygous carriers have no severe clinical symptoms, but they have the so-called "sickle-cell trait," in which the red blood cells collapse into half-moon, or sickle, shapes after 1 to 3 days when sealed under a cover slip on a microscope slide. Homozygous affected persons have "sickle-cell anemia," in which many red blood cells sickle spontaneously while still in the bloodstream, causing severe complications and often death. Why would a genetic disease that is effectively lethal when homozygous be maintained at a frequency of 10 percent or more in a population? Allison noted the correlation between the sickle-cell mutation and malaria and speculated that the sickle-cell trait gives heterozygous carriers some protection against malaria. Key evidence supporting this hypothesis is presented here. Later work showed that the heterozygous carriers have an approximately 15 percent selective advantage as a result of their malaria resistance.

(continues)

THE HUMAN CONNECTION   (continued)

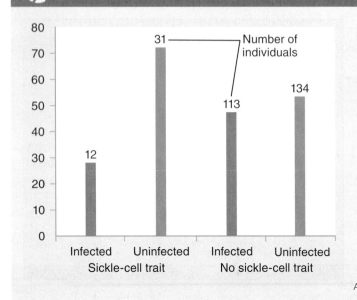

[In 1949 I noticed that] the incidence of the sickle-cell trait was higher in regions where malaria was prevalent than elsewhere, [and] it became imperative to ascertain whether sickle cells can afford some degree of protection against malarial infection. . . [As shown in the bar graph, the incidence of parasites in children with the sickle-cell trait is significantly smaller than in controls.] Children heterozygous for the sickle-cell gene will [therefore] have a selective advantage in regions where malaria is [widespread]. This fact may explain why the sickle-cell gene remains common in these areas in spite of the elimination of genes in patients dying of sickle-cell anemia.

A. A. Allison, *British Medical Journal* 1 (1954): 290–294.

selection. Eventually the population will attain a state of *equilibrium* in which the new mutations exactly balance the selective eliminations. The equilibrium state defines the **mutation–selection balance**. Two important cases, pertaining to a complete recessive and to partial dominance, must be considered. In both cases, equilibrium results from the balance of new mutations against those alleles eliminated by selection.

The allele frequency of a harmful allele maintained at equilibrium depends strongly on whether the allele is completely recessive. Because selection against a complete recessive is so inefficient when the allele is rare, even a small amount of selection against heterozygous carriers results in a dramatic reduction in the equilibrium allele frequency. For example, consider a homozygous-lethal allele that has an equilibrium frequency of 0.01 when completely recessive; if the heterozygous carriers had a relative fitness of 0.99, instead of 1.0, the equilibrium frequency would decrease to 0.001. In other words, a mere 1 percent decrease in the fitness of heterozygous carriers results in a tenfold decrease in the equilibrium frequency. The decrease is so dramatic because there are many more heterozygotes than homozygotes for the rare allele; therefore, a small amount of selection against heterozygotes affects such a relatively large number of organisms that the result is very great.

Occasionally the heterozygote is the superior genotype.

So far, we have considered only cases in which the heterozygote is intermediate in fitness between the homozygotes (or possibly equal in fitness to one homozygote). In these cases, the allele associated with the superior homozygote eventually becomes fixed, unless the selection is opposed by mutation. In this section, we consider the possibility of **heterozygote superiority**, in which the fitness of the heterozygote is greater than that of both homozygotes.

When there is heterozygote superiority, neither allele can be eliminated by selection. In each generation, the heterozygotes produce more offspring than the homozygotes, and the selection for heterozygotes keeps both alleles in the population. Selection eventually produces an equilibrium in which the allele frequencies no longer change. If the relative fitnesses of the genotypes AA, Aa, and aa are $1 - s : 1 : 1 - t$, then it can be shown that at equilibrium, the values of p and q are given by the ratio $p/q = t/s$. This formula makes good intuitive sense, because the greater the selection coefficient against aa (t), the larger the equilibrium ratio of p/q, and vice versa.

Heterozygote superiority does not appear to be a particularly common form of selection in natural populations. However, there are several well-established cases, the best known of which is the sickle-cell hemoglobin mutation (Hb^s) and its relationship to a type of malaria caused by the parasitic protozoan *Plasmodium falciparum* (**FIGURE 14.26**). The mutation in sickle-cell anemia is in the gene for β-globin, and it changes the sixth codon in the coding sequence from the normal 5'-GAG-3', which codes for glutamic acid, into the codon 5'-GUG-3', which codes for valine. In the absence of effective medical care, the Hb^s allele is virtually lethal when homozygous, yet in certain parts of Africa and the Middle East, the allele frequency reaches 10 percent or even higher. The Hb^s allele is

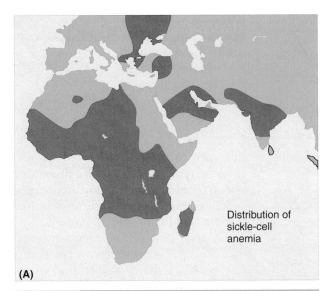

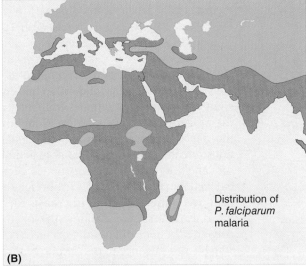

FIGURE 14.26 Geographic distribution of sickle-cell anemia (A) and *P. falciparum* malaria (B) in the 1920s, before extensive malaria-control programs were launched.

maintained because heterozygous persons are less susceptible to malaria, and when they are infected have milder infections, than homozygous-normal persons; the heterozygous genotypes therefore have the highest fitness.

14.9 Some changes in allele frequency are random.

Random genetic drift comes about because populations are not infinitely large, as we have been assuming all along, but finite (limited in size). The breeding organisms in any one generation produce a potentially infinite pool of gametes. Barring differences in fertility, the allele frequencies among gametes would equal the allele frequencies among adults. However, because of the finite size of the population, only relatively few of the gametes participate in fertilization

to form the zygotes of the next generation. In other words, a process of *sampling* takes place from one generation to the next; because there is chance variation among samples, the allele frequencies among gametes and zygotes may differ. Because of sampling variation, some individuals that are present in any generation may contribute no offspring to the following generation, purely because of chance; for this reason, random genetic drift is sometimes called *random failure to breed*.

The concrete example in **FIGURE 14.27** illustrates the essential features of random genetic drift. The graphs in part A show the number of *A* alleles in each of 12 hypothetical subpopulations, each consisting of 8 diploid individuals and all initially containing equal numbers of *A* and *a* alleles. Within each subpopulation, mating is random. For each subpopulation in each generation, a computer program was used to choose 16 gametes from among the pool of gametes produced by the subpopulation in the previous generation. The dispersion of allele frequencies resulting from random genetic drift is apparent. In larger populations than the very small populations illustrated here, the changes in allele frequency would be less pronounced and would require more generations, but the overall effect would be the same. The extent of the dispersion of allele frequency resulting from random genetic drift depends on population size; the smaller the population, the greater the dispersion and the more rapidly it takes place.

In Figure 14.27, part A, the principal effect of random genetic drift is evident even in the first few generations: The allele frequencies begin to spread out over a wider range. By the seventh generation, the spreading is extreme, and the number of *A* alleles ranges from 1 to 15. This spreading out means that the allele frequencies among the subpopulations become progressively more different. In general,

> **KEY CONCEPT**
>
> Random genetic drift causes differences in allele frequency among subpopulations and therefore causes genetic differentiation among subpopulations.

Although allele frequencies among subpopulations spread out over a wide range because of random genetic drift, the *average* allele frequency among subpopulations remains approximately constant. This point is illustrated in Figure 14.27, part B. The average allele frequency stays close to 0.5, its initial value. If an infinite number of subpopulations were being considered instead of only 12 subpopulations, then the average allele frequency would be exactly 0.5 in every generation. This principle implies that in spite of the random drift of allele frequency in individual

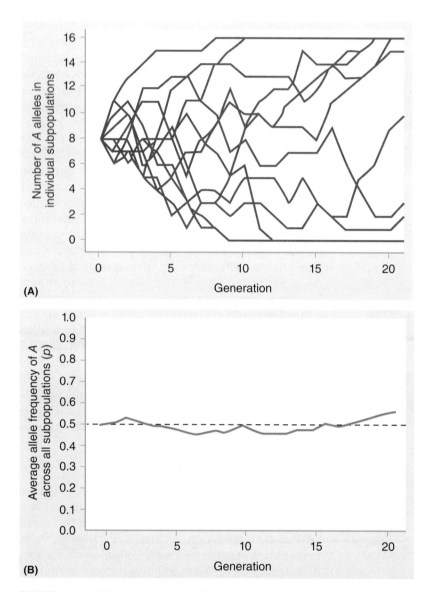

(A)

(B)

FIGURE 14.27 (A) Random genetic drift in 12 hypothetical subpopulations of 8 diploid organisms. (B) Average allele frequency among the subpopulations in part A.

In Figure 14.27, five of the fixed populations become fixed for *A* and three for *a*, which is not very different from the equal numbers expected theoretically with an infinite number of subpopulations.

A real example of random genetic drift in small experimental populations of *Drosophila* that exhibits the characteristics pointed out in connection with Figure 14.27 is illustrated in part A of **FIGURE 14.28**. The figure is based on 107 subpopulations, each initiated with eight *bw^{75}/bw* females (*bw* = brown eyes) and eight *bw^{75}/bw* males and maintained at a constant size of 16 by randomly choosing eight males and eight females from among the progeny of each generation. Note how the allele frequencies among subpopulations spread out because of random genetic drift and how subpopulations soon begin to be fixed for either *bw^{75}* or *bw*. Although the data are somewhat rough because there are only 107 subpopulations, the overall pattern of genetic differentiation has a reasonable resemblance to that expected from the theory based on the binomial distribution (Figure 14.28, part B).

If random genetic drift were the only force at work, all alleles would become either fixed or lost and there would be no polymorphism. On the other hand, many factors can act to retard or prevent the effects of random genetic drift. The most important of these factors are (1) large population size; (2) mutation and migration, which impede fixation because alleles lost by random genetic drift can be reintroduced by either process; and (3) natural selection, particularly those modes of selection that tend to maintain genetic diversity, such as heterozygote superiority.

subpopulations, the average allele frequency among a large number of subpopulations remains constant and equal to the average allele frequency among the original subpopulations.

After a sufficient number of generations of random genetic drift, some of the subpopulations become fixed for *A* and others for *a*. Because we are excluding the occurrence of mutation, a population that becomes fixed for an allele remains fixed thereafter. After 21 generations in Figure 14.27, only four of the populations are still segregating; eventually, these too will become fixed. Because the average allele frequency of *A* remains constant, it follows that a fraction p_0 of the populations will ultimately become fixed for *A* and a fraction $1 - p_0$ will become fixed for *a*. (The symbol p_0 represents the allele frequency of *A* in the initial generation.) Therefore,

Endangered species lose genetic variation.

Restricted population size adversely affects the vigor of individuals and the evolutionary potential of a population for two reasons. First, inbreeding between close or remote relatives is necessary to maintain the population, which results in loss of reproductive fitness through

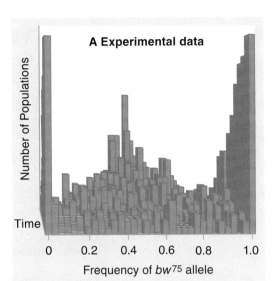

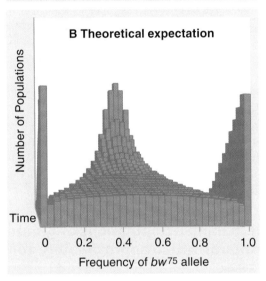

FIGURE 14.28 (A) Random genetic drift in 107 experimental populations of *Drosophila melanogaster*, each consisting of eight females and eight males. (B) Theoretical expectation of the same situation, calculated from the binomial distribution.

(A) Data from P. Buri, *Evolution* 10 (1956): 367–402.

inbreeding depression. Second, and related to the first, random genetic drift can have substantial effects on allele frequencies and the random fixation or loss of alleles. Both processes take place in most agriculturally important plants and animals, as well as in natural or captive populations of rare or endangered species. Endangered species are a focus of **conservation genetics**, which aims to conserve and restore biodiversity by minimizing the loss of genetic variation resulting from inbreeding and random genetic drift.

The genetic diversity of a gene in a population is often measured according to the proportion of heterozygous genotypes. According to the theory of random genetic drift discussed in the previous section, the proportion of heterozygous genotypes is expected to decrease by a fraction $1/(2N)$ in each generation, where N is the number of breeding individuals in the population (technically known as the *effective population size*). The theory applies to an idealized population in which the alleles are selectively neutral and the genes are unlinked.

In the real world, however, experimental studies reveal that the level of genetic diversity maintained in captive populations is often lost more rapidly than expected from the neutral theory and sometimes substantially more rapidly. In actual populations, new mutations are not necessarily selectively neutral. Many new mutations are deleterious, and, as they are eliminated by selection, alleles of linked genes along the chromosome are eliminated, too. For such a region of chromosome, the effective population size is reduced, and the more restricted the level of recombination, the more extensive the effect. Likewise, an occasional new mutation may be beneficial (especially in captive populations whose environment is quite different from that in nature), and as selection drives the beneficial allele to fixation, alleles of linked genes along the chromosome are likely to be fixed, too. Favorable mutations, therefore, also reduce the effective population size in their region of the genome, and, again, the lower the rate of recombination, the wider the effect. Therefore, because of the effect of selection acting on linked genes, different regions of the genome may have different effective population sizes.

Compared with the expected loss of genetic diversity predicted from neutral theory and the average breeding size, actual populations lose genetic diversity some 10 to 30 percent faster. The decrease in diversity is even more rapid in regions of the genome with restricted recombination, including the Y chromosome as well as mitochondrial and chloroplast DNA. A few genes show the opposite effect where genetic diversity is retained longer than expected with neutrality. These genes typically show evidence of heterozygote superiority or selection favoring rare alleles. Hence, while the neutral theory of random genetic drift is generally useful in predicting the loss of genetic diversity in populations of restricted size, it is often a best-case scenario because rates of actual loss can be greater than predicted.

14.10 Mitochondrial DNA is maternally inherited.

Mitochondrial DNA (**mtDNA**) has a number of features that make it useful for studying the genetic relationships among organisms. In higher animals, mitochondria usually show **maternal inheritance**, which means genetic transmission only through the female. The mitochondria are typically maternally inherited because the egg is the major contributor of cytoplasm to the zygote. A typical pattern of maternal inheritance of mitochondria is shown in the human pedigree in **FIGURE 14.29**. When human mitochondrial DNA is

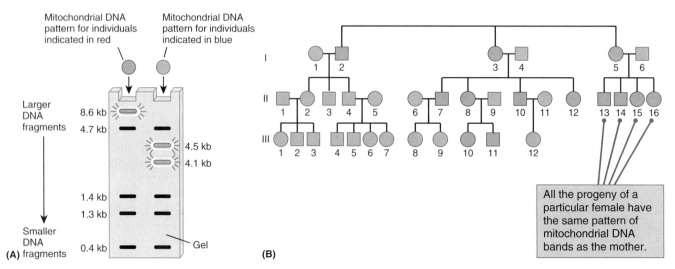

FIGURE 14.29 Maternal inheritance of human mitochondrial DNA. (A) Pattern of DNA fragments obtained when mitochondrial DNA is digested with the restriction enzyme *Hae*II. The DNA type at the left includes a fragment of 8.6 kb (pink). The DNA type at the right contains a cleavage site for *Hae*II within the 8.6-kb fragment, which results in smaller fragments of 4.5 kb and 4.1 kb (blue). (B) Pedigree showing maternal inheritance of the DNA pattern with the 8.6-kb fragment (pink symbols). The mitochondrial DNA type is transmitted only through the mother.

cleaved with the restriction enzyme *Hae*II, the cleavage products include either one fragment of 8.6 kb or two fragments of 4.5 kb and 4.1 kb (Figure 14.29, part A). The pattern with two smaller fragments is typical; it results from the presence of a *Hae*II cleavage site within the 8.6-kb fragment. Maternal inheritance of the 8.6-kb mitochondrial DNA fragment is indicated in part B, because males (I-2, II-7, and II-10) do not transmit the pattern to their progeny, whereas females (I-3, I-5, and II-8) transmit the pattern to all of their progeny. Although the mutation in the *Hae*II site yielding the 8.6-kb fragment is not associated with any disease, a number of other mutations in mitochondrial DNA do cause diseases and have similar patterns of mitochondrial inheritance. Most of these conditions decrease the ATP-generating capacity of the mitochondria and affect the function of muscle and nerve cells, particularly in the central nervous system, leading to blindness, deafness, or stroke. Many of the conditions are lethal in the absence of some normal mitochondria, and there is variable expressivity because of differences in the proportions of normal and mutant mitochondria among affected persons. The condition in which two or more genetically different types of mitochondria are present in the same cell is unusual among animals. For example, a typical human cell contains from 1000 to 10,000 mitochondria, all of them genetically identical.

Human mtDNA evolves changes in sequence at an approximately constant rate.

Another important feature of mitochondrial DNA is that it does not undergo genetic recombination. Hence the DNA molecule in any mitochondrion derives from a single ancestral molecule. Mitochondrial DNA is also a good genetic marker for tracing human ancestry, because it evolves considerably more rapidly than that of nuclear genes. Differences in mitochondrial DNA sequences accumulate among human lineages at a rate of approximately one change per mitochondrial lineage every 3800 years.

Modern human populations originated in subsaharan Africa approximately 200,000 years ago.

Nucleotide differences in mitochondrial DNA have been used to reconstruct the probable historical relationships among human populations. **FIGURE 14.30** shows the gene tree of mtDNA based on the complete mtDNA sequences from 53 persons representing human populations from throughout the world. Because mtDNA undergoes no recombination, the tree is the genetic history of only a single locus; and because mtDNA is maternally inherited, it is a genetic history of females. Nevertheless, the phylogenetic tree shows several remarkable features:

- Much of the mtDNA diversity in African populations is not found among non-Africans; on the average, the mtDNA of Africans shows about twice as much genetic variation as the mtDNA among non-Africans.

- Three of the four major lineages of mtDNA are found only in subsaharan Africans (green); the age of the **most recent common ancestor (MRCA)** of these sequences (green circle) is approximately 170,000 ± 50,000 years, or roughly 200,000 years.

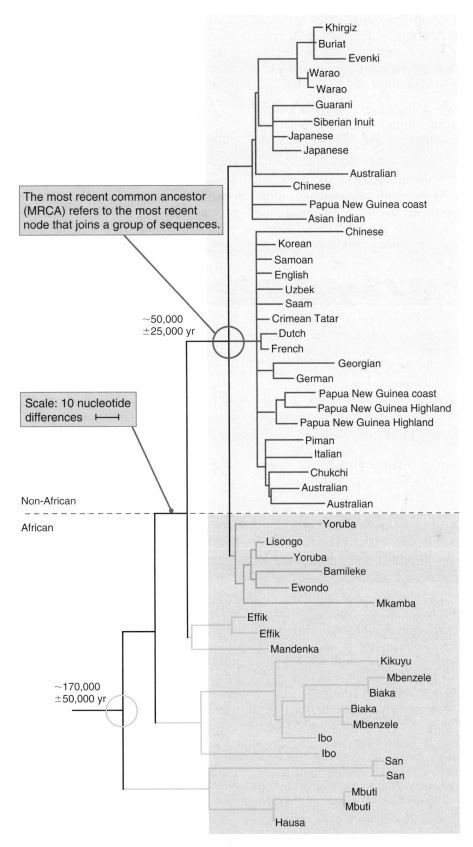

The most recent common ancestor (MRCA) refers to the most recent node that joins a group of sequences.

~50,000 ±25,000 yr

Scale: 10 nucleotide differences

Non-African

African

~170,000 ±50,000 yr

Khirgiz
Buriat
Evenki
Warao
Warao
Guarani
Siberian Inuit
Japanese
Japanese
Australian
Chinese
Papua New Guinea coast
Asian Indian
Chinese
Korean
Samoan
English
Uzbek
Saam
Crimean Tatar
Dutch
French
Georgian
German
Papua New Guinea coast
Papua New Guinea Highland
Papua New Guinea Highland
Piman
Italian
Chukchi
Australian
Australian
Yoruba
Lisongo
Yoruba
Bamileke
Ewondo
Mkamba
Effik
Effik
Mandenka
Kikuyu
Mbenzele
Biaka
Biaka
Mbenzele
Ibo
Ibo
San
San
Mbuti
Mbuti
Hausa

FIGURE 14.30 Phylogenetic tree of human mitochondrial DNA based on analysis of the complete nucleotide sequence of mtDNA from 53 persons.

Data from M. Ingman, et al., *Nature* 408 (2000): 708–713.

- A restricted subset of mtDNA lineages is found among non-Africans (purple and red); these sequences share a more recent MRCA with five African mtDNAs (orange) than these African mtDNAs share with other Africans.

- The age of the MRCA of the mtDNA lineage joining African and non-African populations (blue circle) is approximately 50,000 ± 25,000 years.

These features of the mtDNA tree are consistent with a widely accepted scenario in which all modern human populations are derived from a migration out of Africa that took place approximately 200,000 years ago. There were earlier migrations out of Africa as well, but the descendants of these earlier migrants were largely displaced by those who came later. Although the archaic human populations that overlapped with the ancestors of modern humans later became extinct, they did interbreed to a limited extent with the ancestors of modern humans and left traces of their existence in a small percentage of the nuclear genomes of most modern human populations except those of native Africans. Although evidence of interbreeding is clear from sequences of ancient DNA from bones of the archaic human populations, the mitochondrial DNA in modern human populations all derives from that of a single ancestral human female who lived in Africa about 200,000 years ago.

CHAPTER SUMMARY

- DNA and protein sequences change through time, and the evolutionary relationships among different species can be inferred from comparisons among their DNA and protein sequences.

- Many genes in natural populations are polymorphic; they have two or more common alleles.

- With random mating, the alleles in gametes are combined at random to form the zygotes of the next generation.

- Genetic polymorphisms can be used as genetic markers in pedigree studies and for individual identification (DNA typing).

- Inbreeding means mating between relatives. Relative to the frequencies of genotypes expected with random mating, inbreeding results in an excess of homozygous genotypes.

- Mutation and migration introduce new alleles into populations.

- Natural selection and random genetic drift are the usual causes of change in allele frequency; selection changes allele frequency in a systematic direction, whereas random genetic drift changes allele frequency in an unpredictable direction.

- Analysis of human mitochondrial DNA sequences is consistent with a scenario in which modern humans evolved in southern Africa approximately 200,000 years ago and became the progenitors of all present-day populations existing outside of Africa.

ISSUES AND IDEAS

- What is a gene tree? A species tree? Must they always agree? Explain why or why not.

- Distinguish between orthologous genes and paralogous genes. Which type of duplication provides the raw material for the evolution of new gene functions?

- What does the Hardy–Weinberg principle imply about the relative frequencies of heterozygous carriers and homozygous affected organisms for a rare, harmful recessive allele?

- Traits due to recessive alleles in the X chromosome are usually much more prevalent in males than in females. Explain why this discrepancy is expected with random mating.

- Why are the effects of inbreeding more easily observed with a rare recessive allele than with a rare dominant allele?

- What is the fitness of an organism that dies before the age of reproduction? What is the fitness of an organism that is sterile?

- Many recessive alleles are extremely harmful when homozygous, so there is selection in every generation that tends to reduce the allele frequency. Yet harmful recessive alleles are maintained at a low frequency for almost every gene. What process prevents harmful recessive alleles from being completely eliminated?

- Heterozygote superiority of the type observed with sickle-cell anemia is sometimes called balancing selection. Do you think this is an appropriate term? Explain why or why not.

- What is random genetic drift and why does it occur? Explain why this process implies that in the absence of other forces, the ancestry of all alleles present at a locus in a population can eventually be traced back to a single allele present in some ancestral population.

SOLUTIONS: STEP BY STEP

PROBLEM 1 The DNA sequences of genes *A*, *B*, and *C* are compared in humans (Hu), chimpanzees (Ch), and gorillas (Go). *Gene A* yields the gene tree designated **(1)** in the accompanying illustration, in which the alleles in humans and chimpanzees are more similar to each other than they are to the allele in gorillas. *Gene B* yields the result in **(2)**, in which the allele in humans is more similar to that in gorillas than to that in chimpanzees. And *gene C* yields yet a third pattern indicated in **(3)**. As indicated at the bottom of each gene tree, each result implies a different evolutionary relationship among humans, chimpanzees, and gorillas.

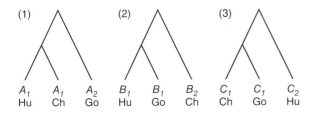

(a) How can such discrepancies between gene trees and species trees originate?
(b) Based on the data from these three genes alone, could you make any judgment about the likely evolutionary relationships among the species?
(c) What additional molecular data might resolve the evolutionary relationships among the species?

SOLUTION. This problem is a typical example of lack of congruence between gene trees and species trees often encountered when comparing genes among closely related species. **(a)** The discrepancies result from the way genetic polymorphisms in the ancestral population become sorted out among the species as they evolve from a common ancestor. The process is known as *lineage sorting*. In this case, the most likely scenario is that *gene A*, *gene B*, and *gene C* were all polymorphic in the common ancestor. The alleles of *gene A* sorted according to pattern **(1)**, those of *gene B* according to pattern **(2)**, and those of *gene C* according to pattern **(3)**. **(b)** If the only data available were data from these three genes, one could not justify any statement about the relative likelihood of any of the species trees. **(c)** Additional molecular data that would clarify the species tree could come from examining many more genes. If one pattern predominated, then the species tree corresponding to that pattern is likely to be correct. In the case of humans, chimpanzees, and gorillas, comparisons of genes across the entire gene give overwhelming support to the species tree depicted in **(1)**.

PROBLEM 2 In human populations, a locus called *secretor* determines whether the carbohydrate A and B antigens of the ABO blood groups are secreted into the saliva and other body fluids. The genotypes *Se/Se* and *Se/se* are secretors, and the genotype *se/se* is not. Among Caucasians known to have blood group A, B, or AB because of the presence of these antigens on the red blood cells, the frequency of the nonsecretor phenotype is about 33 percent.
(a) Assuming random mating, what is the allele frequency of the *se* allele?
(b) Among Caucasians with blood group A, B, or AB, what are the expected proportions of homozygous secretors and heterozygous secretors?

SOLUTION. This is a typical problem that makes use of the Hardy–Weinberg principle for random mating. (a) With Hardy–Weinberg proportions, the frequency of homozygous recessive genotypes equals q^2, which in this case is given as 0.33. Therefore,

$$q = \sqrt{0.33} = 0.574$$

so $p = 1 - q = 0.426$. (b) The expected proportions of *Se/Se* and *Se/se* genotypes are $p^2 = (0.426)^2 = 0.181$ and $2pq = 2(0.426)(0.574) = 0.489$, respectively. Note that approximately half of the genotypes in the population are heterozygous for the *secretor* gene.

PROBLEM 3 Sanfilippo syndrome is an inborn error of metabolism. Affected children develop quite normally in their early years, but the disease grows worse as time goes on. They have difficulties in learning how to speak and exhibit behavioral disorders including temper tantrums, hyperactivity, destructiveness, and aggression. The affected children gradually lose mobility and become unresponsive and usually die in their late teens or twenties. The disorder is an autosomal recessive that affects about 1 in 50,000 newborns when the parents are unrelated.
(a) What is the expected frequency of affected children among the offspring of first-cousin matings?
(b) What is the relative risk (ratio) of the disease among the offspring of first cousins relative to that among the offspring of unrelated individuals?

SOLUTION. First we must calculate the frequency of the recessive allele, q, using the information that the frequency of homozygous recessives is 1 in 50,000. Assuming random mating frequencies,

$$q = \sqrt{(1/50{,}000)} = 0.00447$$

(a) Among the offspring of first-cousin matings, the expected frequency of Sanfilippo syndrome equals

$$q^2(1 - F) + qF,$$

where $F = 1/16$ for the offspring of first cousins. In this case the formula yields $(0.00447)^2 + (0.00447)(1/16) = 0.000298$, or about 1 in 3353. **(b)** Because $q^2 = 0.00002$ is the frequency of affected individuals among the offspring of unrelated individuals, the relative risk due to first-cousin mating is 0.000298/0.00002 = 14.9. In other words, first-cousin mating increases the risk of Sanfilippo syndrome by a factor of almost 15.

CONCEPTS IN ACTION: PROBLEMS FOR SOLUTION

14.1 Given population genotype frequencies $A_1A_1 = 0.20$, $A_1A_2 = 0.30$, and $A_2A_2 = 0.50$, what is the frequency of the A_1 allele in the population?

14.2 Assuming no selection, random genetic drift, migration, or mutation affects the alleles at the locus in problem 14.1, what is the expected frequency of A_2 in the next generation?

14.3 If the genotype AA is an embryonic lethal and the genotype aa is fully viable but sterile, what genotype frequencies would be found in adults in an equilibrium population containing the A and a alleles? Is it necessary to assume random mating?

14.4 Consider a population that, for unknown reasons, is not in Hardy–Weinberg equilibrium for some gene.

(a) Can you calculate the genotype frequencies if you know the allele frequencies?

(b) Can you calculate the allele frequencies if you know the genotype frequencies?

14.5 A trait due to a recessive X-linked allele in a large, randomly mating population affects 1 in 40 males. What is the frequency of heterozygous carrier females? What is the expected frequency of affected females?

14.6 DNA from 100 unrelated people was digested with the restriction enzyme *Hind* III, and the resulting fragments were separated and probed with a sequence for a particular gene. Four fragment lengths that hybridized with the probe were observed, which were of size 5.7, 6.0, 6.2, and 6.5 kb. Each fragment defines a different restriction-fragment allele. The accompanying illustration shows the gel patterns observed among the sample. The number of individuals with each gel pattern is shown across the top. Estimate the allele frequencies of the four restriction-fragment alleles.

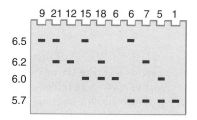

14.7 A condition called nonsyndromic recessive auditory neuropathy results in deafness. One form is caused by a recessive autosomal gene. The frequency of affected individuals due to the mutant gene in one population is 0.002. Assuming Hardy–Weinberg equilibrium, what is the expected incidence of the disorder among the offspring of matings in which both parents are heterozygous carriers?

14.8 Consider a population in which 16 percent of the people exhibit a trait due to a homozygous recessive mutation. What is the allele frequency for the dominant allele if a population is in Hardy–Weinberg equilibrium?

14.9 How does the recessive allele that causes Tay-Sachs disease survive in a population if all affected individuals die before they can reproduce?

14.10 In a population of Austrians, the frequency of alleles determining the ABO blood type groups were estimated as $I^A = 0.20$, $I^B = 0.15$, $I^O = 0.65$. What are the expected ABO genotype frequencies?

14.11 Suppose a randomly mating diploid population has n equally frequent alleles of an autosomal locus. What is the expected frequency of:

(a) Any specified homozygous genotype?

(b) Any specified heterozygous genotype?

(c) All homozygous genotypes together?

(d) All heterozygous genotypes together?

14.12 Which of the following genotype frequencies of AA, Aa, and aa, respectively, satisfy the Hardy–Weinberg principle?

(a) 0.25, 0.50, 0.25

(b) 0.36, 0.55, 0.09

(c) 0.49, 0.42, 0.09

(d) 0.64, 0.27, 0.09

(e) 0.29, 0.42, 0.29

14.13 Suppose that a population has n copies of a mutant allele of a gene in a population of N diploid individuals. If the mutant allele is selectively neutral and mutation and migration can be ignored, what is the probability that the frequency of the mutant allele eventually reaches 1.0 owing to the effects of random genetic drift?

14.14 Galactosemia is an autosomal recessive condition associated with liver enlargement, cataracts, and mental retardation. Among the offspring of unrelated individuals, the frequency of galactosemia is 1×10^{-5}. What is the expected frequency among the offspring of first cousins ($F = 1/16$) and among the offspring of second cousins ($F = 1/64$)?

14.15 A man with normal parents whose brother has phenylketonuria marries a phenotypically normal woman. In the general population, the incidence of phenylketonuria at birth is approximately 1 in 10,000. Assuming Hardy–Weinberg proportions:

(a) What is the probability that the man is a carrier?

(b) What is the probability that the wife is also a carrier?

(c) What is the probability that their first child will be affected?

14.16 Electrophoretic differences in the protein alcohol dehydrogenase in the flowering plant *Phlox drummondii* are determined by codominant alleles of a single gene. In one sample of 35 plants, the following data were obtained:

Genotype AA AB BB BC CC AC
Number   4 10 24 20 10   2

What are the allele frequencies in this sample? With random mating, what are the expected numbers of each of the genotypes?

14.17 Hartnup disease is an autosomal recessive disorder of intestinal and renal transport of amino acids. The frequency of affected newborn infants is about 1 in 14,000. Assuming random mating, what is the frequency of heterozygotes?

14.18 Self-fertilization in the annual plant *Phlox cuspidata* results in an average inbreeding coefficient of $F = 0.66$.

(a) What frequencies of the genotypes for the enzyme phosphoglucose isomerase would be expected in a population with alleles A and B at respective frequencies 0.43 and 0.57?

(b) What frequencies of the genotypes would be expected with random mating?

14.19 The table shown here summarizes the number of synonymous differences in orthologous protein-coding genes in four species denoted A, B, C, and D.

	Species A	Species B	Species C	Species D
Species A	—			
Species B	25	—		
Species C	23	16	—	
Species D	36	35	33	—

(a) What gene-tree topology best fits the data?

(b) If the number of synonymous differences in this gene conforms to a molecular clock, and species B and C diverged 6 million years ago, then what is the estimated divergence time between species A and D?

14.20 Two strains of bacteria, A and B, are placed into direct competition in a chemostat. A is favored over B. If the selection coefficient per generation is constant, what is its value if, in an interval of 100 generations:

(a) The ratio of A cells to B cells increases by 10 percent?

(b) The ratio of A cells to B cells increases by 90 percent?

(c) The ratio of A cells to B cells increases by a factor of 2?

14.21 An allele A undergoes mutation to the allele a at the rate of 10^{-5} per generation. If a very large population is fixed for A (generation 0), what is the expected frequency of A in the following generation (generation 1)? What is the expected frequency of A in generation 2? Deduce the rule for the frequency of A in generation n.

STOP & THINK ANSWERS

ANSWER TO STOP & THINK 14.1

The estimated allele frequency of F is $p = (2 \times 48 + 38)/200 = 0.67$ and that of S is $q = (38 + 2 \times 14)/200 = 0.33$. Assuming random mating, the expected numbers of F/F, F/S, and S/S are, respectively, $(0.67)^2 \times 100 = 44.89$, $2 \times 0.67 \times 0.33 \times 100 = 44.22$, and $(0.33)^2 \times 100 = 10.89$. If you perform a chi-square test for goodness of fit, you will find these numbers in satisfactory agreement with the Hardy–Weinberg principle.

ANSWER TO STOP & THINK 14.2

(a) The DNA samples from the six adults have no bands in common, so the data are consistent with the presence of six unrelated adults. (b) Comparing the bands from the DNA from the juveniles with those from the adults indicates that only male E could have contributed DNA for fragment size 4 in individual G and for fragment size 10 in individuals H and I. Similarly, only female C could have contributed DNA for fragment size 5 in all three juveniles.

(c) Consequently, the remains C are consistent with being those of Tsarina Alexandra, who was evidently homozygous for this STR, and the remains E are consistent with being those of Tsar Nicholas II. The remains of the Tsarevitch Alexis and one of the daughters (thought to be Maria) were found and identified in 2007.

ANSWER TO STOP & THINK 14.3

Across the entire population, the number of affected offspring from first-cousin matings is proportional to $(0.01)(6 \times 10^{-5}) = 6 \times 10^{-7}$ and the number from unrelated parents is proportional to $(1 - 0.01)(1 \times 10^{-6}) = 9.9 \times 10^{-7}$. The probability that an albino offspring has first-cousin parents therefore equals

$(6 \times 10^{-7})/(6 \times 10^{-7} + 9.9 \times 10^{-7}) = 0.38$. This calculation shows that, while first-cousin matings account for only 1 percent of all matings, they account for almost 40 percent of matings that produce homozygous recessive offspring.

ANSWER TO STOP & THINK 14.4

In this case $w = 1 - s = 1 - 0.02 = 0.98$, and therefore $n = [\log(0.1/0.9) - \log(0.8/0.2)]/\log(0.98) = 177.4$ generations, or about twice as long as when $s = 0.04$. The underlying principle is that, when selection is not too strong, the selection coefficient is inversely related to the time required for a specified change in allele frequency.

Complex traits are determined by multiple genetic and environmental factors acting together.
© Yuganov Konstantin/Shutterstock

CHAPTER 15

The Genetic Basis of Complex Traits

LEARNING OBJECTIVES

- To define a complex trait, and distinguish continuous (quantitative), categorical, and threshold traits.

- Among sources of phenotypic variation of a complex trait, to distinguish between variation as a result of genotype, environment, genotype-by-environment interaction, and genotype-by-environment association.

- For a trait in which the phenotype is normally distributed in a population with a given mean and variance, to calculate the range of phenotypes that are expected to include 95 percent of the population and 99 percent of the population.

- For artificial selection of a quantitative trait, to calculate the expected value of the mean of the progeny in the next generation given the mean phenotype in a random-mating population, the threshold value of phenotype to be included among the selected parents, and the narrow-sense heritability of the trait.

- In studies to identify genes affecting a complex trait, to distinguish between a quantitative trait locus (QTL) and a candidate gene, and explain how genome-wide association studies are used to detect QTLs.

Many traits that are important in medical genetics, animal breeding, and plant breeding are influenced by multiple genes as well as by the effects of environment. These are known as **multifactorial traits** because of the multiple genetic and environmental factors implicated in their causation. With a multifactorial trait, a single genotype can have any one of many possible phenotypes (depending on the environment), and similar phenotypes can result from many different genotypes.

Multifactorial traits are often called **complex traits** because each factor that affects the trait contributes, at most, a modest amount to the total variation in the trait observed in the entire population. Most traits that vary in populations of humans and other organisms, including common human diseases that have a genetic component, are complex traits. For a complex trait, the **genetic architecture** consists of a description of all of the genetic and environmental factors that affect the trait, along with the magnitudes of their individual effects and the magnitudes of interactions among the factors. It is, in principle, possible to define the genetic components in terms of Mendelian segregation and locations along a genetic map. Environmental factors are much less easily partitioned into separate factors whose individual effects and interactions can be sorted out. The genetic analysis of complex traits requires special concepts and methods, which are introduced in this chapter.

15.1 Complex traits are determined by multiple genes and the environment.

Complex or multifactorial traits are often called **quantitative traits** because the phenotypes in a population differ in quantity rather than in type. A trait such as height is a quantitative trait. Heights are not found in discrete categories but differ merely in quantity from one person to the next. The opposite of a quantitative trait is a "discrete trait," in which the phenotypes differ in kind—for example, brown eyes versus blue eyes.

Quantitative traits are typically influenced not only by the alleles of two or more genes but also by the effects of the environment. Therefore, with a quantitative trait, the phenotype of an organism is potentially influenced by

- **Genetic factors** in the form of alternative genotypes of one or more genes.

- **Environmental factors** in the form of conditions that are favorable or unfavorable for the development of the trait. Examples include the effect of smoking on the development of lung

cancer in human beings, that of nutrition on the growth rate of animals, and that of fertilizer, rainfall, and planting density on the yield of crop plants.

With some quantitative traits, differences in phenotype result largely from differences in genotype, and the environment plays a minor role. With other quantitative traits, differences in phenotype result largely from the effects of environment, and genetic factors play a minor role. Most quantitative traits fall between these extremes, so both genotype and environment must be taken into account in their analysis.

In a genetically heterogeneous population, many genotypes are formed by the processes of segregation and recombination. Variation in genotype can be eliminated by studying *inbred lines*, which are homozygous for most genes, or the F_1 progeny from a cross of inbred lines, which are uniformly heterozygous for all genes in which the parental inbreds differ (**FIGURE 15.1**). In contrast, complete elimination of environmental variation is impossible, no matter how hard the experimenter may try to render the environment identical for all members of a population. With plants, for example, small variations in soil quality or exposure to the sun will produce slightly different environments, sometimes even for adjacent plants. Similarly, highly inbred *Drosophila* still show variation in phenotype (for example, body size) brought about by environmental differences among animals within the same culture bottle. Therefore, traits that are susceptible to small environmental effects will never be uniform, even in inbred lines.

Important quantitative traits in human genetics include infant growth rate, adult weight, blood pressure, serum cholesterol, and length of life. In plant and animal breeding, traits of key economic importance are often quantitative traits. One economically important quantitative trait in crop plants is yield per unit area—whether it be the yield of corn, tomatoes, soybeans, or grapes. In domestic animals, important quantitative traits include meat quality, milk production per cow, egg laying per hen, fleece weight per sheep, and litter size per sow. In evolutionary studies, fitness is the preeminent quantitative trait. Most quantitative traits cannot be studied by means of the usual pedigree methods, because the effects of segregation of alleles of one gene may be concealed by effects of other genes, and because environmental effects may cause identical genotypes to have different phenotypes. Therefore, individual pedigrees of quantitative traits do not fit any simple pattern of dominance, recessiveness, or X linkage. Nevertheless, genetic effects on quantitative traits can be assessed by comparing the phenotypes of relatives who, because of their familial relationship, must have a certain proportion of their genes in common.

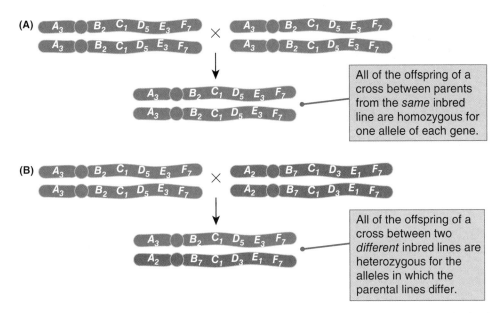

FIGURE 15.1 By definition, a completely inbred line is homozygous for every gene. A population of organisms, all identical in genotype, can be created by crossing inbred parents. (A) If the parents are from the same inbred line, the progeny are all genetically identical and homozygous. (B) If the parents are from different inbred lines, the progeny are all genetically identical but heterozygous for alleles at which the parental inbred lines differ.

Three categories of traits are frequently found to have multifactorial inheritance. They are described in the following section.

Continuous, categorical, and threshold traits are usually multifactorial.

Most phenotypic variation in populations is not manifested in a few easily distinguished categories. Instead, the traits vary continuously from one phenotypic extreme to the other, with no clear-cut breaks in between. Human height is a prime example of such a trait. Other examples include milk production in cattle, growth rate in poultry, yield in corn, and blood pressure in human beings. Such traits are called **continuous traits** because there is a continuous gradation from one phenotype to the next. The range of phenotypes is continuous, from minimum to maximum, with no clear categories. Weight is an example of a continuous trait because the weight of an organism can fall anywhere along a continuous scale of weights, so the number of possible phenotypes is virtually unlimited.

Two other types of quantitative traits are not continuous.

Categorical traits are traits in which the phenotype corresponds to any one of a number of discrete categories. Typically the phenotype corresponds to a count of, for example, the number of skin ridges forming the fingerprints, number of kernels on an ear of corn, number of eggs laid by a hen, number of bristles on the abdomen of a fly, and number of puppies in a litter. An example of a categorical trait is the number of ears on a stalk of corn, which typically has the value 1, 2, 3, or 4 ears on a given stalk.

Threshold traits are traits that have only two, or a few, phenotypic classes, but their inheritance is determined by the effects of multiple genes together with the environment. Examples of threshold traits include twinning in cattle and parthenogenesis (development of unfertilized eggs) in turkeys. In a threshold trait, each organism has an underlying and not directly observable predisposition to express the trait, such as a predisposition for a cow to give birth to twins. If the underlying predisposition is high enough (above a "threshold"), the cow will actually give birth to twins; otherwise, she will give birth to a single calf. Many human diseases with a genetic component are threshold-trait disorders, and the phenotypic classes are "affected" versus "not affected." Examples of such disorders include adult-onset diabetes, schizophrenia, and many congenital abnormalities, such as spina bifida. Threshold traits can be interpreted as continuous traits by imagining that each individual has an underlying risk or *liability* toward manifestation of the condition. A liability above a certain cutoff, or *threshold*, results in expression of the condition; a liability below the threshold results in normality. The liability of an individual to a threshold trait cannot be observed directly, but inferences about liability can be drawn from the incidence of the condition among individuals and their relatives.

The distribution of a trait in a population implies nothing about its inheritance.

The **distribution** of a trait in a population is a description of the population in terms of the proportion of individuals that have each of the possible phenotypes.

TABLE 15.1 Distribution of Height Among British Women

Interval number, i	Height interval (inches)	Midpoint, x_i	Number of women, f_i
1	53–55	54	5
2	55–57	56	33
3	57–59	58	254
4	59–61	60	813
5	61–63	62	1340
6	63–65	64	1454
7	65–67	66	750
8	67–69	68	275
9	69–71	70	56
10	71–73	72	11
11	73–75	74	4
			Total N = 4995

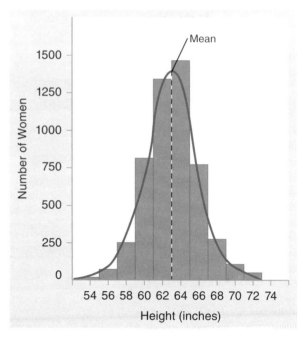

FIGURE 15.2 Distribution of height among 4995 British women and the smooth normal distribution that approximates the data.

FIGURE 15.3 Genetics scholars at the University of Connecticut in Storrs, who have helpfully arranged themselves by height to form a "living histogram."

Peter Morenus/UConn.

Characterizing the distribution of some traits is straightforward because the number of phenotypic classes is small. For example, the distribution of progeny in a certain pea cross may consist of 3/4 green seeds and 1/4 yellow seeds, and the distribution of ABO blood groups among one sample of Greeks may consist of 42 percent O, 39 percent A, 14 percent B, and 5 percent AB. However, with continuous traits, the large number of possible phenotypes makes such summaries impractical. Often, it is convenient to reduce the number of phenotypic classes by grouping similar phenotypes together. Data for an example pertaining to the distribution of height among 4995 British women are given in **TABLE 15.1** and in **FIGURE 15.2**. You can imagine each bar in the graph in Figure 15.2 being built step by step, as each of the women is measured, by placing a small square along the *x*-axis at the location corresponding to the height of each woman. As sampling proceeds, the squares begin to pile up in certain places, leading ultimately to the bar graph shown.

It is worth taking a moment to consider a more vivid example of what a histogram like that in Figure 15.2 really represents. The "real thing" is shown in **FIGURE 15.3**, which is a picture of 162 scholars in genetics at the University of Connecticut, Storrs, who

have arranged themselves in order of their height. The women are in white, the men in blue. This is a "living histogram," in which each building block not merely represents a person but actually *is* a person.

Displaying a distribution completely, either in tabular form, as in Table 15.1, or in graphical form, as in Figure 15.2 (or even Figure 15.3), is always helpful but often unnecessary. In many cases a description of the distribution in terms of two major features is sufficient. These features are the *mean* and the *variance*. To discuss the mean and the variance in quantitative terms, we shall use the data in Table 15.1. The height intervals are numbered from 1 (53 to 55 inches) to 11

(73 to 75 inches). The symbol x_i designates the midpoint of the height interval numbered i; for example, $x_1 = 54$ inches, $x_2 = 56$ inches, and so on. The number of women in height interval i is designated f_i; for example, $f_1 = 5$ women, $f_2 = 33$ women, and so forth. The total size of the sample—in this case 4995—is denoted N. The mean and variance serve to characterize the distribution of height among these women as well as the distribution of many other quantitative traits.

- The **mean**, or average, is the peak of the distribution. The mean of a population is estimated from a sample of individuals from the population as follows:

$$\bar{x} = \frac{\sum f_i x_i}{N} \qquad (15.1)$$

where $\bar{x}$ is the estimate of the mean and Σ symbolizes summation over all classes of data (in this example, summation over all 11 height intervals). In Table 15.1, the mean height in the sample of women is 63.1 inches.

- The **variance** is a measure of the spread of the distribution and is estimated in terms of the squared *deviation* (difference) of each observation from the mean. The variance is estimated from a sample of individuals as follows:

$$s^2 = \frac{\sum f_i (x_i - \bar{x})^2}{N-1} \qquad (15.2)$$

where s^2 is the estimated variance and x_i, f_i, and N are as in Table 15.1. Note that $(x_i - \bar{x})$ is the difference from the mean of each height category, and that the denominator is the total number of individuals minus 1. The variance describes the extent to which the phenotypes are clustered around the mean, as shown in **FIGURE 15.4**. A large value implies that the distribution is spread out, and a small value implies that it is clustered near the mean. From the data in Table 15.1, the variance of the population of British women is estimated as $s^2 = 7.24$ in^2.

A quantity closely related to the variance—the **standard deviation** of the distribution—is defined as the square root of the variance. For the data in Table 15.1, the estimated standard deviation s is obtained from Equation 15.2 as $s = (s^2)^{1/2} = (7.24 \text{ in}^2)^{1/2} = 2.69$ inches. The standard deviation has the useful feature of having the same units of dimension as the mean—in this example, inches.

When the data are symmetrical, or approximately symmetrical, the distribution of a trait can often be approximated by a smooth, bell-shaped curve of the

type shown in Figure 15.4. The bell curve is called the **normal distribution**. Because the normal curve is symmetrical, half of its area is determined by points with values greater than the mean and half by points with values less than the mean; thus the proportion of phenotypes that exceed the mean is 1/2. One important characteristic of the normal distribution is that the entire distribution is completely determined by the value of the mean and the variance.

The mean and standard deviation (square root of the variance) of a normal distribution provide a great deal of information about the distribution of phenotypes in a population, as is illustrated in **FIGURE 15.5**. Specifically, for a normal distribution,

1. Approximately 68 percent of the population have a phenotype within *one* standard deviation of the mean (in the symbols of Figure 15.5, between $\mu - \sigma$ and $\mu + \sigma$).

2. Approximately 95 percent lie within *two* standard deviations of the mean (between $\mu - 2\sigma$ and $\mu + 2\sigma$).

3. Approximately 99.7 percent lie within *three* standard deviations of the mean (between $\mu - 3\sigma$ and $\mu + 3\sigma$).

Applying these rules to the data in Figure 15.2, in which the mean and standard deviation are 63.1 and 2.69 inches, reveals that approximately 68 percent of the women are expected to have heights in the range from $63.1 - 2.69$ inches to $63.1 + 2.69$ inches (that is, 60.4 to 65.8), and approximately 95 percent are expected to have heights in the range from $63.1 - 2 \times 2.69$ inches to $63.1 + 2 \times 2.69$ inches (that is, 57.7 to 68.5).

Real data frequently conform to the normal distribution. Normal distributions are usually the rule when

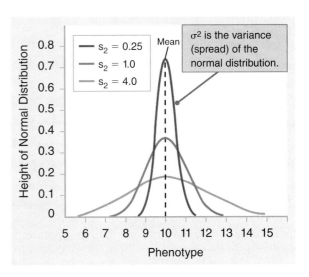

FIGURE 15.4 Graphs showing that the variance of a distribution measures the spread of the distribution around the mean. The area under each curve covering any range of phenotypes equals the proportion of individuals having phenotypes within the range.

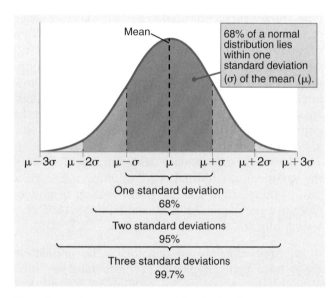

FIGURE 15.5 Features of a normal distribution. The proportions of individuals lying within one, two, and three standard deviations from the mean are approximately 68 percent, 95 percent, and 99.7 percent, respectively. In this normal distribution, the mean is symbolized μ and the standard deviation is σ.

the phenotype is determined by the cumulative effect of many individually small independent factors. This is the case for many multifactorial traits.

STOP & THINK 15.1

A population of wild radish (*Raphanus raphanistrum*) has an average number of flowers per plant that is normally distributed with a mean of 18 and a standard deviation of 3. What proportion of plants is expected to have a flower number between 15 and 21? Between 12 and 24? Less than 12? Greater than 24?

15.2 Variation in a trait can be separated into genetic and environmental components.

In considering the genetics of multifactorial traits, an important objective is to assess the relative importance of genotype versus environment. In some cases in experimental organisms, it is possible to separate genotype and environment with respect to their effects on the mean. For example, a plant breeder may study the yield of a series of inbred lines grown in a group of environments that differ in planting density or amount of fertilizer. It would then be possible to:

1. Compare yields of the same genotype grown in different environments and thereby rank the *environments* relative to their effects on yield, or

2. Compare yields of different genotypes grown in the same environment and thereby rank the *genotypes* relative to their effects on yield.

Such a fine discrimination between genetic and environmental effects is not usually possible, particularly in human quantitative genetics. For example, with regard to the height of the women in Figure 15.2, environment could be considered favorable or unfavorable for tall stature only in comparison with the mean height of a genetically identical population reared in a different environment. This reference population does not exist. Likewise, the genetic composition of the population could be judged as favorable or unfavorable for tall stature only in comparison with the mean of a genetically different population reared in an identical environment. This reference population does not exist, either.

Without such standards of comparison, it is impossible to determine the genetic versus environmental effects on the mean. However, it is still possible to assess genetic versus environmental contributions to the *variance*, because instead of comparing the means of two or more populations, we can compare the phenotypes of individuals within the *same* population. Some of the differences in phenotype result from differences in genotype and others from differences in environment, and it is often possible to separate these effects.

In any distribution of phenotypes, such as the one in Figure 15.2, four sources contribute to phenotypic variation:

1. Genotypic variation

2. Environmental variation

3. Variation due to genotype-by-environment interaction

4. Variation due to genotype-by-environment association

Each of these sources of variation is discussed in the following sections.

The genotypic variance results from differences in genotype.

The variation in phenotype caused by differences in genotype among individuals is called the **genotypic variance** or **genetic variance**. **FIGURE 15.6** illustrates the genotypic variation expected among the F_2 generation from a cross of two inbred lines differing in genotype for three unlinked genes. The alleles of the three genes are represented as A/a, B/b, and C/c, and the genetic variation in the F_2 generation caused

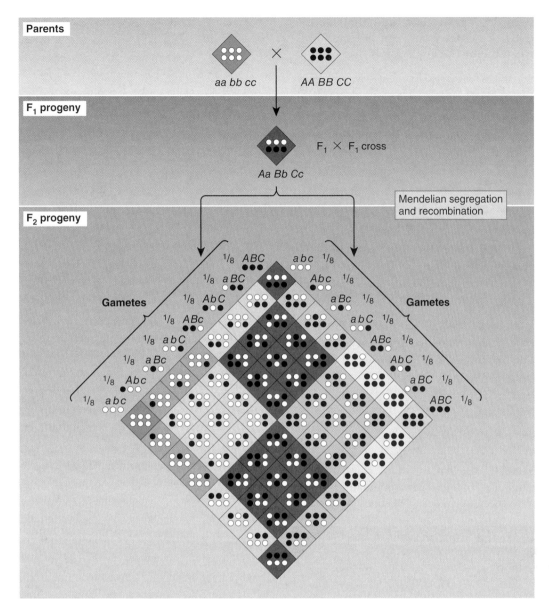

FIGURE 15.6 Segregation of three independent genes affecting a quantitative trait. Each uppercase allele in a genotype contributes one unit to the phenotype.

by segregation and recombination is evident in the differences in color. Relative to a categorical trait (one whose phenotype is determined by counting, such as ears per stalk in corn), if we assume that each uppercase allele is favorable for the expression of the trait and adds one unit to the phenotype, whereas each lowercase allele is without effect, then the *aa bb cc* genotype has a phenotype of 0 and the *AA BB CC* genotype has a phenotype of 6. There are seven possible phenotypes (0 through 6) in the F_2 generation. The distribution of phenotypes in the F_2 generation is shown in the colored bar graph in **FIGURE 15.7**. The normal distribution approximating the data has a mean of 3 and a variance of 1.5. In this case, we are assuming that *all* of the variation in phenotype in

the population results from differences in genotype among the individuals.

Figure 15.7 also includes a bar graph with diagonal lines representing the theoretical distribution when the trait is determined by 30 unlinked genes segregating in a randomly mating population, grouped into the same number of phenotypic classes as the 3-gene case. We assume that 15 of the genes are nearly fixed for the favorable allele and that 15 are nearly fixed for the unfavorable allele. The contribution of each favorable allele to the phenotype has been chosen to make the mean of the distribution equal to 3 and the variance equal to 1.5. Note that the distribution with 30 genes is virtually identical to that with three genes and that both are approximated by the same normal curve. If such distributions were encountered in actual research, the experimenter would not be able to distinguish between them. The key point is that

KEY CONCEPT

Even in the absence of environmental variation, the distribution of phenotypes, by itself, provides no information about the number of genes influencing a trait and no information about the dominance relations of the alleles.

However, the number of genes influencing a quantitative trait is important in determining the potential for long-term genetic improvement of a population by means of artificial selection. For example, in the 3-gene case in Figure 15.7, the best possible genotype would have a phenotype of 6, but in the 30-gene case, the best possible genotype (homozygous for the favorable allele of all 30 genes) would have a phenotype of 60.

In a real example showing the importance of number of genes, a population of the flour beetle *Tribolium* was selected for increased pupa weight over many generations, in each generation choosing as parents those individuals with the greatest pupa weight in the previous generation. At a time when selection no longer was effective in increasing pupa weight, the mean pupa weight in the selected population was 17 standard deviations above the mean of the original population. Determination of traits by a large number of genes implies that

KEY CONCEPT

Selective breeding can create an improved population in which the value of every individual greatly exceeds that of the best individuals that existed in the original population.

This principle may at first seems paradoxical, because in a large enough population, every possible genotype should be created at some low frequency. The explanation of the paradox is that real populations subjected to selective breeding typically consist of a few hundred organisms (at most), and thus many of the theoretically possible genotypes are never formed because their frequencies are much too rare. As selection takes place and the allele frequencies change, these genotypes become more common and allow the selection of superior organisms in future generations.

The environmental variance results from differences in environment.

The variation in phenotype caused by differences in environment among individuals is termed

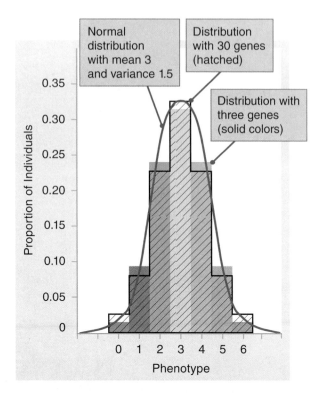

FIGURE 15.7 The distribution of phenotypes determined by the segregation of 3 and 30 independent genes. Both distributions are approximated by the same normal distribution (black curve).

environmental variance. **FIGURE 15.8** is an representation showing the distribution of seed weight in edible beans. The mean of the distribution is 500 mg and the standard deviation is 95 mg. All of the beans in this population are genetically identical and homozygous because they are highly inbred. Therefore, in this population, *all* of the phenotypic variation in seed weight results from environmental variance. A comparison of Figures 15.7 and 15.8 demonstrates the following principle:

KEY CONCEPT

The distribution of a trait in a population provides no information about the relative importance of genotype and environment. Variation in the trait can be entirely genetic, entirely environmental, or a combination of both influences.

Genotypic and environmental variance are seldom separated as clearly as in Figures 15.7 and 15.8, because usually they work together. Their combined effects are illustrated for a simple hypothetical case in **FIGURE 15.9**. At the upper left is the distribution of phenotypes for three genotypes assumed not to be influenced by environment. As depicted, the trait can have one of three distinct and nonoverlapping phenotypes determined by the effects of two additive alleles. The genotypes are in random-mating

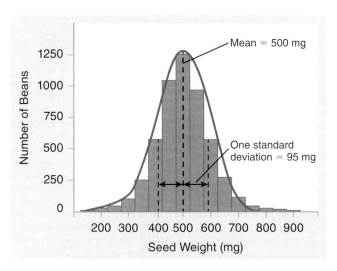

FIGURE 15.8 Distribution of seed weight in a homozygous line of edible beans. All variation in phenotype among individuals results from environmental differences.

proportions for an allele frequency of $1/2$, and the distribution of phenotypes has mean 5 and variance 2. Because it results solely from differences in genotype, this variance is *genotypic variance*, which is symbolized σ_g^2. The three panels at the upper right depict the distribution of phenotypes of each of the three genotypes in the presence of environmental variation. In each case the variance in phenotype that is due to environment alone equals 1. Because this variance results solely from differences in environment, it is *environmental variance*, which is symbolized σ_e^2. When the effects of genotype and environment are combined in the same population, both differences in genotype and differences in environment contribute to variation in the trait, and the distribution shown in the lower part of the figure results. The variance of this distribution is the **total variance** in phenotype, which is symbolized σ_p^2. Because we

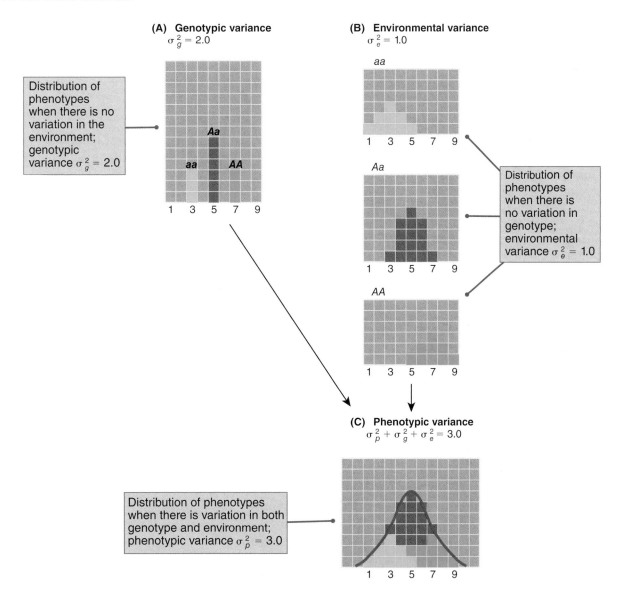

FIGURE 15.9 The combined effects of genotypic and environmental variance. (A) Population affected only by genotypic variance σ_g^2. (B) Populations of each genotype separately showing the effects of environmental variance σ_e^2. (C) Population affected by both genotypic and environmental variance; the total phenotypic variance σ_p^2 equals the sum of σ_g^2 and σ_e^2.

are assuming that genotype and environment have separate and independent effects on phenotype, we expect σ_p^2 to be greater than either σ_g^2 or σ_e^2 alone. In fact,

$$\sigma_p^2 = \sigma_g^2 + \sigma_e^2 \tag{15.3}$$

In words, Equation (15.3) states that

KEY CONCEPT

When genetic and environmental effects contribute independently to phenotype, the total variance equals the sum of the genotypic and environmental variance.

Equation (15.3) is one of the most important relations in quantitative genetics. How it can be used to analyze data will be explained shortly. Although the equation serves as an excellent approximation in very many cases, it is valid in an exact sense only when genotype and environment are independent in their effects on phenotype. The two most important departures from independence are discussed in the next section.

Genotype and environment can interact, or they can be associated.

In the simplest cases, environmental effects on phenotype are additive, and each environment adds (or detracts) the same amount to (or from) the phenotype, independent of the genotype. When this is not true, the environmental effects on phenotype differ according to genotype, and a **genotype-by-environment interaction (G-E interaction)** is said to be present. In some cases, G-E interaction can even change the relative ranking of the genotypes, so a genotype that is superior in one environment may become inferior in another.

An example of genotype-by-environment interaction in maize is illustrated in **FIGURE 15.10**. The two strains of corn are hybrids formed by crossing different pairs of inbred lines, and their overall means, averaged across all of the environments, are approximately the same. However, the strain designated A clearly outperforms strain B in the negative, stressful environments, whereas the performance is reversed when the environment is of high quality. (Environmental quality is judged on the basis of soil fertility, moisture, and other factors.)

Interaction of genotype and environment is common and is very important in both plants and animals. Because of this interaction, no one plant variety outperforms all others in all types of soil and climate, and therefore plant breeders must develop special varieties that are suited to each growing area.

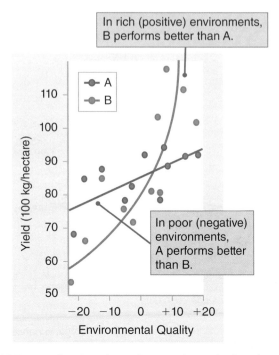

FIGURE 15.10 Genotype-by-environment interaction in maize. Strain A is superior when environmental quality is low (negative numbers), but strain B is superior when environmental quality is high.

Data from W. A. Russell, *Annual Corn & Sorghum Research Conference* 29 (1974): 81.

Another important type of interaction is a **genotype-by-sex interaction**, in which the same genotype results in a different phenotype according to the sex of the organism. Genotype-by-sex interactions are very common in quantitative genetics. One example is seen in the living histogram in Figure 15.3. The distribution of height among the women (dressed in white) is clearly shifted to the left relative to the distribution of height among men (dressed in blue). The averages differ by more than 5 inches (64.8 inches for the women, 70.1 inches for the men). Yet there is no reason to think that the genes that affect height are distributed differently in women and men. The genes simply have somewhat different effects depending on sex—an example of genotype-by-sex interaction.

When different genotypes in a population are not distributed at random among all the possible environments, there is **genotype-by-environment association (G-E association)**. In these circumstances, certain genotypes are preferentially associated with certain environments, which may either increase or decrease the phenotype of these genotypes compared with what would result in the absence of G-E association. An example of deliberate genotype-by-environment association can be found in dairy husbandry, in which some farmers feed each of their cows in proportion to its level of milk production. Because of this practice, cows with superior genotypes with respect to milk production also have a superior environment in that they receive more feed. In plant breeding, genotype-by-environment association can often be eliminated or minimized by appropriate randomization of genotypes within the experimental plots. In other cases, human genetics again being a prime example, the possibility of G-E association cannot usually be controlled.

There is no genotypic variance in a genetically homogeneous population.

Equation (15.3) can be used to separate the effects of genotype and environment on the total phenotypic variance. Two types of data are required:

1. The phenotypic variance of a genetically uniform population, which provides an estimate of σ_e^2 because a genetically uniform population has a value of $\sigma_g^2 = 0$

2. The phenotypic variance of a genetically heterogeneous population, which provides an estimate of $\sigma_g^2 + \sigma_e^2$

An example of a genetically uniform population is the F_1 generation from a cross between two highly homozygous strains, such as inbred lines (Figure 15.1). An example of a genetically heterogeneous population is the F_2 generation from the same cross, as indicated in Figure 15.6. If the environments of both populations are the same, and if there is no G-E interaction, then the estimates may be combined to deduce the value of σ_g^2.

To take a specific numerical illustration, consider variation in the size of the eyes in the cave-dwelling fish *Astyanax*, reared in the same environments (FIGURE 15.11). The variances in eye diameter in the F_1 and F_2 generations from a cross of two highly homozygous strains were estimated as 0.057 and 0.563, respectively. Written in terms of the components of variance, these are

$$F_2: \ \sigma_p^2 = \sigma_g^2 + \sigma_e^2 = 0.563$$
$$F_1: \ \sigma_e^2 = 0.057$$

FIGURE 15.11 Reduced eye size and pigmentation in a cave-dwelling *Astyanax* (below), compared with a surface-dwelling relative (above).

Courtesy of Richard L. Borowsky, New York University.

The estimate of genotypic variance σ_g^2 is obtained by subtracting the second equation from the first; that is,

$$\sigma_g^2 = 0.563 - 0.057 = 0.506$$

because

$$(\sigma_g^2 + \sigma_e^2) - \sigma_e^2 = \sigma_g^2$$

Hence the estimate of σ_g^2 is 0.506, whereas that of σ_e^2 is 0.057. In this example, the genotypic variance is much greater than the environmental variance, but this is not always the case.

🧠 STOP & THINK 15.2

In a cross between two inbred lines of *Drosophila*, the variance in thorax length in the F_1 generation equals 0.0018 mm² and the variance in thorax length in the F_2 generation equals 0.0036 mm². Estimate the environmental variance and the genotypic variance in thorax length.

The broad-sense heritability includes all genetic effects combined.

Separating the genotypic from the environmental variance is worthwhile because it allows one to estimate the relative importance of differences in genotype versus differences in environment to accounting for phenotypic variation in a trait in a population. One measure of the relative importance of differences in genotype makes use of the ratio of the genotypic variance to the total phenotypic variance. This ratio of σ_g^2 to σ_p^2 is called the **broad-sense heritability**, symbolized H^2, and it measures the importance of genetic variation, relative to environmental variation, in causing variation in the phenotype of a trait of interest. Broad-sense heritability is a ratio of variances, specifically

$$H^2 = \frac{\sigma_g^2}{\sigma_p^2} = \frac{\sigma_g^2}{\sigma_p^2 + \sigma_e^2} \qquad (15.4)$$

Substitution of the data for eye diameter in *Astyanax*, in which $\sigma_g^2 = 0.506$ and $\sigma_g^2 + \sigma_e^2 = 0.563$, into Equation (15.4) yields $H^2 = 0.506/0.563 = 0.90$ for the estimate of broad-sense heritability. This value implies that 90 percent of the variation in eye diameter in this population results from differences in genotype among the fish.

Knowledge of heritability is useful in the context of plant and animal breeding because heritability can be used to predict the magnitude and speed of population improvement. The broad-sense heritability defined in Equation (15.4) is used in predicting the outcome of selection practiced among clones, inbred lines, or varieties. Analogous predictions for randomly bred populations utilize another type of heritability (the narrow-sense heritability), which will be discussed in the next section. Broad-sense heritability measures how much of the total variance in phenotype results from differences in genotype. For this reason, H^2 is often of interest in regard to human quantitative traits.

Twin studies are often used to assess genetic effects on variation in a trait.

In human beings, twins would seem to be ideal subjects for separating genotypic and environmental variance because **identical twins**, who arise from the splitting of a single fertilized egg, are genetically identical and are often strikingly similar in such traits as facial features and body build. **Fraternal twins**, who arise from two fertilized eggs, have the same genetic relationship as ordinary siblings, so only half of the genes in either twin are identical with those in the other. Theoretically, the variance between members of an identical-twin pair would be equivalent to σ_e^2, because the twins are genetically identical. The variance between members of a fraternal-twin pair would include not only σ_e^2 but also part of the genotypic variance (approximately $\sigma_g^2/2$ because of the identity of half of the genes in fraternal twins). Consequently, both σ_g^2 and σ_e^2 could be estimated from twin data and combined as in Equation (15.4) to estimate H^2. **TABLE 15.2** summarizes estimates of H^2 based on twin studies of several traits.

Twin studies are subject to several important sources of error, most of which increase the similarity of identical twins, so the numbers in Table 15.2 should be considered approximate and probably too high. One complication of twin studies is that of genotype-by-sex interaction, because half of the fraternal twin pairs are of opposite sex. This problem is usually avoided by studying only same-sex fraternal twins. Here are some other potential sources of error:

1. Genotype-by-environment interaction, which increases the variance in fraternal twins but not in identical twins

2. Frequent sharing of embryonic membranes between identical twins, resulting in a more similar intrauterine environment

3. Greater similarity in the treatment of identical twins by parents, teachers, and peers, resulting in a decreased environmental variance in identical twins

These pitfalls and others imply that data from human twin studies should be interpreted with caution and reservation.

TABLE 15.2 Broad-Sense Heritability, in Percent, Based on Twin Studies

Trait	Heritability, H^2
Longevity	29
Height	85
Weight	63
Amino acid excretion	72
Serum lipid levels	44
Maximum blood lactate	34
Maximum heart rate	84
Verbal ability	63
Numerical ability	76
Memory	47
Sociability index	66
Masculinity index	12
Temperament index	58

15.3 Artificial selection is a form of "managed evolution."

The practice of breeders to choose a select group of organisms from a population to become the parents of the next generation is termed **artificial selection**. When artificial selection is carried out either by choosing the best organisms in a species that reproduces asexually or by choosing the best among subpopulations propagated by close inbreeding (such as self-fertilization), broad-sense heritability permits an assessment of how rapidly progress can be achieved. Broad-sense heritability is important in this context, because with clones, inbred lines, or varieties, superior genotypes can be perpetuated without disruption of favorable gene combinations by Mendelian segregation. An example is the selection of superior varieties of plants that are propagated asexually by means of cuttings or grafts, or of animals that are reproduced by cloning. Because there is no sexual reproduction, each offspring has exactly the same genotype as its parent.

In sexually reproducing populations that are genetically heterogeneous, broad-sense heritability is not relevant in predicting progress resulting from artificial selection, because superior genotypes must necessarily be broken up by the processes of segregation and recombination. For example, if the best genotype is heterozygous for each of two unlinked loci, A/a; B/b, then because of segregation and independent assortment, among the progeny of a cross between parents with the best genotypes—A/a; $B/b \times A/a$; B/b—only 1/4 will have the same favorable A/a; B/b genotype as the parents. The rest of the progeny will be genetically inferior to the parents. For this reason, to the extent that high genetic merit may depend on particular combinations of alleles, each generation of artificial selection results in a slight setback in that the offspring of superior parents are generally not quite so good as the parents themselves. Progress under selection can still be predicted, but the prediction must make use of another type of heritability, the narrow-sense heritability, which is discussed in the next section.

The narrow-sense heritability is usually the most important in artificial selection.

FIGURE 15.12 illustrates a typical form of artificial selection and its result. The organism is *Nicotiana longiflora* (tobacco), and the trait is the length of the corolla tube (*corolla* is a collective term for all the petals of a flower). Part A shows the distribution of phenotypes in the parental generation, and part B shows the distribution of phenotypes in the offspring generation. The parental generation is the population from which the parents were chosen for breeding. The type of selection is called **individual selection**, because each member of the population to be selected is evaluated according to its own individual phenotype. The selection is practiced by choosing some arbitrary level of phenotype—called the **truncation point**—that determines which individuals will be saved for breeding purposes. All individuals with a phenotype above the threshold are randomly mated among themselves to produce the next generation.

In evaluating progress through individual selection, three distinct phenotypic means are important. In Figure 15.12, these means are symbolized as M, M^*, and M', and they are defined as follows:

1. M is the mean phenotype of the entire population in the parental generation, including both the selected and the nonselected individuals.

2. M^* is the mean phenotype among those individuals selected as parents (those with a phenotype above the truncation point).

3. M' is the mean phenotype among the progeny of selected parents.

The relationship among these three means is given by

$$M' = M + h^2(M^* - M) \qquad (15.5)$$

in which the symbol h^2 is the **narrow-sense heritability** of the trait in question.

Later in this chapter, a method for estimating narrow-sense heritability from the similarity in phenotype among relatives will be explained. In Figure 15.12, h^2 is the only unknown quantity, so it can be estimated from the data themselves. Rearranging Equation (15.5) and substituting the values for the means from Figure 15.12, we get

$$h^2 = \frac{M' - M}{M^* - M} = \frac{77 - 70}{81 - 70} = 0.64$$

In a manner analogous to the way in which total phenotypic variance can be split into the sum of the genotypic variance and the environmental variance (Equation 15.3), the genotypic variance can be split into parts resulting from the additive effects of alleles, dominance effects, and effects of interaction between alleles of different genes. The difference between the broad-sense heritability, H^2, and the narrow-sense heritability, h^2, is that H^2 includes all of these genetic contributions to variation, whereas h^2 includes only the additive effects of alleles. In ordinary language, the additive effects of alleles are the genetic effects that are transmissible from parent to offspring; dominance effects are not transmissible because of segregation, and epistatic (interaction) effects are not transmissible because of independent assortment and recombination. It follows that from the standpoint of animal or plant improvement, h^2 is the heritability of interest because

KEY CONCEPT

The narrow-sense heritability, h^2, is the proportion of the variance in phenotype that is transmissible from parents to offspring and that can be used to predict changes in the population mean with individual selection, according to Equation (15.5).

The distinction between the broad-sense heritability and the narrow-sense heritability can be appreciated intuitively by considering a population in which there is a rare recessive gene. In such a case, most homozygous recessive genotypes come from matings between heterozygous carriers. Some such kindreds have more than one affected offspring. Hence affected siblings can resemble each other more than they resemble their parents. For example, if a is a recessive allele, the mating $Aa \times Aa$, in which both parents show the dominant phenotype, may yield two offspring that are aa,

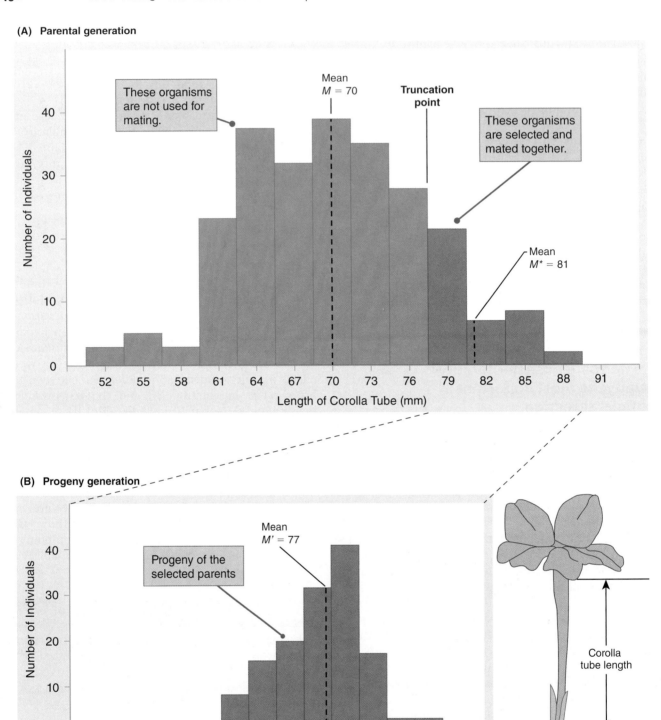

FIGURE 15.12 Selection for increased length of corolla tube in tobacco. (A) Distribution of phenotypes in the parental generation. The symbol M denotes the mean phenotype of the entire population, and M^* denotes the mean phenotype of the organisms chosen for breeding (organisms with a phenotype that exceeds the truncation point). (B) Distribution of phenotypes among the offspring bred from the selected parents. The symbol M' denotes the mean.

which both show the recessive phenotype; in this case, the offspring are more similar to each other than they are to either of their parents. It is the dominance of the wildtype allele that causes this paradox; dominance and epistasis contribute to the broad-sense heritability of the trait but not to the narrow-sense heritability. The narrow-sense heritability includes only those genetic effects that contribute to the resemblance between parents and their offspring, because narrow-sense heritability measures how similar offspring are to their parents.

In general, the narrow-sense heritability of a trait is smaller than the broad-sense heritability. For example, in the parental generation in Figure 15.12, the broad-sense heritability of corolla tube length is $H^2 = 0.82$. The two types of heritability are equal only when the alleles affecting the trait are additive in their effects; *additive effects* means that each heterozygous genotype shows a phenotype that is exactly intermediate between the phenotypes of the respective homozygous genotypes and that the effects are also additive across loci.

Equation (15.5) is of fundamental importance in quantitative genetics because of its predictive value. This can be seen in the following example. The selection in Figure 15.12 was carried out for several generations. After two generations, the mean of the population was 83, and parents having a mean of 90 were selected. By use of the estimate $h^2 = 0.64$, the mean in the next generation can be predicted. The information provided is that $M = 83$ and $M^* = 90$. Therefore, Equation (15.5) implies that the predicted mean is

$$M' = 83 + (0.64)(90 - 83) = 87.5$$

This value is in good agreement with the observed value of 87.9.

🧠 STOP & THINK 15.3

The raised skin ridges on the fingers of each person form a fingerprint pattern that is unique. Each finger can have a ridge pattern in the shape of an arch, a loop, or a whorl, and the number of ridges between the center of a pattern and the edge is the fingerprint ridge count for any finger. The sum of the ridge counts for all 10 fingers constitutes the total fingerprint ridge count, which is normally distributed with a mean of 125 and a standard deviation of 50. Remarkably, whereas the fingerprint pattern differs even among identical twins, the total ridge count is a highly heritable multifactorial trait with a narrow-sense heritability (h^2) of 86 percent. Suppose a man whose fingerprint ridge count is 200 marries a woman whose fingerprint ridge count is 100. What is the expected fingerprint ridge count among their offspring?

There are limits to the improvement that can be achieved by artificial selection.

Artificial selection is analogous to natural selection in that both types of selection cause an increase in the frequency of alleles that improve the selected trait (or traits). Like natural selection, artificial selection is most effective in changing the frequency of alleles that are in an intermediate range of frequency ($0.2 < p < 0.8$). Selection is less effective for alleles with frequencies outside this range, and it is least effective for rare recessive alleles. For quantitative traits, including fitness, the total selection is shared among all the genes that influence the trait, and the selection coefficient affecting each allele is determined by (1) the magnitude of the effect of the allele, (2) the frequency of the allele, (3) the total number of genes affecting the trait, (4) the narrow-sense heritability of the trait, and (5) the proportion of the population that is selected for breeding.

The value of heritability is determined both by the magnitude of effects and by the frequency of alleles. If all favorable alleles were fixed ($p = 1$) or lost ($p = 0$), the heritability of the trait would be 0. Therefore, the heritability of a quantitative trait is expected to decrease over many generations of artificial selection as a result of favorable alleles becoming nearly fixed. For example, 10 generations of selection for less fat in a population of Duroc pigs decreased the heritability of fatness from 73 to 30 percent because of changes in allele frequency resulting from the selection.

Population improvement by means of artificial selection cannot continue indefinitely. A population may respond to selection until its mean is many standard deviations different from the mean of the original population, but eventually the population reaches a **selection limit** at which successive generations show no further improvement. (An exception to this generalization is found in traits that are affected by a very large number of genes and in which selection is carried out in a very large population. In such a case, selective advance can be continued indefinitely because new mutations continue to add genetic variation.) In a few cases, progress under selection ceases because all alleles that affect the trait are either fixed or lost, and so the narrow-sense heritability of the trait becomes 0. On the other hand, fixation or loss of all relevant alleles is rare. The usual reason for a selection limit is that natural selection counteracts artificial selection. Many genes that respond to artificial selection as a result of their favorable effect on a selected trait also have indirect harmful effects on fitness. For example, selection for increased size of eggs in poultry results in a decrease in the number of eggs, and selection for extreme body size (large or small) in most animals results in a decrease in fertility. When one trait (for instance, number of eggs) changes in the course of selection for a different trait (for example, size of eggs), the unselected trait is said to have undergone a **correlated response** to selection. Correlated response of fitness is typical in long-term artificial selection. Each increment of progress in the selected trait is partially offset by a decrease in fitness because of correlated response. Eventually, artificial selection for the trait of interest is exactly balanced by natural selection against the trait, so a selection limit is reached and no further progress is possible without changing the strategy of selection.

Inbreeding is generally harmful, and hybrids may be the best.

Inbreeding can have harmful effects on economically important traits such as yield of grain and egg production. This decline in performance is called **inbreeding depression**, and it results principally from rare harmful recessive alleles becoming homozygous because of inbreeding. **FIGURE 15.13** is an example of inbreeding depression in yield of corn, in which the yield decreases linearly as the degree of inbreeding, measured by the inbreeding coefficient F, increases.

Most highly inbred strains suffer from many genetic defects, as might be expected from deleterious recessive alleles becoming homozygous. One would also expect that if two different inbred strains were crossed, the F_1 would show improved features, because a harmful recessive allele inherited from one parent would likely be concealed due to a normal dominant allele from the other parent. In many organisms, the F_1 generation of a cross between inbred lines is superior to the parental lines, and the phenomenon is called **heterosis**, or **hybrid vigor**. This phenomenon, which is widely used in the production of corn and other agricultural products, yields genetically identical hybrid plants with traits that are sometimes more favorable than those of the ancestral plants from which the inbreds were derived. The features that most commonly distinguish hybrid plants from their

inbred parents are their rapid growth, larger size, and greater yield. Furthermore, the F_1 plants have a fairly uniform phenotype; because the progeny of inbred parents all have the same genotype (Figure 15.1), the genetic variance $\sigma_g^2 = 0$, and so all of the variance in phenotype is due to variation in the environment. Genetically heterogeneous crops with high yields or certain other desirable features can also be produced by traditional plant-breeding programs, but growers often prefer hybrid plants because of this relative uniformity in phenotype. As an example, uniform height and time of maturity facilitate machine harvesting, and plants that all bear fruit at the same time accommodate picking and shipping schedules.

Hybrid varieties of corn are used almost exclusively in the United States for commercial crops. A farmer cannot plant the seeds from his crop because the F_2 generation consists of a variety of genotypes, most of which do not show hybrid vigor. The production of hybrid seeds is a major industry in corn-growing sections of the United States. Since the 1930s, when hybrid corn was first introduced into the United States, average yield has increased about fivefold. About 70 percent of this increase is due to the greater productivity of hybrids, the rest to agricultural practices such as the use of fertilizer and irrigation.

15.4 Genetic variation is revealed by correlations between relatives.

Quantitative genetics relies extensively on similarity among relatives to assess the importance of genetic factors. Particularly in the study of complex behavioral traits in human beings, genetic interpretation of familial resemblance is not always straightforward because of the possibility of nongenetic, but nevertheless familial, sources of resemblance. In plant and animal breeding, the situation is usually less complex because genotypes and environments are under experimental control.

Covariance is the tendency for traits to vary together.

Genetic data about families are frequently pairs of numbers: pairs of parents, pairs of twins, or pairs consisting of a single parent and a single offspring. An important issue in quantitative genetics is the degree to which the numbers in each pair are associated. The usual way to measure the association is to calculate a statistical quantity called the *correlation coefficient* between the members of each pair.

The correlation coefficient among relatives is based on the covariance in phenotype among them. Much as the variance describes the tendency of a set

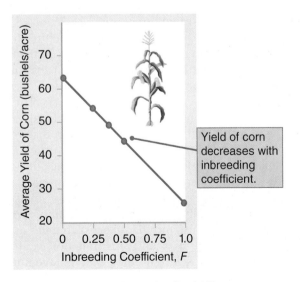

FIGURE 15.13 Inbreeding depression for yield in corn.
Data from N. P. Neal, *Agron. J.* 27 (1935): 666–670.

of measurements to vary [Equation (15.2)], the covariance describes the tendency of pairs of numbers to vary together (co-vary). Calculation of the covariance is similar to that of the variance in Equation (15.2) except that the squared deviation term $(x_i - \bar{x})^2$ is replaced with the product of the deviations of the pairs of measurements from their respective means—that is, $(x_i - \bar{x})(y_i - \bar{y})$.

For instance, $(x_i - \bar{x})$ could be the deviation of a particular father's height from the overall father mean, and $(y_i - \bar{y})$ could be the deviation of his son's height from the overall son mean. In symbols, let f_i be the number of pairs of relatives with phenotypic measurements x_i and y_i. Then the estimated **covariance (Cov)** of the trait among the relatives is

$$Cov(x, y) = \frac{\sum f_i (x_i - \bar{x}_i)(y_i - \bar{y})}{N - 1} \quad (15.6)$$

where N is the total number of pairs of relatives studied.

The **correlation coefficient (r)** of the trait between the relatives is calculated from the covariance as follows:

$$r = \frac{Cov(x, y)}{s_x s_y} \quad (15.7)$$

where s_x and s_y are the standard deviations of the measurements, estimated from Equation (15.2). The correlation coefficient can range from -1.0 to $+1.0$. A value of $+1.0$ means perfect association. When $r = 0$, x and y are not associated.

The additive genetic variance is transmissible; the dominance variance is not.

Covariance and correlation are important in quantitative genetics because the correlation coefficient of a trait between individuals with various degrees of genetic relationship is related fairly simply to the narrow-sense or broad-sense heritability, as shown in **TABLE 15.3**. The table gives the theoretical values of the correlation coefficient for various pairs of relatives; h^2 represents the narrow-sense heritability and H^2 the broad-sense heritability. Considering parent–offspring, half-sibling, or first-cousin pairs, narrow-sense heritability can be estimated directly by multiplication. Specifically, h^2 can be estimated as twice the parent–offspring correlation, four times the half-sibling correlation, or eight times the first-cousin correlation.

With full siblings, identical twins, and double first cousins, the correlation coefficient is related to the broad-sense heritability, H^2, because phenotypic resemblance depends not only on additive effects but

TABLE 15.3 Theoretical Correlation Coefficient in Phenotype Between Relatives

Degree of relationship	Correlation coefficient[*]
Offspring and one parent	$h^2/2$
Offspring and average of parents	$h^2/\sqrt{2}$
Half siblings	$h^2/4$
First cousins	$h^2/8$
Monozygotic twins	H^2
Full siblings	$\sim H^2/2$

[*]Contributions from interactions among alleles of different genes have been ignored. For this and other reasons, H^2 correlations are approximate.

also on dominance. In these relatives, dominance contributes to resemblance because the relatives can share *both* of their alleles as a result of their common ancestry, whereas parents and offspring, half siblings, and first cousins can share at most a single allele of any gene because of common ancestry. Therefore, to the extent that phenotype depends on dominance effects, full siblings can resemble each other more than they resemble their parents.

The most common disorders in human families are multifactorial.

The most common human diseases are caused by genetic and environmental factors acting together. Each of the factors adds to the risk or liability that a person has for manifesting the trait. For example, genetic risk factors for heart disease are revealed in the family history of the disease, and environmental risk factors include being overweight and such behaviors as cigarette smoking. A person's overall risk for manifesting a disease is a multifactorial trait determined by numerous genetic and environmental factors and the interactions among these factors. Such a trait is a *threshold trait*; although the underlying risk itself is not directly observable, the trait will be either present or absent according to whether the risk is above or below a critical (threshold) value.

As with other multifactorial traits, the risk of manifesting a threshold trait has a broad-sense and a narrow-sense heritability that may differ among populations according to the allele frequencies and the distribution of environmental risk factors. The heritabilities cannot be estimated directly, because the risk is not observable directly, but the heritabilities can be inferred from the incidences of the trait among individuals and their relatives. The statistical techniques for doing this are quite specialized, but some of the theoretical results are shown in **FIGURE 15.14**, along with observed data for the most common congenital abnormalities in whites. The x-axis gives the incidence of the trait in the general population, and the y-axis gives the risk of the trait in brothers or sisters of affected persons. The two horizontal lines near the top yield the proportions for simple Mendelian dominance and simple Mendelian recessive inheritance, which are 50 percent (dominance) and 25 percent (recessiveness), respectively. Note that these proportions do not depend on the incidence of the trait in the population. The other curves pertain to threshold traits with the narrow-sense heritabilities of liability as noted. Now the proportion of affected siblings *does* depend on the incidence of the trait in the general population, as well as on the heritability of liability. Consider a trait with a population incidence of 0.05 percent (1 in 2000). If the heritability of liability is 0.40, the proportion of affected siblings among affected persons is a little less than 0.5 percent (1 in 200), or about a 10-fold increase

over that of the general population. If the heritability is 1.00, the proportion of affected siblings among affected persons is a little less than 5 percent (1 in 20), or about a 100-fold increase over that of the general population. Note in Figure 15.14 that the most common traits tend to be threshold traits; with a few exceptions, the proportion of affected siblings tends to be moderate or low, corresponding to the heritabilities indicated.

15.5 Pedigree studies of genetic polymorphisms are used to map loci for quantitative traits.

Genes affecting quantitative traits cannot usually be identified in pedigrees because their individual effects are obscured by the segregation of other genes and by environmental variation. Even so, genes affecting quantitative traits can be localized if they are genetically linked with genetic markers, such as single-nucleotide

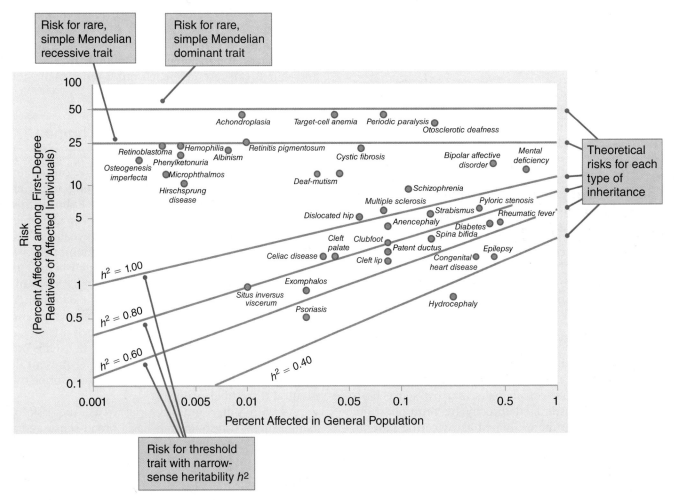

FIGURE 15.14 Risks of common abnormalities in the siblings of affected persons. Diagonal lines are the theoretical risks for threshold traits with the indicated values of the narrow-sense heritability of liability. Horizontal lines are the theoretical risks for simple dominant or recessive traits.

polymorphisms (SNPs). A gene that affects a quantitative trait is a **quantitative-trait locus (QTL)**. (A quantitative-trait locus that affects phenotype through its level of gene expression is called an *expression quantitative-trait locus*, abbreviated *eQTL*.) Locating QTLs in the genome is important to the manipulation of genes in breeding programs and to the identification and study of genes in order to discover their functions.

An illustration of genetic mapping of quantitative trait loci for several quantitative traits in tomatoes is presented in **FIGURE 15.15**. More than 300 highly polymorphic genetic markers have been mapped in the tomato genome, with an average spacing between markers of 5 map units. The chromosome maps in Figure 15.15 show a subset of 67 markers that were segregating in crosses between the domestic tomato and a wild South American relative. The average spacing between the markers is 20 map units. Backcross progeny of the cross $F_1 \times$ domestic tomato were tested for the genetic markers, and the fruits of the backcross progeny were assayed for three quantitative traits—fruit weight, content of soluble solids, and acidity. Statistical analysis of the data was carried out in order to detect marker alleles that were associated with phenotypic differences in any of the traits; a significant association indicates genetic linkage between the marker gene and one or more QTLs affecting the trait. A total of six QTLs affecting fruit weight were detected (green bars), as well as four QTLs affecting soluble solids (blue bars) and five QTLs affecting acidity (dark red bars). Although additional QTLs of smaller effects undoubtedly remained undetected in these types of experiments, the effects of the mapped QTLs are substantial: The mapped QTLs account for 58 percent of the total phenotypic variance in fruit weight, 44 percent of the phenotypic variance in soluble solids, and 48 percent of the phenotypic variance in acidity. The genetic markers linked to the QTLs with substantial effects make it possible to trace the transmission of the QTLs in pedigrees and manipulate them in breeding programs by following the transmission of the linked marker genes. Figure 15.15 also indicates a number of chromosomal regions containing QTLs for two or more of the traits—for example, the QTL regions on chromosomes 6 and 7, which affect all three traits.

Complex traits are usually influenced by many genes, most with small effects.

The fine-scale genetic localization and identification of QTLs affecting complex traits in humans is currently a very active field of genetic research. Most studies rely on **genome-wide association studies (GWAS)**. A typical GWAS involves some number of patients with a disease and a comparable number of controls without the disease, matched with the

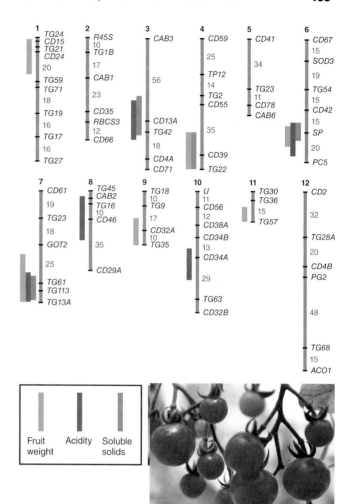

FIGURE 15.15 Location of QTLs for several quantitative traits in the tomato genome. The genetic markers are shown for each of the 12 chromosomes. The numbers in red are distances in map units between adjacent markers, but only map distances of 10 or greater are indicated. The regions in which the QTLs are located are indicated by the bars: green bars, QTLs for fruit weight; blue bars, QTLs for content of soluble solids; and dark red bars, QTLs for acidity (pH). The data are from crosses between the domestic tomato (*Solanum esculentum*) and a wild South American relative with small, green fruit (*Solanum chmielewskii*). The photograph is of fruits of wild tomato. The F_1 generation was backcrossed with the domestic tomato, and fruits from the progeny were assayed for the genetic markers and each of the quantitative traits.

Data from A. H. Paterson, et al., *Nature* 335 (1988): 721–725; Photo © Photos.com.

patients for age, sex, ethnic group, place of residence, and so forth. The sample sizes need to be large, minimally several thousand of both patients and controls, and often tens or hundreds of thousands. Each patient and control is genotyped either by whole-genome sequencing or by means of a microarray that includes oligonucleotides for 500,000 or more SNPs throughout the genome. Then statistical analysis is carried out to ascertain which of the SNP polymorphisms are associated with the disease, and which polymorphic nucleotide increases risk of the disease and which protects against it. The SNP itself may not be causally related to the disease, but the association implies that there is a genetically linked QTL that is causally related. A QTL that increases the risk of a genetic disorder is a *risk factor* for that disorder. Although there is much more to be learned from GWAS, enough has been learned to make some observations about the nature and number of QTLs and the magnitude of their effects.

The first observation is that many genes contribute to variation in complex traits. The number above each bar in **FIGURE 15.16** is the estimated number of genes affecting each of the complex traits. The number differs among traits but is typically more than 50 and often in the hundreds. Each estimate is a minimum because the number of genes that can be detected in an association study depends on the study design and especially on the sample size. The greater the number of individuals in the sample, the more likely a gene affecting the trait is to reach statistical significance and be identified. The largest number in the table is the number 643, the number of genes affecting adult height, which is so large in part because it is based on studies of hundreds of thousands of individuals.

The second observation is that the genes identified as affecting complex traits account collectively for only a modest proportion of genetic variation in the trait.

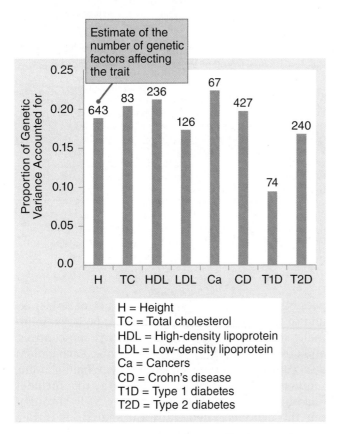

FIGURE 15.16 Estimated number of genes affecting complex traits in humans and the proportion of the total genetic variance explained by these genes.

Data from J.-H. Park, et al., *Proc. Natl. Acad. Sci. USA* 108 (2011): 18026–18031.

In Figure 15.16, the height of each bar corresponds to the proportion of the genetic variance explained by the identified genes, and it ranges from about 10 percent to about 20 percent. If a set of genes were to explain 100 percent of the genetic variation, then these genes would account for all of the heritability of the trait. The shortfall in accounting for the genetic variance by statistically significant genetic factors is called the **missing heritability**. The most common speculation for missing heritability is that genetic factors influencing the trait interact with one another, which they almost certainly do. Gene interactions at the molecular, cellular, and organismal level constitute what is called **physiological epistasis**. Gene interactions at the population level that are statistically significant in association studies constitute **statistical epistasis**. The two kinds of epistasis are detected in different ways and are quite distinct molecularly and statistically. In particular, genes can show a great deal of physiological epistasis yet have undetectable levels of statistical epistasis.

An alternative and more likely explanation for missing heritability is that association studies fail to detect genetic factors with small effects; if these undetected factors were taken into account, most of the missing heritability would no longer be missing. For example, in one association study, genetic effects on height were estimated not gene by gene but chromosome by chromosome, and the proportion of the heritability that could be explained by summing the effects of all the chromosomes was 92 percent.

FIGURE 15.17 gives a closer look at QTLs affecting adult height. Part A is in inferred distribution of relative effects of QTLs based on a sample of approximately 200,000 individuals. The shaded regions indicate the effect sizes that would be statistically significant in a sample of this size. The overall distribution of relative effects is assumed to be normal. On the *x*-axis, relative effect is the magnitude of change in height predicted to result from any one QTL, taking into account simultaneous changes in all the other QTLs that might affect height in the study sample. (Technically, relative effect is the partial regression coefficient of height against genotype for each QTL.)

Another perhaps more easily interpreted illustration of the effects of genes on height is according to how much each QTL contributes to the genetic variation in the trait. This distribution is shown in Figure 15.17, part B. The inset is an expanded version of the right-hand tail of the distribution, with the shaded region again indicating those QTLs that would be statistically significant with a sample size of 200,000. Each statistically significant QTL contributes a relatively small amount to the total genetic variance, usually smaller than 1 percent, which helps explain why the large numbers of genes for the traits in Figure 15.16 account for such a seemingly small proportion of the genetic variation.

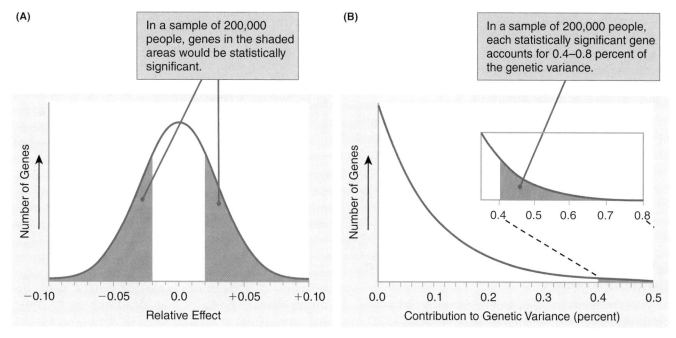

FIGURE 15.17 Quantitative-trait loci (QTLs) affecting adult height. (A) Inferred distribution of QTL effects based on a sample of approximately 200,000 people. (B) Inferred contribution of QTLs to the genetic variance for adult height.

Data from H. L. Allen, et al., *Nature* 467 (2010) 832–838.

Studies such as those summarized in Figure 15.16 and 15.17 support the following observations regarding genes affecting complex traits. We call them observations rather than generalizations because they do not necessarily apply to all complex traits in all organisms.

- Most complex traits are affected by many QTLs.

- The majority of common alleles affecting complex traits account for a relatively small fraction of the genetic variance in the trait. Most genetic disorders appear to be affected by fewer genetic risk factors than the number of genes affecting height; however, as with genes affecting height, most genetic risk factors that have been identified account for a relatively small fraction of the genetic variance in risk. This observation underlines the need for proper interpretation of direct-to-consumer (DTC) or over-the-counter (OTC) genetic testing, because most genetic risk factors contribute only modestly to increase in disease risk.

- QTLs affecting complex traits can affect several complex traits simultaneously; in other words, QTLs often have pleiotropic effects.

- Gene interactions at the molecular, cellular, or organismal levels (physiological epistasis) among QTLs affecting complex traits is likely to be common; however, at the population level, physiological epistasis is often not detectable as statistical epistasis.

- Effects of QTLs are often highly dependent on the environment. Genotype-by-sex interaction is particularly important; in other words, the effect of a QTL can differ dramatically between females and males.

- Some QTLs have effects that are population specific. For example, a QTL in human chromosome 9 leads to a 30 to 40 percent increase in the risk of coronary artery disease in whites but not in blacks.

- The contribution of QTLs to the genetic variance is has an approximately exponential distribution resembling that in Figure 15.17, part B. Most QTLs add small amount to the genetic variance, many fewer add a moderate amount, and very few make a large contribution. Specifically, most QTLs for complex human diseases increase the risk by 10 to 30 percent, and some have effects in the range of 30 to 50 percent. QTLs with larger effects are uncommon.

- Many QTLs are not due to single genes but rather to the combined effects of several closely linked genes. This observation reflects the process of duplication and divergence by which functionally related genes evolve, and it makes sense that functionally related genes would affect the same complex traits.

- Although some genes affecting complex traits might have been predicted based on the biology of the trait and the inferred function of annotated genes in the genomic sequence, many more genes could not have been predicted. Those with known functions have no obvious connection to the trait, and many are genes for which function is unknown.

- Mutations in genes affecting complex traits are sometimes in protein-coding regions and sometimes in regions that affect gene regulation (eQTLs).

- Mutant alleles affecting complex traits may be common (allele frequency greater than 0.05) or rare (allele frequency less than 0.05).

QTLs can also be identified by examining candidate genes.

In many organisms, genotyping hundreds of thousands of individuals for a full GWAS to identify QTLs is prohibitively expensive, and in some organisms obtaining hundreds of thousands of individuals to study is utterly impossible. An alternative to GWAS for the identification of QTLs is to use educated guesswork and an understanding of the biology of a trait to identify candidate genes. A **candidate gene** for a trait is a gene for which there is some *a priori* basis for suspecting that the gene affects the trait. In human behavioral genetics, for example, if a pharmacological agent is known to affect a personality trait, and the molecular target of the drug is known, then the gene that codes for the target molecule and any gene whose product interacts with the drug or with the target molecule are candidate genes for affecting the personality trait.

One example of the use of candidate genes in the study of human behavior genetics is in the identification of a naturally occurring genetic polymorphism associated with serious depression in response to stressful life experiences. The neurotransmitter

 THE HUMAN CONNECTION

Pinch of This and a Smidgen of That

Oliver Smithies (2005)

University of North Carolina, Chapel Hill, NC

Many Little Things: One Geneticist's View of Complex Diseases

This paper was written on the 25th anniversary of Oliver Smithies's first experimental success in introducing exogenous DNA into a chosen site in a mammalian genome. The initial applications were to the genetic analysis of phenotypes in which single genes had a major effect. Here Smithies recounts how he came to realize that the approach also could be used to study complex diseases influenced by many genes, each with a relatively small effect. His idea was to create mouse chromosomes either lacking a gene or having an extra copy. By combining these chromosomes through crosses, he could create mice with zero copies of the gene (if the mouse survived), or one, two, three, or four copies. He could then examine the mice and look changes in phenotype associated with differences in gene dosage. His initial application of this approach was to regulation of blood pressure. This trait was of special interest to

> **❝ My mindset had begun to change . . . to the thought that [hypertension] was more likely to be the result of 'many little things.' ❞**

Smithies because his father had died from complications of high blood pressure, and because he himself required medication to control his own blood pressure. Smithies's insights and approach were a great success and well deserving of his share of the 2007 Nobel Prize in Physiology and Medicine.

My mindset had begun to change from the idea that a few major differences might determine [hypertension] to the thought that it was more likely to be the result of "many little things. . . ." Many quantitative differences have accumulated during human evolution [that have very small effects]. These "many little things" are a joyful source of our individuality. But they are probably also a source of the poorly understood differences in individual susceptibility to [complex diseases].

O. Smithies, *Nat. Rev. Genet.* 6(2005): 419–425.

substance serotonin (5-hydroxytryptamine) is known to influence a variety of psychiatric conditions, such as anxiety and depression. Among the important components in serotonin action is the serotonin transporter protein. Neurons that release serotonin to stimulate other neurons also take it up again through the serotonin transporter. This uptake terminates the stimulation and also recycles the molecule for later use.

The serotonin transporter became a strong candidate gene for depression when it was discovered that the transporter is the target of a class of antidepressants known as selective serotonin reuptake inhibitors. The widely prescribed antidepressant Prozac is an example of such a drug.

Motivated by the strong suggestion that the serotonin transporter might be involved in depression, researchers looked for evidence of genetic polymorphisms affecting the transporter gene in human populations. Such a polymorphism was found in the promoter region of the transporter. About 1 kb upstream from the transcription start site is a series of 16 tandem repeats of a nearly identical DNA sequence of about 15 base pairs. This is the L (*long*) allele, which has an allele frequency of 57 percent among whites. There is also an S (*short*) allele, in which three of the repeated sequences are not present. The S allele has an allele frequency of 43 percent. The genotypes L/L, L/S, and S/S are found in the Hardy–Weinberg proportions of 32 percent, 49 percent, and 19 percent, respectively.

Further studies revealed that the polymorphism does have a physiological effect. In cells grown in culture, L/L cells had approximately 50 percent more mRNA for the transporter than L/S or S/S cells, and L/L cells had approximately 35 percent more membrane-bound transporter protein than L/S or S/S cells. In view of these differences, it is perhaps not surprising that individuals with S/S or S/L genotypes have a much higher risk of severe depression than L/L genotypes. Interestingly, the higher risk shows a genotype-by-environment interaction in that it occurs only among individuals who have experienced three or more stressful life events (threat, defeat, humiliation, or death of a loved one) occurring within the previous 5 years.

The S and L forms of the serotonin transporter also affect drug response. One of the illicit drugs in widespread use is the "club drug" MDMA (3,4-methylenedioxymethamphetamine) also called Ecstasy, which binds to the serotonin transporter and inhibits reuptake. Habitual use of Ecstasy is associated with long-term changes in the release and reuptake of serotonin. This is potentially troubling, as an estimated 1 percent of American young adults between the ages of 18 and 25 are regular users.

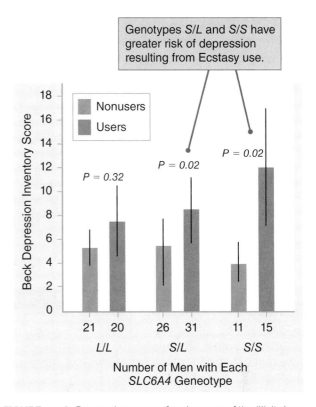

FIGURE 15.18 Depression score of male users of the illicit drug Ecstasy as a function of genotype for the serotonin uptake transporter gene *SLC6A4*. The *P* values are significance levels for each genotype as compared with male nonusers.

Data from J. P. Roiser, et al., *Am. J. Psychiatry* 162 (2005): 609–612.

The effects of the *SLC6A4* polymorphism in habitual Ecstasy users are shown in **FIGURE 15.18**. In this study, male Ecstasy users were classified according to their *SLC6A4* genotype (L/L, S/L, or S/S) and compared with male nonusers with respect to the Beck Depression Inventory score, a standardized rating scale for depression based on a questionnaire covering a variety of different symptoms of depression. A score greater than 9 is regarded as evidence of depression.

Although the Ecstasy users were required to abstain from the drug for at least 3 weeks prior to testing, nevertheless the long-term effects of use are evident in Figure 15.18. Although the difference in the depression score is not statistically significant in the L/L genotype ($P = 0.32$), both the S/L and S/S genotypes have a significantly greater depression score among Ecstasy users than among nonusers ($P = 0.02$). Roughly speaking, there is a twofold greater risk of depression among the S/L genotypes and a 10-fold greater risk of depression among the S/S genotypes. This type of genotype-by-environment interaction is an example of what has been called *pharmacogenomics*, a genotype-specific response to pharmaceutical agents.

CHAPTER SUMMARY

- Complex traits are determined by multiple genetic and environmental factors acting together.

- The relative contributions of genotype and environment to a trait are measured by the variance due to genotype (genotypic variance) and the variance due to environment (environmental variance).

- Correlations between relatives are used to estimate various components of variation, such as genotypic variance, additive variance, and dominance variance.

- Additive variance accounts for the parent–offspring correlation; dominance variance

accounts for the sib–sib correlation over and above that expected from the additive variance.

- Narrow-sense heritability is the ratio of additive (transmissible) variance to the total phenotypic variance; it is widely used in plant and animal breeding.

- Genes that affect quantitative traits (QTLs) can be identified by genetic mapping using highly polymorphic DNA markers, or by specifying candidate genes on the basis of knowledge of the physiology and development of the trait.

- Many complex human behaviors are affected by multiple genetic and environmental factors and the interactions among them.

ISSUES AND IDEAS

- Give an example of a trait in human beings that is affected both by environmental factors and by genetic factors. Specify some of the environmental factors that affect the trait.

- Does the distribution of phenotypes of a trait in a population imply anything about the relative importance of genes and environment in causing differences in phenotype among individuals? Does it imply anything about the number of genes that may affect the trait? Explain.

- In regard to the genotypic variance, what is special about an inbred line or about the F_1 progeny of a cross between two inbred lines?

- The distribution of bristle number on one of the abdominal segments in a population of *Drosophila* ranges from 12 to 26. The narrow-sense heritability of the trait is approximately 50 percent.

Do you think that it would be possible, by practicing artificial selection over a number of generations, to produce a population in which the *mean* bristle number was greater than 26? To produce a population in which the *minimum* bristle number was greater than 26? Explain why or why not.

- Why are correlations between relatives of interest in the study of quantitative traits?

- In the context of complex inheritance, what is a QTL? Does a QTL differ from any other kind of gene? How can QTLs be detected?

- In the context of complex inheritance, what is a candidate gene? Would you regard the human *PAH* gene for phenylalanine hydroxylase as a natural candidate gene for mental retardation? Why or why not?

SOLUTIONS: STEP BY STEP

PROBLEM 1 A random sample of lambs from a large flock was weighed at weaning at 100 days. The weight in kilograms for each of 20 lambs is shown in the accompanying table. Estimate the mean, the variance, and the standard deviation of 100-day weaning weight in this population.

37	37	38	46	39
30	31	35	30	42
43	39	48	27	41
43	41	37	29	26

SOLUTION. The mean and variance of 100-day weaning weight are estimated from the sum of the individual values and the sum of the deviations from the mean, as expressed in Equations (15.1) and (15.2) in the text. Note

that the variance has units of kilograms-squared. The original units of kilograms are restored by taking the square root, which is the standard deviation

$$s = \sqrt{40.16 \text{ kg}^2} = 6.34 \text{ kg}$$

The calculations are shown below.

$$\bar{x} = \frac{\sum x_i}{N} = \frac{739}{20} = 36.95 \text{ kg}$$

$$s^2 = \frac{\sum (x_i - \bar{x})^2}{N - 1} = \frac{762.95}{19} = 40.16 \text{ kg}^2$$

An equivalent formula for the variance, which is much easier to use in calculations, is shown below. The symbol Σx_i^2 means the sum of the squares of the individual values, and $(\Sigma x_i)^2$ means the square of the sum of the individual values.

$$s^2 = \frac{\Sigma x_i^2}{N-1} - \frac{\left(\Sigma x_i\right)^2}{N(N-1)}$$
$$= 1477.32 - 1437.16$$
$$= 40.16 \text{ kg}^2$$

PROBLEM 2 For the flock of sheep in Problem 1, assume that 100-day weaning weight of lambs is normally distributed. What range of weaning weights would be expected to include 68 percent of the lambs? What range of weaning weights would be expected to include 95 percent of the lambs?

SOLUTION. In a normal distribution, 68 percent of the observations are expected to lie within one standard deviation of the mean and 95 percent within two standard deviations of the mean. In this case, the mean weaning weight is estimated as 36.95 kg and the standard deviation as 6.34 kg. The range expected to include 68 percent of the lambs is therefore 36.95 − 6.34 to 36.95 + 6.34 kg, or 30.61 − 43.29 kg. Similarly, the range expected to include 95 percent of the lambs is 36.95 − 2 × 6.34 to 36.95 + 2 × 6.34 kg, or 24.27 − 49.63 kg.

PROBLEM 3 For the flock of sheep in Problem 1, the narrow-sense heritability of 100-day weaning weight is 30 percent. Suppose artificial selection is carried out, choosing as parents adult males and females whose 100-day weaning weight averaged one standard deviation above the current population mean.
(a) What is the expected average weaning weight of the lambs after one generation of selection?
(b) What is the expected average weaning weight of the lambs after five generations of selection? (Assume that the heritability and the standard deviation remain constant over this interval.)

SOLUTION. (a) In the first generation of selection, the mean of the current population is $M = 36.95$ kg and that of the selected parents is $M^* = 36.95 + 6.34 = 43.29$ kg. The narrow-sense heritability h^2 is given as 0.30. The prediction equation

$$M' = M + h^2(M^* - M)$$

applies to this case, where M' is the expected mean of the progeny of the selected parents. Hence $M' = 36.95 + (0.30) \times (43.29 - 36.95) = 38.85$ kg. **(b)** If the selected parents always deviate from the current population mean by one standard deviation, then $M^* - M = 6.34$ in each generation of selection. We are told to assume that h^2 remains at 30 percent. Hence, in each generation of selection, the mean 100-day weaning weight is expected to increase by $0.30 \times 6.34 = 1.90$ kg. After five generations of selection, therefore, the expected 100-day weaning weight is given by $36.95 + 5 \times 0.30 \times 6.34 = 46.46$ kg.

CONCEPTS IN ACTION: PROBLEMS FOR SOLUTION

15.1 In humans, the estimated values of broad-sense heritability for some traits are: maximum heart rate $H^2 = 0.84$, systolic blood pressure $H^2 = 0.57$, and serum lipid level $H^2 = 0.44$. For which of these traits is the phenotypic variance most affected by genetic differences among individuals?

15.2 The following questions pertain to a normal distribution.

  (a) What term applies to the value along the x-axis that corresponds to the peak of the distribution?

  (b) If two normal distributions have the same mean but different variances, which is the broader?

  (c) What proportion of the population is expected to lie within one standard deviation of the mean?

  (d) What proportion of the population is expected to lie within two standard deviations of the mean?

15.3 The mean stem length and variance of stem length in two highly homozygous varieties of roses (I and II) and their F_1 and F_2 progeny are shown below. Calculate the broad-sense heritability.

Variety	Mean Length (cm)	Variance (cm²)
I (short)	35.21	2.365
II (long)	81.56	2.934
F_1	55.59	3.590
F_2	59.79	36.110

15.4 Two varieties of corn, A and B, are field-tested in Indiana and North Carolina. Strain A is more productive in Indiana, but strain B is more productive in North Carolina. What phenomenon in quantitative genetics does this example illustrate?

15.5 Some estimates of broad-sense heritabilities of human traits are 0.85 for adult height, 0.62 for body weight, 0.57 for systolic blood pressure, 0.44 for diastolic blood pressure, 0.50 for twinning, and 0.1 to 0.2 for overall fertility. Which of these characteristics is most likely to "run in families"? If one of your parents and one of your grandparents has high blood pressure, should you be concerned about the likelihood of you having the same problem?

15.6   Two highly inbred strains of mice are crossed. The F_1 generation has a mean tail length of 4.3 cm and a standard deviation of 1.3 cm. The F_2 generation has a mean tail length of 4.3 cm and a standard deviation of 3.5 cm. What are the environmental variance, the genetic variance, and the broad-sense heritability of tail length in this population?

15.7   For the difference between the domestic tomato, *Solanum esculentum*, and its wild South American relative, *Solanum chmielewskii*, the environmental variance accounts for 13 percent of the total phenotypic variance of fruit weight, for 9 percent of the total variance of soluble-solid content, and for 11 percent of the total variance in acidity. What are the broad-sense heritabilities of these traits?

15.8   Two homozygous genotypes of *Drosophila* differ in the number of abdominal bristles. In genotype *AA*, the mean bristle number is 20 with a standard deviation of 2. In genotype *aa*, the mean bristle number is 23 with a standard deviation of 3. Both distributions conform to the normal distribution, in which the proportions of the population with a phenotype within an interval defined by the mean ± 1, ± 1.5, ± 2, and ± 3 standard deviations are 68, 87, 95, and 99.7 percent, respectively.

   (a)   In genotype *AA*, what is the proportion of flies with a bristle number between 20 and 23?

   (b)   In genotype *aa*, what is the proportion of flies with a bristle number between 20 and 23?

   (c)   What proportion of *AA* flies have a bristle number greater than the mean of *aa* flies?

   (d)   What proportion of *aa* flies have a bristle number greater than the mean of *AA* flies?

15.9   The narrow-sense heritability of withers height in a population of quarterhorses is 35 percent. (Withers height is the height at the highest point of the back, between the shoulder blades.) The average withers height in the population is 19 hands. (A "hand" is a traditional measure equal to the breadth of the human hand, now taken to equal 4 inches.) From this population, studs and mares with an average withers height of 16 hands are selected and mated at random. What is the expected withers height of the progeny? How does the value of the narrow-sense heritability change if withers height is measured in meters rather than hands?

15.10   Consider a complex trait in which the phenotypic values in a large population are distributed approximately according to a normal distribution with mean 100 and standard deviation 15. What proportion of the population has a phenotypic value above 130? Below 85? Above 85?

15.11   You are studying fruit weight, seed number, and acid content in tomatoes. From 100 greenhouse plants, you sample a single fruit and measure it for the three traits, obtaining the data shown in the histograms below.

   (a)   Can you infer which of these traits likely has the greatest genetic variance? Explain your reasoning.

   (b)   Can you infer which of these traits has the greatest heritability? Explain your reasoning.

15.12   In an experimental population of the flour beetle *Tribolium castaneum*, the pupal weight is distributed normally with a mean of 2.0 mg and a standard

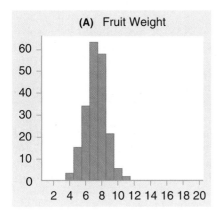

(A) Fruit Weight

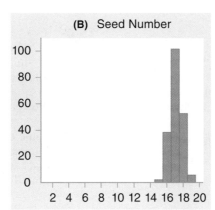

(B) Seed Number

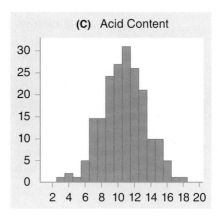

(C) Acid Content

deviation of 0.2 mg. What proportion of the population is expected to have a pupal weight between 1.8 and 2.2 mg? Between 1.6 and 2.4 mg? Would you expect to find an occasional pupa weighing 3.0 mg or more? Explain your answer.

15.13 The F_1 generation of a cross between two inbred lines of tomato has a variance of 12 gm², whereas the F_2 generation has a variance of 30 gm². Estimate the genotypic variance, the environmental variance, and the broad-sense heritability.

15.14 In human beings, the narrow-sense heritability of the total fingerprint ridge count is 90 percent. On the basis of this value, what is the estimated correlation coefficient between first cousins in the total fingerprint ridge count?

15.15 Maternal effects are nongenetic influences on offspring phenotype that derive from the phenotype of the mother. For example, in many mammals, larger mothers have larger offspring, in part because of a maternal effect on birth weight. What result would a maternal effect have on the correlation in birth weight between mothers and their offspring compared with that between fathers and their offspring? How would such a maternal effect influence the estimate of narrow-sense heritability?

15.16 If the correlation coefficient of a trait between first cousins is 0.09, what is the estimated narrow-sense heritability of the trait?

15.17 You are studying sprint speed in lizards. You raise two generations of lizards in the laboratory, measuring speed on a track at age 1 year in both parents and offspring. The resulting data consist of 10 pairs of midparent–offspring values. The midparent speed is the average of the parental speeds.

Pair #	Midparent speed	Offspring speed
1	2.15	1.83
2	2.48	2.91
3	1.29	1.34
4	1.90	2.63
5	2.87	2.95
6	1.91	1.72
7	1.78	1.81
8	1.87	2.62
9	2.18	2.02
10	1.83	1.94

(a) Calculate the mean and variance of both midparent and offspring values.

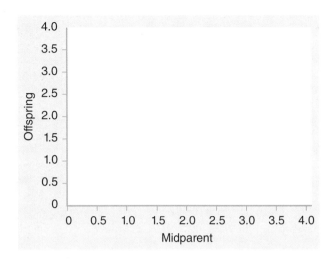

(b) Plot the 10 points on the accompanying graph.

(c) From the scatterplot, would you guess that the trait of sprint speed has high or low heritability in these lizards?

15.18 A mouse population has an average weight gain between ages 3 and 6 weeks of 12 g, and the narrow-sense heritability of the weight gain between 3 and 6 weeks is 20 percent.

(a) What average weight gain would be expected among the offspring of parents whose average weight gain was 16 g?

(b) What average weight gain would be expected among the offspring of parents whose average weight gain was 8 g?

15.19 To estimate the heritability of maze-learning ability in rats, a selection experiment was carried out. From a population in which the average number of trials necessary to learn the maze was 10.8, with a variance of 4.0, animals were selected that managed to learn the maze in an average of 5.8 trials. Their offspring required an average of 8.8 trials to learn the maze. What is the estimated narrow-sense heritability of maze-learning ability in this population?

15.20 The three accompanying graphs show scatterplots of offspring on midparent values for three different traits in the Colorado Blue Columbine (*Aquilegia caerulea*), the official state flower of Colorado. The midparent value is the average of the parental values.

A horticulturalist wishes to increase the value of all these traits because she believes that the resulting flowers would be very successful commercially. However, she needs to produce some impressive flowers quickly if her greenhouse business is not going to fail. As a geneticist, which of these traits would you advise her to concentrate on, and why?

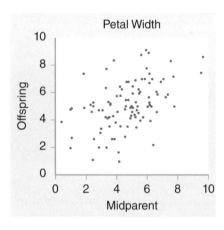

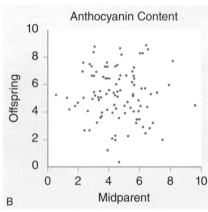

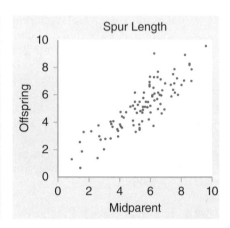

15.21 For a phenotype determined by a single, completely penetrant recessive allele at frequency q in a random-mating population, it can be shown that the narrow-sense heritability is $2q/(1 + q)$.

(a) Calculate the narrow-sense heritability for $q = 1.0, 0.5, 0.1, 0.05, 0.01, 0.005,$ and 0.001.

(b) Note that the narrow-sense heritability goes to 0 as q goes to 0, yet the phenotype is completely determined by heredity. How can this be explained?

-🧠- STOP & THINK ANSWERS

ANSWER TO STOP & THINK 15.1

The range $15-21$ is the mean ± 1 standard deviation, hence approximately 68 percent of the plants are expected to have a flower number in this range. The range $12-24$ is the mean ± 2 standard deviations, hence approximately 95 percent are expected to have a flower number in this range. Because the normal distribution is symmetrical, the expected proportion of plants with ≤ 12 flowers equals $(1 - 0.95)/2 = 0.025 = 2.5$ percent. The same proportion is expected to have a flower number ≥ 24.

ANSWER TO STOP & THINK 15.2

For thorax length, $\sigma_e^2 = 0.0018$ mm² and $\sigma_g^2 + \sigma_e^2 = 0.0036$ mm², so $\sigma_g^2 = 0.0036$ mm² $- 0.0018$ mm² $= 0.0018$ mm². In this case, the genotypic variance and the environmental variance are equal.

ANSWER TO STOP & THINK 15.3

Equation (15.5) $[M' = M + h^2(M^* - M)]$ applies to this situation with $M = 125$, $h^2 = 0.86$, and $M^* = (200 + 100)/2 = 150$. The expected fingerprint ridge count among the offspring is therefore $125 + 0.86 \times (150 - 125) = 146.5$.

READINESS REVIEW

Welcome to your new tool for success in your genetics course! The Readiness Assessment and Review is designed to diagnose your preparation level to learn some of the most important concepts in your course and provide you the help where you most need it, all before you enter the classroom. We identified three skill areas – Mathematics in Genetics, Science Prerequisites, and Thinking Like a Scientist – and created a Readiness Assessment that helps analyze your skill levels and preparedness in each. Our targeted assessment feedback will then point you to the correct section(s) in the Readiness Review where you can review, practice, and improve your readiness using our step-by-step guides and robust practice problem sets.

Once you have completed the Readiness Review, take the assessment again and see how you have improved! It just takes five easy steps to get started:

1. Redeem the access code bound into the front of your book at www.jblearning.com.*
2. Find the Readiness Assessment in your Navigate 2 digital course in the Learning Tools pathway.

LESSONS	**LEARNING TOOLS**	REPORTS & GRADES

⌂ > Courses > Advantage Access for Genetics: Analysis of Genes a...

eBook	Study Center	Resources

3. Take the Readiness Assessment and get your personalized results.

4. Go to the back of your textbook and complete the section of the Readiness Review that corresponds with your results, and practice using the step-by-step guides and problem sets.

Problem 6.3 After 1 year it has divided 12 times. $1\ year = 365\ days \times (24\ hr/day) = 8.67 \times 10^3\ hr$

$$number\ of\ divisions = \frac{8.67 \times 10^3\ hr}{7.20 \times 10^2\ hr} = 1.2 \times 10^1 = 12\ times$$

After 1 year, there will be 4069 cells.

Problem 6.4 The length is 0.5 μm. $\sqrt{0.25} = 0.5$

Problem 6.5 The colony of *Cellularium friendus* is 4.09 mm².

2.03 mm × 2.03 mm or 4.09 mm²

$$0.5\ \mu m \times 4096 \times \frac{10^3\ mm}{\mu m} = 2.0345\ mm$$

5. Retake the assessment to see your improvement!

© Marius Heil/EyeEm/Getty Images

Mathematics in Genetics

OUTLINE

- Introduction
1. Scientific Notation, Significant Figures, and the Metric System
2. Expressing Frequencies: Proportions, Percentages, and Ratios
3. Probability
4. The Power of 2: Exponents
5. Matrices
6. Algebra

Introduction

Genetics is an information science; it is about how information is stored in the DNA, how that information is expressed in the cell, and how the information is transmitted over time to new generations. Math underlies all sciences and genetics is no exception. Good math skills are important to understanding genetics, but do not feel alone if you struggle with mathematics. Many biologists may not expect to use math skills. You have probably already encountered most of the math involved; it's a matter of applying the principles you already know to a new subject matter.

To help you understand the way math is used in genetics, we will review these six topics:

1. Scientific Notation, Significant Figures, and the Metric System
2. Expressing Frequencies: Proportions, Percentages, and Ratios
3. Probability
4. The Power of 2: Exponents
5. Matrices
6. Algebra

Each section includes a brief explanation of the topic followed by the examples of the calculations. There are problems at the end of each section to help you sharpen your skills. Answers to the problems are found at the end, so you can check your work.

TABLE 1 shows the population and area of continents; we will be using this data to provide examples of calculations.

1. Scientific Notation, Significant Figures, and the Metric System

One of the exciting aspects of science is our ability to measure very large or very small lengths and volumes. Scientific notation is a convention that helps us to understand both small and large numbers. One order of magnitude is 10; 10 $= 10^1$, 100 $= 10^2$, 1000 $= 10^3$. You may have noticed that the exponent is the

TABLE 1 Population and Area of Continents		
Continent	**Population**	**Area (km²)**
Asia	4,545,133,094	31,033,131
Africa	1,287,920,518	29,648,481
Europe	742,648,010	22,134,900
North America	587,615,976	21,329,926
South America	428,240,515	17,461,112
Australia/Oceania	41,261,212	8,486,460
Antarctica	0	13,720,000
Total	**7,632,819,325**	**143,814,010**

Data from http://www.worldometers.info/geography/7-continents/, accessed March 3, 2018.

Name _____ Section _____ Date _____

number of zeros. Negative exponents are numbers smaller than 1, so $0.1 = 10^{-1}$, $0.01 = 10^{-2}$, $0.001 = 10^{-3}$, and so on. In this case, the exponent represents the number of places after the decimal point. Search YouTube for a video called Powers of Ten™ (1977) by the Office of Charles and Ray Eames. This video demonstrates the powers of 10 and scientific notation.

In this review, we will use the convention of expressing large (or small) numbers with exponents and two significant figures. For example, the total population of Earth (Table 1) can be expressed as 7.63 billion, or 7.63×10^9 and the area in square km is 1.44×10^8. Calculators and computers use this convention: 7.63e9 and 1.44e8, where "e" stands for exponent and the number that follows "e" is the power of 10. Practice on converting data to scientific notation is provided in Problem 1.1.

The metric system of measurement is used in science and in most countries in the world. The advantage of the metric system is that it is a decimal system, so conversions between different units only require moving the decimal point, essentially changing the order of magnitude. The International System of Units (SI) (**TABLE 2**) demonstrates the relationship between the prefixes used in the metric system.

The prefixes in Table 2 are applicable to computers and to DNA; a byte is the base measurement of computer storage, there are 10^6 bytes in a megabyte. One thousand bases are referred to as a kilobase (kB) and 1 million base pairs (bp) are referred to as a megabase (Mb). Our genomes are 3×10^9 bp, or 3 gigabases (Gb).

One unit missing from the chart above is the ångström (Å), which is defined as 10^{-10} meters; this is another length unit that you may see used in biology.

TABLE 2 Order of Magnitude in the Metric System

Order of Magnitude	Prefix	Symbol	Order of Magnitude	Prefix	Symbol
10^0	–	–	10^0	–	–
10^1	deca	da	10^{-1}	deci	d
10^2	hecto	h	10^{-2}	centi	c
10^3	kilo	k	10^{-3}	milli	m
10^6	mega	M*	10^{-6}	micro	μ
10^9	giga	G*	10^{-9}	nano	n
10^{12}	tera	T*	10^{-12}	pico	p
10^{15}	peta	P*	10^{-15}	femto	f
10^{18}	exa	E*	10^{-18}	atto	a
10^{21}	zetta	Z*	10^{-21}	zepto	z
10^{24}	yotta	Y*	10^{-24}	yocto	y

*These symbols are capitalized to distinguish from other common abbreviations. For example, a ML is a megaliter, 10^6 liters, but mL is a milliliter, 10^{-3} liters.

For example, the bases in DNA are 3.4 Å (0.34 nm) apart. Problem 1.2 provides practice in converting between the factors for length, volume, area, and base pairs.

Molecular genetics deals with structures that are too small to see with the naked eye, so how we understand something that we can't see? One method is microscopy. Light and electron microscopy are two methods commonly used in biology; the units here are fractional units of length. Understanding length measures helps us to understand the order of magnitude of the structures important in genetics. **FIGURE 1** illustrates the sizes of objects and structures from the biological world. When we look at micrographs, it is important to notice the size bar since this provides a ruler to measure the size of structures in the photo. In **FIGURE 2**, the first image is a DNA molecule; note the size bar is 100 nm. The next image is of chromosomes; the size bar indicates that the length is 10 μm, which is 100 times larger than the image of the DNA $\left(\left[(10\,\mu m \times 10^3\,nm\,/\,\mu m)/100\,nm\right] = 100.\right)$

Another example from the text is **FIGURE 3.14** on page 84, indicating that chromosomes are 300 ångströms (Å) (30 nm) wide. Problem 1.3 provides practice to help you with envisioning small structures through microscopy.

One of the challenges in modern science is to understand data that covers several orders of magnitude. In order to condense the information into a more "userfriendly" format, the exponent is used; this is the logarithm (log). An example of where a $\log_{10}$ scale is used is the Richter scale, which is a measure of earthquake intensity based on height of oscillations produced by seismographs. An

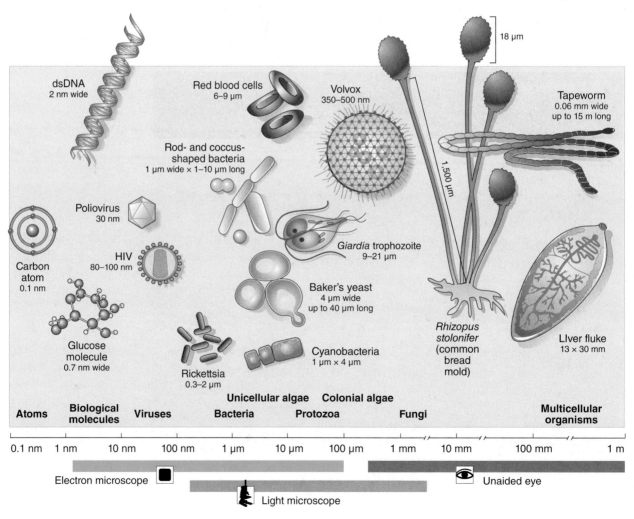

FIGURE RR1.01 Orders of magnitude in biology.

Name _____ Section _____ Date _____

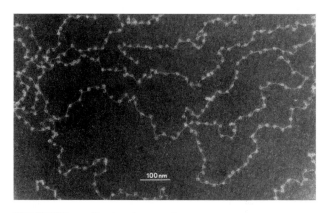 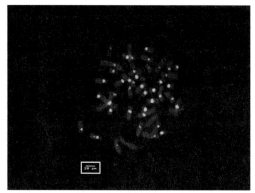

FIGURE RR1.02 Size in micrographs.

(A) Courtesy of Ada L. Olins and Donald E. Olins, University of New England; (B) Courtesy of Paula Coelho and Claudio E. Sunkel Cariola, IBMC.

earthquake that measures 5.0 is 10 times as strong as a 4.0 magnitude earth-quake. A 7.0 magnitude earthquake is 1000 stronger than a 4.0.

Expressing data using exponents is useful when either very small or very large numbers are involved, but there are some tricks to keep in mind when doing calculations. For example, population density is a useful calculation when we want to compare the continents; Table 1 provides the data needed to calculate these numbers. Our units will be people per square kilometer, so we divide the population by the area to get the density.

$$Population\ density = \frac{number\ of\ people}{area\ (km^2)}$$

To complete this calculation with exponents, there is an important rule to keep in mind: In division, the exponents are subtracted. If a calculation involves multiplying two numbers together, the exponents are added. To calculate the overall population density of the Earth, we would divide the population of Earth by the area in km^2.

$$population\ density\ of\ Earth = \frac{7.63 \times 10^9}{1.43 \times 10^8} = 5.34 \times 10^1 = 53.4\ people\ per\ km^2$$

Notice that the answer includes 4/10 of a person. But how can you measure a fraction of a person? Our measurement is whole individuals, it is the division step that creates an illusion of precision; you can't measure a fractional individual. The precision in our answer is limited to the least precise measurement in the data. This can be an issue in genetics; when counting the number of number of offspring produced as a result of a controlled cross, our measurement is whole individuals. One extra digit past the decimal point can help detect sources of rounding error; but more digits are unnecessary and inaccurate since it indicates a level of precision that we do not have. For practice in calculating population density, see Problem 1.4.

Let's ask another question: How many km^2 could an individual own if the continent could be equality divided? In this case, the formula would be the area divided by the number of people. If we do this calculation for everyone on Earth and leave out Antarctica:

$$Area\ per\ person\ on\ Earth = \frac{1.29 \times 10^8}{7.63 \times 10^9} = 0.169 \times 10^{-1} = 0.0169\ km^2$$

We may want to convert into square meters (m²). There are 10^6 m² per km², so when this multiply the number by this factor, the km² cancel out and the answer is in meters:

$$0.169 \times 10^{-1}\, km^2 \times 10^6\, m^2 / km^2 = 0.169 \times 10^5\, m^2$$

In Problem 1.5, you will complete these calculations for all the continents shown in Table 1.

Problems

Problem 1.1 Reformat all the data in Table 1 using scientific notation.

Continent	Population Using Scientific Notation	Area (km²) Using Scientific Notation
Asia		
Africa		
Europe		
North America		
South America		
Australia/Oceania		
Antarctica	-*	
Total	**7.63 × 10⁹**	**1.43 × 10⁸**

*$10^0 = 1$, zero cannot be expressed as an exponent.

Problem 1.2 Convert the following measurements, using scientific notation to express the answer.

1 meter = _____ nm

1 Å = _____ μm

1 L = _____ μL

Name _____ Section _____ Date _____

1 μL = _____ mL

1 km² = _____ m²

1 kilobase (kB) of DNA = _____ bases

1 gigabase (GB) of DNA = _____ kilobases

Problem 1.3 Both of the images (A and B) in **FIGURE** 3 are microbes that represent two extremes of prokaryotic size.

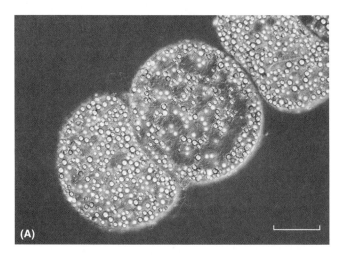

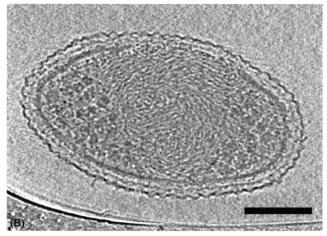

FIGURE RR1.03 (A) A phase microscopy image showing three Thiomargarita namibiensis cells. (Bar = 100 um.) (B) Cryogenic transmission electron microscopy image of an ARMAN cell. (Bar = 100 nm.)

(A) Courtesy of Heide Schulz-Vogt, Max Planck Institute of Marine Microbiology, Germany; (B) Courtesy of Lawrence Berkeley National Lab.

Based on the size bars in the images, which cells are bigger?

_____

What is the difference in the order of magnitude between A and B?

_____

Problem 1.4 Calculate the population density of each continent in people per 10 km² using the data from Problem 1.1.

Continent	Population Density: Persons per 10 km²	Calculation
Asia		
Africa		
Europe		
North America		
South America		
Australia/Oceania		
Antarctica		
Total	**5.34 × 10¹ or 53.4**	

Problem 1.5 How many square <u>meters</u> per person are there in each continent? Use the data in Problem 1.1.

Name _____ Section _____ Date _____

Continent	m² per person	Calculation
Asia		
Africa		
Europe		
North America		
South America		
Australia/Oceania		
Antarctica		
Total	$0.169 \times 10^{-1} = 0.0169\,km^2$	$= \dfrac{1.29 \times 10^8}{7.63 \times 10^9}$

2. Expressing Frequencies: Proportions, Percentages, and Ratios

According to Table 1, there are over 7.6 billion (7.63×10^9) people on Earth; over 587 million (5.87×10^8) live in North America, so the fraction (or proportion) of people who live in North America is:

$$\frac{5.9 \times 10^8 \ human\ population\ in\ North\ America}{7.6 \times 10^9 \ total\ human\ population\ of\ Earth} = 0.77 \times 10^{-1} = 0.077$$

Multiply this by 100 to get the percentage; 7.7% of the world's population resides in North America.

Another way of thinking about this data is as a ratio. What is the ratio of people living in North America when compared to the rest of the world?

5.87×10^8 (*population of North America*): 7.63×10^9 (*population of world*)

Divide both by the population of North America (so it becomes 1):

$$\frac{5.87 \times 10^8}{5.87 \times 10^8} : \frac{7.63 \times 10^9}{5.87 \times 10^8} = 1:12.99$$

This means that 1 out of 13 people in the world live in North America.

You may have noticed that this calculation is the reciprocal of the frequency; $1/13 = 0.077$. Problem 2.1 is practice calculating percentages and ratios from the rest of the continents.

Frequencies are useful to predict expected outcomes. For example, if 10% of M&Ms in a bag are expected to be green, how many green ones will be in a bag of 500? The answer is 10% of 500 or 50. The ratio of green to all colors can be expressed as 1 to 10. See Problem 2.4 for more practice.

Ratios are often used to express frequencies of diseases, genetic conditions, and other attributes in humans: These are estimates based upon sampling populations. For example, the frequency of Turner syndrome, where females only have one X chromosome is estimated to be 1 in 2000 females $(1/2000 = 0.0005 \text{ or } 5.0 \times 10^{-4} \text{ or } 0.05\%)$. In human populations of European descent, approximately 45% have straight hair, 40% have wavy hair, and 15% have curly hair. What is the ratio of individuals with straight to wavy to curly hair?

$$\frac{45}{15} : \frac{40}{15} : \frac{15}{15} = 3.0 : 2.6 : 1.0$$

The genetic basis for hair texture has only been partially determined, but this type of information is critical to find the genes that influence this trait.

As the father of genetics, Gregor Mendel's discovery was that ratios reflect the inheritance pattern. These mathematical relationships discovered by Mendel were key to understanding genetics, but he didn't know about genes or DNA; he used the term "factors." For example, Mendel found that if pea plants with white flowers were crossbred to those with purple flowers, all the new plants in the first generation had purple flowers; was the purple factor masking the white or was the white "disappearing"? When these plants with white flowers were crossbred to each other, the second generation exhibited a 3 : 1 ratio of purple to white, so the white factor was not lost. The same mathematical pattern holds true for a crossbreeding of smooth and wrinkled seeds; all the progeny produce only smooth seeds, but when these smooth seeds are cross a 3 : 1 ratio of smooth to wrinkled is the result in the second generation. If a cross is made between a purple-flowered, smooth-seeded plant and a white-flowered wrinkled-seeded plant, we can predict the ratio of second generation with respect to the two factors by multiplying the ratios together. The mnemonic FOIL (**FIGURE 4**) describes this operation, you multiply the **F**irst numbers, then the **O**utside numbers, then the **I**nside numbers, and finally, the **L**ast numbers.

Another method of calculating ratios for two variables will be discussed in the "Matrices" section.

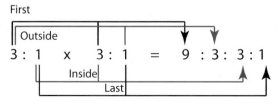

FIGURE RR1.04 F.O.I.L.

Name _____ Section _____ Date _____

Problems

Problem 2.1 Calculate the percentage and ratios of each continent's population to that of the world.

Continent	Proportion (percent)	Calculation	Ratio (Continent : World)	Calculation
Asia				
Africa				
Europe				
North America	0.077 (7.7%)		1:13	
South America				
Australia/ Oceania				
Antarctica				

Problem 2.2 What is the ratio of the population of Asia to North America?

_____

Problem 2.3 What is the ratio of the area of Asia to North America?

_____

Problem 2.4 What is the ratio of the population of Africa to Europe?

_____

Problem 2.5 What is the ratio of the area of Africa to Europe?

_____

Problem 2.6 The Mars candy corporation uses the percentage of each color in a bag as a measure of quality control. This allows us to predict the expected number of each color in each bag. Fill in the following table, calculating the expected number in each bag and the ratio of each color to all colors.

Color	Average % in Bag	Expected Number in a Bag of 500	Calculation	(Ratio Total: Each Color)	Calculation
Brown	30%				
Yellow	20%				
Red	20%				
Orange	10%				
Green	10%				
Blue	10%				

Problem 2.7 A bag of Valentine's Day candy has 50% red hearts and 50% green hearts. Also, 75% of the hearts say "I love you" and 25% say "Be mine."

What is the ratio to red to green hearts?

_____

What is the ratio of hearts that say "I love you" to those that say "Be mine"?

_____

Assuming that the hearts are randomly mixed, what is the expected ratio of red hearts that say "I love you" to red hearts that say "Be mine" to green hearts that say "I love you" to green hearts that say "Be mine"?

_____

Problem 2.8 A pack of 50 baseball trading cards has 40 veteran players and 10 rookie players. In the same deck, there are 45 right-handers and 5 left-handers.

What is the ratio of veterans to rookie players?

_____

Name _____ Section _____ Date _____

What is the ratio of right-handers to left-handers?

_____

Assuming that the number of right-handers and left-handers are randomly distributed among the rookie and veteran players, what is the ratio of right-handed veterans to left-handed veterans to right-handed rookies to left-handed rookies?

_____

Problem 2.9 On the planet Vulcan, there are animals resembling mice; some have tails, while others are tailless, and some have brown coats while others have yellow coats. Vulcan scientists are interested in how these characteristics are inherited, so they crossbred a strain with tails to a strain without tails and all first-generation mice are tailless. When these tailless mice are bred, a ratio of two tailless mice for one tailed mouse was the result. The same holds true for the coat color: A crossbreeding of brown to yellow strains of mice yields first-generation mice that all have yellow coats. Crossbreeding the yellow first generation with each other produces a second generation in which there are two yellow mice for every brown mouse.

If brown mice with tails are crossed to a yellow, tailless strain, what is your prediction for what the first generation will look like?

_____

If these are crossbred, what ratio will you expect in the second generation?

(_2_ : _1_) (_2_ : _1_) = _____ : _____ : _____ : _____

3. Probability

Frequencies are used to estimate probabilities; the more frequent an event is, the more likely that it will occur. Since a coin has two sides, there is a 50% chance $[(1/2) \times 100]$ of a coin landing heads up and 50% landing tails up. Based on the world population data seen earlier, if extraterrestrials visit Earth and decide to randomly beam someone to their ship, there is a 7.7% chance that that person will reside in North America. Similarly, a randomly selected person of European descent has a 15% chance of having curly hair. If you were to randomly select a single M&M from the bag of 500 described earlier, what is the chance that it would be red? It is 0.20 or 20%— the same as the frequency.

Here is a more complicated question: If you were to randomly select two M&Ms from the bag, what is the chance that both would be red? In this case, you multiply the probability for each instance together: $0.20 \times 0.20 = 0.04$ *or* 4.0%. The order in which you draw the M&Ms is irrelevant since there is only one way to make this combination. What is the probability that you will randomly draw a red and a blue M&M in any order? For this calculation, it is necessary to take into account the order in which the M&Ms are drawn, since there are two ways to make the combination: red first followed by blue or blue first followed by red. Both combinations have the same probability $(0.20 \times 0.10 = 0.02$ *or* 2%), and since there are two ways to make the combination, they are independent of each other. The probabilities of the red–blue combination and blue–red combination are added $(0.02 + 0.02 = 0.04$ *or* 4%). If you specify the order in which the colors are drawn the events are dependent since there is only one way to make this combination, so the frequencies are multiplied together $(0.02 \times 0.02 = 0.04)$.

The same rules apply to birth order in families, but it is simplified by the chance of having a boy or girl is equal, 50% or 0.50. The probability of a family consisting of a boy, girl, boy (bgb) family, in that order, is $0.50 \times 0.50 \times 0.50 = 0.125$. Each birth is dependent upon the others, so the probabilities are multiplied together. If we don't care about the order, there are three different ways to have a family that consists of two boys and one girl; bgb, bbg, and gbb. Each of events has a probability of 0.125, but now we are describing events that are independent of each other, so the probabilities are added together, $0.125 + 0.125 + 0.125 = 0.375$.

Problems

Problem 3.1 The table below shows the percentages of each color in a bowl of jelly beans. Fill in the missing frequencies. Use this table for Problems 3.2–3.10.

Color	Number in Bowl	Frequency	Calculation
Brown	112	0.28	
Yellow	88		
Red	80		
Orange	48		
Green	40		
Blue	32		
Total	400	1.00	

Problem 3.2 Referring to the above table, what is the probability of first selecting an orange jelly bean, followed by a red one?

Name _____ Section _____ Date _____

Problem 3.3 What is the probability of first selecting a red jelly bean, followed by an orange one?

_____

Problem 3.4 What is the probability of selecting an orange and a red jelly bean, in any order?

_____

Problem 3.5 What is the probability of first selecting a brown jelly bean, followed by a yellow one?

_____

Problem 3.6 What is the probability of first selecting a yellow jelly bean, followed by a brown one?

_____

Problem 3.7 What is the probability of selecting an orange and a red jelly bean, in any order?

_____

Problem 3.8 What is the probability of first selecting a blue jelly bean, followed by a red one?

_____

Problem 3.9 What is the probability of first selecting a red jelly bean, followed by a blue one?

_____

Problem 3.10 What is the probability of selecting an red and a blue jelly bean, in any order?

_____

TABLE 3 is for Problems 3.11–3.16.

TABLE 3	
Continent	Proportion (percent) of World Population
Asia	0.595 (59.5%)
Africa	0.169 (16.9%)
Europe	0.097 (9.7%)
North America	0.077 (7.7%)
South America	0.056 (5.6%)
Australia/Oceania	0.005 (0.5%)
Antarctica	0.000 (0.0%)

Problem 3.11 What is the probability of extraterrestrials beaming someone up who lives in Asia?

_____

Problem 3.12 What is the probability of extraterrestrials beaming someone up who lives in Asia, followed by a resident of Africa?

_____

Problem 3.13 What is the probability of extraterrestrials beaming someone up who lives in Asia, and a resident of Africa, in any order?

_____

Problem 3.14 What is the probability of the same extraterrestrials beaming someone up from Asia and Europe, in any order?

_____

Name _____ Section _____ Date _____

Problem 3.15 What is the probability of the extraterrestrials beaming someone from North America and then South America?

_____

Problem 3.16 What is the probability of the extraterrestrials beaming someone from North America and then South America in any order?

_____

4. The Power of 2: Exponents

Sports tournaments (such as college basketball) often start with 64 teams. After the first round there are 32, after the second round there are 16, 8 remain after the third round, 4 after the fourth, 2 after the fifth, and finally only 1 team is victorious. This mathematical progression is reverse of what happens when bacteria divide: You start with 1, it divides to produce 2 bacteria, which then divide and 4 bacteria result, and when these divide, there are 8, then 16, 32, and 64. Bacterial division can continue, and the number of bacteria each generation (n) is mathematically described as 2^n. The single bacterium represents generation zero ($2^0 = 1$); after one division, there are $2^1 = 2$, after 2 there are $2^2 = 4$, after 6 divisions there are $2^6 = 64$. This mathematics progress continues exponentially; for example, after 100 generations, there are $2^{100} = 1.27 \times 10^{30}$ bacteria!

The surface area of Earth is about 5×10^{14} m². If a microbe is 0.5 μm long and 0.5 μm wide, how many generations are necessary for it to cover Earth? First, let's convert surface area of Earth from m² into μm²:

$$1\,m = 10^6\,\mu m$$

$$1\,m^2 = 10^{12}\,\mu m^2 (or\ 1e + 12)$$

$$5 \times 10^{14}\,m^2 \times 10^{12}\,\mu m^2\,/\,m^2 = 5 \times 10^{26}\,\mu m^2$$

The surface area of the bacteria is 2.5×10^{-1} μm², as calculated here:

$$0.5\,\mu m \times 0.5\,\mu m = 0.25\,\mu m^2 = 2.5 \times 10^{-1}\,\mu m^2$$

So, for there to be enough to cover Earth, there must be 2.5×10^{26} bacteria (see below for calculation).

$$\frac{5 \times 10^{26}\,\mu m^2}{2.5 \times 10^{-1}\,\mu m^2\,/\,bacteria} = 2.5 \times 10^{27}\,bacteria$$

In order to work backwards and calculate the number of generations, we use the inverse function to exponents, logarithms. In this case it's the $\log_2$.

$$number\ of\ generations = \log_2(number\ of\ bacteria)$$
$$= \log_2(2.5 \times 10^{27}\ of\ bacteria) = 91.0\ generations$$

If a bacterial generation takes 20 minutes, how long would it take for the bacteria to cover Earth?

$$= 91.0 \, generations \times \frac{20 \, minutes}{generation} = 1820 \, minutes$$

$$1820 \, minutes \times \frac{1 \, hour}{60 \, minutes} = 30.3 \, hours$$

So, in less than 2 days, Earth would be covered with bacteria. If you are thinking that there are just too many factors that would prevent this, you are correct. But this exercise provides you with a feeling for the power of 2.

Genetics is all about the power of 2. The power of 2 operates at the molecular genetic level and results in the incredible diversity of life. We can see this in the human family. For example, every human on the planet had two biological parents, and each of those parents had two biological parents, and so on. Each of us has $2^2 = 4$ grandparents, and $2^3 = 8$ great-grandparents. We have half of our mother's DNA and half of our fathers. Going back one more generation, we have 1/4 of a single grandparent's DNA, and we share 1/8 of our DNA with a great-grandparent. Brothers and sisters share some characteristics, but also can be very different from one another. This is because, although all siblings get half of their DNA from the parent, it can be a different half in each sibling. DNA testing for ancestry is becoming more popular and it is possible to identify relatives such as third cousins. How much DNA do you potentially share with a third cousin? To answer this, let's go start with the genetic relationship between parents, and their children and between siblings. These are described as first-degree relationships and all share half (50% or 0.50) of their DNA (2^{-1}). Examples of second-degree relatives are grandchildren, grandparents, uncles, aunts, nephews, and nieces, who share about 0.25 of their DNA (2^{-2}). Cousins are third-degree relatives, so they share about 0.125 of their DNA (2^{-3}). Second cousins are fifth-degree relative and share about 0.03125 of their DNA (2^{-4}). Third cousins are seventh-degree relatives, fourth cousins are ninth-degree relatives, and so on. Problems 4.1–4.7 focus on these issues in humans.

Since we get a set of chromosomes from each of our biological parents, we have two copies of each chromosome, with the notable exception of the X chromosome in males (XY) and in a few other rare conditions. But how does the number of chromosomes stay constant any species overtime? In other words, why doesn't the number of chromosomes double every generation? Humans have 23 pairs of chromosome (46 total), with one of each of the pairs from the mother and the other of the pair from the father. If the chromosomes were not divided up before reproduction, the next generation would have $23 \times 2 = 46$ chromosome pairs, the next would have $23 \times 4 = 92$ pairs. In three generations, the number would double again, and the grandchildren would have $23 \times 8 = 92$ chromosome pairs. The mathematical formula for this progression would be

Number of chromosome pairs $= 23 \times 2^n$, *where* $n = $ # *of generations*

The nucleus of cells would also have to double in size every generation to accommodate the increased amount of DNA. In 12 generations, the number of chromosome pairs would increase 4094 times to 94,208 ($= 23 \times 2^{12} = 23 \times 4096 = 94,208$). This would require an increase in diameter of the nucleus from 10 μm (0.01 mm) to 40 mm ($= 0.01$ mm $\times 4096 = 40.96$ mm); a bit larger than a ping-pong ball (40 mm)! Problems 4.8 and 4.9 continue this line of thought.

The dividing up of the chromosomes so each gamete (egg or sperm) gets one copy of each chromosome takes place during meiosis, a special type of cell division

Name _____ Section _____ Date _____

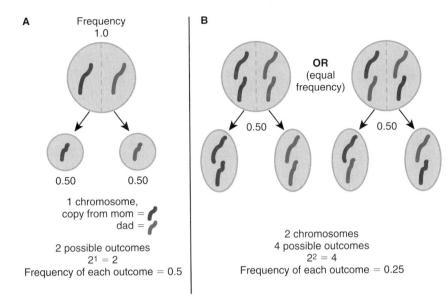

A Frequency 1.0

0.50   0.50

1 chromosome,
copy from mom =
dad =

2 possible outcomes
$2^1 = 2$
Frequency of each outcome $= 0.5$

B

OR
(equal
frequency)

0.50   0.50

2 chromosomes
4 possible outcomes
$2^2 = 4$
Frequency of each outcome $= 0.25$

FIGURE RR1.05 Meiosis for an organism with one (A) or two (B) chromosomes.

where the chromosome pairs line up next to each other and end up in different cells. Meiosis takes place in very special cells in our gonads; only these cells can become the next generation given the right environment. The entire process of meiosis is reviewed in the basic science concepts section of this Readiness Review. Here, we simplify the process and focus on the math in meiosis I; the step that cuts the amount of genetic material in half and also increases genetic diversity in the offspring. If an organism has a single chromosome pair, then each copy goes to a separate cell, forming two different cells (see **FIGURE 5**). If an organism has two chromosome pairs, then there are two different ways for the chromosomes to line up; one way, the maternal chromosomes go into a cell and the paternal chromosomes go into the other cell, or a maternal and paternal copy of each go into the two cells. With two chromosomes, there are four different cells possible (2^2).

In an organism with three chromosomes, there are $2^3 = 8$ different possibilities for cells. See Problems 4.10–4.13 for additional practice.

Problems

Problem 4.1 A seed is planted from a tree, and after the first season, it has only one branch. After the second season, it has two branches, and after three seasons, it has four branches. If it maintains this rate of growth, how many branches will it have after four seasons?

_____

How many will it have after five seasons?

_____

After 10 seasons, how many branches will it have?

_____

Problem 4.2 How many great-great grandparents (that is, three generations back) do each of us have?

_____

Problem 4.3 How much of our DNA is from a great-great grandparent?

_____

Problem 4.4 Going four generations back, how many great-great-great grandparents do each of us have?

_____

Problem 4.5 How much of our DNA is from a great-great-great grandparent?

_____

Problem 4.6 How much DNA do third cousins share?

_____

Problem 4.7 If you could get on a time machine and go back 500 years, what is the maximum* number of relatives would you have, assuming an average life-cycle of 25 years?

_____

* The reason this is a maximum is that it doesn't take into account that some of your relatives may be related to each other; first- and second-cousin matings were common when population numbers were low, and populations were very isolated. This means that one person could be present more than once in your family tree.

Problem 4.8 After 10 generations of reproduction without meiosis, how many chromosomes would humans have?

_____

Problem 4.9 If the nucleus of a cell is 10 μm and it has to double in diameter every generation for 10 generations, what would the diameter be in mm after 10 generations with no meiosis?

_____

Problem 4.10 An organism with three chromosomes will produce eight different combinations. Fill in the figure below, showing all possible combinations.

Name _____ Section _____ Date _____

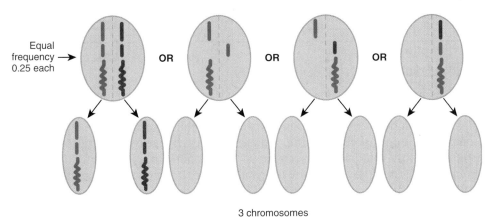

3 chromosomes
8 possible outcomes
$2^3 = 8$
Frequency of each outcome = 0.125

Problem 4.11 If an organism has four different pairs of chromosomes, how many possible combinations could be made?

_____

Problem 4.12 Complete the following figure showing all possible combinations for an organism with four pairs of chromosomes.

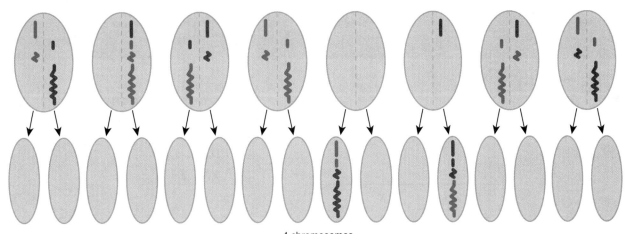

4 chromosomes
16 possible outcomes
$2^4 = 16$
Frequency of each outcome = 0.0625

Problem 4.13 In organisms with 23 pairs of chromosomes (humans), how many different possibilities are possible after meiosis?

5. Matrices

Matrices are a useful way of representing data. There are many types of data matrices, and some are useful to calculate all possible combinations. **TABLE 4** is a data matrix derived from the population data in Table 1; rows and columns are the continents; the proportion of the population of each is shown in red (Antarctica is omitted since there are no permanent human populations residing there.) For example, in order to calculate the probability of beaming up two people from Asia, multiply 0.595 × 0.595 to get 0.354, so 35.4% of random beam-ups would be two people from Asia (see blue arrows). Note that the diagonal (shown in bold) are the probabilities of beaming up two people from the same continent, and that the numbers above the diagonal are transposed (rows swapped for columns) in relation to the numbers below the diagonal. For example, the probability of beaming up one person from Asia and one from Africa is in row 1, column 2 and in row 2, column 1. If the column is selected first, then the row, the probably is 0.101 and if the row is selected first, then the column, the probability is still 0.101. If order doesn't matter, then the probabilities are added together, so there is 0.202 chance of beaming a person from Africa and one from Asia in either order. If we add up all the frequencies in the table, the total will be 1 (allowing for rounding error).

TABLE 4 World Population Frequency Table

	Asia	Africa	Europe	North America	South America	Australia
Asia 0.595	**0.354**	0.101	0.058	0.046	0.033	0.003
Africa 0.169	0.101	**0.286**	0.016	0.013	0.009	0.001
Europe 0.097	0.058	0.016	**0.009**	0.007	0.005	0.0005
North America 0.077	0.046	0.013	0.007	**0.006**	0.003	0.0004
South America 0.056	0.033	0.009	0.005	0.003	**0.005**	0.0003
Australia 0.005	0.003	0.001	0.0005	0.0004	0.0003	**0.0003**

Ratios can also be used in a data matrix. For example, the calculation of (3 : 1) (3 : 1) described above in the example with peas can be derived using a data matrix. Keep in mind that 9 : 3 : 3 : 1 can also be represented as 9/16, 3/16, 3/16, 1/16, or 0.5625, 0.1875, 0.187, 0.0625.

	3 purple	1 white
3 smooth	9 purple, smooth	3 smooth, white
1 wrinkled	3 purple, wrinkled	1 white, wrinkled

This is a Punnett square, named for its developer, Reginald C. Punnett, and uses ratios rather than frequencies. It is a method for predicting the outcome of a breeding experiment, based on a hypothesis about the inheritance pattern. In the computer age, organizing data in matrices is an important starting point when writing program to carry out these types of calculations.

Problems

Problem 5.1 Using Table 4, predict the most common combination of two people that the aliens would expect to beam up.

_____

What is the least common combination?

_____

What is the probability of beaming up a person from Asia and North America, in any order?

_____

Problem 5.2 Fill in the following Punnett square showing the breeding of the yellow and brown, tailed or tailless mice.

	2 yellow	1 brown
2 tailless		
1 tailed		

Problem 5.3 Fill in the following table for M&Ms.

Color	Brown	Yellow	Red	Orange	Green	Blue
Frequency	0.28	0.22	0.2	0.12	0.1	0.08

Color	Frequency					
Brown	0.28	0.0784				
Yellow	0.22	0.0616	0.0484			
Red	0.2	0.056	0.044	0.04		
Orange	0.12	0.0336	0.0264	0.024	0.0144	
Green	0.1	0.028	0.022	0.02	0.012	0.01
Blue	0.08	0.0224	0.0176	0.016	0.0096	0.008

(Blue row last value: 0.0064)

6. Algebra

When there is missing information, algebra can be used to calculate the missing values. Out of a total of 250 snail shells, 220 have a right-handed coil; the number of snails with a left-handed coil can be described by the following equation:

$$250 = 220 + x$$

Rearranging this,

$$x = 250 - 220$$
$$x = 30$$

Algebra is useful in molecular genetics. You may recall from your biology course that DNA has four bases: adenine (A), cytosine (C), guanine (G), and thymine (T). To form the double helix, A pairs with T and C pairs with G. The proportion of G + C base pairs is a useful characteristic. For example, the G + C content of different bacteria varies widely, from 25% to 75%. The bacteria *Wolbachia* that lives inside fruit flies (*Drosophila melanogaster*) has a G + C content of 35.2%, while *Kineococcus radiotolerans* lives inside nuclear waste tanks and has a G + C content of 74.2%. The A + T content of these species is $100 - 35.2 = 64.8\%$ for *Wolbachia* and $100 - 74.2 = 25.8\%$ for *K. radiotolerans*.

Even if we only know the frequency of a single base in DNA, we can calculate the rest. If we know the frequency of A, then we automatically know that T is the same amount since they pair across the double helix. The rest are G + C pairs, with half of the bases C and the other half G.

The relationship between the number of base pairs in DNA that code for a specific amino acid was established in the 1960s to be three bases per amino acid. If you know the length of DNA, you can calculate the maximum number of amino acids for which a DNA fragment can code using the following formula:

$$Number\ of\ amino\ acids = \frac{number\ of\ bases}{3\ bases\ per\ amino\ acid}$$

Alternatively, the number of bases needed to code for a specific size of protein can be calculated with this formula:

$$Number\ of\ bases = Number\ of\ amino\ acids \times 3$$

A 300 base pair section of DNA has the potential to code for 100 amino acids $[= 300\ bases/(3\ bases/amino\ acid)]$. A protein that has 1200 amino acids requires a 3600 base-coding region in the DNA $[1200\ amino\ acids \times 3\ (bases/amino\ acid)]$.

Sometimes it is necessary to do algebra with exponents. For example, if we need to know the length of an edge for a km² plot of land, we first need to understand that the area of a square is calculated by multiplying the length by the width, which are the same, so it can be expressed as an exponent:

$$area\ of\ a\ square\ plot = x^2$$

$$x^2 = 10\ km^2$$

$$\sqrt{x^2} = \sqrt{10\ km^2}$$

The result is

$$x = 3.2\ km$$

Problems

Problem 6.1 The table below lists the characteristic G + C content of bacterial species. Calculate the A + T content of each.

Species	G + C Content %	A + T Content %	Calculation
Lactobacillus brevis	46		
Escherichia coli	50		
Micrococcus luteus	67		
Mycobacterium tuberculosis	75		
Staphylococcus aureus	33		

Problem 6.2 You've discovered a new species of bacteria that lives inside of cell phones, *Cellularium friendus*. The frequency of A is 0.12.

What is the frequency of T? _____

What is the of A + T content? _____

What is the G + C content? _____

What is the frequency of G? _____

Problem 6.3 *Cellularium friendus* is very slow growing, dividing once every 720 hours.

After 1 year, how many times has it divided? _____

Assuming there is no cell death, how many cells are present after 1 year?

_____

Problem 6.4 On micrographs, _C. friendus_ appears to be a perfect cube. If the surface area is 0.25 μm, what is the length of each side?

_____

Problem 6.5 How big would a 1-year old colony of _C. friendus_ be?

_____

Problem 6.6 There are two systems for classifying blood type, the ABO blood system and the Rh system.

The ABO system consists of blood types A, B, AB, and O. In the United States, 42% of people have type B, 10% have type B, and 4% have type AB. What is the percentage of the population with type O blood?

_____

The Rh system is independent of the ABO system and people are either Rh negative (Rh$^-$) or Rh positive (Rh$^+$). If 15% of the population is Rh$^-$, what percentage of the population is Rh$^+$?

_____

Assuming that the the ABO and Rh types are independent of each other, fill in the following table, indicating the percentage of the population with each blood type.

ABO Blood Type	Rh Blood Type	Percentage of Population	Calculation (Assumes that out of 100 people, 42 are type A, 10 are type B, 4 are type AB, and 44 are type O)
A	Positive	35.7%	
A	Negative	6.3%	
B	Positive		
B	Negative		
AB	Positive		
AB	Negative		
O	Positive		
O	Negative		

Problem 6.7 It takes three base pairs of DNA to code of an amino acid, which are the building blocks of protein. Fill in the following table, indicating either the number of bases needed to code for the protein, or the number of amino acids coded for by the number of base pairs.

	DNA Length	Number of Amino Acids	Calculations
Protein 1	3333 bp		
Protein 2	9.63 kb		
Protein 3	1.2 Mb		
Protein 4		4243	
Protein 5		525	
Protein 6		3333	

Answer Key
Problem 1.1

Continent	Population Using Scientific Notation	Area (km²) Using Scientific Notation
Asia	4.55×10^9	3.10×10^7
Africa	1.29×10^9	2.96×10^7
Europe	7.43×10^8	2.21×10^7
North America	5.88×10^8	2.12×10^7
South America	4.28×10^8	1.75×10^7
Australia/Oceania	4.13×10^7	8.50×10^6
Antarctica	–*	1.37×10^7
Total	$\mathbf{7.63 \times 10^9}$	$\mathbf{1.43 \times 10^8}$

* $10^0 = 1$; zero cannot be expressed as an exponent.

Problem 1.2

1 meter = 10^9 (1,000,000,000) nm

1 Å = 10^{-4} (0.0001) μm

1 L = 10^6 (1,000,000) μL

$1\ \mu L = 10^{-3}\ (0.001)\ mL$

$1\ km^2 = 10^6\ (1,000,000)\ m^2$

1 kilobase (kB) of DNA = 10^3 (1000) bases

1 gigabase (GB) of DNA = 10^6 (1,000,000) kilobases

Problem 1.3

Figure A is bigger.

The difference in magnitude is 10^3 or 1000-fold. The size bar in B is in nm and there are 1000 nm per μm, which are the units in A.

Problem 1.4

Continent	Population Density: Persons per km²	Calculation
Asia	1.468×10^2 or 146.8	$= \dfrac{4.55 \times 10^9}{3.10 \times 10^7}$
Africa	4.35×10^1 or 43.5	$= \dfrac{1.29 \times 10^9}{2.96 \times 10^7}$
Europe	3.36×10^1 or 33.6	$= \dfrac{7.43 \times 10^8}{2.21 \times 10^7}$
North America	2.77×10^1 or 27.5	$= \dfrac{5.88 \times 10^8}{2.12 \times 10^7}$
South America	2.45×10^1 or 24.5	$= \dfrac{4.28 \times 10^8}{1.75 \times 10^7}$
Australia/Oceania	0.486×10^{-1} or 4.9	$= \dfrac{4.13 \times 10^7}{8.50 \times 10^6}$
Antarctica	0	$= \dfrac{0}{1.43 \times 10^8}$
Total	5.34×10^1 or 53.4	$= \dfrac{7.63 \times 10^9}{1.43 \times 10^8}$

Problem 1.5

Continent	m² per Person	Calculation
Asia	0.681×10^{-2}	$= \dfrac{3.10 \times 10^7}{4.55 \times 10^9}$
Africa	2.29×10^{-2}	$= \dfrac{2.96 \times 10^7}{1.29 \times 10^9}$
Europe	0.297×10^1	$= \dfrac{2.21 \times 10^7}{7.43 \times 10^8}$
North America	0.361×10^{-1}	$= \dfrac{2.12 \times 10^7}{5.88 \times 10^8}$
South America	2.45×10^1	$= \dfrac{1.75 \times 10^7}{4.28 \times 10^8}$
Australia/Oceania	2.06×10^1	$= \dfrac{8.50 \times 10^6}{4.13 \times 10^7}$
Antarctica	0	*Can't divide by* 0
Total	$0.169 \times 10^{-1} = 0.0169 \; km^2$	$= \dfrac{1.29 \times 10^8}{7.63 \times 10^9}$

Problem 2.1

Continent	Proportion (percent)	Calculation	Ratio (Continent : World)	Calculation
Asia	0.595 (59.5%)	$= \dfrac{4.55 \times 10^9}{7.63 \times 10^9}$	1 : 2	$\dfrac{4.55 \times 10^9}{4.55 \times 10^9} : \dfrac{7.63 \times 10^9}{4.55 \times 10^9}$
Africa	0.169 (16.9%)	$= \dfrac{1.29 \times 10^9}{7.63 \times 10^9}$	1 : 6	$\dfrac{1.29 \times 10^9}{1.29 \times 10^9} : \dfrac{7.63 \times 10^9}{1.29 \times 10^9}$
Europe	0.097 (9.7%)	$= \dfrac{7.43 \times 10^8}{7.63 \times 10^9}$	1 : 10	$\dfrac{7.43 \times 10^8}{7.43 \times 10^8} : \dfrac{7.63 \times 10^9}{7.43 \times 10^8}$
North America	0.077 (7.7%)	$= \dfrac{5.88 \times 10^8}{7.63 \times 10^9}$	1 : 13	$\dfrac{5.88 \times 10^8}{5.88 \times 10^8} : \dfrac{7.63 \times 10^9}{5.88 \times 10^8}$
South America	0.056 (5.6%)	$= \dfrac{4.28 \times 10^8}{7.63 \times 10^9}$	1 : 18	$\dfrac{4.28 \times 10^8}{4.28 \times 10^8} : \dfrac{7.63 \times 10^9}{4.28 \times 10^8}$

Australia/ Oceania	0.005 (0.5%)	$= \dfrac{4.13 \times 10^7}{7.63 \times 10^9}$	1 : 185	$\dfrac{4.13 \times 10^7}{4.13 \times 10^7} : \dfrac{7.63 \times 10^9}{4.13 \times 10^7}$
Antarctica	0.000 (0.0%)	$= \dfrac{0}{7.63 \times 10^9}$	–	$\dfrac{0}{0} : \dfrac{7.63 \times 10^9}{0}$ *Can't divide by 0*

Problem 2.2

$$= \frac{4.55 \times 10^9}{5.88 \times 10^8} : \frac{5.88 \times 10^8}{5.88 \times 10^8} = 7.7 : 1$$

Problem 2.3

$$= \frac{3.10 \times 10^7}{2.12 \times 10^7} : \frac{2.12 \times 10^7}{2.12 \times 10^7} = 1.5 : 1$$

Problem 2.4

$$= \frac{1.29 \times 10^9}{7.43 \times 10^8} : \frac{7.43 \times 10^8}{7.43 \times 10^8} = 1.7 : 1$$

Problem 2.5

$$= \frac{2.96 \times 10^7}{2.21 \times 10^7} : \frac{2.21 \times 10^7}{2.21 \times 10^7} = 1.3 : 1$$

Problem 2.6

Color	Average % in Bag	Expected Number in a Bag of 500	Calculation	(Ratio Total: Each Color)	Calculation
Brown	30%	150	$= 0.30 \times 500$	3 : 1	$\dfrac{500}{150} : \dfrac{150}{150}$
Yellow	20%	100	$= 0.20 \times 500$	5 : 1	$\dfrac{500}{100} : \dfrac{100}{100}$
Red	20%	100	$= 0.20 \times 500$	5 : 1	$\dfrac{500}{100} : \dfrac{100}{100}$
Orange	10%	50	$= 0.10 \times 500$	10 : 1	$\dfrac{500}{50} : \dfrac{50}{50}$
Green	10%	50	$= 0.10 \times 500$	10 : 1	$\dfrac{500}{50} : \dfrac{50}{50}$
Blue	10%	50	$= 0.10 \times 500$	10 : 1	$\dfrac{500}{50} : \dfrac{50}{50}$

Problem 2.7

The ratio of red to green hearts is 1 : 1. Since the percentages are equal, the ratio is 1 red : 1 green.

The ratio of hearts that say "I love you" to those that say "Be mine" is 3 : 1.

$75\% = 0.75$ and $25\% = 0.25$. $\dfrac{0.75}{0.25} : \dfrac{0.25}{0.25} = 3 : 1$

The expected ratio of red hearts that say "I love you" to red hearts that say "Be mine" to green hearts that say "I love you" to green hearts that say "Be mine" is 3 : 1 : 3 : 1.

(1 red : 1 green) × (3 "I love you" : 1 "Be mine") = 3 : 1 : 3 : 1

Problem 2.8

The ratio of veterans to rookie players is 4 : 1 $\left[(40/10):(10/10)\right]$.

The ratio of right-handers to left-handers is 9 : 1 $\left[(45/5):(5/5)\right]$.

The ratio of right-handed veterans to left-handed veterans to right-handed rookies to left-handed rookies is 36 : 4 : 9 : 1.

$$(4:1)\times(9:1)\, 9\times4 = 36:4\times1 = 4:1\times9 = 9:1\times1 = 1$$

Problem 2.9

The first generation will be all yellow tailless.

The second generation will be 4 tailless and yellow : 2 tailless and brown: 2 tailed and yellow: 1 brown and tailed.

Problem 3.1

Color	Number in bowl	Frequency	Calculation
Brown	112	0.28	$= \dfrac{112}{400}$
Yellow	88	0.22	$= \dfrac{88}{400}$
Red	80	0.20	$= \dfrac{80}{400}$
Orange	48	0.12	$= \dfrac{48}{400}$
Green	40	0.10	$= \dfrac{40}{400}$

Blue	32	0.08	$=\dfrac{32}{400}$
Total	400	1.00	$=\dfrac{400}{400}$

Problem 3.2 The probability of first selecting an orange jelly bean, followed by a red one is 0.024.

$0.12 \times 0.20 = 0.024$

Problem 3.3 The probability of first selecting a red jelly bean, followed by an orange one is 0.024.

$0.20 \times 0.12 = 0.024$

Problem 3.4 The probability of selecting an orange and a red jelly bean, in any order is 0.048.

$(0.12 \times 0.20) + (0.20 \times 0.12) = 0.048$

Problem 3.5 The probability of first selecting a brown jelly bean, followed by a yellow one is 0.0616.

$0.28 \times 0.22 = 0.0616$

Problem 3.6 The probability of first selecting a yellow jelly bean, followed by a brown one is 0.0616.

$0.22 \times 0.28 = 0.0616$

Problem 3.7 The probability of selecting an orange and a red jelly bean, in any order is 0.1232.

$(0.28 \times 0.22) + (0.22 \times 0.28) = 0.1232$

Problem 3.8 The probability of first selecting a blue jelly bean, followed by a red one is 0.016.

$0.08 \times 0.20 = 0.016$

Problem 3.9 The probability of first selecting a red jelly bean, followed by a blue one is 0.016.

$0.20 \times 0.08 = 0.016$

Problem 3.10 The probability of selecting a red and a blue jelly bean, in any order is 0.032.

$(0.08 \times 0.20) + (0.20 \times 0.28) = 0.032$

Problem 3.11 The probability of extraterrestrials beaming someone up who lives in Asia is 0.595. Answer is provided in the table of data.

Problem 3.12 The probability of extraterrestrials beaming someone up who lives in Asia, followed by a resident of Africa is 0.101.

$0.595 \times 0.169 = 0.101$

Problem 3.13 The probability of extraterrestrials beaming someone up who lives in Asia and a resident of Africa in any order is 0.202.

$(0.595 \times 0.169) \times 2 = 0.202$

Problem 3.14 The probability of the same extraterrestrials beaming someone from Asia and Europe in any order is 0.115.

$0.595 \times 0.097 \times 2 = 0.115$

Problem 3.15 The probability of the extraterrestrials beaming someone from North America and then South America is 0.004.

$0.077 \times 0.056 = 0.004$

Problem 3.16 The probability of the extraterrestrials beaming someone from North America and someone from South America in any order is 0.009.

$0.077 \times 0.056 \times 2 = 0.009$

Problem 4.1 After four seasons, there will be 8 branches. $2^3 = 8$

After five seasons, there will be 32. $2^5 = 32$

After 10 seasons, there will be 1024. $2^{10} = 1024$

Problem 4.2 We have 16 great-great grandparents. $2^4 = 16$

Problem 4.3 0.0625 of our DNA is from a great-great grandparent. 1/16 or $2^{-4} = 0.0625$

Problem 4.4 We each have 32 great-great-great grandparents. $2^5 = 32$

Problem 4.5 0.03125 of our DNA is from a great-great-great grandparent. 1/32 or 25^5 = 0.03125

Problem 4.6 Third cousins share 0.0078125 DNA. 2^{-7} = 0.0078125. Third cousins are considered seventh-degree relatives.

Problem 4.7 The maximum number of relatives would be 1,048,576.

2^{20} = 1,048,576

$$\frac{500\ years}{25\ years\ per\ generation} = 20\ generations$$

Problem 4.8 After 10 generations of reproduction without meiosis there would be 23,552 chromosomes.

$23 \times 2^{10} = 23 \times 1024 = 23,552$

Problem 4.9 The diameter would be 23.552 mm after 10 generations with no meiosis.

$$23.552\ mm = 10\,\mu m \times \frac{10^{-3}\ mm}{per\ \mu m} \times 23,552$$

Problem 4.10

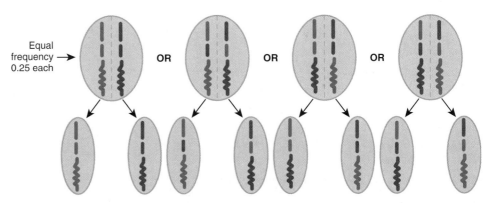

Equal frequency → 0.25 each

OR   OR   OR

3 chromosomes
8 possible outcomes
2^3 = 8
Frequency of each outcome = 0.125

Problem 4.11 16 combinations could be made. $16 = 2^4$

Problem 4.12

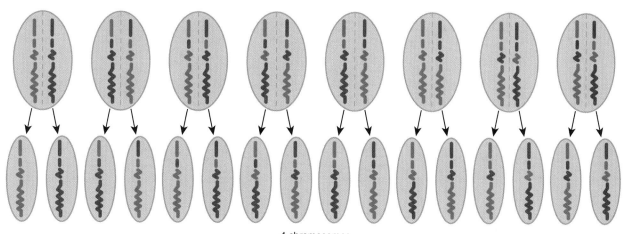

4 chromosomes
16 possible outcomes
$2^4 = 16$
Frequency of each outcome = 0.0625

Problem 4.13 There are 8,388,608 different possibilities after meiosis. $2^{23} = 8,388,608$

Problem 5.1 The most common combination is two people from Asia. The least common combination is two people from Australia, where the probability is 0.092. $0.046 + 0.046 = 0.092$

Problem 5.2

	2 yellow	1 brown
2 tailless	4 yellow, tailless	1 brown, tailless
1 tailed	2 yellow, tailed	1 brown, tailed

Problem 5.3

Color	Brown	Yellow	Red	Orange	Green	Blue
Frequency	0.28	0.22	0.2	0.12	0.1	0.08

Color	Frequency						
Brown	0.28	0.0784	0.0616	0.056	0.0336	0.028	0.0224
Yellow	0.22	0.0616	0.0484	0.044	0.0264	0.022	0.0176
Red	0.2	0.056	0.044	0.04	0.024	0.02	0.016
Orange	0.12	0.0336	0.0264	0.024	0.0144	0.012	0.0096
Green	0.1	0.028	0.022	0.02	0.012	0.01	0.008
Blue	0.08	0.0224	0.0176	0.016	0.0096	0.008	0.0064

Note on solution: All the answers are already in the table. For example, yellow × brown is 0.0616, the same as brown × yellow.

Problem 6.1

Species	G + C Content %	A + T Content %	Calculation
Lactobacillus brevis	46	54	100 − 46 = 54
Escherichia coli	50	50	100 − 50 = 50
Micrococcus luteus	67	33	100 − 67 = 33
Mycobacterium tuberculosis	75	25	100 − 75 = 25
Staphylococcus aureus	33	67	100 − 33 = 67

Problem 6.2 The frequency of T is 0.12. Since A pairs with T, Freq A = Freq T.

The A + T content is 0.24. $Freq\ A + T = 0.12 + 0.12$

The G + C content is 0.76.

$$(Freq\ G + C) + (Freq\ A + T) = 1$$
$$(Freq\ G + C) = 1 - (Freq\ A + T)$$
$$(Freq\ G + C) = 1 - 0.24 = 0.76$$

The frequency of G is 0.38. $0.76 \div 2 = 38$

Problem 6.3 After 1 year it has divided 12 times. $1\ year = 365\ days \times (24\ hr/day) = 8.67 \times 10^3\ hr$

$$number\ of\ divisions = \frac{8.67 \times 10^3\ hr}{7.20 \times 10^2\ hr} = 1.2 \times 10^1 = 12\ times$$

After 1 year, there will be 4069 cells.

Problem 6.4 The length is 0.5 μm. $\sqrt{0.25} = 0.5$

Problem 6.5 The colony of *Cellularium friendus* is 4.09 mm².

2.03 mm × 2.03 mm or 4.09 mm²

$$0.5\ \mu m \times 4096 \times \frac{10^3\ mm}{\mu m} = 2.0345\ mm$$

Problem 6.6 The percentage of the population with type O blood is 44%. $1 - (0.42 + 0.10 + 0.04) = 1 - 0.56 = 0.44\ or\ 44\%$

The percentage of the population that is Rh⁺ is 85%. $1 - 0.15 = 0.85\ or\ 85\%$

ABO Blood Type	Rh Blood Type	Percentage of Population	Calculation (Assumes that out of 100 people, 42 are type A, 10 are type B, 4 are type AB, and 44 are type O)
A	Positive	35.7%	$= 42 \times 0.85 = 35.7\%$
A	Negative	6.3%	$= 42 \times 0.15 = 6.3\%$
B	Positive	8.5%	$= 10 \times 0.85 = 8.5\%$
B	Negative	1.5%	$= 10 \times 0.15 = 1.5\%$
AB	Positive	3.4%	$= 4 \times 0.85 = 3.4\%$
AB	Negative	0.6%	$= 4 \times 0.15 = 0.6\%$
O	Positive	37.4%	$= 44 \times 0.85 = 37.4\%$
O	Negative	6.6%	$= 44 \times 0.15 = 6.6\%$

ANSWER KEY

Problem 6.7

	DNA Length	Number of Amino Acids	Calculations
Protein 1	3333 bp	1111	$\dfrac{3333\ bases}{3\ bases\ per\ amino\ acid} = 1111$
Protein 2	9.63 kb	4320	$\dfrac{9630\ bases}{3\ bases\ per\ amino\ acid} = 3210$
Protein 3	1.2 Mb	400,000	$\dfrac{1.2 \times 10^6\ bases}{3\ bases\ per\ amino\ acid} = 400{,}000$
Protein 4	12,729	4243	$4243 \times 3 = 12{,}729$
Protein 5	1575	525	$525 \times 3 = 1575$
Protein 6	9999	3333	$3333 \times 3 = 9999$

UNIT 2 READINESS REVIEW

© Witthaya Prasongsin/Getty Images

Science Prerequisites

OUTLINE

- Introduction
1. Atoms, Molecules, and What Holds Them Together
2. Biological Molecules
3. Cell Structure and Function
4. Cellular Metabolism
5. The Cell Cycle, Mitosis, and Meiosis

Introduction

It has been said that physics is applied mathematics, chemistry is applied physics, and biology is applied chemistry. Keep in mind that biology is the level that deals with living organisms and incorporates all of these levels. In this section, we will cover some important science prerequisites.

1. Atoms, Molecules, and What Holds Them Together
2. Biological Molecules
3. Cell Structure and Function
4. Cellular Metabolism
5. The Cell Cycle, Mitosis, and Meiosis

Each section includes a brief explanation of the topic followed by the examples of the calculations. There are questions at the end of each section to help you sharpen your skills. Answers to the questions are found at the end so you can check your work.

1. Atoms, Molecules, and What Holds Them Together

The smallest unit of matter is the atom. There are three parts to atoms called subatomic particles: electrons, protons, and neutrons. The nucleus of an atom contains neutrons, which have no charge, and positively charged protons. The negatively charged electrons orbit around the atomic nucleus. **FIGURE 1** is a simplified version of an atom.

An element, such as carbon or hydrogen, is the smallest unit of matter that retains physical properties. The periodic table (**FIGURE 2**) is a listing of the elements organized into groups. Each element has a characteristic number of protons unique to that element; this is the atomic number. The atomic mass is the number of protons plus neutrons. In order for the element to have no charge, the number of electrons must equal the number of protons. There are variants of elements called isotopes that contain a different number of neutrons than the number shown on the periodic table; these may be unstable and emit radioactive energy.

CHNOPS is an acronym for the six most common elements in biological material: carbon, hydrogen, nitrogen, oxygen, phosphorus, and sulfur. These elements make up 97% of biological tissue.

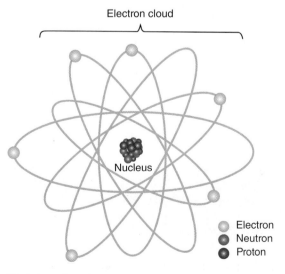

FIGURE RR2.01 Simplified view of an atom.

Name _____ Section _____ Date _____

	Group IA																		Group 0
Period 1	**1** **H** **1.008**	Group IIA											Group IIIB	Group IVB	Group VB	Group VIB	Group VIIB		2 He 4.003
Period 2	3 Li 6.941	4 Be 9.012												5 B 10.81	**6** **C** **12.01**	7 N 14.01	8 O 16.00	9 F 19.00	10 Ne 20.18
Period 3	11 Na 22.99	12 Mg 24.31	Group IIIA	Group IVA	Group VA	Group VIA	Group VIIA	Group VIII			Group IB	Group IIB		13 Al 26.98	14 Si 28.09	**15** **P** **30.97**	**16** **S** **32.06**	17 Cl 35.45	18 Ar 39.95
Period 4	19 K 39.10	20 Ca 40.08	21 Sc 44.96	22 Ti 47.90	23 V 50.94	24 Cr 52.00	25 Mn 54.94	26 Fe 55.85	27 Co 58.93	28 Ni 58.70	29 Cu 63.55	30 Zn 65.38	31 Ga 69.72	32 Ge 72.59	33 As 74.92	34 Se 78.96	35 Br 79.90	36 Kr 83.80	
Period 5	37 Rb 85.47	38 Sr 87.62	39 Y 88.91	40 Zr 91.22	41 Nb 92.91	42 Mo 95.94	43 Tc (98)	44 Ru 101.1	45 Rh 102.9	46 Pd 106.4	47 Ag 107.9	48 Cd 112.4	49 In 114.8	50 Sn 118.7	51 Sb 121.8	52 Te 127.6	53 I 126.9	54 Xe 131.3	
Period 6	55 Cs 132.9	56 Ba 137.3	57 La 138.9	72 Hf 178.5	73 Ta 180.9	74 W 183.9	75 Re 186.2	76 Os 190.2	77 Ir 192.2	78 Pt 195.1	79 Au 197.0	80 Hg 200.6	81 Tl 204.4	82 Pb 207.2	83 Bi 209.0	84 Po (209)	85 At (210)	86 Rn (222)	
Period 7	87 Fr (223)	88 Ra (226.0)	89 Ac (227)	104 Unq	105 Unp	106 Unh	107 Uns		109 Une										

Lanthanides (rare earth metals)

58 Ce 140.1	59 Pr 140.9	60 Nd 144.2	61 Pm (145)	62 Sm 150.4	63 Eu 152.0	64 Gd 157.3	65 Tb 158.9	66 Dy 162.5	67 Ho 164.9	68 Er 167.3	69 Tm 168.9	70 Yb 173.0	71 Lu 175.0

Actinides

90 Th 232.0	91 Pa (231)	92 U 238.0	93 Np (244)	94 Pu (242)	95 Am (243)	96 Cm (247)	97 Bk (247)	98 Cf (251)	99 Es (252)	100 Fm (257)	101 Md (258)	102 No (259)	103 Lr (260)

KEY
16 ——— Atomic number
S ——— Symbol of element
32.06 ——— Atomic mass

☐ Metals   ☐ Metalloids
☐ Nonmetals   ☐ Noble Gases

FIGURE RR2.02 Periodic Table of Elements with the six most common elements in living tissue shown in boxes outlined in red.

The electrons that orbit the nucleus are organized into shells; the shell closest to the nucleus can hold only 2 electrons. The second shell can hold a maximum of 8, while the third shell can hold 18 electrons. This is diagrammed in **FIGURE 3** for elements common in biological material.

Elements are connected together by chemical bonds to form molecules, which can have very different characteristics than the elements from which they are composed. For example, chlorine is a poisonous gas, while sodium is a highly reactive metal that reacts violently when exposed to water. However, when ionically bonded together, they form table salt. Ionic bonds involve the transfer of electrons; in this case, sodium losses an electron to chlorine (see **FIGURE 4**).

Covalent bonding occurs when electrons are shared equally (**FIGURE 5**). These types of bonds are stronger than ionic. Covalent bonds can also be polar; this occurs when the electrons are closer to one atom than the others but are still shared. This is the case in water (**FIGURE 6**).

Water is absolutely necessary for life. Since water is composed of two hydrogens bonded to one oxygen, it can be called dihydrogen monoxide (DHMO); you may

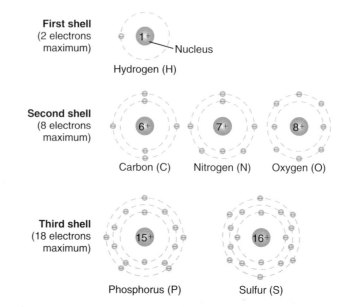

First shell
(2 electrons maximum)

Nucleus

Hydrogen (H)

Second shell
(8 electrons maximum)

Carbon (C)   Nitrogen (N)   Oxygen (O)

Third shell
(18 electrons maximum)

Phosphorus (P)   Sulfur (S)

FIGURE RR2.03 The orbitals for six elements commonly found in biological material.

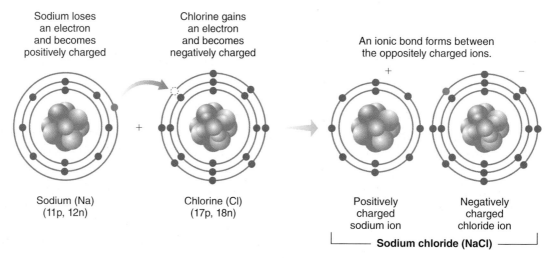

Sodium loses an electron and becomes positively charged

Chlorine gains an electron and becomes negatively charged

An ionic bond forms between the oppositely charged ions.

+   −

+

Sodium (Na)
(11p, 12n)

Chlorine (Cl)
(17p, 18n)

Positively charged sodium ion

Negatively charged chloride ion

Sodium chloride (NaCl)

FIGURE RR2.04 Ionic bonds are the result of electrons being captured by other elements.

1e⁻

2e

FIGURE RR2.05 Equal sharing of electrons results in a covalent bond.

be aware of an internet hoax that claims that DHMO is a dangerous substance about which you should be concerned. More commonly known as H_2O, the oxygen atom is covalently bonded to the two hydrogen atoms (Figure 6a). However, unlike covalent bonds where the electrons are shared equally, the electrons in water are held closer to the oxygen, giving the oxygen partial negative charge. As a result, they hydrogens are slightly positively charged. These partial charge differences are designated as (δ^-) and (δ^+) (Figure 6b). Because opposite charges attract, the

Name _____ Section _____ Date _____

oxygen of one molecule becomes associated with the hydrogen of another molecule through what are called hydrogen bonds. The molecules form a lattice structure that accounts for the property of liquid water to stick to itself, a property known as cohesion. This is one of the unique properties that make water a critical component of living systems; about two-thirds of your body is water.

Water can dissociate into its component hydrogen and hydroxide ions, as shown below in **FIGURE 7**. In pure water, this occurs in 10^{-7} molecules. Since the concentration of hydrogen ions [H$^+$] in water is 10^{-7}, the $-\log_{10}$ of [H$^+$] is 7; this is the pH of pure water. Since the pH scale is based upon a log scale, each unit represents 10-fold change (**FIGURE 8**). Acids have a pH lower than 7, while bases have a pH above 7. Any molecule that adds [H$^+$] to a liquid lowers pH and is acidic, while any substance that absorbs [H$^+$] is basic.

The ability of liquid water to adhere to itself through the polar covalent bonds as shown in Figure 6 is relevant to life. Hydrogen bonds are the result of partial charges; these are also very important in DNA structure.

The weakest bonds are van der Waals (also known as van der Waals interactions). These temporary interactions take place when molecules come close to each other. An example of van der Waals bonds is the ability of geckos to walk up slick surfaces such as glass. van der Waals and hydrogen bonds are important in temporary association of molecules that are required during the chemical reactions that make up metabolic pathways.

Functional groups are molecules that are attached to others that can change the way the molecule works. The most common functional groups in biological systems are hydroxyl ($-OH$), methyl ($-CH_3$), carbonyl ($-CO$), carboxyl ($-COOH$), amino ($-NH_2$), phosphate ($-PO_4$), and sulfhydryl ($-SH$). In **TABLE 1**, R represents the molecule to which the functional group is attached.

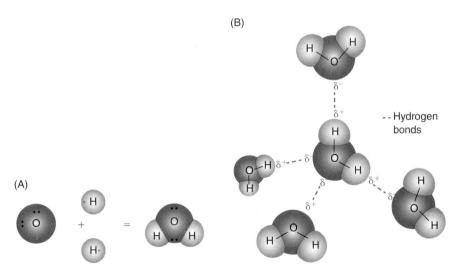

FIGURE RR2.06 The special properties of water are due to polar covalent bonds. (a) Oxygen in water is a bit more negative (δ^-) because electrons and the hydrogen atoms are slightly positive, (δ^+). (b) This polarity results in a lattice of water molecules in liquid water.

$$H_2O \quad \leftrightarrow \quad H^+ \quad + \quad OH^-$$

water   hydrogen ion   hydroxide ion

FIGURE RR2.07 Water can undergo reversible disassociation into a hydrogen ion and a hydroxide ion.

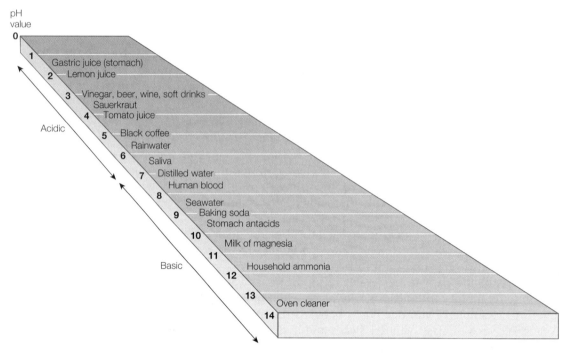

FIGURE RR2.08 The pH scale as a measure of the acidic or basic nature of a liquid.

TABLE 1 Functional Groups and Structures		
Functional Group	**Structure**	**Notes**
Hydroxyl (OH)	R—O—H	
Methyl (CH₃)	R—C—H (with H above and H below)	Can function as a mark on DNA
Amino (NH₂)	R—N with H and H	Present in amino acids
Carboxyl (COOH)	R—C(=O)—O—H	Present in amino acids
Carbonyl (CO)	R—C(=O)—R	
Phosphate (PO₄)	R—O—P(=O)(OH)—OH	Part of the DNA backbone
Sulfhydryl (SH)	R—S—H	In a specific amino acid

Name _____ Section _____ Date _____

Problems

Problem 1.1 What does the atomic number represent?

_____

True or False
Problem 1.2 The atomic mass is the number of protons and electrons in the atomic nucleus.

_____

Problem 1.3 Electrons orbit the atomic nucleus.

_____

Problem 1.4 The number of electrons is equal to the number of protons, which ensures the element has a neutral charge.

_____

Problem 1.5 The special properties of water are due to ionic bonding of oxygen and hydrogen.

_____

Problem 1.6 Hydrogen bonds are the strongest bonds in biological material.

_____

Problem 1.7 Name the six most common elements in biological systems.

_____

_____

Fill in the Blank
Problem 1.8 _____ are weak interactions that occur when molecules are in close proximity.

Problem 1.9 Water adheres to itself because of _____

_____ between molecules.

Problem 1.10 The pH of a liquid is based upon the _____

_____ of the [H^+].

Problem 1.11 Which functional group(s) contain carbon?

_____

Problem 1.12 Which functional group(s) contain phosphorus?

_____

Problem 1.13 Which functional group(s) contain nitrogen?

_____

Matching

Problem 1.14 Match each of these functional groups to the correct structure below.

Methyl _____

Hydroxyl _____

Amino _____

Phosphate _____

Carboxyl _____

Carbonyl _____

(A)

(B)

R—O—H

(C)

(D)

(E)

(F)

2. Biological Molecules

It's a life-eats-life world. When you think about it, all your food was alive at some time (or at least it should have been). Take a look at any nutritional facts label and you will see these three categories: total fats, total carbohydrates, and protein (**FIGURE 9**). These are categories of biological molecules. But there are four categories of biological molecules, so what is missing and why?

Hydrates of carbon, better known as carbohydrates, are sugars and starches. They have a general formula of CH_2O and the building blocks are called saccharides (**FIGURE 10**). In plants, animals, and microbes, energy can be extracted from the six-carbon monosaccharide glucose. Starches, which are long chains of sugar molecules known as polysaccharides, store energy for later use in plants. Table sugar is sucrose, a disaccharide composed of glucose plus fructose; both are five-carbon sugars. The five-carbon sugar ribose is present in the backbone of DNA and in RNA. In nutritional labels, DNA and RNA are included in the carbohydrate category. They are considered separate biological molecules in the category nucleic acids because of their special properties.

Name _____ Section _____ Date _____

SIDE-BY-SIDE COMPARISON

Original Label

Nutrition Facts
Serving Size 2/3 cup (55g)
Servings Per Container About 8

Amount Per Serving

Calories 230   Calories from Fat 72

% Daily Value*

Total Fat 8g	12%
Saturated Fat 1g	5%
Trans Fat 0g	
Cholesterol 0mg	0%
Sodium 160mg	7%
Total Carbohydrate 37g	12%
Dietary Fiber 4g	16%
Sugars 1g	
Protein 3g	

Vitamin A	10%
Vitamin C	8%
Calcium	20%
Iron	45%

* Percent Daily Values are based on a 2,000 calorie diet. Your daily value may be higher or lower depending on your calorie needs.

	Calories:	2,000	2,500
Total Fat	Less than	65g	80g
Sat Fat	Less than	20g	25g
Cholesterol	Less than	300mg	300mg
Sodium	Less than	2,400mg	2,400mg
Total Carbohydrate		300g	375g
Dietary Fiber		25g	30g

New Label

Nutrition Facts
8 servings per container
Serving size 2/3 cup (55g)

Amount per serving
Calories **230**

% Daily Value*

Total Fat 8g	10%
Saturated Fat 1g	5%
Trans Fat 0g	
Cholesterol 0mg	0%
Sodium 160mg	7%
Total Carbohydrate 37g	13%
Dietary Fiber 4g	14%
Total Sugars 12g	
Includes 10g Added Sugars	20%
Protein 3g	

Vitamin D 2mcg	10%
Calcium 260mg	20%
Iron 8mg	45%
Potassium 235mg	6%

* The % Daily Value (DV) tells you how much a nutrient in a serving of food contributes to a daily diet. 2,000 calories a day is used for general nutrition advice.

Note: The images above are meant for illustrative purposes to show how the new Nutrition Facts label might look compared to the old label. Both labels represent fictional products. When the original hypothetical label was developed in 2014 (the image on the left-hand side), added sugars was not yet proposed so the "original" label shows 1g of sugar as an example. The image created for the "new" label (shown on the right-hand side) lists 12g total sugar and 10g added sugar to give an example of how added sugars would be broken out with a % Daily Value.

FIGURE RR2.09 Nutrition facts label. Fats, carbohydrates, and protein are classifications of biological molecules. But where are the nucleic acids?

Courtesy of U.S. Food and Drug Administration.

FIGURE RR2.10 Examples of saccharides, the building blocks of carbohydrates. The numbers indicate the numbering conventions to refer to specific carbons. The numbers are referred to 1-prime (1'), 2', etc. (a) The monosaccharide glucose. (b) The disaccharide sucrose. (c) The monosaccharide ribose.

Nucleic acids are so named because they were first detected in the nucleus and they are acidic. They are modified carbohydrates that contain a ribose sugar. One modification is the attachment of a nitrogenous base to the 1′ base of the ribose sugar (**FIGURE 11**). The ribose molecules are attached though phosphate groups that are attached to the 5′ and 3′ carbons that have either one ring (pyrimidines) or two rings (purines).

DNA (deoxyribose nucleic acid) and RNA (ribonucleic acid) are composed of nucleotides (**FIGURE 12**). In the case of DNA, the OH on the 2′ carbon is missing from the ribose sugar, hence the name deoxyribose. DNA forms a stable

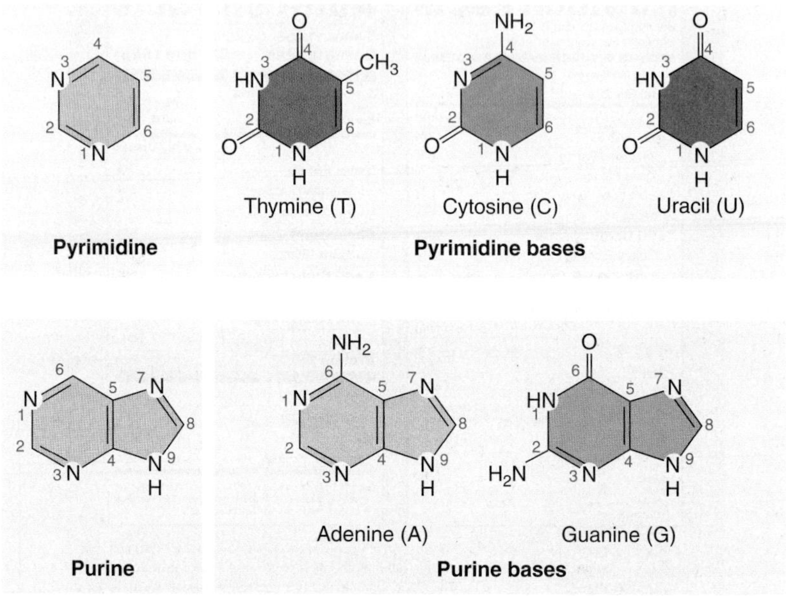

FIGURE RR2.11 Nitrogenous bases that are attached to 1′ carbon of ribose sugars in nucleic acids.

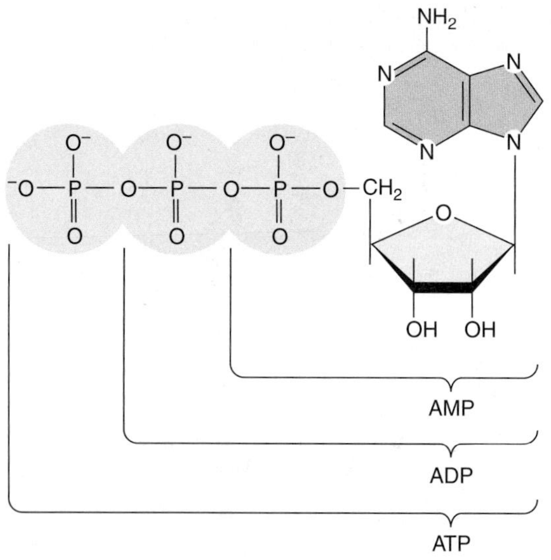

FIGURE RR2.12 Naming conventions for nucleotides, the building blocks of nucleic acids. This example is for the base adenosine; with three phosphate groups, it is called adenosine triphosphate (ATP), with two phosphate groups it is called adenosine diphosphate (ADP), and with only one phosphate group, it is called adenosine monophosphate (AMP).

Name _____ Section _____ Date _____

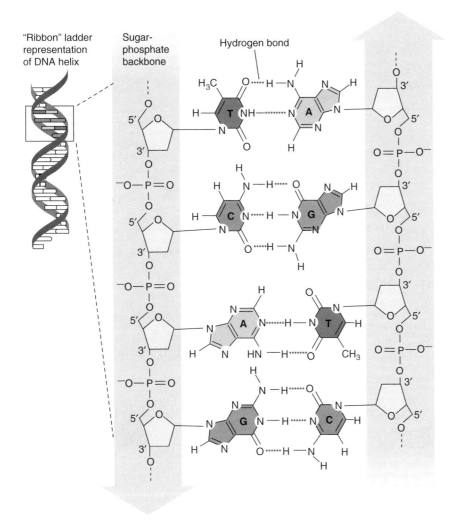

FIGURE RR2.13 The double helix of DNA indicating that A pairs with T and C with G through hydrogen bonding.

double-stranded molecule (**FIGURE 13**), with A pairing with T and C pairing with G; the bases are held together across the helix by hydrogen bonds. The sugar-phosphate backbone is oriented in opposite directions on the two strands. RNA is usually found in single-stranded form (**FIGURE 14**), although it is capable of forming regions that are double stranded. RNA is less stable than DNA and is quickly degraded by enzymes in the cytoplasm. In genetics, the messenger (RNA) is killed.

Fats are also known as lipids, which are important components of membranes in all organisms. In animals, lipids are used for long-term energy storage. Cholesterol is a lipid that helps stabilize cell membranes in animals, but when there is too much in our diets it can be deposited in the arteries and cause atherosclerosis, where blood flow is restricted to the heart. Lipids are covalently bonded chains of carbon and hydrogen. The building blocks of lipids are fatty acids, which have a hydrophobic (water-hating) tail and a hydrophilic (water-loving) head (**FIGURE 15**). The reason that oil and water don't mix is because

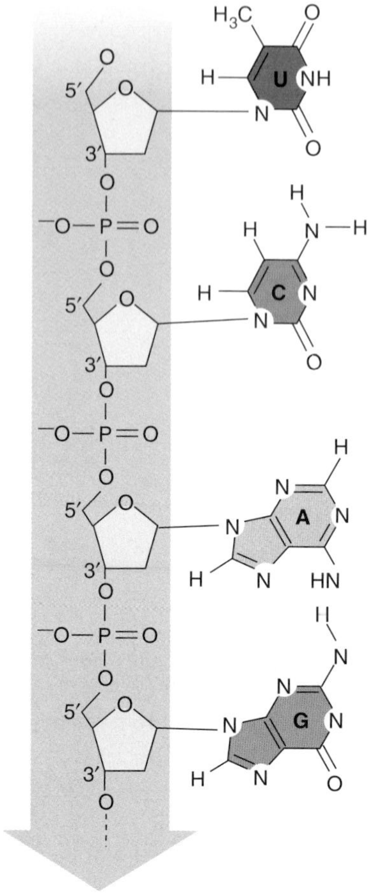

FIGURE RR2.14 RNA is often single stranded.

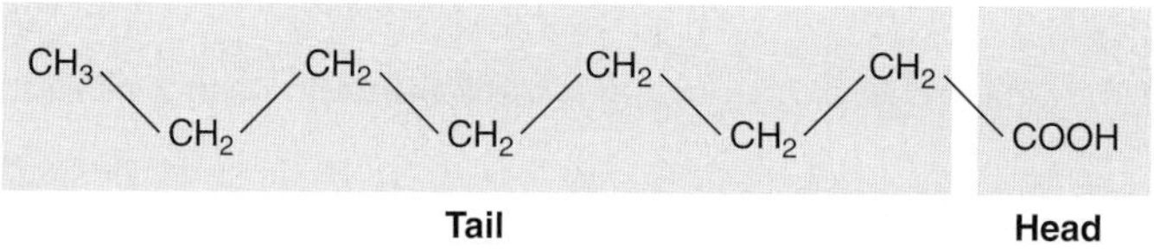

Tail **Head**

FIGURE RR2.15 The basic structure of a lipid. The head is a carboxyl group that is hydrophilic, while the tail is hydrophobic.

of this hydrophobic tail, which repels water. Lipids are built through a series of enzyme-catalyzed reactions.

The building blocks of proteins are amino acids. All 20 amino acids have a central carbon that is attached to a carboxyl group, an amine group, and a hydrogen. The central carbon is attached to an R group, which is different for each amino acid. There are 11 amino acids that can be classified as polar (**FIGURE 16**) and 9 that are non polar (**FIGURE 17**). These categories can be further classified depending upon the characteristics of the R group.

Examples of proteins include keratin, which is found in our hair and nails. It is composed of about 312 amino acids. Myosin and actin are proteins found in muscles. Enzymes are proteins that catalyze the steps in metabolism and end in the suffix -ase (e.g., amylase). The linear order of amino acids in proteins is coded for by DNA.

Name _____ Section _____ Date _____

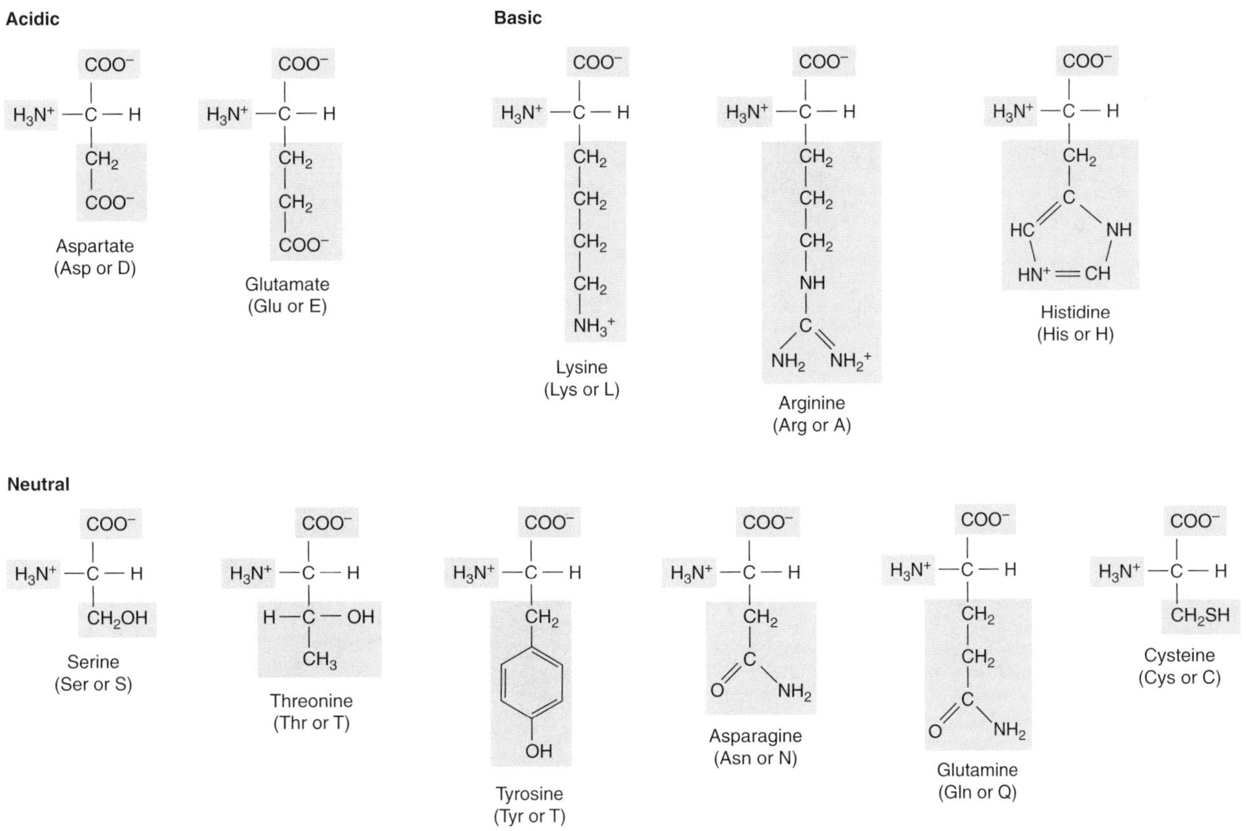

Acidic

Basic

Aspartate
(Asp or D)

Glutamate
(Glu or E)

Lysine
(Lys or L)

Arginine
(Arg or A)

Histidine
(His or H)

Neutral

Serine
(Ser or S)

Threonine
(Thr or T)

Tyrosine
(Tyr or T)

Asparagine
(Asn or N)

Glutamine
(Gln or Q)

Cysteine
(Cys or C)

FIGURE RR2.16 Polar amino acids.

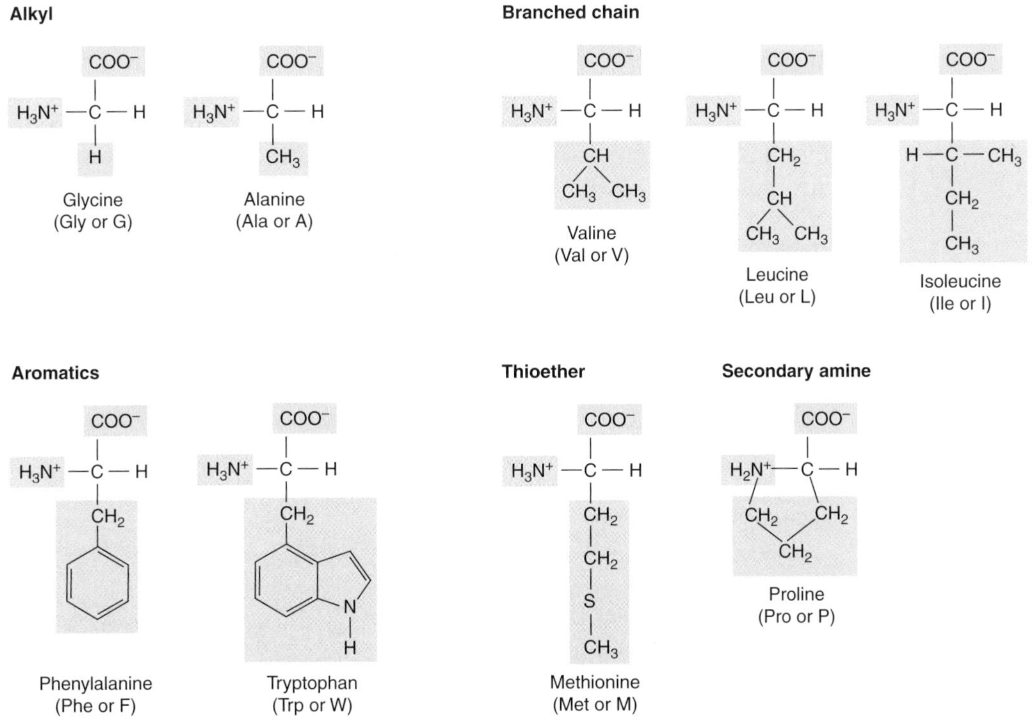

Alkyl

Branched chain

Glycine
(Gly or G)

Alanine
(Ala or A)

Valine
(Val or V)

Leucine
(Leu or L)

Isoleucine
(Ile or I)

Aromatics

Thioether

Secondary amine

Phenylalanine
(Phe or F)

Tryptophan
(Trp or W)

Methionine
(Met or M)

Proline
(Pro or P)

FIGURE RR2.17 Non polar amino acids.

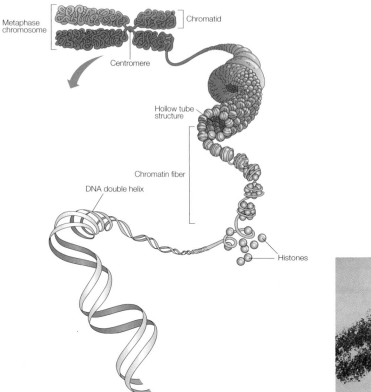

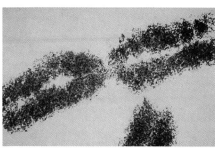

FIGURE RR2.18 Chromosomes are composed of both DNA and proteins.

Photo © Science VU/Visuals Unlimited.

The biological molecules can be combined to form hybrid molecules with new functions. The ABO blood groups are glycoproteins—sugar groups covalently bonded to proteins. Lipoproteins are proteins combined with lipid molecules; these can transport fats in the blood and include high-density lipoprotein (HDL) and low-density lipoprotein (LDL). Chromosomes are another example of a hybrid molecule. When double-stranded DNA is complexed with histone protein, it is referred to as chromatin. The chromatin fiber is wrapped tightly to form chromosomes (**FIGURE 18**).

Problems

Problem 2.1 Complete the following table.

Molecule	Building Blocks	Function(s)	Examples
			Glucose, sucrose, ribose
			Cholesterol
			Keratin (hair and nails), myosin and actin (muscle), amylase
			DNA and RNA

Name _____ Section _____ Date _____

Problem 2.2 Which molecule(s) contain both phosphorus and nitrogen?

_____

Problem 2.3 Which molecules are coded for by DNA?

_____

Problem 2.4 Which class of molecules have both a hydrophilic and a hydrophobic end?

_____

Problem 2.5 Which class of molecules are an important source of energy in plants, animals, and bacteria?

_____

Problem 2.6 What type of molecules always end with the suffix -ase?

_____

Problem 2.7 If everything we eat was alive, why aren't nucleic acids listed on nutritional facts labels?

_____

_____

Problem 2.8 Chromatin is composed of what two different molecules?

_____

3. Cell Structure and Function

There are two broad categories of cell types: prokaryotic and eukaryotic. Prokaryotic cells are distinguished by a cell wall, a lack of a nucleus, and no internal membrane-bound compartments (FIGURE 19). The DNA is found in a structure called the nucleoid, which is not bound by a membrane.

As eukaryotes, both animal (FIGURE 20) and plant cells (FIGURE 21) have extensive membrane-bound compartments, including the nucleus, mitochondria, and the Golgi apparatus. The nucleus is surrounded by a double membrane that contains pores for the exchange of material with the cytoplasm, including messenger RNA. Unlike animal cells, plant cells have cell walls that support the cells. Animal cell membranes contain cholesterol, which helps to support the other lipids in the membrane (FIGURE 20). Plant cells contain chloroplasts, which convert light energy into chemical energy that is stored in carbohydrates (FIGURE 21). The functions of cell structures are listed in **TABLE 2**.

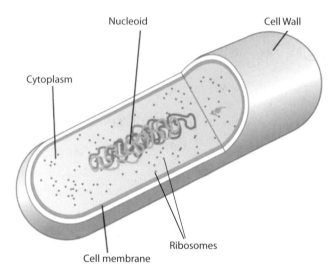

Nucleoid

Cell Wall

Cytoplasm

Ribosomes

Cell membrane

FIGURE RR2.19 Prokaryotic cell structure.

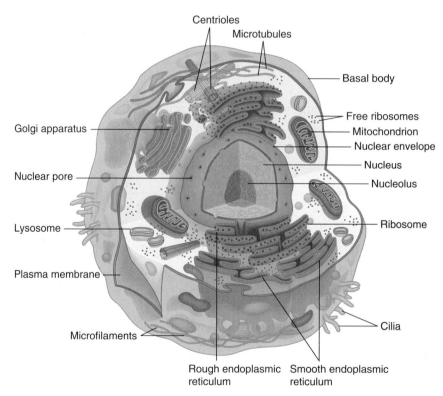

Centrioles

Microtubules

Basal body

Golgi apparatus

Free ribosomes

Mitochondrion

Nuclear envelope

Nuclear pore

Nucleus

Nucleolus

Lysosome

Ribosome

Plasma membrane

Cilia

Microfilaments

Rough endoplasmic reticulum

Smooth endoplasmic reticulum

FIGURE RR2.20 Animal cell.

Name _____ Section _____ Date _____

TABLE 2 Overview of Cell Organelles

Organelle	Structure	Function
Nucleus	Round or oval body; surrounded by nuclear envelope.	Contains the genetic information necessary for control of cell structure and function; DNA contains hereditary information.
Nucleolus	Round or oval body in the nucleus consisting of DNA and RNA.	Produces ribosomal RNA.
Endoplasmic reticulum	Network of membranous tubules in the cytoplasm of the cell. Smooth endoplasmic reticulum contains no ribosomes. Rough endoplasmic reticulum is studded with ribosomes.	Smooth endoplasmic reticulum (SER) is involved in the production of phospholipids and has many different functions in different cells; round endoplasmic reticulum (RER) is the site of the synthesis of lysosomal enzymes and proteins for extracellular use. Generates new membranes for certain organelles and the cell membrane.
Ribosomes	Small particles found in the cytoplasm; made of RNA and protein.	Aid in the production of proteins on the RER and polysomes.
Polysome	Molecule of mRNA bound to ribosomes.	Site of protein synthesis.
Golgi complex	Series of flattened sacs usually located near the nucleus.	Sorts, chemically modifies, and packages proteins produced on the RER.
Secretory vesicles	Membrane-bound vesicles containing proteins produced by the RER and repackaged by the Golgi complex; contain protein hormones or enzymes.	Store protein hormones or enzymes in the cytoplasm awaiting a signal for release.
Food vacuole	Membrane-bound vesicle containing material engulfed by the cell.	Stores ingested material and combines with lysosome.
Lysosome	Round, membrane-bound structure containing digestive enzymes.	Combines with food vacuoles and digests materials engulfed by cells.
Mitochondria	Round, oval, or elongated structures with a double membrane. The inner membrane is thrown into folds.	Complete the breakdown of glucose, producing NADH and ATP.
Cytoskeleton	Network of microtubules and microfilaments in the cell.	Gives the cell internal support, helps transport molecules and some organelles inside the cell, and binds to enzymes of metabolic pathways.
Cilia	Small projections of the cell membrane containing microtubules; found on a limited number of cells.	Propel materials along the surface of certain cells.
Flagella	Large projections of the cell membrane containing microtubules; found in humans only on sperm cells.	Provide motive force for sperm cells.
Centrioles	Small cylindrical bodies composed of microtubules arranged in nine sets of triplets; found in animal cells, not plants.	Help organize spindle apparatus necessary for cell division.

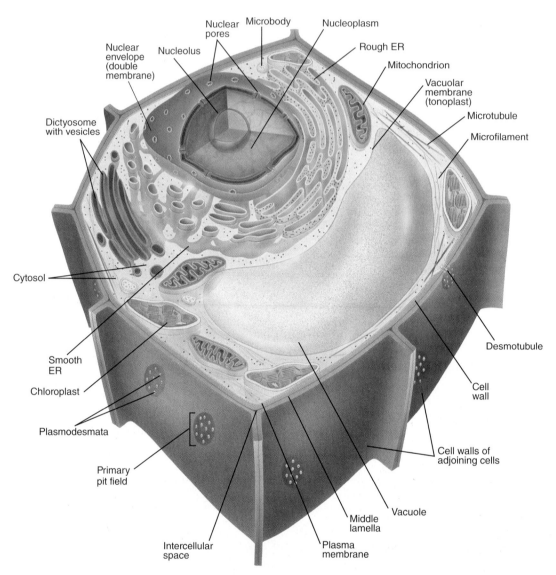

FIGURE RR2.21 Plant cell structure.

Problems

Fill in the Blank

Problem 3.1 _____ cells lack a membrane-bound nucleus.

Problem 3.2 _____ and _____ cells have cell walls.

Problem 3.3 Polysomes are ribosomes complexed with _____ .

Name _____ Section _____ Date _____

Problem 3.4 The _____ is composed of microtubules and microfilaments.

Problem 3.5 _____ store molecules for eventual release out of the cell.

Matching
Match the structure on the left with the function on the right.

Problem 3.6   Cytoskeleton _____   a. Modifies proteins

Problem 3.7   Ribosomes _____   b. Organize spindle fibers

Problem 3.8   Golgi complex _____   c. Provides internal support

Problem 3.9   Mitochondria _____   d. Produces ribosomal RNA

Problem 3.10   Nucleolus _____   e. Site of protein synthesis

Problem 3.11   Centrioles _____   f. Breakdown of glucose

4. Cellular Metabolism

Metabolism is the sum of all chemical reactions in living material and can be divided into two categories: catabolic and anabolic. When you eat, the food is broken down by your digestive system into its component molecules; these are catabolic reactions. The breakdown of glucose to extract energy and the digestion of protein into amino acids are specific examples. Anabolic reactions involve the building of molecules; protein synthesis is an example. You may have heard the term "anabolic steroids"; these are synthetic molecules that can help build muscle, but also have negative effects on the brain and reproductive system. Anabolic steroid use is illegal without a prescription and is banned by most sports. A metabolic pathway is the series of steps (anabolic and catabolic) necessary to transform molecules. Each step involves substrate(s), catalysts (enzymes), and product(s). In **FIGURE 22**, products from the first step become the substrate for the next step. Pathways can also run backwards but are usually catalyzed by a different set of enzymes. The substrate and products can be carbohydrates, proteins, lipids, nucleic acids, or any combination of molecules, but the enzymes are proteins and are therefore coded by DNA.

FIGURE 23 demonstrates the role of enzymes as catalysts to lower the activation energy required for a reaction to occur. In an uncatalyzed reaction, the activation energy (E_a) necessary for the reaction to proceed is too high, which prevents the reaction from taking place. In a catalyzed reaction, the activation energy is lowered, so the reaction can proceed more easily. The double peak in Figure 23 represents a two-step reaction, where the enzyme and substrate first

FIGURE RR2.22 A simplified metabolic pathway. Substrate 1 is acted on by enzyme 1 forming a product that then becomes the substrate for another reaction.

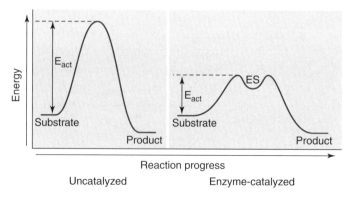

FIGURE RR2.23 Activation energies are lowered by enzymes. The enzyme catalyzed reaction can involve multiple steps. ES = enzyme plus substrate.

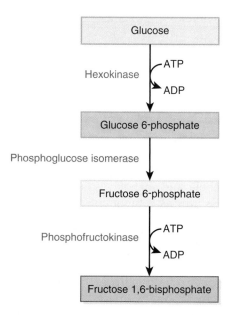

FIGURE RR2.24 Example of a pathway that utilizes the energy from ATP.

bind, then a chemical reaction occurs, releasing the enzyme and products. One way that activation energies are lowered by the enzyme is by bringing substrates closer together, so the reaction can proceed. Enzymes are not altered or consumed in this process, so once the reaction is complete, they can catalyze more reactions.

When additional energy and/or a functional group is needed for a step, it can be coupled to a step that extracts the necessary energy from a cofactor, such as ATP. **FIGURE 24** shows the first few steps of the metabolic pathway that extracts energy from glucose. In this case, a phosphate functional group is removed from the ATP and added to glucose.

The central dogma of molecular biology is a complicated pathway that describes the steps necessary for the synthesis of proteins in the cell. It is summarized in **FIGURE 25**. Protein synthesis consumes most of the energy in a cell. During cell division, the DNA is replicated (see cell cycle, below). Transcription is the process of copying the information in DNA into an RNA molecule. Some of these RNA molecules, known as messenger RNA (mRNA), find their way to ribosomes in the cytoplasm and are translated into proteins.

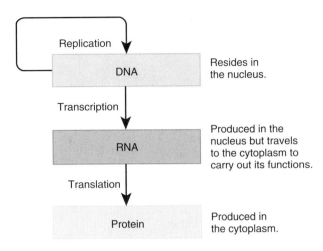

FIGURE RR2.25 Overview of the central dogma of molecular biology.

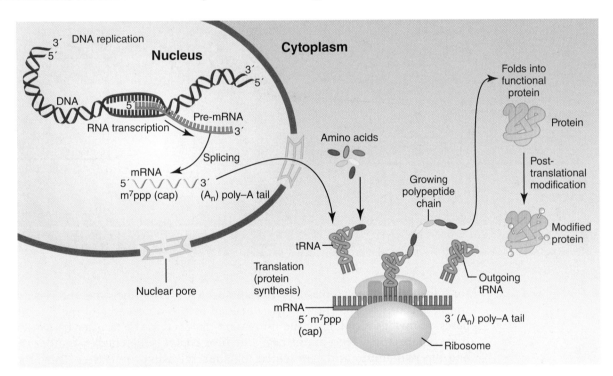

FIGURE RR2.26 The process of protein synthesis in a eukaryotic cell.

The replication of DNA takes place during the cell cycle before cells divide. Transcription of regions of the DNA that contain genes copies the nucleic acid sequence into an RNA molecule. This is analogous to reading written information out loud. The information is still in the same language, but the format is changed from words to sound. During translation, genetic information is changed into a completely different language—that of proteins.

Since prokaryotic cells have no distinct nucleus, transcription and translation take place in the same compartment. Sometimes translation can begin on an RNA while it is still being transcribed. In eukaryotic cells, transcription and translation take place in different compartments of the cell, the nucleus and the cytoplasm, respectively (**FIGURE 26**). Once mRNA is transcribed, it is processed in the nucleus

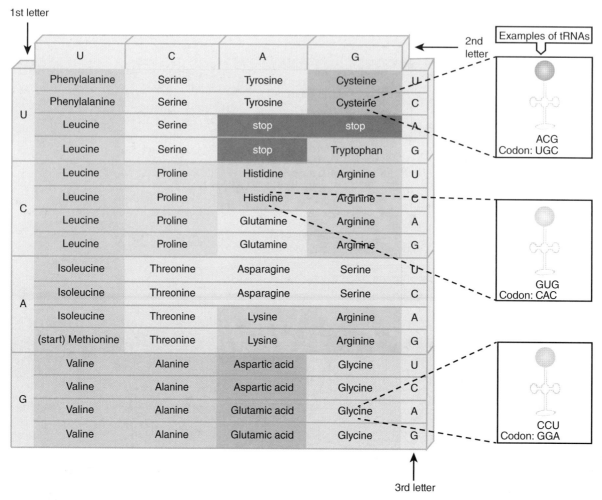

1st letter

2nd letter

Examples of tRNAs

	U	C	A	G	
U	Phenylalanine	Serine	Tyrosine	Cysteine	U
	Phenylalanine	Serine	Tyrosine	Cysteine	C
	Leucine	Serine	stop	stop	A
	Leucine	Serine	stop	Tryptophan	G
C	Leucine	Proline	Histidine	Arginine	U
	Leucine	Proline	Histidine	Arginine	C
	Leucine	Proline	Glutamine	Arginine	A
	Leucine	Proline	Glutamine	Arginine	G
A	Isoleucine	Threonine	Asparagine	Serine	U
	Isoleucine	Threonine	Asparagine	Serine	C
	Isoleucine	Threonine	Lysine	Arginine	A
	(start) Methionine	Threonine	Lysine	Arginine	G
G	Valine	Alanine	Aspartic acid	Glycine	U
	Valine	Alanine	Aspartic acid	Glycine	C
	Valine	Alanine	Glutamic acid	Glycine	A
	Valine	Alanine	Glutamic acid	Glycine	G

ACG
Codon: UGC

GUG
Codon: CAC

CCU
Codon: GGA

3rd letter

FIGURE RR2.27 The genetic code. The first, second, and third letter of the codons are specified on the left, top, and right, respectively.

Reproduced from Karp G, Patton JG. *Cell and molecular biology: Concepts and experiments.* John Wiley & Sons; 2013. Reproduced with permission of John Wiley.

for transport to the cytoplasm. When the mRNA finds a ribosome, the process of translation can begin. The mRNA is aligned to the ribosome and the first three bases, known as a codon, in the mRNA are matched to the transfer RNA (tRNA) that contains the correct amino acid. The next codon is matched to another tRNA and the two amino acids are joined together, releasing the tRNAs. This process is repeated until a stop codon is encountered and translation is terminated. The correspondence between codons and amino acids is shown in **FIGURE 27**. Once a protein is formed, it may undergo modifications to become fully functional.

Problems

Fill in the Blank

Problem 4.1 _____ is the breakdown of molecules during metabolism.

Problem 4.2 Protein synthesis is an example of a(an) _____ _____ process.

Name _____ Section _____ Date _____

Problem 4.3 _____ are catalysts that are not consumed by the reactions in which they participate.

Problem 4.4 Enzymes reduce the _____ of reactions.

Problem 4.5 _____ describe the steps necessary to transform biological molecules.

Problem 4.6 If more energy is needed to carry out a step, the reaction can be coupled to a _____, such as ATP.

Problem 4.7 The metabolic pathway that described the process of protein synthesis is called the _____.

Problem 4.8 Replication is the synthesis of DNA necessary prior to cell division, _____ is the process of copying the information from DNA into RNA, and _____ is process where proteins are synthesized.

Problem 4.9 In prokaryotes, _____ of a RNA molecule can begin before _____ is complete because these processes take place in the same compartment.

Problem 4.10 A(an) _____ is the three-base combination that codes for a specific amino acid.

Problem 4.11 Amino acids are brought to the ribosome by _____ _____ molecules.

5. The Cell Cycle, Mitosis, and Meiosis

FIGURE 28 is a diagram of major steps in the cell cycle. In the G1 (gap 1) phase, the cell carries out its usual metabolic functions; when it time to divide, it moves into the S or synthesis phase. This is when the DNA is replicated. During the G2 phase, the cell prepares to divide. Cell division include the M phase, which is mitosis and cytokinesis. The chromosomes are divided equally during mitosis and the cytoplasm is divided during cytokinesis, forming two identical daughter cells.

In the S phase, the DNA is replicated and the chromosomes go from consisting of a single chromatid to having two identical chromatids (called sister chromatids) held together by the centromere. Mitosis consists of four stages (**FIGURE 29**). During prophase, the structure of the nucleus changes and the chromosomes condense; these become visible in the light microscope. During metaphase, the chromosomes line up end to end. Anaphase involves the separation of the sister

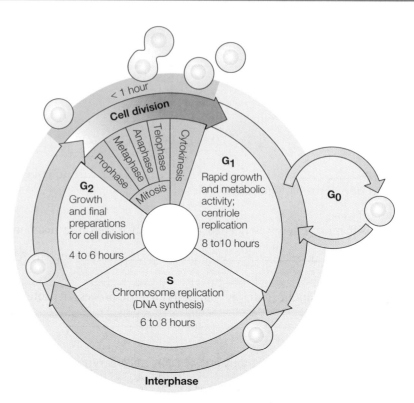

FIGURE RR2.28 The cell cycle, in which two identical daughter cells are produced from a single cell.

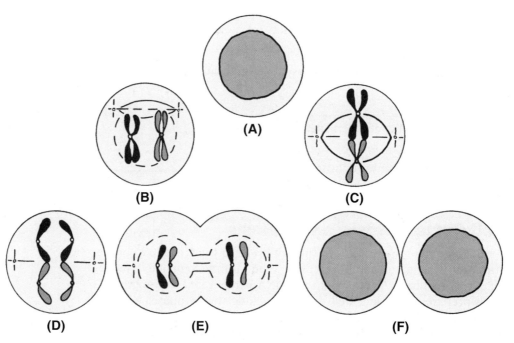

FIGURE RR2.29 Mitosis results in two identical daughter cells. The stages are: (a) G2, (b) prophase, (c) metaphase, (d) anaphase, (e) telophase, and (f) cytokinesis.

chromatids so each daughter cell will get a copy of each chromosome. In telophase, the chromosomes decondense and the nucleus resumes its usual structure. Cell division is complete once cytokinesis occurs, which is the division of the cytoplasm and the organelles. Mitosis insures that the genetic makeup of the daughter cells is identical.

The cells in our bodies that have the potential to become the next generation are called germ cells, while the rest of the cells are called somatic cells. The term *germ* is derived from the Latin term for seed or sprout, and does not refer to microbes, which we commonly refer to as "germs." Meiosis is the process that takes place in our germline cells (a.k.a., gonads) and results in either egg or sperm. It is characterized by homologous chromosomes lining up next to each other during meiosis I (**FIGURE 30**). We get two copies each chromosome, one from each biological parent. The pairs of chromosomes are called homologues. Most of our cells have two copies of each chromosome; this condition is called diploid. Ploidy levels indicate how many copies of each chromosome a cell (or an individual organism) contains. There are a few special cells in our bodies that contain only one copy of each chromosome; these are called haploid cells. In females, these are the eggs or ovum; in males these are the sperm. Meiosis involves two consecutive cell divisions: meiosis I and meiosis II (Figure 30). The major difference between mitosis and meiosis is that in metaphase of meiosis I (metaphase I), the homologous chromosomes line up next to each other rather than end to end. These alignments are called synapses and allow the chromosomes to cross over so that the genetic material is rearranged. This process of recombination is one

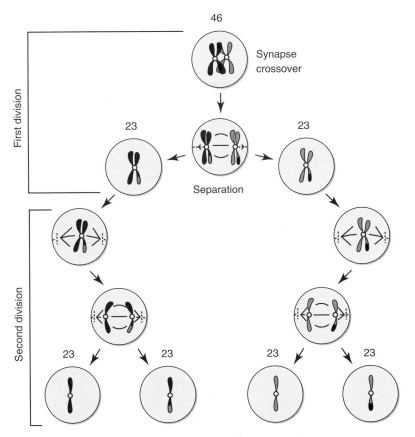

FIGURE RR2.30 The process of meiosis results in four different haploid products.

way that the genetic deck is shuffled every generation. New combinations are genes that are produced. After meiosis I, the amount of information is cut in half since each cell now contains only one copy of each chromosome, so this is called the reduction division. During meiosis II, the chromosomes line up end to end, just like in meiosis and the sister chromatids go to separate cells. Four different haploid cells are the result of meiosis. The difference are due to the of segregation of chromosomes into different cells and to crossing-over during metaphase I. In contrast, mitosis results in two identical daughter cells.

Gametogenesis is the process that produces egg and sperm (**FIGURE 31**). Meiosis I occurs in the primary oocytes and spermatocytes to produce the secondary oocytes and secondary spermatocytes. The secondary spermatocytes go through meiosis II and eventually become sperm (spermatozoa). Secondary oocytes also go through meiosis II, but only one matures to become the ovum (egg). Polar bodies are the products of meiosis II that do not become the egg but help nourish the ovum by filling it the necessary biological molecules for embryonic development. Once a haploid egg is fertilized by a haploid sperm, the diploid condition is re-established, an embryo is formed, and development can proceed.

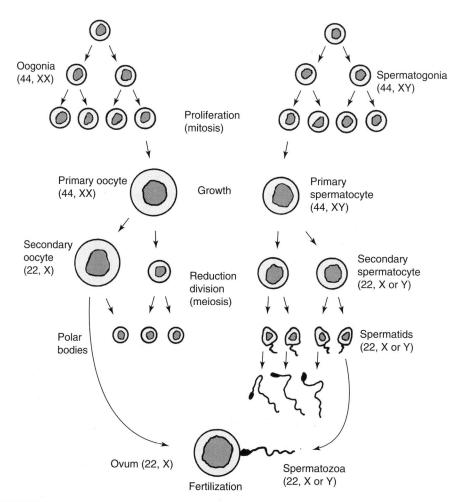

FIGURE RR2.31 Gametogenesis followed by fertilization is necessary to produce progeny. Upon fertilization, the original number of chromosomes is produced, which is 23 pairs in humans (22 plus XX or XY).

Name _____ Section _____ Date _____

Problems

True or False

Problem 5.1 In the cell cycle, G1 is followed by the S phase, then M, and lastly G2.

_____

Problem 5.2 DNA is replicated during the S phase.

_____

Problem 5.3 Cytokinesis results in the final generation of two cells during the cell cycle.

_____

Problem 5.4 Mitosis generates two genetically diverse daughter cells.

_____

Problem 5.5 During mitosis and meiosis II, chromosomes line up end to end.

_____

Problem 5.6 During the S-phase, the DNA is replicated and sister chromatids are the result.

_____

Problem 5.7 Gametes have two copies of each chromosome and are therefore diploid.

_____

Problem 5.8 Gametes formed by mitosis are genetically diverse.

_____

Problem 5.9 Complete the following table.

Comparison of mitosis and meiosis		
	Mitosis	**Meiosis**
Number of cells produced		
Number of sets of chromosomes (ploidy)		
Genetic content of cells		
Cells that can undergo each process		

Answer Key

Problem 1.1
The number of protons.

Problem 1.2
False (The atomic mass is the number of protons and neutrons in the atomic nucleus.)

Problem 1.3
True

Problem 1.4
True

Problem 1.5
False (The special properties of water are due to polar covalent bonds between oxygen and hydrogen.)

Problem 1.6
False (Hydrogen bonds are weaker than ionic and polar, but stronger than van der Waals forces.)

Problem 1.7
carbon, hydrogen, oxygen, nitrogen, phosphorus, and sulphur

Problem 1.8
van der Waals

Problem 1.9
polar covalent or hydrogen bonds

Problem 1.10
$-\log_{10}$

Problem 1.11
Methyl (CH_3), carboxyl (COOH), and carbonyl (CO)

Problem 1.12
Phosphate (PO_4)

Problem 1.13
Amino (NH_2)

Problem 1.14
Methyl → d
Hydroxyl → b
Amino → c
Phosphate → e
Carboxyl → f
Carbonyl → a

Problem 2.1 Complete the following table.

Molecule	Building Blocks	Function(s)	Examples
Carbohydrates	Saccharides	Energy	Glucose, sucrose, ribose
Lipids	Fatty acids	Long-term energy storage, membrane function	Cholesterol
Proteins	Amino acids	Structural molecules, enzymes	Keratin (hair and nails), myosin and actin (muscle), amylase
Nucleic acids	Nucleotides	Information storage and transmission	DNA and RNA

Problem 2.2
RNA and DNA (or nucleic acids)

Problem 2.3
Proteins

Problem 2.4
Lipids

Problem 2.5
Carbohydrates

Problem 2.6
Enzymes

Problem 2.7
Nucleic acids are specialized carbohydrates, so they are included in this class.

Problem 2.8
DNA and protein (histone)

Problem 3.1
Prokaryotic

Problem 3.2
Prokaryotic (or bacterial); plant

Problem 3.3
messenger RNA (mRNA)

Problem 3.4
cytoskeleton

Problem 3.5
Secretory vesicles

Problem 3.6
c

Problem 3.7
e

Problem 3.8
a

Problem 3.9
f

Problem 3.10
d

Problem 3.11
b

Problem 4.1
Catabolism

Problem 4.2
anabolic

Problem 4.3
Enzymes

Problem 4.4
activation energy

Problem 4.5
Metabolic pathways

Problem 4.6
cofactor

Problem 4.7
central dogma

Problem 4.8
Transcription; translation

Problem 4.9
translation; transcription

Problem 4.10
codon

Problem 4.11
transfer RNA (tRNA)

Problem 5.1
False (The correct order is G1, S, G2, M.)

Problem 5.2
True

Problem 5.3
True

Problem 5.4
False (Mitosis generates identical daughter cells.)

Problem 5.5
True

Problem 5.6
True

Problem 5.7
False (Gametes only have one copy of each chromosome and are haploid.)

Problem 5.8
False (Gametes are formed by meiosis, not mitosis.)

Problem 5.9

	Mitosis	Meiosis
Number of cells produced	2	4
Number of sets of chromosomes (ploidy)	2	1
Genetic content of cells	Identical	All different
Cells that can undergo each process	Any cell in the body (somatic)	Only cells destined to become egg or sperm (germ cells)

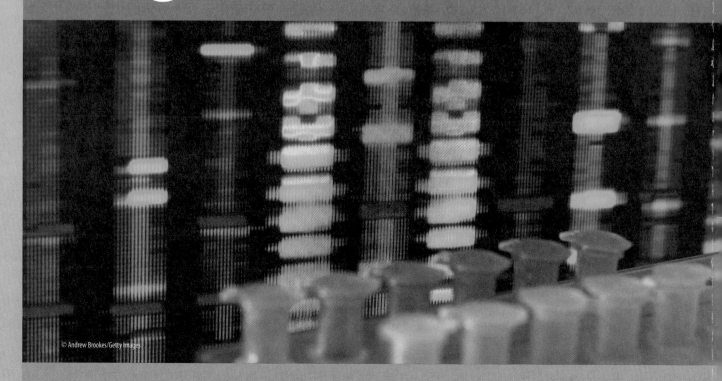

© Andrew Brookes/Getty Images

Thinking Like a Scientist

OUTLINE

- Introduction
  1. Logic
  2. Critical Thinking
  3. The Scientific Method

Name _____ Section _____ Date _____

Introduction

In order to understand any scientific discipline, it is necessary to understand scientific thinking. Thinking like a scientist not only requires education and discipline but also understanding our own biases and fallibility. When a scientist identifies a problem of interest, the first step is to learn as much as possible about the all aspects of the issue. This often takes even older scientists into new disciplines; being a scientist is about lifelong learning. Understanding how logic, critical thinking, and the scientific method are related and contribute to the current state of sciences such as genetics will help with your comprehension of this rapidly-advancing field.

Each section includes a brief explanation of the topic followed by the examples. There are questions at the end of each section to help you sharpen your skills. Answers to the questions are found at the end so you can check your work.

1. Logic

What kind of person do you envision when you think of a scientist? Someone who is logical? What is logic? Vulcans on the science fiction series, *Star Trek*, are renowned for their logic. It is a line of reasoning that leads to conclusions that can be validated. A way to sharpen your logical skills is to evaluate a series of three statements, the first two are true, the last is for your consideration: Does it logically follow? Can you think of exceptions to the last statement that make it false? Consider these statements:

Living trees have green leaves.
Leaves contain chlorophyll.
Therefore, everything green contains chlorophyll.

The first two statements are true, but the third is false; there are lots of green items that don't contain chlorophyll. A green light at an intersection is an example. When exceptions to the third statement are easily found, then an error in reasoning has occurred. This is a logical fallacy. This is the way Vulcans would express this (if they were on Earth):

Living trees have green leaves.
Leaves contain chlorophyll.
Therefore, chlorophyll is responsible for green color.

The change in the last statement eliminates the fallacy by specifying that greenness of leaves occurs because they contain chlorophyll. There is adequate scientific evidence for this last statement and it follows logically.

If we were to travel to another planet and saw green organisms growing in the sunlight, we might be tempted to assume the same series of statements logically apply. A scientist would change the logic to reflect the potential for a different explanation:

These organisms are green and living in sunlight.
Plant leaves are green.
Therefore, these organisms *might* contain chlorophyll.

The last statement is now a hypothesis, which can be tested. The next logical statement will be, "let's test these organisms for the presence of chlorophyll," leading to a period of critical thinking necessary to develop an experimental protocol.

It can be tricky to decide when statements contain logical fallacies:

All dogs are canines.
All dogs are mammals.
Therefore, mammals are canines.

For someone who is unfamiliar with biological taxonomy, it might not be immediately clear that the last statement is backwards. Dogs belong to the taxonomic family called canines, and all canines are mammals.

Let's consider a common misconception exemplified by these statements:

Bacteria are everywhere.
Bacteria can cause disease.
Therefore, all bacteria are dangerous.

Bacteria are everywhere; this is why extreme precautions are necessary to sterilize surfaces for medical procedures. There are many bacteria that can cause disease, but not all bacteria are dangerous; some are actually necessary for our survival. We rely on populations of microbes in our bodies to maintain health. Our "microbiome," as it's called, is a collection of bacteria in our own bodies upon which we rely. For example, the bacterial community in our lower intestines is important in digestion and nutrition. The overuse of antibiotics creates problems when the target of the antibiotic gains resistance and the good bacteria are destroyed by the same antibiotic. Sometimes a result of taking antibiotics is diarrhea, because the bacteria that aid in digestion are killed along with the disease-causing bacteria. Fortunately, this effect is only temporary, and the digestive system can return to normal. You may be familiar with products that contain *probiotics*, which are living bacteria that aid in digestion; why would you take such a product if all bacteria are dangerous? The diversity of microbes is just one example of the diversity of life. It can be tempting to make logical associations based on a limited sample size, but when exceptions are found because only a limited set of organisms were considered, such logic is found to be flawed.

Of concern in genetics is that a statement may be true in some cases but not in all cases due to genetic diversity. For example, a statement might be true for a percentage of the population that carries a particular disease variant of a gene (known as an allele), but not for the rest of the population. Here is an example:

People can be genetically predisposed to allergies.
Peanuts can trigger allergies.
Therefore, peanuts should be avoided by everyone.

Allergies are attributable to an overreaction by the extremely complex immune system. Since our genes code for the proteins that make up the immune system, there is a great deal of diversity among peoples' immune systems. Also, there are environmental triggers for allergies. So, it is a fallacy to assume that everyone will react the same.

As our knowledge of genetics increases, we are identifying the specific genes that contribute to how we respond to environmental cues, including the taste of food. A recent example is the identification of genetic differences that are linked to how we respond to the taste of cilantro.

The herb cilantro is used in gourmet cooking.
I like the taste of cilantro.
Therefore, everyone likes cilantro.

It turns out cilantro tastes bad to some people, and a specific change in the DNA is correlated to how a person responds to the taste. The exact nature of how this change influences our sense of taste is under investigation.

Consider this example regarding the rules for chromosomes that determine gender.

Males have an X and a Y chromosome.
Females have two X chromosomes.
Therefore, males inherit their Y chromosome from their fathers.

Except for extremely rare genetic changes, this is true for humans and for mammals in general. However, it isn't true for birds, some insects, and other taxonomic groups. What are the exceptions in humans? Females who have only one X chromosome (XO) have Turner syndrome, so not all females are XX. In addition, there are some very rare sex-reversal syndromes, where one can appear male but have two X chromosomes or appear female and be XY. Not only are these rare, but these individuals are sterile and cannot reproduce, so the third statement remains true. Here is a way to more accurately express this relationship:

Most human males have an X and a Y chromosome (XY).
Most human females have two X chromosomes (XX).
Therefore, males get their Y chromosome from their fathers.

There is a common misconception regarding evolution that is exemplified in these statements.

Chimpanzees can walk upright and resemble humans.
Humans walk upright and share over 90% similarities at the protein level.
Therefore, humans evolved from chimpanzees.

Humans and chimpanzees share many similarities, including over 95% of our DNA sequence, but they evolved from a common ancestor instead of one evolving from the other. Both species we observe today have been changing over time, so that common ancestor only exists in the fossil record.

As science progresses and more advanced tools are developed, it becomes necessary to re-evaluate logical statements. Neanderthals are a branch of humans that are known from the fossil record; they are named because they were first discovered in the Neander Valley in Germany. Populations of Neanderthals coexisted with human populations in Europe and Asia from about 430,000 to 40,000 years ago. Prior to 2010, there was no evidence for the breeding of Neanderthals with modern humans, so this line of logic was accepted among archeologists:

Neanderthals coexisted with modern humans in Europe and Asia.
There is no archeological evidence for social interactions that could lead to breeding.
Therefore, Neanderthals and modern humans did not breed.

Advances in DNA sequencing have allowed geneticists to obtain DNA sequences from Neanderthal bones. By 2010, it was confirmed that some people of European and/or Asian ancestry contain sequences that can be directly traced to the Neanderthals. This indicates that there was some interbreeding between these species. Some commercial genetic tests can estimate the amount of Neanderthal ancestry that an individual has inherited.

Logic is important in both critical thinking and the scientific method; as you will see below, it helps scientists to develop testable hypotheses that advance our knowledge of the world.

Problems

Find the fallacy. In these statements, the first two statements are true, but there is a problem with the third. Explain the fallacy in each and provide an example.

Problem 1.1 Snakes have scales.

Fish have scales.

Therefore, snakes and fish must be closely related taxonomically.

_____

_____

_____

Problem 1.2 Females with Turner syndrome only have one X chromosome (XO).

Most males only have one X chromosome (XY).

Therefore, individuals with Turner syndrome must get the X from their mother.

_____

_____

_____

Problem 1.3 All bacteria are single-celled organisms.

Bacteria cause disease.

Therefore, all single-celled organisms cause disease.

_____

_____

_____

Problem 1.4 Mushrooms are a fungus.

Mushrooms can be purchased in a grocery store.

Therefore, all mushrooms are edible.

_____

_____

_____

Problem 1.5 The common cold is caused by viruses.

The flu is caused by viruses.

Name _____ Section _____ Date _____

Therefore, all viruses cause illness.

_____

_____

_____

Problem 1.6 Redheads are relatively rare (about 1%).

Brunettes are more common (estimated 75–97%).

Redheads are not as healthy as brunettes.

_____

_____

_____

Problem 1.7 Most humans (about 85%) are right-handed.

Other primates do not exhibit this preference; they are ambidextrous (use both hands with equal frequency).

Humans do not share a common ancestor with other primates.

_____

_____

_____

Problem 1.8 There is archeological evidence that most Neanderthals were right-handed.

Most humans are right-handed, but primates are ambidextrous.

Therefore, handedness is not a genetically determined trait.

_____

_____

_____

2. Critical Thinking

When hear the term "critical thinking" you may imagine someone critiquing or criticizing something; for example, saying, "those shoes are ugly." But critical thinking actually refers to thinking about an issue from multiple perspectives without including your own biases. Thinking that a pair of shoes is ugly is a judgment based on personal tastes: "Beauty is in the eye of the beholder." Thinking critically about a pair of shoes can include the questions: Are they comfortable? Is there arch support? What are they made from? If there is an emergency, can I run in these shoes? Will the color complement clothes in my wardrobe? Are they well made? How long will they last? Critical thinking is global thinking—considering an issue from all perspectives.

It takes training to become a critical thinker. It can be difficult to recognize our own biases or to realize that we have misconceptions. Have you ever heard that we use only 10% of our brains? Rather than accept this, a critical thinker asks questions such as: Who calculated this? How was it done? Can I see the data? It turns out that this is a myth (a.k.a. urban legend). The source of this estimate is unknown, and it has been attributed to a number of different people. If it were true, then we could lose 90% of our brain without effect. But damage to even a small part of the brain has serious repercussions; the problems with multiple concussions in athletes is an example. Despite the findings of modern neurobiology that have debunked this myth, it does seem to persist.

There is an Internet site devoted to the dangers of dihydrogen monoxide (DHMO), also known as H_2O (water). There are those who have been fooled by this idea. But a careful reading of this information reveals that the claims made on this site aren't outright false, but key information is omitted. For example, DHMO is a major component of acid rain, but it's the rain part, not the acid. Since you can drown in water, the claim that DHMO is a health threat is technically true.

An Internet search for old cigarette ads reveals some surprising advertising methods. It was known in the 1930s that cigarette smoking decreased appetite, so the promise of staying thin was used as a method to get people to smoke. Advertisements including medical doctors recommending brands of cigarettes to smoke. Some even claimed that their brand prevented coughing and irritation associated with smoking. The association of coughing with smoking was understood in the 1930 and no brand of cigarettes prevented coughing. It took decades for the association of cigarette smoking with disease to be proven and the claims of cigarette manufacturers shown to be false.

Through the 1940s to the 1960s, studies began accumulating that smoking was hazardous to health. In 1965, the following label was mandated on all cigarettes in the U.S.: "Caution: Cigarette Smoking May Be Hazardous to Your Health." In 2003, the European Union adopted the warning labels "Smoking kills" and "Smoking seriously harms you and others around you." The statement must cover 40% of the pack's surface.

The Tobacco Master Settlement Agreement (MSA) in 1998 required tobacco companies to limit the advertising of cigarettes and to pay into a fund that helps states pay for tobacco-related health costs. In 2008, World Health Organization (WHO) called tobacco "the single most preventable cause of death in the world today."

What caused this change? The discovery that chemicals in cigarette smoke can damage DNA and directly contribute to cancer and other diseases. There are many scientists responsible for this shift who thought critically about cigarette smoking and used scientific methods to collect evidence.

Problems

Evaluate the following statements as examples of critical thinking or opinion based on personal bias and provide a brief explanation for your answer.

Problem 2.1 I don't like beans because they give me gas.

Problem 2.2 Beans cause gas due to the presence of a saccharide called raffinose, which is not digestible by humans, so it is broken down by bacteria in the intestine. This produces the gas.

Name _____ Section _____ Date _____

Problem 2.3 The trees are such a beautiful shade of green.

_____

Problem 2.4 Leaves are green because they contain chlorophyll, which allows them to harvest energy from the sun.

_____

Problem 2.5 Wow, this highway looks like a parking lot!

_____

Problem 2.6 There must be an accident ahead, this traffic is backed up.

_____

Problem 2.7 Bumblebees shouldn't be able to fly because they are not aerodynamic.

_____

Problem 2.8 Bumblebees can fly because they have special flight muscles that are very efficient at using energy.

_____

Problem 2.9 This house is perfect for you and your family!

_____

Problem 2.10 This house is 2000 square feet, with two bedrooms and one bath; it has a yard, and the metro train tracks are directly behind the property.

_____

3. The Scientific Method

It took several decades to arrive at the consensus that smoking is hazardous to health and to stop the advertisement of cigarettes as "healthy." The scientific method involves cycles of asking questions, developing potential explanations, known as hypotheses (singular: hypothesis), and hypothesis testing (**FIGURE 1**). Review of results by experts in the field is required before publication in scientific journals; this allows for fact checking. This peer-review process makes science self-correcting. As more evidence accumulates in the scientific literature, a consensus is reached among scientists and a theory is developed. Sometimes the term *theory* is confused with *hypothesis*; keep in mind that a scientific theory is composed of substantiated hypotheses. As new evidence accumulates in the literature, theories are subject to refinement.

Using logic and critical thinking, let's start with this series of statements:

People who smoke cough a lot.
Coughing is associated with lung and throat irritation.
Therefore, smoking causes lung and throat irritation.

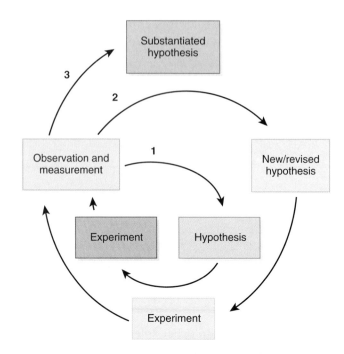

FIGURE RR3.F01 The scientific method involves cycles of hypothesis testing. Substantiated hypotheses are those subjected to peer-review and publication.

The last statement is a testable hypothesis; we can survey smokers and non-smokers to determine how much each group coughs and also determine how often each group suffers from either throat or lung irritation. A survey of medical records will indicate the number of smokers and nonsmokers who are admitted to hospitals for respiratory problems. The collection of data in the years even before the 1930s clearly support the assertion that smoking can cause lung and throat irritation.

Once a hypothesis is developed, it is then tested. This requires scientists to use their critical-thinking skills to establish the experimental design. An important task for the scientist is to clearly define control and experimental groups. The control group helps to establish a baseline and the experimental group is when they test the hypothesis. For example, the initial observation that smokers cough may be true, but how much do nonsmokers cough? A control group of nonsmokers is necessary to compare to smokers, who are the experimental group. Just measuring the number of hospital visits for respiratory distress for smokers isn't enough, because the numbers don't mean anything without a group to which to compare the results.

The progress toward identifying, verifying, and curing the effect of smoking on human health occurred over several decades and involves many scientists, including students training for scientific careers. Scientists who were also smokers have to ignore any personal convictions regarding the effect of smoking on their own health. Other scientists are inspired to try to solve a problem affecting family members and friends. It was important that scientists critically analyze advertisements that made claims regarding health benefits of smoking. One difficult position for a scientist is when funding for research is obtained from a company with financial interests in the results. There is a way to detect where there may be undue influence by a funding source in peer-reviewed literature. Funding sources must be revealed so those who review the findings as part of peer review can evaluate the potential for any undue influence.

Name _____ Section _____ Date _____

Here is an example of the importance of carefully controlled studies relating to human genetics. In the early 1960s, it became possible to determine the chromosomal make-up of individuals, known as a *karyotype*. **FIGURE 2** is an example of a karyotype for a XY male.

Soon after the karyotyping technique was developed, it was discovered that there is variation in the human population for the number of chromosomes. The first example of a male with two Y chromosome (XYY) was described in 1961. Soon after, a study was published that indicated that 9 of 315 males at a hospital for the developmentally disabled had this karyotype; this was considered a high frequency by some. These individuals shared characteristics such as being tall and having acne problems, as well as developmental disabilities. Surveys among males in prison for violent crimes indicated what appeared to be a surprisingly high number of XYY males. There are recorded cases where men were erroneously assumed to be XYY based on height, acne problems, and violent history, including the mass murderer Richard Speck. The press promoted the idea that the XYY chromosome was "supermale" syndrome despite the lack of evidence from peer-reviewed studies. The Y chromosome was dubbed the "criminal" chromosome by the popular press. Many of these studies did not consider the frequency of XYY outside the prison system. Carefully controlled studies that involved screening newborns for XYY and following their progress indicates that the frequency of XYY is not significantly different in non institutionalized populations, and that many males with an extra Y are not violent and have normal complexions.

The logical fallacy in this story can be summed up as follows:

Some males have an extra Y chromosome.
Some males in prisons for violent behavior have an extra Y chromosome.
Therefore, the extra Y chromosome is the cause of violent behavior.

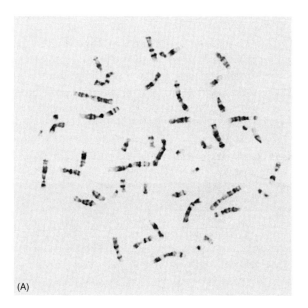

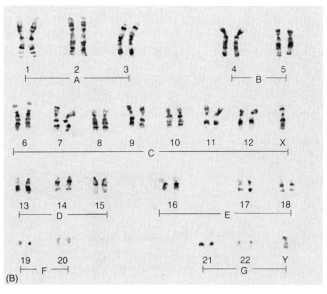

FIGURE RR3.F02 A. Metaphase chromosomes from white blood cells, processed and spread out for karyotyping analysis. B. Chromosomes organized by groupings; homologous are next to each other. There is only one X and there is a Y, so this is an XY male.

(A) Courtesy of Patricia A. Jacobs, Wessex Regional Genetics Laboratory, Salisbury District Hospital; (B) Courtesy of Patricia A. Jacobs, Wessex Regional Genetics Laboratory, Salisbury District Hospital.

Controlled studies disproved the last statement. Here is another logical fallacy exemplified by this story:

Men with XYY syndrome are found in prisons and institutions for the developmentally disabled.

These men have bad complexions, are unusually tall, and have violent dispositions.

Therefore, men who are tall, have bad complexions, and exhibit violent behavior are XYY.

The problem here is that the association of traits with a chromosomal abnormality must be supported by testing a large number of individuals who share the traits but do not share the abnormality. When we think critically about this issue, there is another side that emerges. Is it possible that unusually tall adolescents with bad complexions face social pressures, such as bullying, that can lead them to be more violent, irrespective of genetic makeup? The scientific method can test this hypothesis with carefully controlled studies.

Scientists must step back from deeply held beliefs when generating and testing hypotheses. A hypothesis must be falsifiable, so if the results don't support the hypothesis, an alternative can be formulated. It can be difficult for anyone (including scientists) to admit he or she wrong, but it is a keystone of the scientific method. Disproved hypotheses can lead to great discoveries when critical-thinking skills and logic are used to identify the reasons that data may not support the original hypothesis.

How can you identify non falsifiable hypotheses? They often include phases that explain all exceptions. Or they require experimental protocols that are not available. Here are two examples of non falsifiable hypotheses:

Earth is 6000 years old; anything that seems older was a result of divine intervention designed to test our faith.

Aliens are responsible for seeding Earth with the components from which life arose.

In the first statement, all exceptions are explained by divine intervention, so exceptions can never be identified. In the second statement, until we have solid evidence of aliens with the technology to seed a planet, it cannot be tested.

As data accumulates on a specific topic, such as the effect of smoking on health, scientists arrive at a consensus that can then be called a theory. Scientific theories are not merely conjectures based on personal biases, but are collections of substantiated hypotheses that explain phenomena. It is not necessary for the results of every experiment to support a theory, but a consensus is necessary. It may seem ironic that theories are subject to constant verification, but this is important in the scientific method. As techniques advance we may discover the underlying reasons for validated observations is not what we expect; this leads to new discoveries.

Problems

Fill in the Blank

Problem 3.1 A(n) _____ is a statement that can be scientifically verified or disproved by observation and experimentation.

Problem 3.2 When enough evidence is collected on a topic, a(n) _____

_____ can be formulated.

Problem 3.3 The _____ includes cycles of observation and measurement, followed by hypothesis development, then experimentation, which often leads to revised a hypothesis, that is tested, until a substantiated hypothesis can be developed.

Problem 3.4 A hypothesis is _____ if all exceptions are explained as part of the hypothesis.

Problem 3.5 A(n) _____ group provides a measurement of those not subjected to a treatment to which the experimental group is compared.

Identify Each Hypothesis as Testable or Un Testable.

Problem 3.6 Different breeds of dogs and wolves share a common wolf-like ancestor.

_____

Problem 3.7 Cancer is caused by a lack of faith.

_____

Problem 3.8 People with high blood pressure are at risk for heart disease.

_____

Problem 3.9 Redheads are more prone to skin cancer because they have light skin.

_____

Problem 3.10 Too much time on a smartphone can cause repetitive motion syndrome.

_____

Answer Key

Problem 1.1
The presence of a structure like scales doesn't necessarily mean the organisms are closely related. The last statement is vague; what constitutes closely related? Moths and butterflies also have scales but are not closely related to snakes or fish.

Problem 1.2
There is no evidence in the statement to preclude individuals with Turner syndrome from getting their mother's *or* father's X.

Problem 1.3
There are lots of single-celled organisms that do not cause disease; for example, baker's yeast (*Saccharomyces cerevisiae*) is a single-celled eukaryote that makes bread rise and produces alcohol. The bacteria in probiotics do not cause disease.

Problem 1.4

There are many examples of poisonous mushrooms; only very specific species of mushrooms are sold for food.

Problem 1.5

Colds and flu are health problems of humans and other mammals. There are viruses that infect plants and bacteria, but these do not affect humans.

Problem 1.6

The frequency of hair color does not predict health; lifestyle and overall genetic background are important.

Problem 1.7

This observation can just as easily be used to argue that humans and other primates share a common ancestor and this trait arose after the human lineage branched off.

Problem 1.8

The first two statements don't address the genetics of handedness. The origin of handedness has yet to be determined.

Problem 2.1

This is an opinion. Some people aren't bothered by having gas.

Problem 2.2

This is critical thinking. It provides a scientific explanation.

Problem 2.3

This is an opinion; someone with red-green colorblindness would have a different perception and opinion on green.

Problem 2.4

This is critical thinking. It provides an explanation base on scientific evidence.

Problem 2.5

This is an opinion, comparing a parking lot to a traffic back-up doesn't address the problem.

Problem 2.6

This is critical thinking. It provides an explanation.

Problem 2.7

This can be considered critical thinking, even if we know that bees can fly. Applying the rules of aerodynamics to bees is a critical-thinking exercise and it forces us to consider why bees can fly.

Problem 2.8
This is critical thinking. It provides a scientific explanation.

Problem 2.9
Without any reasons why the house it perfect, this is an opinion.

Problem 2.10
This is critical thinking.

Problem 3.1
hypothesis

Problem 3.2
theory

Problem 3.3
scientific method

Problem 3.4
non falsifiable

Problem 3.5
control

Problem 3.6
Testable. There is both archeological evidence and genetic evidence for this hypothesis.

Problem 3.7
Untestable. The first problem is faith in what? Even if "faith" is clearly defined, this still leaves the problem of measuring faith. Since faith requires deeply held beliefs, it can never be measured or tested scientifically.

Problem 3.8
Testable. This is a validated hypothesis.

Problem 3.9
Testable. There is support for this hypothesis. There is a variant of a gene involved in pigment production that is responsible for red hair and also results in pale skin. Pigments in our skin provide protection against UV damage from the sun, so the less pigment, the higher the risk of cancer.

Problem 3.10
Testable. There are studies that suggest that there is a correlation with specific repetitive motion syndromes correlated with the types of motions used on touch screens.

Answers for Even-Numbered Problems

Chapter 1

1.2 It was generally believed that the genetic material must be a very complex molecule. Proteins were chemically the most complex macromolecules known at the time. DNA was thought to be a monotonous polymer composed of a simple repeating unit.

1.4 RNA differs from DNA in that the sugar-phosphate backbone contains ribose rather than deoxyribose. RNA contains the base uracil (U) instead of thymine (T). RNA usually exists as a single strand (although any particular molecule of RNA may contain short regions of complementary base pairs that can come together to form duplexes).

1.6 Percent G = 23.6%, so percent C = 23.6% also. Percent A + T = 100 − 47.2 = 52.8%, but because percent A = percent T, it must be that percent A = 26.4%.

1.8 3′-CA-5′, because the dinucleotide 3′-CA-5′ pairs with 5′-GT-3′, so where either strand contains 5′-GT-3′, the other contains 3′-CA-5′.

1.10 5′-ATGAC-3′

1.12 5′-TGTCGTATTTGCAAG-3′

1.14 Cys–Arg–Ile–Cys–Lys or, using the single-letter abbreviations, CRICK.

1.16 The amino acid Pro could be specified by either 5′-CCU-3′ or 5′-CCC-3′; the former has an expected frequency of $3/4 \times 3/4 \times 1/4 = 9/64$ and the latter of $3/4 \times 3/4 \times 3/4 = 27/64$, totaling 9/16. The amino acid Phe could be specified by 5′-UUU-3′ or 5′-UUC-3′ in this random polymer, and so Phe would have an expected frequency of $(1/4 \times 1/4 \times 1/4) + (1/4 \times 1/4 \times 3/4) = 1/64 + 3/64 = 1/16$.

1.18 **(a)** Met–Ser–Thr–Ala–Val–Leu–Glu–Asn–Pro–Gly. **(b)** The mutation changes the initiation codon into a noninitiation codon, so translation will not start with the first AUG; translation will start either with the next AUG farther along the mRNA or, if this is too distant, not at all. **(c)** Met–Ser–Thr–Ala–Val–Leu–Glu–Asn–Pro–Gly; there is no change, because both 5′-UCC-3′ and 5′-UCG-3′ code for serine. **(d)** Met–Ser–Thr–Ala–Ala–Leu–Glu–Asn–Pro–Gly; there is a Val-to-Ala amino acid replacement because 5′-GUC-3′ codes for Val, whereas 5′-GCC-3′ codes for Ala. **(e)** Met–Ser–Thr–Ala–Val–Leu; translation is terminated at UAA because 5′-UAA-3′ is a "stop" (termination) code.

1.20 **(a)** Y, Z and W missing, X in excess; **(b)** Z and W missing, Y in excess; **(c)** W missing, Z in excess.

Chapter 2

2.2 2/3. Since the chosen seed is round, it cannot be homozygous recessive (*ww*). The pattern on the gel demonstrates codominance, and so the bands reveal that the DNA must have come from a heterozygous genotype, and the ratio of homozygous *WW* to heterozygous *Ww* among round F_2 seeds is 1/3 : 2/3.

2.4 The birth of an affected offspring implies that parents are at risk of another affected offspring. Once the genotypes of the parents have been deduced from their own phenotypes and the fact that they have an affected offspring, the recurrence risk is calculated as the probability of an affected offspring in the next birth. **(a)** The mating must be $Aa \times aa$, and so the risk of an *Aa* offspring in any birth (and, therefore, the recurrence risk) is 1/2. **(b)** The mating must be $Aa \times Aa$, and so the risk of an *aa* offspring in any birth is 1/4. **(c)** The mating must be $Aa \times aa$, and so the risk of an *aa* offspring in any birth is 1/2.

2.6 The DNA fragment from a_1 with the 2-kb insertion results in a band at 3 kb + 2 kb = 5 kb, and the DNA fragment from a_2 with the 1-kb deletion results in a band at 3 kb − 1 kb = 2 kb. DNA from each homozygous genotype produces only one band, whereas that from each heterozygous genotype produces 2 bands.

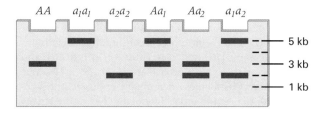

2.8 **(a)** II-2 is the affected person, and so the genotype must be *aa*. **(b)** I-I and I-2 are the parents of the affected person, and because neither is affected, their genotypes must both be *Aa*. **(c)** II-I and II-3 are siblings of the affected person. Because they are not affected, and because they result from the mating $Aa \times Aa$, their possible genotypes are *AA* and *Aa*. **(d)** From the mating $Aa \times Aa$, the ratio of *AA* : *Aa* is 1 : 2, hence the probability that II-3 is a carrier (genotype *Aa*) equals 2/3.

2.10 If IV-3 is affected, III-2 and III-3 must be heterozygous. A cross of heterozygotes, $Aa \times Aa$ produces 1 *AA* : 2 *Aa* : 1 *aa*. Unaffected offspring from this cross cannot be *aa*, so the genotypes and their relative proportions are 1/3 *AA* : 2/3 *Aa*. The probability that at least one of IV-1, IV-2, and IV-4 is a carrier is 1 minus the probability that all three are homozygous *AA*, which equals $1 − (1/3)^3 = 26/27 = 0.96$.

2.12 The existing data enable us to group the mutants into three complementation groups as follows: {*a, c, d*}, {*b, f*}, and {*e*}. The missing entries are shown in the accompanying table.

	a	*b*	*c*	*d*	*e*	*f*
a	⊖	+	−	⊖	+	⊕
b		⊖	⊕	⊕	⊕	−
c			⊖	−	⊕	⊕
d				⊖	⊕	⊕
e					⊖	+
f						⊖

2.14 Since the two genes are in different chromosomes, we can consider them separately. If the woman has a daughter, the chance that the child receives the PKU allele is 1/2 and the chance that she receives the CF allele is also 1/2. The chance of receiving exactly one of the recessives is $1/2 \times 1/2 + 1/2 \times 1/2 = 1/2$. And the chance of receiving both is $1/2 \times 1/2 = 1/4$. If the woman has a granddaughter, the chance of the grandchild receiving both recessives is $1/4 \times 1/4 = 1/16$. In this expression, the first 1/4 is the chance that the grandchild's mother received both recessive alleles, and the second 1/4 is the chance that the grandchild receives both recessive alleles when the mother carries them.

2.16 Because of Mendelian segregation, the expected numbers are **(a)** 0; **(b)** 100; **(c)** 200; **(d)** 100; **(e)** 0; **(f)** 0.

2.18 (a) In problems of this kind, it is sometimes simplest to calculate the probability of the complement of the outcomes in question and then subtract this probability from 1. In this case, note that sibships with at least one boy include all sibships except those consisting only of girls. All-girl sibships of size 4 have a probability of $(1/2)^4 = 1/16$, and, hence, sibships with one or more boys have a probability of $1 - 1/16 = 15/16$. **(b)** There are 6 birth orders constituting two girls and two boys, namely, GGBB, GBGB, GBBG, BGGB, BGBG, and BBGG. Each of these has a probability of $(1/2)^4 = 1/16$, and because the six possibilities are mutually exclusive, the probabilities add. Hence, the probability of any one of the birth orders is 6/16 = 3/8. **(c)** In this case, because a particular birth order is specified, the answer is $(1/2)^4 = 1/16$.

2.20 In the functional female gametes, the ratio of *A* : *a* is 1/2 : 1/2 because of Mendelian segregation. In males, the products of meiosis in an *Aa* individual also consist of *A* + *A* + *a* + *a*, but as stated in the problem, half of the *A*-bearing products are nonfunctional. Hence, each male meiosis produces, on the average, three functional products, namely *A* + *a* + *a*. The ratio of *A* : *a* among functional male gametes is, therefore, 1 : 2 or, converting to proportions, 1/3 *A* : 2/3 *a*. The Punnet square shown here indicates that the F_2 ratio of *A* : *Aa* : *aa* is 1/6 : 3/6 : 2/6 (or, reducing the fractions, 1/6 : 1/2 : 1/3).

		Eggs	
		1/2 *A*	1/2 *a*
Functional	1/3 *A*	1/6 *AA*	1/6 *Aa*
pollen	2/3 *a*	2/6 *Aa*	2/6 *aa*

Chapter 3

3.2 After one cell cycle carried out in the presence of colchicine, a chimpanzee cell would be expected to have $48 \times 2 = 96$ chromosomes.

3.4 (a) $2/3 \times 2/3 \times 1/4 = 1/9$; **(b)** 0; **(c)** $1 \times 1/2 = 1/2$.

3.6 9 *R– S–* : 6 *rr S–* or *R– ss*: 1 *rr ss*.

3.8 Panel A is anaphase I of meiosis, because the homologous chromosomes are paired. Panel B is anaphase II of meiosis, because the chromosome number has been reduced by half. Panel C is anaphase of mitosis, because the homologous chromosomes are not paired.

3.10 The easiest approach to this problem is to calculate the fraction of fertilized eggs that die and subtract this fraction from 1. The expected fraction of XXX fertilized eggs is $1/6 \times 1/2 = 1/12$, and the expected fraction of YY fertilized eggs is $1/6 \times 1/2 = 1/12$. Altogether 1/6 of the fertilized eggs are expected to die; hence, 5/6 of the fertilized eggs are expected to survive.

3.12 (a) $1 - [1/2 \times 1/2] = 3/4$; **(b)** $1/2 \times 1/2 = 1/4$.

3.14 The mating between relatives, denoted by the double horizontal line in the pedigree, suggests an autosomal recessive. This deduction is also supported by the observations that affected individuals have unaffected parents and that both males and females are affected.

3.16 (c) The *P*-value is the probability of a chi-square value as great or greater than that observed, given that the hypothesis is true.

3.18 The cutoffs for a *P*-value of $P = 0.05$ are 3.84 for 1 df, 5.99 for 2 df, 7.81 for 3 df, 9.48 for 4 df, and 11.07 for 5 df. (These values are from direct computation; values read from a graph will not be so accurate.) For the goodness-of-fit type of chi-square test in this chapter, the degrees of freedom 1–5 correspond to two, three, four, five, and six classes of data, respectively. Perhaps surprisingly, the increasing chi-square value needed for statistical significance does not imply that one is less likely to obtain a statistically significant chi-square with increasing degrees of freedom. The probability of obtaining a statistically significant chi-square value, given that the genetic hypothesis is true, is 5%, irrespective of the number of degrees of freedom.

3.20 The possible birth orders could all be listed, but the list is very long. There are exactly $8! / (3! \times 5!) = 56$ birth orders with 3 boys and 5 girls. Note that this expression is of the form $n!/(s!\ t!)$ as in Equation 3.1, which enumerates the number of possible combinations of *n* events (in this case births) of which *s* are of one kind (boys) and $t = n - s$ are of another kind (girls).

Chapter 4

4.2 Females produce gametes of types *A B*, *a b*, *A b*, and *a B* in the proportions $(1 - 0.08)/2 = 0.46$ and $(1 - 0.08)/2 = 0.46$ for each of the nonrecombinant types, and $0.08/2 = 0.04$ and $0.08/2 = 0.04$ for each of the nonrecombinant types. Owing to the absence of crossing-over in male *Drosophila*, males produce gametes only of types *A B* and *a b* in the proportions 0.50 and 0.50.

4.4 Only **(b)** is true; all the others are false.

4.6 This is a testcross, so the phenotypes of the progeny reveal the ratio of meiotic products from the doubly heterozygous parent. The expected progeny are 82 percent parental (41 percent wildtype and 41 percent purple, crossveinless) and 18 percent recombinant (9 percent purple and 9 percent crossveinless).

4.8 The data include $49 + 59 = 108$ progeny with either both mutations or neither, and $51 + 41 = 92$ with one mutation or the other. The first of these groups consists of parental chromosomes and the second of recombinant chromosomes, and so the appropriate chi-square test compares the ratio 108 : 92 against an expected 100 : 100 under the hypothesis of no linkage. The chi-square value equals 1.28 and there is one degree of freedom, from which P equals approximately 0.26. There is no evidence of linkage even though both genes are in the X chromosome.

4.10 (a) The mutant genes are not alleles because they do not show segregation; if they were alleles, all the progeny would be resistant to one insecticide or the other. **(b)** The genes are not linked. The chi-square of 177 parental : 205 recombinant types against 191 : 191 equals 2.05 with one degree of freedom, which is not significant ($P = 0.15$). **(c)** We can deduce that the two genes in the chromosome must be far apart.

4.12 (a) The parental types are evidently $v^+ \, pr \, bm$ and $v \, pr^+ \, bm^+$ and the double-recombinant types $v^+ \, pr^+ \, bm$ and $v \, pr \, bm$. This puts v in the middle. The $pr-v$ recombination frequency is $(59 + 71 + 36 + 41)/1000 = 20.7$ percent, and the $v-bm$ recombination frequency is $(153 + 185 + 36 + 41)/1000 = 41.5$ percent. The expected number of double crossovers equals $0.207 \times 0.415 \times 1000 = 85.9$, so the coincidence is $(36 + 41)/85.9 = 0.90$. The interference is, therefore, $1 - 0.90 = 0.10$. **(b)** The true map distances are certainly larger than 20.7 and 41.5 cM. The frequencies of recombination between these genes are so large that there are undoubtedly many undetected double crossovers in each region.

4.14 (a) The three-allele hypothesis predicts that the matings will be $FS \times FS$ and should yield $FF : FS : SS$ offspring in a ratio of 1 : 2 : 1. **(b)** These ratios are not observed, and furthermore, some of the progeny show the "null" pattern. **(c)** One possibility is that the F and S bands are from unlinked loci and that there is a "null" allele of each, says, f and s. Because f and s are common and F and S are rare, most persons with two bands would have the genotype $Ff \, Ss$. The cross should, therefore, yield 9/16 $F- \, S-$ (two bands), 3/16 $F- \, ss$ (fast band only), 3/16 $ff \, S-$ (slow band only), and 1/16 $ff \, ss$ (no bands). **(d)** The data are consistent with this hypothesis ($\chi^2 = 2.67$ with three degrees of freedom, P value approximately 0.50).

4.16 In these types of problems, it is necessary to start by calculating the number of double-crossover gametes that would be observed, given the interference. The number of observed doubles equals the number of expected doubles times the coincidence, or $0.15 \times 0.20 \times 0.20 = 0.006$, yielding among 1000 gametes, an expected 3 each of $o \, ++$ and $+ \, ci \, p$. The single recombinants in the $o-ci$ interval would therefore number $0.15 \times 1000 - 6 = 144$, for an expected 72 each of $o \, + \, p$ and $+ \, ci \, +$. The single recombinants in the $ci-p$ interval would number $0.20 \times 1000 - 6 = 194$, for an expected 97 each of

$o \, ci \, p$ and $+++$. The remaining 656 gametes are nonrecombinant, yielding an expected 328 each of $o \, ci \, +$ and $++ \, p$.

4.18 The gene–centromere map distance equals 1/2 the frequency of second-division segregation, which also equals the frequency of crossing over in the region. In this problem, it is easiest to answer the questions by taking the cases out of order, considering the second-division segregation at the beginning. **(a)** The frequency of second-division segregation of $cys\text{-}1$ must be 14 percent, because the map distance is 7 cM. Because of the complete interference, a crossover on one side of the centromere precludes a crossover on the other side, so these asci must have first-division segregation for $pan\text{-}2$. **(b)** Similarly, the frequency of second-division segregation of $pan\text{-}2$ must be 6 percent, because the map distance is 3 cM; these asci must have first-division segregation for $cys\text{-}1$. **(c)** Because of the complete interference, second-division segregation of both markers is not possible. **(d)** The only remaining possibility is first-division segregation of both markers, which must have a frequency of $1 - 0.14 - 0.06 = 80\%$. **(e)** First-division segregation of both markers yields a PD tetrad, and second-division segregation for one of the markers yields a TT tetrad. Because there are no double crossovers, there can be no NPD tetrads. Hence, the frequencies are PD = 80% and TT = 20%.

4.20 (a) To solve this type of problem, you should first deduce the genotype of the gamete contributed by the triply heterozygous parent to each class of progeny. In this case, the gamete types are, from top to bottom, $+++$, $++ \, unc$, $+ \, dor \, unc$, $dpy \, ++$, $dpy \, dor \, +$, $dpy \, dor \, unc$. To determine the order of the genes, you must identify which gene is in the middle. You can do this by comparing the parental type gametes with the double crossovers. The two parental types are $dpy \, dor \, +$ and $++ \, unc$, since these are the most frequent. In these data, there are no observed double crossovers (which means that interference across the region is complete). The missing classes of progeny correspond to the double crossovers. The missing phenotypes of progeny are normal body, deep orange eye, coordinated, corresponding to the genotype $+ \, dor \, +$, and dumpy body, red eye, uncoordinated, corresponding to the genotype $dpy \, + \, unc$. Comparing the parental types with the inferred double crossover types reveals that dpy^+ and dpy are interchanged relative to the alleles of the other two genes. Hence, the genotype of the F₁ heterozygous female is: $dor \, dpy \, unc^+/dor^+ \, dpy^+ \, unc$ or $unc^+ \, dpy \, dor/unc \, dpy^+ \, dor^+$.

(b) Analysis of the single recombinants indicates $(96 + 110)/1000 = 20.6$ percent recombination between dor and dpy, and $(75 + 65)/1000 = 14.0$ percent recombination between dpy and unc. The resulting genetic map is:

or, equivalently,

(c) Because there are no double recombinants observed, the interference is complete, which means interference = 1.

Chapter 5

5.2 The genetic consequence of the obligatory crossover is that alleles in the pseudoautosomal region in the X chromosome can be interchanged with their homologous alleles in the pseudoautosomal region in the Y chromosome, yielding a pattern of inheritance in pedigrees that is indistinguishable from that of ordinary autosomal inheritance.

5.4 The mosaic females are heterozygous; in some skin-cell lineages the X chromosome with the wildtype allele is expressed, and in other skin-cell lineages the X chromosome with the mutant allele is expressed.

5.6 The observation is quite unexpected, because a 47, XXX female would be expected to produce many XX-bearing eggs, and a 47, XYY male would be expected to produce many XY-bearing sperm. Apparently, the extra X chromosome in 47, XXX females, and the extra Y chromosome in 47, XYY males, are eliminated from the nucleus prior to meiosis.

5.8 The calico male has the karyotype XXY, and, because one of the X-chromosomes carries the allele associated with black fur, the calico male arose from sex-chromosome nondisjunction in the father.

5.10 The overlaps between the deletions are w and x between 1 and 2, and v between 2 and 3. There is no overlap between 1 and 3. The gene order that can be inferred from these data is, therefore, z (w x) y v u (or the reverse), where the parentheses mean that the gene order cannot be determined from these data. The gene order could be completed by trying to isolate additional deletions in the region. Any deletion that uncovers either w or x (but not both), plus at least one other marker on either side, would provide the information we need to complete the ordering. Alternatively, the gene order could be determined by carrying out a three-point cross with z (w x) or (w x) y.

5.12 The mother has a Robertsonian translocation that joins the long arm of chromosome 21 with the long arm of another acrocentric chromosome. The affected child has 46 chromosomes. This karyotype differs from the usual karyotype for trisomy 21 because the extra chromosome 21 is not a free chromosome.

5.14 The deletion evidently results from unequal crossing over between the duplications, which yields only one copy of the duplication and a deletion of the intervening chromosomal material containing the male-fertility genes. The observation implies that the duplications are oriented in the same direction, rather than inverted.

5.16 **(a)** In females homozygous for the inversion, there is no impediment to crossing-over. Hence, for a map distance of 12 map units, one should expect to observe 12 percent recombination, because over this length of genetic interval, multiple crossing-over can be neglected. **(b)** In females heterozygous for the inversion, the products of recombination would not be recovered, and if the entire region between the genes were involved in the inversion, the frequency of recombination would be 0. Because only 1/3 of the interval is inverted in this case, the recombination frequency is expected to be (2/3) × 12 percent = 8 percent.

5.18 Species A (with a haploid chromosome number of $n = 6$) hybridizes with species B (also with $n = 6$). The F_1 progeny will have $n = 12$ chromosomes and be sterile. The sterility can be overcome by endoreduplication in an F_1 organism, which creates a fertile new species with a chromosome number of $2n = 24$. This new species (with a haploid chromosome number of $n = 12$) hybridizes with a third species, C (haploid number $n = 6$), yielding another sterile F_1 progeny with 18 chromosomes. Endoreduplication in one of these sterile F_1 organisms gives rise to a fully fertile new species with $2n = 36$ chromosomes. This scenario is very similar to that which produced hexaploid wheat.

5.20 a (c e) g b h d f or the reverse. The order of the genes c and e cannot be determined from these data.

Chapter 6

6.2 **(a)** 5′-ACGCGT-3′. **(b)** 5′-ATCGAT-3′. **(c)** 5′-AGCT-3′. **(d)** 5′-GATATC-3′.

6.4 The 3′-to-5′ exonuclease activity of DNA polymerase has the critically important proofreading function. If this error-correcting mechanism is incapacitated, the frequency of mutations resulting from the incorporation of wrong nucleotides will increase by a factor of about 100.

6.6 A replication fork has an asymmetrical structure because of the inability of the DNA polymerase to carry out synthesis in the 3′-to-5′ direction and because of the antiparallel orientation of the strands in the DNA double helix. Both daughter strands are synthesized in the 5′-to-3′ direction; therefore, one strand (the leading strand) is synthesized continuously, whereas the other strand (the lagging strand) requires the transient existence of the short pieces of DNA known as precursor fragments that serve as replication intermediates. The precursor fragments are joined together by a type of DNA ligase.

6.8 The molecule is 10^7 base pairs in length, and because replication is bidirectional, the time required for replication is the time needed to replicate 5×10^6 base pairs. **(a)** At a rate of 1500 nucleotide pairs per second, the time required is 56 minutes. **(b)** At a rate of 50 nucleotide pairs per second, the time required is 28 hours. Replication of eukaryotic chromosomes of this size actually takes less time because there are multiple origins of replication.

6.10 Evidently, DNA in this species is replicated by a conservative mechanism, in which the parental strands stay together and the daughter strands stay together.

6.12 Samples 1 and 2 are made up of deoxyribonucleotides and, therefore, are DNA samples. Sample 3 is composed of ribonucleotides and contains U; therefore, it is a sample of RNA. Because of Watson-Crick base pairing, double-stranded DNA must have equal proportions of dATP and dTTP as well as dGTP and dCTP. Only sample 1 has this feature (ignoring the slight discrepancies due to experimental error) and is, therefore, likely to be double-stranded DNA. Sample 2 is single-stranded DNA, and sample 3 is single-stranded RNA.

6.14

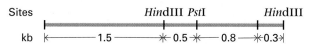

6.16 The difference in sizes of the restriction fragment could be due either to the insertion of DNA between the two restriction sites or to the elimination of one of the two flanking *Xho*I restriction sites. The pattern of hybridization of the 10-kb fragment to multiple locations along the polytene chromosomes, which differs between strains, suggests not only that the mutation is caused by insertion of DNA but also that the inserted DNA is a transposable element. Transposable elements are often present in multiple copies in a genome, and because of the lack of specificity for insertion sites that many transposable elements display, one would not expect to find elements inserted at the same genomic locations in two independent wildtype isolates.

6.18 The sequence, read from left to right, is

3'-CTTGACCGATACACAACTGAGACT-5'.

6.20

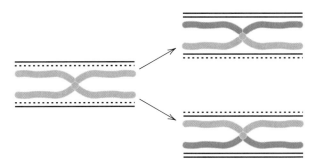

Chapter 7

7.2 For a mean of 1 phage per cell, the expected proportion of uninfected cells is given by the Poisson distribution as $e^{-1} = 0.368$. If you are not familiar with the Poisson distribution, you can work out the answer as follows. The probability that a particular bacteria is infected by a particular phage is a 1×10^{-6}, hence, the probability of a particular bacteria escaping a particular phage is 0.999999. The probability of a particular bacteria escaping all 1,000,000 phages is, therefore, $(0.999999)^{1000000} = 0.368$, which equals the expected proportion of uninfected cells.

7.4 The mating pairs should be plated on medium supplied with kanamycin (counterselective marker) and lacking methionine (selective marker). The Hfr strain is prototrophic for methionine production, and the F⁻ strain is a methionine auxotroph.

7.6 You wish to have 50 cells in 0.1 ml, so the concentration in the final dilution should be 5×10^2 viable cells per ml. Dilution of the original 5×10^7 to a final 5×10^2 would require a total dilution of 10^5, which could be accomplished with two dilutions of 100-fold each and one dilution of 10-fold.

7.8 Because the genes are so distant in the DNA, their cotransformation is independent. Hence, the expected frequency of cotransformation is $(6 \times 10^{-4} \times 1 \times 10^{-3}) = 6 \times 10^{-7}$.

7.10 Apparently *h* and *tet* are closely linked, so recombinants that contain the *h⁺* allele of the Hfr also tend to contain the *tet-s* allele of the Hfr, and these recombinants are eliminated by the counterselection for *tet-r*.

7.12 The attachment site *att* must be between genes *f* and *g*.

7.14 The three possible orders are (1) *pur – pro– his*, (2) *pur – his – pro*, and (3) *pro – pur – his*. The predictions of the three orders are as follows: (1) Virtually all *pur⁺ his⁻* transductants should be *pro⁻*, but this is not true. (2) Virtually all *pur⁺ pro⁻* transductants should be *his⁻*, but this is not true. (3) Some *pur⁺ pro⁻* transductants will be *his⁻*, and some *pur⁺ his⁻* transductants will be *pro⁻* (depending on the locations of the exchanges). Therefore, order (3) is the only one that is not contradicted by the data.

7.16 Any phage transduces one small, contiguous piece of DNA. Cotransduction, therefore, indicates very close linkage. Hence, *G* and *H* are close and *G* and *I* are close, but *H* and *I* are not close (because they do not cotransduce), so the order of these three genes must be *I G H* (or the reverse). The location of the other three genes can be deduced similarly: *A* is close to *H*, *B* is close to *I*, and *T* must be close to *A* but not close to *H*. Hence, the gene order is *B I G H A T* or, equivalently, *T A H G I B*.

7.18

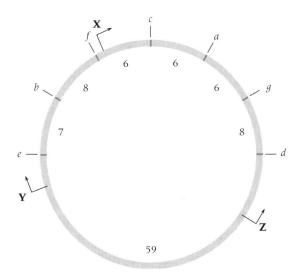

7.20 The times of entry correspond to the *x*-intercepts in the accompanying graph. For the genes *a, b, c,* and *d* the times of entry are 10, 15, 20, and 30 minutes, respectively.

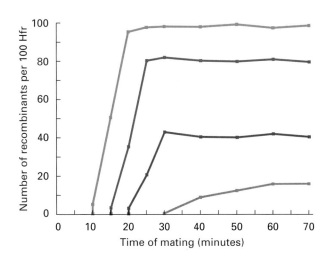

Chapter 8

8.2 (a) The original DNA duplex must have the sequence 5'-TGG-3'/3'-ACC-5', where the bottom strand (written to the right of the slash) is the one transcribed; the inversion has the sequence 5'-CCA-3'/3'-GGT-5', where again the bottom strand is transcribed. The mRNA, therefore, contains 5'-CCA 3' (the "reverse complement" of the original codon), which specifies Pro. **(b)** None of the reverse-complement codons are synonymous. This answer can be seen by checking the reverse complement of each codon in turn, or more easily by noting that there are no synonymous codons with a complementary base in the second position. **(c)** The chain-termination codons are 5'-UAA-3', 5'-UAG-3', and 5'-UGA-3', which have reverse complements 5'-UUA-3' (Leu), 5'-CUA-3' (Leu), and 5'-UCA-3' (Ser), respectively. Hence, inversion of any of the reverse-complement codons would yield a chain-termination codon. **(d)** The chain-termination codons are 5'-UAA-3', 5'-UAG-3', and 5'-UGA-3', which have reverse complements 5'-UUA-3' (Leu), 5'-CUA-3' (Leu), and 5'-UCA-3' (Ser), respectively.

8.4 The mRNA sequence is 5'-AAAAUGGCCUUAAUCU-CAGCGUCCUAC-3' and the first AUG is the initiation codon; so the resulting amino acid sequence is Met–Ala–Leu–Ile–Ser–Ala–Ser–Tyr or in the single-letter abbreviations M–A–L–I–S–A–S–Y.

8.6 (a) Methionine has 1 codon, histidine 2, and proline 4, so the total number is $1 \times 2 \times 4 = 8$. **(b)** Because arginine and serine each have six codons, the possible number of sequences coding for Met–Arg–Ser is $1 \times 6 \times 6 = 36$.

8.8 The first reading frame translates into polyaspartic acid (Asp), the second into polymethionine (Met). There is no product from the third reading frame because 5'-UGA-3' is a stop codon.

8.10 The alternating polymer UGUGU encodes the alternating polypeptide Cys–Val.

8.12

3'-GGCUGACCGAGACCAAAGGAAGUGCAA-5'.

8.14

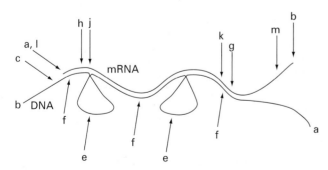

8.16 (a) DNA double helix

	ACTTGATACGCAACCCGTATCGGT
	TGAACTATGCGTTGGGCATAGCCA
mRNA	UGAACUAUGCGUUGGGCAUAGCCA
tRNA anticodon	ACUUGAUACGCAACCCGUAUCGGU

(b) The top strand is transcribed.

(c) Reading frames

```
UGA ACU AUG CGU UGG GCA UAG CCA
 *  Thr Met Arg Trp Ala  *  Pro

U GAA CUA UGC GUU GGG CAU AGC CA
  Glu Leu Cys Val Gly His Ser

UG AAC UAU GCG UUG GGC AUA GCC A
   Asn Tyr Ala Leu Gly Ile Ala
```

The * designates a stop codon.

(d) In a Southern blot hybridization, double-stranded DNA is denatured and transferred to a filter, so the probe can be complementary to either strand. Thus, either strand of the DNA molecule may be used. (The mRNA could also be used, though this is not usually done in practice because RNA is easily degraded.)

(e) A Northern blot hybridization detects mRNA immobilized on a membrane. Thus, the probe should be a strand complementary to the mRNA and should, therefore, be the top strand of the DNA as written here.

8.18 There are three possible reading frames for the alternating polymer GUCGUC, each of which encodes a different repeating polymer. One has repeating Val (GUC), another repeating Ser (UCG), and the third repeating Arg (CGU).

8.20 (a) For an error rate of 10^{-3}, the probability of no errors in 300 amino acids is $(1 - 10^{-3})^{300} = 0.74$. (Alternatively you can use the zero term of the Poisson distribution, noting that the mean number of errors would be $300 \times (1 - 10^{-3}) = 0.30$, and, hence, the probability of no error equals $e^{-0.30} = 0.74$.) **(b)** The probability of no errors in a polypeptide chain of 1000 amino acids equals $(1 - 10^{-3})^{1000} = 0.368$ (or, alternatively, $e^{-1} = 0.368$); hence, the probability of no errors in a complete tetramer is $(0.368)^4 = 0.018$. (In other words, less than 2 percent of tetramers are completely error free.)

Chapter 9

9.2 In the presence of glucose, bacterial cells have very low levels of cAMP. Without cAMP, the binding of the CRP protein cannot take place, transcription occurs only at a low level, and few *lac* proteins are made. Lack of induction is true even if lactose is present. The system ensures that preferred carbon sources (in this case glucose) are exhausted before the cell expends energy for the synthesis of new enzymes to utilize unpreferred carbon sources (in this case lactose).

9.4 The hemoglobin messenger RNA is very stable and is not broken down.

9.6 The mutant gene should bind more of the activator protein or bind it more efficiently. Therefore, the mutant gene should be induced with lower levels of the activator protein or expressed at higher levels than the wildtype gene.

9.8 In the first cross, active repressor is already present in the recipient cells, so no activation of *lac* genes is possible. In the second cross, when the *lac* genes enter the recipient, no repressor is present, so *lac* mRNA is produced and β-galactosidase is synthesized. The effect is transient, because after a short period of time, repressor is also made using the donor *lacI*⁺ gene, and transcription of *lac* genes is repressed.

9.10 (a) The phenotype of cells with the mutant GAL4p would be noninducible; the mutant gene would be recessive because the wildtype GAL4p can still function normally. **(b)** The phenotype of cells with the mutant GAL80p would be constitutive; the mutant gene would be recessive, since wildtype GAL80p would bind GAL4p in the normal way.

9.12 (a) Mutation A is an O^c mutation and is cis-dominant to the wildtype allele. **(b)** The O^c mutant operator does not bind with the mutant repressor protein. **(c)** Mutation B is an I^- mutation and is recessive to the wildtype allele. **(d)** The original mutation was an I^S (super-repressor) mutation that binds the wildtype operator even in the presence of lactose; the I^- mutation in the same gene knocks out production of any functional repressor.

9.14 (a) The phenotype would be constitutive because deletion of the red A causes a frameshift that switches the reading frame from AUG–AUG–AUG and so forth to UGA–UGA–UGA and so forth. UGA is a stop codon, so the attenuator will not be translated and transcription will never be attenuated. **(b)** The phenotype will be Met$^-$ because deletion of the red A and the underlined A changes the reading frame from AUG–AUG–AUG and so forth to GAU–GAU–GAU and so forth. GAU is an Asp codon. Therefore, translation of these codons will occur even when the concentration of methionine is very low, and transcription will terminate prematurely. **(c)** The phenotype is expected to be wildtype because a deletion of all three A's does not change the reading frame.

9.16 (a) Design a repressor–operator pair that work in a fashion analogous to those in the *lac* operon. When low levels of lead are present, the repressor protein binds an operator sequence and prevents transcription. When high levels of lead are present, the repressor binds lead ions, which causes it to dissociate from the operator and allows transcription to proceed. **(b)** A mutation in the repressor that prevents it from binding the operator. Expression of *Pb* would be constitutive. **(c)** A mutation in the operator that prevents the repressor from binding. Expression of *Pb* would be constitutive.

9.18 (a) Yes; the repressor is functional, and the presence of lactose activates transcription of the *lac* genes. **(b)** and **(d)** Yes; at 42°C, the repressor cannot bind the operator, which means that the *lac* operon is transcribed whether or not the inducer is present. **(c)** No; at this temperature, the repressor functions normally, and because lactose is absent, the *lac* operon is repressed.

9.20 Recessive mutations will fail to complement if they are in the same exon, or if one mutation is in an exon that is shared among all of the alternatively spliced transcripts. However, recessive mutations will complement if they are in different exons and these are not shared among all of the alternatively spliced transcripts. **(a)** The complementation matrix is as shown in the accompanying diagram. **(b)** This matrix is unusual in that some mutations are shared between complementation groups. In particular, mutations 1, 2, 3, and 4 fail to complement in all combinations, and mutations 3, 4, 5, and 6 fail to complement in all combinations. One would, therefore, expect mutations 1 and 2 to fail to complement mutations 5 and 6, but in fact these mutations do complement. In other words, mutations 3 and 4 appear to belong to two different complementation groups.

	1	2	3	4	5	6
1	−	−	−	−	+	+
2		−	−	−	+	+
3			−	−	−	−
4				−	−	−
5					−	−
6						−

Chapter 10

10.2 No, one of the fragments could have been ligated with the other in an inverted orientation.

10.4 (a) The *tet-r* gene is not cleaved with *Bgl*I, so addition of tetracycline to the medium requires that the colonies be tetracycline-resistant (Tet-r) and, hence, contain the plasmid. **(b)** Cells with the phenotypes Tet-r Kan-r or Tet-r Kan-s will form colonies. **(c)** Colonies with the phenotype Tet-r Kan-s contain inserts within the cleaved *kan-r* gene.

10.6 Brown with black spots.

10.8 No, they are not compatible. *Acc*651 produces a 5′ overhang whereas *Kpn*I produces a 3′ overhang.

10.10 The 3.1-kb fragment has inverted repeats at the ends; it is probably a DNA transposable element.

10.12 The E-B fragment contains a repressor-binding site, and the H-S fragment contains a repressor-binding site; these act cooperatively to produce full repression.

10.14 The results are consistent with both of the mirror-image restriction maps shown here.

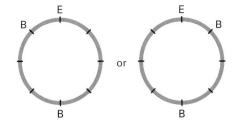

10.16 The hint says that if the genome were represented x times in the library, the probability that a particular sequence would be missing is e^{-x}, which we want to equal 0.05. Hence, $x = 3.0$. Because one haploid representation of the genome equals $(1.6 \times 10^{10})/2 = 8 \times 10^9$ base pairs, and the average insert size is 1×10^4 base pairs, one requires $[(8 \times 10^9)/(1 \times 10^4)] \times 3.0 = 2.4 \times 10^6$ clones.

10.18 Interpretation in terms of level of expression in the experimental sample relative to the control sample: **(a)** pronounced overexpression, **(b)** pronounced underexpression, **(c)** approximately equal expression, **(d)** moderate overexpression, **(e)** moderate underexpression.

10.20 Genes encoding proteins that are used in the synthesis of amino acids, nucleosides and nucleotides, vitamins, and other small molecules that are not present in minimal medium but are present in complete medium.

Chapter 11

11.2 In a loss-of-function mutation, the genetic information in a gene is not expressed in some or all cells. In a gain-of-function mutation, the genetic information is expressed at inappropriate times or in inappropriate cells. Both mutations can occur in the same gene: Some alleles may be loss-of-function and others gain-of-function. However, a particular allele must be one or the other (or neither).

11.4 The first experiment suggests that the genetic program of cell death is already activated at the time of transplantation and cannot be turned off. The second experiment shows that intercellular signaling must be responsible for inducing the cell-death fate, because when cells are transplanted to a new site prior to receiving this signal, they do not die.

11.6 **(a)** and **(b)** are true; **(c)** is false, because mutations in gap genes result in contiguous segments that are missing.

11.8 Distinct sets of genes are activated in the different target cells.

11.10 A loss-of-function mutation in a gene that is necessary for a developmental fate to be expressed prevents the fate from being realized; such a mutation is expected to be recessive.

11.12 The loss-of-function mutation results in the floral phenotype sepal-sepal-carpel-carpel; the gain-of-function mutation results in the floral phenotype petal-petal-stamen-stamen. The results imply that Ap3/Pi is necessary for petal development in whorl 2 and for stamen development in whorl 3. In a wildtype genetic background, Ap3/Pi is sufficient for petal development in whorl 1 and for stamen development in whorl 4.

11.14 **(a)** A loss-of-function mutation in *X* eliminates sepals and petals from the first two whorls, which are replaced with carpels in the first whorl and stamens in the second whorl. **(b)** A loss-of-function mutation in *Y* eliminates petals from the second whorl and stamens from the third whorl, which are replaced with sepals in the second whorl and carpels in the third whorl. **(c)** A loss-of-function mutation in *Z* eliminates stamens from the third whorl and carpels from the fourth whorl, which are replaced with petals in the third whorl and sepals in the fourth whorl.

11.16 The observations imply that gap genes act earlier in the hierarchy than pair-rule genes.

11.18 The 20 mutations need not be mutations of 20 different genes; indeed, many of them proved to be alleles of one or another of a small set of genes.

11.20 Owing to the presence of *mal*$^+$ gene product in the oocyte, the *mal*/*mal* female and *mal*/Y male progeny have enough XDH activity to produce a wildtype eye color.

Chapter 12

12.2 Such people are somatic mosaics, a condition that can be explained by somatic mutations in the pigmented cells of the iris of the eye or in their precursor cells.

12.4 $1/(2 \times 10^{-4}) = 5000$. The incidence of the condition is much smaller than expected because of low penetrance.

12.6 There are nine different mutant codons: (1) UGC → CGC (Arg), (2) UGC → AGC (Ser), (3) UGC → GGC (Gly), (4) UGC → UUC (Phe), (5) UGC → UCC (Ser), (6) UGC → UAC (Tyr), (7) UGC → UGU (Cys), (8) UGC → UGA (Stop), and (9) UGC → UGG (Trp). Number 7 is a silent mutation, number 8 is a termination codon, and the rest are missense mutations.

12.8 The serine codon must be UCN; otherwise it could not change into an alanine in one step. The alanine codon must be GCN. Hence the amino acid change results from a U → G change in the codon or from an A → C base change in the transcribed strand of the *psbA* gene. It is, therefore, a transversion.

12.10 **(a)** 1/2 females and 1/2 males (the females are heterozygous); **(b)** all female (the males die).

12.12 Transposition is an ongoing process, and an element that is inserted into a gene by transposition can also be cut out of it and pasted elsewhere; the original gene can remain unaltered once the cut DNA strands are repaired.

12.14 At 1 Sv the total mutation rate will be two times the spontaneous mutation rate. An increase of 50 percent would require 1/2 Sv, and an increase of 20 percent would require 1/5 Sv.

12.16 The lysine codons are AAA and AAG and the glutamic acid codons are GAG and GAA. Hence, an A → G transition in the first codon position will change a codon for lysine into one for glutamic acid.

12.18 The high temperature knocks out *X* function and the low temperature knocks out *Y* function. **(a)** Development would proceed to A at both high and low temperature. **(b)** Development would proceed to B at both high and low temperature. **(c)** Development would proceed to A at high temperature and B at low temperature. **(d)** Development would proceed to B at high temperature and A at low temperature.

12.20 **(a)** Acsi of types (4) and (5) show typical 3 : 1 patterns of segregation expected with gene conversion, which strongly suggests that mismatch repair does take place. The ratios of *m* : + in these asci are 6 : 2 and 2 : 6, respectively. Asci (2) and (3) also show evidence of mismatch repair because the segregation ratios are, respectively, 5 : 3 and 3 : 5. **(b)** In asci of types (1), (2), and (3), adjacent pairs of ascospores, which result from postmeiotic mitosis, do not have the same genotype; this result implies that the immediate product of meiosis was a heteroduplex, resulting in genetically different ascospores after the mitotic division. The persistence of the heteroduplexes implies that mismatch repair is not 100 percent efficient.

Chapter 13

13.2 It is a "genetic disease" in that genetic mutations take place in cells and make them undergo uncontrolled growth and division. However, most cancers are caused by somatic mutations in the cells of affected individuals; hence, they are not inherited from the parents (familial) but, rather, are due to new somatic mutations.

13.4 Ras-GDP is activated by cellular growth factors to produce Ras-GTP, which stimulates cellular growth. A gain-of-function mutation such as G19V promotes constitutive growth; hence, *Ras* is a proto-oncogene.

13.6 Because p53 plays a key role in monitoring DNA damage through transcriptional activation of the genes for p21, GADD, and 14-3-3σ and in activating the pathway of apoptosis through transcriptional activation of the genes for Bax and Apaf1.

13.8 The accompanying diagram shows how mitotic recombination can cause loss of heterozygosity. The diagram also shows that all genetic markers distal to (toward the telomere side of) the site of recombination also lose heterozygosity.

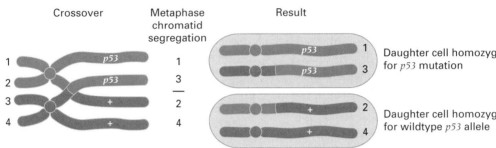

13.10 The cell-cycle checkpoints highlighted in this chapter are a DNA damage checkpoint that responds to double-stranded breaks or other damage to DNA; a centrosome duplication checkpoint that is activated unless the centrosome has duplicated and the products have moved into position; and a spindle checkpoint that is activated when the spindle is not completely assembled or when one or more chromosomes are unattached to the spindle or improperly aligned on the spindle.

13.12 **(a)** The smaller band should be associated with the mutant *RB1* allele, the larger with the nonmutant allele. **(b)** The loss of heterozygosity is independent in different tumors; hence, the mechanism need not be the same in both cases. **(c)** In this case, the wildtype allele has undergone a new mutation (or perhaps a small deletion or insertion), which inactivates the gene but does not detectably change the size of the band.

13.14 3 tumors/(4 × 10^6 cells) = 0.75 × 10^{-6} = 7.5 × 10^{-7} per cell. This is not exactly a "mutation" rate, because it includes such events as chromosome loss, mitotic recombination, and so forth, which are not, strictly speaking, mutations.

13.16 Loss of heterozygosity will generally proceed by different mechanisms in opposite eyes, or in different tumors in the same eye, because the events that trigger cancer progression occur independently.

13.18 **(a)** The cells should stop dividing at the restrictive temperature because they would no longer sense and propagate a growth-stimulatory signal (probably the presence of nutrients rather than growth factors in this example). **(b)** The mutation would be recessive; the Ras protein produced from the wildtype gene in the same cell would still propagate the signal. **(c)** The mutant cell would continue to divide even when nutrients are exhausted but would die when starved rather than cease growing. **(d)** The mutation would be dominant; the mutant Ras would remain in the GTP-bound form regardless of the presence of the wildtype Ras and would ceaselessly propagate a signal to grow.

13.20 For generations I and II, the band at position *c* is associated with the disease. In generation III, individuals 2 and 7 are at risk owing to the parent 5 and individuals 21 and 22 are at risk owing to the parent 20. Individual 13, however, is not at risk because the parent 14 with a band at position *c* is not affected; band *c* in this individual evidently originates from a nonmutant allele.

Chapter 14

14.2 The expected frequency of an A_2 allele in the next generation equals the frequency of this allele in the current generation, 0.66.

14.4 **(a)** No; **(b)** Yes.

14.6 Since the total number of individuals equals 100, the total number of alleles equals 200. The estimated allele frequencies are as follows: "15.7," (6 + 7 + 5 + 2 × 1)/200 = 0.10; "6.0," (15 + 18 + 2 × 6 + 5)/200 = 0.25; "6.2," (21 + 2 × 12 + 18 + 7)/200 = 0.35; and "6.5," (2 × 9 + 21 + 15 + 6)/200 = 0.30.

14.8 The frequency of homozygous recessives is q^2 is 0.16, which implies that $q = 0.4$. The frequency of the dominant allele is, therefore, $p = (1 − q) = 0.6$, or 60%.

14.10 $I^A I^A = (0.20)^2 = 0.04$; $I^A I^B = 2 (0.20) (0.15) = 0.06$; $I^B I^B = (0.15)^2 = 0.022$; $I^A I^O = 2(0.20) (0.65) = 0.26$; $I^B I^O = 2(0.15) (0.65) = 0.20$; $I^O I^O = (0.65)^2 = 0.422$. (Note: The proportions do not add to exactly 1.0 because of round-off error.)

14.12 If the genotype frequencies satisfy the Hardy–Weinberg principle, they should be in the proportions p^2, $2pq$, and q^2. In each case, p equals the frequency of *AA* plus one-half the frequency of *Aa*, and $q = 1 − p$. The allele frequencies and expected genotype frequencies with random mating are **(a)** $p = 0.50$, expected: 0.25, 0.50, 0.25; **(b)** $p = 0.635$, expected: 0.403, 0.464, 0.133; **(c)** $p = 0.7$, expected: 0.49, 0.42, 0.09; **(d)** $p = 0.775$, expected: 0.601, 0.349, 0.051; and **(e)** $p = 0.5$, expected: 0.25, 0.50, 0.25. Therefore, only **(a)** and **(c)** fit the Hardy–Weinberg principle. [Another approach to this problem is to note that, with Hardy–Weinberg frequencies, the square of the proportion of heterozygous genotypes ($2pq × 2pq$) equals four times the product of the frequencies of the homozygous genotypes ($4p^2q^2$). This equality is true only for **(a)** and **(c)**.]

14.14 The allele frequency is the square root of the frequency of affected offspring among unrelated individuals, or $(1 × 10^{-5})^{1/2} = 3.2 × 10^{-3}$. With inbreeding the expected frequency of homozygous recessives is $q^2 (1 − F) + qF$, in which F is the inbreeding coefficient. When $F = 1/16$, the expected frequency of homozygous recessives is $(3.2 × 10^{-3})^2 (1 − 1/16) + (3.2 × 10^{-3}) (1/16) = 2.1 × 10^{-4}$; when $F = 1/64$, the value is $6 × 10^{-5}$.

14.16 The numbers of alleles are as follows: A, $8 + 10 + 2 = 20$; B, $10 + 48 + 20 = 78$; and C, $20 + 20 + 2 = 42$. The total number of alleles is 140, and so the allele frequencies in the sample are as follows: A, $20/140 = 0.14$; B, $78/140 = 0.56$; and C, $42/140 = 0.30$. Among 70 plants, the expected numbers are

Genotype	AA	AB	BB	BC	CC	AC
Expected	1.42	11.14	21.72	23.40	6.30	6.00

14.18 (a) Here, $F = 0.66$, $p = 0.43$, and $q = 0.57$. The expected genotype frequencies are as follows: AA, $(0.43)^2 (1 - 0.66) + (0.43)(0.66) = 0.347$; AB, $2(0.43)(0.57) (1 - 0.66) = 0.167$; and BB, $(0.57)^2(1 - 0.66) + (0.57) (0.66) = 0.487$. (b) With random mating, the expected genotype frequencies are as follows: AA, $(0.43)^2 = 0.185$; AB, $2(0.43)(0.57) = 0.490$; and BB, $(0.57)^2 = 0.325$.

14.20 To obtain the relative fitness of B relative to A use, the equation $p_n/q_n = (p_0/q_0)(1/w)^n$. In this example, $n = 100$ and p_n is the frequency of A and q_n is the frequency of B in generation n. The selection coefficient is $s = 1 - w$. (a) This result implies that $p_n/q_n = 1.10 \times p_0/q_0$, and so the equation to solve is $1.10 = (1/w)^{100}$. Rearranging and taking logarithms, we find that $w = 0.99905$, or $s = 1 - w = 0.00095$. (b) In this case, $w = 0.9936$ and $s = 0.0064$. (c) Here $w = 0.99309$ and $s = 0.00691$.

Chapter 15

15.2 (a) The mean. (b) The one with the larger variance. (c) 68%. (d) 95%.

15.4 Genotype-by-environment interaction.

15.6 All individuals in F_1 generation are genetically identical, and all of the phenotypic variation in tail length in this generation results from environmental variance (σ_e^2). In the F_2 generation, both the genetic (σ_g^2) and the environmental variance (σ_e^2) contribute to the total phenotypic variance [$\sigma_p^2 = (\sigma_g^2 + \sigma_e^2)$]. In this example, $\sigma_e^2 = (1.3 \text{ cm})^2 = 1.69 \text{ cm}^2$ and $\sigma_p^2 = (3.5 \text{ cm})^2 = 12.25 \text{ cm}^2$. Hence, $\sigma_g^2 = 12.25 \text{ cm}^2 - 1.69 \text{ cm}^2 = 10.56 \text{ cm}^2$, and $H^2 = 10.56 \text{ cm}^2/12.25 \text{ cm}^2 = 0.862$.

15.8 (a) In the genotype AA population, 23 is the mean plus 1.5 standard deviations. The problem states that 87 percent of the population falls within 1.5 standard deviations, and in the AA genotype the standard deviation is 2 bristles. Hence, 87 percent of the population is expected to have between $20 - 1.5 \times 2 = 17$ and $20 + 1.5 \times 2 = 23$ bristles. Since the normal distribution is symmetrical around the mean, half of the 87 percent will be in the range 17–20 bristles and the other half in the range 20–23 bristles. Hence, the answer is that $87/2 = 43.5$ percent of the population is expected to have between 20 and 23 bristles. (b) In the aa genotype, 20 is the mean minus 1 standard deviation, and since the standard deviation of the aa genotype is 3 bristles, then we expect 68

percent of the population to be in the range $23 \pm 3 = 20$ to 26 bristles. This distribution is symmetrical around 23, so 34 percent of the population is expected to have between 20 and 23 bristles. (c) The mean of AA flies is 1.5 standard deviations smaller than the mean of aa flies; because we know from part (a) that 87 percent of AA flies have a bristle number in the range 17–23, and the distribution is symmetrical, then $(1 - 0.87)/2 = 6.5$ percent have fewer than 17 bristles and 6.5 percent have more than 23 bristles. Hence, 6.5 percent of the AA flies have more bristles than the mean of aa flies. (d) The mean of aa flies is 1 standard deviation greater than the mean of AA flies. Because we know from part (b) that 68 percent of aa flies have a bristle number in the range 20 to 26 and the distribution is symmetrical, then $(1 - 0.68)/2 = 16$ percent have a bristle number greater than 26 and 16 percent have a bristle number smaller than 20. Hence, the proportion of aa flies with a bristle number greater than 20 (the mean of the AA flies) is $0.68 + 0.16 = 84$ percent.

15.10 A phenotypic value of 130 is 2 standard deviations above the mean, so the proportion greater equals $(1 - 0.95)/2 = 2.5$ percent. A value of 85 is 1 standard deviation below the mean, so the proportion smaller equals $(1 - 0.68)/2 = 16$ percent. The proportion above 85, therefore, equals $1 - 0.16 = 84$ percent; however, this value can be calculated alternatively as $0.50 + 0.68/2 = 0.84$.

15.12 The range 1.8 to 2.2 mg is the mean ± 1 standard deviation, and 68% of the population is expected to fall within this range. Similarly, the range 1.6 to 2.4 mg is the mean ± 2 standard deviations, and 95% of the population is expected to fall within this range. A beetle with a pupal weight of 3.0 mg or more deviates by more than $+5$ standard deviations from the mean. Because only 0.15% of the population deviates by more than $+3$ standard deviations, animals with pupae weighing more than 3.0 mg are expected to be exceedingly rare and, in practice, are not found. (The expected proportion weighing ≥ 3.0 mg actually equals 2.9×10^{-7}, or roughly 1 animal in 3 million.)

15.14. The narrow-sense heritability equals 8 times the first-cousin correlation coefficient, so the first cousin correlation coefficient for total fingerprint ridge count is $0.90/8 = 0.112$.

15.16 $h^2 = 8 \times 0.09 = 0.72$.

15.18 (a) $12 + 0.20 \times (16 - 12) = 12.8$ g. (b) $12 + 0.20 \times (8 - 12) = 11.2$ g.

15.20 Spur length. Visually, the offspring-midparent scatterplots show the strongest relationship in the third scatterplot (spur length). This means that the heritability for spur length is higher than the heritability of the other two traits. Because heritability determines the speed of the response to selection [from Equation 15.6, $M' - M = h^2(M^* - M)$] spur length should respond the fastest to the horticulturalist's selective breeding.

Word Roots
Prefixes, Suffixes,
and Combining Forms

Roots and Prefixes	Meaning	Example
a-, an-	*absence or lack*	acentric, lack of a centromere
ab-	*departing from, away from*	abnormal, departing from normal
ac-, acro-	*extreme or extremity, peak*	acrocentric, centromere near the end of a chromosome
allel-	*of one another*	alleles, alternative forms of a gene
allo-	*other*	allosteric protein, a protein possessing a distinct binding site for each of two different molecules
amphi-	*on both sides, of both kinds*	amphidiploid, an organism containing diploid genomes from two different organisms
ana-	*apart, up, again*	anaphase of mitosis, when the chromosomes separate (move apart)
ant-, anti-	*opposed to, preventing or inhibiting*	antibiotic, preventing or inhibiting life
ante-	*preceding, before*	antedate, preceding a date
apo-	*former, from*	aporepressor, precursor to repressor
aut-, auto-	*self*	autogenous, self-generated
bi-	*two*	bidirectional, going in two directions
bio-	*life*	biology, the study of life and living organisms
blast-	*bud or germ*	blastoderm, structure formed in early development
carcin-	*cancer*	carcinogen, a cancer-causing agent
cata-	*down*	catabolism, chemical breakdown
caud-	*tail*	caudal (directional term)
chiasm-	*crossing*	chiasma, the cross-shaped figure that occurs at the site of crossing-over between homologous chromosomes
chrom-	*colored*	chromosome (staining body), so named because they stain darkly
circum-	*around*	circumnuclear, surrounding the nucleus
co-, con-, com-	*together*	codominant, expression of both alleles in a heterozygote
cyt-	*cell*	cytology, the study of cells
de-	*undoing, reversal, loss, removal*	deoxy, lacking an oxygen atom
di-	*twice, double*	dicentric, having two centromeres
dys-	*difficult, faulty, painful*	dysfunctional, disturbed function
ec-, ex-, ecto-	*out, outside, away from*	excise, to cut away
ectop-	*displaced*	ectopic, expression that occurs in the wrong tissue or cell type
endo-	*within, inner*	endonuclease, cleaving a nucleic acid in the interior, not the end, of a molecule
epi-	*over, above*	episome, a genetic element over (beyond) the core genome
eu-	*good, well*	eukaryote, a cell with a good or true nucleus
exo-	*outside, outer layer*	exonuclease, an enzyme that digests nucleic acids beginning at the ends of the molecule
extra-	*outside, beyond*	extracellular, outside the body cells of an organism
flagell-	*whip*	flagellum, the tail of a sperm cell
gam-, gamet-	*married, spouse*	gametes, the sex cells
gene	*beginning, origin*	genetics
gon-, gono-	*seed, offspring*	gonads, the sex organs
haplo-	*single*	haploid, gametic chromosome number
hema-, hemato-, hemo-	*blood*	hemoglobin, blood protein
hemi-	*half*	hemimethylated, methylated on one DNA strand only
hetero-	*different or other*	heterosexuality, sexual desire for a person of the opposite sex
holo-	*whole*	holoenzyme, form of an enzyme containing all subunits
hom-, homo-	*same*	homozygous, having the same allele of a gene in homologous chromosomes
homeo-	*similar*	homeotic, related structurally

Roots and Prefixes	Meaning	Example
hyper-	*excess*	hypermorphic, state in which a gene is expressed at levels greater than normal
hypo-	*below, deficient*	hypomorphic, state in which a gene is expressed at levels less than normal
in-	*in, into*	induce, to lead into (a new state)
inter-	*between*	intercellular, between the cells
intercal-	*insert*	intercalated dyes, dyes that insert between adjacent base pairs in DNA
intra-	*within, inside*	intracellular, inside the cell
iso-	*equal, same*	isothermal, equal, or same, temperature
juxta-	*near, close to*	juxtapose, place near or next to
karyo-	*kernel, nucleus*	karyotype, the set of the nuclear chromosomes
kin-, kines-	*move*	kinetic energy, the energy of motion
lact-	*milk*	lactose, milk sugar
lumen	*light*	lumen, center of a hollow structure
lys-	*dissolution or loosening*	lysis, disruption by dissolution
macro-	*large*	macromolecule, large molecule
mal-	*bad, abnormal*	malfunction, abnormal functioning of an organ
mater-	*mother*	maternal, pertaining to the mother
mega-	*large*	megabase, a million base pairs
meio-	*less*	meiosis, nuclear division that halves the chromosome number
mero-	*partial*	merodiploid, partial diploid (in bacteria)
meta-	*beyond, between, transition*	metaphase, the stage of mitosis or meiosis in which chromosomes are located between the poles of the spindle
micro-	*small*	microscope, an instrument used to make small objects appear larger
mito-	*thread, filament*	mitochondria, small, filament like structures located in cells
mono-	*single*	monohybrid, heterozygous for one allelic pair
morpho-	*form*	morphology, the study of form and structure of organisms
multi-	*many*	multinuclear, having several nuclei
muta-	*change*	mutation, change in the base sequence of DNA
nano-	*dwarf*	nanometer, one billionth of a meter
nucle-	*pit, kernel, little nut*	nucleus
nulli-, nullo-	*none*	nullisomic, having no copies of a particular chromosome
oligo-	*few*	oligonucleotide, a nucleic acid molecule containing a few nucleotides
onco-	*a mass*	oncology, study of cancer
oo-	*egg*	oocyte, precursor of female gamete
org-	*living*	organism
ov-, ovi-	*egg*	ovum, oviduct
oxy-	*oxygen*	oxygenation, the saturation of a substance with oxygen
para-	*near, beside*	paracentric, near the centromere
pater-	*father*	paternal, pertaining to the father
patho-	*pathogen*	disease-causing
pent-	*five*	pentose, a 5-carbon sugar
per-	*through*	permease, a protein that carries a small molecule through the cell membrane
peri-	*around*	pericentric, around the centromere
phago-	*eat*	bacteriophage, a virus that infects bacteria
pheno-	*show, appear*	phenotype, the physical appearance of an individual
pili	*hair*	arrector pili muscles of the skin, which make the hairs stand erect
poly-	*multiple*	polymorphism, multiple forms
post-	*after, behind*	posterior, places behind (a specific) part
pre-, pro-	*before, ahead of*	prenatal, before birth
proto-	*first or original*	prototroph, having the nutritional requirements of the wild type
pseudo-	*false*	pseudodominant, appears but isn't dominant
re-	*back, again*	reinfect
retro-	*backward, behind*	retrovirus, to move "backward" from RNA into DNA
sanguin-	*blood*	consanguineous, indicative of a genetic relationship between individuals
se-	*apart*	segregate, to set apart

Roots and Prefixes	Meaning	Example
semi-	*half*	semicircular, having the form of half a circle
septum	*fence*	septum, membrane or wall between cells
soma-	*body*	somatic cell
sub-	*under*	subunit
super-	*above, over*	supercoil
telo-	*end*	telomere, the end of a chromosome arm
tetra-	*four*	tetraploid, having four sets of chromosomes
thermo-	*heat*	thermophile, heat-loving, able to grow at high temperatures
topo-	*locale, local*	topoisomerase, altering the local state
toti-	*wholly*	totipotent, having the ability to generate all cell types
tra-, trans-	*across, through*	transgenic, placement of novel DNA into an organism
tri-	*three*	triploid, three complete sets of chromosomes in a cell
ultra-	*beyond*	ultraviolet radiation, beyond the band of visible light
vita-	*life*	vital, alive
vitre-	*glass*	in vitro, in "glass"—the test tube
viv-	*live*	in vivo
zyg-	*a yoke, twin*	zygote

Suffixes	Meaning	Example
-able	*able to, capable of*	viable, ability to live or exist
-ac	*referring to*	cardiac, referring to the heart
-age	*action, process*	cleavage, process of cleaving
-al	*relating to, pertaining to*	chromosomal
-ary	*associated with, relating to*	coronary, associated with the heart
-bryo	*swollen*	embryo
-cide	*destroy or kill*	germicide, an agent that kills germs
-ell, -elle	*small*	organelle
-emia	*condition of the blood*	anemia, deficiency of red blood cells
-gen	*an agent that initiates*	pathogen, any agent that produces disease
-gram	*data that are systematically recorded, a record*	autoradiogram, a record of where radioactive atoms decayed
-ic	*having the character of*	acidic
-logy	*the study of*	pathology, the study of changes in structure and function brought on by disease
-lysis	*loosening or breaking down*	hydrolysis, chemical decomposition of a compound into other compounds as a result of taking up water
-oid	*like, resembling*	cuboid, shaped as a cube
-oma	*tumor*	lymphoma, a tumor of the lymphatic tissues
-ory	*referring to, of*	auditory, referring to hearing
-pathy	*disease*	osteopathy, any disease of the bone
-phil, -philo	*like, love*	hydrophilic, water-attracting (e.g., molecules)
-phobia	*fear*	acrophobia, fear of heights
-phragm	*partition*	diaphragm, which separates the thoracic and abdominal cavities
-plasm	*form, shape*	cytoplasm
-scope	*instrument used for examination*	microscope, instrument used to examine small things
-some	*body*	chromosome
-stasis	*arrest, fixation, stand*	epistasis, "stand upon," the genotype at one locus affects the phenotypic expression of the genotype at a second locus
-troph	*nutrition*	prototroph, nutritional requirements of wild-type organism
-ula, -ule	*diminutive*	blastula, little "bud," sphere of cells in early development
-zyme	*ferment*	enzyme

Glossary

acentric chromosome A chromosome with no centromere.

acridine A chemical mutagen that intercalates between the bases of a DNA molecule, causing single-base insertions or deletions.

acrocentric chromosome A chromosome with the centromere near one end.

adapter molecule Any molecule covalently bonded to a target molecule that permits specific chemical interaction with another substrate. Typical examples include adaptors that permit binding with preselected PCR primers or adaptors that interact with receptors on natural or synthetic surfaces.

addition rule The principle that the probability that any one of a set of mutually exclusive events is realized equals the sum of the probabilities of the separate events.

adenine (A) A nitrogenous purine base found in DNA and RNA.

adjacent segregation Type of segregation from a heterozygous reciprocal translocation in which a structurally normal chromosome segregates with a translocated chromosome. In adjacent-1 segregation, homologous centromeres go to opposite poles of the first-division spindle; in adjacent-2 segregation, homologous centromeres go to the same pole of the first-division spindle.

albinism Absence of melanin pigment in the iris, skin, and hair of an animal; absence of chlorophyll in plants.

alkaptonuria A recessively inherited metabolic disorder in which a defect in the breakdown of tyrosine leads to excretion of homogentisic acid (alkapton) in the urine.

alkylating agent An organic compound capable of transferring an alkyl group to other molecules.

allele Any of the alternative forms of a given gene.

allele frequency The relative proportion of all alleles of a gene that are of a designated type.

allopolyploid A polyploid formed by hybridization between two different species.

alpha satellite DNA Highly repetitive DNA sequences associated with mammalian centromeres.

alternate segregation Segregation from a heterozygous reciprocal translocation in which both parts of the reciprocal translocation separate from both nontranslocated chromosomes in the first meiotic division.

Ames test A bacterial test for mutagenicity; also used to screen for potential carcinogens.

amino acid Any one of a class of organic molecules that have an amino group and a carboxyl group; 20 different amino acids are the usual components of proteins.

aminoacylated tRNA A tRNA covalently attached to its amino acid; charged tRNA.

A (aminoacyl) site The tRNA-binding site on the ribosome to which each incoming charged tRNA is initially bound.

aminoacyl-tRNA synthetase The enzyme that attaches the correct amino acid to a tRNA molecule.

amino terminus The end of a polypeptide chain at which the amino acid bears a free amino group ($-NH_2$).

amniocentesis A procedure for obtaining fetal cells from the amniotic fluid for the diagnosis of genetic abnormalities.

anaphase The stage of mitosis or meiosis in which chromosomes move to opposite ends of the spindle. In anaphase I of meiosis, homologous centromeres separate; in anaphase II, sister centromeres separate.

anaphase-promoting complex (APC/C) A ubiquitin-protein ligase that targets proteins whose destruction enables a cell to transition from metaphase into anaphase.

aneuploid A cell or organism in which the chromosome number is not an exact multiple of the haploid number; more generally, aneuploidy is a condition in which particular genes or chromosomal regions are present in extra or fewer copies compared with wildtype.

antibiotic-resistant mutant A cell or organism that carries a mutation conferring resistance to an antibiotic.

antibody A blood protein produced in response to a specific antigen and capable of binding with the antigen.

anticodon The three bases in a tRNA molecule that are complementary to the three-base codon in mRNA.

antigen A substance able to stimulate the production of antibodies.

antiparallel The chemical orientation of the two strands of a double-stranded nucleic acid molecule; the 5′-to-3′ orientations of the two strands are opposite one another.

antisense RNA An RNA molecule complementary in nucleotide sequence to all or part of a messenger RNA.

antiterminator A sequence in RNA that allows transcription to continue through the gene.

AP endonuclease An endonuclease that cleaves a DNA strand at any site at which the deoxyribose lacks a base.

AP repair Means by which restoration of apurinic and apyrimidinic sites occur.

apoptosis Genetically programmed cell death.

aporepressor A protein converted into a repressor by binding with a particular molecule.

Archaea One of the three major classes of organisms; also called archaebacteria, they are unicellular microorganisms, usually found in extreme environments, that differ as much from bacteria as either group differs from eukaryotes. *See also* **Bacteria**.

artificial selection Selection, imposed by a breeder, in which organisms of only certain phenotypes are allowed to breed.

ascospore *See* **ascus**.

ascus A sac containing the spores (ascospores) produced by meiosis in certain groups of fungi, including *Neurospora* and yeast.

asexual polyploidization Formation of a polyploid through the fusion of normal gametes followed by endoreduplication of the chromosome sets in the hybrid.

attached-X chromosome A chromosome in which two X chromosomes are joined to a common centromere; also called a compound-X chromosome.

attenuation *See* **attenuator**.

attenuator A regulatory base sequence near the beginning of an mRNA molecule at which transcription can be terminated; when an attenuator is present, it precedes the coding sequences.

autonomous determination Cellular differentiation determined intrinsically and not dependent on external signals or interactions with other cells.

autopolyploid An organism whose cells contain more than two sets of homologous chromosomes.

autoregulation Regulation of gene expression by the product of the gene itself.

autosomes All chromosomes other than the sex chromosomes.

auxotroph A mutant microorganism that is unable to synthesize a compound required for its growth but is able to grow if the compound is provided.

BAC *See* **bacterial artificial chromosome**.

backcross The cross of an F$_1$ heterozygote with a partner that has the same genotype as one of its parents.

Bacteria One of the major kingdoms of living things; includes most bacteria. *See also* **Archaea**.

bacterial artificial chromosome (BAC) A plasmid vector with regions derived from the F plasmid that contains a large fragment of cloned DNA.

bacteriophage A virus that infects bacterial cells; commonly called a phage.

basal transcription factors Transcription factors that are associated with transcription of a wide variety of genes.

base Single-ring (pyrimidine) or double-ring (purine) component of a nucleic acid.

base analog A chemical so similar to one of the normal bases that it can be incorporated into DNA.

base excision repair A mechanism of DNA repair initiated by removal of a mismatched or damaged base from its associated deoxyribonucleotide sugar.

base pair A pair of nitrogenous bases, most commonly one purine and one pyrimidine, held together by hydrogen bonds in a double-stranded region of a nucleic acid molecule; commonly abbreviated bp, the term is often used interchangeably with the term *nucleotide pair*. The normal base pairs in DNA are A–T and G–C.

base-substitution Incorporation of an incorrect base into a DNA duplex.

β-galactosidase An enzyme that cleaves lactose into its glucose and galactose constituents; produced by a gene in the *lac* operon.

biochemical pathway A diagram showing the order in which intermediate molecules are produced in the synthesis or degradation of a metabolite in a cell.

bioinformatics An interdisciplinary field combining computer science, engineering, statistics, and mathematics to analyze and interpret biological data.

bivalent A pair of homologous chromosomes, each consisting of two chromatids, associated in meiosis I.

blastoderm Structure formed in the early development of an insect larva; the syncytial blastoderm is formed from repeated cleavage of the zygote nucleus without cytoplasmic division; the cellular blastoderm is formed by migration of the nuclei to the surface and their inclusion in separate cell membranes.

block in a biochemical pathway Stoppage in a reaction sequence due to a defective or missing enzyme.

blunt ends Ends of a DNA molecule in which all terminal bases are paired; the term usually refers to termini formed by a restriction enzyme that does not produce single-stranded ends.

bootstrapping Analysis of multiple data sets formed by random sampling with replacement from an actual data set in order to estimate a degree of confidence in a particular branch or branching pattern in a gene tree.

broad-sense heritability The ratio of genotypic variance to total phenotypic variance.

cAMP–CRP complex A regulatory complex consisting of cyclic AMP (cAMP) and the CAP protein; the complex is needed for transcription of certain operons.

cancer Any of a large number of diseases characterized by the uncontrolled proliferation of cells.

candidate gene A gene proposed to be involved in the genetic determination of a trait because of the role of the gene product in the cell or organism.

cap A complex structure at the 5′ termini of most eukaryotic mRNA molecules, having a 5′-5′ linkage instead of the usual 3′-5′ linkage.

carbon-source mutant A cell or organism that carries a mutation preventing the use of a particular molecule or class of molecules as a source of carbon.

carboxyl terminus The end of a polypeptide chain at which the amino acid has a free carboxyl group (–COOH).

carrier A heterozygote for a recessive allele.

cassette In bacterial genetics, a circular molecule of duplex DNA containing a target sequence for a site-specific recombinase (integrase) and usually one or more protein-coding sequences. In yeast genetics, either of two sets of inactive mating-type genes that can become active by relocating to the *MAT* locus.

categorical trait A complex trait in which each possible phenotype can be classified into one of a number of discrete categories. Also called a meristic trait.

cDNA *See* **complementary DNA**.

cell cycle The growth cycle of a cell; in eukaryotes, it is subdivided into G$_1$ (gap 1), S (DNA synthesis), G$_2$ (gap 2), and M (mitosis).

cell cycle arrest A blockage in the progression of the cell cycle.

cell division cycle (*cdc*) mutant A mutant whose phenotype is to arrest the cell cycle at a specific and reproducible point.

cell fate The pathway of differentiation that a cell normally undergoes.

cell lineage The ancestor–descendant relationships of a group of cells in development.

cell senescence A normal process in which mammalian cells in culture cease dividing after about 50 doublings.

cellular oncogene A gene coding for a cellular growth factor whose abnormal expression predisposes to malignancy. *See also* **oncogene**.

centimorgan A unit of distance in the genetic map equal to 1 percent recombination; also called a map unit.

central dogma The concept that genetic information is transferred from the nucleotide sequence in DNA to the nucleotide sequence in an RNA transcript to the amino acid sequence of a polypeptide chain.

centriole In animal cells, one of a pair of particulate structures composed of an array of microtubules around which the spindle is organized.

centromere The region of the chromosome that is associated with spindle fibers and that participates in normal chromosome movement during mitosis and meiosis.

centrosome A localized region of clear cytoplasm found near the nucleus of nondividing cells, which in dividing cells duplicates to form the centers around which the spindle is organized. *See also* **centriole**.

centrosome duplication checkpoint A mechanism that arrests the cell cycle while the centrosome (the spindle-organizing center) remains undivided.

chain elongation The process of addition of successive amino acids to the growing end of a polypeptide chain.

chain initiation The process by which polypeptide synthesis is begun.

chain termination The process of ending polypeptide synthesis and releasing the polypeptide from the ribosome; a chain-termination mutation creates a new stop codon, resulting in premature termination of synthesis of the polypeptide chain.

chaperone A protein that assists in the three-dimensional folding of another protein.

charged tRNA A tRNA molecule to which an amino acid is linked; acylated tRNA.

checkpoint Any mechanism that arrests the cell cycle until one or more essential processes are completed.

chiasma The cytological manifestation of crossing-over; the cross-shaped exchange configuration between nonsister chromatids of homologous chromosomes that is visible in prophase I of meiosis. The plural can be either *chiasmata* or *chiasmas*.

chimeric gene A gene produced by recombination, chromosome rearrangement, or genetic engineering that is a mosaic of DNA sequences from two or more different genes.

ChIP *See* **chromatin immunoprecipitation**

chi-square (χ^2) A statistical quantity calculated to assess the goodness of fit between a set of observed numbers and the theoretically expected numbers.

chromatid Either of the longitudinal subunits produced by chromosome replication.

chromatid interference In meiosis, the effect that crossing-over between one pair of nonsister chromatids may have on the probability that a second crossing-over in the same chromosome will involve the same or different chromatids; chromatid interference does not generally occur.

chromatin The aggregate of DNA and histone proteins that makes up a eukaryotic chromosome.

chromatin immunoprecipitation Method in which antibodies specific to proteins that bind with chromatin are used to precipitate and study the DNA sequences with which they bind.

chromatin remodeling complex Any of a number of complex protein aggregates that reorganizes the nucleosomes of chromatin in preparation for transcription.

chromomere A tightly coiled, bead-like region of a chromosome most readily seen during the pachytene substage of meiosis; the beads are in register in a polytene chromosome, resulting in the banded appearance of the chromosome.

chromosome In eukaryotes, a DNA molecule that contains genes in linear order to which numerous proteins are bound and that has a telomere at each end and a centromere; in prokaryotes, the DNA is associated with fewer proteins, lacks telomeres and a centromere, and is often circular; in viruses, the chromosome is DNA or RNA, single-stranded or double-stranded, linear or circular, and often free of bound proteins.

chromosome complement The set of chromosomes in a cell or organism.

chromosome map A diagram showing the locations and relative spacing of genes along a chromosome.

chromosome painting Use of differentially labeled, chromosome-specific DNA strands for hybridization with chromosomes to label each chromosome with a different color.

chromosome territory The three-dimensional region occupied by a particular chromosome in the nucleus of an interphase or noncycling cell.

chromosome theory of heredity The theory that chromosomes are the cellular objects that contain the genes.

cis configuration The arrangement of linked genes in a double heterozygote in which both mutations are present in the same chromosome—for example, $a_1 a_2/+ +$; also called coupling.

cis-dominant Of or pertaining to a mutation that affects the expression of only those genes on the same DNA molecule.

cis heterozygote *See* **cis configuration**.

cistron A DNA sequence specifying a single genetic function as defined by a complementation test; a nucleotide sequence coding for a single polypeptide.

ClB method A genetic procedure used to detect X-linked recessive lethal mutations in *Drosophila melanogaster*; so named because one X chromosome in the female parent is marked with an inversion (*C*), a recessive lethal allele (*l*), and the dominant allele for Bar eyes (*B*).

clone A collection of organisms derived from a single parent and, except for new mutations, genetically identical to that parent; in genetic engineering, the linking of a specific gene or DNA fragment to a replicable DNA molecule, such as a plasmid or phage DNA.

cloned DNA fragment A DNA fragment inserted into a vector and transformed into a host organism.

cloned gene A DNA sequence incorporated into a vector molecule capable of replication in the same or a different organism.

cloning The process of producing cloned genes.

CNV *See* **copy number variation**

coding region The part of a DNA sequence that codes for the amino acids in a protein.

coding sequence A region of a DNA strand with the same sequence as is found in the coding region of a messenger RNA, except that T is present in DNA instead of U.

codominant Refers to phenotypes in which the presence of both alleles in heterozygous genotypes can be detected.

codon A sequence of three adjacent nucleotides in an mRNA molecule, specifying either an amino acid or a stop signal in protein synthesis.

coefficient of coincidence An experimental value obtained by dividing the observed number of double recombinants by the expected number calculated under the assumption that the two events take place independently.

cohesive ends Single-stranded regions at the ends of otherwise double-stranded DNA molecules that are complementary in base sequence.

cointegrate A DNA molecule, usually circular and formed by recombination, that joins two replicons.

colchicine A chemical that prevents formation of the spindle in nuclear division.

colinearity The linear correspondence between the order of amino acids in a polypeptide chain and the corresponding sequence of nucleotides in the DNA molecule.

colony A visible cluster of cells formed on a solid growth medium by repeated division of a single parental cell and its daughter cells.

colony hybridization assay A technique for identifying colonies that contain a particular cloned gene; many colonies are transferred to a filter, lysed, and exposed to radioactive DNA or RNA complementary to the DNA sequence of interest, after which colonies that contain a sequence complementary to the probe are located by autoradiography.

color blindness In human beings, the usual form of color blindness is X-linked red–green color blindness. Unequal crossing-over between the adjacent red and green opsin pigment genes results in chimeric opsin genes that cause mild or severe green-vision defects (deuteranomaly or deuteranopia, respectively) and mild or severe red-vision defects (protanomaly or protanopia, respectively).

combinatorial control Strategy of gene regulation in which a relatively small number of time- and tissue-specific positive and negative regulatory elements are used in various combinations to control the expression of a much larger number of genes.

comparative genomics One of the most powerful strategies for identifying genetic elements in the human genome and those of model organisms. The process of comparing the genome sequences of groups of related species that have a graded series of divergence times.

compartment A region of the nucleus in which a subset of topologically associating domains are present.

complementary base pairing Regions of nucleic acid molecules whose nucleotides can undergo Watson-Crick base pairing.

complementary DNA (cDNA) A DNA molecule made by copying RNA with reverse transcriptase.

complementation The phenomenon in which two recessive mutations with similar phenotypes result in a wildtype phenotype when both are heterozygous in the same genotype; complementation means that the mutations are in different genes.

complementation group A group of mutations that fail to complement one another.

complementation test A genetic test to determine whether two mutations are alleles (are present in the same functional gene).

complete medium Culture medium containing all required nutrients to support growth and cell division.

complex trait A multifactorial trait influenced by multiple genetic and environmental factors, each of relatively small effect, and their interactions.

computational genomics The use of computers in the management and interpretation of genomic data.

conditional mutation A mutation that results in a mutant phenotype under certain (restrictive) environmental conditions but results in a wildtype phenotype under other (permissive) conditions.

configuration *See* cls configuration.

conjugation A process of DNA transfer in sexual reproduction in certain bacteria; in *E. coli*, the transfer is unidirectional, from donor cell to recipient cell. Also, a mating between cells of *Paramecium*.

conjugative plasmid A plasmid encoding proteins and other factors that make possible its transmission between cells.

consensus sequence A generalized base sequence derived from closely related sequences found in many locations in a genome or in many organisms; each position in the consensus sequence consists of the base found in the majority of sequences at that position.

conservation genetics Application of genetic principles to the preservation and restoration of biodiversity.

conserved sequence A base or amino acid sequence that changes very slowly in the course of evolution.

constant antibody region The part of the heavy and light chains of an antibody molecule that has the same amino acid sequence among all antibodies derived from the same heavy-chain and light-chain genes.

constitutive mutant A mutant in which synthesis of a particular mRNA molecule (and the protein that it encodes) takes place at a constant rate independent of the presence or absence of any inducer or repressor molecule.

contact inhibition A phenomenon in normal mammalian cells in culture whereby cells cease to grow and divide when they are in close physical proximity.

continuous trait A trait in which the possible phenotypes have a continuous range from one extreme to the other rather than falling into discrete classes.

coordinate gene Any of a group of genes that establish the basic anterior–posterior and dorsal–ventral axes of the early embryo.

coordinate regulation Control of synthesis of several proteins by a single regulatory element; in prokaryotes, the proteins are usually translated from a single mRNA molecule.

copy number variation Differences in which a substantial portion of the human genome is duplicated or deleted in submicroscopic chunks ranging from 1 kb to 1 Mb.

core particle The aggregate of histones and DNA in a nucleosome, without the linking DNA.

co-repressor A small molecule that binds with an aporepressor to create a functional repressor molecule.

correlated response Change of the mean in one trait in a population accompanying selection for another trait.

correlation coefficient (r) A measure of association between pairs of numbers, equaling the covariance divided by the product of the standard deviations.

cotransduction Transduction of two or more linked genetic markers by one transducing particle.

cotransformation Transformation in bacteria of two genetic markers carried on a single DNA fragment.

counterselected marker A mutation used to prevent growth of a donor cell in an Hfr × F⁻ bacterial mating.

coupled transcription–translation In prokaryotes, the translation of an mRNA molecule before its transcription is completed.

coupling configuration *See* **cis configuration**.

covariance (*Cov*) A measure of association between pairs of numbers that is defined as the average product of the deviations from the respective means.

CRISPR-Cas9 The term applied to a method for precise cleavage and alteration of a specific DNA sequence based on components originally discovered in bacteria that serve as a defense against invading viral or plasmid DNA.

crossing-over A process of exchange between nonsister chromatids of a pair of homologous chromosomes that results in the recombination of linked genes.

cut-and-paste transposition A mechanism of transposition in which the transposable element is not replicated in the process of transposition, but is cleaved ("cut") from an existing site in the genome and inserted ("pasted") at a new target site.

C-value paradox The observation that among eukaryotes, the DNA content of the haploid genome (C-value) bears no consistent relationship to the metabolic, developmental, or behavioral complexity of the organism.

cyclin One of a group of proteins that participates in controlling the cell cycle. Different types of cyclins interact with the p34 kinase subunit and regulate the G_1/S and G_2/M transitions. The proteins are called cyclins because their abundance rises and falls rhythmically in the cell cycle.

cyclin–CDK complex Protein complex formed by the interaction between a cyclin and a cyclin-dependent protein kinase (CDK).

cyclin-dependent protein kinase (CDK) Any of a number of proteins that are activated by combining with a cyclin and that regulate the cell cycle by phosphorylation of other proteins.

cytokinesis Division of the cytoplasm.

cytological map Diagrammatic representation of a chromosome.

cytosine (C) A nitrogenous pyrimidine base found in DNA and RNA.

daughter strand A newly synthesized DNA or chromosome strand.

deficiency *See* **deletion**.

degeneracy *See* **redundancy**.

degrees of freedom An integer that determines the significance level of a particular statistical test. In the goodness-of-fit type of chi-square test in which the expected numbers are not based on any quantities estimated from the data themselves, the number of degrees of freedom is one less than the number of classes of data.

deletion Loss of a segment of the genetic material from a chromosome; also called deficiency.

denaturation The separation of DNA strands.

Denisovans An extinct species or subspecies of archaic humans in the genus Homo.

deoxyribonucleic acid *See* **DNA**.

deoxyribose The five-carbon sugar present in DNA.

depurination Removal of purine bases from DNA.

diakinesis The substage of meiotic prophase I, immediately preceding metaphase I, in which the bivalents attain maximum shortening and condensation.

dicentric chromosome A chromosome with two centromeres.

dideoxyribose A deoxyribose sugar that lacks the 3′ hydroxyl group; when incorporated into a polynucleotide chain, it blocks further chain elongation.

dideoxy sequencing method Procedure for DNA sequencing in which a template strand is replicated from a particular primer sequence and terminated by the incorporation of a nucleotide that contains dideoxyribose instead of deoxyribose; the resulting fragments are separated by size via electrophoresis.

dihybrid Heterozygous at each of two loci; progeny of a cross between true-breeding, or homozygous, strains that differ genetically at two loci.

diploid A cell or organism with two complete sets of homologous chromosomes.

diplotene The substage of meiotic prophase I, immediately following pachytene and preceding diakinesis, in which pairs of sister chromatids that make up a bivalent (tetrad) begin to separate from each other and chiasmata become visible.

direct repeat Copies of an identical or very similar DNA or RNA base sequence in the same molecule and in the same orientation.

direct-to-consumer (DTC) genetic testing Genetic testing services advertised by television, radio, print, or online and sold directly to the consumer who provides a tissue or blood sample to the purveyor.

distance matrix A matrix showing the amount of sequence divergence between all possible pairs of a set of protein or nucleic acid sequences.

distribution In quantitative genetics, the mathematical relation that gives the proportion of members in a population that have each possible phenotype.

D loop A DNA structure where the two strands of a double-stranded DNA molecule are separated for a stretch and held apart by a third strand of DNA.

DNA Deoxyribonucleic acid, the macromolecule, usually composed of two polynucleotide chains in a double helix, that is the carrier of the genetic information in all cells and many viruses.

DNA chip An array of tiny dots ("microarray") of DNA molecules immobilized on glass or on another solid support used for hybridization with a probe of fluorescently labeled nucleic acid.

DNA cloning See cloned gene.

DNA damage bypass A mechanism of DNA repair in which a damaged region of DNA is skipped over during replication and replaced later by a region of DNA excised from the equivalent strand in the daughter duplex.

DNA editing Use of CRISPR-Cas9 or comparable methods to alter a specific genomic sequence in a predetermined manner.

DNA damage checkpoint A mechanism that arrests the cell cycle while damaged DNA remains unrepaired.

DNA ligase An enzyme that catalyzes formation of a covalent bond between adjacent 5′-P and 3′-OH termini in a broken polynucleotide strand of double-stranded DNA.

DNA looping A mechanism by which enhancers that are distant from the immediate proximity of a promoter can still regulate transcription; the enhancer and promoter, both bound with suitable protein factors, come into indirect physical contact by means of the looping out of the DNA between them. The physical interaction stimulates transcription.

DNA methylase An enzyme that adds methyl groups (–CH$_3$) to certain bases, particularly cytosine.

DNA microarray An array of tiny dots of DNA molecules immobilized on glass or on another solid support used for hybridization with a probe of fluorescently labeled nucleic acid.

DNA polymerase Any enzyme that catalyzes the synthesis of DNA from deoxynucleoside 5′-triphosphates, using a template strand.

DNA repair Any of several different processes for restoration of the correct base sequence of a DNA molecule into which incorrect bases have been incorporated or whose bases have been chemically modified.

DNA replication The semiconservative copying of a DNA molecule.

DNA transposon A type of transposable element that undergoes transposition without its genetic information passing through an RNA intermediate.

DNA typing Electrophoretic identification of individual persons by the use of DNA probes for highly polymorphic regions of the genome, such that the genome of virtually every person exhibits a unique pattern of bands; sometimes called DNA fingerprinting.

DNA uracil glycosylase An enzyme that removes uracil bases when they occur in double-stranded DNA.

DTC genetic testing *See* **direct-to-consumer genetic testing**

domain Contiguous region of a polypeptide chain or protein molecule that folds into a characteristic structure relatively independently of other such regions, often contributing unique structural characteristics or binding properties to the molecule as a whole.

dominant trait Refers to an allele whose presence in a heterozygous genotype results in a phenotype characteristic of the allele.

dosage compensation A mechanism regulating X-linked genes such that their activities are equal in males and females; in mammals, random inactivation of one X chromosome in females results in equal amounts of the products of X-linked genes in males and females.

double-stranded DNA A DNA molecule consisting of two antiparallel strands that are complementary in nucleotide sequence.

double-Y syndrome The clinical features of the karyotype 47, XYY.

Down syndrome The clinical features of the karyotype 47,+21 (trisomy 21).

duplex DNA A double-stranded DNA molecule.

duplication A chromosome aberration in which a chromosome segment is present more than once in the haploid genome; if the two segments are adjacent, the duplication is a tandem duplication.

dynamic mutation Mutations in certain genetically unstable tandem repeats that increase or decrease in repeat number at a relatively high rate.

ectopic recombination Genetic exchange in which crossing over takes place between DNA sequences located at positions other than corresponding sites in homologous chromosomes.

editing function The activity of DNA polymerases that removes incorrectly incorporated nucleotides; also called the proofreading function.

electrophoresis A technique used to separate molecules on the basis of their different rates of movement in response to an applied electric field, typically through a gel.

elongation (protein synthesis) Addition of amino acids to a growing polypeptide chain.

embryoid A small mass of dividing cells formed from haploid cells in anthers that can give rise to a mature haploid plant.

embryonic stem cells Cells in the blastocyst that give rise to the body of the embryo.

endonuclease An enzyme that breaks internal phosphodiester bonds in a single- or double-stranded nucleic acid molecule; usually specific for either DNA or RNA.

endoreduplication Doubling of the chromosome complement because of chromosome replication and centromere division without nuclear or cytoplasmic division.

enhancer A base sequence in eukaryotes and eukaryotic viruses that increases the rate of transcription of nearby genes; the defining characteristics are that it need not be adjacent to the transcribed gene and that the enhancing activity is independent of orientation with respect to the gene.

environmental variance The part of the phenotypic variance that is attributable to differences in environment.

enzyme A protein that catalyzes a specific biochemical reaction and is not itself altered in the process.

epigenetic Persistent changes in chromatin structure or gene expression usually resulting from chemical modification of histones or DNA rather than from changes in DNA sequence.

episome A DNA element that can persist in the cell by undergoing autonomous replication or by becoming incorporated into the genome.

epistatic gene In developmental genetics, any gene whose mutant phenotype masks that of another gene. *See* **hypostatic gene.**

epistasis A term referring to an interaction between nonallelic genes in their effects on a trait. Generally, *epistasis* means any type of interaction in which the genotype at one locus affects the phenotypic expression of the genotype at another locus. In a more restricted sense, it refers to a situation in which the genotype at one locus determines the phenotype in such a way as to mask the genotype present at a second locus.

eQTL Expression quantitative trait locus; a quantitative-trait locus that affects phenotype according to the level of its expression.

equational division Term applied to the second meiotic division because the haploid chromosome complement is retained throughout.

euchromatin A region of a chromosome that has normal staining properties and undergoes the normal cycle of condensation; relatively uncoiled in the interphase nucleus (compared with condensed chromosomes), it apparently contains most of the genes.

Eukarya One of the major kingdoms of living organisms, in which the cells have a true nucleus and divide by mitosis or meiosis.

eukaryote A cell with a true nucleus (DNA enclosed in a membranous envelope) in which cell division takes place by mitosis or meiosis; an organism composed of eukaryotic cells.

euploid A cell or an organism having a chromosome number that is an exact multiple of the haploid number.

evolution Cumulative change in the genetic characteristics of a species through time.

excisionase An enzyme that is needed for prophage excision; works together with an integrase.

excision repair Type of DNA repair in which segments of a DNA strand that are chemically damaged are removed enzymatically and then resynthesized, using the other strand as a template.

E (exit) site The tRNA-binding site on the ribosome that binds each uncharged tRNA just prior to its release.

exon The sequences in a gene that are retained in the messenger RNA after the introns are removed from the primary transcript.

exon shuffle The theory that new genes can evolve by the assembly of separate exons from preexisting genes, each coding for a discrete functional domain in the new protein.

exonuclease An enzyme that removes a terminal nucleotide in a polynucleotide chain by cleavage of the terminal phosphodiester bond; nucleotides are removed successively, one by one; usually specific for either DNA or RNA and for either single-stranded or double-stranded nucleic acids. A 5′-to-3′ exonuclease cleaves successive nucleotides from the 5′ end of the molecule; a 3′-to-5′ exonuclease cleaves successive nucleotides from the 3′ end.

factorial Of a number, the product of all integers from 1 through the number itself; 0! is defined to equal 1.

familial Tending to be present in multiple generations of a pedigree.

F_1 generation The first generation of descent from a given mating.

F_2 generation The second generation of descent from a given mating, produced by intercrossing or self-fertilizing F_1 organisms.

first-division segregation Separation of a pair of alleles into different nuclei in the first meiotic division; happens when there is no crossing-over between the gene and the centromere in a particular cell.

first meiotic division The meiotic division that reduces the chromosome number; sometimes called the reduction division.

fitness A measure of the average ability of organisms with a given genotype to survive and reproduce.

5′ end The end of a DNA or RNA strand that terminates in a free phosphate group not connected to a sugar farther along.

5′ untranslated region The initial part of a messenger RNA, which does not code for protein.

fixed allele An allele whose allele frequency equals 1.0.

F^- cell A cell, typically of *Escherichia coli*, that lacks the F plasmid.

flower ABC model A model of floral determination in which a unique combination of gene activities present in each whorl of the floral meristem results in the differentiation of a distinct organ in the mature flower.

forward mutation A change from a wildtype allele to a mutant allele.

F factor A bacterial plasmid—often called the fertility factor or sex plasmid—that is capable of transferring itself from a host (F^+) cell to a cell not carrying an F factor (F^- cell); when an F factor is integrated into the bacterial chromosome (in an Hfr cell), the chromosome becomes transferrable to an F^- cell during conjugation.

F^+ cell A cell, typically of *Escherichia coli*, that contains the F plasmid.

F′ plasmid An F plasmid that contains genes obtained from the bacterial chromosome in addition to plasmid genes; formed by aberrant excision of an integrated F factor, taking along adjacent bacterial DNA.

fractal globule Proposed structure in which chromatin is packed that allows accessibility while minimizing the formation of tangles and knots.

fragile-X chromosome A type of X chromosome containing a site toward the end of the long arm that tends to break in cultured cells that are starved for DNA precursors; causes fragile-X syndrome.

frameshift mutation A mutational event caused by the insertion or deletion of one or more nucleotide pairs in a gene, resulting in a shift in the reading frame of all codons following the mutational site.

fraternal twins Twins that result from the fertilization of separate ova and are genetically related as siblings; also called dizygotic twins.

frequency of cotransduction The proportion of transductants carrying a selected genetic marker that also carry a nonselected genetic marker.

frequency of recombination The proportion of gametes carrying combinations of alleles that are not present in either parental chromosome.

functional genomics The use of DNA microarrays and other methods to study the coordinated expression of many genes simultaneously.

gain-of-function mutation Mutation in which a gene is overexpressed or inappropriately expressed.

gamete A mature reproductive cell, such as a sperm or egg in animals.

gametophyte In plants, the haploid part of the life cycle that produces the gametes by mitosis.

gap gene Any of a group of genes that control the development of contiguous segments or parasegments in *Drosophila* such that mutations result in gaps in the pattern of segmentation.

gel electrophoresis *See* **electrophoresis**.

gene The hereditary unit defined experimentally by the complementation test. At the molecular level, a region of DNA containing genetic information, usually transcribed into an RNA molecule that is processed and either functions directly or is translated into a polypeptide chain; a gene can mutate to various forms called alleles.

gene amplification A process in which certain genes undergo differential replication either within the chromosome or extrachromosomally, increasing the number of copies of the gene.

gene cloning *See* **cloned gene**.

gene conversion The phenomenon in which the products of a meiotic division in an *Aa* heterozygous genotype are in some ratio other than the expected 1*A* : 1*a*—for example, 3*A* : 1*a*, 1*A* : 3*a*, 5*A* : 3*a*, or 3*A* : 5*a*.

gene dosage Number of gene copies.

gene expression The multistep process by which a gene is regulated and its product synthesized.

gene fusion A new gene created by the joining of DNA from two preexisting genes. *See also* **chimeric gene**.

gene library A large collection of cloning vectors containing a complete (or nearly complete) set of fragments of the genome of an organism.

gene pool The totality of genetic information in a population of organisms.

gene product A term used for the polypeptide chain translated from an mRNA molecule transcribed from a gene; if the RNA is not translated (for example, ribosomal RNA), the RNA molecule is the gene product.

generalized transducing phage *See* **transducing phage**.

general transcription factor A protein molecule needed to bind with a promoter before transcription can proceed; transcription factors are necessary, but not sufficient, for transcription, and they are shared among many different promoters.

gene regulation Processes by which gene expression is controlled in response to external or internal signals.

gene targeting Disruption or mutation of a designated gene by homologous recombination.

gene therapy Deliberate alteration of the human genome for alleviation of disease.

gene tree A diagram showing the real or estimated ancestral relationships among a set of protein or nucleic acid sequences.

genetic analysis Use of genetic methods to identify the genes whose products participate in pathways of metabolism, development, behavior, or other biological processes.

genetic architecture Of a complex trait, specification of the genetic and environmental factors that contribute to the trait, and their interactions.

genetic code The set of 64 triplets of bases (codons) that correspond to the 20 amino acids in proteins and the signals for initiation and termination of polypeptide synthesis.

genetic engineering The linking of two DNA molecules by *in vitro* manipulations for the purpose of generating a novel organism with desired characteristics.

genetic map *See* **linkage map**.

genetic marker Any pair of alleles whose inheritance can be traced through a mating or through a pedigree.

genetic variance *See* **genotypic variance**

genetics The study of biological heredity.

genome The total complement of genes contained in a cell or virus; commonly used to refer to all genes present in one complete haploid set of chromosomes in eukaryotes.

genome annotation The process of identifying the locations of genes and all of the coding regions in a genome and determining what those genes do.

genome-wide association study A population study in which genotypes from whole-genome sequencing or a large number of genetic markers across the genome (typically single-nucleotide polymorphisms) are matched with phenotypes for a complex trait to identify genes or genetic markers that are correlated with phenotypic expression.

genomic imprinting A process of DNA modification in gametogenesis that affects gene expression in the zygote; one probable mechanism is the methylation of certain bases in the DNA.

genomics The systematic study of the genome using large-scale DNA sequencing, gene-expression analysis, or computational methods.

genotype The genetic constitution of an organism or virus, typically with respect to one or a few genes of interest, as distinguished from its appearance, or phenotype.

genotype-by-sex interaction A type of genotype-by-environment interaction in which the phenotypic effects of genes differ according to the sex of the individual.

genotype-by-environment (G-E) association The condition in which genotypes and environments are not in random combinations.

genotype-by-environment (G-E) interaction The condition in which genetic and environmental effects on a trait are not additive.

genotype frequency The proportion of members of a population that are of a prescribed genotype.

genotypic variance The magnitude of the phenotypic variance in a population that can be attributed to differences in genotype among individuals in the population.

germ cell A cell that gives rise to reproductive cells.

germ-line mutation A mutation that takes place in a reproductive cell.

goodness of fit The extent to which observed numbers agree with the numbers expected on the basis of some specified genetic hypothesis.

G protein One of a family of signaling proteins that is activated by binding to a molecule of guanosine triphosphate (GTP).

G_1 period *See* **cell cycle**.

G_1 restriction point The "start" point at which cells are arrested in the G_1 phase until they become committed to division.

G_1/S transition Transition from the first "growth" phase of the cell cycle to DNA synthesis.

G_2/M transition Transition between the second "growth" phase of the cell cycle to mitosis.

G_2 period *See* **cell cycle**.

GWAS *See* genome-wide association study

guanine (G) A nitrogenous purine base found in DNA and RNA.

guide RNA The RNA template present in telomerase.

gyrase A type of topoisomerase II that cleaves and rejoins both strands of a DNA duplex to relieve torsional stress.

haploid A cell or organism of a species containing the set of chromosomes normally found in gametes.

haplotype The allelic form of each of a set of linked genes present in a single chromosome.

Hardy–Weinberg principle The genotype frequencies expected with random mating.

helicase A protein that separates the strands of double-stranded DNA.

hemophilia A One of two X-linked forms of hemophilia; patients are deficient in blood-clotting factor VIII.

heritability A measure of the degree to which a phenotypic trait can be modified by selection. *See also* **broad-sense heritability** and **narrow-sense heritability**.

heterochromatin Chromatin that remains condensed and heavily stained during interphase; commonly present adjacent to the centromere and in the telomeres of chromosomes. Some chromosomes are composed primarily of heterochromatin.

heteroduplex All or part of a double-stranded nucleic acid molecule in which the two strands have different hereditary origins; produced either as an intermediate in recombination or by the *in vitro* annealing of single-stranded complementary molecules.

heterokaryon A cell containing two or more nuclei of differing genetic constitutions.

heterosis The superiority of hybrids over either inbred parent with respect to one or more traits; also called hybrid vigor.

heterozygote superiority The condition in which a heterozygous genotype has greater fitness than either of the homozygotes.

heterozygous Carrying dissimilar alleles of one or more genes; not homozygous.

hexaploid A cell or organism with six complete sets of chromosomes.

H4 histone *See* **histone**.

Hfr cell An *E. coli* cell in which an F plasmid is integrated into the chromosome, making possible the transfer of part or all of the chromosome to an F⁻ cell.

histone Any of the small basic proteins bound to DNA in chromatin; the five major histones are designated H1, H2A, H2B, H3, and H4. Each nucleosome core particle contains two molecules each of H2A, H2B, H3, and H4. The H1 histone forms connecting links between nucleosome core particles.

histone tail Region at the amino end of histone subunits that is susceptible to chemical modification affecting gene activity.

Holliday junction A cross-shaped configuration of two DNA duplexes formed as an intermediate in recombination.

Holliday junction-resolving enzyme An enzyme that catalyzes the breakage and rejoining of two DNA strands in a Holliday junction to generate two independent duplex molecules.

homeobox A DNA sequence motif found in the coding region of many regulatory genes; the amino acid sequence corresponding to the homeobox has a helix-loop-helix structure.

homeotic (*Hox*) gene Any of a group of genes in which a mutation results in the replacement of one body structure by another body structure.

homogentisic acid Substance excreted in the urine of alkaptonurics that turns black upon oxidation.

homologous Genes derived from a common ancestral gene, or proteins whose genes derive from a common ancestral gene. *See also* **orthologous** and **paralogous**.

homologous chromosomes Chromosomes that pair in meiosis and have the same genetic loci and structure; also called homologs.

homothallism The capacity of cells in certain fungi to undergo a conversion in mating type to make possible mating between cells produced by the same parental organism.

homozygous Having the same allele of a gene in homologous chromosomes.

horizontal transmission Transfer of genes from one species to another, as by transmissible bacterial plasmids.

hotspot With respect to mutation, a site in a DNA molecule at which the mutation rate is much higher than the rate for most other sites; with respect to recombination, a site in a DNA molecule at which recombination is much more likely than at most other sites.

hotspots of recombination Regions in a genome that exhibit elevated rates of recombination relative to a neutral expectation.

housekeeping gene A gene that is expressed at the same level in virtually all cells and whose product participates in basic metabolic processes.

Human Genome Project A worldwide project to map genetically and sequence the human genome.

Huntington disease Dominantly inherited degeneration of the neuromuscular system, with onset in middle age.

hybrid An organism produced by the mating of genetically unlike parents; also, a duplex nucleic acid molecule produced of strands derived from different sources.

hybrid vigor *See* **heterosis**.

hydrogen bond A weak, noncovalent linkage between two negatively charged atoms in which a hydrogen atom is shared.

hypermorphic mutation A mutation in which the wildtype gene function is overexpressed or overactive.

hypomorphic mutation A mutation in which the wild-type gene function is underexpressed or only partially active.

hypostatic gene In developmental genetics, any gene whose mutant phenotype is masked by that of another gene. *See* **epistatic gene**.

identical twins Twins developed from a single fertilized egg that splits into two embryos at an early division; also called monozygotic twins.

imaginal disk Structures present in the body of insect larvae from which the adult structures develop during pupation.

immunity A general term for resistance of an organism to specific substances, particularly agents of disease.

imprinting A process of DNA modification in gametogenesis that affects gene expression in the zygote; a probable mechanism is the methylation of certain bases in the DNA.

inborn error of metabolism A genetically determined biochemical disorder, usually in the form of an enzyme defect that produces a metabolic block.

inbreeding Mating between relatives.

inbreeding coefficient (F) A measure of the genetic effects of inbreeding in terms of the proportionate reduction in heterozygosity in an inbred organism compared with the heterozygosity expected with random mating.

inbreeding depression A phenomenon in which the average value of a quantitative trait in a population undergoes progressive deterioration as the level of inbreeding increases.

incomplete dominance Condition in which the phenotype of a heterozygous genotype is intermediate between the phenotypes of the homozygous genotypes.

independent assortment Random distribution of unlinked genes into gametes, as with genes in different (nonhomologous) chromosomes or genes that are so far apart on a single chromosome that the recombination frequency between them is 1/2.

individual selection Selection based on each organism's own phenotype.

induced mutation A mutation formed under the influence of a chemical mutagen or radiation.

inducer A small molecule that inactivates a repressor, usually by binding to it and thereby altering the ability of the repressor to bind to an operator.

inducible transcription Transcription of a gene, or a group of genes, only in the presence of an inducer molecule.

induction Activation of an inducible gene; prophage induction is the derepression of a prophage that initiates a lytic cycle of phage development.

initiation (protein synthesis) The process by which mRNA binds with ribosomes and other factors and protein synthesis begins.

inosine (I) One of a number of unusual bases found in transfer RNA.

insertion sequence A DNA sequence capable of transposition in a prokaryotic genome; such sequences usually code for their own transposase.

insulator A region of DNA that acts as a barrier preventing enhancers and silencers on one side from affecting the expression of genes on the other side.

integrase An enzyme that catalyzes a site-specific exchange between two DNA sequences.

integrated phage A state in which the phage DNA molecule is inserted intact into the bacterial chromosome; the integrated phage is called a prophage.

integron In bacteria, a DNA structure consisting of a promoter, a flanking target site for a site-specific recombinase (integrase), a coding sequence for the integrase, and usually one or more protein-coding cassettes that have been "captured" by site-specific recombination.

interference The tendency for crossing-over to inhibit the formation of another crossover nearby.

interphase The interval between nuclear divisions in the cell cycle, extending from the end of telophase of one division to the beginning of prophase of the next division.

interrupted-mating technique In an Hfr × F⁻ cross, a technique by which donor and recipient cells are broken apart at specific times, allowing only a particular length of DNA to be transferred.

intervening sequence *See* **intron**.

intron A noncoding DNA sequence in a gene that is transcribed but is then excised from the primary transcript in forming a mature mRNA molecule; found primarily in eukaryotic cells. *See also* **exon**.

inversion A structural aberration in a chromosome in which the order of several genes is reversed from the normal order. A pericentric inversion includes the centromere within the inverted region, and a paracentric inversion does not include the centromere.

inversion loop Loop structure formed by synapsis of homologous genes in a pair of chromosomes, one of which contains an inversion.

inverted repeat Either of a pair of base sequences present in the same molecule that are identical or nearly identical but are oriented in opposite directions; often found at the ends of transposable elements.

ion torrent sequencing A method of sequencing by synthesis in which each newly added nucleotide is detected by a change in pH in the solution in a tiny chamber in a semiconductor chip.

ionizing radiation Electromagnetic or particulate radiation that produces ion pairs when dissipating its energy in matter.

IS element *See* **insertion sequence**.

J (joining) region Any of multiple DNA sequences that code for alternative amino acid sequences of part of the variable region of an antibody molecule; the J regions of heavy and light chains are different.

karyotype The chromosome complement of a cell or organism; often represented by an arrangement of metaphase chromosomes according to their lengths and the positions of their centromeres.

kilobase (kb) Unit of length of a duplex DNA molecule; equal to 1000 base pairs.

kinetochore The cellular structure, formed in association with the centromere, to which the spindle fibers become attached in cell division.

kissing complex The bipartite structure formed by base pairing between a small regulatory RNA and a messenger RNA.

Klinefelter syndrome The clinical features of human males with the karyotype 47,XXY.

knockout mutations A genetic technique in which one of an organism's genes is made inoperative; often referred to as targeted deletions.

lactose permease An enzyme responsible for transport of lactose from the environment into bacteria.

lagging strand The DNA strand whose complement is synthesized in short fragments that are ultimately joined together.

leader polypeptide A short polypeptide encoded in the leader sequence of some operons coding for enzymes in amino acid biosynthesis; translation of the leader polypeptide participates in regulation of the operon through attenuation.

leading strand The DNA strand whose complement is synthesized as a continuous unit.

leptotene The initial substage of meiotic prophase I during which the chromosomes become visible in the light microscope as unpaired, thread-like structures.

leukocyte Any of several classes of mature white blood cells.

liability Risk, particularly toward a threshold type of quantitative trait.

library *See* **gene library**.

ligand The molecule that binds to a specific receptor.

lineage diagram A diagram of cell lineages and their developmental fates.

LINE element A type of transposable element lacking long terminal repeats that undergoes transposition via an RNA intermediate; the acronym LINE stands for long interspersed element.

linkage The tendency of genes located in the same chromosome to be associated in inheritance more frequently than expected from their independent assortment in meiosis.

linkage group The set of genes present together in a chromosome.

linkage map A diagram of the order of genes in a chromosome in which the distance between adjacent genes is proportional to the rate of recombination between them; also called a genetic map.

linker DNA In genetic engineering, synthetic DNA fragments that contain restriction enzyme cleavage sites used to join two DNA molecules. *See also* **nucleosome**.

lnc RNA *See* long noncoding RNA.

local population A group of organisms of the same species occupying an area within which most individual members find their mates; synonymous terms are *deme* and *Mendelian population*.

locus The site or position of a particular gene on a chromosome.

long noncoding RNA RNA transcripts of greater than 200 nucleotides that help regulate transcription, splicing, translation, and epigenetic modifications.

loss-of-function mutation A mutation that eliminates gene function; also called a null mutation.

loss of heterozygosity Loss of the presence of the wild-type allele, or loss of its function, in a heterozygous cell, enabling the phenotype of a recessive mutant allele to be expressed; mechanisms for loss of heterozygosity include chromosome loss, gene conversion, and mutation.

lost allele An allele no longer present in a population; its frequency is 0.

LTR retrotransposon A type of transposable element that transposes via an RNA intermediate and that has long terminal repeats (LTRs) in direct orientation at its ends.

LUCA Acronym for last universal common ancestor of all living things on Earth.

lysis Breakage of a cell caused by rupture of its cell membrane and cell wall.

lysogen Clone of bacterial cells that have acquired a prophage.

lysogenic cycle In temperate bacteriophage, the phenomenon in which the DNA of an infecting phage becomes part of the genetic material of the cell.

lytic cycle The life cycle of a phage, in which progeny phage are produced and the host bacterial cell is lysed.

major groove In B-form DNA, the larger of two continuous indentations running along the outside of the double helix.

map distance The genetic distance between two marker genes expressed as the sum of the length in map units across of a set of small, nonoverlapping intervals between the marker genes; corresponds to one-half of the average number of chiasmata between the genes multiplied by 100.

mapping function The mathematical relation between the genetic map distance across an interval and the observed percentage of recombination in the interval.

map unit A unit of distance in a linkage map that corresponds to a recombination frequency of 1 percent. Technically, the map distance across an interval in map units equals one-half the average number of crossovers in the interval, expressed as a percentage. Map units are sometimes called centimorgans (cM).

maternal-effect gene A gene that influences early development through its expression in the mother and the presence of the gene product in the oocyte.

maternal inheritance Extranuclear inheritance of a trait through cytoplasmic factors or organelles contributed by the female gamete.

mating-type interconversion Phenomenon in homothallic yeast in which cells switch mating type as a result of the transposition of genetic information from an unexpressed cassette into the active mating-type locus.

MCS Multiple cloning site. *See* **polylinker**.

mean The arithmetic average.

megabase (Mb) Unit of length of a duplex nucleic acid molecule; equal to 1 million base pairs.

meiocyte A germ cell that undergoes meiosis to yield gametes in animals or spores in plants.

meiosis The process of nuclear division in gametogenesis or sporogenesis in which one replication of the chromosomes is followed by two successive divisions of the nucleus to produce four haploid nuclei.

Mendelian genetics The mechanism of inheritance in which the statistical relations between the distribution of traits in successive generations result from (1) particulate hereditary determinants (genes), (2) random union of gametes, and (3) segregation of unchanged hereditary determinants in the reproductive cells.

meristem The mitotically active growing point of plant tissue.

messenger RNA (mRNA) An RNA molecule transcribed from a DNA sequence and translated into the amino acid sequence of a polypeptide. In eukaryotes, the primary transcript undergoes elaborate processing to become the mRNA.

metabolic pathway A set of chemical reactions that take place in a definite order to convert a particular starting molecule into one or more specific products.

metabolism The totality of chemical and physical processes that take place in cells or organisms.

metabolite Any small molecule that serves as a substrate, an intermediate, or a product of a metabolic pathway.

metacentric chromosome A chromosome with its centromere about in the middle so that the arms are equal or almost equal in length.

metaphase In mitosis, meiosis I, or meiosis II, the stage of nuclear division in which the centromeres of the condensed chromosomes are arranged in a plane between the two poles of the spindle.

metaphase plate Imaginary plane, equidistant from the spindle poles in a metaphase cell, on which the centromeres of the chromosomes are aligned by the spindle fibers.

microRNA Small double-stranded RNA molecules that repress translation of mRNAs containing complementary sequences.

migration Movement of organisms among subpopulations; also, the movement of molecules in electrophoresis.

minimal medium A growth medium consisting of simple inorganic salts, a carbohydrate, vitamins, organic bases, essential amino acids, and other essential compounds; its composition is precisely known. Minimal medium contrasts with complex medium or broth, which is an extract of biological material (vegetables, milk, meat) that contains a large number of compounds and the precise composition of which is unknown.

minor groove In B-form DNA, the smaller of two continuous indentations running along the outside of the double helix.

mitotic spindle Formed at the beginning of metaphase. The spindle is an elongated, football-shaped array of spindle fibers consisting primarily of microtubules formed by polymerization of the protein tubulin.

mismatch repair Removal, from duplex DNA, of a single-stranded region in which a nucleotide pair does not form proper hydrogen bonds, followed by replacement with a region of newly synthesized DNA using the intact strand as a template.

missense mutation An alteration in a coding sequence of DNA that results in an amino acid replacement in the polypeptide.

missing heritability The proportion of the heritability of a complex trait that cannot be accounted for by genetic factors identified as statistically significant in genome-wide association tests.

mitosis The process of nuclear division in which the replicated chromosomes divide and the daughter nuclei have the same chromosome number and genetic composition as the parent nucleus.

mobile DNA Alternative term for transposable elements.

molecular clock A condition in which a protein or nucleic acid molecule has the same probability of change per unit time in every branch of a gene tree.

molecular evolution The study of how (and why) the sequences of macromolecules change through time

molecular genetics The branch of genetics concerned with the chemistry of DNA and the molecules that participate in its replication, function, mutation, and repair.

molecular phylogenetics A group of statistical methods for estimating gene trees and often, by inference, the evolutionary relationships among the taxa of which the genes are representative.

monohybrid A genotype that is heterozygous for one pair of alleles; the offspring of a cross between genotypes that are homozygous for different alleles of a gene.

monoploid The basic chromosome set that is reduplicated to form the genomes of the species in a polyploid series; the smallest haploid chromosome number in a polyploid series.

monosomy Condition of an otherwise diploid organism in which one member of a pair of chromosomes is missing.

mosaic An organism composed of two or more genetically different types of cells.

most recent common ancestor (MRCA) In a phylogenetic tree, the most recent node that unites a particular subset of sequences, characters, or species.

M period *See* **cell cycle**.

mRNA *See* **messenger RNA**.

mtDNA Mitochondrial DNA.

multifactorial trait A trait determined by the combined action of many factors, typically some genetic and some environmental.

multiple alleles The presence, in a population, of more than two alleles of a gene.

multiple cloning site *See* **polylinker**.

multiplication rule The principle that the probability that all of a set of independent events are realized simultaneously equals the product of the probabilities of the separate events.

mutagen An agent that is capable of increasing the rate of mutation.

mutant Any heritable biological entity that differs from wildtype, such as a mutant DNA molecule, mutant allele, mutant gene, mutant chromosome, mutant cell, mutant organism, or mutant heritable phenotype; also, a cell or organism in which a mutant allele is expressed.

mutant screen A type of genetic experiment in which the geneticist seeks to isolate multiple new mutations that affect a particular trait.

mutation A heritable alteration in a gene or chromosome; also, the process by which such an alteration happens. Used incorrectly, but with increasing frequency, as a synonym for *mutant*, even in some excellent textbooks.

mutation rate The probability of a new mutation in a particular gene, either per gamete or per generation.

mutation-selection balance A state of equilibrium in a population at which the elimination of deleterious alleles of a gene by selection is equal to the creation of new deleterious alleles of the same gene by mutation.

nanopore sequencing A method of single-molecule DNA sequencing that uses a motor protein to drive a strand of DNA through a transmembrane protein, identifying reach nucleotide as it passes through according to the magnitude of the change in membrane conductance.

narrow-sense heritability The fraction of the phenotypic variance revealed as resemblance between parents and offspring; technically, the ratio of the additive genetic variance to the total phenotypic variance.

natural selection The process of evolutionary adaptation in which the genotypes genetically best suited to survive and reproduce in a particular environment give rise to a disproportionate share of the offspring and so gradually increase the overall ability of the population to survive and reproduce in that environment.

Neanderthals An extinct species of human that was widely distributed in ice-age Europe.

negative regulation Regulation of gene expression in which mRNA is not synthesized until a repressor is removed from the DNA of the gene.

neighbor joining A method for estimating a gene tree in which pairs of taxa are joined sequentially according to which pair are separated by the shortest distance.

nick A single-strand break in a DNA molecule.

nitrous acid HNO_2, a chemical mutagen.

noncomplementation Failure of two mutations to compensate each another's defects when present in the same cell or individual.

nondisjunction Failure of chromosomes to separate (disjoin) and move to opposite poles of the division spindle; the result is loss or gain of a chromosome.

nonhomologous end joining Mechanism of DNA repair in which broken ends of DNA are joined together despite lacking sequence homology.

non-LTR retrotransposon A type of transposable element that transposes via an RNA intermediate and that lacks terminal repeats at its ends.

nonparental ditype (NPD) An ascus containing two pairs of recombinant spores.

nonselective medium A growth medium that allows growth of wildtype and of one or more mutant genotypes.

nonsense mutation A mutation that changes a codon specifying an amino acid into a stop codon, resulting in premature polypeptide chain termination; also called a chain termination mutation.

nonsynonymous substitution A coding sequence that does result in an amino acid replacement.

normal distribution A symmetrical bell-shaped distribution characterized by the mean and the variance; in a normal distribution, approximately 68 percent of the observations are within 1 standard deviation of the mean, and approximately 95 percent are within 2 standard deviations.

nuclease An enzyme that breaks phosphodiester bonds in nucleic acid molecules.

nucleic acid A polymer composed of repeating units of phosphate-linked five-carbon sugars to which nitrogenous bases are attached. *See also* **DNA** and **RNA**.

nucleic acid hybridization The formation of duplex nucleic acid from complementary single strands.

nucleolus (*pl.* nucleoli) Nuclear organelle in which ribosomal RNA is made and ribosomes are partially synthesized; usually associated with the nucleolar organizer region. A nucleus may contain several nucleoli.

nucleoside A purine or pyrimidine base covalently linked to a sugar.

nucleosome The basic repeating subunit of chromatin, consisting of a core particle composed of two molecules each of four different histones around which a length of DNA containing about 145 nucleotide pairs is wound, joined to an adjacent core particle by about 55 nucleotide pairs of linker DNA associated with a fifth type of histone.

nucleotide A nucleoside phosphate.

nucleotide excision repair A mechanism of DNA repair in which nucleotides with mismatched or damaged bases are cleaved from a DNA strand and replaced using the complementary DNA strand as a template.

nutritional mutation A mutation in a metabolic pathway that creates a need for a substance to be present in the growth medium or that eliminates the ability to utilize a substance present in the growth medium.

oligonucleotide primer A short, single-stranded nucleic acid synthesized for use in DNA sequencing or as a primer in the polymerase chain reaction.

oncogene A gain-of-function mutation in a cellular gene, called a proto-oncogene, whose normal function is to promote cellular proliferation or inhibit apoptosis; oncogenes are often associated with tumor progression.

one gene–one enzyme hypothesis The theory that each gene directly produces a single enzyme.

open reading frame (ORF) In the coding strand of DNA or in mRNA, a region containing a series of codons uninterrupted by stop codons and therefore capable of coding for a polypeptide chain.

operator A regulatory region in DNA that interacts with a specific repressor protein in controlling the transcription of adjacent structural genes.

operon A collection of adjacent structural genes regulated by an operator and a repressor.

operon model The genetic regulatory mechanism of the lac system first explained by François Jacob and Jacques Monod.

ORF *See* **open reading frame**.

orthologous Genes that share a common ancestral gene through the process of speciation. *See also* **paralogous**.

OTC genetic testing *See* **over-the-counter genetic testing**

over-the-counter (OTC) genetic testing Self-administered genetic tests based on diagnostic kits purchased over the counter without prescription.

PAC *See* **P1 artificial chromosome**.

pachytene The middle substage of meiotic prophase I, in which the homologous chromosomes are closely synapsed.

pair-rule gene Any of a group of genes active early in *Drosophila* development that specifies the fates of alternating segments or parasegments. Mutations in pair-rule genes result in loss of even-numbered or odd-numbered segments or parasegments.

paracentric inversion An inversion that does not include the centromere.

paralogous Genes that share a common ancestral gene through the process of gene duplication within a species. *See also* **orthologous**.

parasegment Developmental unit in *Drosophila* consisting of the posterior part of one segment and the anterior part of the next segment in line.

parental combination Alleles present in an offspring chromosome in the same combination as that found in one of the parental chromosomes.

parental ditype (PD) An ascus containing two pairs of nonrecombinant spores.

parental strand In DNA replication, the strand that served as the template in a newly formed duplex.

partial diploid A cell in which a segment of the genome is duplicated, usually in a plasmid.

Pascal's triangle Triangular configuration of integers in which the nth row gives the binomial coefficients in the expansion of $(x + y)^{n-1}$. The first and last numbers in each row are 1, and the others equal the sum of the adjacent numbers in the row immediately above.

pathogenicity island A horizontally transferred region of a bacterial DNA molecule containing genes causing disease.

pattern formation The creation of a spatially ordered and differentiated embryo from a seemingly homogeneous egg cell.

PCR *See* **polymerase chain reaction**.

pedigree A diagram representing the familial relationships among relatives.

penetrance The proportion of organisms having a particular genotype that actually express the corresponding phenotype. If the phenotype is always expressed, penetrance is complete; otherwise, it is incomplete.

peptide bond A covalent bond between the amino group $(-NH_2)$ of one amino acid and the carboxyl group $(-COOH)$ of another.

peptidyl transferase In translation of messenger RNA, the enzyme that creates the peptide bond between the growing polypeptide chain and the incoming amino acid.

pericentric inversion An inversion that includes the centromere.

permissive condition An environmental condition in which the phenotype of a conditional mutation is not expressed; contrasts with the nonpermissive or restrictive condition.

personalized medicine Choosing appropriate medical practices, interventions, or treatments based on an individual patient's likelihood of disease or success of response.

p53 transcription factor A key protein that helps regulate a mammalian cell's response to stress, especially to DNA damage.

P_1 generation The parents used in a cross, or the original parents in a series of generations; also called the P generation if there is no chance of confusion with the grandparents or more remote ancestors.

phage *See* **bacteriophage**.

phage repressor Regulatory protein that prevents transcription of genes in a prophage.

phenotype The observable properties of a cell or an organism, which result from the interaction of the genotype and the environment.

phenylalanine hydroxylase (PAH) The enzyme, deficient in phenylketonuria, that converts phenylalanine into tyrosine.

phenylketonuria (PKU) A hereditary human condition resulting from the inability to convert phenylalanine into tyrosine; causes severe mental retardation unless treated in infancy and childhood via a low-phenylalanine diet.

phenylthiocarbamide (PTC) A type of organosulfur thiourea containing a phenyl ring that has either a very bitter taste or no detectable taste according to the genotype of the individual doing the tasting.

phosphodiester bond In nucleic acids, the covalent bond formed between the 5'-phosphate group (5'-P) of one nucleotide and the 3'-hydroxyl group (3'-OH) of the next nucleotide in line; these bonds form the backbone of a nucleic acid molecule.

phylogenetic tree A diagram showing the genealogical relationships among a set of genes or species.

physiological epistasis Gene interactions at the molecular, cellular, or organismic levels such that the phenotypic effects of the genes acting together cannot be predicted based on the phenotypic effects of the individual genes.

plaque A clear area in an otherwise turbid layer of bacteria growing on a solid medium, caused by the infection and killing of the cells by a phage; because each plaque results from the growth of one phage, plaque counting is a way of counting viable phage particles. The term is also used for animal viruses that cause clear areas in layers of animal cells grown in culture.

plasmid An extrachromosomal genetic element that replicates independently of the host chromosome; it may exist in one or many copies per cell and may segregate in cell division to daughter cells in either a controlled or a random fashion. Some plasmids, such as the F factor, may become integrated into the host chromosome.

pleiotropic effect Any phenotypic effect that is a secondary manifestation of a mutant gene.

pleiotropy The condition in which a single mutant gene affects two or more distinct and seemingly unrelated traits.

point mutation Mutations that replace one base pair with a different base pair or that add or delete a single base pair in DNA.

polarity The 5'-to-3' orientation of a strand of nucleic acid.

pole cell Any of a group of cells, set off at the posterior end of the *Drosophila* embryo, from which the germ cells are derived.

Pol II holoenzyme A large protein complex containing the type of RNA polymerase used in transcribing most protein-coding genes.

poly-A tail The sequence of adenines added to the 3' end of many eukaryotic mRNA molecules in processing.

polycistronic mRNA An mRNA molecule from which two or more polypeptides are translated; found primarily in prokaryotes.

polylinker A short DNA sequence that is present in a vector and that contains a number of unique restriction sites suitable for gene cloning.

polymerase chain reaction (PCR) Repeated cycles of DNA denaturation, renaturation with primer oligonucleotide sequences, and replication, resulting in exponential growth in the number of copies of the DNA sequence located between the primers.

polymorphic gene A gene for which there is more than one relatively common allele in a population.

polymorphism The presence, in a population, of two or more relatively common forms of a gene, chromosome, or genetically determined trait.

polynucleotide chain A polymer of covalently linked nucleotides.

polypeptide *See* **polypeptide chain.**

polypeptide chain A polymer of amino acids linked together by peptide bonds.

polyploidy The condition of a cell or organism with more than two complete sets of chromosomes.

polysome A complex of two or more ribosomes associated with an mRNA molecule and actively engaged in polypeptide synthesis; a polyribosome.

polytene chromosome A giant chromosome consisting of many identical strands laterally apposed and in register, exhibiting a characteristic pattern of transverse banding.

population A group of organisms of the same species.

population genetics Application of Mendel's laws and other principles of genetics to entire populations of organisms.

positional information Developmental signals transmitted to a cell by virtue of its position in the embryo.

positive regulation Mechanism of gene regulation in which an element must be bound to DNA in an active form to allow transcription. Positive regulation contrasts with negative regulation, in which a regulatory element must be removed from DNA.

postreplication mismatch repair DNA repair that takes place via recombination in nonreplicating DNA or after the replication fork is some distance beyond a damaged region.

precision medicine *See* **personalized medicine**

P (peptidyl site) The tRNA-binding site on the ribosome to which the tRNA bearing the nascent polypeptide becomes bound immediately after formation of the peptide bond.

precursor fragment *See* **Okazaki fragment.**

primary transcript An RNA copy of a gene; in eukaryotes, the transcript must be processed to form a translatable mRNA molecule.

primer In nucleic acids, a short RNA or single-stranded DNA segment that functions as a growing point in polymerization.

primosome The enzyme complex that forms the RNA primer for DNA replication in eukaryotic cells.

principle of epistasis In the genetic analysis of linear developmental-switch pathways, the principle that the epistatic gene in a double mutant genotype acts downstream in the pathway from the hypostatic gene.

probe A radioactive DNA and RNA molecule used in DNA–RNA and DNA–DNA hybridization assays.

processivity Refers to the number of consecutive nucleotides in a template strand of nucleic acid that are traversed before a DNA polymerase or an RNA polymerase detaches from the template.

product molecule The end result of a biochemical reaction or a metabolic pathway.

progenitor cell An adult stem cell.

programmed cell death Cell death that happens as part of the normal cellular response to damage or as part of the normal developmental process. *See also* **apoptosis**.

prokaryote An organism that lacks a nucleus; prokaryotic cells divide by fission.

promoter A DNA sequence at which RNA polymerase binds and initiates transcription.

promoter fusion Joining of the promoter region of one gene with the protein-coding region of another.

proofreading function *See* **editing function**.

prophage The form of phage DNA in a lysogenic bacterium; the phage DNA is repressed and is usually integrated into the bacterial chromosome, but some prophages are in plasmid form.

prophage induction Activation of a prophage to undergo the lytic cycle.

prophase The initial stage of mitosis or meiosis, beginning after interphase and terminating with the alignment of the chromosomes at metaphase; often absent or abbreviated between meiosis I and meiosis II.

proteome The set of all proteins encoded in a genome.

proteomics Study of the complement of proteins present in a cell or organism in order to identify their localization, functions, and interactions.

proto-oncogene A eukaryotic gene that functions to promote cellular proliferation or inhibit apoptosis, in which gain-of-function mutations (oncogenes) are associated with cancer progression.

prototroph Microbial strain capable of growth in a defined minimal medium that ideally contains only a carbon source and inorganic compounds. The wildtype genotype is usually regarded as a prototroph.

pseudoautosomal region In mammals, a small region of the X and Y chromosome containing homologous genes.

pseudogene A DNA sequence that is not functional because of one or more mutations but that has a functional counterpart in the same organism; pseudogenes are regarded as mutated forms of ancient gene duplications.

P transposable element A *Drosophila* transposable element used for the induction of mutations, germ-line transformation, and other types of genetic engineering.

Punnett square A cross-multiplication square used for determining the expected genetic outcome of a mating.

purine An organic base found in nucleic acids; the predominant purines are adenine and guanine.

pyrimidine An organic base found in nucleic acids; the predominant pyrimidines are cytosine, uracil (in RNA only), and thymine (in DNA only).

pyrimidine dimer Two adjacent pyrimidine bases, typically a pair of thymines, in the same polynucleotide strand, between which chemical bonds have formed; the most common lesion formed in DNA by exposure to ultraviolet light.

quantitative trait A trait—typically measured on a continuous scale, such as height or weight—that results from the combined action of several or many genes in conjunction with environmental factors.

quantitative trait locus (QTL) A locus segregating for alleles that have different, measurable effects on the expression of a quantitative trait.

random genetic drift Fluctuation in allele frequency from generation to generation resulting from restricted population size.

random mating System of mating in which mating pairs are formed independently of genotype and phenotype.

random spore analysis In fungi, the genetic analysis of spores collected at random rather than from individual tetrads.

rate of mutation The probability that a unit length of DNA (generally a base pair) mutates with time.

reading frame The phase in which successive triplets of nucleotides in mRNA form codons; depending on the reading frame, a particular nucleotide in an mRNA could be in the first, second, or third position of a codon. The reading frame actually used is defined by the AUG codon that is selected for chain initiation.

recessive trait Refers to an allele, or the corresponding phenotypic trait, expressed only in homozygotes.

reciprocal cross A cross in which the sexes of the parents are the reverse of those in another cross.

reciprocal translocation Interchange of parts between nonhomologous chromosomes.

recombinant A chromosome that results from crossing-over and that carries a combination of alleles differing from that of either chromosome participating in the crossover; the cell or organism that contains a recombinant chromosome.

recombinant DNA A DNA molecule composed of one or more segments from other DNA molecules.

recombination Exchange of parts between DNA molecules or chromosomes; recombination in eukaryotes usually entails a reciprocal exchange of parts, but in prokaryotes it is often nonreciprocal.

recruitment The process in which a transcriptional activator protein interacts with one or more components of the transcription complex and attracts it to the promoter.

red–green color blindness *See* **color blindness**.

reductional division Term applied to the first meiotic division because the chromosome number (counted as the number of centromeres) is reduced from diploid to haploid.

redundancy The feature of the genetic code in which an amino acid corresponds to more than one codon; also called degeneracy.

release phase The termination of polypeptide synthesis.

release (protein synthesis) The process by which the completed polypeptide chain is freed from the ribosome.

relative fitness The fitness of a genotype expressed as a proportion of the fitness of another genotype.

renaturation The reconstruction of a protein or nucleic acid (such as DNA) to their original form especially after denaturation.

replica plating Procedure in which a particular spatial pattern of colonies on an agar surface is reproduced on a series of agar surfaces by stamping them with a template that contains an image of the pattern; the template is often produced by pressing a piece of sterile velvet upon the original surface, which transfers cells from each colony to the cloth.

replication *See* **DNA replication; θ replication**.

replication fork In a replicating DNA molecule, the region in which nucleotides are added to growing strands.

replication origin The base sequence at which DNA synthesis begins.

replication slippage The process in which the number of copies of a small tandem repeat can increase or decrease during replication.

reporter gene A gene whose expression can readily be monitored.

repressible transcription A regulatory process in which a gene is temporarily rendered unable to be transcribed.

repressor A protein that binds specifically to a regulatory sequence adjacent to a gene and blocks transcription of the gene.

repulsion configuration *See* ***trans* configuration**.

restriction endonuclease A nuclease that recognizes a short nucleotide sequence (restriction site) in a DNA molecule and cleaves the molecule at that site; also called a restriction enzyme.

restriction enzyme *See* **restriction endonuclease**.

restriction fragment A segment of duplex DNA produced by cleavage of a larger molecule by a restriction enzyme.

restriction fragment length polymorphism (RFLP) Genetic variation in a population associated with the size of restriction fragments that contain sequences homologous to a particular probe DNA; the polymorphism results from the positions of restriction sites flanking the probe, and each variant is essentially a different allele.

restriction map A diagram of a DNA molecule showing the positions of cleavage by one or more restriction endonucleases.

restriction site The base sequence at which a particular restriction endonuclease makes a cut.

restrictive condition A growth condition in which the phenotype of a conditional mutation is expressed.

retinoblastoma An inherited cancer caused by a mutation in the tumor-suppressor gene located in chromosome band *13q14*. Inheritance of one copy of the mutation results in multiple malignancies in retinal cells of the eyes in which the mutation becomes homozygous—for example, through a new mutation or mitotic recombination.

retinoblastoma protein Any of a family of proteins found in animal cells that functions to hold cells at the G_1/S restriction point ("start") by binding to and sequestering a transcription factor that initiates the cell cycle.

retrovirus One of a class of RNA animal viruses that cause the synthesis of DNA complementary to their RNA genomes on infection.

reverse genetics Procedure in which mutations are deliberately produced in cloned genes and introduced back into cells or the germ line of an organism.

reverse transcriptase An enzyme that makes complementary DNA from a single-stranded RNA template.

reverse transcriptase PCR (RT-PCT) Amplification, using an RNA template, of a duplex DNA molecule originally produced by reverse trascriptase.

reversion Restoration of a mutant phenotype to the wildtype phenotype by a second mutation.

RFLP *See* **restriction fragment length polymorphism**.

R group Refers to the side chain connected to the alpha carbon that differs for each amino acid.

ribonucleic acid *See* **RNA**.

ribose The five-carbon sugar in RNA.

ribosomal RNA (rRNA) RNA molecules that are components of the ribosomal subunits; in eukaryotes, there are four rRNA molecules—5S, 5.8S, 18S, and 28S; in prokaryotes, there are three—5S, 16S, and 23S.

ribosome The cellular organelle on which the codons of mRNA are translated into amino acids in protein synthesis. Ribosomes consist of two subunits, each composed of RNA and proteins. In prokaryotes, the subunits are 30S and 50S particles; in eukaryotes, they are 40S and 60S particles.

ribosome-binding site The base sequence in a prokaryotic mRNA molecule to which a ribosome can bind to initiate protein synthesis; also called the Shine–Dalgarno sequence.

ribosome tRNA-binding sites The tRNA-binding sites on the ribosome to which tRNA molecules are bound. The aminoacyl site receives the incoming charged tRNA, the peptidyl site holds the tRNA with the nascent polypeptide chain, and the exit site holds the outgoing uncharged tRNA.

riboswitch A 5′ RNA leader sequence that, according to whether it is bound with a small molecule, can adopt either of two configurations, one of which permits transcription and the other of which terminates transcription.

ribozyme An RNA molecule able to catalyze one or more biochemical reactions.

risk factor Any gene, epigenetic alteration, or environmental circumstance that, acting together with other such agents, increases the likelihood of abnormality or disease.

RNA Ribonucleic acid, a nucleic acid in which the sugar constituent is ribose; typically, RNA is single-stranded and contains the four bases adenine, cytosine, guanine, and uracil.

RNA interference (RNAi) The ability of small fragments of double-stranded RNA to silence genes whose transcripts contain homologous sequences.

RNA polymerase An enzyme that makes RNA by copying the base sequence of a DNA strand.

RNA polymerase holoenzyme Any of several large protein complexes that includes RNA polymerase among its constituents.

RNA processing The conversion of a primary transcript into an mRNA, rRNA, or tRNA molecule; includes splicing, cleavage, modification of termini, and (in tRNA) modification of internal bases.

RNA-seq Assay of gene-expression levels by massively parallel sequencing of cDNA from RNA transcripts.

RNA splicing Excision of introns and joining of exons.

Robertsonian translocation A chromosomal aberration in which the long arms of two acrocentric chromosomes become joined to a common centromere.

rolling-circle replication A mode of replication in which a circular parent molecule produces a linear branch of newly formed DNA.

R plasmid A bacterial plasmid that carries drug-resistance genes; commonly used in genetic engineering.

rRNA *See* **ribosomal RNA**.

RT-PCR *See* **reverse transcriptase PCR**.

Sanger sequencing A method for sequencing DNA that makes use of dideoxynucleotides to terminate strand elongation during synthesis.

satellite DNA Eukaryotic DNA that forms a minor band at a different density from that of most of the cellular DNA in equilibrium density gradient centrifugation; consists of short sequences repeated many times in the genome (highly repetitive DNA) or of mitochondrial or chloroplast DNA.

scaffold A protein-containing material in chromosomes, believed to be responsible in part for the compaction of chromatin.

scanning The process in which the eukaryotic translational initiation complex moves along the mRNA until the first start codon AUG is encountered.

second-division segregation Segregation of a pair of alleles into different nuclei in the second meiotic division, the result of crossing-over between the gene and the centromere of the pair of homologous chromosomes.

second meiotic division The meiotic division in which the centromeres split and the chromosome number is not reduced; also called the equational division.

segment Any of a series of repeating morphological units in a body plan.

segmentation gene Any of a group of genes that determines the spatial pattern of segments and parasegments in *Drosophila* development.

segment-polarity gene Any of a group of genes that determines the spatial pattern of development within the segments of *Drosophila* larvae.

segregation Separation of the members of a pair of alleles into different gametes in meiosis.

selected marker A genetic mutation that allows growth in selective medium.

selection In evolution, intrinsic differences in the ability of genotypes to survive and reproduce; in plant and animal breeding, the choosing of organisms with certain phenotypes to be parents of the next generation; in mutation studies, a procedure designed in such a way that only a desired type of cell can survive, as in selection for resistance to an antibiotic.

selection coefficient The amount by which relative fitness is reduced or increased.

selection limit The condition in which a population no longer responds to artificial selection for a trait.

selectively neutral mutation A mutation that has no (or negligible) effects on fitness.

selective medium A medium that allows growth only of cells with particular genotypes.

selfish DNA DNA sequences that do not contribute to the fitness of an organism but are maintained in the genome through their ability to replicate and transpose.

semiconservative replication The usual mode of DNA replication, in which each strand of a duplex molecule serves as a template for the synthesis of a new complementary strand, and the daughter molecules are composed of one old (parental) strand and one newly synthesized strand.

semisterility A condition in which a significant proportion of the gametophytes produced by a plant or of the zygotes produced by an animal are inviable, as in the case of a translocation heterozygote.

sequencing by synthesis A method of automated DNA sequencing in which the templates have been amplified by the polymerase chain reaction.

sex chromosome A chromosome, such as the human X or Y, that plays a role in the determination of sex.

sexual polyploidization Formation of a polyploid through the fusion of unreduced gametes.

short tandem repeat (STR) A genetic polymorphism resulting from a tandemly repeated short DNA sequence.

sib *See* **sibling**.

sibling A brother or sister, each having the same parents.

sibship A group of brothers and sisters.

sickle-cell anemia A severe anemia in human beings, inherited as an autosomal recessive and caused by an amino acid replacement in the β-globin chain; heterozygotes tend to be more resistant to falciparum malaria than are normal homozygotes.

significant *See* **statistically significant**.

silencer A nucleotide sequence that binds with certain proteins whose presence prevents gene expression.

silent substitution A mutation that has no phenotypic effect.

simple Mendelian trait A phenotypic attribute whose principal cause is a mutant gene inherited according to Mendel's laws, and whose expression, under normal conditions, is minimally affected by environment.

simple sequence repeat (SSR) A DNA polymorphism in a population in which the alleles differ in the number of copies of a short, tandemly repeated nucleotide sequence.

simple tandem repeat (STR) A DNA polymorphism in a population in which the alleles differ in the number of copies of a short, tandemly repeated nucleotide sequence.

SINE element A type of transposable element lacking long terminal repeats that undergoes transposition via an RNA intermediate; the acronym SINE stands for short interspersed element.

single-active-X principle In mammals, the genetic inactivation of all X chromosomes but one in each cell lineage, except in the very early embryo.

single-molecule sequencing A method of automated DNA sequencing that uses individual DNA molecules as the sequencing template.

single-nucleotide polymorphism (SNP) A site in the DNA occupied by a different nucleotide pair among a significant fraction of the individuals in a population.

single-stranded DNA A DNA molecule that consists of a single polynucleotide chain.

single-stranded DNA binding protein A protein able to bind and stabilize single-stranded DNA.

sister chromatids Chromatids produced by replication of a single chromosome.

site-specific recombinase An enzyme that catalyzes intermolecule recombination between two duplex DNA molecules at the site of a target sequence that they have in common.

small interfering RNA (siRNA) Small cleavage products of double-stranded RNA used to target RNAs containing complementary sequences for destruction or for inhibition of their function.

small ribonucleoprotein particles Small nuclear particles that contain short RNA molecules and several proteins.

They are involved in intron excision and splicing and in other aspects of RNA processing.

snRNP Any of several classes of small ribonucleoprotein particles involved in RNA splicing.

somatic cell Any cell of a multicellular organism other than the gametes and the germ cells from which gametes develop.

somatic mutation A mutation arising in a somatic cell.

Southern blot A nucleic acid hybridization method in which, after electrophoretic separation, denatured DNA is transferred from a gel to a membrane filter and then exposed to radioactive DNA or RNA under conditions of renaturation; the radioactive regions locate the homologous DNA fragments on the filter.

species In sexual organisms, a group of organisms that can interbreed and produce fully viable and fertile offspring; in asexual organisms, a group of organisms more closely related to one another than they are to members of other such groups.

specialized transducing phage *See* **transducing phage**.

species tree A real or estimated ancestral history of a group of species.

S period *See* **cell cycle**.

spindle A structure composed of fibrous proteins on which chromosomes align during metaphase and move during anaphase.

spindle assembly checkpoint A mechanism that arrests the cell division cycle until the spindle is properly deployed.

spliceosome An RNA-protein particle in the nucleus in which introns are removed from RNA transcripts.

spontaneous mutation A mutation that happens in the absence of any known mutagenic agent.

sporadic An instance of a disease that is solitary, lacking other affected members in the same pedigree.

spore A unicellular reproductive entity that becomes detached from the parent and can develop into a new organism upon germination; in plants, spores are the haploid products of meiosis.

sporophyte The diploid, spore-forming generation in plants, which alternates with the haploid, gamete-producing generation (the gametophyte).

standard deviation The square root of the variance.

start codon An mRNA codon, usually AUG, at which polypeptide synthesis begins.

statistical epistasis That component of the genetic variance in a population that remains after allocating as much of the genetic variance as possible to the additive and dominance effects of genes.

statistically significant Said of the result of an experiment or study that has only a small probability of happening by chance on the assumption that some hypothesis is true. Conventionally, if results as bad or worse would be expected less than 5 percent of the time, the result is said to be statistically significant; if less than 1 percent of the time, the result is called **statistically highly significant;** both outcomes cast the hypothesis into serious doubt.

stem cell Undifferentiated cell able to divide indefinitely and to differentiate into any of a number of different cell types.

sticky ends The single-stranded ends of a DNA fragment produced by certain restriction enzymes, each capable of reannealing with a complementary sequence in another molecule.

stochastic noise In genetics, random perturbations in molecular interactions that affect levels of gene expression.

stop codon One of three mRNA codons—UAG, UAA, and UGA—at which polypeptide synthesis stops.

STR *See* **simple tandem repeat**

STRP *See* **simple tandem repeat polymorphism**.

STS *See* **sequence-tagged site**.

subfunctionalization Evolutionary change in a gene that results in loss of one or more of its functional or regulatory motifs.

submetacentric chromosome A chromosome whose centromere divides it into arms of unequal length.

subpopulation Any of the breeding groups within a larger population between which migration is restricted.

substrate molecule A substance acted on by an enzyme.

synapsis The pairing of homologous chromosomes or chromosome regions in the zygotene substage of the first meiotic prophase.

synonymous substitution A change in a coding region that alters the nucleotide sequence of a codon without changing the amino acid that is specified.

synteny group A group of genes present in a continuous region of chromosome in two or more species.

synthetic guide RNA (sgRNA) In DNA editing by CRISPR, the synthesized RNA that is complementary to the DNA sequence of the target DNA to be cleaved.

TAD *See* **topologically associating domain**

tandem duplication A pair of identical or closely related DNA sequences that are adjacent and in the same orientation.

TATA-box binding protein (TBP) A protein that binds to the TATA motif in the promoter region of a gene.

TATA box The base sequence 5′-TATA-3′ in the DNA of a promoter.

taxon A population, species, or other group of organisms of which a protein or nucleic acid sequence, or a set of such sequences, is regarded as representative.

TBP-associated factor (TAF) Any protein found in close association with TATA binding protein.

***T* DNA** Transposable element found in *Agrobacterium tumefaciens*, which produces crown gall tumors in a wide variety of dicotyledonous plants.

telomerase An enzyme that adds specific nucleotides to the tips of the chromosomes to form the telomeres.

telomere The tip of a chromosome, containing a DNA sequence required for stability of the chromosome end.

telophase The final stage of mitotic or meiotic nuclear division.

temperature-sensitive mutation A conditional mutation that causes a phenotypic change only at certain temperatures.

template A strand of nucleic acid whose base sequence is copied in a polymerization reaction to produce either a complementary DNA or an RNA strand.

template-directed gap repair Mechanism of DNA repair in which a gap in a DNA strand is filled by means of replication using a homologous chromosome, sister chromatid, or duplicate gene as a template.

terminator A sequence in RNA that halts transcription.

testcross A cross between a heterozygote and a recessive homozygote, resulting in progeny in which each phenotypic class represents a different genotype.

testis-determining factor (TDF) Genetic element on the mammalian Y chromosome that determines maleness.

tetrad The four chromatids that make up a pair of homologous chromosomes in meiotic prophase I and metaphase I; also, the four haploid products of a single meiosis.

tetraploid A cell or organism with four complete sets of chromosomes; in an autotetraploid, the chromosome sets are homologous; in an allotetraploid, the chromosome sets consist of a complete diploid complement from each of two distinct ancestral species.

tetratype An ascus containing spores of four different genotypes—one each of the four genotypes possible with two alleles of each of two genes.

θ replication Bidirectional replication of a circular DNA molecule, starting from a single origin of replication.

30-nm fiber The level of compaction of eukaryotic chromatin resulting from coiling of the extended, nucleosome-bound DNA fiber.

three-point cross Cross in which three genes are segregating; used to obtain unambiguous evidence of gene order.

3′ end The end of a DNA or RNA strand that terminates in a sugar and so has a free hydroxyl group on the number-3′ carbon.

3′ untranslated region The terminal portion of a messenger RNA, following the stop codon, which does not code for protein.

threshold trait A trait with a continuously distributed liability or risk; organisms with a liability greater than a critical value (the threshold) exhibit the phenotype of interest, such as a disorder.

thymine (T) A nitrogenous pyrimidine base found in DNA.

thymine dimer *See* **pyrimidine dimer**.

time of entry In an Hfr × F⁻ bacterial mating, the earliest time that a particular gene in the Hfr parent is transferred to the F⁻ recipient.

Ti plasmid A plasmid that is present in *Agrobacterium tumefaciens* and is used in genetic engineering in plants.

topologically associating domain A region of DNA in the nucleus that is folded in such a way that many contacts affecting gene expression can occur within the region but that is largely separate from other such regions.

total variance Summation of all sources of genetic and environmental variation.

trait Any aspect of the appearance, behavior, development, biochemistry, or other feature of an organism.

***trans* configuration** The arrangement in linked inheritance in which a genotype heterozygous for two mutant sites has received one of the mutant sites from each parent—that is, $a_1 + / + a_2$.

transcript An RNA strand that is produced from, and is complementary in base sequence to, a DNA template strand.

transcription The process by which the information contained in a template strand of DNA is copied into a single-stranded RNA molecule of complementary base sequence.

transcription complex An aggregate of RNA polymerase (consisting of its own subunits) along with other polypeptide subunits that makes transcription possible.

transcription factor A molecule, usually a protein, that binds to a specific sequence in DNA and affects the rate of transcription of a gene.

transcriptional activator protein Positive control element that stimulates transcription by binding with particular sites in DNA.

transducing phage A phage type capable of producing particles that contain bacterial DNA (transducing particles). A specialized transducing phage produces particles that carry only specific regions of chromosomal DNA; a generalized transducing phage produces particles that may carry any region of the genome.

transduction The carrying of genetic information from one bacterium to another by a phage.

transfer RNA (tRNA) A small RNA molecule that translates a codon into an amino acid in protein synthesis; it has a three-base sequence, called the anticodon, complementary to a specific codon in mRNA, and a site to which a specific amino acid is bound.

transformation Change in the genotype of a cell or organism resulting from exposure of the cell or organism to DNA isolated from a different genotype; also, the conversion of an animal cell, whose growth is limited in culture, into a tumor-like cell whose pattern of growth is different from that of a normal cell.

transforming rescue The ability of an introduced DNA fragment to correct a gametic defect in a mutant organism.

transgenic organism An animal or plant in which novel DNA has been incorporated into the germ line.

***trans* heterozygote** *See* ***trans* configuration**.

transition mutation A mutation resulting from the substitution of one purine for another purine or that of one pyrimidine for another pyrimidine.

translation The process by which the amino acid sequence of a polypeptide is synthesized on a ribosome according to the nucleotide sequence of an mRNA molecule.

translocation Interchange of parts between nonhomologous chromosomes; also, the movement of mRNA with respect to a ribosome during protein synthesis. *See also* **reciprocal translocation**.

transmembrane receptor A receptor protein containing amino acid sequences that span the cell membrane.

transmission genetics The processes by which genes are passed from one generation to the next.

transposable element A DNA sequence capable of moving (transposing) from one location to another in a genome.

transposase Protein necessary for transposition.

transposition The movement of a transposable element.

transposon A transposable element that contains bacterial genes—for example, for antibiotic resistance; also used loosely as a synonym for *transposable element*.

transversion mutation A mutation resulting from the substitution of a purine for a pyrimidine or that of a pyrimidine for a purine.

trinucleotide repeat A tandemly repeated sequence of three nucleotides; genetic instability in some trinucleotide repeats is the cause of a number of human hereditary diseases.

triplet code A code in which each codon consists of three bases.

triploid A cell or organism with three complete sets of chromosomes.

trisomic A diploid organism with an extra copy of one of the chromosomes.

trisomy-X syndrome The clinical features of the karyotype 47,XXX.

trivalent Structure formed by three homologous chromosomes in meiosis I in a triploid or trisomic chromosome when each homolog is paired along part of its length with first one and then the other of the homologs.

tRNA *See* **transfer RNA**.

trombone model A model of DNA synthesis in which the active polymerase complex of the lagging strand is connected to the active polymerase complex of the leading strand by means of a protein clamp, which helps coordinate the replication of both strands of double-stranded DNA.

true-breeding Refers to a strain, breed, or variety of organism that yields progeny like itself; homozygous.

truncation point In artificial selection, the value of the phenotype that determines which organisms will be retained for breeding and which will be culled.

tumor-suppressor gene A gene that normally controls cell proliferation or that activates the apoptotic pathway, in which loss-of-function mutations are associated with cancer progression.

Turner syndrome The clinical features of human females with the karyotype 45,X.

two-hybrid analysis A method for detecting protein-protein interactions that makes use of two fused (hybrid) proteins, one including the DNA-binding domain and the other the transcriptional activation domain of a trancriptional activator protein. If the polypeptide chains attached to these components interact within the nucleus, then the interaction brings the domains together and transcription of a reporter gene takes place.

unequal crossing-over Crossing-over between nonallelic copies of duplicated or other repetitive sequences—for example, in a tandem duplication, between the upstream copy in one chromosome and the downstream copy in the homologous chromosome.

univalent Structure formed in meiosis I in a monoploid or a monosomic when a chromosome has no pairing partner.

uracil (U) A nitrogenous pyrimidine base found in RNA.

uridylate A nucleotide in RNA in which the base is uridine.

variable antibody region The portion of an immunoglobulin molecule that varies greatly in amino acid sequence among antibodies in the same subclass.

variable expressivity Differences in the severity of expression of a particular genotype.

variance A measure of the spread of a statistical distribution; the mean of the squares of the deviations from the mean.

vector A DNA molecule, capable of replication, into which a gene or DNA segment is inserted by recombinant DNA techniques; a cloning vehicle.

viral oncogene A class of genes found in certain viruses that predispose to cancer. Viral oncogenes are the viral counterparts of cellular oncogenes. *See also* **cellular oncogene; oncogene**.

V–J joining DNA splicing that unites one of the V (variable) regions of an antibody light-chain gene with one of the J (joining) regions of the same gene to create a unique antibody sequence.

Watson–Crick base pairing Base pairing in DNA or RNA in which A pairs with T (or U in RNA) and G pairs with C.

wildtype The most common phenotype or genotype in a natural population; also, a phenotype or genotype arbitrarily designated as a standard for comparison.

wobble The acceptable pairing of several possible bases in an anticodon with the base present in the third position of a codon.

X chromosome A chromosome that plays a role in sex determination and that is present in two copies in the homogametic sex and in one copy in the heterogametic sex.

xeroderma pigmentosum An inherited defect in the repair of ultraviolet-light damage to DNA, associated with extreme sensitivity to sunlight and multiple skin cancers.

X inactivation In mammals, the genetic inactivation of all X chromosomes but one in each cell lineage, except in the very early embryo.

X-linked gene A gene located in the X chromosome; X-linked inheritance is usually evident from the production of nonidentical classes of progeny from reciprocal crosses.

YAC *See* **yeast artificial chromosome**.

Y chromosome The sex chromosome present only in the heterogametic sex; in mammals, the male-determining sex chromosome.

zygote The product of the fusion of a female gamete and a male gamete in sexual reproduction; a fertilized egg.

zygotene The substage of meiotic prophase I in which homologous chromosomes synapse.

zygotic gene Any of a group of genes that control early development through their expression in the zygote.

Index

Note: Page numbers followed by *f* or *t* indicate material in figures or tables, respectively.